www.brookscole.com

brookscole.com is the World Wide Web site for BrooksCole/Thomson Learning and is your direct source to dozens of online resources.

At *brookscole.com* you can find out about supplements, demonstration software, and student resources. You can also send email to many of our authors and preview new publications and exciting new technologies.

brookscole.com
Changing the way the world learns®

Books in the Brooks/Cole Biology Series

Oceanography

An Invitation to Marine Science

FOURTH EDITION

TOM GARRISON
Orange Coast College
University of Southern California

BROOKS/COLE
TM
THOMSON LEARNING

Australia • Canada • Mexico • Singapore • Spain • United Kingdom • United States

BROOKS/COLE
THOMSON LEARNING

Earth Science Editor: *Keith Dodson, Nina Horne*
Development Editor: *Mary Arbogast*
Assistant Editor: *Mark Andrews*
Editorial Assistant: *Rebecca Eisenman*
Marketing Manager: *Tom Ziolkowski*
Marketing Assistant: *Maureen Griffin*
Project Manager: *Teri Hyde*
Print/Media Buyer: *Karen Hunt*
Permissions Editor: *Joohee Lee*
Advertising Project Manager: *Laura Hubrich*
Technology Project Manager: *Sherill Meaney*

Production Service: *Joan Keyes, Dovetail Publishing Services*
Text Designer: *Adriane Bosworth and Joan Keyes*
Photo Researcher: *Myrna Engler, Tom Garrison*
Copy Editor: *Cathy Baehler*
Illustrators: *Precision Graphics, Graphic World*
Cover Designer: *Roy R. Neuhaus*
Cover Image: *Tim Thompson/Stone*
Indexer: *Nancy Ball*
Cover Printer: *Phoenix Color Corp.*
Compositor: *Graphic World*
Printer: *R. R. Donnelley & Sons, Willard*

Printed in United States of America
2 3 4 5 6 7 05 04 03 02

Library of Congress Cataloging-in-Publication Data

Garrison, Tom
    Oceanography: an invitation to marine science / Tom Garrison.—4th ed.
        p.   cm.
    Includes bibliographical references and index.
    ISBN 0-534-37557-X
    1. Oceanography.
    GC11.2 .G37   2002
    551.46—dc21   2001035127

Wadsworth/Thomson Learning
511 Forest Lodge Road
Pacific Grove, CA 93950
USA

For more information about our products, contact us:
Thomson Learning Academic Resource Center
1-800-423-0563
http://www.brookscole.com

International Headquarters
Thomson Learning
International Division
290 Harbor Drive, 2nd Floor
Stamford, CT 06902-7477
USA

UK/Europe/Middle East/South Africa
Thomson Learning
Berkshire House
168-173 High Holborn
London WC1V 7AA
United Kingdom

Asia
Thomson Learning
60 Albert Street, #15-01
Albert Complex
Singapore 189969

Canada
Nelson Thomson Learning
1120 Birchmount Road
Toronto, Ontario M1K 5G4
Canada

To my family and my students:
My hope for the future.

Brief Contents

Detailed Contents

Preface for Students and Instructors

This book was written to provide an *interesting*, clear, current, and reasonably comprehensive overview of the marine sciences. It was designed for college and university students who are curious about Earth's largest feature, but who may have little formal background in science. Oceanography is broadly interdisciplinary; students are invited to see the *connections* between astronomy, economics, physics, chemistry, history, meteorology, geology, and ecology—areas of study they once considered separate. It's no surprise that oceanography courses have become increasingly popular in the last decade.

Students bring a natural enthusiasm to their study of this field. Even the most indifferent reader will perk up when presented with stories of encounters with huge waves, photos of giant squids, tales of exploration under the best and worst of circumstances, evidence that vast chunks of Earth's surface slowly move, news of Earth's past battering by asteroids, micrographs of glistening diatoms, and data showing the growing economic importance of seafood and marine materials. If pure spectacle is required to generate an initial interest in the study of science, oceanography wins hands down!

In the end, however, it is subtlety that triumphs. Studying the ocean reinstills in us the sense of wonder we felt as children when we first encountered the natural world. There is much to tell. The story of the ocean is a story of change and chance—its history is written in the rocks, the water, and the genes of the millions of organisms that have evolved here.

The Fourth Edition

My aim in writing this book was to produce a text that would enhance students' natural enthusiasm for the ocean. My students have been involved in this book from the very beginning—indeed, it was their request for a readable, engaging, and thorough text that initiated the project a long time ago. Through the 20 years I have been writing textbooks, my enthusiasm for oceanic knowledge has increased (if that is possible), forcing my patient reviewers and editors to weed out an excessive number of exclamation points, but enthusiasm does shine through. One student reading the final manuscript of the first edition commented, "At last, a textbook that does not read like stereo instructions." Good!

This fourth edition builds on its predecessors. As before, a great many students have participated alongside professional marine scientists in the writing and review process. In response to their recommendations, as well as those of instructors who have adopted the book and the many specialists and reviewers who contributed suggestions for strengthening the earlier editions, we have

- Emphasized the *process* of science throughout. Underlying assumptions and limitations are discussed as appropriate in all sections of the book.
- Rearranged the chapters on the physics and chemistry of seawater to present the physical properties before discussing chemistry.
- Enhanced the visual program for increased clarity and accuracy, adding and modifying illustrations to make ideas easier to grasp.
- Expanded discussions of physical oceanography, marine geology, environmental issues, and the ocean's contribution to world economics.
- Covered communities more thoroughly, emphasizing community ecology in the sections of the text on marine biology.
- Refined every chapter to reflect current thought and recent research.
- Produced an interactive CD-ROM to be included with each copy of the text. Students can see oceanography in action.
- Generated a new Web site specific to this textbook. The Web site is open to all without cost or subscription.

The Plan

The plan of the book is straightforward: Because all matter on Earth—except hydrogen and some helium—was generated in stars, our story of the ocean starts with stars. Have oceans evolved elsewhere? We continue with a brief look at the history of marine science (with additional historical information sprinkled through later chapters). The theories of Earth structure and plate tectonics are presented next as a base on which to build the explanation of bottom features that follow. A survey of ocean physics and chemistry prepares us for discussions of atmospheric circulation, classical physical oceanography, and coastal processes. Our look at marine biology begins with an overview of the problems and benefits of living in seawater, continues with a discussion of the production and consumption of food, and ends with taxonomic *and* ecological surveys of marine organisms. The last chapters address marine resources and environmental concerns.

Connections between disciplines are emphasized throughout. Marine science draws on several fields of study integrating the work of specialists into a unified whole. For example, a geologist studying the composition of marine sediments on the deep seabed must be aware of the biology and life histories of the organisms in the water above, the chemistry that affects the shells and skeletons of the creatures as they fall to the ocean floor, the physics of particle settling and water density and ocean currents, and the age and underlying geology of the study area. This book is organized to make those connections from the first.

Organization

A broad view of marine science is presented in 18 chapters, each free-standing (or nearly so) to allow an instructor to assign chapters in any order he or she finds appropriate. Each chapter begins with a **vignette**—a short illustrated tale, observation, or sea story to whet your appetite for the material to come—followed by **What to Watch For,** a survey of the chapter's content. Some vignettes spotlight scientists at work; others describe the experiences of people or animals in the sea.

The chapters are written in an **engaging style.** Terms are defined and principles developed in a straightforward manner. Some complex ideas are initially outlined in broad-brush strokes, then discussed again in greater depth after you have a clear view of the overall situation. When appropriate to their meanings, the derivations of words are shown. **Measurements** are given in both metric (S.I.) and American systems. At the request of a great many students, the units are spelled out (that is, we write *kilometer* rather than *km*) to avoid ambiguity and for ease of reading.

The photos, charts, graphs, and paintings in the **extensive illustration program** have been chosen for their utility, clarity, and beauty. **Boxes** in each chapter present commentaries of special interest on unique topics or controversies. **Internet icons** are provided at nearly all of the subheads, boxes, and essays, indicating that text-specific Web links provide additional information.

Concluding each chapter is a **Questions from Students** section. These questions are ones that students have asked me over the years. This material is an important extension of the chapters and occasionally contains key words and illustrations. Each chapter ends with an array of study materials for students, beginning with a new feature called **Key Concepts to Review.** Important **Terms** are listed next. These are also defined in an extensive **Glossary** in the back of the book. **Study Questions** are also included in each chapter. Writing the answers to these questions will cement your understanding of the concepts presented. The **annotated bibliography**

at the book's Web site will be helpful when you wish to know more about a particular topic.

Appendixes will help you master measurements and conversions, geological time, latitude and longitude, chart projections, the mathematics of the Coriolis effect and tidal forces, and the taxonomy of marine organisms. In case you'd like to join us in our life's work, the last appendix discusses **jobs in marine science.**

This book has been thoroughly **student tested.** You need not feel intimidated by the concepts—this material has been mastered by students just like you. Read slowly and go step-by-step through any parts that give you trouble. Your predecessors have found the ideas presented here to be useful, inspiring, and applicable to their lives. Best of all, they have found the subject to be *interesting!*

The Web Site

This book is fully Internet integrated. You'll find its dedicated Web site at:

www.brookscole.com/product/053437557xs

Internet icons at each head and subhead, and at the ends of Vignettes and Boxes, indicate text-specific information is available on the site. References, bibliography entries, chapter outlines, sample quizzes, flashcards, links, exercises, animations, and tutorial assistance are also accessible at the site. We will periodically update the Web site to add features and announce important advances in marine science. *The site is open to anyone without cost or subscription.*

Suggestions for Using This Book

1. *Begin with a preview.* Scout the territory ahead: Read "What to Watch For" after each vignette, flip through the assigned pages reading only the headings and subheadings, look at the figures and read captions that catch your attention.

2. *Keep a pen and paper handy.* Jot down a few questions—any questions—that this quick glance stimulates. *Why* is the deep ocean cold if the inside of the Earth is so hot? *What* makes storm conditions like those seen in 1997–98? *Where* did sea salt come from? *Will* global warming actually be a problem? *Does* anybody still hunt whales? Writing questions will help you focus when you start studying.

3. *Now read in small but concentrated doses.* Each chapter is written in a sequence and tells a story. The logical progression of ideas is going somewhere. Follow the organization of the chapter. Stop occasionally to review what you've learned. Flip back and forth to review and preview.

4. *Strive to be actively engaged.* Write marginal notes, underline occasional passages (underlining whole sections is seldom useful), write more questions, draw on the diagrams, check off subjects as you master them, make

flashcards while you read (if you find them helpful), *use your book!*

5. *Monitor your understanding.* If you start at the beginning of the chapter, you will have little trouble understanding the concepts as they unfold. But if you find yourself at the bottom of the page having only scanned (rather than understood) the material, stop there and start that part again. Look ahead to see where we're going. Remember, students have been here before, and I have listened to their comments to make the material as clear as I can. This book was written for you.

6. *Use the Internet sites.* We have placed more than 1,500 Internet links throughout the text. If you have access to a computer, fire it up and scan the designated site as you read the associated passage of the book.

7. *Enjoy the journey.* Your instructor would be glad to share his or her understanding and appreciation of marine science with you—you need only ask. Students, instructors, and authors all work together toward a common goal—an appreciation of the beauty and interrelationships a growing understanding of the ocean can provide.

Acknowledgments

Jack Carey at Thomson Learning, the grand master of college textbook publishing, willed the first edition of this book into being. His suggestions have been combined with those of more than 750 undergraduate students and 118 reviewers to contribute to my continuously growing understanding of marine science. Donald Lovejoy, Stanley Ulanski, Richard Yuretich, Ronald Johnson, and John Mylroie deserve special recognition for many years of patient direction. For the fourth edition, I have especially depended on the expert advice of Ernest Knowles at North Carolina State University, Hans-Peter Bunge of Princeton, William Cochlan of San Francisco State University, James Ingle at Stanford, Richard Murray at Boston University, Morris Sotonoff at Chicago State University, Kent Syverson of the University of Wisconsin, Michelle Kominz at Western Michigan University, Lawrence Krissek at Ohio State, Scott King at Purdue, Brian McAdoo at Vassar, James Stratton of Eastern Illinois University, Michael Lyle at Tidewater Community College, and Bruce Fouke at the University of Illinois, Urbana.

My long-suffering departmental colleagues Dennis Kelly, Jay Yett, and Robert Profeta again should be awarded medals for putting up with me, answering hundreds of my questions, and being so forbearing through the book's lengthy gestation period. Thanks again to our dean, Stanley Johnson, and our college president, Margaret Gratton, for supporting this project and encouraging our faculty to teach, conduct research, and be involved in community service. Our department teaching assistants deserve praise as well, especially Timothy Riddle and Matthew Bollen, the Internet wizards responsible for the Web site and extensive links.

Yet another round of gold medals should go to my family for being patient (well, *relatively* patient) during those years of days and nights when dad was holed up in his dark reference-littered cave, listening to really loud Glenn Gould Bach recordings, working late on The Book. Thank you Marsha, Jeanne, Greg, and John for your love and understanding.

The people who provided pictures and drawings have worked miracles to obtain the remarkable images in these pages. To mention just a few: Gerald Kuhn sent classics taken by his late SIO colleague Francis Shepard, Vincent Courtillot of the University of Paris contributed the remarkable photo of the Aden Rift, Catherine Devine at Cornell provided time-lapse graphics of tsunami propagation, Jian Lin of Woods Hole sent a photo of Iceland's Thingvellir Graben, Robert Headland of the Scott Polar Research Institute in Cambridge searched out prints of polar subjects, Charles Hollister at Woods Hole kindly provided seafloor photos from his important books, Andreas Rechnitzer and Don Walsh recalled their exciting days with *Trieste,* and Bruce Hall, Pat Mason, Ron Romanoski, Ted Delaca, William Cochlan, Christopher Ralling, Mark McMahon, John Shelton, Alistair Black, Howard Spero, Eric Bender, Ken-ichi Inoue, and Norman Cole contributed beautiful slides. Seran Gibbard provided the highest-resolution images yet made of the surface of Titan, and Michael Malin forwarded truly beautiful images of erosion on Mars. Herbert Kawainui Kane again allowed us to reprint his magnificent paintings of Hawaiian subjects. Deborah Day and Cindy Clark at Scripps Institution, Jutta Voss-Diestelkamp at the Alfred Wegener Institut in Bremerhaven, and David Taylor at the Centre for Maritime Research in Greenwich dug through their archives one more time. Don Dixon, William Hartmann, Ron Miller, and William Kaufmann provided paintings, Dan Burton sent photos, and Andrew Goodwillie printed customized charts. Wim van Egmond contributed striking photomicrographs of diatoms, forams, and copepods. Bill Haxby at Lamont provided truly beautiful seabed scans. Karen Riedel helped with DSDP core images. James Ingle offered me a desk at Stanford whenever I needed it. NOAA, NASA, USGS, the Smithsonian Institution, the Royal Geographical Society, the U.S. Navy, and the U.S. Coast Guard came through time and again, as did private organizations like Alcoa Aluminum, Cunard, Shell Oil, The Maersk Line, Grumman Aviation, Breitling-SA, CNN, Associated Press, Oakley, and the *Los Angeles Times.* The Woods Hole team was also generous: Philip Richardson, William Schmitz, Susumu Honjo, Doug Webb, James Broda, Albert Bradley, and Kathy Patterson all provided photographs, diagrams, and advice. Individuals with special expertise have also been willing to share: Hank Brandli processed satellite digital images of storms, Peter Sloss at the National Geophysical Data Center helped me sort through computer-generated seabed images, Steven Grand of the University of Texas provided a descending deep-slab image, Hans-Peter Bunge of Princeton patiently explained mantle-core dynamics, Michael Gentry again mined the archives of the Johnson Space Center for Earth images, Jurrie van der Woulde at JPL helped with images of oceans elsewhere, John Maxtone-Graham of New York's Ocean Liner Museum found me a rogue wave picture, Ed Ricketts, Jr., contributed a portrait of his father, and professor Lynton Land of the University of Texas sent a rare photo of a turbidity current.

The Brooks/Cole team performed the customary miracles. The charge was again led from her redwood forest keep by Joan Keyes, production editor and champion of every known means of digital communication. The text was polished by Mary Arbogast, the best developmental editor in the Orion arm of the galaxy, and Cathy Baehler and Mary Roybal the copy editors who saved me from many errors. Myrna Engler worked tirelessly on photo research and permissions. Project manager Teri Hyde kept us all running in the same direction. And Nina Horne, my excellent editor, gave birth to this book and a baby daughter simultaneously. What skill!

My thanks to all.

A Goal and a Gift

The goal of all this effort: *To allow you to gain an oceanic perspective.* "Perspective" means being able to view things in terms of their relative importance or relationship to one an-other. An oceanic perspective lets you see this misnamed planet in a new light, and helps you plan for its future. You will see that water, continents, seafloors, sunlight, storms, seaweeds, and society are connected in subtle and beautiful ways.

The ocean's greatest gift to humanity is intellectual—the constant challenge its restless mass presents. Let yourself be swept into this book and the class it accompanies. Give yourself time to ponder: "Meditation and water are wedded forever," wrote Herman Melville in Moby Dick. Take pleasure in the natural world. Ask questions of your instructors and teaching assistants, read the references, try your hand at the questions at the ends of the chapters.

Be optimistic. Take pleasure in the natural world. Please write to me when you find errors or if you have comments. Above all, *enjoy yourself!*

Tom Garrison
Orange Coast College
University of Southern California
tsgarri@attglobal.net

Reviewers

ERNEST ANGINO, *University of Kansas*
M. A. ARTHUR, *Pennsylvania State University*
HENRY A. BART, *La Salle University*
STEVEN R. BENHAM, *Pacific Lutheran University*
LATSY BEST, *Palm Beach Community College*
EDWARD BEUTHER, *Franklin and Marshall College*
WILLIAM L. BILODEAU, *California Lutheran University*
JULIE BRIGHAM-GRETTE, *University of Massachusetts at Amherst*
LAURIE BROWN, *University of Massachusetts*
KEITH A. BRUGGER, *University of Minnesota, Morris*
ZANNA CHASE, *Lamont-Doherty Earth Observatory, Columbia University*
KARL M. CHAUFF, *St. Louis University*
WILLIAM COCHLAN, *San Francisco State University*
JAMES E. COURT, *City College of San Francisco*
RICHARD DAME, *University of South Carolina, Columbia*
DAVID DARBY, *University of Minnesota at Duluth*
ROBERT J. FELLER, *University of South Carolina, Columbia*
L. KENNETH FINK, JR., *University of Maine*
BRUCE FOUKE, *University of Illinois, Urbana*
DIRK FRANKENBURG, *University of North Carolina, Chapel Hill*
ROBERT R. GIVEN, *Marymount College, Rancho Palos Verdes*
WILLIAM GLEN, *U. S. Geological Survey*
KAREN GROVE, *San Francisco State University*
BARRON HALEY, *West Valley College*
JACK C. HALL, *University of North Carolina at Wilmington*
WILLIAM HAMNER, *University of California, Los Angeles*
WILLIAM B. HARRISON III, *Western Michigan University*
DAVID HASTINGS, *University of British Columbia*
TED HERMAN, *West Valley College*
JOSEPH HOLLIDAY, *El Camino College*
ANDREA HUVARD, *California Lutheran University*
JAMES C. INGLE, JR., *Stanford University*
RONALD E. JOHNSON, *Old Dominion University*
SCOTT KING, *Purdue University*
JOHN A. KLASIK, *California Polytechnic University, Pomona*
ERNEST C. KNOWLES, *North Carolina State University, Raleigh*
EUGENE KOZLOFF, *Friday Harbor, WA*
MICHELLE KOMINZ, *Western Michigan University*
LAWRENCE KRISSEK, *Ohio State University*
ALBERT M. KUDO, *University of New Mexico*
FRANK T. KYTE, *University of California, Los Angeles*
LYNTON S. LAND, *University of Texas, Austin*
RICHARD W. LATON, *Western Michigan University*

RUTH LEBOW, *University of California, Los Angeles Extension*
DOUGLAS R. LEVIN, *Bryant College*
LARRY LEYMAN, *Fullerton College*
TIMOTHY LINCOLN, *Albion College*
DONALD L. LOVEJOY, *Palm Beach Atlantic College*
MICHAEL LYLE, *Tidewater Community College*
JAMES MACKIN, *State University of New York, Stony Brook*
DAVID C. MARTIN, *Centralia College*
ELLEN MARTIN, *University of Florida*
BRIAN MCADOO, *Vasser College*
JAMES MCWHORTER, *Miami-Dade Community College, Kendall Campus*
CHRIS METZLER, *Mira Costa College*
RICHARD W. MURRAY, *Boston University*
JOHN E. MYLROIE, *Mississippi State University*
CONRAD NEWMAN, *University of North Carolina, Chapel Hill*
JAMES G. OGG, *Purdue University*
B. L. OOSTDAM, *Millersville University*
JAN PECHENIK, *Tufts University*
BERNARD PIPKIN, *University of Southern California*
MARK PLUNKETT, *Bellevue Community College*
K. M. POHOPIEN, *Covina, CA*
RICHARD G. ROSE, *West Valley College*
JUNE R. P. ROSS, *Western Washington University*
WENDY L. RYAN, *Kutztown University*
ROBERT J. SAGER, *Pierce College, Washington*
ROBERT F. SCHMALZ, *Pennsylvania State University*
DON SEAVY, *Olympic College*
SAM SHABB, *Highline Community College*
WILLIAM G. SIESSER, *Vanderbilt University*
RALPH SMITH, *University of California, Berkeley*
SCOTT W. SNYDER, *East Carolina University*
MORRIS L. SOTONOFF, *Chicago State University*
JAMES F. STRATTON, *Eastern Illinois University*
KENT SYVERSON, *University of Wisconsin*
J. COTTER THARIN, *Hope College*
STANLEY ULANSKI, *James Madison University*
J. J. VALENCIC, *Saddleback College*
RAYMOND E. WALDNER, *Palm Beach Atlantic College*
JILL M. WHITMAN, *Pacific Lutheran University*
P. KELLY WILLIAMS, *University of Dayton*
BERT WOODLAND, *Centralia College*
JOHN H. WORMUTH, *Texas A&M University*
RICHARD YURETICH, *University of Massachusetts at Amherst*
MEL ZUCKER, *Skyline College*

An Ocean World 1

A MARINE POINT OF VIEW

I believe Earth is misnamed. From space our planet is brilliantly blue, white in places with clouds and ice, sometimes swirling with storms. Dominating its surface is a single great ocean of liquid water. This ocean affects and moderates temperature and dramatically influences weather. From its floor is pumped about one-third of the world's supply of petroleum and natural gas. The ocean borders most of the planet's largest cities. It is a primary shipping and transportation route and a major recreational resource. The dry land on which nearly all of human history has unfolded is hardly visible from space, for nearly three-quarters of the planet is covered by water. Oceanus would surely be a better name for our watery home.

Earth, its ocean, and its organisms have changed together over the ages. Life originated in the ocean, developing and flourishing there for more than 3 billion years before venturing onto the unwelcoming continents. All our planet's living things—terrestrial as well as aquatic—carry an ocean within themselves. Their blood, their eggs, and the fluids that bathe their cells are all saline. We ourselves are made largely of water with about the same relative proportion of salts as the sea. The first nine months of human life are spent in a water world, a warm supportive ocean that cradles from shock and provides a stable and weightless environment for the complex processes of growth and development. After birth, we view the universe through an ocean—the fluid behind the corneas of our eyes is similar to seawater. In a sense we see everything from a marine point of view.

We have been profoundly influenced by the ocean. It separates enemies but fosters trade; it inspires our spirits of adventure and imagination. Its

NASA

A view of Earth. Most of its surface is covered by a liquid-water ocean averaging 3,796 meters (12,451 feet) deep. The tail of the space shuttle *Discovery* is visible at the top.

waters and floor provide important physical and biological resources. Studies of the ocean have extended our scientific understanding of Earth and the universe. Nearly half of the planet's 6 billion inhabitants now live within 150 kilometers (93 miles)[1] of a coastline. The ocean gives gainful employment to millions of these people, provides transportation for more than 90% of their commerce, and silently hides their wastes. Best of all, it gives us hope. We look to the ocean for solitude and inspiration. Mariners, divers, surfers, and anyone else wishing time for quiet reflection know the special peace that the ocean can bring. To fly for hours over the vast Pacific brings a new perspective to anyone content with the poor name Earth.

1-1

---

[1] Throughout this book, metric measurements precede American measurements. For a brief review of metric units and their relation to American units, please turn to Appendix I.

WHAT TO WATCH FOR IN CHAPTER 1

Think of oceanography as a story. In this chapter, the main character—the world ocean—is introduced in broad brush strokes, and we begin our investigation of the ocean with questions about its origin. As you read, keep in mind these questions: Where did the ocean come from? How do we know?

You may be startled to discover that most of the atoms that make up Earth, its ocean, and its inhabitants were formed within stars billions of years ago. Stars spend their lives changing hydrogen and helium to heavier elements. As they die, some stars eject these elements into space during cataclysmic explosions. The sun and the planets, including Earth, probably condensed from a cloud of dust and gas enriched by the recycled remnants of exploded stars.

Our ocean is not a remnant of that cloud, however. Most of the ocean formed later, as water vapor trapped in Earth's outer layers escaped to the surface through volcanic activity during the planet's youth. The vapor cooled and condensed to form an ocean. Life originated there soon after. Life and Earth have grown old together.

How do we know? By using the scientific method, a systematic way of asking and answering questions about the natural world. Science assumes that nature "plays fair," that rules don't capriciously change, and that answers to questions are ultimately knowable as our powers of observation improve. Using the tools of science to investigate nearby worlds, we have discovered that our beautiful liquid ocean may not be entirely unique.

Traditionally, we have divided the ocean into artificial compartments called oceans and seas, using the boundaries of continents and imaginary lines such as the equator. In fact there are few dependable natural divisions, only one great mass of water. The Pacific and Atlantic Oceans and the Mediterranean and Baltic Seas, so named for our convenience, are in reality only temporary features of a single **world ocean.**[2] In this book we refer to the world ocean, or simply the **ocean,** *as a single entity,* with subtly different characteristics at different locations but with very few natural partitions. Such a view emphasizes the interdependence of ocean and land, life and water, atmospheric and oceanic circulation, and natural and man-made environments. The ocean may be defined as the vast body of saline water that occupies the depressions of Earth's surface. Over 97% of the water on or near Earth's surface is contained in the ocean; less than 3% is held in land ice, groundwater, and all the freshwater lakes and rivers (see **Figure 1.1**).

On a *human* scale the ocean is impressively large. It covers 361 million square kilometers (139 million square miles) of Earth's surface. If Earth's elevations were leveled to a smooth ball, the ocean would cover it to a depth of 2,686 meters (8,810 feet). Average ocean depth is 4½ times greater than average land elevation. The Pacific Ocean is Earth's most prominent single feature. **Figure 1.2a** summarizes some basic characteristics of the world ocean.

On a *planetary* scale, however, the ocean itself is insignificant. Its average depth is a tiny fraction of Earth's radius—the blue ink representing the ocean on an 8-inch paper globe is proportionally thicker. The ocean accounts for only slightly more than 0.02% of Earth's mass, or 0.13% of its volume. There is much more water chemically trapped within Earth's hot interior than there is in its ocean and atmosphere.

For Internet sites related to topics in the text, go to the Web site and click on the numbered icon.

1-2

MARINE SCIENCE, OCEANOGRAPHY, AND THE NATURE OF SCIENCE

Marine science (or **oceanography**) is the process of discovering unifying principles in data obtained from the ocean, its associated life forms, and the bordering lands. It draws on several disciplines, integrating the fields of geology, physics, biology, chemistry, and engineering as they apply to the ocean and its surroundings. *Marine geologists* focus on questions such as the composition of the inner Earth, the mobility of the crust, the characteristics of seafloor sediments, and the history of Earth's climate. Some of their work touches on areas of intense scientific and public concern, including earthquake prediction and the distribution of valuable resources. *Physical oceanographers*

---

[2] When an important new term is introduced and defined, it is printed in boldface type. These terms are listed at the end of the chapter and defined in the glossary.

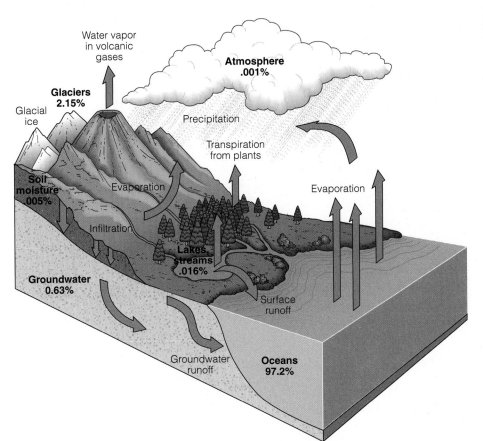

Water vapor
in volcanic
gases

**Glaciers
2.15%**

Glacial
ice

**Atmosphere
.001%**

Precipitation

Transpiration
from plants

**Soil
moisture
.005%**

Evaporation

Infiltration

Evaporation

**Lakes
streams
.016%**

**Groundwater
0.63%**

Surface
runoff

Groundwater
runoff

**Oceans
97.2%**

Figure 1.1 The relative amount of water in various locations on or near Earth's surface. More than 97% of the water lies in the ocean. Ice on land contains about 1.9%, groundwater 0.6%, rivers and lakes 0.02%, and the atmosphere 0.001%. The cycling of water between locations is shown in this diagram. The movement is driven by radiant energy from the sun.

study and observe wave dynamics, currents, and ocean–atmosphere interaction. Their predictions of long-term climate trends are becoming increasingly important as pollutants change Earth's atmosphere. *Marine biologists* work with the nature and distribution of marine organisms, the impact of oceanic and atmospheric pollutants on the organisms, the isolation of disease-fighting drugs from marine species, and the yields of fisheries. *Chemical oceanographers* study the ocean's dissolved solids and gases, and the relationships of these components to the geology and biology of the ocean as a whole. *Marine engineers* design and build oil platforms, ships, harbors, and other structures that enable us to use the ocean wisely. Other marine specialists study weather forecasting, ways to increase the safety of navigation, methods to generate electricity, and much more. Virtually all marine scientists specialize in one area of research, but they also must be familiar with related specialties and appreciate the linkages between them. **Figure 1.3** shows marine scientists in action.[3]

Marine scientists today are asking some critical questions about the origin of the ocean, the age of its basins, and the nature of the life forms it has nurtured. We are fortunate to live at a time when scientific study may be able to answer some of those questions. **Science** is a systematic *process* of asking questions about the observable world and

then testing the answers to those questions. Scientists gather and study information (data), but the information itself is not science. Science interprets raw information by constructing a general explanation with which the information is compatible.

Scientists start with a question—a desire to understand something they have observed or measured. The process of science begins with an informed guess called a working **hypothesis,** a speculation about the natural world that can be tested and verified or disproved by further observations and controlled experiments. (An **experiment** is a test that simplifies observation in nature or in the laboratory by manipulating or controlling the conditions under which observations are made.) Hypotheses consistently supported by observation or experiment are advanced to the status of **theory,** a statement of relationship accepted by most scientists. The largest constructs, known as **laws,** are principles explaining events in nature that have been observed to occur with unvarying uniformity under the same conditions.

Theories and laws in science do not arise fully formed or all at once. Scientific thought progresses as a continuing chain of questioning, testing, and matching theories to observations. A theory is strengthened if new facts support it. If not, the theory is modified or a new explanation is sought. The power of science lies in the ability of the process to operate *in reverse*, that is, in the use of a theory or law to make predictions and anticipate new facts to be observed.

---

[3] Would you like to join us? Appendix VIII discusses careers in oceanography.

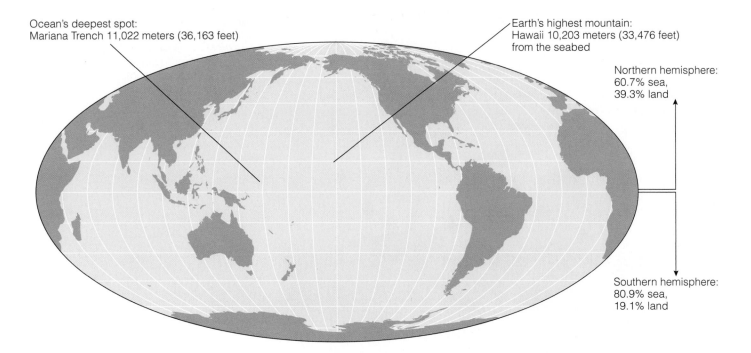

Ocean's deepest spot:
Mariana Trench 11,022 meters (36,163 feet)

Earth's highest mountain:
Hawaii 10,203 meters (33,476 feet)
from the seabed

Northern hemisphere:
60.7% sea,
39.3% land

Southern hemisphere:
80.9% sea,
19.1% land

Area: 361,100,000 square kilometers (139,400,000 square miles)
Volume: 1,370,000,000 cubic kilometers (329,000,000 cubic miles)
Average depth: 3,796 meters (12,451 feet)
Average temperature: 3.9°C (39.0°F)
Average salinity: 34,482 grams per kilogram (0.56 ounce per pound), 3.4%
Average land elevation: 840 meters (2,772 feet)
Age: About 4 billion years
Future: Uncertain

Figure 1.2 (**a**) The proportion of sea versus land, shown on an equal-area projection of Earth. An equal-area projection is a map drawn to represent areas in their correct relative proportions. The average depth of the ocean is 4½ times greater than average land elevation. Note the extent of the Pacific Ocean, Earth's most prominent single feature. (**b**) Light, air, wind, water, and life—a typical scene on the misnamed Earth.

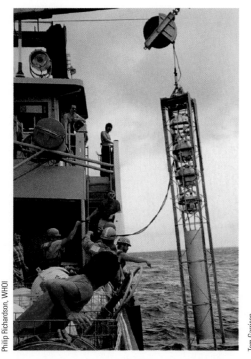

Philip Richardson, WHOI

a

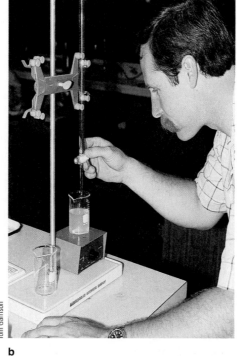

Tom Garrison

b

Dan Burton Photography

c

Jay Dickman

d

Figure 1.3 Marine scientists at work. (**a**) A sound research float being launched from the Woods Hole Oceanographic Institution's research ship *Oceanus*. The probe will drop to a depth of 3,500 meters (11,500 feet) and produce a low-frequency tone once each day for tracking. (**b**) A marine chemist tests a water sample to find the concentrations of dissolved elements. (**c**) Three researchers peer from a viewing port in the *Aquarius Habitat* off Key Largo, Florida. *Aquarius* rests on the seabed 15 meters (49 feet) below the surface and serves as a base from which divers can study the area. (**d**) The nuclear submarine USS *Hawkbill* surfaces at the North Pole during a mission to collect water samples and probe undersea mountain ranges.

This procedure, often called the **scientific method,** is an orderly process by which theories are verified or rejected. It is based on the assumption that nature "plays fair"—that the rules governing natural phenomena do not change capriciously as our powers of questioning and observing improve. We believe that the answers to our questions about nature are *ultimately knowable.*

There is no one scientific method. Some researchers observe, describe, and report on some subject and leave it to others to hypothesize. Scientists don't have one single method in common—the general method they employ is a critical attitude about being *shown* rather than being *told* and taking a logical approach to problem solving. **Figure 1.4** summarizes the main points.

Nothing is ever proved absolutely true by the scientific method. Theories may change as our knowledge and powers of observation change; thus all scientific understanding is tentative. Science is not a democratic process or a popularity contest. The conclusions about the natural world that we reach by the process of science may not always be comfortable, easily understood, or immediately embraced, but if those conclusions consistently match observations, they may be considered true.

Solitude   BOX 1.1

The ocean has always challenged the human spirit. Meeting that challenge alone is a supreme triumph of humanity over the uncontrollable and unpredictable forces of nature. Whatever the reasons for their voyages, all who travel by sea alone have experienced profound solitude, loneliness, helplessness, and—if things went well—the exultation of success.

The first man to sail alone around the world was Joshua Slocum, a Massachusetts sailor who first went to sea when he was 14. He was 51 when he began his solitary voyage. He started from Boston on 24 April 1895 in *Spray*, an 11-meter (37-foot) sloop he had rebuilt from a derelict hulk. More than three years and 74,000 kilometers (46,000 miles) later, on 3 July 1898, he tied the boat to the cedar spike driven into the bank that had held her when she was first launched. "I could bring her no nearer home," he said.

In his book, *Sailing Alone Around the World*, Slocum wrote about the peace and heightened awareness that intimate contact with the ocean can bring.

> The fog lifted just before night, and I was afforded a look at the sun just as it was touching the sea. I watched it go down and out of sight. Then I turned my face eastward, and there, apparently at the very end of the bowsprit, was the smiling full moon rising out of the sea. Neptune himself coming over the bows could not have startled me more. "Good evening, sir," I cried; "I'm glad to see you." Many a long talk since then I have had with the man in the moon; he had my confidence on the voyage.

> About midnight the fog shut down again denser than ever before. One could almost "stand on it." It continued so for a number of days, the wind increasing to a gale. The waves rose high, but I had a good ship. Still, in the dismal fog I felt myself drifting into loneliness, an insect on a straw in the midst of the elements. I lashed the helm, and my vessel held her course, and while she sailed I slept.

> During these days a feeling of awe crept over me. My memory worked with startling power. The ominous, the insignificant, the great, the small, the wonderful, the commonplace—all appeared before my mental vision in magical succession. Pages of my history were recalled which had been so long forgotten that they seemed to belong to a previous existence. I heard all the voices of the past laughing, crying, telling what I had heard them tell in many corners of the Earth. The loneliness of my state wore off when the gale was high and I found much work to do. When fine weather returned, then came the sense of solitude, which I could not shake off. . . .

The ocean is as vast, as quiet, as furious, as inspiring as it was in Slocum's day, but our understanding of it has grown immensely. I think he would have enjoyed the insights that nearly a century of progress in marine science has added to the things he saw, heard, and felt.

Source: Joshua Slocum, *Sailing Alone Around the World* (New York: Sheridan House, 1972). Originally published in 1899.

Joshua Slocum, shown here rounding the tip of Cape Horn in his 37-foot sloop, *Spray*, was the first person to sail around the world alone. After a difficult voyage of more than three years and 74,000 kilometers (46,000 miles), he returned to Newport, Rhode Island, and tied *Spray* to the same cedar spike driven into the bank that had held her when she was first launched. "I could bring her no nearer home."

© 1972 Sheridan House

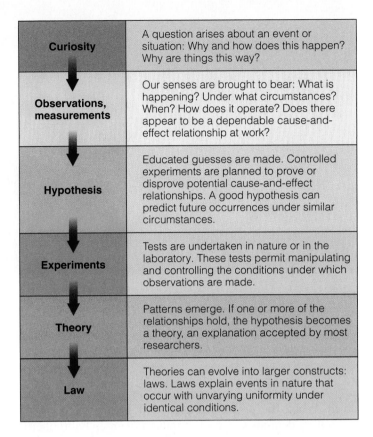

Curiosity	A question arises about an event or situation: Why and how does this happen? Why are things this way?
Observations, measurements	Our senses are brought to bear: What is happening? Under what circumstances? When? How does it operate? Does there appear to be a dependable cause-and-effect relationship at work?
Hypothesis	Educated guesses are made. Controlled experiments are planned to prove or disprove potential cause-and-effect relationships. A good hypothesis can predict future occurrences under similar circumstances.
Experiments	Tests are undertaken in nature or in the laboratory. These tests permit manipulating and controlling the conditions under which observations are made.
Theory	Patterns emerge. If one or more of the relationships hold, the hypothesis becomes a theory, an explanation accepted by most researchers.
Law	Theories can evolve into larger constructs: laws. Laws explain events in nature that occur with unvarying uniformity under identical conditions.

Figure 1.4 An outline of the scientific method, a systematic process of asking questions about the observable world and testing the answers to those questions. There is not a single scientific method. Science depends on a critical attitude about being shown rather than being told, and taking logical approaches to problem solving. Application of a scientific method leads us to truth based on the observations and measurements that have been made—a work in progress, never completed. The external world, not internal conviction, must be the testing ground for scientific beliefs.

This book shows some of the results of the scientific process as it has been applied to the world ocean. It presents facts, interpretations of facts, examples, stories, and some of the crucial discoveries that have led to our present understanding of the ocean and the world on which it formed. As the results of science change, so will the ideas and interpretations presented in books like this one.

THE ORIGIN OF THE EARTH

We have always wondered about our origins—how Earth was formed, how the ocean arose, and how life came to be. In the last 50 years researchers using the scientific method have determined a tentative age for the ocean, Earth, and the universe. They have developed hypotheses about how matter is assembled, how stars and planets are formed, and even how life may have arisen. Many of the details are still sketchy, of course, but these hypotheses have predicted some important recent discoveries in subatomic physics and molecular biology. Perhaps the most dramatic discoveries in natural science

in the past century have been those dealing with the origin and history of the universe.

The universe apparently had a beginning. The **big bang,** as that event is modestly named, probably occurred about 13 billion years ago. All of the mass and energy of the universe is thought to have been concentrated at a geometric point at the beginning of time, the moment when the expansion of the universe began. We don't know what initiated the expansion, but it continues today and will probably continue for billions of years, perhaps forever.

The very early universe was unimaginably hot, but as it expanded it cooled. About a million years after the big bang, temperatures fell enough to permit the formation of atoms from the energy and particles that had predominated up to that time. Most of these atoms were hydrogen, then as now the most abundant form of matter in the universe. About a billion years after the big bang, this matter began to congeal into the first galaxies and stars.

Galaxies and Stars

A **galaxy** is a huge rotating aggregation of stars, dust, gas, and other debris held together by gravity. There are perhaps 200 billion galaxies in the universe and 200 billion stars in each galaxy. Our galaxy is named the **Milky Way galaxy** (*galaktos* = milk).[4] A galaxy very similar to our own is shown in **Figure 1.5.**

The **stars** that make up a galaxy are massive spheres of incandescent gases. They are usually intermingled with diffuse clouds of gas and debris. In spiral galaxies like the Milky Way, the stars are arrayed in spiral arms radiating from the galactic center. Other galaxies are elliptical or irregular in shape. Our part of the Milky Way is populated with many stars, but distances within a galaxy are so huge that the star nearest the sun is about 42 trillion kilometers (26 trillion miles) away. Astronomers tell us there are as many galaxies in the universe as there are stars in our own galaxy, and more stars in the Milky Way than grains of sand on a small beach!

Our sun is a typical star (**Figure 1.6**). The sun and its family of planets, called the **solar system,** is located about three-quarters of the way out from the galaxy's center, in a spiral arm. We orbit the galaxy's brilliant core, taking about 230 million years to make one orbit—even though we are moving at about 280 kilometers per second (half a million miles an hour). Earth has made about 20 circuits of the galaxy since the ocean formed.

The Lives of Stars

As we will see, most of the substance of Earth and its ocean was formed by stars.

Stars form in **nebulae,** clouds of dust and gas within galaxies. With the aid of telescopes and infrared-sensing

---

[4]Because they can be useful as well as interesting, the derivations of words are sometimes included.

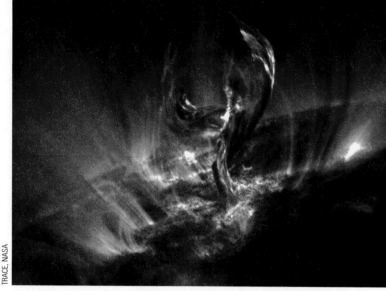

TRACE, NASA

ESO

Figure 1.5 This spiral galaxy in the constellation Coma Berenices is very similar in size and structure to our own Milky Way galaxy. The galaxy, photographed in 1999 by the Hubble Space Telescope, is 62 million light-years away and about 56,000 light-years in diameter. The few stars in the foreground are in our galaxy. Note the clouds of dust and gas that obscure our view of some parts of the spiral arms. Planets and oceans are made of such material.

Figure 1.6 A filament of hot gas erupts from the face of our sun. Like all normal stars, our sun is powered by nuclear fusion—the welding together of small atoms to make larger ones. These violent reactions generate the heat, light, matter, and radiation that pour from stars into space. The entire Earth could easily fit into this filament's outstretched arms.

satellites, astronomers have observed such clouds in our own and other galaxies. They have seen stars in different stages of development and have inferred a sequence in which these stages occur. The **condensation theory,** a theory based on this inference, explains how stars and planets are believed to form.

The life of a star begins when a diffuse area of a nebula begins to shrink under the influence of its own weak gravity. Gradually, the cloudlike sphere condenses into a knot of gases called a **protostar** (*protos* = first). The original diameter of the protostar may be many times the diameter of our solar system, but gravitational energy causes it to contract. As the

protostar shrinks, the compression raises its internal temperature. When the protostar reaches a temperature of about 10 million degrees Celsius (18 million degrees Fahrenheit), nuclear fusion begins; that is, hydrogen atoms begin to fuse together to form helium, a process that liberates even more energy. This rapid release of energy, which marks the transition from protostar to star, stops the young star's shrinkage. (The process is shown in **Figure 1.7.**)

After fusion reactions begin, the star becomes stable—neither shrinking nor expanding, and burning its hydrogen fuel at a steady rate. This stable phase does not last forever. After a long and productive life, the star has converted a large percentage of its hydrogen to atoms as heavy as carbon or oxygen. When a medium-mass star (like our sun) begins to consume these heavier atoms, its energy output slowly rises and its body swells to a stage aptly named *red giant* by as-

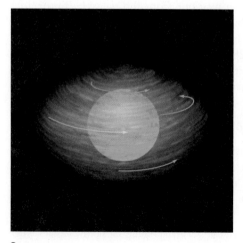

a

b

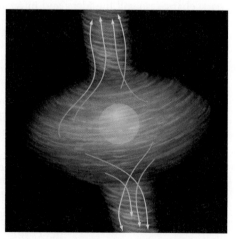

c

tronomers. The dying giant slowly pulsates, throwing off concentric shells of light gas enriched with these heavy elements. But most of the harvest of carbon and oxygen is forever trapped in the cooling ember at the star's heart.

The life history of a star depends on its initial mass. Much more massive stars have shorter but more interesting lives. They, too, fuse hydrogen to form atoms as heavy as carbon and oxygen, but being larger and hotter their internal nuclear reactions consume hydrogen at a much faster rate. Higher core temperatures permit the formation of atoms up to the mass of iron.

The dying phase of a massive star's life begins when its depleted core collapses in on itself. This rapid compression causes the star's internal temperature to soar. When the infalling material can no longer be compressed, the energy of the inward fall is converted to a cataclysmic expansion called a **supernova.** The explosive release of energy in a supernova is so sudden that the star is blown to bits and its shattered mass is accelerated outward at nearly the speed of light. The explosion lasts only about 30 seconds, but in that short time the nuclear forces holding apart individual atomic nuclei are overcome—and atoms heavier than iron are formed. The gold of your rings, the mercury in a thermometer, and the uranium in nuclear power plants were all created during such a brief and stupendous flash. The atoms produced by a star through millions of years of orderly fusion, *and* the heavy atoms generated in a few moments of unimaginable chaos, are sprayed into space (**Figure 1.8**). Every chemical element heavier than hydrogen—most of the atoms that make up the planets, the ocean, and living creatures—was manufactured by the stars.

The Formation of the Solar System

Earth and its ocean are the indirect result of a supernova explosion. The thin cloud, or **solar nebula,** from which our sun and its planets formed was probably struck by the shock wave and some of the matter of an expanding supernova remnant. Indeed, the turbulence of the encounter may have caused the condensation of our solar system to begin. The solar nebula was affected in at least two important ways: First, the shock wave caused the condensing mass to spin; second, the nebula absorbed some of the heavy atoms from the passing supernova remnant. In other words, a massive star had to live its life (constructing elements in the process) and then undergo explosive disintegration in order to seed heavy elements back into the nebular nursery of dust and gas from which our solar system arose. The planets are made mostly of matter assembled in a star (or stars) that disappeared billions of years ago. We are made of that stardust. Our bones and brains are composed of ancient atoms constructed by stellar fusion long before the solar system existed.

By about 5 billion years ago, the solar nebula was a rotating, disk-shaped mass of about 75% hydrogen, 23% helium, and 2% other material (including heavier elements, gases, dust, and ice). Like a spinning skater bringing in her arms, the nebula spun faster as it condensed. Material concentrated near its center became the protosun. Much of the outer material eventually became **planets,** the smaller bodies that orbit a star and do not shine by their own light (see again Figure 1.7).

The new planets formed in the disk of dust and debris surrounding the young sun through a process known as **accretion**—the clumping of small particles into large masses (**Figure 1.9**). Bigger clumps with stronger gravity pulled in most of the condensing matter. Near the protosun, where temperatures were highest, the first materials to solidify were substances with high boiling points, mainly metals and certain rocky minerals. The planet Mercury, closest to the sun, is mostly iron because iron is a solid at high temperatures. Somewhat farther out, in the cooler regions, magnesium, silicon, water, and oxygen condensed. Methane and ammonia accumulated in the frigid outer zones. Earth's array of water, silicon–oxygen compounds, and metals results from its middle position within that accreting cloud. The planets of the outer solar system—Jupiter, Saturn, Uranus, and Neptune—are composed mostly of methane and ammonia ices because those gases can congeal only at cold temperatures.

d

e

Figure 1.7 The origin of a star in the spiral arm of a galaxy. As a cloud of gas and dust contracts under the influence of gravity (**a**), the spinning mass forms a disk with a warm, thick center, a protostar (**b**). The center contracts under its own gravity and continues to heat up while the gas and dust in the surrounding disk continue to fall in. As the protostar begins to radiate more heat, it ejects some matter outward from its poles (**c**). When fusion begins in the star's core (**d**), the shock waves will disperse much of the remaining matter that has not coalesced into planets, moons, or comets around the periphery. Our solar system (**e**) was formed in this way about 5 billion years ago. (Source: *The Universe Explained*, Colin A. Ronan, ed. New York: H. Holt. © 1994.)

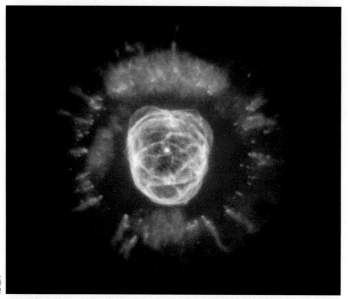

Figure 1.8 The end of a solar system? The glowing gas in this beautiful nebula once formed the outer layers of a sunlike star that exploded only 10,000 years ago. The inner loops are being ejected by a strong wind of particles from the remnant central star. If planets orbited this star, their shattered remnants are contained in the outward-rushing orange filaments at the periphery. Perhaps 8 billion years from now, observers 5,000 light-years away would see a similar sight as our sun reaches the end of its life.

a

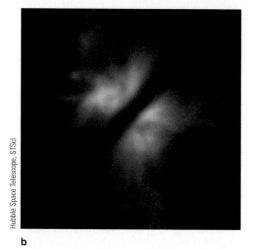

b

Figure 1.9 (**a**) In this artist's conception, our sun begins to shine at the center of a spinning disk of dust, ice, gases, and rock. The accretion phase is in progress—the bits will condense into planets and other bodies. (**b**) A solar nebula about the same size as the one that formed our solar system is seen in this 1998 image by the Hubble Space Telescope. The new star is hidden from direct view by a thick, dark disk crossing the center of the image. (The same sort of disk is represented in Figures 1.7d and 1.9a.) The disk—seen here edge-on—is about 130 billion kilometers (80 billion miles) in diameter—about 15 times the diameter of Neptune's orbit around our sun. Dark clouds and bright wisps above and below the disk suggest it is still growing from infalling dust and gas.

The period of accretion lasted perhaps 50 to 70 million years. The protosun became a star when its internal temperature became high enough to fuse atoms of hydrogen into helium. The violence of these nuclear reactions sent radiation sweeping past the inner planets, clearing the area of excess particles and ending the period of rapid accretion. Gases like those we now see on the giant outer planets may once have surrounded the inner planets, but this rush of solar energy and particles stripped them away.

This process might not be rare. As you'll learn in **Box 1.2,** 67 planets have been discovered orbiting other stars as of April, 2001. One of them is shown in **Figure 1.10.**

EARTH AND OCEAN

The young Earth, formed by the accretion of particles, was probably homogeneous throughout. During the accretion phase, Earth's surface was heated by the impact of meteors and more infalling debris. This heat, combined with gravitational compression and heat from the decay of radioactive elements accumulating within the newly assembled planet, caused Earth to partially melt. Gravity pulled most of the iron inward to form the planet's core. The sinking iron released huge amounts of gravitational energy, which, through friction, heated Earth even more. At the same time, a slush of lighter minerals—silicon, magnesium, aluminum, and oxygen-bonded compounds—migrated toward the surface,

a

Hubble Space Telescope, STScI

Figure 1.10 The first image of a possible planet outside our solar system. The bright object lies at the end of a filament of dust and gas extending from a newly formed sun.

forming Earth's crust (**Figure 1.11**). This important process, called **density stratification,** lasted perhaps 100 million years.

Then Earth began to cool. Its first surface is thought to have formed about 4.6 billion years ago. That surface did not remain undisturbed for long. A planetary body somewhat larger than Mars smashed into the young Earth (**Figure 1.12**). The rocky mantle of the impactor was ejected to form a ring of debris around Earth, and its metallic core fell into Earth's core and joined with it. The newly formed moon, still glowing from heat generated by the kinetic energy of infalling objects, is depicted in **Figure 1.13.** Could a similar cataclysm happen today? The issue is addressed in Box 13.1, page 327.

Radiation from the energetic young sun then stripped away our planet's outermost layer of gases, its first atmosphere; soon, however, gases that had been trapped inside the forming planet burped to the surface to form a second atmosphere. This volcanic venting of volatile substances—including water vapor—is called **outgassing** (**Figure 1.14a**). As the hot vapors rose, they condensed into clouds in the cool upper atmosphere. Recent research suggests that millions of tiny, icy comets colliding with Earth may also have contributed to the accumulating mass of water vapor, this ocean-to-be (**Figure 1.14b**).

Earth's surface was so hot that no water could settle there and no sunlight could penetrate the thick clouds. A visitor approaching from space 4.5 billion years ago would have seen a vapor-shrouded sphere blanketed by lightning-stroked clouds. After millions of years the upper clouds cooled enough for some of the outgassed water to form droplets. Hot rains fell toward Earth, only to boil back into the clouds again.

b

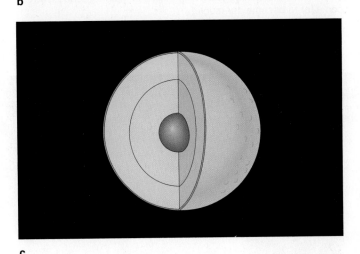

c

Figure 1.11 A representation of the formation of Earth. (**a**) The planet grew by the aggregation of particles. Meteors and asteroids bombarded the surface, heating the new planet and adding to its growing mass. At the time, Earth was composed of a homogeneous mixture of materials. (**b**) Earth lost volume because of gravitational compression. High temperatures in the interior turned the inner Earth into a semi-solid mass; dense iron (red drops) fell toward the center to form the core, while less dense silicates moved outward. Friction generated by this movement heated the Earth even more. (**c**) The result of *density stratification* is evident in the formation of an inner and outer core, a mantle, and the crust.

Figure 1.12 An early stage in the formation of the moon. A planetary body somewhat larger than Mars smashed into the young Earth about 4.4 billion years ago. The rocky mantle of the impactor was ejected to form a ring of debris around the Earth, and its metallic core fell into Earth's core and joined with it. Rocks brought from the lunar surface by Apollo astronauts suggest that the ejected material condensed soon after to become our moon.

As the surface became cooler, water collected in basins and began to dissolve minerals from the rocks. Some of the water evaporated, cooled, and fell again. The world ocean was gradually accumulating.

These heavy rains may have lasted about 25 million years. Large amounts of water vapor and other gases continued to escape through volcanic vents during that time and for millions of years thereafter. The ocean grew deeper. Evidence suggests that Earth's crust grew thicker as well, perhaps in part from chemical reaction with oceanic compounds.

The physical expanse and distribution of the early ocean is a matter of some controversy. Most researchers hold that masses of rock have always protruded through the ocean surface to form continents. However, some recent studies suggest that water may have covered Earth's entire surface for some 200 million years before the continents emerged. Although most of the ocean was in place about 4 billion years ago, ocean formation continues very slowly even today: About 0.1 cubic kilometer (0.025 cubic mile) of new water is added to the ocean each year, mostly as steam flowing from volcanic vents and in the form of microscopic cometary fragments.

The composition of the early atmosphere (often called the *primitive atmosphere*) was much different from today's (**Figure 1.15**). Geochemists believe it may have been rich in carbon dioxide, nitrogen, and water vapor, with traces of ammonia and methane. Beginning about 3.5 billion years ago, this mixture began a gradual alteration to its present composition, mostly nitrogen and oxygen. At first this change was brought about by carbon dioxide dissolving in seawater to form carbonic acid, then combining with crustal rocks. The chemical breakup of water vapor by sunlight high in the at-

Figure 1.13 Within a thousand years of the giant impact, our moon (foreground) was forming. In this painting the sky is dominated by a red-hot Earth, recently reshaped and melted by the moon-forming impact. The ring of debris will eventually fall to Earth or be captured by the still-growing molten moon.

mosphere also played a role. Then, about 1.5 billion years later, the ancestors of today's green plants produced—by photosynthesis—enough oxygen to oxidize minerals dissolved in the ocean and surface sediments. Oxygen began to accumulate in the atmosphere. (This monumental event in Earth's history is called the *oxygen revolution.* You'll find more about it in Chapter 15.)

THE ORIGIN OF LIFE

Life, at least as we know it, would be inconceivable without large quantities of water. Water has the ability to retain heat, moderate temperature, dissolve many chemicals, and suspend nutrients and wastes. These characteristics make it a mobile stage for the intricate biochemical reactions that allowed life to begin and prosper on Earth.

As early as 1929, biologist J.B.S. Haldane proposed that exposing a primitive atmosphere to ultraviolet radiation or lightning might produce some of the same carbon compounds found in living things. Building on this idea in a classic 1953 experiment, Stanley Miller and Harold Urey

Figure 1.14 Sources of the ocean. (**a**) Outgassing. Volcanic gases emitted by fissures add water vapor, carbon dioxide, nitrogen, and other gases to the atmosphere. Volcanism was a major factor in altering Earth's original atmosphere; the action of photosynthetic plants was another. (**b**) Comets may have delivered some of Earth's surface water. Intense bombardment of the early Earth by large bodies—comets and asteroids—probably lasted until about 3.8 billion years ago.

© William K. Hartmann

a

b

© Don Dixon

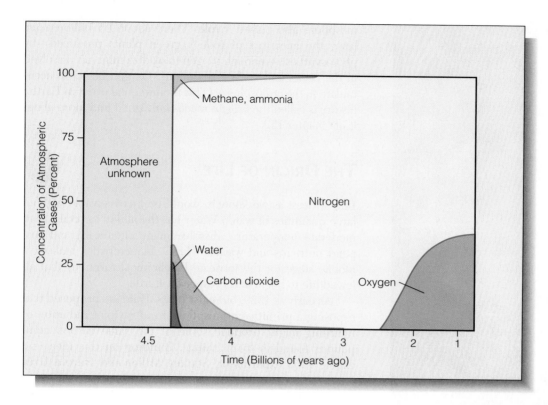

Figure 1.15 The evolution of our atmosphere. The early atmosphere had high concentrations of water and carbon dioxide with traces of methane and ammonia. Much of the carbon dioxide dissolved in seawater to form carbonic acid, then combined with crustal rocks. Nitrogen became the dominant gas. After the emergence of photo-synthetic organisms, oxygen began to accumulate in the atmosphere. (Source: C.J. Allègre and S. H. Schneider. "The Evolution of the Earth," *Scientific American,* October 1994. Reprinted by permission of Ian Worpole.)

mixed together water vapor, ammonia, methane, and hydrogen—gases that researchers then thought were abundant in the early atmosphere of Earth—and passed an electric spark through them for a week. In that short time, Miller's apparatus produced a number of different amino acids and other organic compounds (**Figure 1.16**). The electric discharge had provided energy for the formation of these simple molecules.

Recent analysis casts doubt on this hypothesis. The violent collisions now believed to have occurred early in Earth's formation would have prevented the formation of large quantities of methane and ammonia. Where did the candidate carbon compounds come from? There is growing consensus that these organic materials were transported to Earth by the comets, asteroids, meteors, and interplanetary dust particles that crashed into our planet during its birth.

Since Miller's and Urey's first experiments, other mixtures thought to reflect more accurately the early atmosphere of Earth have been tested, with similar results. When exposed to ultraviolet light, heat, and an electrical spark, these mixtures produce simple sugars and most of the biologically important amino acids. They even produce small proteins and nucleotides (components of the molecules that transmit genetic information between generations). The main chemical requirement seems to be the absence (or near absence) of free oxygen, a compound that can disrupt any unprotected large molecule.

Did *life* form in these experiments? No, the compounds that formed are only building blocks of life. But the experiments do tell us something about the commonality and unity of life on Earth. The facts that the crucial compounds can be synthesized so easily and are present in virtually all living forms are probably not coincidental. Those compounds are "permitted" by physical laws and by the chemical composition of this planet. The experiments also underscore the special role of water in life processes. The fact that all life, from a jellyfish to a dusty desert weed, depends on saline water within its cells to dissolve and transport chemicals is certainly significant. It strongly suggests that simple, self-replicating—living—molecules arose somewhere in the early ocean.

The early steps in the evolution of living organisms from simple organic building blocks, a process known as **biosynthesis,** are still speculative (see **Figure 1.17**). One idea, popularized in the 1950s, suggests that life may have originated in shallow tidal pools at the ocean's edge. Evaporation of water from these pools would have concentrated the amino acid and nucleotide building blocks into a rich organic soup. Grains of sand or tiny bubbles would have provided handy surfaces on which larger chemical combinations could be assembled. Sunlight would have supplied the energy for these reactions. Accumulating in protected pools, reacting aggregates could have become progressively more complex, eventually evolving into biochemical systems capable of reproducing.

Unfortunately for this hypothesis, the sunlight needed to energize the reactions might not have been present; even if it had been, strong sunlight has harmful effects on unprotected large molecules. Planetary scientists now suggest that the sun was faint in its youth, putting out so little heat that the ocean may have been frozen to a depth of around 300 meters (1,000 feet). The ice would have formed a blanket that kept most of

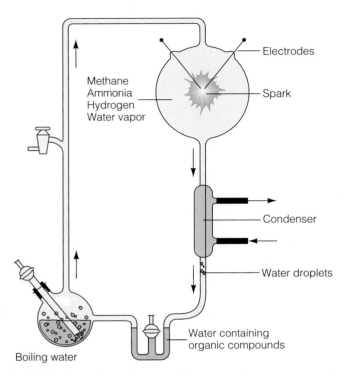

Methane
Ammonia
Hydrogen
Water vapor

Electrodes

Spark

Condenser

Water droplets

Water containing
organic compounds

Boiling water

Figure 1.16 An apparatus constructed by Stanley Miller in 1953 for producing organic molecules from a mixture of gases rich in methane and ammonia, substances thought at the time to have been abundant in Earth's primitive atmosphere. Biochemists have since learned that the early atmosphere was rich in carbon dioxide, nitrogen, water vapor, and hydrogen, with little methane and ammonia. But passing an electric spark through either mixture produces simple sugars and most of the biologically important amino acids. The substances that form in these experiments are not alive but are the building blocks of life.

the ocean fluid and relatively warm. Periodic fiery impacts by asteroids, comets, and meteor swarms could have thawed the ice, but between batterings it would have re-formed. In 1994, chemists Jeffrey Bada and Stanley Miller suggested that organic material may have formed and then been trapped beneath the ice—protected from the atmosphere, which contained chemical compounds capable of shattering the complex molecules. The first self-sustaining molecules might have arisen deep below the layers of surface ice, on clays or pyrite crystals near warm hydrothermal vents on the ocean floor. In 2000, organic chemist Günther Wächtershäuser suggested energy-releasing iron–sulfur chemical pathways that would have allowed early life to thrive deep in sediments or rocks.

There are other hypotheses, but most researchers agree that, whatever the details, some quiet corner of the ocean was the likely place where life began.

How long ago might life have begun? The oldest fossils yet found, from northwestern Australia, are between 3.4 and 3.5 billion years old (**Figure 1.18**). They are remnants of fairly complex bacteria-like organisms, indicating that life must have originated even earlier, probably only a few hundred million years after a stable ocean formed. Evidence of an even more ancient beginning has been found in the form of

Tom Garrison

a

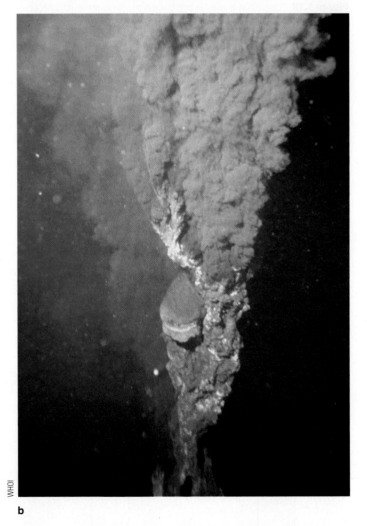

WHOI

b

Figure 1.17 Environments for biosynthesis? Some scientists believe that life could have originated in coastal tidal pools (**a**). Weak sunlight and unstable conditions at Earth's surface, however, may have favored the origin of life near deep-ocean hydrothermal vents (**b**).

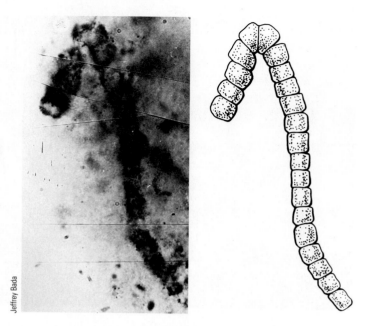

Jeffrey Bada

Figure 1.18 Fossil of a bacteria-like organism (with an artist's reconstruction) that photosynthesized and released oxygen into the atmosphere. Among the oldest fossils ever discovered, this microscopic filament from northwestern Australia is about 3.5 billion years old.

carbonaceous residues in some of the oldest rocks on Earth, from Akilia Island near Greenland. These 3.85-billion-year-old specks of carbon bear a chemical fingerprint that researchers feel could only have come from a living organism. Life and Earth have grown old together; each has greatly influenced the other.

THE DISTANT FUTURE OF EARTH?

Our descendants may enjoy another 5 billion years of Earth as we know it today. But then our sun, like any other star, will begin to die. The sun is not massive enough to become a supernova, but its red giant phase will engulf the inner planets perhaps 6 billion years from now. Its fiery atmosphere will expand to a radius greater than the orbit of Earth. The ocean and atmosphere, all evidence of life forms, the crust, and perhaps the whole planet will be recycled into component atoms and hurled by shock waves into space (as in Figure 1.8). Our successors, if any, will have perished or fled to safer worlds. Its fuel exhausted and its energies spent, the sun will cool to a glowing ember and ultimately to a dark cinder. Perhaps a new system of stars and planets will form from the debris of our remains.

The history of past and future Earth is shown as a graph in **Figure 1.19.**

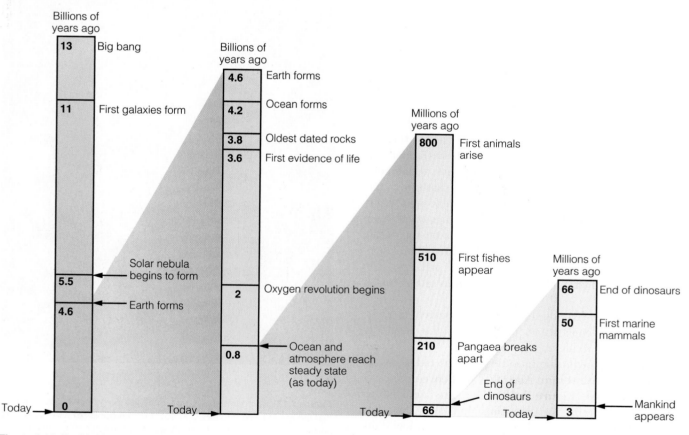

Figure 1.19 Earth's history—past and future. (Source: From Hartman & Miller, *The History of Earth*, pp. 22–23. New York: Workman Publishing. © 1991.)

Planets with liquid on or near their surfaces may not be as rare as once thought. Water itself is not scarce. In the solar system, for example, Jupiter has hundreds of times more water than Earth does, nearly all of it in the form of ice. In 1998, ice was discovered in deep craters near our moon's south pole. Astronomers have even located water molecules drifting in free space. But *liquid* water is unexpected. Although it might be a bit premature to consider "comparative oceanography" as a career choice, researchers are increasingly certain that liquid water exists (or existed quite recently) on at least two other bodies in our solar system. We can begin to compare and contrast them.

Europa and Ganymede

The spacecraft *Galileo* passed close to Europa—a moon of Jupiter—in early 1997. Photos sent to Earth revealed a cracked, icy crust covering what appears to be a slushy mix of ice and water (**Figures a–c**). The jigsaw-puzzle pattern of ice pieces appears to have formed when the ice crust cracked apart, moved slightly, and then froze together again. In another pass early in 2000, *Galileo* detected a distinctive magnetic field, the signature of a salty liquid-water ocean below the ice.

The volume of this ocean is astonishing. Though Europa is slightly smaller than our own moon, the depth of the ocean averages about 160 kilometers (100 miles) deep. The amount of water in its ocean is perhaps 40 times that in Earth's! Europa's ocean is probably kept liquid by heat escaping Europa's interior and by gravitational friction of tidal forces generated by Jupiter itself. Though the surface of the ice is about 8 kilometers (5 miles) thick and as cold as the surface of Jupiter, the liquid interior of the ocean, cradled deep in rocky basins, may be warm enough to sustain life. No continents emerge from this alien sea. A mission to the surface of Europa is being planned for launch in 2008.

Ganymede, another satellite of Jupiter, was surveyed by *Galileo* in May 2000. In January 2001, researchers released photographs that showed structures strikingly similar to those on Europa. Again, magnetometer data suggested a salty ocean beneath a moving, icy crust.

Mars

Europa may have an icy ocean now, but Mars, a much nearer neighbor, probably had an ocean in the distant past. An ocean probably occupied the low places of the northern hemisphere of Mars between 3.2 billion and 1.2 billion years ago when conditions were warmer (**Figure d**). In 1998, cameras aboard *Mars Global Surveyor* sent photos from the surface that may show evidence that water once flowed from

rock layers below the surface onto the bottom of a crater, leaving sediments that look suspiciously marine (see Figure 5.21). Where is the water now? Mars has become much colder in the past billion years, perhaps because of the loss of greenhouse gases in the atmosphere. If a large quantity of water is present, most of it probably lies beneath the surface in the form of permafrost.

Could wet pockets exist today? In June 2000 Michael Malin and Kenneth Edgett released photographs from *Mars Global Surveyor* that showed clear evidence of recently

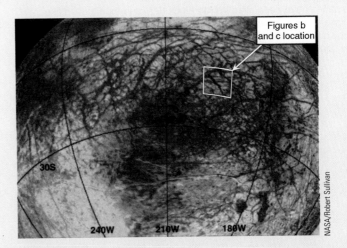

a A view of Jupiter's moon Europa showing location of detailed Figures **b** and **c**.

b Close-up view of some of Europa's ice fractures. The darker bands indicate recent breaks between the brighter platelike masses of ice.

c Photographic reconstruction of ice plates in Figure b. Fracture spaces have been filled in with black (left) and then moved like jigsaw-puzzle pieces to represent the original surface. Europa's icy crust is moving, with fresh ice filling the fractures. This suggests a liquid-water ocean beneath.

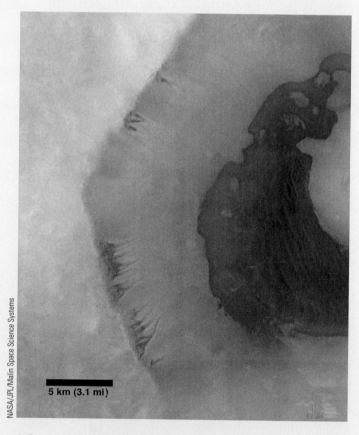

5 km (3.1 mi)

d Current models suggest that, early in its history, Mars had a thick, carbon dioxide–rich atmosphere, much like the atmosphere of the early Earth. Carbon dioxide is a "greenhouse" gas—it traps the sun's heat like the glass panels of a greenhouse. The atmosphere kept Mars warm and allowed water to flow freely, and an ocean may have existed on Mars about 3.2 billion years ago. Over the eons, rocks on the Martian surface absorbed the carbon dioxide, and the atmosphere grew thin and cold. The ocean disappeared, its water binding to rocks or freezing beneath the planet's surface.

flowing water (**Figures e** and **f**). Images of crater walls show channels that end in fan-shaped "aprons"—debris fields similar to the alluvial fans deposited by water flowing rapidly over arid landscapes on Earth. The surfaces of these aprons are not marked by craters or the dark dust that covers nearly all Martian features. These gullies appear young.

How might liquid water form channels on such a cold planet? Warmed by volcanic heat within Mars, water might seep from the cliffs, then freeze to form ice dams at the tops of these gullies. Stressed by growing pressure from the warm water rising behind it, the dams might give way in a gully-forming flood. Confirmation must await a visit by planetary oceanographers.

e Narrow gullies (small valleys) possibly carved by recently flowing water line the north wall of a crater on Mars. Dirt and rocks carried downhill form fingerlike deposits at the bottom of the crater.

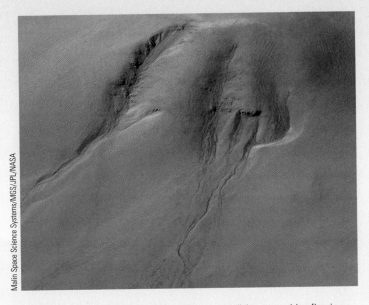

Malin Space Science Systems/MGS/JPL/NASA

f Young gullies on Mars. These valleys, possibly caused by flowing water, are about 3 kilometers (2 miles) wide and line a south-facing wall of a large crater. Their recent formation is indicated by the lack of surface craters.

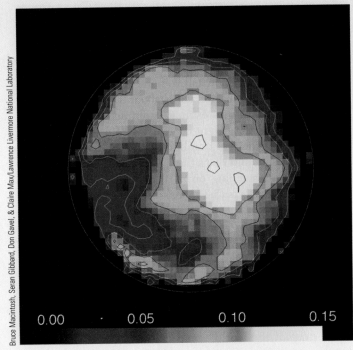

Bruce Macintosh, Seran Gibbard, Don Gavel, & Claire Max/Lawrence Livermore National Laboratory

| 0.00 | 0.05 | 0.10 | 0.15 |

g In 1999 astronomers were able to peer through the haze on Saturn's moon Titan and glimpse an ocean of liquefied natural gas. The surface temperature is a chilly −180°C. (Source: Gibbard, et al., *ICARUS*, Lawrence National Lab, CA)

Titan

Must an ocean consist of liquid water? Hydrocarbons have been seen on the surface of Titan, Saturn's largest moon. In 1999 scientists using the huge W. M. Keck telescope in Hawaii detected cold, dark infrared-absorbing organic matter surrounding a bright area about the size of Australia (**Figure g**). The evidence suggests a cold liquid ocean of methane, ethane, and other hydrocarbons. The "continent" is probably made of water ice. More information may be obtained in 2004 by the *Cassini* spacecraft when it drops a probe into Titan's surface.

Extrasolar Planets

Sixty-seven planets had been discovered outside our solar system as of April, 2001. Most of these planets were found by watching the wobbling path a star takes through space when influenced by the gravity of a massive companion planet. In June 1996 astronomers at the University of Arizona announced the discovery of water vapor, methane, and ammonia in the atmosphere of one of the largest of these planets. The high temperature of its surface (as high as 700°C, or 1,300°F) would prevent the formation of a liquid-water ocean, but smaller and cooler planets with similar atmospheres have been found around two stars nearer to the sun. Researchers at San Francisco State University have estimated surface temperatures on these planets at 85°C (185°F) and 44°C (112°F). These planets could have water in both vapor and liquid form—rain and oceans are possible.

Life and Oceans?

Could the presence of life be a clue to the existence—or past existence—of an ocean on a planet or moon? Since CO_2 is the "normal" composition for the atmosphere of a terrestrial planet, a large quantity of atmospheric oxygen would be unexpected—after all, oxygen is one of the most reactive of gases. Any oxygen in an atmosphere is likely to react with other materials. The red, rust-colored rocks typical of Mars almost certainly resulted from the oxidation (rusting) of iron-containing minerals. If a planet is found with lots of free oxygen, something is probably replenishing that oxygen.

That "something" is probably life. The action of photosynthetic organisms (including plants) produces excess oxygen. Without photosynthesis, Earth's atmosphere would be all but oxygen-free. As noted in the text of this chapter, life, at least on Earth, probably originated in the ocean. If the atmosphere of distant planets contains significant quantities of oxygen, an ocean and life might be possible. Scientists are close to technologies that will allow them to detect chemical signatures in the atmospheres of planets orbiting other stars.

1-11

1. How far away is the horizon when I stand on the beach and look out to sea?

That depends on how tall you are. If you're about 6 feet tall, the distance to the horizon is about 3 miles (4.8 kilometers). For a more precise estimate, subtract 4 inches (a third of a foot) to find the height of your eyes (in feet) above the ground, then divide that number by 0.5736. Take the square root of the result, and that's the number of miles to the horizon. Lifeguards standing on a 10-foot tower see a horizon roughly 8.5 kilometers (5.3 miles) distant.

2. Why do we call this planet Earth?

We are terrestrial organisms, so the name is understandable. The modern English word *Earth* is derived from the Anglo-Saxon word *eorthe,* which itself comes from the ancient German word for Earth, *Erde.* The earliest German tribes were landlocked people who had no knowledge of the true extent of the world ocean. We seem to be stuck with Earth.

3. Where did the word *ocean* come from?

Ocean derives from the Greek word *okeanos* (oceanus), a word meaning "outer sea" (in contrast to the Greeks' "inner sea," the Mediterranean). The term connotes a large moving river. Okeanos was also a mythical Greek titan who was god of the sea before Poseidon and father of the Oceanids, or ocean nymphs. The later Latin name for the ocean was *oceanus.* The Latin word evolved into the Middle English term *occean,* in which the double *c* was pronounced as a *k.* This later became the English word *ocean* (which, until the twentieth century, was pronounced with a soft *c*, as in the word *celery*). Perhaps the best name for Earth would be Oceanus, but so far I've not had much luck in changing the name!

4. You wrote that "Nothing is ever proven absolutely true by the scientific method." What good is it then—can't we depend on the process of science?

One philosopher of science has described truth as a liquid—it flows around ideas and is hard to grasp. The progressive improvement in our understanding of nature is subject to the limitations inherent in our observations. As our observations become more accurate, so do our conclusions about the natural world. But because observation (and interpretation of observation) is never perfect, truth can never be absolute. In the 1920s, for example, astronomers assumed that the universe was limited to our own Milky Way galaxy.

---

[5]Each chapter ends with a few questions students have asked me after a lecture or reading assignment. These questions and their answers may be interesting to you, too.

Observations made with a large new telescope on Mt. Wilson in California by Harlow Shapley and Edwin Hubble allowed them to measure more distant objects. Galaxies were discovered in profusion, "like grains of sand on a beach" in Shapley's words. Thus truth changed its shape. In a more directly marine example, one of the *Challenger* expedition's responsibilities was investigating Edinburgh professor Edward Forbes's contention that life below about 550 meters (1,800 feet) was impossible. Consistent sampling below that depth was not practical until the advent of *Challenger*'s steam winch. Sure enough, when observations were made, life was there, and truth changed. We learn as we go. We depend on the underlying assumption that nature "plays fair"—that is, is consistent and does not capriciously change the rules as our powers of observation grow. What we have learned so far is of inestimable practical and aesthetic value, and we have only scratched the surface.

5. How do scientists know how old Earth, the solar system, and the galaxy are? How can they calculate the age of the universe?

The age estimates presented in this chapter are derived from interlocking data obtained by many researchers using different sources. One source is meteorites, chunks of rock and metal formed at about the same time as the sun and planets and out of the same cloud. Some meteorites contain gases thought to be remnants of the original solar nebula. Many have fallen to Earth in comparatively recent times. We know from signs of radiation within these objects how long it has been since they were formed. That information, combined with the rate of radioactive decay of unstable atoms in meteorites, moon rocks, and the oldest rocks on Earth, allows astronomers to make a reasonably accurate estimate of how long ago these objects formed.

The light from stars and planets can be analyzed to discover their composition. A spectroscope breaks starlight into component colors by passing it through a device that works like a prism. Careful study of the character and distribution of this light yields information on the star's temperature and composition. The elements present in stars and the distribution of types of stars in the sky suggest that most stars go through a predictable life cycle based on their beginning mass.

As for the age of the universe itself, by June 1999 astronomers had obtained very accurate measurements of its rate of expansion. By calculating backward, they found the universe would have begun its expansion between 12 and 13.4 billion years ago.

6. Life appears to have arisen on Earth soon after the formation of a stable surface. Could life have formed on other planets?

We have no evidence, direct or indirect, of life on other planets around our sun or elsewhere in the universe. Yet it seems provincial to assume that life could have arisen only here. The formation of organic molecules from simple chemicals receiving energy from lightning, heat, ultraviolet light, and other sources may be quite common, and increasing complexity in these compounds may be a universal phenomenon.

7. Would life on other planets resemble life on Earth?

Organisms elsewhere might be very different. Recall that life on this planet probably arose in the ocean, and all life forms here carry an ocean of sorts within their bodies. On a planet without water, the organisms would be much different.

For example, on a hypothetical planet with an ammonia ocean, life would not have a structure of cells surrounded by lipid membranes. Lipid membranes are the sheets of fatty molecules that keep the inside of a cell separate from the environment, and ammonia prevents them from forming. Without membranes, cells as we know them are not possible. Notwithstanding this argument, life need not be confined to planets with water. Other life forms may exist, based on other "brews."

8. You mentioned two hypotheses for the origin of life. Are there others?

Yes, indeed. Divine creation is one, but it is permanently impossible to test that idea using the scientific method. Another is *panspermia*, which proposes that life formed (or was created) in space and was then distributed throughout the galaxy. Still another involves intelligences "seeding" planets with DNA (or molecules of similar function), later to return and see what useful creatures have resulted. Neither of these ideas can be tested, at least for now. Moreover, both in a sense beg the question; that is, they assume that some extraterrestrial life already existed, but they don't deal with how that life may have arisen.

In 1996 Stanford chemistry professor Richard N. Zare proposed another interesting hypothesis for the origin of life on Earth. Violent collisions have blasted chunks of the surface of Mars to Earth. These meteorites are linked to Mars by traces of the Martian atmosphere they contain. There may be some interchange of material—perhaps living material—between planets. "Who's to say that we are not all Martians, that Mars was the place where life on Earth first started?"

9. What happened to Joshua Slocum after his single-handed circumnavigation?

He continued to work periodically as a master mariner, but most of his time was spent in writing, lecturing, exhibiting at expositions and fairs, selling curios, farming (a vocation he did not enjoy), and reading. In 1905 he set sail, again in *Spray*, for the West Indies. After two more trips to the same

islands, he departed in 1909, at age 65, in his well-traveled boat to explore South America's Orinoco and Amazon Rivers. No trace of him or his boat was ever found.

KEY CONCEPTS TO REVIEW

- The world ocean may be considered as a single entity with temporary partitions. It covers about 71% of Earth's surface.

- Marine science—oceanography—is the process of discovering unifying principles in data obtained from the ocean.

- Science is the *process* of asking questions about the observable world and then testing the answers to those questions. The external world, not internal conviction, is the testing ground for scientific beliefs.

- The objects comprising our solar system condensed about 5 billion years ago from a thin cloud that had been enriched by heavy elements made in exploding stars.

- Earth is density-stratified. During its formation heavy materials fell inward to form the core, and lighter substances rose to become the outer layers. Ocean and atmosphere are the least dense of these layers.

- Earth first had a solid surface about 4.6 billion years ago. The ocean formed from clouds of steam and water vapor when Earth's surface became cool enough to allow liquid water to rest on the surface.

- Life probably arose on Earth shortly after its formation about 4 billion years ago. Life may have arisen in the deep ocean.

- Other planets and moons may have—or may have had—oceans.

INTERNET STUDY RESOURCES

The Web site for this book contains helpful study aids. Each vignette, box, and first- and second-level head has a number that corresponds to information at the Web site. Log on to:

www.brookscole.com/product/053437557xs

and click on the Chapter-by-Chapter area. Choose Chapter 1 and select a resource:

- **Flash Cards** allows you to test your mastery of the Terms and Concepts to Remember for this chapter.
- **Tutorial Quizzes** provides a multiple-choice practice quiz.

- **Student Guide to InfoTrac® College Edition** will lead you to Critical Thinking Projects that use InfoTrac College Edition as a research tool.

TERMS AND CONCEPTS TO REMEMBER

accretion	oceanography
big bang	outgassing
biosynthesis	planet
condensation theory	protostar
density stratification	science
experiment	scientific method
galaxy	solar nebula
hypothesis	solar system
law	star
marine science	supernova
Milky Way galaxy	theory
nebula	world ocean
ocean	

STUDY QUESTIONS

Review Questions

1. Which hemisphere contains the greatest percentage of ocean? Is most of Earth's water in the ocean?

2. Can the scientific method be applied to speculations about the natural world that are not subject to test or observation?

3. What element makes up most of the detectable mass in the universe?

4. Briefly trace the life of a typical star.

5. How are planets and stars related?

6. How are light elements converted into heavy ones?

7. Will all stars end their lives as supernovas? What happens to the heavy elements made by small stars?

8. What is density stratification?

9. Are the ocean and present atmosphere "leftovers" from the original atmosphere of Earth?

10. What is biosynthesis? Where do researchers think it might have occurred on our planet? Could it happen again today?

Critical Thinking Questions

1. Marine biologists sometimes say that all life forms on Earth, even desert lizards and alpine plants, are marine. Why?

2. Where did Earth's surface water come from?

3. Do you think water planets are common in the galaxy? What about planets in general?

4. How might Earth be different if an ocean had not formed on its surface?

5. How do we know what happened so long ago?

6. **InfoTrac College Edition Project** Life as we know it on Earth apparently requires water in some form. The planet Mars appears not to have liquid water on its surface, but evidence suggests that it may have had water at one time. Could Mars also have hosted Earth-like life? Might life still exist there? Do some research using InfoTrac College Edition to support your answer.

 Systems Today CD Question

Take a few minutes to install and become familiar with **Earth Systems Today,** the interactive CD-ROM that accompanies this text. At the end of most of the chapters in this book are questions based on this CD-ROM. Answering these questions using the CD-ROM will help you understand basic oceanographic principles.

1. Although there are no questions specific to Chapter 1 on the CD-ROM, go to "Geologic Time" in the CD-ROM's main menu, and click on "Changing Earth." Begin the animation. Notice that the span of time included in this animation is only 750 million years (0.75 billion years), but Earth is around 4,600 million (4.6 billion) years old. The slow movement of continents—about as fast as your fingernails grow—combined with the immensity of geological time, gives an appreciation of Earth's great age.

RESOURCES FOR FURTHER READING AND RESEARCH

The Web site for this book contains many ideas for further reading and research. Log on to:

www.brookscole.com/product/053437557xs

and click on the Chapter-by-Chapter area. Choose Chapter 1 and select a resource:

- **References** lists the major books and articles consulted in writing this chapter, along with comments from the author about their content and reading level.
- **Hypercontents** takes you to an extensive list of sites with news, research, and images related to individual sections of the chapter.

- In **Student Guide to InfoTrac College Edition,** scroll down to Suggested Readings from InfoTrac College Edition for brief descriptions of the articles listed and search hints for finding them in InfoTrac.
- **Regional InfoTrac College Edition** articles are organized into East Coast, West Coast, and Gulf Coast regions, allowing you to study oceanography on a more local level. You can also access Regional InfoTrac College Edition at www.localocean.com.

 For additional readings, go to InfoTrac College Edition, your online research library, at:

http://infotrac.thomsonlearning.com/

A History of Oceanography 2

MAKING MARINE HISTORY

History is made a day at a time. Sometimes those who make history realize the significance of a single day's effort, but more often the days blend into weeks and months of hard work punctuated only occasionally by scientific or artistic insights. Still, individual days are important, and none is more important to a discoverer than the day that *first* piqued his or her involvement in a field that becomes the subject of lifelong study. No matter what follows, that day, spent perhaps with a good teacher and good friends, will remain unique in memory.

Student oceanographers get cold, and wet, and often seasick when they make their first study cruises to sample the ocean they've been learning about

The *Alvin* team prepares to make history. *Alvin* is a small research submarine capable of transporting a pilot and two researchers to a depth of 4,000 meters (13,120 feet).

Dan Dion

in books and classes. They may take samples with sophisticated computerized probes or with buckets and thermometers lowered on a line. They may work from outboard-powered boats or well-equipped research vessels. They struggle with buckets of sediment, cut fragrant piles of frozen fish food for aquarium animals, stand watch in howling storms, steady their microscopes, learn arcane computer commands, memorize the interior of a shark, and are the last ones out of the library at night. They write, they talk, and they party. A few of them experience a flash of commitment and decide to continue. That is the moment they will remember most vividly—the day they start to make their own history.

With few exceptions, we have no knowledge of how the travelers and scientists described in this chapter became personally interested in advancing marine science, but we do know that, as before, some of today's student oceanographers will contribute to the oceanography texts of tomorrow. Progress in marine science depends on them, and perhaps on you.

WHAT TO WATCH FOR IN CHAPTER 2

The early history of marine science is closely associated with the history of voyaging. The first marine studies had practical aims: travel, trade, warfare, and facilitating the search for food. Later the search for new knowledge became a goal in itself. The first part of this chapter focuses on the mastery of marine knowledge for voyaging; it describes some of the voyagers and their journeys, the inventions that made their adventures possible, and some of the discoveries they made. The second part covers voyaging for science, the rise of oceanographic institutions, and the present thrust of ocean science.

As you read you'll notice a subtle progression from the involvement of whole human populations, to the contribution of extraordinary individuals, to epic scientific expeditions, to the rise of national and academic centers of learning. Oceanographers now spend more time ashore than at sea; some work exclusively with data sent by space-based sensors or robot sea probes. All would agree with Isaac Newton's observation that history is important: "If I have seen farther, it is because I have stood on the shoulders of giants."

Ours is a restless and inquisitive species. Despite the vast ocean, we have populated nearly every habitable place on Earth. This fact was aptly illustrated when European explorers set out to "discover" the world, only to be met by native peoples at almost every landfall! Clearly, the ocean did not prevent the spread of humanity.

Consider for a moment the difficulty people must have had in migrating from eastern Africa, the probable place of human origin, to the rest of the world. Generations of nomads spread slowly into the Near East and central Europe, and on across the plains of central Asia. They wandered north to the shores of the Bering Sea and, later, over a land or ice bridge to North America, reaching the southern tip of South America about 11,000 years ago. Though most traveled on foot, some people undoubtedly used rafts or boats. Later travelers even colonized remote Pacific islands.

Ocean transportation offers people the benefits of mobility and greater access to food supplies. Any coastal culture skilled at raft building or small-boat navigation had economic and nutritional advantages over its less adept competitors. From the earliest period of human history, then, understanding and appreciating the ocean and its life forms benefited those patient enough to learn.

VOYAGING BEGINS

The first direct evidence we have of **voyaging,** traveling on the ocean for a specific purpose, comes from records of trade in the Mediterranean Sea. The Egyptians organized shipborne commerce on the Nile River, but the first regular ocean traders were probably the Cretans—or the Phoenicians, who inherited maritime supremacy in the Mediterranean after the Cretan civilizations were destroyed by earthquakes and political instability around 1200 B.C. Skilled sailors, Phoenicians carried their wares through the Straits of Gibraltar to markets as distant as Britain and the west coast of Africa. Considering the simple ships they used, this was quite an achievement.

The Greeks began to explore outside the Mediterranean into the Atlantic Ocean around 900–700 B.C. (**Figure 2.1**). Early Greek seafarers noticed a current running from north to south beyond Gibraltar. Believing that only rivers had currents, they decided that this great mass of water, too wide to see across, was part of an immense flowing river. The Greek name for this river was *okeanos*. Our word *ocean* is derived from **oceanus,** a Latin variant of that root. Phoenician sailors were also very much at home in this "river," but like the Greeks they rarely ventured out of sight of land.

As they went about their business, early mariners began to record information to make their voyages easier and safer—the location of rocks in a harbor, landmarks and the sailing times between them, the direction of currents. These first **cartographers** (chart makers) were probably Mediterranean traders who made routine journeys from producing areas to markets. Their first charts (drawn about 800 B.C.) were merely notes to jog their memory for obvious features along the route. Today's **charts** are graphic representations

Figure 2.1 A Greek ship from about 500 B.C. Such ships were used for trade and to continue the exploration of the Atlantic, outside the Mediterranean.

that primarily depict water and water-related information. (*Maps* primarily represent land.) For more on maps and charts, please see Appendix IV.

In this early time other cultures also traveled on the ocean. The Chinese began to engineer an extensive system of inland waterways, some of which connected with the Pacific Ocean to make long-distance transport of goods more convenient. The Polynesian peoples had been moving easily among islands off the coasts of Southeast Asia and Indonesia since 3000 B.C. and were beginning to settle the mid-Pacific islands. Though none of these civilizations had contact with the others, each developed methods of charting and navigation. All these early travelers were skilled at telling direction by the stars and by the position of the rising or setting sun.

Curiosity and commerce encouraged adventurous people to undertake ever more ambitious voyages. But these voyages were possible only with the coordination of astronomical direction-finding (and knowledge of the shape and size of Earth), advanced shipbuilding technology, accurate graphic charts (not just written descriptions), and, perhaps most important, a growing understanding of the ocean itself. Marine science, the organized study of the ocean, began with the technical studies of voyagers.

SCIENCE FOR VOYAGING

Progress in applied marine science began at the **Library of Alexandria,** in Egypt. Founded in the third century B.C. by Alexander the Great, the library could be considered the first university in the world. Scholars worked and researched there, and students came from around the Mediterranean to study. Written knowledge of all kinds was warehoused around its leafy courtyards. This library constituted history's greatest accumulation of ancient writings. When any ship entered the harbor, the books (actually scrolls) it contained were by law removed and copied; the *copies* were returned to the owner and the originals kept for the library. Caravans arriving overland were also searched. The characteristics of nations, trade, natural wonders, artistic achievements, tourist sights, investment opportunities, and other items of interest to seafarers were cataloged and filed. Manuscripts describing the Mediterranean coast were of great interest. Traders quickly realized the competitive benefit of this information.

Yet marine science was only one of the library's many research areas. For 600 years, it was the greatest repository of wisdom of all kinds and the most influential institution of higher learning in the ancient world. Here, perhaps, was the first instance of cooperation between a university and the commercial community, a partnership that has paid dividends for both science and business ever since.

Alexandrian Contributions

The second librarian at Alexandria (235 B.C.–192 B.C.) was the Greek astronomer, philosopher, and poet **Eratosthenes of Cyrene.** This remarkable man was the first to calculate the circumference of Earth. The Greek Pythagoreans had realized Earth was spherical by the sixth century B.C., but Eratosthenes was the first to estimate its true size.

Eratosthenes had heard from travelers returning from Syene (now Aswan, site of the great Nile dam) that at noon on the longest day of the year the sun shone directly onto the waters of a deep vertical well. In Alexandria, he noticed that a vertical pole cast a slight shadow on that day. He measured the shadow angle and found it to be a bit more than 7°, about $\frac{1}{50}$ of a circle. He correctly assumed that the sun is a great distance from Earth; so the sun's rays would approach Syene and

Alexandria in essentially parallel lines. If the sun was directly overhead at Syene but not directly overhead at Alexandria, then the surface of Earth would have to be curved. But what was the *circumference* of Earth?

By studying the reports of camel caravan traders, he estimated the distance from Alexandria to Syene at about 785 kilometers (491 miles). Eratosthenes now had the two pieces of information needed to derive the circumference of the Earth by geometry. **Figure 2.2** shows his method. The precise size of the units of length (*stadia*) Eratosthenes used is thought to have been 555 meters (607 yards), and historians estimate that his calculation, made in about 230 B.C., was accurate to within about 8% of the true value. Within a few hundred years most people in the West who had contact with the library or its scholars knew Earth's approximate size.

Even without the contributions of Eratosthenes, the significance of the Library of Alexandria to marine science is immense. In Alexandria, traders, explorers, scholars, and students had a place to conduct research and exchange information (and rumors) about the seas. Library researchers invented the astronomical, geometric, and mathematical base for **celestial navigation,** the technique of finding one's position on Earth by reference to the apparent positions of heavenly bodies.

Cartography flourished. The first workable charts representing a spherical surface on a flat sheet were developed by Alexandrian scholars. Latitude and longitude, systems of imaginary lines dividing the surface of Earth, were invented by Eratosthenes. **Latitude** lines were drawn parallel to the equator, and **longitude** lines ran from pole to pole (see **Box 2.1**). Eratosthenes placed the lines through prominent landmarks and important places, creating a convenient though irregular grid. Our present regular grid of latitude and longitude was invented by Hipparchus (c. 165–c. 127 B.C.), a librarian who divided the surface of the Earth into 360 degrees. A later Egyptian–Greek, Claudius Ptolemy (A.D. 90–168), *oriented* charts by placing east to the right and north at the top. Ptolemy's division of degrees into minutes and seconds of arc is still used by navigators.

Ptolemy also introduced an "improvement" to Eratosthenes's surprisingly accurate estimate of Earth's circumference. Unfortunately, Ptolemy wrongly depended on flawed calculations of the effects of atmospheric refraction. He publicized a too low estimate of the size of Earth—about 70% of the true value. This error, coupled with his mistake of overestimating the size of Asia, greatly reduced the apparent width of the unknown part of the world between the Orient and Europe. More than 1,500 years later, these mistakes made it possible for Columbus to convince people he could reach Asia by sailing west.

Though it weathered the dissolution of Alexander's empire, the Library of Alexandria did not survive the subsequent period of Roman rule. The last librarian was Hypatia, the first notable woman mathematician, philosopher, and scientist. In Alexandria she was a symbol of science and knowledge, concepts the early Christians identified with pagan practices. The mission of the library, as personified by the last librarian, antagonized the governors and citizens of the city of Alexandria.

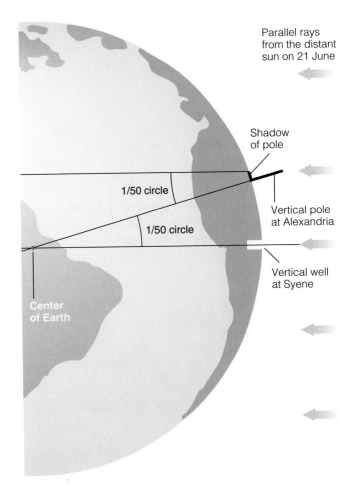

Figure 2.2 A diagram showing Eratosthenes's method for calculating the circumference of Earth. As is described in the text, he used simple geometric reasoning based on the assumptions that Earth is spherical and that the sun is very far away. Using this method, he was able to discover the circumference of Earth to within about 8% of its true value. (The diagram is not drawn to scale.)

After years of rising tensions, in A.D. 415 a mob brutally murdered her and burned the library with all its contents. Most of the community of scholars dispersed, and Alexandria ceased to be a center of learning in the ancient world. The academic loss was incalculable, and trade suffered because shipowners no longer had a clearinghouse for updating the nautical charts and information they had come to depend on. All that remains of the library today is a remnant of an underground storage room. We will never know the true extent and influence of its collection of over 700,000 irreplaceable scrolls.

Western intellectual development slackened during the so-called Dark Ages that followed the fall of the Roman Empire in A.D. 476. For almost 1,000 years, until the European Renaissance, much of the progress in medicine, astronomy, philosophy, mathematics, and other vital fields of human endeavor was made by the Arabs or imported by them from Asia. For example, the Arabs used the Chinese-invented compass for navigating caravans over seas of sand, and their understanding of the Indian Ocean's periodic winds—the monsoons—allowed an Arabian navigator to guide Vasco da

A sphere has no edges, no beginnings or ends, so what should we use as a frame of reference for positioning and navigation? The question was first successfully addressed by geographers at the Library of Alexandria, in Egypt. In the third century B.C., Eratosthenes drew latitude and longitude lines through important places (**Figure a**). The Alexandrian perception of the world is reflected in the size of the continents and the central position of Alexandria.

A later Alexandrian scholar divided Earth into an orderly grid based on 360 increments, or degrees (*degre* = step). The equator was a natural dividing point for the north–south (latitude) positioning grid, but there was no natural dividing point for the east–west (longitude) grid. Not surprisingly, Alexandria was arbitrarily selected as the first "zero longitude" and a regular grid laid out east and west of that city.

The general scheme has withstood the test of time, but there has been controversy. Though use of the equator as "zero latitude" has never been in question, each seafaring country wanted the prestige of having the world's longitude centered on its capital. For centuries, maritime nations issued charts with their own longitude "zeros." After much political disagreement, nations agreed in 1884 that the Greenwich meridian near London would be the world's "zero longitude" (**Figure b–d**). Given the accuracy of that meridian's known position and the long history and success of British navigation and timekeeping, Greenwich was an excellent choice.

To read more on latitude, longitude, time, and navigation, please turn to Appendix III.

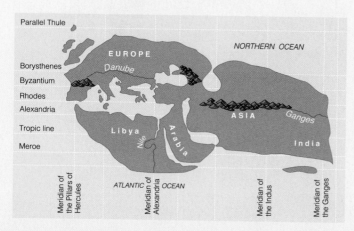

a The world, according to a chart from the third century B.C. Eratosthenes drew latitude and longitude lines through important places rather than spacing them at regular intervals as we do today. The Alexandrian perception of the world is reflected in the size of the continents and the central position of Alexandria.

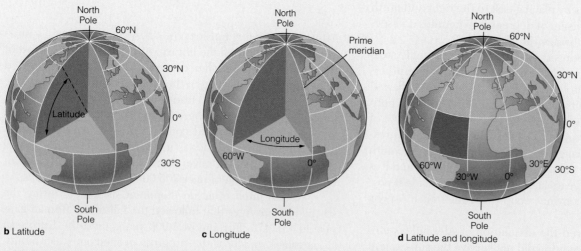

b Latitude   **c** Longitude   **d** Latitude and longitude

Latitude and longitude are measured as angles between lines drawn from the center of Earth to the surface. **b** Latitude is measured as the angle between a line from Earth's center to the equator and a line from Earth's center to the measurement point. **c** Longitude is measured as the angle between a line from Earth's center to the measurement point and a line from Earth's center to the prime (or Greenwich) meridan, which is a line drawn from the north pole to the south pole passing through Greenwich, England. **d** Lines of latitude are always the same distance apart while the distance between two lines of longitude varies with latitude.

Gama from East Africa to India in 1498. Earlier, at the height of the Dark Ages, Vikings raided and explored to the south and west. And half a world away the Polynesians continued some of the most extraordinary voyages in history.

Voyages of the Oceanian Peoples

In the history of human migration, no voyaging saga is more inspiring than that of the **Polynesian** colonizations, the peopling of the central and eastern Pacific islands. A profound knowledge of the sea was required for these voyages, and the story of the Polynesians is a high point in our chronology of marine science applied to travel by sea.

The Polynesians are one of four cultures inhabiting some 10,000 islands scattered across nearly 26 million square kilometers (10 million square miles) of open Pacific Ocean (**Figure 2.3**). The Southeast Asian or Indonesian ancestors of the Oceanian peoples, as these cultures are collectively called,

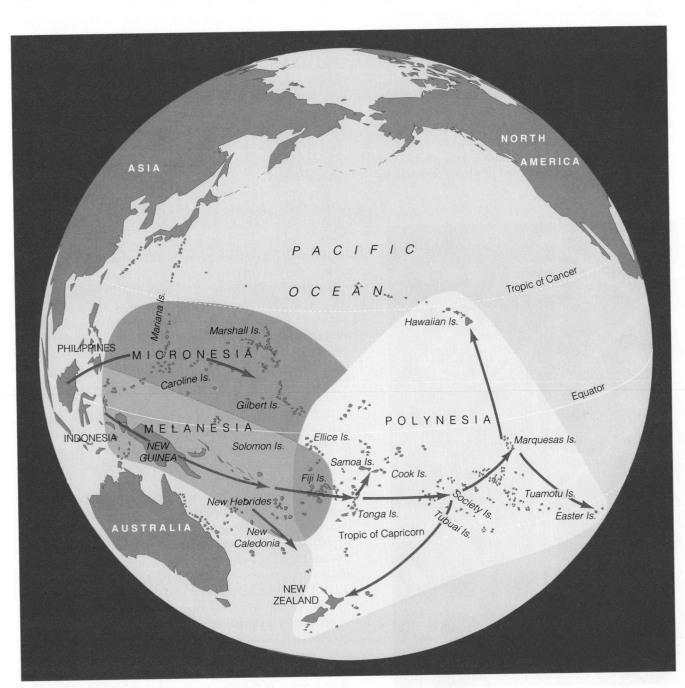

Figure 2.3 The Polynesian Triangle. Ancestors of the Polynesians had spread from Southeast Asia or Indonesia to New Guinea and the Philippines by about 20,000 years ago. The mid-Pacific islands have been colonized for about 2,500 years, but the explosive dispersion that led to the settlement of Hawaii occurred about A.D. 450–600. Arrows show a possible direction and order of settlement.

spread eastward in the distant past. Although experts vary in their estimates, there is some consensus that by 30,000 years ago New Guinea was populated by these wanderers and that by 20,000 years ago the Philippines were occupied. By around 500 B.C. the so-called cradle of Polynesia—Tonga, Samoa, the Marquesas, and the Society Islands—was settled. Oceanian navigators may already have been using shells attached to a bamboo grid to represent the positions of their islands. (A Micronesian stick chart from recent times is shown in **Figure 2.4.**)

For a long and evidently prosperous period the Polynesians spread from island to island until the easily accessible islands had been colonized. Eventually, however, overpopulation and depletion of resources became a problem. Politics, intertribal tensions, and religious strife shook society. Groups of people scattered in all directions from some of the "cradle" islands during a period of explosive dispersion. Between A.D. 300 and 600, Polynesians successfully colonized nearly every habitable island within the vast triangular area shown in Figure 2.3. Easter Island was found against prevailing winds and currents, and the remote islands of Hawaii were discovered and occupied. These were among the last places on Earth to be populated.

Figure 2.4 A modern Micronesian stick chart. Knots or shells tied at the junctions between bamboo sticks represent islands. Straight strips represent patterns of regular waves; bent strips depict waves curving around islands.

How did these risky voyages into unexplored territory come about? Religious warfare may have been the strongest stimulus to colonization. If the losers of a religious war were banished from the home islands under penalty of death, their only hope for survival was to reach a distant and hospitable new land.

Seafaring had long been a tradition in the home islands, but such trips called for radical new technology. Great dual-hulled sailing ships, some capable of transporting up to 100 people, were designed and built. New navigation techniques were perfected that depended on the positions of stars barely visible to the north. New ways of storing food, water, and seeds were devised. Whole populations left their home islands in fleets designed specially for long-distance discovery (**Figure 2.5**). In some cases, fire was nurtured on board in case of landfall on an island that lacked volcanic flame. But a new island was only a possibility, a dream. Their gods may have promised the voyagers safe deliverance to new lands, but how many fleets set out from the troubled homelands only to fall victim to storms, thirst, or other dangers?

In that anxious time the Polynesians practiced and perfected their seafaring knowledge. To a skilled navigator, a change in the rhythmic set of waves against the hull could indicate an island out of sight over the horizon. The flight tracks of birds at dusk could suggest the direction of land. The positions of the stars told stories, as did the distant clouds over an unseen island. The smell of the water, or its temperature, or salinity, or color, conveyed information—as did the direction of the wind relative to the sun, and the type of marine life clustering near the boat. The sunrise colors, the sunset colors, the hue of the moon—every nuance had meaning, every detail had been passed in ritual from father to son. The greatest Polynesian minds were navigators, and reaching Hawaii was their greatest achievement.

Of all the islands colonized by the Polynesians, Hawaii is farthest away, across an ocean whose guide stars were completely unknown to the southern navigators. The Hawaiian Islands are isolated in the northern Pacific. There are no islands of any significance for more than 2,000 miles to the south. Moreover, Hawaii lies beyond the equatorial doldrums, a hot and often windless stretch across which these pioneers must somehow have paddled. And yet some fortunate and knowledgeable people colonized Hawaii sometime between A.D. 450 and 600. Try to imagine their feelings of relief and justification upon reaching a promised paradise under a new night sky. Think of that first approach to the high islands of Hawaii, the first unlimited drink of fresh water, the first solid earth after months of uncertainty!

Within a hundred years of their first arrival, Hawaiian navigators were routinely piloting vessels on regular return trips to the Marquesas and the Society Islands (Tahiti and others). Some of the trips were undertaken to import needed food species to the newly found islands, but others were made to recruit new citizens and leaders to "green-clad Hawaii."

At a time when seafarers of other civilizations sailed beside the comforting bulk of a charted coast, Polynesians

Figure 2.5 A Polynesian voyaging canoe similar to the ones used for long-range colonization and the discovery of Hawaii. Note the double-hulled design and the small deckhouse for shelter. The crew is seen paddling through the doldrums, a relatively windless area near the equator. Water would have been in short supply, the voyage tense and tiresome.

looked to the open sea for sustenance, deliverance, and hope. Their great knowledge of the ocean protected them.

Meanwhile, Back in Europe . . .

The Dark Ages were periodically punctuated by the raids of **Vikings,** bands of Scandinavian adventurers and treasure seekers whose remarkably fast, strong, and stable ships (**Figure 2.6**) enabled them to row up rivers faster than a horse and rider could spread warning. Danish and Norwegian Vikings swept down the coast of Europe; they methodically pillaged Paris, robbed monasteries in Ireland, and looted Britain. The Swedish Vikings foraged as far away as Kiev and Constantinople! In A.D. 859 Vikings spent a week or so ashore in Morocco, rounding up prisoners for sale as slaves or to hold for ransom. Sixty-two Viking ships participated, a spectacular display of technology, sea power, seamanship, and navigation.

At first the Europeans were powerless against these marauders, but eventually the need for common defense overcame provincial hostility and xenophobia. One of the causes of the Renaissance in Europe may have been the experience of banding together for protection against these northern raiders.

As the French, Irish, and British defenses became more effective, the Norwegian Vikings began to look west.

Iceland and Greenland had been discovered by ships blown off course during storms. Iceland was colonized by about A.D. 700, Greenland by 995. In an early voyage from Norway to Greenland in 986, a commuter named Bjarni Herjulfsson was blown past his goal by unfavorable winds. For about five days he sailed up and down the coast of a new land (which was, in fact, North America) without landing or making charts. His sketchy reports kindled a real-estate fever; Leif, son of Eric the Red, purchased Herjulfsson's ship and returned. His party found salmon-filled lakes, vines and grapes, and fodder for cattle in what was probably the northeastern tip of Newfoundland. With a bit of advertising overstatement, he called the place *Vinland* ("wine-land").

By A.D. 1000 the Norwegians had colonized Vinland. The settlements were modest, and at first relationships with the natives were encouraging. Unlike the Spanish who landed in the New World 500 years later, the Norwegian colonists tried to cooperate with the locals, to learn from them, and to help them in a mutual pact of assistance. Unfortunately misunderstandings arose, battles ensued, and the weather turned colder. The colony had to be abandoned in 1020. The Norwegians lacked the numbers, the weapons, and the trading goods to make the colony a success.

Lars Karl Ingason

Figure 2.6 *Islendingur*, an accurate replica of a Viking ship, in trials before crossing the North Atlantic in the summer of 2000. Fast, stable ships like *Islendingur* were used to transport explorers and colonists from what is now Iceland, Denmark, and Norway to the New World about 500 years before Columbus set sail.

Chinese Contributions

As the Dark Ages continued in Europe, Chinese navigators became more skilled, and their vessels grew larger and more seaworthy. They then set out to explore the other side of the world. Between 1405 and 1433, admiral Zheng He commanded the greatest fleet the world had ever known. At least 317 ships and 37,000 men undertook seven missions to explore the Indian Ocean, Indonesia, and around the tip of Africa into the Atlantic. Their aim: to display the wealth and power of the young Ming dynasty and to show kindness to people of distant places. The largest ship in the fleet, with nine masts and a length of 134 meters (440 feet) (**Figure 2.7**), was a huge treasure ship carrying objects of the finest materials and craftsmanship. The mission of the fleet was not to accumulate such treasure, but to give it away! Indeed, the primary purpose of these expeditions was to convince all nations with which the fleet had contact that China was the only truly civilized state and beyond any imaginable need for knowledge or assistance.

Many technical innovations had been required to make such an ambitious undertaking possible. In addition to the compass, the Chinese invented the central rudder, watertight compartments, and sophisticated sails on multiple masts, all of which were critically important for the successful operation of large sailing vessels. Until Europeans adopted the rudder around A.D. 1100, long-distance voyaging in a Western ship large enough to be stable in rough seas was usually difficult. Early Mediterranean traders and, later, the Polynesians and the Vikings had used specialized steering oars held against the right side (steer-board = starboard side) of their boats. While this system worked well in protected waters, the small area of the steering oar (and the exposed position of the steersman) made it difficult to hold a course on long ocean passages. The centrally mounted submerged rudder solved that problem. Also, dividing the ship into separate compartments below the waterline meant that flooding due to hull damage could be confined to a relatively small area of the ship, and the vessel could then be repaired and saved from sinking. Since sails provided the power to move, advances in sail design could drastically influence the success of any voyage. The Chinese fitted their trapezoidal or triangular sails with battens (pieces of bamboo inserted into stitched seams running the width of the sail) and placed the sails on multiple masts. The sails resembled venetian blinds covered with cloth. It was not necessary for Chinese sailors to climb the masts to unfurl the sails every time the wind changed; everything could be done from the deck with windlasses and lines. The shape of the sails made it easier to sail close to the wind in confined seaways.

Figure 2.7 The treasure ship, largest in a vast Chinese fleet whose purpose was to show kindness to people of distant places. The fleet sailed the Pacific and Indian Oceans between 1405 and 1433.

Despite enjoying these technical advances, the Chinese intentionally abandoned oceanic exploration in 1433. The political winds had changed, and the cost of the "reverse tribute" system was judged too great. In all, until late in the twentieth century, the Chinese made very few contributions to our understanding of the ocean. Still, their voyaging technology filtered into the West and made subsequent discoveries possible.

The Age of Discovery: From Prince Henry to Magellan

Having been jolted by internal awakening and external reality, Renaissance Europeans set out to explore the world by sea. They did not undertake exploration for its own sake, however; any voyage had to have a material goal. Trade between east and west had long been dependent on arduous and insecure desert caravan routes through the central Asian and Arabian deserts. This commerce was cut off in 1453 when the Turks captured Constantinople, and an alternate ocean route was needed.

A European visionary who thought ocean exploration held the key to great wealth and successful trade was **Prince Henry the Navigator,** third son of the royal family of Portugal (**Figure 2.8**). Prince Henry established a center at Sagres for the study of marine science and navigation ". . . through all the watery roads." Although he personally was not well traveled (he went to sea only twice in his life), captains

under his patronage explored from 1451 to 1470, compiling detailed charts wherever they went. Henry's explorers pushed south into the unknown and opened the west coast of Africa to commerce. He sent out small, maneuverable ships designed for voyages of discovery and manned by well-trained crews. For navigation, his mariners used the **compass**—an instrument (invented in China in the fourth century B.C.) that points to magnetic north. Although Arab traders had brought the

Figure 2.8 Prince Henry of Portugal, the Navigator. In the mid-1400s, Henry established a center at Sagres for the study of marine science and navigation ". . . through all the watery roads."

compass from China in the twelfth century, navigators still considered it a magical tool. They concealed the compass in a special box (predecessor of today's binnacle) and consulted it out of view of the crew. Henry's students knew Earth was round (but because of the errors publicized by Claudius Ptolemy they were wrong in their estimation of its size).

A master mariner (and skilled salesman), **Christopher Columbus,** "discovered" the New World quite by accident. Native Americans had been living on the continent for about 11,000 years, and the Norwegian Vikings had made about two dozen visits to a functioning colony on the continent 500 years before his noisy arrival; yet Columbus gets the credit. Why? Because his interesting souvenirs, exaggerated stories, inaccurate charts, and promises of vast wealth excited the imagination of royal courts. Columbus made North America a media event without ever sighting it!

Columbus wasn't trying to discover new lands. His intention was to pioneer a sea route to the rich and fabled lands of the East, made famous more than 200 years earlier in the overland travels of Marco Polo. As "Admiral of the Ocean Sea," Columbus was to have a financial interest in the trade routes he blazed. He was familiar with Prince Henry's work and, like all other competent contemporary navigators, knew Earth was spherical. By sailing west he believed he could come close to his eastern destination, the latitude of which he thought he knew. Because of wishful thinking and dependence on Ptolemy's data, however, Columbus made the *smallest* estimate of the size of Earth by any navigator in modern history; he assumed Earth to be only about half its actual size!

Not surprisingly, Columbus mistook the New World for his goal of India or Japan. He thought that the notable absence of wealthy cities and well-dressed inhabitants resulted from striking the coast too far north or south of his desired latitude. He made three more trips to the New World but went to his grave believing that he had found islands off the coast of Asia. He never saw the mainland of North America and never realized the size and configuration of the continents whose future he had so profoundly changed.

Other explorers quickly followed, and Columbus's error was soon corrected. Charts drawn as early as 1507 included the New World (**Figure 2.9**). Such charts perhaps inspired **Ferdinand Magellan** (**Figure 2.10**), a Portuguese navigator in the service of Spain, to believe that he could open a westerly trade route to the Orient. Unfortunately, the chart makers estimated the Americas and the Pacific Ocean to be much smaller than they actually are. (Compare Figure 2.9 with Magellan's route, shown in **Figure 2.11**.) In the Philippines Magellan was killed, and his men decided to continue sailing west around the world. Only 34 of the original crew of 260 survived, returning to Spain three years after they had set out. But they had proved it was possible to circumnavigate the globe.

The Magellan expedition's return to Spain in 1522 marks the end of the Age of European Discovery. An unpleasant era of exploitation of the human and natural resources of the Americas followed. Native empires were destroyed, and objects of priceless cultural value were melted into coin to fund European warfare and greed.

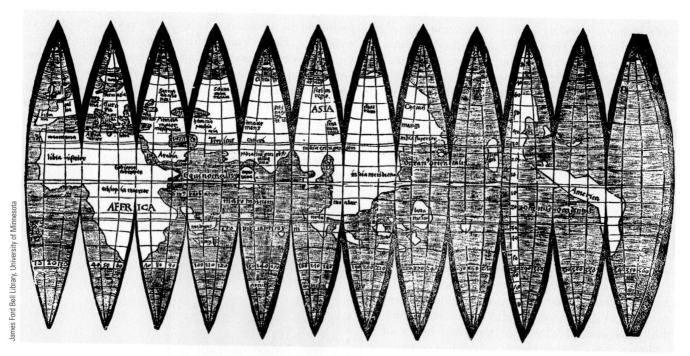

James Ford Bell Library, University of Minnesota

Figure 2.9 The Waldseemüller Map, published in 1507—the first map to name America and to show the New World as separate from Asia. The deep gores are designed to form a globe about 10 centimeters (4 inches) in diameter.

VOYAGING FOR SCIENCE

British sea power arose after the Age of Discovery to compete with the colonial aspirations of France and Spain. Sailing ships require dependable supply and repair stations, especially in remote areas. The great powers sent out expeditions to claim appropriate locations, preferably inhabited by friendly natives eager to help provision ships half a globe from home. The French sent Admiral de Bougainville into the South Pacific in the mid-1760s. His 1768 claim for France of what is now called French Polynesia opened the area to the powerful European nations. The British followed immediately.

James Cook

Scientific oceanography begins with the departure from Plymouth Harbor in 1768 of HMS *Endeavour* under the capable command of **James Cook** of the British Royal Navy (**Figure 2.12**). An intelligent and patient leader, Cook was also a skillful navigator, cartographer, writer, artist, diplomat, sailor, scientist, and dietician. The primary reason for the voyage was to assert the British presence in the South Seas, but the expedition had numerous scientific goals as well. First, Cook conveyed several members of the Royal Society (a scientific research group) to Tahiti to observe the transit of Venus across the disk of the sun. Their measurements verified calculations of planetary orbits made earlier by Edmund Halley (later of comet fame) and others. Then Cook turned south into unknown territory to search for a hypothetical southern continent, which some philosophers believed had to exist to balance the land-

The Granger Collection, New York

Figure 2.10 Ferdinand Magellan, a Portuguese navigator in service to Spain whose expedition was first to circumnavigate the world.

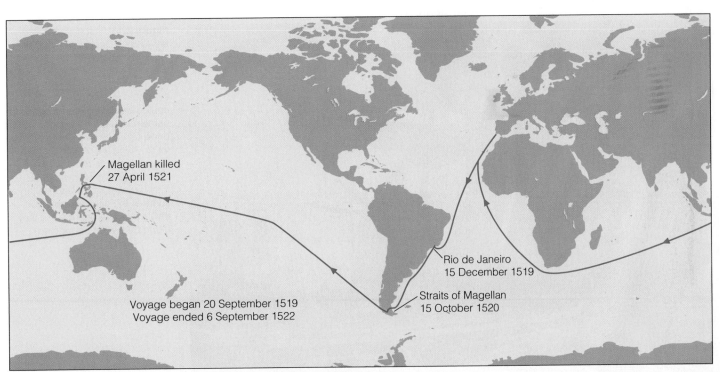

Magellan killed
27 April 1521

Voyage began 20 September 1519
Voyage ended 6 September 1522

Rio de Janeiro
15 December 1519

Straits of Magellan
15 October 1520

Figure 2.11 Track of the Magellan expedition, the first voyage around the world. Magellan himself did not survive the voyage; only 34 out of 260 sailors managed to return after three years of dangerous travel.

Figure 2.12 Captain James Cook, Royal Navy, painted in 1776 by Nathaniel Dance, shortly before embarking on his third, and fatal, voyage. Cook is a fully matured, self-confident captain who has twice circled the globe, penetrated into the Antarctic, and charted coastlines from Newfoundland to New Zealand.

mass of the Northern Hemisphere. Cook and his men found and charted New Zealand, mapped Australia's Great Barrier Reef, marked the positions of tens of small islands, made notes on the natural history and human habitation of these distant places, and initiated friendly relations with many chiefs. Cook survived an epidemic of dysentery contracted by the ship's company while ashore in Batavia (Djakarta) and sailed home to England around the world in 1771. Because of his insistence on cleanliness and ventilation (and because his provisions included cress, sauerkraut, and citrus extracts), his sailors avoided scurvy—a vitamin-C deficiency disease that for centuries had decimated crews on long voyages.

The Admiralty was deeply impressed. Cook was promoted to the rank of commander and in 1772 was given command of the ships *Resolution* and *Adventure,* in which he embarked on one of the great voyages in scientific history.[1] On this second voyage he charted Tonga and Easter Islands, and discovered New Caledonia in the Pacific and South Georgia in the Atlantic. He was first to circumnavigate the world at high latitudes. Though he sailed to 71°S latitude, he never sighted Antarctica. He returned home again in 1775.

Promoted to the rank of captain, Cook set off in 1776 on his third, and last, expedition, in *Resolution* and *Discovery.* His commission was to find a northwest passage around Canada and Alaska or a northeast passage above Siberia. He "discovered" the Hawaiian Islands (Hawaiians were there to greet him, of course, as shown in **Figure 2.13**) and charted

---

[1] Sailing master in *Resolution* was the 21-year-old William Bligh, later the object of the famous mutiny aboard HMS *Bounty.*

Figure 2.13 First contact! Captain James Cook, commanding HMS *Resolution,* off the Hawaiian island of Kauaʻi, wrote in 1778: "It required but little address to get them to come along side, but we could not prevail upon any one to come on board; they exchanged a few fish they had in the canoes for anything we offered them, but valued nails, or iron above every other thing; the only weapons they had were a few stones in some of the canoes and these they threw overboard when they found they were not wanted."

the west coast of North America. After searching unsuccessfully for a passage across the top of the world, Cook retraced his route to Hawaii to provision his ships for departure home. On 14 February 1779, after an elaborate farewell dinner with the chief of the island of Hawaii, Cook and his officers prepared to return to *Resolution,* anchored in Kealakekua Bay. The Englishmen somehow angered the Hawaiians and were beset by the crowd. Cook, among others, was killed in the fracas. The tracks of Cook's voyages are shown in **Figure 2.14.**

Cook deserves to be considered a scientist as well as an explorer because of his accuracy and thoroughness and the completeness in his descriptions. He and the scientists aboard took samples of marine life, land plants and animals, the ocean floor, and geological formations; they also reported the characteristics of these samples in their logbooks and journals. His navigation was outstanding, and his charts of the Pacific were accurate enough to be used by the Allies in World War II invasions of the Pacific islands. He drew accurate conclusions, did not exaggerate his findings in his reports, and

Figure 2.14 The tracks of Cook's three voyages exploring the Pacific.

opened friendly diplomatic relations with many native populations. Cook recorded and successfully interpreted events in natural history, anthropology, and oceanography. Unlike many captains of his day, he cared for his men. He was a thoughtful and clear writer. This first marine scientist peacefully changed the map of the world more than any explorer or scientist in history.

The Longitude Problem

How did Cook (or Columbus, or any ocean explorer) know where he was? Unless explorers could record position accurately on a chart, exploration was essentially useless. They could not find their way home, nor could they or anyone else find the way back to the lands they had discovered.

At night, Columbus and his European predecessors used the stars to find latitude and, as a consequence, knew their position north or south of home. You can do this, too. In the Northern Hemisphere, take a simple protractor and measure the angle between the horizon, your eye, and the north polar star. The protractor reads approximately in degrees of latitude. To find the Indies, for example, Columbus dropped south to a line of latitude and followed it west. But to pinpoint a location you need both latitude *and* the east–west position of longitude. (For more about latitude, longitude, and their use in marine navigation, see Appendix III.)

You can find longitude with a clock. First, determine local noon by observing the path of the shadow of a vertical shaft—it is shortest at noon—and set your clock accordingly. After traveling some distance to the west, you will notice that noon according to your *clock* no longer marks the time when the shadow of the *shaft* is shortest at your new location. If "clock" noon occurs 3 hours before "shaft" noon, you can do some simple math to see how far west of your starting point you have come. Earth turns toward the east, making one rotation of 360° in 24 hours, so its rotation rate is 15° per hour (360°/24 hours = 15°/hour). The 3-hour difference between "clock" noon and "shaft" noon puts you 45° west of your point of origin (3° × 15° = 45°). The more accurate the clock (and the measurement of the shaft's shadow), the more accurate your estimate of westward position. In Columbus's time no clocks were accurate enough to make this calculation practical after a few days at sea. How such clocks were invented is a fascinating story.

In 1707 a British battle fleet led by Sir Clowdisley Shovell ran aground in the Scilly Islands because they had lost track of their longitude after many weeks at sea; four ships and 2,000 men were lost. This was the most serious in a long string of English maritime disasters involving inaccurate estimation of longitude. In 1714, after lengthy study, the British government offered a £20,000 prize (equivalent to more than $12 million in modern currency) for a method of determining longitude to within $\frac{1}{2}°$ after 30 days at sea.

The time method would work in theory, but existing clocks were too delicate and too inaccurate to meet these standards. The key to the problem was inventing a sturdy clock that ran at a constant rate under any circumstance, even the changeable conditions of a ship at sea.

In 1728 **John Harrison,** a Yorkshire cabinetmaker, began working on such a clock. His radical new timepiece, called a **chronometer,** was governed not by a pendulum (which would be useless in a rolling ship) but by a spring escapement. His first version (**Figure 2.15a**) was tested at sea in 1736, and Harrison was awarded £500 as encouragement to continue his efforts. Over the next 25 years he built three more clocks, culminating in 1760 in his Number Four (**Figures 2.15b–c**), perhaps the most famous timekeeper in the world.

A sea trial of Number Four was begun in HMS *Deptford* in 1761. Harrison, too old and infirm to accompany the chronometer, sent his son and collaborator to tend the instrument. *Deptford* crossed the Atlantic from England to Jamaica and made a near-perfect landfall. After taking the clock's known error rate into account—its "rate of going" was $2\frac{2}{3}$ seconds a day—the clock was found to be only 5 seconds slow. This would have meant an error in longitude of only 2.3 kilometers (1.4 miles), an astonishing achievement by then-current standards of long-distance navigation.

Technically, Harrison had won the prize—Number Four had more than met the criteria set by the British Board of Longitude—but he was granted only part of the promised reward. Understandably, the officials would not hand over the money until it had been determined that the clock's secrets could be applied to quantity production. Just as understandably, Harrison did not wish his life's work compromised without compensation. He feared (correctly, as it turned out) that once the clockwork was examined by a competent watchmaker, his ideas would be copied. Finally, in 1769, a single copy was made; its success clinched Harrison's achievement. Captain Cook took a chronometer on his last two voyages, but Harrison received the balance of the prize only in 1773 (when he was 80) and then only through the direct intervention of King George III.

All four of Harrison's chronometers are on view still functioning in Britain's National Maritime Museum at Greenwich, in eastern London. Greenwich is an ideal site for the museum; in 1884 the Greenwich meridian, a longitude line at the naval observatory there, became "longitude zero" for the world (**Figure 2.16**). Not since Eratosthenes's selection of Alexandria as the first "longitude zero" had Western nations recognized a common base for positioning.

The Sampling Problem

Marine science advances by the analysis of samples. The chronometer permitted investigators to determine the precise location at which they collected samples of water, bottom sediments, and marine life; but first they had to overcome the difficulty of actually *obtaining* a sample. Sampling of floor sediments or bottom water is not an easy task in the deep ocean. The line used to suspend the sampling device snakes back and forth as currents strike it, and the weight of the line itself makes it difficult to tell when the sampler has hit bot-

a

b

c

Figure 2.15 The first chronometers. (a) Harrison's Number One time-keeper, a clock built to prove principles but unfit for use at sea. (b) The face of Harrison's Number Four timekeeper, which won the £20,000 award offered by the British Board of Longitude. (c) The intricate workmanship is clearly visible in this view of the Number Four clockwork and protective case. Both are functioning and are on display at the National Maritime Museum in Greenwich, England.

tom. Deploying and recovering the line are laborious and time consuming, and sometimes the sampling device does not work properly. Early bottom-sampling devices (such as those used by Cook) were simple wax-covered lead weights lowered to shallow bottoms to pick up sediments and test the suitability of anchorages. Later devices took deep-water samples, extracted cores from the sediments, grabbed samples of the bottom, or scooped biological specimens from the ocean floor.

The first researchers to attack the deep-sampling problem successfully were British explorers Sir John Ross and his nephew Sir James Clark Ross (**Figure 2.17**). During an expedition to scout the Northwest Passage in 1818, Sir John Ross obtained a bottom sample from 1,919 meters (6,294 feet) near Greenland by using a clamping sampler to trap the specimen. Sir James Clark Ross, discoverer of the Ross Sea and the area of Antarctica known as Victoria Land, obtained depth **soundings** (depth measurements) of 4,433 meters (14,545 feet) and 4,893 meters (16,054 feet) in the South Atlantic.

Sampling techniques improved through the century. Using a sounding method perfected in the late 1840s by a U.S. Navy midshipman, American commodore Matthew Maury used a long lightweight line and lead weight to discover the Mid-Atlantic Ridge, an important hidden range of mountains. Fridtjof Nansen perfected the deep-water sampling bottle bearing his name near the end of the century. Even today, in spite of modern advances, deep sampling remains difficult. Remotely operated vehicles can work at great depths and return samples and pictures to the surface, but their electronic complexity makes them delicate and expensive to operate.

Figure 2.16 The prime meridian transit circle, Greenwich, England. Zero longitude is marked by the brass rail extending from the instrument in the building toward the camera.

Tom Garrison

© National Maritime Museum, London

Figure 2.17 Sir James Clark Ross, the first researcher to conduct extensive deep soundings.

THE FIRST SCIENTIFIC EXPEDITIONS

Great as his contributions undoubtedly were, Cook's three voyages (and those of the Rosses) were not purely scientific expeditions. These men were British naval officers engaged in Crown business, concerned with charting, foreign relations, and natural phenomena as they applied to Royal Navy matters. The first genuine *only-for-science* expedition may well have been the British *Challenger* Expedition of 1872–76, but the United States got into the act first with a hybrid expedition in 1838.

The United States Exploring Expedition

After a 10-year argument over its potential merits, the **United States Exploring Expedition** was launched in 1838. It was primarily a naval expedition, but its captain was somewhat freer in maneuvering orders than Cook had been. The work of the scientists aboard the flagship USS *Vincennes* and the expedition's five other vessels helped establish the natural sciences as reputable professions in America. Had it not been for the combative and disagreeable personality of its leader, Lt. Charles Wilkes (**Figure 2.18**), this expedition might have become as famous as those of Cook or the upcoming *Challenger* voyage.

The expedition departed on a four-year circumnavigation. Its goals included showing the flag, whale scouting, mineral gathering, charting, observing, and pure exploration. One unusual goal was to disprove a peculiar theory that Earth was hollow and could be entered through huge holes at either pole.

Wilkes's team explored and charted a large sector of the east Antarctic coast and made observations that confirmed the landmass as a continent. A map of the Oregon Territory produced in 1841, one of 241 maps and charts drawn by members of the expedition, proved especially valuable when connected to the map of the Rocky Mountains prepared the following year by Captain John C. Fremont. Hawaii was thoroughly explored, and Wilkes led an ascent of Mauna Loa, one of the two highest peaks of Hawaii's largest island. The expedition returned with many scientific specimens and artifacts, which formed the nucleus of the collection of the newly established Smithsonian Institution in Washington, D.C. No evidence of polar holes was found.

Upon their return in 1842, Wilkes and his "scientifics" prepared a final report totaling 19 volumes of maps, text, and illustrations. The report is a landmark in the history of American scientific achievement.

United States Naval Academy Museum

Figure 2.18 Lt. Charles Wilkes soon after his return from the United States Exploring Expedition in 1842.

Library of Congress

Figure 2.19 Matthew Fontaine Maury, perhaps the first person for whom oceanography was a full-time occupation. This photograph was probably taken in 1853.

The Work of Matthew Maury

2-16

At about the time the Wilkes expedition returned, **Matthew Maury** (**Figure 2.19**), a Virginian and fellow U.S. naval officer, became interested in exploiting winds and currents for commercial and naval purposes. After being crippled in a stagecoach accident, in 1842 Maury was given charge of the Navy's Depot of Charts and Instruments. There he studied a huge and neglected treasure trove of ships' logs, with their many regular readings of temperature and wind direction. By 1847 Maury had assembled much of this information into coherent wind and current charts. Maury began to issue these charts free to mariners in exchange for logs of their own new voyages.

Slowly a picture of planetary winds and currents began to emerge. Maury himself was a compiler, not a scientist, and he was vitally interested in the promotion of maritime commerce. His understanding of currents was built on the work of **Benjamin Franklin.** Nearly 100 years earlier, Franklin had noticed the peculiar fact that the fastest ships were not always the fastest ships; that is, hull speed did not always correlate with out-and-return time on the European run. Franklin's cousin, a Nantucket merchant named Tim Folger, noted Franklin's puzzlement and provided him with a rough chart of the "Gulph Stream" that he (Folger) had worked out. By staying within the stream on the outbound leg and adding its speed to their own, and by avoiding it on their return, captains could

traverse the Atlantic much more quickly. It was Franklin who published, in 1769, the first chart of any current (**Figure 2.20**).

But it was Maury who was the first person to sense the worldwide pattern of surface winds and currents. Based on his analysis, he produced a set of directions for sailing great distances more efficiently. Maury's sailing directions quickly attracted worldwide notice: He had shortened the passage for vessels traveling from the North American East Coast to Rio de Janeiro by 10 days, and to Australia by 20. His work became famous in 1849 during the California gold rush—his directions made it possible to save 30 days on the voyage around Cape Horn to California. Applicable U.S. charts still carry the inscription "Founded on the researches of M.F.M. while serving as a lieutenant in the U.S. Navy." His crowning achievement, *The Physical Geography of the Seas,* a book explaining his discoveries, was published in 1855.

Maury, considered by many to be the father of physical oceanography, was perhaps the first man to undertake the systematic study of the ocean as a full-time occupation.

Charles Darwin and HMS *Beagle*

2-17

Though he was later to become famous for his explanation of biological evolution through natural selection, **Charles Darwin**'s first contributions to science were in the field of

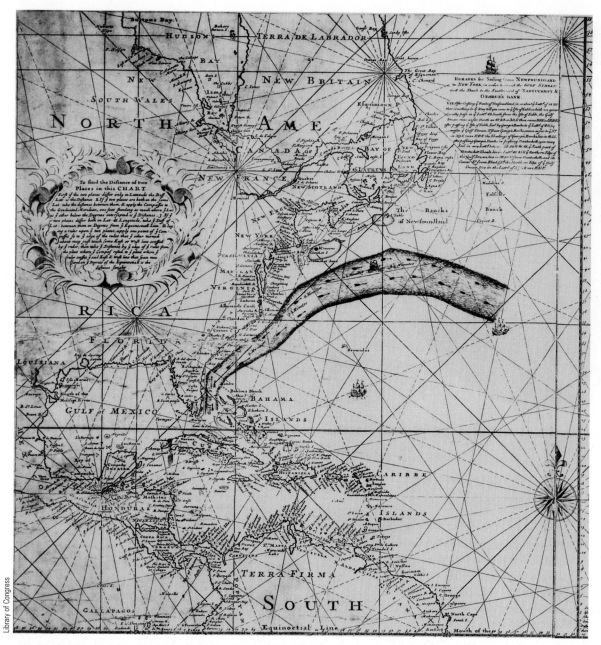

Figure 2.20 Benjamin Franklin's 1769 chart of the Gulf Stream system. His cousin, Tim Folger, discovered that Yankee whalers had learned to use the Gulf Stream to their advantage. Others, especially English shipowners, were slower to learn. Folger, himself a sea captain, wrote that Nantucket whalers ". . . in crossing it have sometimes met and spoke with those packets who were in the middle of and stemming it. We have informed them that they were stemming a current that was against them to the value of three miles an hour and advised them to cross it, but they were too wise to be counseled by simple American fishermen."

marine biology (**Figure 2.21**). Darwin became interested in natural history during his student days in Cambridge, where he was studying to become a minister of the Church of England. Because of the influence of John Henslow, a professor of botany, he was selected in 1831 to accompany HMS *Beagle* on a voyage to South America and some Pacific islands as an unpaid naturalist. Nearly all of Darwin's subsequent work stemmed directly from observations he made on this 4½-year voyage. In his first major work, *Struc-*

ture and Distribution of Coral Reefs (1842), Darwin correctly argued that atolls and reefs such as those he observed could only have resulted from subsidence of the seafloor, with corals growing upward as their bases dropped. His monographs on barnacle biology, volcanic islands, and fossils are overshadowed by his remarkable work *On the Origin of Species,* published in 1859. More on the contributions of Charles Darwin may be found in Chapter 13 on pages 326 and 329.

Figure 2.21 Charles Darwin late in life pictured between a marine iguana (left) and a land iguana (right), its presumed evolutionary antecedent. Darwin studied marine iguanas during his voyage aboard HMS *Beagle*.

The *Challenger* Expedition

The first sailing expedition devoted completely to marine science was conceived by Charles Wyville Thomson, a professor of natural history at Scotland's University of Edinburgh, and his Canadian-born student of natural history, John Murray. Stimulated by their own curiosity and by the inspiration of Charles Darwin's voyage in HMS *Beagle*, they convinced the Royal Society and the British government to provide a Royal Navy ship and trained crew for a prolonged and arduous voyage of exploration across the oceans of the world. Thomson and Murray even coined a word for their enterprise: **oceanography.** Though the term literally implies only marking or charting, it has come to mean the science of the ocean. The government and the Royal Society agreed to the endeavor provided that a proportion of any financial gain from discoveries was handed over to the Crown. This arranged, the scientists made their plans.

HMS *Challenger*, a 2,306-ton steam corvette (**Figure 2.22**), set sail on 21 December 1872 on a four-year voyage around the world, covering 127,600 kilometers (79,300 miles). Although the captain was a Royal Navy officer, the six-man scientific staff directed the course of the voyage. *Challenger*'s track is shown in **Figure 2.23**.

One important mission of the **Challenger Expedition** was to investigate Edinburgh professor Edward Forbes's contention that life below 549 meters (1,800 feet) was impossible because of high pressure and lack of light. The steam winch on board made deep sampling practical, and samples from depths as great as 8,185 meters (26,850 feet) were collected off the Philippines. Through the course of 492 deep soundings with mechanical grabs and nets at 362 stations (including 133 dredgings), Forbes was proved resoundingly wrong. With each hoist, animals new to science were strewn on the deck; in all, staff biologists discovered 4,717 new species! **Figure 2.24** shows one of the large trawl nets used in making some of these discoveries.

The scientists also took salinity, temperature, and water-density measurements during these soundings. Each reading contributed to a growing picture of the physical structure of the deep ocean. They completed at least 151 open-water trawls and stored 77 samples of seawater for detailed analysis ashore. The expedition collected new information on ocean currents, meteorology, and the distribution of sediments. The locations and profiles of coral reefs were charted. Thousands of pounds of specimens were brought to British museums for study. Manganese nodules, brown lumps of mineral-rich sediments, were discovered on the seabed, sparking interest in deep-sea mining.

This first pure oceanographic investigation was an unqualified success. The discovery of life in the depths of the oceans stimulated the new science of marine biology. The scope, accuracy, thoroughness, and attractive presentation of the researchers' written reports made this expedition a high point in scientific publication. The *Challenger Report*, the record of the expedition, was published between 1880 and 1895 by Sir John Murray in a well-written and magnificently

Figure 2.22 Lt. Pelham Aldrich, first lieutenant of HMS *Challenger*, kept a detailed journal of the *Challenger* Expedition. With accuracy and humor he kept this record in good weather and bad, and he had the patience and skill to include watercolors of the most exciting events. This is part of the first page of his journal.

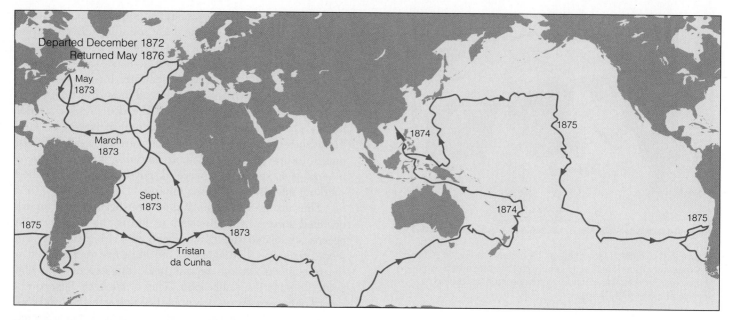

Figure 2.23 HMS *Challenger's* track, December 1872–May 1876.

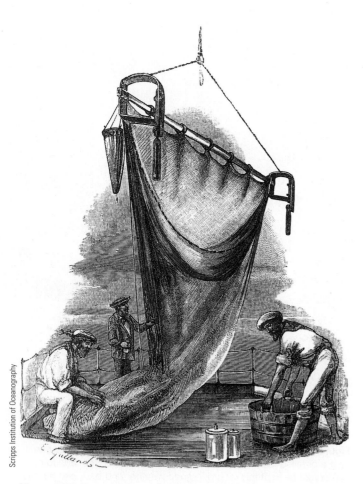

Scripps Institution of Oceanography

Figure 2.24 Emptying a trawl net; an engraving from the *Challenger Report*.

illustrated 50-volume set; it is still used today. Indeed, it was the 50-volume *Report*, rather than the cruise itself, that provided the foundation for the new science of oceanography. The expedition's many financial spin-offs indicated that pure research was a good investment, and the British government realized quick profits from the exploitation of newly discovered mineral deposits on islands. The *Challenger* Expedition remains history's longest continuous scientific oceanographic expedition.

With successes like these, the pace of exploration accelerated. American naturalist Alexander Agassiz, sailing in 1877 in the U.S. Coast and Geodetic Survey ship *Blake*, collected data corroborating the *Challenger* material at 355 deep-sea stations. The distribution of manganese nodules was found to be widespread. Further work by Agassiz and his students around the turn of the twentieth century in the survey ship *Albatross* helped train a generation of influential American marine biologists. In 1886 the Russians entered the field of marine exploration with the three-year cruise of *Vitiaz* under the leadership of S. O. Makarov; their main contribution was a careful analysis of the salinity and temperature of North Pacific waters.

The Rise of Modern Sea Power

Marine science is also applied to military interests. **Sea power** is the means by which a nation extends its military capacity onto the ocean. History has been greatly influenced by sea power—for example, the defeat of the Persian fleet by the Greeks at Salamis in 480 B.C. and the triumph of British admiral Horatio Nelson over French forces at Trafalgar in 1850 led to eras of cultural and economic supremacy by both nations.

In 1892 **Alfred Thayer Mahan,** an American naval officer and historian, published *The Influence of Sea Power upon History, 1660–1783.* Based on his studies of the rise and fall of nation–states, this book was to have profound consequences for the development of the modern world. Mahan stressed the interdependence of military and commercial control of seaborne commerce and the ability of safe lines of transportation and communication to influence the outcomes of conflicts. Coming at a time of unprecedented technological improvements in shipbuilding, Mahan's work was read avidly in Great Britain, Germany, and the United States. For better or worse, the naval hardware, strategy, and tactics of the last century's greatest wars—along with their outcomes—was influenced by his clear analysis.

Interest in marine science is maintained today by the U.S. Navy through the Office of Naval Research and through the Oceanographer of the Navy.

VOYAGES FOR SCIENCE IN THE TWENTIETH CENTURY

In the twentieth century, oceanographic voyages became more technically ambitious and expensive. Scientist–explorers sought out and investigated places that once had been too difficult to reach. Though the deep ocean floor was coming

into reach, it was the forbidding polar ocean that attracted their first attentions.

Polar Exploration

Polar oceanography began with the pioneering efforts of Fridtjof Nansen (**Figure 2.25**). Nansen courageously allowed his specially designed ship *Fram* to be trapped in the Arctic ice, where he and his crew of 13 drifted with the pack for nearly four years (1893–96), exploring to 85°57'N, a record for the time. The 1,650-kilometer (1,025-mile) drift of *Fram* proved that no Arctic continent existed. Nansen's studies of the drift, of meteorological and oceanographic conditions, of life at high latitudes, and of deep sounding and sampling techniques form the underpinnings of modern polar science.

Living up to her name—*Fram* means "forward" in Norwegian—Nansen's ship continued to play a pivotal role in exploration (**Figure 2.26**). In 1910 Roald Amundsen, a student of Nansen's, set out in the sturdy little vessel for the coast

Alda Amundsen, SPRI

Figure 2.26 Nansen's 123-foot schooner *Fram* ("forward"). With 13 men, *Fram* sailed on 22 June 1893 to the high Arctic with the specific purpose of being frozen into the ice. *Fram* was designed to slip up and out of the frozen ocean, and drifted with the pack ice to within about 4° of the North Pole. The whole harrowing adventure took nearly four years. The ship's 1,650-kilometer (1,025-mile) drift proved no Arctic continent existed beneath the ice. Living conditions aboard can be sensed from this recently discovered photograph.

Hulton Deutsch Collection Limited

Figure 2.25 Fridtjof Nansen, pioneering Norwegian oceanographer and polar explorer, looking every inch the Viking. In 1908 Nansen became the first professor of oceanography, a post created for him at Christiania University.

of Antarctica, the first leg of a journey to the South Pole. Nansen himself settled down to a long and distinguished career as an oceanographer, inventor, zoologist, artist, statesman, and professor. He was awarded the Nobel Peace Prize in 1922 for his unstinting work in worldwide humanitarian causes.

Scientific curiosity, national pride, new ideas in shipbuilding, advances in nutrition, and great personal courage led in the early years of the twentieth century to the golden age of polar exploration. After some heroic attempts by a number of explorers to reach the poles, an American naval officer, Robert E. Peary, accompanied by his African-American assistant Matthew Henson and four Inuit (Eskimos), reached the vicinity of the North Pole in April 1909. A party of five men led by Roald Amundsen of Norway achieved the South Pole in December 1911.

Peary's and Amundsen's scientific contributions were limited to a few meteorological observations, but around the same time another Antarctic explorer, Captain Robert Falcon Scott of the British Royal Navy, led a scientific expedition to the South Pole. Chief scientist of the British expedition was Dr. Edward A. Wilson (**Figure 2.27**), a Cambridge-trained physician and naturalist with a special interest in ornithology. In 1911, before the assault on the pole, Wilson and three other scientists conducted the first extensive scientific investigations at high latitude. Laboring in a cramped expedition hut at Cape Evans, the researchers tended continuous-reading barometers and temperature gauges. They noted the direction of winds at the surface (and aloft by observing the smoke plume from Mt. Erebus, a nearby active volcano), measured atmospheric radioactivity, and recorded displays of auroras. They also made weekly observations of the strength and direction of Earth's magnetic field, trawled for plankton, trapped fish beneath the ice, made soundings through cracks in the pack ice, analyzed water samples from various depths, collected samples of life from various freshwater lakes and pools ashore, explored nearby dry valleys and glaciers, and analyzed the ecological relationships among species. Ice received their full attention; using methods pioneered by Nansen, the team made the first detailed on-site studies of the formation, deposition, temperature, and thermal conductivity of sea ice.

Although the expedition was a scientific success, the polar journey itself was a tragic failure. Under Scott's leadership, members of the expedition arrived at the pole four weeks after Amundsen, but they died on the return trip to base camp. There were many reasons for the polar party's inability to complete the 2,000-kilometer (1,300-mile) round-trip (Scott, Wilson, and a companion died just 18 kilometers—11 miles—from their final food depot). Among these reasons must surely be counted the team's unwillingness to abandon their scientific notes, their measurements and records, and 35 pounds of geological specimens collected en route. It was not geographical victory but courage in adversity that captures the imagination.

Modern technology has eased the burden of high-latitude travel. In 1958, under the command of Capt. William Anderson, the U.S. nuclear submarine *Nautilus* sailed beneath the North Pole during a submerged transit beneath the Arctic pack from Point Barrow, Alaska, to the Norwegian Sea. As is apparent in Box 4.1c, a warm, strong, and stable nuclear submarine is a nearly ideal platform for conducting oceanographic research at high latitudes.

Scott Polar Research Institute

Figure 2.27 Dr. Edward Adrian Wilson, physician and ornithologist, chief scientist of the 1910–12 British Antarctic Expedition. Wilson and three other scientists conducted the first extensive scientific investigations at high latitude. In this photograph, he is completing a watercolor (actually, a "gin" color, as water would have frozen) showing the positions of ice-crystal halos around the sun.

Figure 2.28 USS *Pargo* surfaced at the edge of the Arctic ice pack during the 38-day 1993 SCICEX cruise. A scientific staff of 7 was embarked along with a full operating crew; 39 shore-side researchers participated in the analysis of samples and data taken during the cruise. A superbly strong, fast, quiet, powerful (and pleasantly warm) nuclear submarine makes an ideal platform for oceanographic research at high northern latitudes. USS *Pargo,* a *Sturgeon*-class boat specially strengthened for operation beneath the northern ice cap, is 89 meters (292 feet) long, 9.75 meters (32 feet) wide. Her speed and depth capability are classified.

At the request of oceanographic researchers, the U.S. Navy initiated a series of research cruises in the northern polar ocean. Called SCICEX (for Scientific Ice Exercise), and funded by governmental agencies, the cruises began in 1993 and continued through the year 2000. **Figure 2.28** shows USS *Pargo,* a fast and powerful *Sturgeon*-class submarine hardened for surfacing through ice, during 1993 operations in the Arctic. A staff of seven scientists spent 45 days operating in the Arctic Ocean during a 1996 research cruise aboard USS *Pogy,* another *Sturgeon*-class submarine. More than 12,800 kilometers (9,000 miles) of underway data—including bathymetry, gravity anomaly, temperature, salinity, ice draft, and images of the underside of the ice—were collected; more than 1,500 water samples were taken for biological and chemical analysis; and four long-term observation buoys were placed. The 1998 and 1999 expeditions aboard USS *Hawkbill* made use of a sub-bottom profiler to provide the first images of the shallow strata of the Arctic Ocean floor. The Navy hosted annual 45- to 60-day research cruises aboard *Sturgeon*-class submarines through the year 2000.

Other Twentieth-Century Voyages

2-22

In 1925 the German *Meteor* **Expedition,** which crisscrossed the South Atlantic for two years, introduced modern optical and electronic equipment to oceanographic investigation. Its most important innovation was use of an **echo sounder,** a device that bounces sound waves off the ocean bottom, to study the depth and contour of the seafloor (see Figure 4.2). The echo sounder revealed to *Meteor* scientists a varied and often extremely rugged bottom profile rather than the flat floor they had anticipated.

Atlantis, launched in 1931, was the first U.S. research ship built specifically for ocean studies. Investigations by her scientists confirmed Matthew Maury's findings of a mid-Atlantic ridge and helped discover its extent. The 32-meter (104-foot) schooner *E.W. Scripps,* under the direction of Harald Sverdrup, began a wide-ranging program of chemical, biological, and geophysical exploration off the coast of southern California in 1937. These voyages led to publication in 1942 of *The Oceans,* the first modern reference work on all phases of marine science.

In October 1951 a new HMS *Challenger* began a two-year voyage that would make precise depth measurements in the Atlantic, Pacific, and Indian Oceans and in the Mediterranean Sea. With echo sounders, measurements that would have taken the crew of the first *Challenger* nearly four hours to complete could be made in seconds. *Challenger II*'s scientists discovered the deepest part of the ocean's deepest trench, naming it *Challenger Deep* in honor of their famous predecessor. In 1960 U.S. Navy lieutenant Don Walsh and Jacques Piccard descended into the Challenger Deep in

Much can be gained from the study of underwater artifacts. Objects protected beneath soft marine sediment can age more gracefully than objects exposed to eroding water, winds, or frost. As you would expect, study of well-preserved objects from the seabed can provide insight into ship construction or the layout of a town, but there is always more to learn—information about a culture and people in the form of personal objects, fragments of clothing, remnants of cargo, even scraps of written material or paintings. The resting place of the RMS *Titanic* was discovered in 1985, and artifacts brought ashore for study have stimulated our imagination (and provided the story line for one of the most profitable movies ever made).

With luck, underwater archaeology can also solve mysteries. The loss of the CSS *Hunley* in February 1864 has bewildered generations of sailors and historians. The first submarine to sink a ship in battle, *Hunley* was also the first application of stealth technology to naval warfare. Designed to destroy Union ships blockading Charleston harbor during the American Civil War, the unstable 12-meter- (39-foot-) long sub was powered by eight sailors turning a crank attached to the propeller. The crew had enough air for 150 minutes—long enough to attach a bomb to the hull of the USS *Housatonic* and send it to the bottom.

Their mission accomplished, the captain turned *Hunley* toward shore. They flashed the "success" signal as they re-turned—one blue light from a lantern. Then they disappeared. No one knew *Hunley*'s location until 1995 when a research team located the hull using side-scan sonar. In 1999 *Hunley* was raised on two massive trusses fitted with nylon slings and inflated foam pillows (**Figure a**). The trusses and sub were loaded aboard a barge and taken to the Charleston naval base for a restoration process expected to last about 10 years. The privately funded *Hunley* foundation expects the cost of recovery and conservation to exceed $10 million.

What do historians hope to learn? Currents have filled the hull with fine sediment. When X-ray study suggests the best way to open the hull, years of sifting and digging will begin. It may be that the crew is sitting at their posts inside the sub, their skeletons held by their heavy winter uniforms. Around them in this time capsule may be artifacts of their world, their mission, their life and loves, their hopes for the future. Was a critical through-hull fitting left open in the euphoria of victory? Were valves stuck or handles jammed? Did the explosion fatally damage the hull? Who exactly was aboard? There will be much to learn, most of it unanticipated. "This isn't about Yankees or Confederates," says Robert Neyland, the U.S. Navy's chief underwater archaeologist. "It's about science."

2-23

The Confederate submarine *Hunley* is raised from a barge by crane on 8 August 2000, at the former Charleston Naval Base in North Charleston, South Carolina. The *Hunley*, which sank with its crew of nine on 17 February 1864 after blowing up the Union blockade ship *Housatonic* with a charge of black powder, was lifted from its watery grave about four miles off Sullivans Island, secured on a barge, and transported to a conservation lab at the old base.

AP/Wide World Photos

Trieste, a Swiss-designed, blimplike bathyscaphe (that dramatic story opens Chapter 4).

But in many ways the last voyage to be discussed here is the most portentous of all. In 1968 the drilling ship *Glomar Challenger* (**Figure 2.29**) set out to test a controversial hypothesis about the history of the ocean floor. It was capable of drilling into the ocean bottom beneath more than 6,000 meters (20,000 feet) of water and recovering samples of seafloor sediments. These long and revealing plugs of seabed rock provided confirming evidence for seafloor spreading and plate tectonics. (The wonderful details are found in Chapter 3.) In 1985 deep-sea drilling duties were taken over by the much larger and more technologically advanced ship *JOIDES Resolution* (see Figure 5.17).[2] The ship contains equipment capable of drilling in water 8,100 meters (27,000 feet) deep and houses the most completely equipped geological laboratories ever put to sea.

---

[2] *JOIDES* is the acronym for Joint Oceanographic Institutions for Deep Earth Sampling.

THE RISE OF OCEANOGRAPHIC INSTITUTIONS

The demands of scientific oceanography have become greater than the capability of any single voyage. Oceanographic institutions, agencies, and consortia evolved in part to ensure continuity of effort. The first of these coordinating bodies was founded by Prince Albert I of Monaco, who endowed his country's oceanographic laboratory and museum in 1906. The most famous alumnus of Albert's Musée Océanographique is Jacques Cousteau, coinventor in 1943 of the scuba underwater breathing system. Monaco also became the site of the International Hydrographic Bureau, founded in 1921 as an association of maritime nations. This bureau published one of the first general charts of the ocean showing bottom contours.

In the United States, the three preeminent oceanographic institutions are the Woods Hole Oceanographic Institution on Cape Cod, founded in 1930 (and associated with the Massachusetts Institute of Technology and the neighboring

Deep Sea Drilling Program

Figure 2.29 *Glomar Challenger,* operated by the Deep Sea Drilling Project from 1968 to 1983. The 122-meter (400-foot) ship used computers to maintain her position with the precision needed to complete the drilling of cores up to 1.5 kilometers (about 1 mile) long. A few of these cores—from several sites in the South Atlantic—yielded samples of sediments down to the solid oceanic crust. The oldest sediments, and thus the oceanic crust immediately below, were shown to be surprisingly young—only about 180 million years old. Yet the oldest continental crust had been dated at more than 3.8 billion years. What could explain this dramatic discrepancy? Chapter 3 tells the story.

Marine Biological Laboratory, founded in 1888); the Scripps Institution of Oceanography, founded in La Jolla, California, and affiliated with the University of California in 1912 (see **Figure 2.30**); and the Lamont–Doherty Earth Observatory of Columbia University, founded in 1949.[3]

The U.S. government has been active in oceanographic research. Within the Department of the Navy are the Office of Naval Research, the Office of the Oceanographer of the Navy, the Naval Oceanic and Atmospheric Research Laboratory, and the Naval Ocean Systems Command. These agencies are responsible for oceanographic research related to national defense. The National Oceanic and Atmospheric Administration (NOAA), founded within the Department of Commerce in 1970, seeks to facilitate commercial uses of the ocean. NOAA includes the National Ocean Service, the National Weather Service, the National Marine Fisheries Service, and the Office of Sea Grant.

---

[3] There are, of course, other prominent institutions involved in studying the ocean. A list of some of them and thoughts on how a student might enter the field appear in Appendix VIII, "Working in Oceanography."

SATELLITE OCEANOGRAPHY

The National Aeronautics and Space Administration (NASA), organized in 1958, has become an important institutional contributor to marine science. For four months in 1978, NASA's **Seasat,** the first oceanographic satellite, beamed oceanographic data to Earth. More recent contributions have been made by satellites beaming radar signals off the sea surface to determine wave height, variations in sea-surface contour and temperature, and other information of interest to marine scientists.

The first of a new generation of oceanographic satellites was launched in 1992 as a joint effort of NASA and the Centre National d'Études Spatiales (the French space agency). The centerpiece of **TOPEX/Poseidon,** as the project is known, is a satellite orbiting 1,336 kilometers (835 miles) above Earth in an orbit that allows coverage of 95% of the ice-free ocean every 10 days (**Figure 2.31**). The satellite's *TOPography EXperiment* uses a positioning device that allows researchers to determine its position to within 1 centimeter (½ inch) of Earth's center. The radars aboard can then

a

b

Figure 2.30 (**a**) The Woods Hole Oceanographic Institution, Woods Hole, Massachusetts. Marine science has been an important part of this small Cape Cod fishing community since Spencer Fullerton Baird, then assistant secretary of the Smithsonian Institution, established the U.S. Commission of Fish and Fisheries there in 1871. The Marine Biological Laboratory was founded in 1888, the Oceanographic Institution in 1930. (**b**) The Scripps Institution of Oceanography, La Jolla, California. Begun in 1892 as a portable laboratory-in-a-tent, Scripps was founded by William Ritter, a biologist at the University of California. Its first permanent buildings were erected in 1905 on a site purchased with funds donated by philanthropic newspaper owner E. W. Scripps and his sister, Ellen.

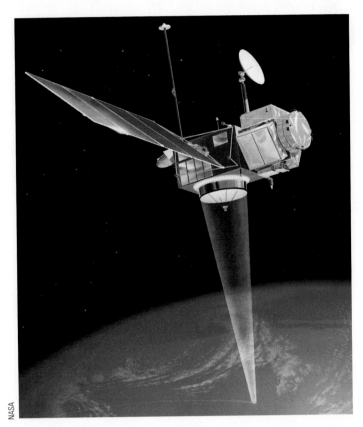

Figure 2.31 The joint U.S.–French *TOPEX/Poseidon* satellite orbits 1,336 kilometers (835 miles) above Earth in an orbit that allows coverage of 95% of the ice-free ocean every 10 days. The satellite was launched in 1992 and is supplied with a positioning device that allows researchers to determine its position to within 1 centimeter (½ inch) of Earth's center. Such accuracy makes possible very accurate determination of sea-surface height by radar transmitters on board. For examples of the results of *TOPEX/ Poseidon* studies, see Figures 9.7b, and 9.18.

determine the height of the sea surface with unprecedented accuracy. Other experiments in this five-year program include sensing water vapor over the ocean, determining the precise location of ocean currents, and determining wind speed and direction.

SEASTAR, launched by NASA in 1997, carries a color scanner called SeaWiFS (sea-viewing wide-field-of-view sensor). This device measures the distribution of chlorophyll at the ocean surface, a measure of marine productivity.

Still to come is NASA's ambitious *Jason-1,* a follow-on to *TOPEX/Poseidon.* Scatterometers will measure ocean surface winds, a radiometer will sense water vapor, and sea-surface height will be reported by even more accurate radar altimeters. A five-year mission is planned.

The U.S. Department of Defense has built a satellite system you can use every day called the **GPS (Global Positioning System)**, a "constellation" of 24 satellites (21 active, 3 spare) in orbit 10,600 miles above Earth. The satellites are spaced so at least four satellites will be above the horizon from any point on Earth. Each satellite contains a computer, an atomic clock, and a radio transmitter. On the ground, any GPS receiver contains a computer that calculates its own po-

sition by information from at least three of the satellites. The result is provided in the form of a geographic position— longitude and latitude—that is accurate to less than 1 meter (depending on the type of equipment used). Handheld GPS receivers can be purchased for less than $100. Their use in marine navigation and positioning has revolutionized the process of data collection at sea.

Satellite oceanography is an important frontier, and discoveries made by satellites are discussed in later chapters.

ALPHABET OCEANOGRAPHY

Oceanography has become big science. Governments and institutions now cooperate in funding research to describe large-scale marine processes, not just to describe the nature of a small aspect of the ocean. The largest programs are known by their acronyms—words formed from the first letters of their titles. Here are a few of the most important:

- CLIVAR. The CLImate VARiability and Predictability program continues the work of TOGA (the Tropical Ocean and Global Atmosphere program) that officially ended in 1994 and WOCE (the World Ocean Circulation Experiment) that ended in 1998. It is scheduled to become the largest scientific program ever attempted by physical oceanographers. Its three goals: (1) A study of seasonal climate variability and the dynamics of the global ocean–atmosphere–land system; (2) a study of climate variability and predictability in the 10- to 100-year range; and (3) detection of human-induced changes to atmospheric temperature and circulation.
- CoOP. The interdisciplinary Coastal Ocean Processes program seeks to increase understanding of processes that occur at the edges of ocean basins. Continental shelves are a special focus for the investigations.
- GLOBEC. Global Ocean Ecosystem Dynamics will increase our understanding of the causes of variations in the populations of marine organisms resulting from global climate change. It will also address issues of biological diversity.
- JGOFS. Scientists in the Joint Global Ocean Flux Study have concentrated on the ocean's chemical, physical, and biological processes to increase understanding of the ocean's carbon cycle. Their goals are to determine what processes control the movement of carbon between the ocean and the atmosphere and to improve our ability to make global-scale predictions of the likely response of the ocean and atmosphere to human activities.
- ODP. The Ocean Drilling Program involves about 20 countries and expends more than $50 million each year. The successor to the DSDP (Deep Sea Drilling Project), ODP cores the seabed using the drillship *JOIDES Resolution.* ODP's goal is to discover the geological histories of the ocean basins and their margins. ODP's discoveries figure prominently in the history of the ocean floor discussed in Chapters 3 and 4.

Table 2.1   Time Line for the History of Marine Science

Date	Event	Date	Event
4000 B.C.	Egyptian trade on Nile.	1758	Carolus Linnaeus publishes tenth edition of *Systema Naturae*, in which biological nomenclature is formalized (see **Chapter 13**).
3800 B.C.	First maps showing water (river charts).		
1200 B.C.	Phoenicians trade from Mediterranean to Britain and West Africa.	1760	John Harrison's Number Four chronometer (see **Chapter 2**).
1000 B.C.	Polynesians first inhabit Tonga, Samoa.	1768	James Cook's first voyage of discovery (see **Chapter 2**).
900 B.C.	Greeks first use the term *okeanos*, root of our word *ocean*.	1769	Benjamin Franklin publishes first chart showing an ocean current (see **Chapter 2**).
800 B.C.	First graphic aids to marine navigation.	1779	James Cook dies in Hawaii.
600 B.C.	Greek Pythagoreans assume a spherical Earth.	1818	John Ross takes first deep-water and sediment samples.
325 B.C.	Pytheas voyages to Britain, links tides to movement of the moon (see **Chapter 11**); Chinese invent the compass.	1831	Charles Darwin departs on five-year voyage aboard HMS *Beagle* (see **Chapter 13**).
300 B.C.	Library founded at Alexandria.	1835	Gaspard Coriolis publishes first papers on an object's horizontal motion across Earth's surface (see **Chapter 8**).
230 B.C.	Eratosthenes calculates circumference of Earth, invents latitude and longitude.		
127 B.C.	Hipparchus arranges latitude and longitude in regular grid by degrees.	1836	William Harvey devises a taxonomy of seaweeds (see **Chapter 14**).
A.D. 150	Claudius Ptolemy errs in estimating Earth's circumference.	1838	Departure of the United States Exploring Expedition.
A.D. 415	Alexandrian library destroyed.	1847	Hans Christian Oersted observes plankton (see **Chapter 14**).
A.D. 500	Hawaii colonized by Polynesians.	1855	Matthew Maury publishes *Physical Geography of the Seas* (see **Chapter 2**).
A.D. 780	Viking raids begin.		
1000	Norwegian colonies in North America.	1859	Darwin's *Origin of Species* published (see **Chapter 13**).
1460	Prince Henry the Navigator dies.		
1492	Columbus's first voyage.	1872	Departure of *Challenger* Expedition.
1522	Magellan's crew completes first circumnavigation.	1877	Alexander Agassiz begins research in *Blake*.
1609	Hugo Grotius publishes *Mare Liberum*, the foundation for all modern law of the sea (see **Chapter 18**).	1880	William Dittmar determines major salts in seawater (see **Chapter 7**).
1687	Isaac Newton's publication of *Principia Mathematica*, which includes an explanation of the operation of gravity (see **Chapter 11**).	1888	Marine Biological Laboratory founded at Woods Hole, Massachusetts.
1742	Anders Celsius invents the centigrade temperature scale (see **Chapter 6**).	1890	Alfred Thayer Mahan completes *The Influence of Sea Power upon History*.

- RIDGE. The Ridge Interdisciplinary Global Experiments (and its international component Inter-RIDGE) are directed at the dynamics of mid-ocean spreading centers. As you will learn in the next chapter, ridges are the boundaries along which new oceanic crust is created. Remotely piloted research vehicles are playing an important role in these studies. The U.S. contribution is about $6 million per year.

HISTORY IN PROGRESS

As we have seen, hundreds of marine scientists and their students are using an impressive array of equipment to probe the world ocean. *Marine science is by necessity a field science:*

Ships and distant research stations are essential to its progress. The business of operating these ships and staffing the research stations is costly and sometimes dangerous, yet these researchers are willing to meet the daily challenge. They feel a sense of continuity within their separate specialties because oceanographers must be familiar with the scientific literature, the written history of their fields. History is not an abstract area of interest, but rather a part of their daily lives. Some of the important milestones in the history of marine science are shown in **Table 2.1**.

One cannot help but feel that the Alexandrian scholars would have appreciated the 2,000-year effort to understand the ocean. Application of the scientific method to oceanographic studies has yielded many benefits in that time, not the least of which is the satisfaction of knowing a small frac-

Table 2.1 Time Line for the History of Marine Science *(continued)*

Date	Event	Date	Event
1891	Sir John Murray and Alphonse Renard classify marine sediments (see **Chapter 5**).	1962	Rachel Carson's book *Silent Spring* initiates the U.S. environmental movement.
1893	Fridtjof Nansen in Arctic in *Fram* (see **Chapter 2**).	1968	*Glomar Challenger* returns first cores, indicating the age of Earth's crust. The cores support theories of plate tectonics (see **Chapters 3** and **5**).
1900	Richard D. Oldham identifies P and S waves on seismograph (see **Chapter 3**).	1969	Santa Barbara, California, oil well blowout captures national attention (see **Chapter 18**).
1906	Prince Albert I of Monaco establishes the Musée Océanographique.	1970	National Oceanic and Atmospheric Administration (NOAA) established.
1907	Bertram Boltwood calculates age of Earth by radioactive decay (see **Chapters 1** and **3**).	1970	John Tuzo Wilson writes brief history of the tectonic revolution in geology in *Scientific American* (see **Chapter 3**).
1911	Roald Amundsen first at South Pole.		
1912	Alfred Wegener's Frankfurt lectures on continental drift (see **Chapter 3**).	1974	Project FAMOUS (French-American Mid-Ocean Undersea Study) maps and samples the Mid-Atlantic Ridge, a zone of seafloor spreading (see **Chapter 3**).
1912	Scripps Institution allied with the University of California.	1977	*Alvin* finds hydrothermal vents in the Galápagos rift (see **Chapters 3, 4,** and **16**).
1918	Vilhelm Bjerknes formulates theory of atmospheric fronts (see **Chapter 8**).	1978	*Seasat*, the first satellite dedicated to ocean studies, is launched.
1921	International Hydrographic Bureau founded.	1985	*JOIDES Resolution* replaces *Glomar Challenger* in Deep Sea Drilling Project (see **Chapters 2** and **3**).
1925	Departure of *Meteor* Expedition; first echo sounder in operation (see **Chapters 2** and **3**).	1985	R. D. Ballard locates wreck of *Titanic*.
1930	Woods Hole Oceanographic Institution founded.	1987	Observations of supernova 1987A confirm theories of the origin of elements (see **Chapter 1**).
1931	*Atlantis* launched.		
1937	*E. W. Scripps* launched.	1991	JOI researchers bore to a depth of 2 kilometers (1.24 miles) beneath the seafloor near the Galápagos Islands (see **Chapter 3**).
1942	*The Oceans*, first modern reference text, published.		
1943	Jacques Cousteau and Emile Gagnan invent the scuba regulator and tank combination, the "aqualung."	1992	U.S.-French *TOPEX/Poseidon* satellite launched.
1949	Maurice Ewing forms the Lamont-Doherty Geological Observatory (see **Chapters 2** and **3**).	1995	*Keiko*, a small remotely controlled Japanese submersible, sets a new depth record: 10,978 meters (36,008 feet) in the Challenger Deep.
1958	U.S. nuclear submarine *Nautilus* makes first submerged transit of the Arctic ice pack, passes through North Pole (see **Chapter 2**).	1998	*Galileo* spacecraft finds possible evidence of an ocean on Jupiter's moon Europa (see **Chapter 1**).
1960	Bathyscaphe *Trieste* carrying Jacques Piccard and Don Walsh reaches bottom of deepest trench at 10,915 meters (35,801 feet) (see **Chapter 4**).	2000	*Mars Global Surveyor* photographs channels perhaps carved by flowing water (see **Chapter 1**).

tion of the story of Earth and its ocean. The lure of voyaging has been behind many of these discoveries. Scientists in the oceanographic research ships, laboratories, and libraries of the world go on collecting knowledge today. With luck and support, their efforts will continue into the distant future.

QUESTIONS FROM STUDENTS

1. **How do we know when people reached certain locations? How do researchers know New Guinea was inhabited by 30,000 years ago, the tip of South America by 11,400 years ago, and Hawaii by A.D. 450–600?**

Researchers use various methods to date the human artifacts they find. The **radiometric dating** technique, for example, depends on the slow and predictable decrease in radioactivity of naturally radioactive materials. One commonly used form of radiometric dating measures the amount of radioactive carbon in once-living material such as wood or bone. A small amount of the carbon dioxide in the atmosphere contains carbon atoms that have been made radioactive by natural processes. This carbon dioxide is made into glucose and then into cellulose (and wood) by plants. The analysis assumes a fixed ratio of radioactive to nonradioactive carbon dioxide in the air through time. By knowing how the present proportion of radioactive carbon in wood differs from the presumed initial proportion, scientists know how long ago the wood ceased taking up CO_2, and therefore when it stopped living. Animal remains may also be dated this way because animals derive nutrients from photosynthesizing plants. All organic material up to 40,000 years old—including human bones, tools with wood or bone pieces, and food remains—can be dated using this method. More on the topic of radiometric dating can be found in Box 3.2, page 70.

2. It would be difficult for humans to walk from Siberia to Alaska today. How was it possible in the past?

The great migrations to the Americas took place at the end of the last ice age, about 13,000 years ago. At that time the large amount of water trapped as ice on the continents caused sea level to fall 100–125 meters (300–400 feet) lower than it is today. The lower sea level exposed land in the Bering Sea and the Aleutian arc between Siberia and Alaska. Lower sea level, combined with the jam of pack ice against the islands themselves, made a passage for migrating game. People followed the animals for food. Both ended up in the "New World" in the process.

3. What's this about a chronometer not having to keep perfect time? I thought you had to know exactly what time it is to be able to calculate your longitude.

Yes, you need accurate time. But a chronometer is valuable not because it necessarily keeps perfect time, but because it loses or gains time at a constant, known rate. Each day the navigator multiplies the number of seconds the clock is known to gain (or lose) in a day by the number of days since the clock was last set—and then adds the total to the time shown on the chronometer's face to obtain the real time. The value of a chronometer lies entirely in its consistency.

4. How do modern navigators find their position at sea?

Very dull story. They push a few buttons on a box and read their latitude and longitude directly on a screen. This is accomplished by observation of the accelerations experienced by a "black box" since leaving a starting point (inertial navigation) or by analysis of radio transmissions from land stations or satellites. The satellite system is rapidly becoming affordable: For less than $100, you can now buy a small, handheld portable receiver capable of receiving Global Positioning System satellite signals. The GPS system can be accurate to a few centimeters! None of these methods is nearly as much fun as the old-fashioned sextant-and-chronometer method, but I suspect that Captain Cook would be *very* impressed by our new tools.

5. Speaking of James Cook, what was his motivation for those extraordinary voyages?

One could say simply that he was a serving Royal Naval Officer and was ordered to go. But Beaglehole, Hough, and other biographers suggest that the story is more complex. How did a relatively unschooled man become leader of one of the

first scientific oceanographic expeditions? Cook had the usual attributes of a successful person—intelligence, strength of character, meeting the right people at the right time, health, focus, luck—but he also had a driving intellectual curiosity and a rare (for that era) tolerance and respect for alien cultures. As Hough (1994) writes, "Cook stood out like a diamond amidst junk jewelery. . . ." It was no surprise that the Lords of the Admiralty settled upon this unique man to lead the adventure.

And the atmosphere of that adventure? Another excerpt from Hough's 1994 biography is instructive: ". . . we can still stand back in wonder at the dimensions of these orders, at the sheer effrontery of asking nearly one hundred men to embark in a wooden vessel, scarcely more than one hundred feet in length, and dependent for its mobility upon the whim of winds, and currents and tides, to sail to the other side of a world only dimly charted, there to carry out an exacting observation, and then to discover a completely uncharted continent, survey its coastline and take note of its characteristics. Perhaps an even greater wonder was that men of experience and wisdom were prepared to take on this undertaking with pleasure and excitement."

6. Did Columbus discover North America?

No. He never saw North America.

KEY CONCEPTS TO REVIEW

- The ocean did not prevent the spread of humanity to virtually all habitable areas.

- The origins of marine science lie in voyaging—traveling on the ocean for a specific purpose. Technological advances made during voyaging and later marine exploration led to the rise of scientific oceanography.

- Earth's shape and circumference were accurately estimated around 230 B.C. at the Library of Alexandria, Egypt. The use of latitude and longitude for positioning and navigation also began there.

- Captain James Cook, Royal Navy, was perhaps the first ocean observer careful enough to be considered a marine scientist.

- The HMS *Challenger* Expedition was the first purely scientific voyage of oceanic exploration.

- Polar studies greatly advanced marine science at the beginning of the twentieth century.

- Now, nearly all research is conducted not by individuals but by teams of specialists working in large, nationally funded oceanographic or military institutions.

- The tools of modern oceanography include satellites, manned and remotely controlled vehicles, and computer modeling.

The Web site for this book contains helpful study aids. Each vignette, box, and first- and second-level head has a number that corresponds to information at the Web site. Log on to:

 www.brookscole.com/product/053437557xs

and click on the Chapter-by-Chapter area. Choose Chapter 2 and select a resource:

- **Flash Cards** allows you to test your mastery of the Terms and Concepts to Remember for this chapter.
- **Tutorial Quizzes** provides a multiple-choice practice quiz.
- **Student Guide to InfoTrac College Edition** will lead you to Critical Thinking Projects that use InfoTrac College Edition as a research tool.

TERMS AND CONCEPTS TO REMEMBER

cartographer
celestial navigation
Challenger Expedition
chart
chronometer
Columbus, Christopher
compass
Cook, James
Darwin, Charles
echo sounder
Eratosthenes of Cyrene
Franklin, Benjamin
GPS (Global Positioning System)
Harrison, John
Jason-1
latitude
Library of Alexandria
longitude

Magellan, Ferdinand
Mahan, Alfred Thayer
Maury, Matthew
Meteor Expedition
oceanography
oceanus
Polynesia
Prince Henry the Navigator
radiometric dating
sea power
Seasat
SEASTAR
sounding
TOPEX/Poseidon
United States Exploring Expedition
Vikings
voyaging

STUDY QUESTIONS

Review Questions

1. What features would be most useful to include in a nautical chart? Why?

2. What were the stimuli to Polynesian colonization? How were the long voyages accomplished? How were Polynesian voyages different from (and similar to) those of the Vikings?

3. What were the main stimuli to European voyages of exploration during the Age of Discovery? Why did it end?

4. What were the oceanographic contributions of Prince Henry the Navigator, Benjamin Franklin, Matthew Maury, and Charles Darwin?

5. How can you find your approximate latitude and longitude at sea?

6. What were the goals and results of the United States Exploring Expedition? What U.S. institution greatly benefited from its efforts?

7. What was the first purely scientific oceanographic expedition, and what were some of its accomplishments? What contributions did the earlier, hybrid expeditions make?

Critical Thinking Questions

1. How could you convince a 10-year-old that Earth is round? What evidence would a child offer that it's flat? How can you counter those objections?

2. How did the Library of Alexandria contribute to the development of marine science? What happened to most of the information accumulated there? Why do you suppose the residents of Alexandria became hostile to the librarians and the many achievements of the library?

3. How did Eratosthenes calculate the approximate size of Earth? Which of his assumptions was the "shakiest"?

4. If Columbus did not discover North America, then who did?

5. Imagine that you set your watch at local noon in Kansas City on Monday and then fly to the Coast on Tuesday. You stick a pole into the ground on a sunny day at the beach, wait until its shadow is shortest, and look at your watch. The watch says 10:00 A.M. Are you on the East Coast or the West Coast?

6. Sketch briefly the major developments in marine science since 1900. Do individuals, separate voyages, or institutions figure most prominently in this history?

7. **InfoTrac College Edition Project** Navigation technology has become highly sophisticated in the past 30 years. Ships' crews now rely on satellite transmissions and computer systems to tell them their location, often with accuracy within a few meters. Companies are promoting small Global Positioning System (GPS) devices for use by automobile travelers and hikers. What are the pros and cons of increased reliance on such devices? Research this question using InfoTrac College Edition.

The Web site for this book contains many ideas for further reading and research. Log on to:

www.brookscole.com/product/053437557xs

and click on the Chapter-by-Chapter area. Choose Chapter 2 and select a resource:

- **References** lists the major books and articles consulted in writing this chapter, along with comments from the author about their content and reading level.
- **Hypercontents** takes you to an extensive list of sites with news, research, and images related to individual sections of the chapter.

- In **Student Guide to InfoTrac College Edition,** scroll down to Suggested Readings from InfoTrac College Edition for brief descriptions of the articles listed and search hints for finding them in InfoTrac.
- **Regional InfoTrac College Edition** articles are organized in East Coast, West Coast, and Gulf Coast regions, allowing you to study oceanography on a more local level. You can also access Regional InfoTrac College Edition at www.localocean.com.

For additional readings, go to InfoTrac College Edition, your online research library, at:

http://infotrac.thomsonlearning.com/

Plate Tectonics 3

A DANGEROUS BREAKTHROUGH

Powerful forces inside our planet are continually forming the major features of its surface. These forces determine the outlines and locations of the continents and ocean floors. They build and destroy mountains and seabeds, raise islands, power volcanoes, form deep trenches, and, through earthquakes, influence the lives of millions of people. Few discoveries in marine science are as exciting to researchers as the recent breakthroughs in our understanding of how these forces work. But intellectual breakthroughs are one thing, physical breakthroughs quite another.

In October 1987 the scientists and crew of the Scripps Institution research vessel *Melville* were present for a breakthrough of sorts—the eruption of an undersea volcano directly beneath their ship! *Melville* was on an expedition to collect rock and water samples from the MacDonald Seamount, a submerged volcano in French Polynesia, located 1,100 kilometers (700 miles) west of Pitcairn Island. When the research team arrived on station they noticed large patches of greenish brown water containing fine particles of volcanic ash, which suggested recent volcanic activity. The crew was lowering their gear to the top of the seamount, which rises to within 40 meters (130 feet) of the surface, when huge bubbles of gas and steam suddenly engulfed the ship, making "horrendous clangs and clamors" as they burst against *Melville*'s hull. Chocolate-colored water containing steaming lava balls too hot to hold in bare hands streamed to the surface. The ship's depth recorder and hull-mounted water

Lava meets the ocean on the active island of Hawaii.

temperature sensor broke down, but no one was injured. The eruption lasted about five minutes.

The MacDonald Seamount is the last in a chain of submerged and emergent volcanoes stretching 2,000 kilometers (1,200 miles) across the floor of the South Pacific. Here magma (molten rock) rises from deep inside Earth. The composition of gases within the young rock would tell *Melville* researchers about the forces and conditions that cause island chains to form and something about the history of the Pacific floor itself. Samples they obtained that morning helped to confirm recent theories about oceanic *hot spots,* themselves a verification of the theory of plate tectonics. Seldom do researchers have the good fortune to be in exactly the right place at exactly the right time; scientific breakthroughs are generally less threatening to life and property!

3-1

WHAT TO WATCH FOR IN CHAPTER 3

This chapter describes the inner structure of Earth and how that structure forms the ocean floor and the continents. You may be surprised to discover that the rigid, brittle surface of our planet floats on a hot, deformable layer of partially melted rock. Over long spans of time, movement in and below this layer carries continents across the face of Earth, splits ocean basins, and shatters cities. Much of what you'll learn in the next few chapters depends on an understanding of this movement and its consequences.

As you read this chapter you should be especially alert to the idea of *density*—mass per unit of volume. How does density affect the way Earth's inner layers are stacked? How can such heavy layers move? How did people figure out what's inside Earth without having been there or having deep samples to study?

Answers to these questions are found in a set of interrelated ideas known as plate tectonics. As you read about this theory, notice the contributions of the major players and the scientific logic they used. As you may have already done, they began by wondering about the jigsaw-puzzle fit of Africa and South America.

The chapter ends with descriptions of some of the evidence researchers have uncovered in support of plate tectonics. Did you ever wonder why oceans and continents are where they are? Here are some answers. Mysteries remain, but tectonic theory explains (and predicts) many of Earth's features and activities. For me, the most interesting bit of information from studies of plate tectonics is an explanation of why our planet has such young seafloors, the oldest of which is only as old as the dinosaurs—that is, about $\frac{1}{23}$ the age of Earth.

A LAYERED EARTH

It might seem easy to satisfy our curiosity about the nature of the inner Earth by digging or drilling for samples. The deepest hole drilled so far is being bored by researchers on the Kola Peninsula in Russia. In 1992 they reached a depth of 12,063 meters (7.5 miles), where temperatures are 245°C (473°F) and pressure caused the hole casing to collapse. Drilling has also been conducted at sea. The oceanic drilling record is held by *JOIDES Resolution.* In 1991, after 12 years of intermittent effort, a drill aboard the ship penetrated 2 kilometers (1.2 miles) of seafloor beneath 2.5 kilometers (1.6 miles) of seawater.

No matter where investigators drill, the samples of rock they recover are not exceptionally dense. But the deepest probes have penetrated less than 1/500 of the radius of Earth. Should we assume that Earth consists of lightweight (less dense) rock all the way through?

Thanks to studies of Earth's orbit begun in the late 1700s, we know Earth's total mass. This mass is much greater than would have been predicted from even the deepest rocks ever collected. Earth's interior must therefore contain heavier (denser) substances than the rocks from the deep drill holes. We might expect materials from the interior to get denser gradually with increasing depth, but geologists have shown that the density of the materials increases abruptly at specific depths. Thus, they are convinced that Earth has *distinct interior layers,* somewhat resembling the inside of a peach (see **Figure 3.1**).

Density is an expression of the relative heaviness of a substance; it is defined as the mass per unit volume, usually expressed in grams per cubic centimeter (g/cm^3). The density of pure water is 1 g/cm^3. Granite rock is about 2.7 times denser, at 2.7 g/cm^3. As we saw in Chapter 1, Earth was formed by accretion from a cloud of dust, gas, and stellar debris. Gravity later sorted the components by density, stratifying Earth into layers (see again Figure 1.11). Because each deeper layer is denser than the layer above, we say Earth is **density stratified.**

Although researchers have never directly collected samples from below the outermost layer of Earth, they have indirect evidence about the chemical composition, density, temperature, and thickness of each layer.

Classifying the Layers by Composition

Early in this century researchers pieced together a view of Earth's interior based on measurements of volcanic gases, earthquake shocks, and variations in the pull of gravity. The first useful classification of Earth's interior emphasized chemical composition. Geologists named the layers *crust, mantle,* and *core.*

The **crust** is the thin, relatively lightweight outermost layer. It accounts for only 0.4% of Earth's total mass and less than 1% of its volume. The crust beneath the ocean differs in thickness, composition, and age from the crust of the continents. The thin **oceanic crust** is mostly **basalt**—a heavy,

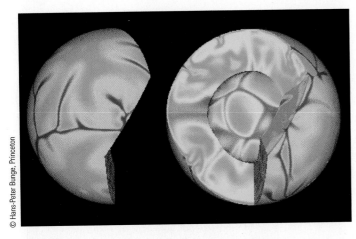

Figure 3.1 Earth in cross section, showing zones of hot and cooler material. The outer core is seen beneath the mantle. Heat generated within Earth causes its layers to move slowly, powering the drift of continents and the spreading of seafloors.

dark-colored rock composed mostly of oxygen, silicon, magnesium, and iron. Its density is about 2.9 g/cm³. By contrast, the most common material in the thicker **continental crust** is **granite,** a familiar light-colored rock composed mainly of oxygen, silicon, and aluminum. Its density is about 2.7 g/cm³.

The **mantle,** the layer beneath the crust, comprises 68.1% of Earth's mass and 83% of its volume. Mantle materials are thought to contain mainly silicon and oxygen with iron and magnesium. Its average density is about 4.5 g/cm³. The mantle is about 2,900 kilometers (1,800 miles) thick.

The **core,** Earth's innermost layer, consists mainly of iron (90%) and nickel, along with silicon, sulfur, and heavy elements. Its average density is about 13 g/cm³, and its radius is approximately 3,470 kilometers (2,160 miles). The core accounts for about 31.5% of Earth's mass and about 16% of its volume.

Classifying the Layers by Physical Properties

Subdividing Earth on the basis of chemical composition does not necessarily reflect the *physical* properties and behavior of rock materials in these layers. Different conditions of temperature and pressure prevail at different depths, and these conditions influence the physical properties of the materials subjected to them.

Slabs of Earth's relatively cool and solid crust and upper mantle float—and move—independently of one another over the hotter, pliable (deformable) mantle layer directly below. Physical properties are more important than chemical ones in determining this movement, so geologists have devised a classification based on physical properties. These are shown in **Figure 3.2:**

- The **lithosphere** (*lithos* = rock)—Earth's cool, rigid outer layer—is 100 to 200 kilometers (60 to125 miles) in thickness. It is comprised of the continental and oceanic crusts *and* the uppermost cool and rigid portion of the mantle.

- The **asthenosphere** (*asthenes* = weak) is the hot, partially melted, slowly flowing layer of upper mantle below the lithosphere extending to a depth of about 350 to 650 kilometers (220 to 400 miles).
- The **mantle** extends to the core. The asthenosphere and the mantle below the asthenosphere (the lower mantle) have a similar chemical composition. Although it is hotter, mantle material below the asthenosphere does not melt because of rapidly increasing pressure. As a result it is denser and flows much more slowly.
- The **core** has two parts. The outer core is a dense, viscous liquid. The inner core is a solid with a maximum density of about 16 g/cm³, nearly six times the density of granite rock. Both parts are extremely hot, with an average temperature of about 5,500°C (9,900°F). Recent evidence indicates that the inner core may be as hot as 6,600°C (12,000°F) at its center, hotter than the surface of the sun!

Figure 3.2 shows the lithosphere and asthenosphere in detail. *Note that the rigid sandwich of crust and upper mantle— the lithosphere—floats on (and is supported by) the denser deformable asthenosphere.* Note also that the structure of oceanic lithosphere differs from that of continental lithosphere. Because the thick granitic continental crust is not exceptionally dense, it can project above sea level. In contrast, the thin, dense basaltic oceanic crust is almost always submerged.

Isostatic Equilibrium

Why do large regions of continental crust stand high above sea level? If the asthenosphere is nonrigid and deformable, why don't mountains sink because of their mass and disappear? Another look at Figure 3.2 will help to explain the situation. The mountainous parts of continents have "roots" extending into the asthenosphere. The continental crust and the rest of the lithosphere "float" on the denser asthenosphere. The situation involves buoyancy, the principle that explains why ships float.

Buoyancy is the ability of an object to float in a fluid by displacing a volume of that fluid equal in weight to the floating object's own weight. *A steel ship floats because it displaces a volume of water equal in weight to its own weight plus the weight of its cargo.* An empty containership displaces a smaller volume of water than the same ship fully loaded (**Figure 3.3**). The water supporting the ship is not *strong* in the mechanical sense—water does not support a ship the same way a steel bridge supports the weight of a car. Buoyancy, rather than mechanical strength, supports the ship and her cargo.

Any part of a continent that projects above sea level is supported in the same way. Consider the continent containing Mount Everest, highest of Earth's mountains at 8.84 kilometers (29,007 feet) above sea level. Mount Everest and its neighboring peaks are not supported by the *mechanical* strength of the materials within Earth; nothing in our world is that strong. Over a long period, and under the tremendous weight of the overlying crust, the asthenosphere behaves like

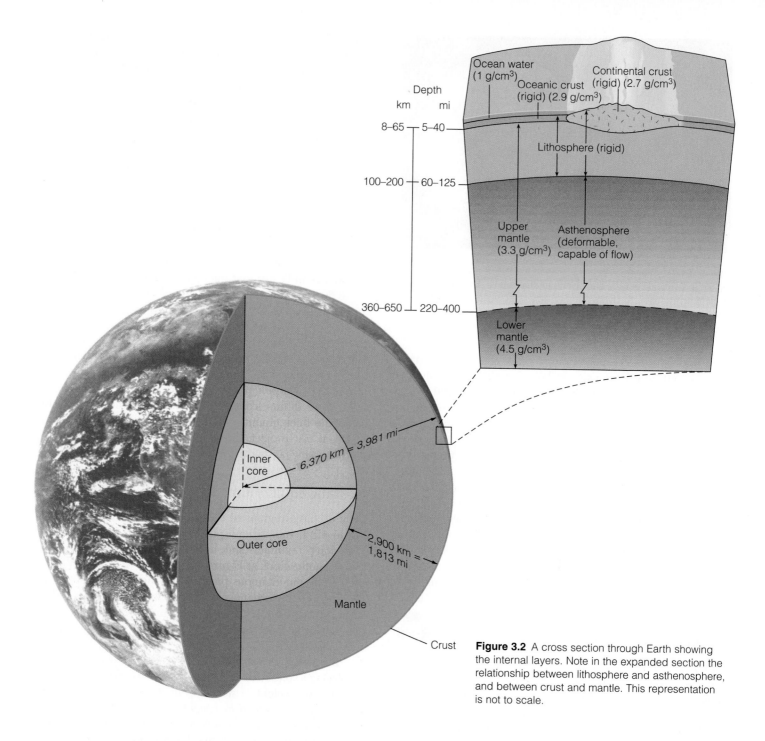

Figure 3.2 A cross section through Earth showing the internal layers. Note in the expanded section the relationship between lithosphere and asthenosphere, and between crust and mantle. This representation is not to scale.

a dense, viscous, slowly moving fluid. *The continent's mountains float high above sea level because the lithosphere gradually sinks into the deformable asthenosphere until it has displaced a volume of asthenosphere equal in mass to their mass.* The mountains stand at great height, nearly in balance with their subterranean underpinnings but susceptible to rising or falling as erosion or crustal stresses dictate. Lower regions are supported by shallower roots. In a slow-motion version of a ship floating in water, the entire continent stands in **isostatic equilibrium** (*isos* = equal, *stasis* = standing).

What happens when a mountain erodes? In much the same way a ship rises when cargo is removed, Earth's crust will rise in response to the reduced load. Ancient mountains that have undergone millions of years of erosion often expose rocks that were once embedded deep within their roots. This kind of isostatic readjustment results in the thinning of the continental crust beneath the mountains, and subsidence beneath areas of deposited sediments. This process is shown in **Figure 3.4.**

Unlike the asthenosphere on which lithosphere floats, crustal rock does not slowly flow at normal surface temperatures. A ship reacts to any small change in weight with a correspondingly small change in vertical position in the water, but an area of continent or ocean floor cannot react to every small weight change because the underlying rock is *not* liquid,

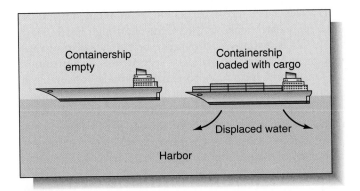

Figure 3.3 The principle of buoyancy. A ship sinks until it displaces a volume of water equal in weight to the weight of the ship and its cargo.

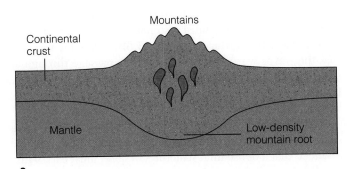

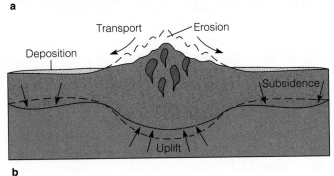

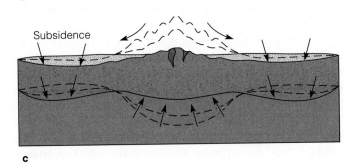

Figure 3.4 Erosion and isostatic readjustment can cause continental crust to become thinner in mountainous regions. As mountains are eroded, isostatic uplift causes their roots to rise. The same thing happens when a ship is unloaded. Further erosion exposes rocks that were once embedded deep within the peaks. Deposition of sediments away from the mountains often causes nearby crust to sink.

the deformation does not occur rapidly, and the edges of the continent or seabed are mechanically bound to adjacent crustal masses. When the force of uplift or downbending exceeds the mechanical strength of the adjacent rock, the rock will fracture along a plane of weakness—a **fault.** The adjacent crustal fragments will move vertically in relation to each other. This sudden adjustment of the crust to isostatic forces by fracturing, or faulting, is one cause of earthquakes. The lithosphere does not always behave as a rigid, brittle solid, however. Where forces are applied slowly enough, some of this material may deform without breaking.

Sources of Internal Heat

Heat from within Earth keeps the asthenosphere pliable and the lithosphere in motion. Studies of the sources and effects of Earth's internal heat have contributed to our understanding of the planet's internal construction. In the late 1800s the British mathematician and physicist William Thomson (Lord Kelvin) calculated that Earth was about 80 million years old, an estimate based on the rate at which the planet would have cooled from an original molten mass. Geologists tried to explain mountain building based on Lord Kelvin's assumption of progressive cooling. In their drying-fruit model, Earth was considered to have shrunk as it cooled. Mountains were thought to be shrinkage wrinkles—like those seen as a grape transforms into a raisin—and earthquakes were thought to be caused by jerks during this wrinkling. Earth's true age was not then known, and the "wrinkle" theory depended on rapid cooling of a relatively young Earth. When further research showed that Earth is about 4.6 billion years old (see Chapter 1), calculations of heat flow indicated that our planet should have cooled almost completely by now. Earthquakes, active volcanoes, and hot springs clearly indicated it had not! There must be other sources of heat energy besides the trapped ancient heat of formation. Better explanations for mountain building and earthquakes were needed.

An important source of heat that was not recognized in Lord Kelvin's time is **radioactive decay.** Though most atoms are stable and do not change, some unstable forms of elements are unstable and give off heat when their nuclei break apart (decay). Radioactive particles are ejected in the process. As we also saw in Chapter 1, radioactive decay within the newly formed Earth released heat that contributed to the melting of the original mass. Most of the melted iron sank toward the core, releasing huge amounts of energy. This residual heat combines with the much greater heat given off by the continuous decay of radioactive elements within the crust and upper mantle (primarily potassium, uranium, and thorium).

Some of Earth's internal heat journeys toward the surface by **conduction,** a process analogous to the slow migration of heat along a skillet's handle. Some heat also rises by **convection** in the asthenosphere. Convection occurs when a fluid is heated, expands and becomes less dense, and rises. (Convection causes air to rise over a warm radiator—see, for example, Figure 8.6.)

So, even after 4.6 billion years, heat continues to flow out from within Earth. As we shall see, this heat, not raisinlike global shrinkage, builds mountains and volcanoes, causes earthquakes, moves continents, and shapes ocean basins.

THE EVIDENCE FOR LAYERING

It has been known since the mid-1800s that low-frequency waves can travel through the interior of Earth. The forces that cause **earthquakes** generate low-frequency waves called **seismic waves** (*seismos* = earthquake). Some of these waves radiate through Earth, reflecting or bending as they travel, and eventually reappear at the surface. Careful study of the time and location of their arrival at the surface, along with changes in the frequency and strength of the waves themselves, has revealed the information about the nature of Earth's interior discussed earlier in this chapter. (We use the same kind of analysis to select a ripe watermelon. If we tap the outside and hear a *tick*, we suspect the melon isn't ripe. A *thunk* indicates a winner.)

Primary and Secondary Seismic Waves

Two kinds of seismic waves are particularly useful for analyzing Earth's interior structure. One kind of wave, the **P wave** (or primary wave), is a compressional wave similar in behavior to a sound wave. Rapidly pushing and pulling a very flexible spring (like a Slinky) generates P waves. The **S wave** (or secondary wave) is a transverse wave like that seen in a rope shaken side to side. Both kinds of seismic waves are shown in **Figure 3.5.**

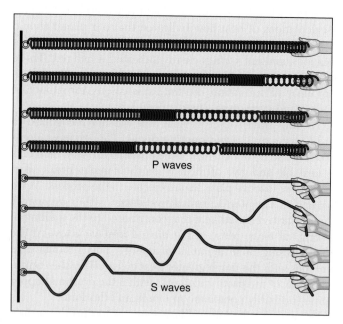

Figure 3.5 P waves (primary waves) are compressional waves like those seen in a Slinky that is alternately stretched and compressed. S waves (secondary waves) are side-to-side waves like those seen in a shaken rope. Both kinds of waves are associated with earthquakes. (Source: From *The Earth: An Introduction to Physical Geology,* 4/e by Edward J. Tarbuck and Frederick K. Lutgens. © 1993 Macmillan College Publishing. Reprinted by permission of Pearson Education, Inc. Upper Saddle River, NJ 07458.)

P waves and S waves are generated simultaneously at the source of an earthquake. P waves, travel through Earth nearly twice as fast as S waves, so they arrive first at the **seismograph** (*seismos* = earthquake, *graphos* = to write), an instrument that senses and records earthquakes. Liquids are unable to transmit the side-to-side S waves but do propagate compressional P waves. Solid rock transmits both kinds of waves. Analysis of the characteristics of seismic waves returning to Earth's surface after passage through the interior suggests which parts of the interior are solid, liquid, or partially melted.

Shadow Zones

In 1900 the English geologist Richard Oldham first identified P and S waves on a seismograph. If Earth were perfectly homogeneous, seismic waves would travel at constant speeds from an earthquake, and their paths through the interior would be straight lines (**Figure 3.6a**). Oldham's investigations, however, showed that seismic waves were arriving *earlier* than expected at seismographs far from the quake. This meant that the waves must have traveled *faster* as they went down into Earth. They must also have been refracted—bent back toward the surface (**Figure 3.6b**). Oldham reasoned that the waves were being influenced by passage through areas of Earth with different density and elastic properties than those seen at the surface. Thus it was found that Earth is not homogeneous and that its properties vary with depth.

In 1906 Oldham made a critical discovery: No S waves survived deep passage through Earth (**Figure 3.6c**). Oldham deduced that a dense fluid structure, or core, must exist within Earth to absorb the S waves (**Box 3.1**). He further predicted that a **shadow zone,** a wide band from which S waves were absent, would be found on the side of Earth opposite the location of the earthquake. The existence of the shadow zone and liquid core was verified by seismographic analysis in 1914.

What about the P waves—the compressional waves that can pass through liquid? Oldham found that P waves arrived at a seismograph farthest away from an earthquake (that is, on the opposite side of the globe) much more slowly than expected. They had been deflected but not stopped by Earth's core (**Figure 3.6d**). Working from this information, later researchers were able to calculate the mantle–core boundary to be 2,900 kilometers (1,800 miles) below the surface.

More sensitive seismographs were developed in the 1930s. In 1935 the Danish seismologist Inge Lehmann suggested that the very faint, very low frequency P waves discovered opposite the earthquake site had speeded up as they passed through an inner core, indicating that it was a solid (see again Figure 3.6d). Measurements of subtle differences in the pull of gravity, plus a more accurate estimate of Earth's mass (derived from precise timings of the orbits of artificial satellites), gave further clues to the layered structure of Earth. By the early 1960s another new generation of sensitive seismographs stood ready to provide geologists with an even better understanding of Earth's inner configuration. To confirm their theories, scientists needed data from a very large earthquake.

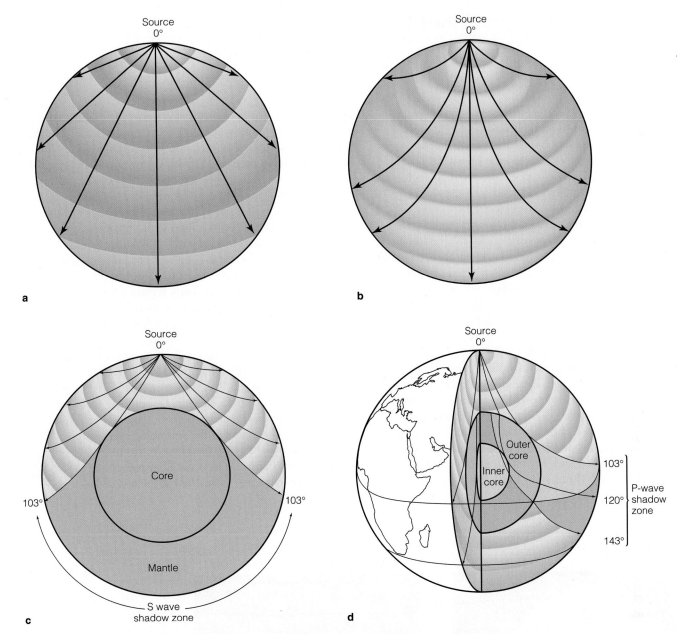

Figure 3.6 How earthquakes contributed to our model of the layered Earth. (**a**) Earthquake waves passing through a homogeneous planet would not be reflected or refracted (bent). The waves would follow linear paths. (**b**) In a planet that gets progressively denser and more rigid with depth, the waves would bend along curved paths. (**c**) Our Earth has a liquid outer core through which the side-to-side S waves cannot penetrate. A shadow zone extends about halfway around the world from an earthquake's focus. (**d**) P waves (compressional waves) can penetrate the liquid outer core but are bent in transit. A P wave shadow band forms between 103° and 143° from an earthquake's focus. Very sensitive seismographs can sometimes detect weak P wave signals reflecting off the solid inner core (see Box 3.1). (Source: From *The Earth's Dynamic Systems*, 9/e, by Hamblin & Christiansen. Reprinted by permission of Pearson Education, Inc., Upper Saddle River, NJ 07458.)

Alaska, 27 March 1964

They did not have to wait long. On the last Friday of March 1964, at 1736 (5:36 P.M.), one of the largest earthquakes ever recorded struck 144 kilometers (90 miles) east of Anchorage, Alaska (**Figure 3.7**). The release of energy tore the surface of Earth for 800 kilometers (500 miles) between the small port of Cordova in the east and Kodiak Island in the west. In some places the vertical movement of the crust was 3.7 meters (12 feet); one small island was lifted 11.6 meters (38 feet). Horizontal movement caused the greatest damage: 65,000 square kilometers (25,000 square miles) of land abruptly moved west. In 4½ minutes of violent shaking, Anchorage had moved sideways 2 meters (6.6 feet), and the town of Seward had moved

By carefully studying low-frequency pulses of energy generated by the forces that cause earthquakes, researchers have been able to discover much about the nature of Earth's interior, much as a tap on a melon can tell the buyer if it is ripe. In one of the twentieth century's nicest pieces of geological detective work, a similar analysis allowed scientists to discover the concentric cores within our planet.

In 1906, British geologist Richard Oldham published a paper showing that some types of earthquake-related low-frequency pulses—called seismic waves—do not penetrate straight through Earth. He found that other types of seismic waves arrive at measuring instruments farthest away from an earthquake (that is, on the opposite side of the globe) much more slowly than expected. Oldham thought that a dense fluid structure, or core, must exist within Earth to slow and absorb the waves. He further predicted that a shadow zone, a wide band from which seismic waves would be nearly absent, would encircle the side of Earth opposite the location of the earthquake. Such a shadow zone could be formed by reflection and refraction of waves by a liquid core. The existence of the shadow zones and liquid core was verified in 1914.

A subtler feature was discovered by Inge Lehmann, a Danish seismologist. Denmark is well situated for recording waves that pass through the core of Earth from large earthquakes around the southern Pacific rim. By 1935 a new generation of very sensitive detection instruments was available, and Lehmann used them to sense the very faint, very low frequency waves reaching Earth's surface opposite an earthquake. The waves had speeded up as they passed through an inner core, indicating that it was a solid. Delicate secondary reflections suggested some of the characteristics of this material. She had discovered a rigid white-hot mass about the size of the moon suspended within Oldham's liquid core.

Excellent teachers had prepared both researchers to learn from the ghostly echoes of the cores. While Oldham's educational background was fairly typical for a Victorian gentleman of the era, Inge Lehmann's personal history was not. Educated at the first coeducational school in Denmark, Lehmann decided to pursue a scientific career, partly because "no difference between the intellect of boys and girls was recognized [at the school], a fact that brought me disappointment later in life when I had to realize that this was not the general attitude." A few years after graduating from the University of Copenhagen in mathematics and physical sciences, she was appointed chief of the seismological department of the Royal Danish Geodetic Institute in Copenhagen, a post she held until her retirement in 1953. She passed away in 1993 at the age of 105.

3-11

a Richard Oldham, discoverer of Earth's core.

National Portrait Gallery, London

b Inge Lehmann, discoverer of Earth's inner core.

© National Survey & Cadastre, Denmark

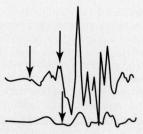

Sverdlosk

c Traces from Lehmann's 1936 paper showing the arrival in Russia of seismic waves from a New Zealand earthquake on 16 June 1929. The core wave arrivals are marked by arrows. (Source: B. Bolt, "Earthquakes and Geological Discovery." Scientific American Books, © 1993.)

14 meters (46 feet)! A seismic (earthquake-generated) sea wave destroyed two harbors. Over 75% of the state's commerce was disrupted, and thousands of people were made homeless. Damage exceeded $750 million, and, considering the violence of the earthquake, it is a wonder that only 115 lives were lost.

Seismic stations over much of the world saw extraordinarily large P waves arrive from Alaska. Many of the 800 seismographs online worldwide were physically damaged. The arrival times of the P and S waves at each station were carefully noted. When correlated with the frequency, intensity, and phase characteristics of the waves, this information helped to confirm the models of Earth layering. The shaken citizens of Anchorage probably didn't derive much comfort from the knowledge gained from "their" earthquake.

TOWARD A NEW UNDERSTANDING OF EARTH

Earth's layered internal structure—its brittle lithosphere floating on the hot and viscous asthenosphere—ensures that its surface will be geologically active. Earthquakes and volcanoes attest to the commotion within. But how do internal layering and heat contribute to mountain building, the arrangement of continents, the nature of the seafloors, and the wealth of seemingly randomly distributed geological features found everywhere? Are there patterns and order in this apparent chaos? Let us see how we achieved our current understanding of the answers to these questions.

The Age Debate

Geologists read Earth's history in its surface rocks and features. In the past the effort to understand geological processes was especially hindered by an incomplete understanding of Earth's age. Advances in geology had to wait—quite literally—for the time to be right.[1]

As you saw in Chapter 1, interlinking lines of evidence now tell us the 4.6-billion-year age of Earth can be accepted with confidence. But before appropriate tools were developed, researchers were frustrated by tradition and seemingly contradictory data. At the end of the eighteenth century most European natural scientists believed in a young Earth, one that had formed only about 6,000 years ago. This age had been determined not by an analysis of rocks, but through the genealogy of the Bible's Old Testament. Careful reading of the book of Genesis in 1654 had convinced Irish bishop James Ussher that the Creation had taken place on 26 October 4004 B.C.

James Hutton—a Scottish physician with an interest in geology—decided that the biblical account of the Creation was incorrect because it implied that the landscape was mostly stable and unchanging. Hutton, however, had measured geological changes: the rate at which streambeds eroded, the distribu-

---

[1] For information on geologic time and its divisions, please see Appendix II.

Figure 3.7 Much of the Anchorage suburb of Turnagain Heights was destroyed by landslides and ground liquefaction in the 27 March 1964 Alaska earthquake. The earthquake's magnitude was 9.2—the greatest ever recorded in the United States.

Wilbur Garrett/NGS Image Collection

tion of sediments by rivers, and the patterns of rocks in the Scottish countryside. From his observations, Hutton concluded that the rate of geological change today is not greatly different from the rate of change in the past. His principle of **uniformitarianism,** formalized in 1788, suggested that all of Earth's geological features and history could be explained by processes identical to ones acting today, and that these processes must have been at work for a very long time (see **Figure 3.8**).

A few scientists agreed with Hutton. His detractors, however, asked some painful questions: If Earth is very old, and if erosive forces have continued uniformly through time, why isn't Earth's surface eroded flat? Why isn't the ocean brimming with sediment? What post-Creation forces could build mountains?

Believers in another school of thought, **catastrophism,** were able to answer these objections by interpreting the biblical account of the Creation literally. The catastrophists were convinced that Earth was very young and that the biblical flood was responsible for the misleading appearance of Earth's great age. The flood, they maintained, had folded and exposed strata, toppled mountains, filled shallow ocean basins with sediments, and caused many plants and animals to become extinct. This theory had the additional benefit of explaining how seashell fossils could be present on mountaintops.

A further complication was introduced in the late 1850s when Charles Darwin and Alfred Russell Wallace proposed a rational mechanism, natural selection, by which new kinds of living things might come about. Because natural selection required long periods of time to generate the overwhelming variety of life forms on Earth, biological evidence also suggested an ancient Earth. The arguments intensified.

Scientists attempting to prove or disprove uniformitarianism, catastrophism, and evolution made a wealth of new discoveries during the last half of the nineteenth century. As

Figure 3.8 Rocks at Siccar Point, Scotland, that helped convince James Hutton of Earth's great age. Both layers of rock, now at right angles, were laid down horizontally on the seafloor at different times. After the first layer was deposited, Earth movements tilted it vertically and uplifted it to form mountains. Erosion wore down those mountains. When the lower layer was once again submerged, the upper layers were deposited horizontally on top. Now both sets of rock layers have been uplifted and eroded. Clearly this process required time—and lots of it.

we've seen, they improved the seismograph and discovered long-distance earthquake (seismic) waves. Others probed the ocean floors, drew more accurate charts, collected mineral samples from great heights and depths, measured the flow of heat from within Earth, and identified patterns in the worldwide distribution of fossils. All the theories were reassessed. *The evidence from these explorations convinced most researchers that Earth was truly of great age.* The stage was now set for a revolution in geology, the development of what we know today as the theory of plate tectonics. The first steps toward the theory were tentative, however, and some of its proponents were dismissed as lunatics.

A Puzzling Fit

As Leonardo da Vinci noticed on early charts, in some regions the continents looked as if they would fit together like jigsaw-puzzle pieces if the intervening ocean were removed. In 1620 Francis Bacon also wrote of a "certain correspondence" between shorelines on either side of the South Atlantic. In 1885 Edward Suess, a respected German scientist, elaborated on suggestions that the Southern Hemisphere's continents might once have been a single large land mass. He based his belief in part on the similarities of fossils found on these continents, especially fossils of the fern *Glossopteris*. Suess was not taken seriously by his colleagues because he could not explain how

the continents had moved. Still, the fit of South America to Africa was striking (**Figure 3.9**).

As they probed the submerged edges of the continents, marine scientists found that the ocean bottom nearly always sloped gradually out to sea for some distance and then dropped steeply to the deep-ocean floor. They realized that these shelflike continental edges were extensions of the continents themselves. In the few locations where they had measurements, researchers found that the fit between South America and Africa, impressive at the shoreline, was even better along the submerged edges of the continents.

Though so accurate a fit almost certainly could *not* have occurred by chance, no one had yet proposed a mechanism that could separate whole continents into moving pieces. If such a mechanism did exist, its gradual operation would surely require a great deal of time.

Continental Drift

Into the fray stepped **Alfred Wegener,** a busy German meteorologist and polar explorer (**Figure 3.10**). In a lecture in 1912 he proposed a startling and original theory, **continental drift.** Wegener suggested that all Earth's land had once been joined into a single supercontinent surrounded by an ocean. He called the land mass **Pangaea** (*pan* = all, *gaea* = Earth, land) and the surrounding ocean **Panthalassa** (*pan* = all, *thalassa* = ocean). Wegener thought Pangaea had broken into pieces about 200 million years ago. Since then, he said, the pieces had moved to their present positions and were still moving.

Wegener's evidence included the apparent shoreline fit of continents across the North and South Atlantic and new information on offshore contours obtained by contemporary oceanographic expeditions. He pointed to Suess's *Glossopteris* fossils; to areas of erosion apparently caused by the same glacier in tropical areas now widely separated (South Africa, India, and Australia); and to Ernest Shackleton's 1908 discovery of coal, the fossilized remains of tropical plants, in frigid Antarctica. Wegener even suggested that volcanic activity was powered by the friction of continental movement.

Unlike anyone before him, Wegener also proposed a mechanism to account for the hypothetical drift. He believed that the heavy continents were slung toward the equator on the spinning Earth by a centrifugal effect. This inertia, coupled with the tidal drag on the continents from the combined effects of sun and moon, would account for the phenomenon of drifting continents, he thought.

Wegener was dismissed as a crank. His detractors claimed, with some justification, that he had carefully selected only those data supporting his hypothesis, ignoring contrary evidence. Where, for instance, were the wakes or tracks through old seabed that the migrating continents would leave? But a few geologists sided with Wegener. These "drifters" were hesitant to embrace the centrifugal force theory, yet they were unable to propose an alternative power source that could move the massive granitic continents.

The greatest block to the acceptance of continental drift lay in geologists' view of Earth's mantle. The available evi-

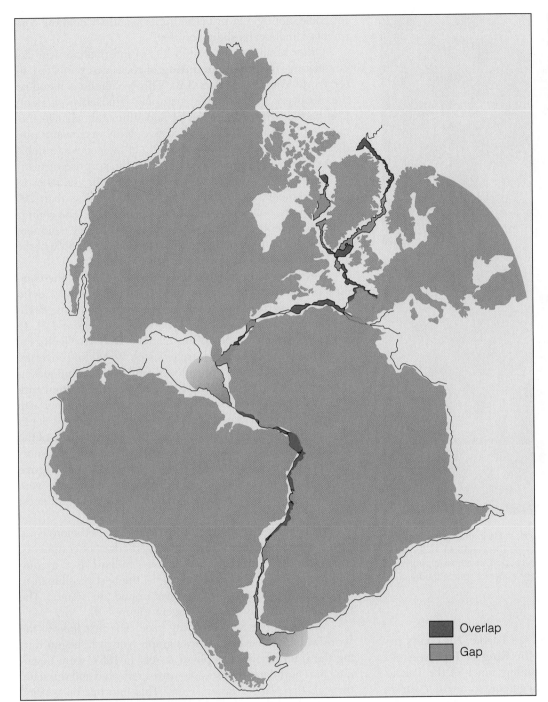

Figure 3.9 The fit of all the continents around the Atlantic at a water depth of about 137 meters (450 feet), as calculated by Sir Edward Bullard at the University of Cambridge in the 1960s. This graphic was an effective stimulus to the tectonic revolution.

■ Overlap
■ Gap

dence seemed to suggest that a deep, solid mantle supported the crust and mountains *mechanically* (not *isostatically*) from below. Drift would be impossible with this kind of subterranean construction. A few perceptive seismic researchers then noticed that the upper mantle reacted to earthquake waves as if it were a deformable mass, not a rigid solid. Perhaps such a layer would resemble a slug of iron heated in a blacksmith's forge; it would deform with pressure and even flow slowly. Established geologists, however, dismissed this interpretation, saying that the mountains would simply fall over or sink without rigid underpinnings. By 1926 the "drifters"

were in full retreat. When Wegener died on an expedition across Greenland in 1930, his theory was already in eclipse.

The Idea Transformed

The concept of continental drift refused to die—those neatly fitted continents provided a haunting reminder of Wegener to anyone looking at an Atlantic chart. In 1935 a Japanese scientist, Kiyoo Wadati, speculated that earthquakes and volcanoes near Japan might be associated with continental drift. In 1940 seismologist Hugo Benioff plotted the locations of

Figure 3.10 Alfred Lothar Wegener writes in his journal at the beginning of what would be his last expedition to Greenland, in 1930. His remarkable book *The Origin of Continents and Oceans* was published in 1915. In it, he outlined interdisciplinary evidence for his theory of continental drift.

deep earthquakes at the edges of the Pacific. His charts revealed the true extent of the **Pacific Ring of Fire,** a circle of violent geologic activity surrounding much of the Pacific Ocean. Seismographs were now beginning to reveal a worldwide pattern of earthquakes and volcanoes. Deep earthquakes did not occur randomly over Earth's surface but were concentrated in zones that extended in lines along Earth's surface.

Benioff, Wadati, and others wondered what could cause such an orderly pattern of deep earthquakes. Many of the lines corresponded with a worldwide system of oceanic ridges, the first of which was plotted in 1925 by oceanographers aboard the research ship *Meteor* working in the middle of the North Atlantic. **Figure 3.11** is a plot of about 30,000 earthquakes. Notice the odd pattern they form—almost as if Earth's lithosphere is divided into sections! Benioff's sensitive seismographs also began to gather strong evidence for a de-

formable, nonrigid layer in the upper mantle. Could the continents somehow be sliding on that?

Other seemingly unrelated bits of information were accumulating. **Radiometric dating** of rocks was perfected after World War II (see **Box 3.2**). This technique is based on the discovery that unstable, naturally radioactive elements lose particles from their nuclei and ultimately change into new stable elements. The radioactive decay occurs at a predictable rate, and measuring the ratio of radioactive to stable atoms in a sample provides its age. To the surprise of many geologists, the maximum age of the ocean floor and its overlying sediments was radiometrically dated to less than 200 million years, only about 4% of the age of Earth. The centers of the continents are much older; some parts of the continental crust are more than 3.9 billion years old, about 85% of the age of Earth. *Why was oceanic crust so young?*

Attention had turned to the deep-ocean floors, the complex profiles of which were now being revealed by **echo sounders,** devices that measure depth by bouncing high-frequency sound waves off the bottom (see Figure 4.3). In particular, scientists aboard the Lamont–Doherty Geological Observatory deep-sea research vessel *Vema* (a converted three-masted schooner) invented deep survey techniques as they went. After World War II they probed the bottom with powerful echo sounders and looked beneath sediments with reflected pressure waves generated by surplus Navy depth charges dropped gingerly overboard. The overall shape of the Mid-Atlantic Ridge was slowly revealed. The ridge's conformance to shorelines on either side of the Atlantic (**Figure 3.12**) raised many eyebrows. Ocean floor sediments were shown to be thickest at the edge of the Atlantic and thinnest near this mid-ocean ridge.

Vema and other research ships also compiled more complete, accurate charts of the submerged edges of continents. At Cambridge University, Sir Edward Bullard used a computer to process these data to achieve the best possible fit of the continental jigsaw-puzzle pieces around the Atlantic. The fit was astonishingly good (see again Figure 3.9).

Mantle studies were keeping pace. The first links in the Worldwide Standardized Seismograph Network, begun during the International Geophysical Year in 1957, were beginning to report data from seismic waves reflected and refracted through the planet's inner layers. This information verified the existence of a layer in the upper mantle that caused a decrease in the velocity of seismic waves. This finding strongly suggested that the layer was deformable. Perhaps the lithosphere was isostatically balanced in this partially melted layer, and perhaps continents could move around in it *if* a suitable power source existed.

The Breakthrough: From Seafloor Spreading to Plate Tectonics

In 1960 Professor Harry Hess of Princeton University and Robert Dietz of the Scripps Institution of Oceanography proposed a radical idea to explain the features of the ocean floor

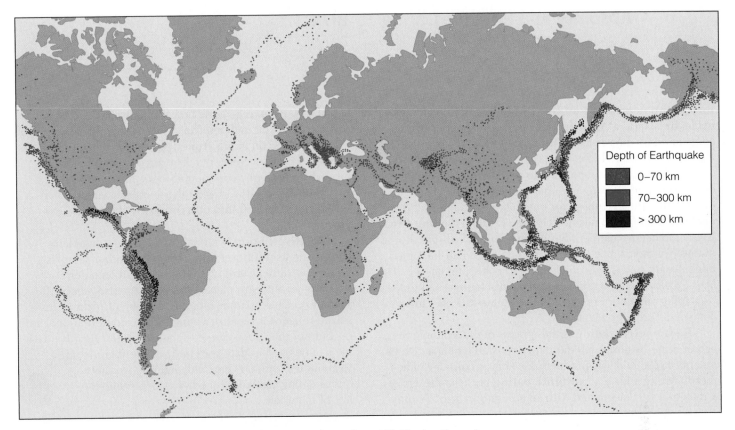

Figure 3.11 Seismic events worldwide, January 1977 through December 1986. The locations of about 10,000 earthquakes are colored red, green, and blue to represent event depths of 0 to 70 kilometers, 70 to 300 kilometers, and below 300 kilometers, respectively.

Depth of Earthquake
- 0–70 km
- 70–300 km
- > 300 km

and the fit of the continents. They suggested that new seafloor develops at the Mid-Atlantic Ridge (and the other newly discovered ocean ridges) and then spreads outward from this line of origin. Continents would be pushed aside by the same forces that cause the ocean to grow. This motion could be powered by **convection currents,** slow-flowing circuits of material within the mantle.

Seafloor spreading, as the new theory was called, pulled many loose ends together. If the mid-ocean ridges were **spreading centers** and sources of new ocean floor rising from the asthenosphere, they should be hot. They were. If the new oceanic crust cooled as it moved from the spreading center, it should shrink in volume and become denser, and the ocean should be deeper farther from the spreading center. It was. Sediments at the edges of the ocean basin should be thicker than those near the spreading centers. They were, and they were also older.

Did this mean that Earth was continuously expanding? Because there was no evidence for a growing Earth, the creation of new crust at spreading centers would have to be balanced by the destruction of crust somewhere else. Then researchers discovered that the crust plunges down into the mantle along the periphery of the Pacific. The process is known as **subduction,** and these areas are called

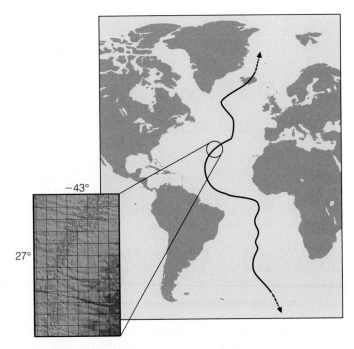

Figure 3.12 The Mid-Atlantic Ridge, showing its conformance to the coastlines of the adjacent continents. The inset shows a detail of the ridge—the central rift is clearly visible.

Radioactive decay is the process by which unstable atomic nuclei break apart. As we have seen, radioactive decay is accompanied by the release of heat, some of which warms Earth's interior and helps drive the processes of plate tectonics. Although it is impossible to predict exactly when any one unstable nucleus in a sample will decay, it is possible to discover the time required for one-half of all the unstable nuclei in a sample to decay. This time is called the **half-life**. Every radioactive element has its own unique half-life. For example, a radioactive form of uranium has a half-life of 4.5 billion years, and a radioactive form of potassium has a half-life of 8.4 billion years. During each half-life, one-half of the remaining amount of the radioactive element decays to become a different element.

Radiometric dating is the process of determining the age of rocks by observing the ratio of unstable radioactive elements to stable decay products. Geologists consider radiometric dating a form of **absolute dating** because the age of a rock may be determined with an accuracy of 1 to 2% of its

actual age. **Figure a** shows how samples may be dated by this means. Using radiometric dating, researchers have identified small zircon grains from western Australian sandstone that are 4.2 billion years old. The zircons were probably eroded from nearby continental rocks and deposited by rivers. (Older crust is now unidentifiable, having been altered and converted into other rocks by geological processes.)

Relative dating is a method of dating a sample by comparing its position to the positions of other samples. Younger sediments are typically laid down on older deposits—events are placed in their proper sequence. Researchers can determine whether a group of rocks or fossilized remains is older or younger than a different one close by, but not the actual age of the assemblages. Relative dating can be used to date rock and sediment layers up to about 600 million years old, and the two methods of dating can work together to determine the age of materials.

3-18

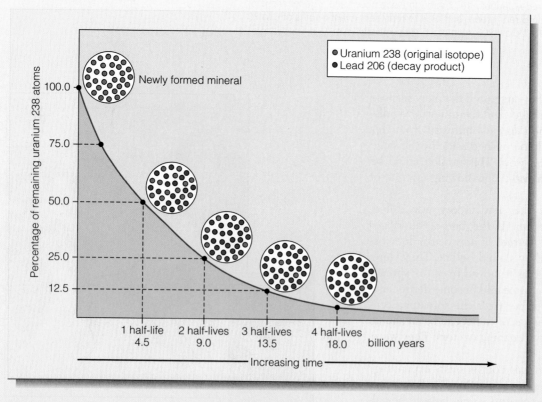

a The rate of radioactive decay of a radioactive form of uranium (^{238}U) into lead. The half-life of ^{238}U is 4.5 billion years. During each half-life one-half of the remaining amount of the radioactive element decays to become a different element. The assumption is made that the system has remained closed—no radioactive atoms or decayed products have been added or removed from the sample. Ages obtained from absolute and relative dating of continental rocks and rocks and sediments from the seabeds coincide with ages predicted by plate tectonic theory.

subduction zones (or Wadati–Benioff zones in honor of their discoverers). The zones of concentrated earthquakes (see again Figure 3.11) were found in regions of crustal formation (spreading centers) and crustal destruction (subduction zones).

In 1965 the ideas of continental drift and seafloor spreading were integrated into the overriding concept of **plate tectonics** (*tekton* = builder; our word *architect* has the same root), primarily by the work of **John Tuzo Wilson,** a geophysicist at the University of Toronto. In this theory Earth's outer layer consists of about a dozen separate major lithospheric **plates** floating on the asthenosphere. When heated from below, the deformable asthenosphere expands, becomes less dense, and rises. It turns aside when it reaches the lithosphere, however, and it drags the plates laterally until it turns under again to complete the circuit. The large plates include both continental and oceanic crust. The major plates jostle about like huge flats of ice on a warming lake. Plate movement is slow in human terms, averaging about 5 centimeters (2 inches) a year. The plates interact at converging, diverging, or slipping boundaries, sometimes forcing one another below the surface or wrinkling into mountains.

Plate movement appears to be caused by a combination of three forces:

- The outward push of new seabed formed at spreading centers. Plates slide off the raised ridges.
- Friction of mantle convection currents against the bottom of a plate.
- The downward pull of a descending plate's dense leading edge.

We now know *through the great expanse of geologic time, this slow movement remakes the surface of Earth, expands and splits continents, and forms and destroys ocean basins.* The less dense, ancient granitic continents ride high in the lithospheric plates, rafting on the slowly moving asthenosphere below. This process has progressed since Earth's crust first cooled and solidified.

Figure 3.13 presents an overview of the tectonic system. Literally and figuratively, it all fits; a cooling, shrinking, raisin-like wrinkling is no longer needed to explain Earth's surface features. This twentieth-century understanding of the ever-changing nature of Earth has given fresh meaning to historian Will Durant's warning: "Civilization exists by geological consent, subject to change without notice."

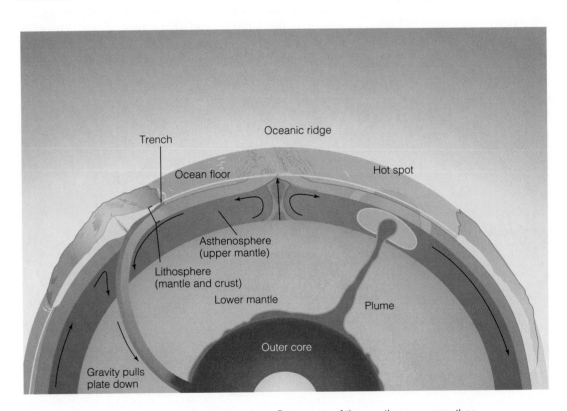

Figure 3.13 The tectonic system is powered by heat. Some parts of the mantle are warmer than others, and convection currents form when warm mantle material rises and cool material falls. Above the mantle floats the cool, rigid lithosphere, which is fragmented into plates. Large convection currents in the partially melted, deformable asthenosphere drag the plates away from one another (along the ocean ridges), toward one another (at subduction zones or areas of mountain building), or past one another (as at California's San Andreas Fault). Smaller localized convection currents form cylindrical plumes that rise to the surface to form hot spots (like the Hawaiian Islands). Note that the whole mantle is involved in thermal convection currents (see again Figure 3.1). (Source: From *The Earth's Dynamic Systems,* 9/e, by Hamblin & Christiansen, Fig. 2.8. Reprinted by permission Pearson Education, Inc., Upper Saddle River, NJ 07458.)

After a series of raucous scientific meetings in 1966 and 1967, the revolution in geology entered a period of rapid consolidation. In 1968 *Glomar Challenger* drilled her first deep-ocean crustal cores and provided the confirmation of plate tectonics. Researchers found supporting data from many sources that tended to confirm Wilson's surprising synthesis. Every scientist had to reexamine his or her specialty in light of this new information. Zoologists found new explanations for the unusual animals of Australia. Biologists discovered a new cause of the isolation required for the formation of new species by natural selection. Paleontologists found an explanation for similar fossils on different continents. Resource specialists could at last explain why coal deposits were buried in Antarctica. Some geologists were pleased, some were skeptical. All were eager to explore further to prove or disprove this new theory.

This historical overview is useful in transmitting the sense of discovery and excitement surrounding the twentieth-century revolution in geology. Now we can investigate the workings of plate tectonics in more detail.

PLATE BOUNDARIES

The lithospheric plates and their margins are shown in **Figure 3.14**. The plates float on a dense, deformable asthenosphere and are free to move relative to each other. Plates interact with neighboring plates along their mutual boundaries. In **Figure 3.15**, movement of Plate A to the left (west) requires it to slide along its north and south margins. An overlap is produced in front (to the west), and a gap is created behind (to the east). Different places on the margins of Plate A experience separation and extension, convergence and compression, and transverse movement (shear).

The three types of plate boundaries that result from these interactions are called *divergent, convergent, and transform boundaries*, depending on their sense of movement. **Figure 3.16** provides an overview of the plate boundary interactions we will study next.

Divergent Plate Boundaries: Forming Ocean Basins

The spreading center at the Mid-Atlantic Ridge is a **divergent plate boundary,** a line along which two plates are moving apart. Oceanic crust forms along divergent plate boundaries. The formation of the Atlantic is shown in **Figure 3.17.** Heat from the lower crust and mantle accumulates beneath the continents. About 210 million years ago this heat caused the asthenosphere to expand and rise, lifting and fracturing the lighter, solid lithosphere above. The plate and its embedded continent split in two, and a new plate boundary—a rift valley—was formed between the pieces. As

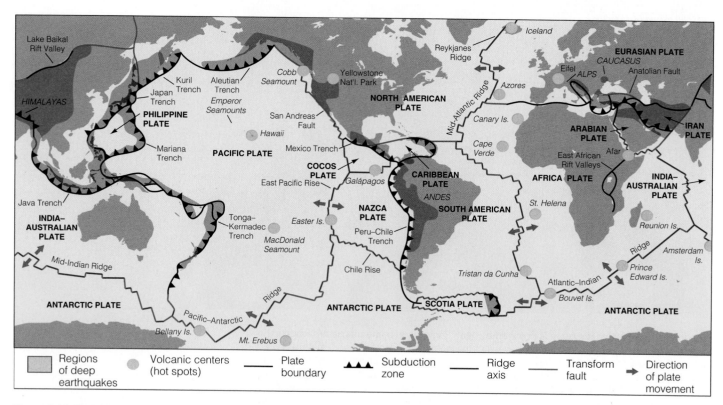

Figure 3.14 The major lithospheric plates, showing their directions of relative movement and the location of the principal hot spots. Note the correspondence of plate boundaries and earthquake locations—compare this figure to Figure 3.11. Most of the million or so earthquakes and volcanic events each year occur along plate boundaries.

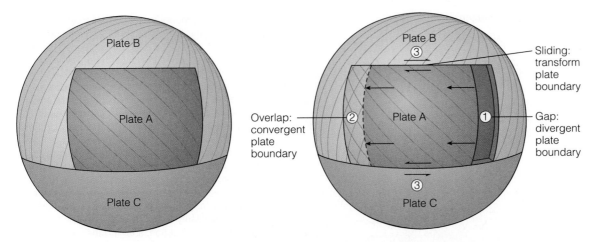

Figure 3.15 Plate boundaries in action. As Plate A moves to the left (west), a gap forms behind it ① and an overlap with Plate B forms in front ②. Sliding occurs along the top and bottom sides ③. The margins of Plate A experience the three types of interactions: at ①, extension characteristic of divergent boundaries; at ②, compression characteristic ofconvergent boundaries; and at ③ the shear characteristic of transform plate boundaries. Figure 3.16 shows these kinds of boundaries in more detail.

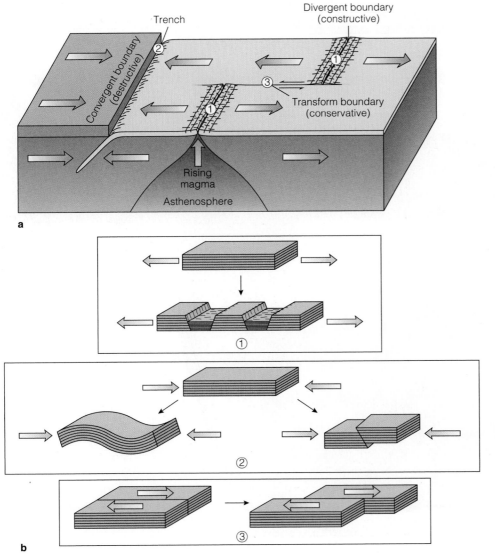

Figure 3.16 (a) Plate boundaries on the surface of a sphere. The divergent ①, convergent ②, and transform ③ margins match those of the margins of Plate A in Figure 3.15. (b) Extension of divergent boundaries causes splitting and rifting ①, compression at convergent boundaries produces buckling and shortening ②, and translation at transform boundaries causes shear ③.

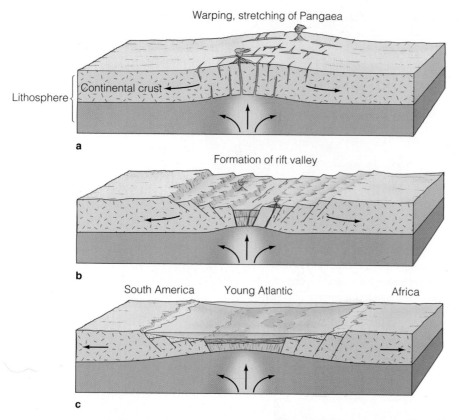

Warping, stretching of Pangaea

Lithosphere {

Continental crust

a

Formation of rift valley

b

South America   Young Atlantic   Africa

c

Figure 3.17 A model for the formation of a new plate boundary: the breakup of Pangaea and the formation of the Atlantic beginning about 210 million years ago. (**a**) As the lithosphere began to crack, a rift formed beneath the continent, and molten basalt from the asthenosphere began to rise. (**b**) As the rift continued to open, the two new continents were separated by a growing ocean basin. Volcanoes and earthquakes occur along the active rift area, which is the mid-ocean ridge. The East African Rift Valley currently resembles this stage. (**c**) A new ocean basin (shown in green) forms beneath a new ocean. The Red Sea (**d**) currently resembles this stage. Note the remarkable sawtooth configuration of the peaks on the horizon, and their similarity to the diagram. (**e**) The South Atlantic in cross section, showing the mid-ocean ridge in the middle of the growing basin.

V. Courtillot

d

|←—Mid-ocean ridge—→|

South America   Rift   Atlantic   Africa

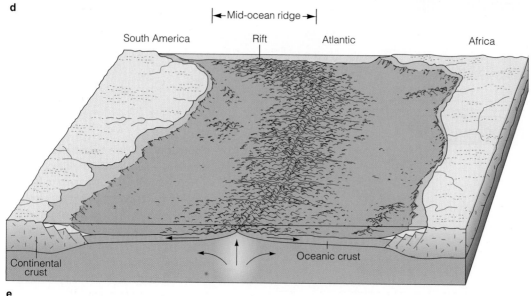

Continental crust

Oceanic crust

e

the broken plate separated at this new spreading center, molten rock called **magma** rose into the crustal fractures. (Magma is called *lava* when found above ground.) Some of the magma solidified in the fractures; some erupted from volcanoes on the seafloor. Together these processes produced new oceanic crust. The South Atlantic, a large new ocean basin formed between the diverging plates, is shown in Figure 3.17e. A long mid-ocean ridge divided by a central rift valley traverses the ocean floor roughly equidistant from the shorelines in both the North and South Atlantic, terminating north of Iceland.[2]

Plate divergence is not confined to the Atlantic, nor has it been limited to the last 200 million years. As may also be seen in Figure 3.14, the Mid-Atlantic Ridge has counterparts in the Pacific and Indian Oceans. The Pacific floor, for example, diverges along the East Pacific Rise and the Pacific–Antarctic Ridge, spreading centers that form the eastern and southern boundaries of the great Pacific Plate. In East Africa rift valleys have formed relatively recently as plate divergence begins to separate another continent. As happened in the Red Sea (Figure 3.17d), the ocean will invade when the rift becomes deep enough. **Figure 3.18** shows how divergence has formed other ocean basins. About 20 cubic kilometers (4.8 cubic miles) of new ocean crust forms each year.

Convergent Plate Boundaries: Recycling Crust, Building Island Arcs and Continents

Because Earth is not getting larger, divergence in one place must be offset by convergence in another. Oceanic crust is destroyed at **convergent plate boundaries,** regions of violent geologic activity where plates are pushing together. South America, embedded in the westward-moving South American Plate, encounters the Pacific's Nazca Plate as it moves eastward. The relatively thick and light continental lithosphere of South America rides up and over the heavier oceanic lithosphere of the Nazca Plate, which is subducted along the deep trench that parallels the west coast of South America. **Figure 3.19** is a cross section through these plates.

The subducting plate's periodic downward lurches cause earthquakes. Some of the oceanic crust and its sediments will melt as the plate plunges downward, forming a magma rich in water and carbon dioxide. In places this magma rises through overlying layers to the surface and causes volcanic eruptions. The active volcanoes of Central America and South America's Andes Mountains are a

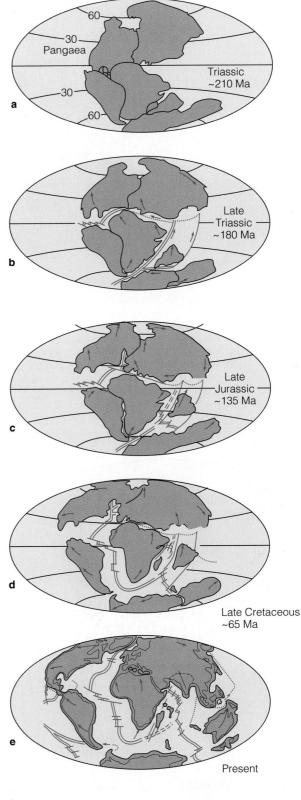

Ma = *mega-annum*, indicating millions of years ago

Figure 3.18 The breakup of Pangaea shown in five stages beginning about 210 million years ago. Inferred motion of lithospheric plates is indicated by arrows. Spreading centers (mid-ocean ridges) are shown in red.

---

[2]The Atlantic's rate of spreading is about 5 centimeters (2 inches) a year. This young ocean was about 25 meters (82 feet) narrower when Columbus sailed it than it is today.

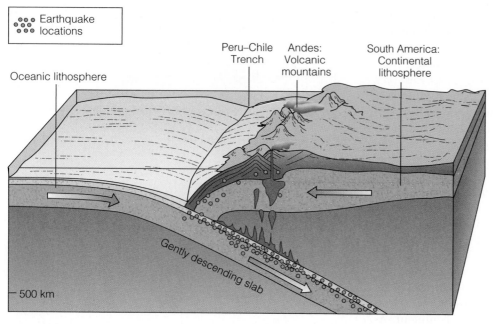

Earthquake locations

Oceanic lithosphere

Peru–Chile Trench

Andes: Volcanic mountains

South America: Continental lithosphere

Gently descending slab

500 km

Figure 3.19 A cross section through the west coast of South America, showing the convergence of a continental plate and an oceanic plate. The subducting oceanic plate becomes denser as it descends, its downward slide propelled by gravity. Starting at a depth of about 100 kilometers (60 miles), heat drives water and other volatile components from the subducted sediments into the overlying mantle, lowering its melting point. Masses of the melted material, rich in water and carbon dioxide, rise to power Andean volcanoes.

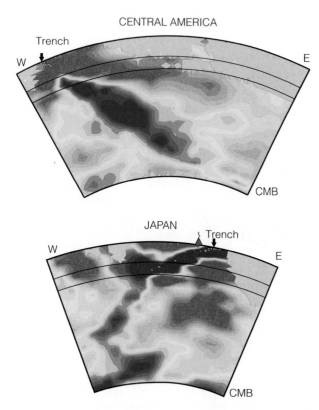

CENTRAL AMERICA

Trench

W   E

CMB

JAPAN   Trench

W   E

CMB

Figure 3.20 Vertical slices through Earth's mantle beneath Central America and Japan. Colder material is shown in blue, warmer in red. The distribution of colder material suggests that the subducting slabs beneath both areas have penetrated to the core–mantle boundary at a depth of 2,900 kilometers (1,800 miles).

product of this activity, as are the area's numerous earthquakes. The North American Cascade volcanoes, including Mount St. Helens, result from similar processes. Most of the subducted crust mixes with the mantle. As shown in **Figure 3.20,** some of it continues downward through the mantle, eventually reaching the mantle–core boundary 2,800 kilometers (1,700 miles) beneath the surface! Subduction at converging oceanic plates was responsible for the great Alaska earthquake of 1964. Plate convergence (and divergence) is faster in the Pacific than in the Atlantic, in a few places reaching a rate of 18 centimeters (7 inches) a year. You can now clearly see the source of the Pacific Ring of Fire.

In the previous example continental crust met oceanic crust. What happens when two *oceanic* plates converge? One of the colliding plates will usually be older, and therefore cooler and denser, than the other. Pulled by gravity, this heavier plate will slip steeply below the lighter one into the asthenosphere. The ocean bottom is distorted in these areas to form deep trenches, the ocean's greatest depths. Water and carbon dioxide trapped with the melting rock of the subducting plate rise into the overlying mantle, lowering its melting point. This fluid mix of magma and subducted material forms a relatively light magma that powers vigorous volcanoes, but the volcanoes emerge from the seafloor rather than from a continent. These volcanoes appear in patterns of curves on the overriding oceanic crust; when they emerge above sea level they form curving arcs of islands (see **Figure 3.21**).

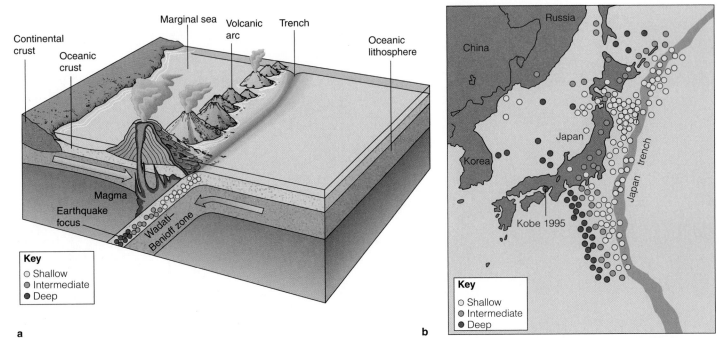

Figure 3.21 (**a**) The formation of an island arc along a trench as two oceanic plates converge. The volcanic islands form as masses of magma reach the seafloor. The Japanese islands were formed in this way. (**b**) The distribution of shallow, intermediate, and deep earthquakes for part of the Pacific Ring of Fire in the vicinity of the Japan trench. Note that earthquakes occur only on one side of the trench, the side on which the plate subducts. The site of the catastrophic 1995 Kobe subduction earthquake is marked.

Convergent margins are vast "continent factories" where materials from the surface descend and are heated, compressed, partially liquefied, separated, mixed with surrounding materials, and recycled to the surface. Relatively light continental crust is the main product, and it is produced at a rate of about 1 cubic kilometer (0.24 cubic mile) per year. Some geophysicists believe all of Earth's continental crust may have originated from granitic rock produced in this way. The island arcs may have coalesced to form larger and larger continental masses.

Two plates bearing continental crust can also converge. The most spectacular example of such a collision, between the India–Australian and Eurasian Plates some 45 million years ago, formed the Himalayas. Because both plates are of approximately equal density, neither plate edge is being subducted; instead, both are compressed, folded, and uplifted, as **Figure 3.22** shows. The lofty top of Mount Everest is made of rock formed from sediments deposited long ago in a shallow sea!

Transform Plate Boundaries: Fracturing Crust

Remember, movement of lithospheric plates over the mantle is occurring on the surface of a sphere, not on a plane. The axis of spreading is not a smoothly curving line, but rather a jagged trace abruptly offset by numerous faults. These features are called **transform faults** (see **Figure 3.23,** and look for these features on the mid-ocean ridges of Figure 3.14). Transform faults are named from the fact that the relative motion is changed, or transformed, along them. We will discuss transform faults in more detail in Chapter 4 in our discussion of the mid-ocean ridge system (see, for example, Figure 4.23 on page 109), but the concept is important in our discussion of plate boundaries because lithospheric plates shear laterally past one another at **transform plate boundaries.** Crust is neither produced nor destroyed at this type of junction.

The potential for earthquakes at transform plate boundaries can be great as the plate edges slip past each other. The eastern boundary of the Pacific Plate is a long transform fault

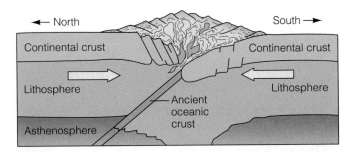

Figure 3.22 A cross section through southern China, showing the convergence of two continental plates. Neither plate is dense enough to subduct; instead, their compression and folding uplift the plate edges to form the Himalayas. Notice the massive supporting "root" beneath the emergent mountain needed for isostatic equilibrium.

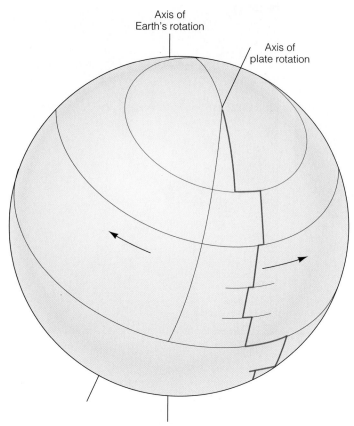

Figure 3.23 Transform faults (red) form because the axis of seafloor spreading on the surface of a sphere cannot follow a smoothly curving line. The motion of two diverging lithospheric plates (arrows) rotates about an imaginary axis extending through Earth. The plates shear laterally past one another at transform faults. Look for these features worldwide on Figure 3.14.

system. As you can see in Figure 3.14, California's San Andreas Fault (**Figure 3.24**) is merely the most famous of the many faults marking the junction between the Pacific and North American Plates. The Pacific Plate moves steadily, but its movement is stored elastically at the North American Plate boundary until friction is overcome. Then the Pacific Plate lurches in abrupt jerks to the northwest along much of its shared border with the North American Plate, an area that includes the major population centers of California. These jerks cause California's famous earthquakes. Because of this movement, coastal southwestern California is gradually sliding north along the rest of North America; some 50 million years from now, it will encounter the Aleutian Trench.

Plate Interactions: A Summary

3-23

There are, then, two kinds of plate divergences: divergent oceanic crust (such as that in the mid-Atlantic) and divergent continental crust (as in the Rift Valley of East Africa). And there are three kinds of plate convergences: oceanic crust toward continental crust (west coast of South America), oceanic crust toward oceanic crust (northern Pacific), and continental crust toward continental crust (Hi-

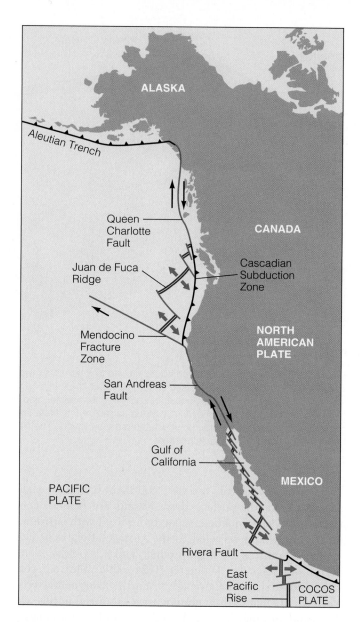

Figure 3.24 A long transform plate boundary, which includes California's San Andreas Fault.

malayan mountains). Transform boundaries mark locations at which crustal plates move past one another (San Andreas Fault). Each of these movements produces a distinct topography, and each zone contains potential dangers for its human inhabitants.

The characteristics of plate boundaries are summarized in **Table 3.1**.

THE CONFIRMATION OF PLATE TECTONICS

The theory of plate tectonics has had the same effect on geology that the theory of evolution has had on biology. In each case a catalog of seemingly unrelated facts was unified by a

Table 3.1   Characteristics of Plate Boundaries

Plate Boundary		Plate Movement	Seafloor	Events Observed	Example Locations
Divergent plate boundaries	Ocean–ocean	Apart	Forms by seafloor spreading.	Ridge forms at spreading center. Ocean basin expands, plate area increases. Many small volcanoes and/or shallow earthquakes.	Mid-Atlantic Ridge, East Pacific Rise
	Continent–continent		New ocean basin may form as continent splits.	Continent spreads, central rift collapses, ocean fills basin.	East African Rift Valley, Red Sea
Convergent plate boundaries	Ocean–continent	Together	Destroyed at subduction zones.	Dense oceanic lithosphere plunges beneath less dense continental. Earthquakes trace path of downmoving plate as it descends into asthenosphere. A trench forms. Subducted plate partially melts. Magma rises to form continental volcanoes.	Western South America, Cascade Mountains in western U.S.
	Ocean–ocean			Older, cooler, denser crust slips beneath less dense crust. Strong quakes. Deep trench forms in arc shape. Subducted plate heats in upper mantle, magma rises to form curving chains of volcanic islands.	Aleutians, Marianas
	Continent–continent	N.A.		Collision between masses of granitic continental lithosphere. Neither mass is subducted. Plate edges are compressed, folded, uplifted; one may move beneath the other.	Himalayas, Alps
Transform plate boundaries		Past each other	Neither created nor destroyed.	A line (fault) along which lithospheric plates move past each other. Strong earthquakes along fault.	San Andreas Fault; South Island, New Zealand

powerful central idea. As we will see, many discoveries contributed to our present understanding of plate tectonics, but the most compelling evidence is locked within the floors of the young ocean basins themselves.

Paleomagnetism

3-25

A compass needle points toward the magnetic north pole because it aligns with Earth's magnetic field (**Figure 3.25**). Tiny particles of iron-bearing magnetic minerals are found in basaltic magma. These minerals act like miniature compass needles; as they cool to form new seafloor, their magnetic fields align with Earth's magnetic field. Thus the orientation of Earth's magnetic field at that particular time becomes frozen in the rock as it solidifies. Any later change in the strength or direction of Earth's magnetic field will not significantly change the characteristics of the field trapped within the solid rocks. The "fossil," or remanent, magnetic field of a rock is known as **paleomagnetism** (*palaios* = ancient).

A **magnetometer** measures the amount and direction of residual magnetism in a rock sample. In the late 1950s geophysicists towed sensitive magnetometers just above the ocean floor to detect the weak magnetism frozen in the rocks. When plotted on charts, the data revealed a pattern of symmetrical magnetic stripes or bands on both sides of a spread-

ing center (**Figure 3.26a**). The magnetized minerals contained in the rocks of some stripes add to Earth's present magnetic orientation to enhance the strength of the local magnetic field, but the magnetism in rocks in adjacent stripes weakens it. What could cause such a pattern?

In 1963 geologists Drummond Matthews, Frederick Vine, and Lawrence Morley proposed a clever interpretation. They knew that Earth's magnetic field reverses at irregular intervals of a few hundred thousand years. In a time of reversal a compass needle would point south instead of north, and any particles of magnetic material in fresh seafloor basalt at a spreading center would be imprinted with the reversed field. The alternating magnetic stripes represent rocks with alternating magnetic polarity—one band having normal polarity (magnetized in the same direction as today's magnetic field direction), the next band having reversed polarity (opposite from today's direction). These researchers realized that the pattern of alternating weak and strong magnetic fields was symmetrical because freshly magnetized rocks born at the ridge are spread apart and carried away from the ridge by plate movement (**Figure 3.26b**). Similar magnetic patterns have been found on land and independently dated by other means.

By 1974, scientists had compiled charts showing the paleomagnetic orientation—and the age—of the seafloors of the eastern Pacific and the Atlantic for about the last 200 million

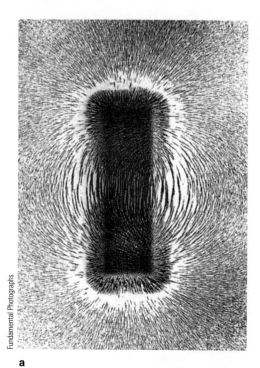

a

a

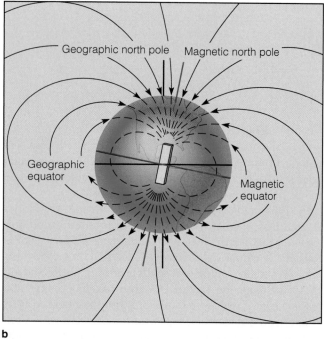

b

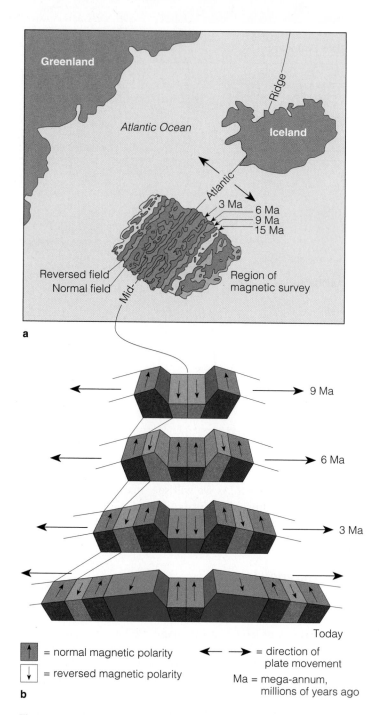

= normal magnetic polarity

↓ = reversed magnetic polarity

→← = direction of plate movement

Ma = mega-annum, millions of years ago

b

Figure 3.25 (**a**) The magnetic field pattern of a simple bar magnet is revealed by iron filings that align themselves along the lines of magnetic force. (**b**) Earth's magnetic field is thought to arise from fluid motions and electric currents in the outer core. The likely energy source is heat from the solid inner core causing convection currents in the outer core. These flowing currents of liquid iron, coupled with the rotation of Earth, form the magnetic field in a process analogous to the way a power-station generator produces electricity. The axis of Earth's magnetic field is tilted about 11° from the axis of the geographic North and South Poles. Compass needles point to magnetic north, not to geographic (true) north.

Figure 3.26 Patterns of paleomagnetism and their explanation by plate tectonic theory. (**a**) When scientists conducted a magnetic survey of a spreading center, the Mid-Atlantic Ridge, they found bands of weaker and stronger magnetic fields frozen in the rocks. (**b**) The molten rocks forming at the spreading center take on the polarity of the planet when they are cooling, and then move slowly in both directions from the center. When Earth's magnetic field reverses, the polarity of newly formed rocks changes, creating symmetrical bands of opposite polarity.

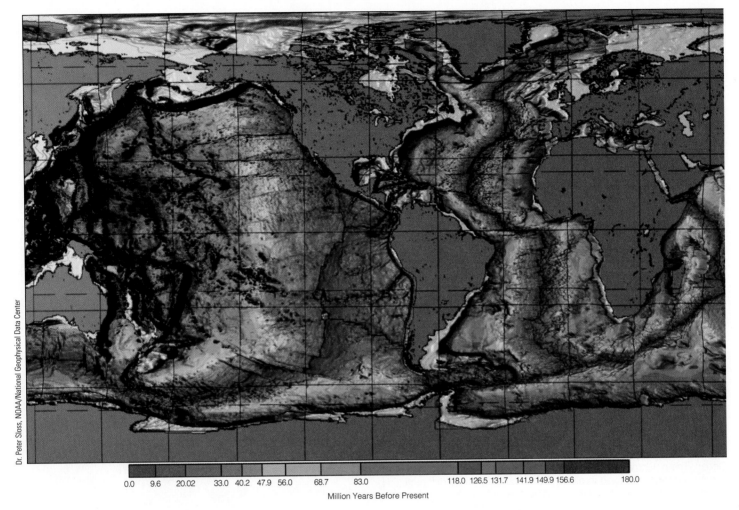

| 0.0 | 9.6 | 20.02 | 33.0 | 40.2 | 47.9 | 56.0 | 68.7 | 83.0 | 118.0 | 126.5 | 131.7 | 141.9 | 149.9 | 156.6 | 180.0 |

Million Years Before Present

Figure 3.27 The age of the ocean floors. The colors seen here represent an expression of seafloor spreading over the last 200 million years as revealed by paleomagnetic patterns.

years (see **Figure 3.27**). Plate tectonics beautifully explains these patterns, and the patterns themselves are among the most compelling of all arguments for the theory.

Paleomagnetic data have recently been used to measure spreading rates, to calibrate the geologic time scale, and to reconstruct continents. Paleomagnetism has been among the most productive specialties in geology for the past three decades, and other lines of paleomagnetic investigation have also shed light on the process of plate tectonics.

Apparent Polar Wandering

Geologists already knew that periodic magnetic field reversals were not the only unusual feature of Earth's magnetic field. A plot of the apparent position of the north magnetic pole, as measured by the magnetic orientation of rocks in North America, South America, Europe, and Africa, showed that the magnetic pole seemed to have moved—it appeared to

have migrated to its present position from much farther south, from a point in the Pacific Ocean.

Because *actual* pole wandering was a very remote scientific possibility, and having more than one north magnetic pole at the same time would be impossible, the pattern strongly suggested the poles probably stayed put and the continents might have moved in relation to the pole and to one another.

Geophysicists examined this possibility. If these continents had once been united and had drifted over Earth's surface together, they should have identical paths of apparent polar wandering terminating in coincidence at the North Pole's present position (see **Figure 3.28**). Armed with this information, geologists quickly noticed a fault through Scotland's Caledonian Mountains joined with the Cabot Fault extending from Newfoundland to Boston. This and other evidence strongly suggested that the continents have indeed drifted apart, carrying their magnetized rocks with them.

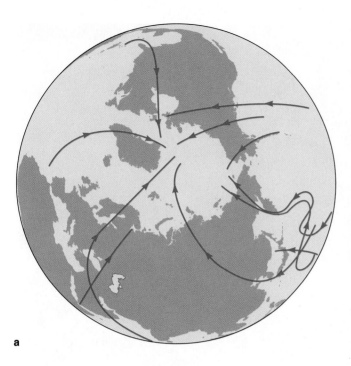

a

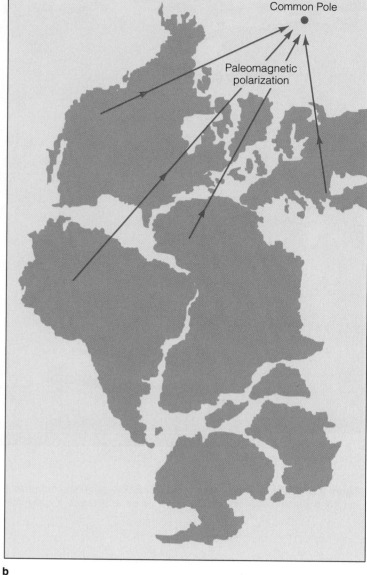

b

Figure 3.28 Polar wandering? (**a**) The magnetic properties of rocks in North America, South America, Europe, and Africa suggest that the north magnetic pole has apparently migrated to its present position from much farther south, possibly from a point in the Pacific Ocean. The paleomagnetic evidence also suggests that, if the continents had remained fixed in place, Earth would have had different magnetic poles at the same time—an impossibility. (**b**) The problem can be solved if the pole has remained fixed while the continents have moved. For example, if Europe and North America were previously joined, the paleomagnetic fields preserved in their rocks would indicate a single pole location until they drifted apart. Evidence from South America and Africa supports this interpretation. (Source: From *The Earth's Dynamic Systems,* 8/e, by Hamblin & Christiansen. © 1998. Reprinted by permission of Pearson Education, Inc., Upper Saddle River, NJ 07458.)

Hot Spots

Hot spots represent surface expressions of plumes of magma rising from stationary sources of heat in the mantle (see again Figure 3.13). Hot spots are not always located at plate boundaries, and no one knows why their source of heat is localized or what anchors them in place. As lithospheric plates slide over these fixed locations, they are weakened from below by rising heat and magma. A volcano can form over the hot spot; but because the plate is moving, the volcano is carried away from its source of magma after a few million years and becomes inactive. It is replaced at the hot spot by a new volcano a short distance away. A chain of volcanoes and volcanic islands results (**Figure 3.29**). The MacDonald Seamount, which gave researchers aboard *Melville* such an exciting morning (see the beginning of this chapter), is being formed by a hot spot.

Figure 3.30 shows the most famous of these hot spot chains, which extends from the old eroded volcanoes of the Emperor Seamounts to the still-growing island of Hawaii. In fact, the abrupt bend in the chain was caused by a change in direction of the Pacific Plate, from a largely northward to a more westward movement, about 40 million years ago. The next Hawaiian island that will come into being—already named Loihi—is building on the ocean floor to the southeast. Now about 1,000 meters (3,300 feet) beneath the surface, Loihi will break the surface about 30,000 years from now.

There are other hot spots in the Pacific. The island chains formed by their activity also jog in the Hawaiian pattern, indicating that they are positioned on the same lithospheric plate. Chains of undersea volcanoes in the Atlantic, centered on the Mid-Atlantic Ridge, suggest that a similar process is at work there. Hot spots can exist beneath continental crust as well; Yellowstone National Park is believed to be over a hot spot beneath

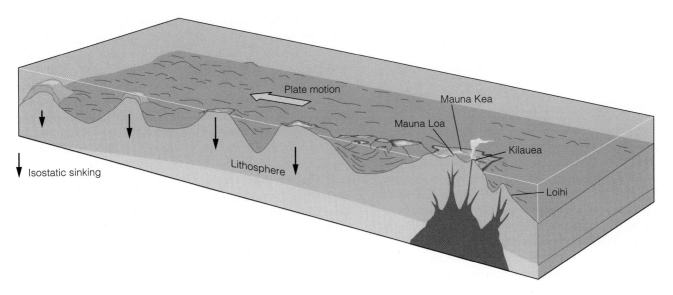

Figure 3.29 Formation of a volcanic island chain by the movement of an oceanic plate over a stationary mantle plume and hot spot. The age of the islands increases toward the left. New islands—like Loihi—will continue to form over the hot spot.

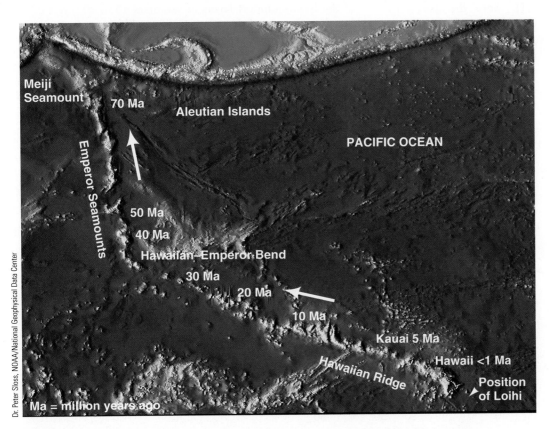

Figure 3.30 The Hawaiian chain, islands formed one by one as the Pacific Plate slid over a hot spot. The oldest known member of the chain, the Meiji Seamount, formed about 70 million years ago (70 Ma), and the bend in the chain shows that the plate changed direction about 40 million years ago. The island of Hawaii still has active volcanism, but the next island in the chain, Loihi, has begun building on the ocean floor. The upper arrow shows the initial direction of plate motion; the lower arrow shows the present direction.

the westward-moving North American Plate. Look again at Figure 3.14 to see the locations of hot spots around the world. The configuration and length of these chains of volcanoes and geothermal sites are consistent with the theory of plate tectonics.

Atolls and Guyots

Atolls are ring-shaped islands of coral reefs and reef-derived sediment centered over submerged inactive volcanoes. The coral animals that build atolls can live only in the upper sunlit layer of seawater. How, then, did their skeletal remains end up at a depth of 1,280 meters (4,200 feet) within Eniwetok Atoll? This surprising discovery was made in 1954 when bore holes were being drilled in preparation for the first hydrogen bomb tests. Plate tectonics suggests an answer. Coral animals can build atop the skeletons of their dead predecessors at a rate of about 1 centimeter (½ inch) each year. Coral animals living on a volcano's flanks can grow upward as the crust beneath the volcano slowly cools and contracts (sinks) during its movement away from the warm spreading center where it formed. A deep column of coral skeletons can accumulate if the rate of sinking is less than about 1 centimeter per year. Thus, the coral record traces plate subsidence for millions of years into the past, supporting the plate tectonics theory. (This process is shown in Figure 12.27.)

Guyots (pronounced ghee-OH) were discovered by Harry Hess during his service as commander of a U.S. Navy transport in the Pacific during World War II. With the ship's echo sounder he found chains of odd, flat-topped, submerged volcanic mountains; he named them after Princeton's first professor of geology, Arnold Guyot. More than 500 guyots have been located, and plate tectonics neatly explains the formation of most of them. Like atolls, they were once volcanic peaks (oceanic islands) standing above sea level. As the plate on which they were riding cooled, contracted, and was carried away from the spreading center, they became inactive and stopped growing. They were eroded flat by wave action as they sank beneath the ocean surface. Evidence of ancient beaches along their rims suggests that this hypothesis is correct. **Figure 3.31** shows a sinking progression of guyots. They never formed atolls, either because they began in water too cold for coral or because they moved and sank too rapidly for coral growth to keep up. Indeed, an atoll can become a guyot if its rate of subsidence increases, coral growth slows, or plate motion takes it into colder water.

The Age and Distribution of Sediments

If the ocean basins are genuinely ancient, and if the processes that produce sediments have been operating for most or all of that time, both the thickness and the age of sediments on the ocean floor should be great. They are not. The young spreading ridges are almost free of sediment, and the oldest edges of the basins support layers of sediment 15 to 20 times thinner than the age of the ocean itself would suggest.

Powerful low-frequency echo sounders have probed the depth and structure of these sediments in many locations. Core

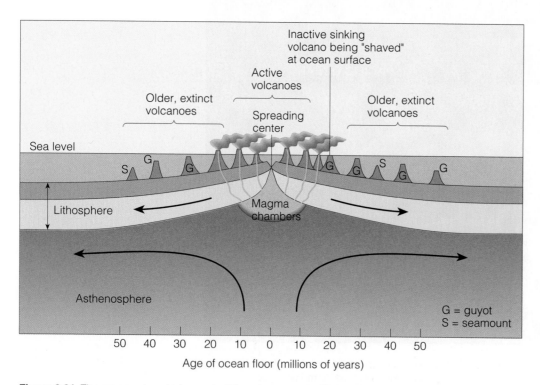

Figure 3.31 The process by which guyots (G) and seamounts (S) form. Guyots have flat tops because they have grown tall enough to be eroded by waves at the ocean's surface. Seamounts have a similar origin but retain their more pointed volcano shape because they never reached the surface.

sampling has shown that sediments resting directly on the basalt floor are almost always youngest near the spreading centers and oldest near subduction zones. The oldest sediments of the ocean basins are rarely more than 180 million years old (see Figure 5.20). These data are consistent with the plate tectonics idea that ocean basins are continuously created and destroyed.

The Oceanic Ridges

The location and configuration of the oceanic ridges are clear evidence of past events. The volcanic nature of ridge islands like Iceland, the shape of the longitudinal rifts splitting the ridge tops, and the sinking of the seabed as new oceanic crust cools and travels outward are all consistent with the theory of plate tectonics. The distribution of transform faults and fracture zones along the oceanic ridges (features you'll learn about in Chapter 4) also supports plate tectonics theory, as do on-the-spot geological observations made by researchers in deep submersibles.

Terranes

Buoyant continental and oceanic plateaus, island arcs, and fragments of granitic rock and sediments can be rafted along with a plate and scraped off onto a continent when the plate is subducted. This process is similar to what happens when a sharp knife is scraped across a table top to remove pieces of cool candle wax. The wax accumulates and wrinkles on the knife blade in the same way land masses and ocean sediments accumulate against the face of a continent as the lithosphere in which they are embedded reaches a plate boundary. Plateaus, isolated segments of seafloor, ocean ridges, ancient island arcs, and parts of continental crust that are squeezed and sheared onto the face of a continent are called **terranes.** The thickness and low density of terranes prevent their subduction. A simplified account of terrane accumulation is diagrammed in **Figure 3.32.**

Terranes are surprisingly common. New England, much of the Pacific Northwest of North America, and most of Alaska appear to be composed of this sort of crazy-quilt assemblage of material, some of which has evidently arrived from thousands of kilometers away. For example, western Canada's Vancouver Island has moved north some 3,500 kilometers (2,200 miles) in the last 75 million years.

Curiously, terranes can also contain fragments of dense oceanic crust. Roughly 0.001% of oceanic lithosphere is not subducted, but rather *obducted*—scraped off—onto the edges of continents. The heavy wrinkled rocks contain pillow basalts and material derived from the upper mantle. Called **ophiolites** (*ophion* = snake) because of their sinuous shape, these assemblages can contain metallic ores similar to those known to exist at the mid-ocean-ridge spreading centers. Ophiolites are found on all continents, and as might be expected those near the edges of the present continents are generally younger than those embedded deep within continental interiors. Some of these ancient reminders of past episodes of plate tectonics are more than 1.2 billion years old.

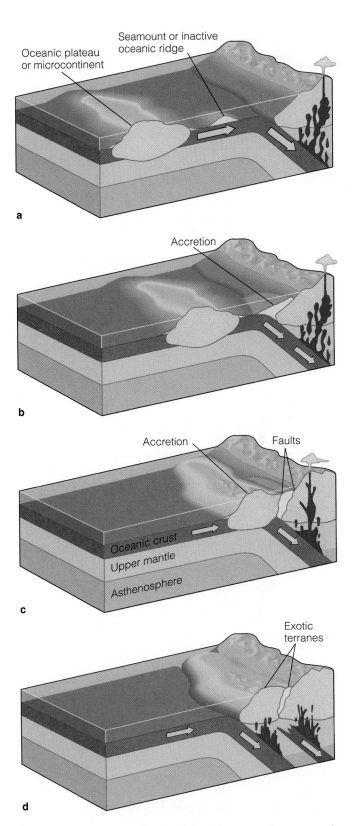

Figure 3.32 Terrane formation. Oceanic plateaus usually composed of relatively low density rock are not subducted into the trench with the oceanic plate. Instead, they are "scraped off," causing uplifting and mountain building as they strike a continent (**a–d**). Though rare, assemblages of subducting oceanic lithosphere can also be scraped off (obducted) onto the edges of continents. These dense, mineral-rich assemblages are known as ophiolites after their serpentlike shapes. Rich ore deposits are sometimes found in them.

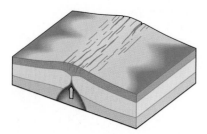

Stage 1: **Conception**
Motion: Crustal upwarp
Features: Elevation of continental crust, beginning of rifting and volcanism
Example: No good example today (perhaps West Africa)

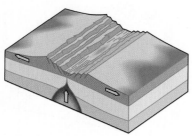

Stage 2: **Embryonic**
Motion: Uplift
Features: Complex system of rift valleys and lakes on continent
Example: East African rift valleys

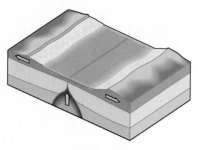

Stage 3: **Juvenile**
Motion: Divergence (spreading)
Features: Narrow sea with matching coasts. Oceanic ridge formed.
Example: Red Sea

Stage 4: **Mature**
Motion: Divergence (spreading)
Features: Ocean basin with continental margins. Ocean continues to widen at oceanic ridge.
Example: Atlantic Ocean, Arctic Ocean

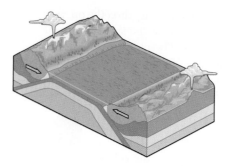

Stage 5: **Declining**
Motion: Convergence (subduction)
Features: Subduction begins. Island arcs and trenches form around basin edge.
Example: Pacific Ocean

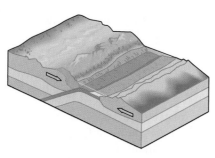

Stage 6: **Terminal**
Motion: Convergence, collision, and uplift
Features: Oceanic ridge subducted. Narrow, irregular seas with young mountains.
Example: Mediterranean Sea

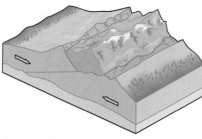

Stage 7: **Suturing**
Motion: Convergence and uplift
Features: Mountains form as two continental crust masses collide, are compressed, and override.
Example: India–Eurasia collision. Himalaya Mountains

Figure 3.33 The Wilson Cycle, named in honor of John Tuzo Wilson's synthesis of plate tectonics. Over great spans of time, ocean floors form and are destroyed. Mountains erode, sediments subduct, and continents rebuild. Seawater moves from basin to basin.

An Overview, Problems, and Implications

The theory of plate tectonics reveals much about the nature of Earth's surface. **Figure 3.33** summarizes surface tectonic activity.

In case you feel geophysicists have all the answers, however, consider just a few of the theory's unsolved problems:

- Why should long *lines* of asthenosphere be any warmer than adjacent areas?
- If plate movement depends on drag within mantle convection cells, why should the partially melted mantle material flow parallel to the plate bottoms for long distances instead of cooling and sinking near the spreading center?
- Is the increasing density of the leading edge of a subducting plate more important than convective friction in making the plate move?
- Why do mantle plumes form?
- How far do most plates descend? Recent evidence suggests much of the material spans the entire mantle, reaching the core.
- Has seafloor spreading always been a feature of Earth's surface? Has a previously thin crust become thicker with time, permitting plates to function in the ways described here?
- There is evidence of tectonic movement prior to the breakup of Pangaea. Will the process continue indefinitely, or are there cycles within cycles?

Though there is clearly much to learn, plate tectonics is already an especially powerful predictive theory. Discoveries and insights made by the researchers mentioned in this chapter, and hundreds of others, have borne out the intuition of Alfred Wegener. Our understanding of the process will evolve as more data become available, but there seems very little chance that geologists will ever return to the dominant pre-1960 view of a stable and motionless crust. Plate tectonic theory shows us the picture of an actively cycling Earth and an ever-changing surface, with *a single world ocean* changing shape and shifting positions as the plates slowly move.

The configuration of the ocean basins—discussed in the next chapter—is the result of plate tectonic activity. The variety of these features will make more sense now that you are armed with an understanding of the theory.

QUESTIONS FROM STUDENTS

1. How far beneath Earth's surface have people actually ventured?

Gold miners in South Africa have excavated ore 3,777 meters (12,389 feet, 2.35 miles) beneath the surface, where rock temperature is 55°C (131°F). The miners work in pairs, one chipping the rock and the other aiming cold air at his partner. After a time they trade places.

2. What is the difference between crust and lithosphere?

Lithosphere includes crust (oceanic and continental) and rigid upper mantle down to the asthenosphere. The velocity of seismic waves in the crust is much different from that in the mantle. This suggests differences in chemical composition or crystal structure, or both. The lithosphere and asthenosphere have different physical characteristics: The lithosphere is generally rigid, but the asthenosphere is capable of slow movement. Asthenosphere and lithosphere also transmit seismic waves at different speeds.

3. What are the most abundant elements comprising Earth?

You may be surprised to learn that oxygen accounts for about 46% of the mass of Earth's crust. On an atom-for-atom basis the proportion is even more impressive: Of every 100 atoms of Earth's crust, 62 are oxygen. Most of this oxygen is not present as the gaseous element but is combined with other atoms into oxides and other compounds. Most of the familiar crustal rocks and minerals are oxides of aluminum, silicon, and iron (rust, for example, is iron oxide).

In Earth as a whole, iron is the most abundant element, making up 35% of the mass of the planet, while oxygen accounts for 30% overall. Remember, most of the mass of the universe is hydrogen gas. Oxygen and iron are abundant on Earth only because fusion reactions in stars can transform light elements like hydrogen into heavy ones.

4. How do geologists determine the location and magnitude of an earthquake?

They use the time difference in the arrival of the P and S waves at their instruments to determine the distance to an earthquake. At least three seismographs in widely separated locations are needed to get a fix on the location.

The strength of the waves, adjusted for the distance, is used to calculate the earthquake's magnitude. Earthquake magnitude is often expressed on the **Richter scale.** Each full step on the Richter scale represents a 10-fold change in surface wave amplitude and a 32-fold change in energy release. Thus an earthquake with a Richter magnitude of 6.5 releases about 32 times more energy than an earthquake with a magnitude of 5.5, and about 1,000 times that of a 4.5-magnitude quake. People rarely notice an earthquake unless the Richter magnitude is 3.2 or higher, but the energy associated with a magnitude-6 quake may cause significant destruction.

The energy released by the 1964 Alaska earthquake was over a billion times greater than that released by the smallest earthquakes felt by humans, an energy release equal to about twice the energy content of world coal and oil production for an entire year. Very low or very high Richter magnitudes are not easy to measure accurately. The Alaska earthquake's magnitude was initially calculated between 8.3 and 8.6 on the

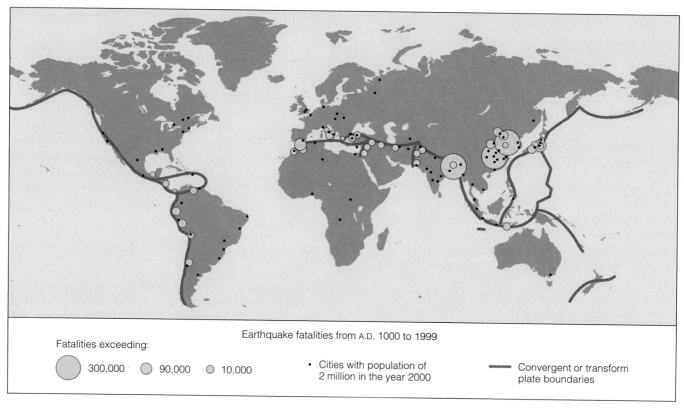

Figure 3.34 Population centers and earthquake areas. Large cities near plate boundaries in developing countries are likely to have the highest fatalities in a massive earthquake. (Source: Reprinted by permission of Dr. Roger Billham, University of Colorado.)

Earthquake fatalities from A.D. 1000 to 1999

Fatalities exceeding:
- 300,000
- 90,000
- 10,000
- Cities with population of 2 million in the year 2000
- Convergent or transform plate boundaries

Richter scale, but recent reassessment has yielded an extraordinary magnitude of 9.2. Earth rang like a great silent bell for ten days after the earthquake.

5. What's the potential for serious loss of life and property damage due to tectonic plate movement?

Relatively great. About 40% of the world's largest cities lie within 160 kilometers (100 miles) of a plate boundary. By the year 2000, about 290 million people lived in high-risk areas. About 80% of those at risk live in developing nations, where seismic safety is not a high priority in building design. Another 210 million people are also threatened by active volcanoes; most live along the subduction zones of the Pacific Ring of Fire, where 75% of Earth's 850 active volcanoes are located. Since 1600, there have been approximately 262,000 deaths from volcanic eruptions, about 76,000 in the twentieth century. See **Figure 3.34** for a graphic assessment of the risk.

6. How common are large earthquakes?

About every two days, somewhere in the world, there's an earthquake of from 6 to 6.9 on the Richter scale—roughly equivalent to the quake that shook Northridge and the rest of southern California in January 1994; or Kobe, Japan, in January 1995. Once or twice a month, on average, there's a 7 to 7.9 quake somewhere. There is about one 8 to 8.9 earthquake—similar in magnitude to the 1964 earthquake in Alaska or the 1812 New Madrid quake—each year.

Northridge- and Kobe-size quakes are moderate in size and occur fairly infrequently. Large losses of life and property can occur when these earthquakes occur in populated areas. Damage estimates from the Northridge earthquake exceeded $40 billion. In Kobe, more than 5,000 people died, and more than 26,000 were injured. Some 56,000 buildings were destroyed; estimates of the cost of reconstruction exceeded $400 billion.

7. Do other Earth-like planets or moons have hot internal layers and slowly drifting lithospheric plates?

The two bodies studied, the moon and Mars, appear to have much thicker crusts than Earth. The crusts of the moon and Mars make up about 10% of their planetary masses, while the crust of Earth constitutes only about 0.4% of its mass. The crust of the moon appears to be contiguous; one researcher has called it a "single-plate" planet.

Although plate tectonics, at least in the terrestrial sense, almost certainly does not occur on the moon, there is some controversial evidence that the surface of Mars may at one time have consisted of two plates. Mars is pocked by giant volcanoes and marked by gigantic lava plains; it probably had an active interior at some time in the not-too-distant past. The size of Martian volcanoes is so huge that some researchers believe hot spots may exist beneath a stationary lithosphere. The case for Martian plate tectonics was bolstered in May 1999 when *Mars Global Explorer* reported evidence of paleomagnetic stripes on Mars. The stripes are about ten times wider than those seen on Earth.

KEY CONCEPTS TO REVIEW

- Earth is composed of concentric spherical layers, with the least dense layer on the outside and the densest as the core. The lithosphere, the outermost solid shell that includes the crust, floats on the hot, deformable asthenosphere. The mantle is the largest of the layers.

- Large regions of Earth's continents are held above sea level by isostatic equilibrium, a process analogous to a ship floating in water.

- Some seismic waves—energy associated with earthquakes—can pass through Earth. Analysis of how these waves are changed, and the time required for their passage, has told researchers much about conditions inside Earth.

- Plate tectonics theory suggests that Earth's surface is not a static arrangement of continents and ocean, but a dynamic mosaic of jostling segments of lithosphere called lithospheric plates. The plates have collided, moved apart, and slipped past one another since Earth's crust first solidified.

- Plate motion is driven by slow, heat-generated (convection) currents flowing in the mantle. Most of the heat is generated by the decay of radioactive elements within Earth.

- Most of the large-scale features seen at Earth's surface may be explained by the interactions of plate tectonics. Plate tectonics also explains why our ancient planet has surprisingly young seafloors, the oldest of which is only as old as the dinosaurs, that is, about $\frac{1}{23}$ the age of Earth.

- The confirmation of plate tectonics rests on diverse scientific studies from many disciplines. Among the most convincing is the study of paleomagnetism, the orientation of Earth's magnetic field frozen into rock as it solidifies.

INTERNET STUDY RESOURCES

The Web site for this book contains helpful study aids. Log on to:

www.brookscole.com/product/053437557xs

and click on the Chapter-by-Chapter area. Choose Chapter 3 and select a resource:

- **Flash Cards** allows you to test your mastery of the Terms and Concepts to Remember for this chapter.
- **Tutorial Quizzes** provides a multiple-choice practice quiz.
- **Student Guide to InfoTrac College Edition** will lead you to Critical Thinking Projects that use InfoTrac College Edition as a research tool.

TERMS AND CONCEPTS TO REMEMBER

absolute dating	mantle
asthenosphere	oceanic crust
atoll	ophiolite
basalt	P wave (primary wave)
buoyancy	Pacific Ring of Fire
catastrophism	paleomagnetism
conduction	Pangaea
continental crust	Panthalassa
continental drift	plate
convection	plate tectonics
convection current	radioactive decay
convergent plate boundary	radiometric dating
core	relative dating
crust	Richter scale
density	S wave (secondary wave)
density stratification	seafloor spreading
divergent plate boundary	seismic waves
earthquake	seismograph
echo sounder	shadow zone
fault	spreading center
granite	subduction
guyot	subduction zone
half-life	terrane
hot spot	transform fault
isostatic equilibrium	transform plate boundary
lithosphere	uniformitarianism
magma	Wegener, Alfred
magnetometer	Wilson, John Tuzo

STUDY QUESTIONS

Review Questions

1. How are Earth's internal layers classified? Which method is more useful in explaining plate tectonics?

2. How is crust different from lithosphere?

3. Would the most violent earthquakes be associated with spreading centers or with subduction zones? Why?

4. Describe the mechanism that powers the movement of the lithospheric plates.

5. Why are the continents about 20 times older than the oldest ocean basins?

Critical Thinking Questions

1. Seawater is more dense than fresh water. A ship moving from the Atlantic into the Great Lakes goes from seawater to fresh water. Will the ship sink farther into the water during the passage, stay at the same level, or rise slightly?

2. Some earthquakes are linked to adjustments of isostatic equilibrium. How can this occur? Where would you be likely to experience such an earthquake?

3. Where are the youngest rocks in the ocean crust? The oldest? Why?

4. What biological evidence supports the theory of plate tectonics?

5. What evidence can you cite to support the theory of plate tectonics? What questions remain unanswered? Which side would you take in a debate?

6. **InfoTrac College Edition Project** Continental drift illustrations sometimes give the impression that the continents have always looked the same. However, plate tectonic research has shown that the areas of land we now identify as continents may have had interesting transformations during the long periods of geologic time. North and South America, for example, contain rocks and structures in common with regions now found at much greater distances. Do some research using InfoTrac College Edition to discover some of the connections between the American continents and other land masses. Write an essay on your findings and speculate on how these connections can be explained—or not explained—by plate tectonic theory.

Earth Systems Today CD Questions

1. Go to "Geologic Time" in the CD-ROM's main menu, and click on "Changing Earth." Begin the animation of plate motion. When it arrives at "present day," predict plate movement 250 million years into the future. Sketch your answer, then play the rest of the animation to see if you are correct.

2. Go to "Geologic Time," then "Relative Dating" and "Absolute Dating." After reviewing the contents of Box 3.2 (Absolute and Relative Dating), try your hand at some of the experiments. Pay particular attention to the concept of half-life. See if you can correctly answer the questions dealing with the age of layers in "Layer Study."

3. Go to "Earth's Layers," then to "P and S Waves." Before activating the animation, can you predict the type of movement for each wave?

4. Go to "Earth's Layers," then to "Core Studies." Put yourself in the place of Inge Lehmann and Richard Oldham (Box 3.1), and perform some experiments in all three sections of "Core Studies." Would you have reached the same cnclusions they did?

5. Go to "Plate Tectonics," then "Plate Locations." With the "Tectonic Activity" boxes toggled off, predict the positions of recent earthquakes and volcanoes based on the plate lo-

cations shown. Toggle the "Tectonic Activity" boxes back on, and gauge your success.

6. Go to "Plate Tectonics," then "Plate Boundaries." Note there are three types of boundaries: divergent, convergent, and transform. Note also there are three settings for convergent boundaries (ocean–ocean, ocean–continent, and continent–continent). Before animating the frames, write a brief prediction of the resulting geological activity for each of the combinations of boundaries and settings. Now animate the frames, and see if your choices were appropriate.

7. Go to "Plate Tectonics," then "Triple Junctions and Seafloor Study." Take the VR (virtual reality) tour of a segment of the East Pacific Risa. Before you move and zoom (using your mouse), can you predict the kinds of features you will encounter?

8. Go to "Volcanism," then "Distribution of Volcanism." Toggle off the volcano types. Now looking at the upper chart showing spreading plate boundaries, directions of plate motion, and the location of subduction zones, predict where subduction zone volcanoes, rifting site volcanoes, and hot spots might be expected. Now toggle the volcanoes on to see how accurate your predictions were.

RESOURCES FOR FURTHER READING AND RESEARCH

The Web site for this book contains many ideas for further reading and research. Log on to:

www.brookscole.com/product/053437557xs

and click on the Chapter-by-Chapter area. Choose Chapter 3 and select a resource:

- **References** lists the major books and articles consulted in writing this chapter, along with comments from the author about their content and reading level.

- **Hypercontents** takes you to an extensive list of sites with news, research, and images related to individual sections of the chapter.

- In **Student Guide to InfoTrac College Edition** scroll down to Suggested Readings from InfoTrac College Edition for brief descriptions of the articles listed and search hints for finding them in InfoTrac College Edition.

- **Regional InfoTrac College Edition** articles are organized into East Coast, West Coast, and Gulf Coast regions, allowing you to study oceanography on a more local level. You can also access Regional InfoTrac College Edition at www.localocean.com.

For additional readings, go to InfoTrac College Edition, your online research library, at:

http://infotrac.thomsonlearning.com/

Continental Margins and Ocean Basins

4

TRIESTE AT THE BOTTOM

Studying deep-ocean basins has always presented daunting obstacles. The most difficult problem is to reach extreme depths, but, amazingly, scientists have visited the bottom of the deepest ocean basin. On 23 January 1960 U.S. Navy lieutenant Don Walsh and Dr. Jacques Piccard descended to a depth of 11,022 meters (6.85 miles) into the Challenger Deep, an area of the Mariana Trench discovered in 1951 by the British oceanographic research vessel *Challenger II*.[1] The vehicle used in the descent was the **bathyscaphe** *Trieste* (**Figures a, b,** and **c**), a deep-diving submersible designed like a blimp with a very strong and thick (and cramped) steel crew sphere suspended below. A blimp uses helium gas for buoyancy, but a gas would be compressed by water pressure; so gasoline, which is relatively incompressible, was used to provide lift.

First, *Trieste* made a series of increasingly deep practice dives; then the big day arrived. Don Walsh wrote of the experience in 1979:

> At about 600 feet we entered a zone of deepening twilight where colors faded into gray. By 1,000 feet the light had gone completely. We turned out the lights in the sphere to watch for the luminescent creatures that are sometimes visible at this level. We saw very few. Eventually we turned the cabin lights back on and briefly tested the forward lights that throw a beam in front of the observation window. Formless plankton streamed past, giving us a sensation of great speed.
>
> There were minor incidents such as the small leak that always developed in one of the hull connectors—a place where wires from lights and instruments on the outside of the sphere pass through the hull. The leak started at about 10,000 feet. It was an old friend, a tiny drip, drip, drip. I timed the drips and found no change from before, which meant that it had not become more serious. We expected it to disappear at about

Don Walsh

a The bathyscaphe *Trieste* before launching. Note the crew sphere suspended below the gasoline-filled flotation hull. *Trieste* is 15.7 meters (59.5 feet) long.

---

[1] The position of the Challenger Deep is marked in Figure 4.28, and its depth in relation to Mt. Everest is shown in Figure 4.29a.

91

Trieste's steel personnel sphere. Its thickness varies from 13 to 18 centimeters (5 to 7 inches) in thickness. The small, thick window is cone-shaped to withstand the pressure of the deepest spot in the world ocean—11 million kilograms per square meter (8 tons per square inch)!

We checked our speed to half a foot a second and continued. At that rate, time and distance pass very slowly, and I think for the first time in the dive both of us had the feeling of awe that comes from exploring the totally unknown.

I did not take my eyes off the fathometer, and Jacques never stopped watching out of the tiny porthole with its weak probe of light. No bottom was in sight. . . . Soon Jacques could see a difference in the effect of our light in the water, as the rays reflected off the bottom. As we approached the floor I called the fathometer readings to Jacques in fathoms: "Thirty . . . twenty . . . ten. . . ." At eight he called that he could see the bottom.

At 1:10 P.M. we sank gently onto the soft floor. A great cloud of silt rose around us. The fifteen-man, Navy civilian/military team had set a record that could not be broken.

The voyagers took temperature readings and saw a pair of bottom fish and some crustaceans before ascending. With the possible exception of the moon, no place visited by humans has been more hazardous to explore.

4-1

WHAT TO WATCH FOR IN CHAPTER 4

The principles you learned in Chapter 3 are put to use here—tectonic forces and erosion have built and shaped the seabed. First, though, consider how difficult it is to see the signature of these forces. Light doesn't penetrate far in the ocean, so "seeing" must be by other means. For me, the most unlikely advance in remote sensing is the use of orbiting satellites to sense seabed contours!

Tectonic history divides the seabed into two regions: continental margins and deep-ocean basins. The continental margin, which is the relatively shallow ocean floor nearest the shore, shares the structure and composition of the adjacent continent. The dense deep-ocean floor away from land has a much different tectonic origin, history, and crustal character.

The continental margin consists of the continental shelf and the continental slope, with prominent secondary features that include submarine canyons and continental rises. Features of the deep-ocean basins include rugged oceanic ridges, flat abyssal plains, occasional deep trenches, and curving chains of volcanic islands. Most of the ocean floor is blanketed with sediment, the subject of Chapter 5.

As you read, try to imagine what it might be like to see these features as we see mountains or canyons on land. How does seabed topography compare to the familiar landforms around us?

15,000 feet, when the water pressure packed the plastic sealer in more tightly—and it did.

At 27,000 feet we checked our rate of descent to 2 feet per second by dumping some shot ballast. We were not too sure of the underwater currents here, and we did not want to go crashing into a wall of the trench by mistake. As we neared 30,000 feet I started thinking about the changes we had planned to make when we got to within 1,000 feet of the bottom—which we now were expecting to find only another 3,500 feet below us. I was running through a mental checklist when I heard and felt a powerful crack. The sphere rocked as though we were on land and going through a mild earthquake.

We waited anxiously for what might happen next. Nothing did. We flipped off the instruments and the underwater telephone so that we could hear better. Still nothing happened. We switched the instruments back on and studied the dials that would tell us if something critical had occurred. No, we had our equilibrium and were descending exactly as before.

BATHYMETRY

In February 2000 the *Mars Global Surveyor* spacecraft completed a map of the surface of our planetary neighbor. The photos from orbit were clear and detailed, and objects about 2 meters (7 feet) across could be seen almost anywhere on the planet. There was no ocean, and few storms to spoil the view.

Mapping Earth is much more difficult because water and clouds hide more than three-quarters of the surface. Until surprisingly recently we have known more about the global contours of the moon and the inner planets than we knew about our own home. Thanks to modern bathymetry, our view is clearing.

The discovery and study of ocean floor contours is called **bathymetry** (*bathy* = deep + *meter* = measure). The earliest known bathymetric studies were carried out in the Mediterranean by a Greek named Posidonius in 85 B.C. He and his crew let out nearly 2 kilometers (1.25 miles) of rope until a stone tied to the end of the line touched bottom. Bathymetric technology had not improved by the time Sir James Clark Ross obtained soundings of 4,893 meters (16,054

Archive, Scripps Institution of Oceanography

Figure 4.1 An illustration from the *Challenger Report* (1880). Seamen are handling the steam winch used to lower a weight on the end of a line to the seabed to find ocean depth.

feet) in the South Atlantic in 1818. In the 1870s the researchers aboard HMS *Challenger* added the innovation of a steam-powered winch to raise the line and weight, but the bathymetric method was the same (**Figure 4.1**). The *Challenger* crew made 492 bottom soundings and confirmed Matthew Maury's earlier discovery of the Mid-Atlantic Ridge.

The sinking of the RMS *Titanic* in 1912 stimulated research that finally ended slow, laborious weight-on-a-line efforts. By April 1914, Reginald A. Fessenden, a former employee of Thomas Edison, had developed an "Iceberg Detector and Echo Depth Sounder." The detector worked by directing a powerful sound pulse ahead of a ship and then listening for an echo from the submerged portion of an iceberg. It was easy to direct the beam downward to sense the distance to the bottom. It might take most of a day to lower and raise a weighted line, but echo sounders could take many bottom recordings in a minute.

When World War I intervened, Fessenden's talents were turned to enemy submarine detection devices. But when peace returned, he continued his echo sounder research. In June 1922, an echo sounder based on his designs made the first continuous profile across an ocean basin aboard the USS *Stewart*, a U.S. Navy vessel. Using an improved echo sounder based on his design, the German research vessel *Meteor* made 14 profiles across the Atlantic from 1925 to 1927. The wandering path of the Mid-Atlantic Ridge was revealed, and its obvious coincidence with coastlines on both sides of the Atlantic stimulated the discussions that culminated in our present understanding of plate tectonics.

Echo sounding wasn't perfect, as **Figure 4.2a** shows. The ship's exact position was sometimes uncertain. The speed of sound through seawater varies with temperature, pressure, and salinity, and those variations made depth readings slightly inaccurate. Simple depth sounder images (shown in **Figure 4.2b**) were also unable to resolve the fine detail that oceanographers needed to explore seabed features. Even so, researchers using depth sounder tracks had painstakingly compiled the first comprehensive charts of the ocean floor by 1959. (A portion of one of those hand-drawn charts is shown in Figure 4.22a.)

Two new techniques—multibeam systems and satellite altimetry—made possible by improved sensors and high-speed computers have been perfected to minimize inaccuracies and speed the process of bathymetry. The features discussed in this chapter have been studied using these (and other) systems, any of which is surely an improvement over lowering rocks into the ocean!

Multibeam Systems

Like an echo sounder, a multibeam system bounces sound off the seafloor to measure ocean depth. Unlike a simple echo sounder, a multibeam system may have as many as 121 beams radiating from a ship's hull. Fanning out at right angles to the direction of travel, these beams can cover a 120° arc (**Figure 4.3a**). Typically, a pulse of sound energy is sent toward the seabed every 10 seconds. Listening devices record sounds

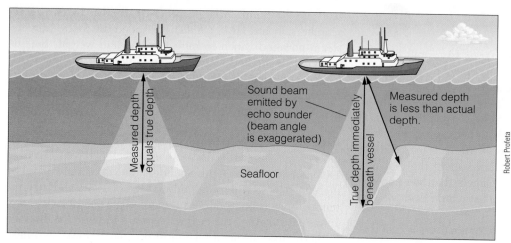

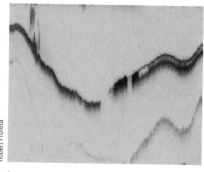

a

b

Figure 4.2 (**a**) The accuracy of an echo sounder can be affected by water conditions and bottom contours. The pulses of sound energy, or "pings," from the sounder spread out in a narrow cone as they travel from the ship. When depth is great, the sounds reflect from a large area of seabed. Because the first sound of the returning echo is used to sense depth, measurements over deep depressions are often inaccurate. (**b**) An echo sounder trace. A sound pulse from a ship is reflected off the seabed and returns to the ship. Transit time provides a measure of depth. For example, it takes about 2 seconds for a sound pulse to strike the bottom and return to the ship when water depth is 1,500 meters (4,900 feet). Bottom contours are revealed as the ship sails a steady course. In this trace, the horizontal axis represents the course of the ship, and the vertical axis represents water depth. The ship has sailed over a small submarine canyon.

reflected from the bottom, but only from the narrow corridors corresponding to the outgoing pulse. Successive observations build a continuous swath of coverage beneath the ship. By "mowing the lawn"—moving the ship in a coverage pattern similar to cutting grass—researchers can build a complete map of an area (**Figure 4.3b**). (Further processing can yield remarkably detailed images, including those of Figures 4.12, 4.13, and 4.15.) Fewer than 200 research vessels are equipped with multibeam systems. At the present rate, charting the entire seafloor in this way would require more than 125 years.

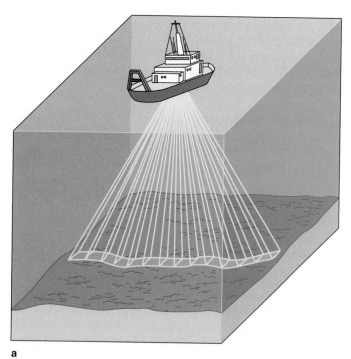

a

b

Figure 4.3 (**a**) A multibeam echo sounder uses as many as 121 beams radiating from a ship's hull. Fanning out at right angles to the direction of travel, these beams can cover a 120° arc and measure a swath of bottom about 3.4 times as wide as the water is deep. Typically, a "ping" is sent toward the seabed every 10 seconds. Listening devices record sounds reflected from the bottom, but only from the narrow corridors corresponding to the outgoing pulse. Thus a multibeam system is much less susceptible to contour error than the single-beam device shown in Figure 4.2a. (**b**) A multibeam reading of a fragment of seafloor near the East Pacific Rise south of the tip of Baja California, Mexico. The uneven coverage reflects the path of the ship across the surface. Detailed analysis requires sailing a careful pattern, rather like the pattern for mowing a lawn evenly. Computer processing provides extraordinarily detailed images, such as those in Figures 4.12, 4.13, and 4.15.

Satellite Altimetry

4-4

Satellites cannot measure ocean depths directly, but they can measure small variations in the elevation of surface water. Using about a thousand radar pulses each second, the U.S. Navy's *Geosat* satellite (**Figure 4.4a**) can measure its distance from the ocean surface to within 0.03 meter (1 inch)! Because the precise position of the satellite can be calculated, the average height of the ocean surface can be known with great accuracy.

Disregarding waves or tides or currents, the ocean surface can vary from the ideal smooth (ellipsoid) shape by as much as 200 meters (660 feet). This is because the pull of gravity varies across the surface of Earth depending on the nearness (or distance away) of massive parts of Earth. A mountain or ridge "pulls" water toward it from the sides, forming a mount of water over itself (**Figure 4.4b**). For example, a typical undersea volcano with a height of 2,000 meters (6,600 feet) above the seabed and a radius of 20 kilometers (32 miles) would produce a 2-meter (6.6-foot) rise in the ocean surface. (This mound cannot be seen with the unaided eye because the slope of the surface is very gradual.) The large features of the seabed are amazingly and accurately reproduced in the subtle standing irregularities of the sea surface (**Figure 4.4c**).

Geosat and its successors, *TOPEX/Poseidon* and *Jason*, have allowed the rapid mapping of the world ocean floor from space. Hundreds of previously unknown features have been discovered through the data they have provided.

THE TOPOGRAPHY OF OCEAN FLOORS

4-5

Most people think an ocean basin is shaped like a giant bathtub. They imagine that the continents drop off steeply just beyond the surf zone and that the ocean is deepest somewhere out in the middle. As is clear in **Figure 4.5,** bathymetric studies have shown that this is not the case.

U.S. Department of the Navy

a

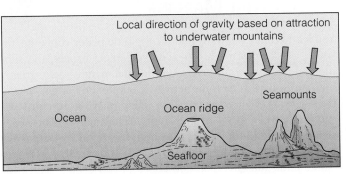

b

Figure 4.4 (**a**) *Geosat,* a U.S. Navy satellite that operated from 1985 through 1990, provided measurements of sea surface height from orbit. Moving above the ocean surface at 7 kilometers (4 miles) a second, *Geosat* bounced 1,000 pulses of radar energy off the ocean every second. Height accuracy was within 0.03 meter (1 inch)! (**b**) Distortion of the sea surface above a seabed feature occurs when the extra gravitational attraction of the feature "pulls" water toward it from the sides, forming a mound of water over itself. (**c**) This view of the complex system of ridges, trenches, and fracture zones on the South Atlantic seafloor east of the tips of South America (SA) and Antarctica's Palmer Peninsula (ANT) was derived in large part from *Geosat* data declassified by the U.S. Navy in 1995.

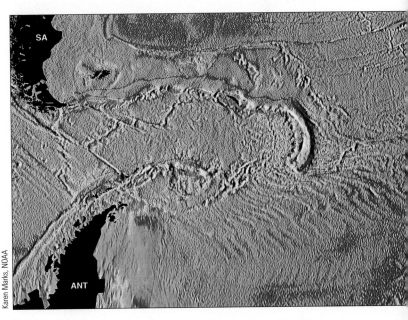

Karen Marks, NOAA

c

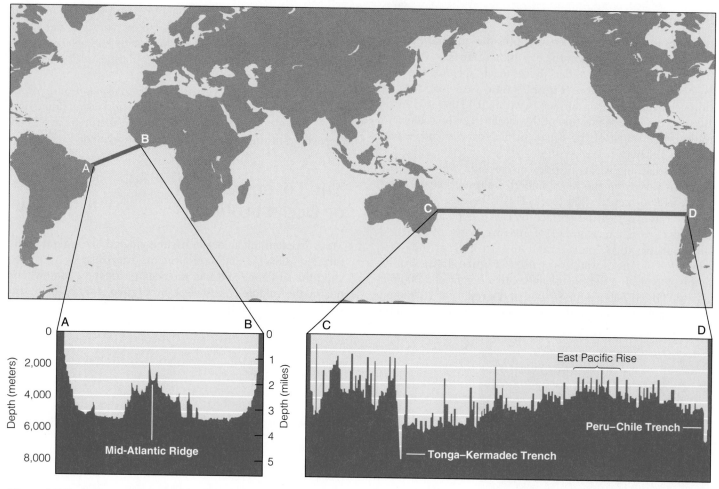

Figure 4.5 Typical cross sections of the Atlantic and Pacific ocean basins. The vertical exaggeration is 100:1.

Why? As you read in the last chapter, plate tectonics theory suggests that Earth's surface is not a static arrangement of continents and ocean, but a dynamic mosaic of jostling lithospheric plates. The lighter continental lithosphere floats in isostatic equilibrium above the level of the heavier lithosphere of the ocean basins. The great density of the seabed partly explains why more than half of Earth's solid surface is at least 3,000 meters (10,000 feet) below sea level (**Figure 4.6**). The continental crust is thicker than oceanic crust, and there is a difference in density between continental lithosphere and oceanic lithosphere. Therefore, the less dense lithosphere containing the continents floats in isostatic equilibrium above the level of the denser lithosphere containing ocean basins. The shape of the curve results from the isostatic position of continental crust relative to oceanic crust and the volume of water available to fill the basins. **Figure 4.7** shows the relative amounts of Earth's surface composed of continental and oceanic structures and the important subdivisions of each.

Notice in **Figure 4.8** the transition between the thick (and less dense) granitic rock of the continents and the relatively thin (and denser) basalt of the deep-sea floor. Near

shore the features of the ocean floor are similar to those of the adjacent continents because they share the same granitic basement. The transition to basalt marks the *true* edge of the continents and divides ocean floors into two major provinces. The submerged outer edge of a continent is called the **continental margin.** The deep-sea floor beyond the continental margin is properly called the **ocean basin.**

CONTINENTAL MARGINS

You learned in Chapter 3 that lithospheric plates converge, diverge, or slip past each other. As you might expect, the submerged edges of continents—continental margins—are greatly influenced by this tectonic activity. Continental margins facing the edges of *diverging* plates are called **passive margins** because relatively little earthquake or volcanic activity is now associated with them. They surround the Atlantic, so passive margins are sometimes referred to as *Atlantic-type* margins. Continental margins near the edges of *converging* plates (or near places where plates are slipping past each other) are called

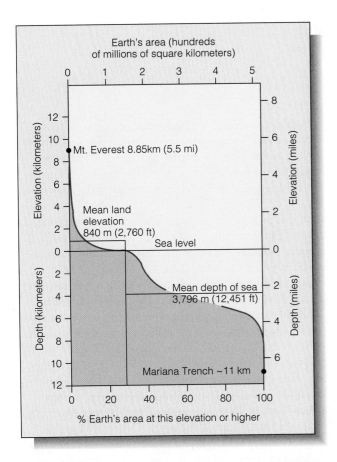

Figure 4.6 A graph showing the distribution of elevations and depths on Earth. This graph is not a land-to-sea profile of Earth, but rather a plot of the area of Earth's surface above any given elevation or depth below sea level. Note that more than half of Earth's solid surface is at least 3,000 meters (10,000 feet) below sea level. The average depth of the ocean is much greater than the average elevation of the continents: The average depth of the world ocean (3,796 meters or 12,451 feet) is much greater than the average height of the continents (840 meters or 2,760 feet).

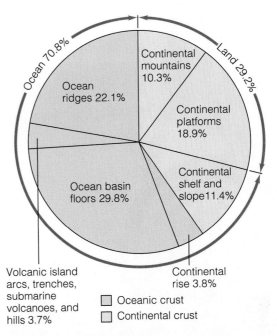

Figure 4.7 Features of Earth's solid surface shown as percentages of the planet's total surface.

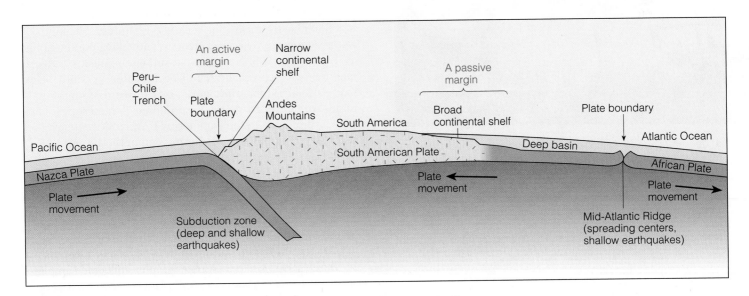

Figure 4.8 Typical continental margins bordering the leading (tectonically active) and trailing (passive) edges of a moving continent. (The vertical scale has been exaggerated.)

Technological advances have permitted researchers to explore ocean basins in person. The sophistication and cost of deep-diving devices increase with their depth capability: Scuba outfits cost only a few hundred dollars, but a human-carrying vehicle suitable for a trip to the abyssal plains costs millions.

With the proper mix of gases and adequate training, a scuba diver can descend to about 60 meters (200 feet). With special mixtures of breathing gases and dry suits inflated with argon to avoid heat loss, professional divers have reached a depth of 300 meters (984 feet). Below that depth a small research submarine is better equipped to withstand the cold and pressure. One of the newest and most promising is the maneuverable, compact, one-person *DeepWorker 2000* (**Figure a**) developed by Nuytco Research, Ltd. The sub is "flown" by a pilot seated in a comfortable chair and surrounded by a strong plastic bubble allowing 360° visibility. Using foot pedals and a control stick, *DeepWorker 2000* can descend to 610 meters (2,000 feet) and remain on station for up to 12 hours. A small number of these quick and maneuverable subs will explore the continental shelf and upper continental slope for five years in association with the National Geographic Society's Sustainable Seas Expeditions.

For descents deeper than mid-slope, however, stronger and more costly submersibles like *Alvin* (**Figure b**) must be used. *Alvin* was commissioned in 1964 and is operated jointly by the Woods Hole Oceanographic Institution, the National Science Foundation, the Office of Naval Research, and the National Oceanic and Atmospheric Administration. The most famous of the research submersibles, *Alvin* has explored the Mid-Atlantic Ridge near the Azores at 2,700 meters (9,000 feet), measuring rock temperatures, collecting water samples for chemical analysis, and taking photographs.

A modern U.S. Navy nuclear submarine makes a superb manned research vehicle. The USS *Pargo* has recently been made available to marine scientists (**Figure d**). Operating under Arctic ice, *Pargo* made an unprecedented series of measurements in the summer of 1993. USS *Pogy*, a sister ship, completed a similar mission in the summer of 1996.

Remotely operated robots are becoming more popular with researchers. Known as ROVs (remotely operated vehicles), they're cheaper to operate and safer than people-carrying vehicles. The deepest-diving of these, *Kaiko* (**Figure e**), descended to a measured depth of 10,914 meters (35,798 feet) near the bottom of the Challenger Deep on 24 March 1995. *Kaiko* took photographs, collected samples of sediments, and planted a small plaque to commemorate the event.

NUYTCO Research

a *DeepWorker 2000,* among the newest and most capable of human-carrying research vehicles. Described by its inventor as ". . . an underwater sports car," *DeepWorker 2000* and its sister vessels embarked in 1999 on a five-year exploration of the continental shelf and upper continental slope. Equipped with high-resolution sonar, video cameras, and two robotic arms that can extend 3.7 meters (12 feet) from the vehicle, the maneuverable little sub can take a researcher to a depth of 610 meters (2,000 feet) for up to 12 hours.

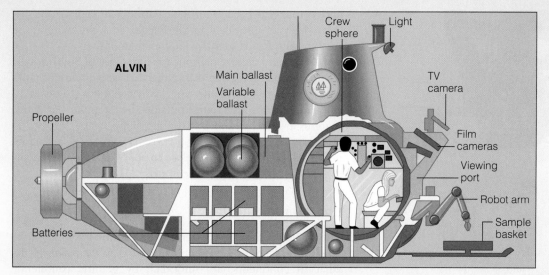

ALVIN

Crew sphere
Light
Main ballast
Variable ballast
Propeller
TV camera
Film cameras
Viewing port
Robot arm
Sample basket
Batteries

b *Alvin,* the best-known and oldest of the six deep-diving manned research submarines now in operation, made its 3,000th dive in September 1995. *Alvin* carries three people and is capable of diving to 4,000 meters (13,120 feet).

JAMSTEC

c The newest deep-diving vehicle capable of carrying a human crew is Mitsubishi Heavy Industries' *Shinkai 6500.* Now the deepest-diving submersible in operation, *Shinkai 6500* safely descended to a depth of 6,527 meters (21,409 feet) on 11 August 1989 and thus can reach all but the deepest trench floors, or about 98% of the world ocean bottom.

d Very fast, strong, and silent, a nuclear submarine makes an ideal platform for oceanographic research. The U.S. Navy made one of its nuclear submarines available to marine scientists during the summer of 1993. The research cruise aboard USS *Pargo,* a *Sturgeon*-class submarine strengthened for operation in ice, began in Groton, Connecticut, and ended in Bergen, Norway. More than 9,000 kilometers (5,600 miles) of underway data were collected in the deep Arctic Ocean beneath the pack ice. Water samples were taken for chemical and biological analysis (as seen here), current probes were released and tracked, and buoys were placed to monitor weather and surface water conditions. The data from the cruise will be placed in the public domain and results published in unclassified literature. USS *Pogy,* a sister ship, completed a similar mission in the summer of 1996. USS *Hawkbill* followed in 1998.

Ted Delaca, University of Alaska

active margins because of their earthquake and volcanic activity. They are prevalent in the Pacific, so active margins are sometimes referred to as *Pacific-type* margins.

Active and passive margins west and east of South America are shown in **Figure 4.9.** Note that active margins coincide with plate boundaries but passive margins do not. Passive margins are also found outside the Atlantic, but active margins are confined mostly to the Pacific.

Continental margins have two main divisions: a shallow, nearly flat continental *shelf* close to shore and a more steeply sloped continental *slope* to seaward.

Continental Shelves

The shallow, submerged extension of a continent is called the **continental shelf.** Continental shelves, an extension of the adjacent continents, are underlain by granitic continental crust. They are much more like the continent than like the deep-ocean floor, and they may have hills, depressions, sedimentary rocks, and mineral and oil deposits similar to those on the dry land nearby. Earth's continental shelves are shown in **Figure 4.10.** Taken together, the continental shelves comprise 7.4% of Earth's ocean area.

Figure 4.11 shows a passive-margin continental shelf characteristic of Atlantic Ocean edges. The broad shelf extends far from shore in a gentle incline, typically 1.7 meters per kilometer (0.1°, or about 9 feet per mile), much less than the slope of a well-drained parking lot. Shelves along the margin of the Atlantic Ocean often reach 350 kilometers (220 miles) in width and end at a depth of about 140 meters (460 feet), where a steeper drop-off begins.

The passive-margin shelves of the Atlantic Ocean formed as the fragments of Pangaea were carried away from each other by seafloor spreading. The continental lithosphere, thinned during initial rifting, cooled and contracted as it moved away from the spreading center, submerging the trailing edges of the continents and forming the shelves.

Most of the material comprising a shelf comes from erosion of the adjacent continental mass. Rivers assist in passive shelf building by transporting huge amounts of sediments to the shore from far inland. In some places the sediments accumulate behind natural dams formed by ancient reefs or ridges of granitic crust (see again Figure 4.9). The weight of the sediment isostatically depresses the continental edges and allows the sediment load to grow even thicker. Sediment at the outer edge of a shelf can be up to 15 kilometers (9 miles) thick and 150 million years old.

The width of a shelf is usually determined by proximity to a plate boundary. You can see in Figure 4.8 that the shelf at the *passive* margin (east of South America) is broad but the shelf at the *active* margin (west of South America) is very narrow. The widest shelf, 1,280 kilometers (800 miles) across, lies north of Siberia in the tectonically quiet Arctic Ocean. Shelf width depends not only on tectonics, but also on marine processes: Fast-moving ocean currents can sometimes prevent sediments from accumulating. For example, the east coast of Florida has a very narrow shelf because there is no

e *Kaiko,* the deepest-diving vehicle presently in operation, descended to a measured depth of 10,914 meters (35,798 feet) near the bottom of the Challenger Deep on 24 March 1995. The small ROV (remotely operated vehicle) sends information back to operators in the mothership by fiber-optic cable. *Kaiko* is operated by JAMSTEC, a Japanese marine science consortium.

Perhaps the most imaginative new technology is **telepresence,** the extension of a person's senses by remote manipulators. A scientist might wear a helmet containing small stereo television screens and earphones and place his hands in special gloves equipped with tactile feedback units. His head and hand movements would be duplicated by a robot on the seafloor, and sensations "felt" by the robot would be relayed back to the scientist through the TVs, earphones, and gloves. He would thus have the sensation of being on the seafloor and could take samples, manipulate tools, or just look around. Other researchers could watch or participate at distant locations via the information superhighway. Oceanographers have set a course for interesting times!

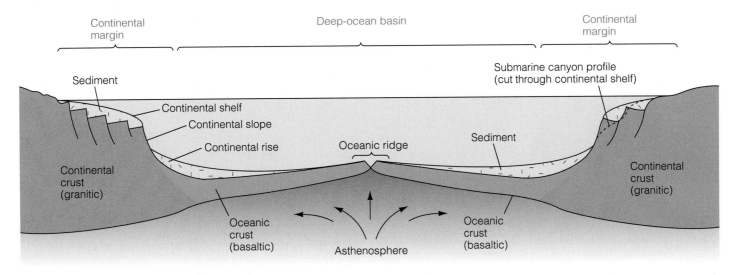

Figure 4.9 Cross section of a typical ocean basin flanked by passive continental margins. (The vertical scale has been greatly exaggerated to emphasize the basin contours.)

natural offshore dam formed by ridges of granitic crust, and because the swift current of the nearby Gulf Stream scours surface sediment away. The west coast, however, has a broad shelf (**Figure 4.12**) with a steep terminating slope.

A broad shelf results as sediment accumulates along the sheltered edges of the Gulf of Mexico. In **Figure 4.13,** sediment eroded from the Rocky Mountains has been carried by the Mississippi River and deposited south of Texas and Louisiana. This sediment overlies salt deposited about 180 million years ago when much of the water in the gulf evaporated. The weight of the overlying sediment causes domes of salt to rise, spread out, and dissolve to create the pockmarked bottom characteristic of the area.

The shelves of the active Pacific margins are generally not as broad and flat as Atlantic shelves. An example is the abbreviated shelf off the west coast of South America, where the steep western slope of the Andes Mountains continues nearly uninterrupted beneath the sea into the depths of the Peru–Chile Trench (see again Figure 4.8). Active-margin shelves have more varied topography than passive-margin shelves. The character of continental shelves at an active margin may be determined more by faulting, volcanism, and tectonic deformation than by sedimentation.

A few Pacific shelves are broad, however. As in the Atlantic, natural offshore dams trap sediments and form shelves. But in the Pacific the sediment-trapping dams are

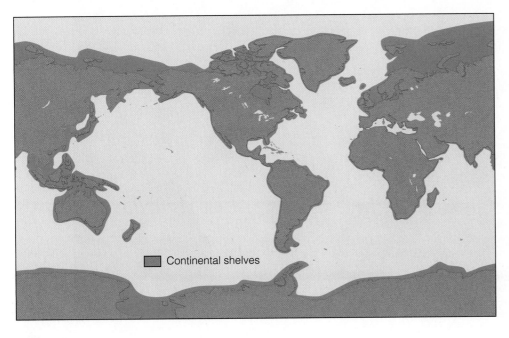

Figure 4.10 The worldwide distribution of continental shelves. The continental islands of Ireland and Great Britain are part of the continent of Europe, and Alaska connects to Siberia.

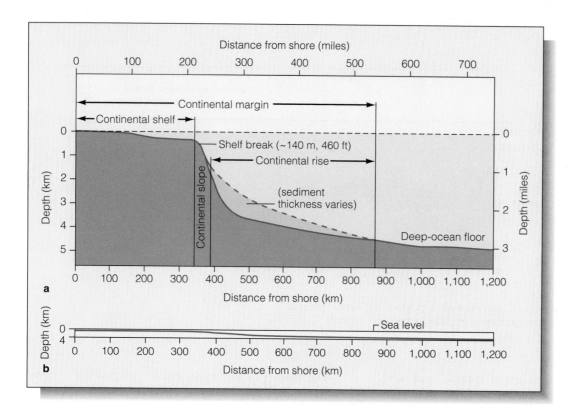

Distance from shore (miles)

Continental margin

Continental shelf

Shelf break (~140 m, 460 ft)

Continental slope

Continental rise

(sediment thickness varies)

Deep-ocean floor

Depth (km)

Depth (miles)

Distance from shore (km)

a

Sea level

Depth (km)

Distance from shore (km)

b

Figure 4.11 The features of a passive continental margin. (**a**) Vertical exaggeration 50:1. (**b**) No vertical exaggeration.

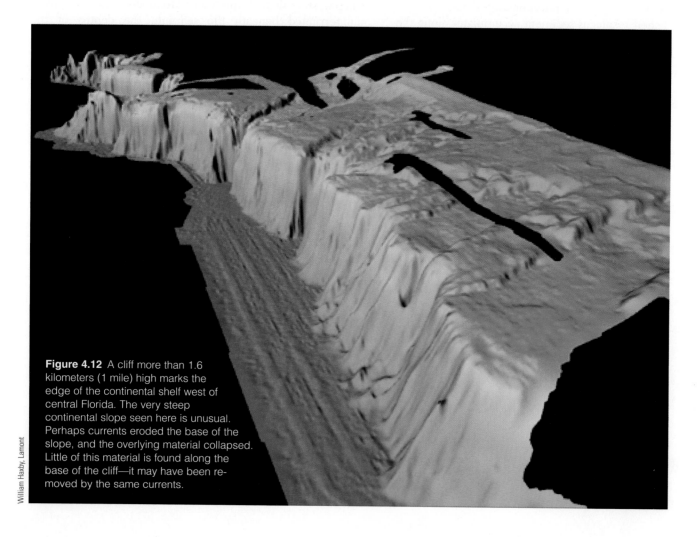

Figure 4.12 A cliff more than 1.6 kilometers (1 mile) high marks the edge of the continental shelf west of central Florida. The very steep continental slope seen here is unusual. Perhaps currents eroded the base of the slope, and the overlying material collapsed. Little of this material is found along the base of the cliff—it may have been removed by the same currents.

William Haxby, Lamont

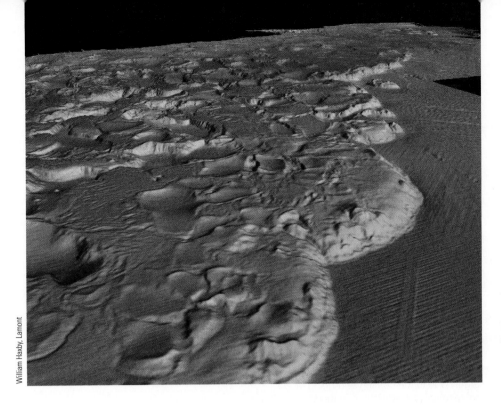

Figure 4.13 The edge of the Gulf of Mexico south of Texas and Louisiana. Sediments carried by the Mississippi have overlain salt deposits. The salt squeezed to the surface in places where it dissolved, leaving pockmarks.

more commonly offshore chains of volcanoes or lines of coral reefs. Volcanic activity east of China and Southeast Asia has formed a broad basin that is filling with sediments and is one of the largest shelves in the Pacific.

Because of their gentle slope, continental shelves are greatly influenced by changes in sea level. Around 18,000 years ago—at the height of the last **ice age** (periods of extensive glaciation)—massive ice caps covered huge regions of the continent. The water that formed the thick ice sheets was derived from the ocean, and sea level fell about 125 meters (410 feet) below its present position (**Figure 4.14**). The continental shelves were almost completely exposed, and the surface area of the continents was about 18% greater than it is today. Rivers and waves cut into the sediments that had accumulated during periods of higher sea level, and they transported some coarse sediments to their present locations at the shelves' outer edges. Sea level began to rise again when the ice caps melted, and sediments again began to accumulate on the shelves. (More on the history and effects of sea level change can be found in the discussion of coasts in Chapter 12.)

The continental shelves have been the focus of intense exploration for natural resources. Because shelves are the submerged margins of continents, any deposits of oil or minerals along a coast are likely to continue offshore. Water depth over shelves averages only about 75 meters (250 feet), so large areas of the shelves are accessible to mining and drilling activities. Many of the techniques used to find and exploit natural resources on land can also be used on the continental shelves. Resource development requires intense scientific investigations, and our understanding of the geology of the shelves has benefited greatly from the search for offshore

oil and natural gas. (The economic and legal implications of the exploitation of these riches are touched on in Chapter 17 in the discussion of the Law of the Sea and the establishment of exclusive economic zones.)

Continental Slopes

The **continental slope** is the transition between the gently descending continental shelf and the deep-ocean floor. Continental slopes are formed of sediments that reach the built-out edge of the shelf and are transported over the side. At active margins a slope may also include marine sediments scraped off a descending plate during subduction (**Figure 4.15**). The inclination of a typical continental slope is about 4° (or 70 meters per kilometer, 370 feet per mile), slightly steeper than the steepest road slope allowed on the interstate highway system. As Figure 4.11b implies, even the steepest of these slopes is not precipitous: A 25° slope is the greatest incline yet discovered. In general, continental slopes at active margins are steeper than those at passive margins. Continental slopes average about 20 kilometers (12 miles) in width and end at the continental rise, usually at a depth of about 3,700 meters (12,000 feet). The bottom of the continental slope is the true edge of a continent.

The **shelf break** marks the abrupt transition from continental shelf to continental slope. The depth of water at the shelf break is surprisingly constant—about 140 meters (460 feet) worldwide—but there are exceptions. The great weight of ice on Antarctica, for example, has isostatically depressed that continent, and the depth at the shelf break is 300 to 400 meters (1,000 to 1,300 feet). The shelf break in Greenland is similarly depressed.

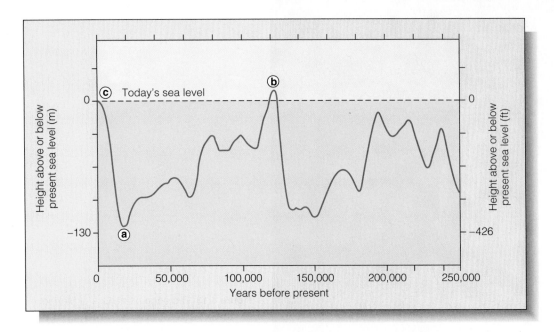

Figure 4.14 Changes in sea level over the last 250,000 years, as traced by data taken from ocean-floor cores. The rise and fall of sea level are largely due to the coming and going of ice ages—periods of increased and decreased glaciation, respectively. Water that formed the glaciers during these periods came from the ocean, and this caused sea level to drop. Point **a** indicates a low stand of −125 meters (−410 feet) at the climax of the last ice age some 18,000 years ago. Point **b** indicates a high stand of +6 meters (+19.7 feet) during the last interglacial period about 120,000 years ago. Point **c** shows the present sea level. The rate of sea-level rise has nearly leveled off, but sea level is rising slowly as we continue to emerge from the last period of glaciation.

Note that sea level has been considerably below its present position for nearly all of the last quarter-million years. Estuaries would have been less prevalent, and continental shelves much narrower. Consider the implications for coastal civilizations. During the time between 18,000 years ago and 8,000 years ago, sea level rose over 100 meters (330 feet) at a rate of about 1 centimeter (1/2 inch) a year, over 0.5 meter in a human lifetime. Could this have given rise to the flood legends common to many religions?

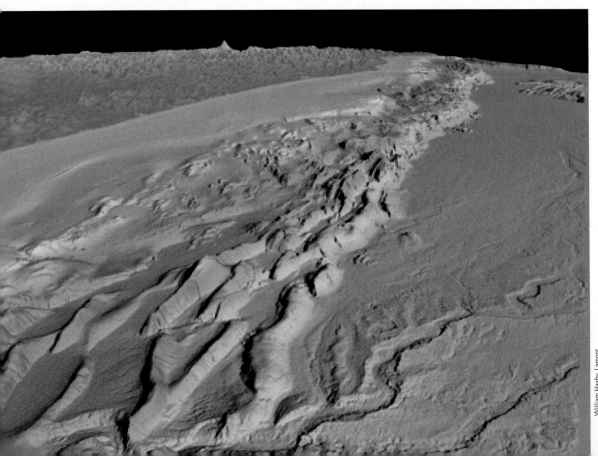

Figure 4.15 Folded ridges of sediment cover the ocean floor west of Oregon. The crinkles result from the collision between the North American Plate and the Juan de Fuca Plate. Like a bulldozer, the North American Plate scrapes sediments from the subducting Juan de Fuca Plate and piles it into folds. To the south (the upper right), part of the Juan de Fuca Plate breaks through the sediment.

William Haxby, Lamont

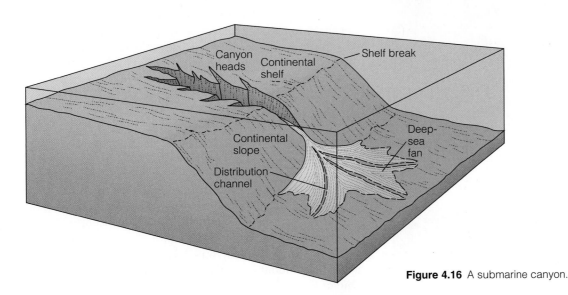

Figure 4.16 A submarine canyon.

Submarine Canyons

Submarine canyons cut into the continental shelf and slope, often terminating on the deep-sea floor in a fan-shaped wedge of sediment (**Figure 4.16**). More than a hundred submarine canyons nick the edge of nearly all of Earth's continental shelves. The canyons generally trend at right angles to the shoreline (and shelf edge), sometimes beginning very close to shore. Congo Canyon actually extends into the African continent as a deep estuary at the mouth of the Congo River. These enigmatic features are quite large. In fact, submarine canyons similar in size and profile to Arizona's Grand Canyon have been discovered!

Hudson Canyon, a typical large canyon on a passive margin, is shown in **Figure 4.17.** Like many submarine canyons, Hudson Canyon is located just offshore of the mouth of a river or stream—in this case, New York's Hudson River. Because of their similarity to canyons on land, submarine canyons appear to have been created by erosion, so marine geologists initially thought the canyons were carved into the shelves by stream erosion at times of lower sea level. But most researchers agree that sea level has never fallen more than 200 meters (660 feet) below its present level in the last 600 million years. Stream erosion could account for the shape of the uppermost parts of the canyons. However, since submarine canyons can be traced to depths in excess of 3,000 meters (10,000 feet) below sea level, stream erosion could not have played a direct role in cutting their lower depths.

What, then, caused the submarine canyons to form? As is so often the case in the history of marine science, the answer to this riddle was associated with a peculiar event.

A sharp earthquake struck the Grand Banks south of Newfoundland at 1531 (3:31 P.M.) Greenwich time on 18 November 1929. The quake's **epicenter,** the point on the surface of Earth directly above the focus of an earthquake, was located on the continental slope at a water depth of about 2,200 meters (7,200 feet). Ordinarily such an earthquake would not attract much attention, but a number of important transatlantic communication cables lay on the ocean floor south of the earthquake area. One by one these cables failed. A segment of the Western Union cable nearest the epicenter broke almost immediately, and a French cable about 160 kilometers (100 miles) south of the Western Union cable parted at 1835 (6:35 P.M.). Another Western Union cable to the south snapped early the next morning, and cables even farther south were damaged later that day. Officials had expected an occasional random failure, but a *sequence* of cable failures, from north to south along 480 kilometers (300 miles) of seafloor in the mid-Atlantic, was baffling.

Local landslides or sediment liquefaction triggered by the earthquake probably caused the first cable breaks, and although no one thought of it right away, an abrasive underwater "avalanche" almost certainly caused the more distant breaks. These avalanche-like sediment movements, called **turbidity currents,** are caused when turbulence mixes sediments into water

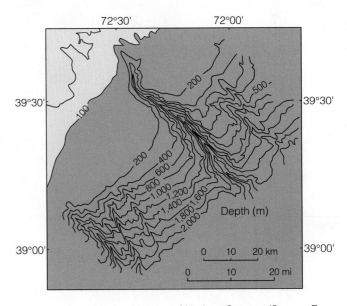

Figure 4.17 A topographic map of Hudson Canyon. (Source: From Taylor & Francis International Scientific and Educational Publishers.)

CONTINENTAL MARGINS AND OCEAN BASINS • • • •

Society for Sedimentary Geology

Figure 4.18 A turbidity current flowing down a submerged slope off the island of Jamaica. The propeller of a submarine caused the turbidity current by disturbing sediment along the slope.

above a sloping bottom. The sediment-filled water is denser than the surrounding water, so the thick muddy fluid runs down the slope at speeds of up to 27 kilometers (17 miles) per hour. **Figure 4.18** is a rare photograph of a turbidity current.

What is the connection between turbidity currents and submarine canyons? Most geologists believe that the canyons have been formed by abrasive turbidity currents plunging down the canyons. Small amounts of debris may cascade continuously down the canyons (**Figure 4.19**), but earthquakes

U.S. Department of the Navy

Figure 4.19 A continuous cascade of sediment at the head of San Lucas submarine canyon (off the coast of Baja California, Mexico), which may be eroding the narrow gorge in conjunction with occasional turbidity currents.

can shake loose huge masses of sediment that rush down the edge of the shelf, scouring the canyon deeper as they go. In this way the canyons can be cut to depths far below the reach of streams even during the low sea levels of the ice ages.

Continental Rises

Along passive margins the oceanic crust at the base of the continental slope is covered by an apron of accumulated sediment called the **continental rise** (see again Figure 4.9). Sediments from the shelf slowly descend to the ocean floor along the whole continental slope, but most of the sediments that form the continental rise are transported to the area by turbidity currents. The width of the rise varies from about 100 to 1,000 kilometers (63 to 630 miles), and its slope is gradual—about one-eighth that of the continental slope. One of the widest and thickest continental rises has formed in the Bay of Bengal at the mouths of the Ganges–Brahmaputra River, the most sediment-laden of the world's great rivers.

Deep-ocean currents are an important factor in shaping continental rises, especially along the western boundaries of most ocean basins. Deep boundary currents held against the continental slopes by the Coriolis effect (news of which awaits you in Chapter 8) pick up volcanic debris and sediments transported by turbidity currents and sweep them along the ocean floor. As the currents are forced around bends and across depressions, they slow and deposit the suspended material on the rises in the form of ridges and mud waves (**Figure 4.20**).

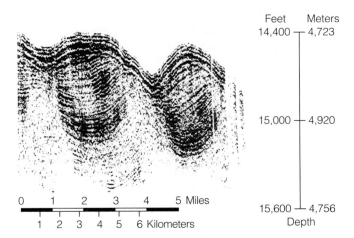

	Feet	Meters
	14,400	4,723
	15,000	4,920
	15,600	4,756
		Depth

0 1 2 3 4 5 Miles

1 2 3 4 5 6 Kilometers

Figure 4.20 Large stationary waves of sediment deposited at the base of the continental rise in the western North Atlantic. Deep boundary currents pick up sediments and sweep them across the ocean floor. As the currents are forced around bends and across depressions, they slow and deposit the suspended material at the base of the rises in the form of ridges or mud waves.

DEEP-OCEAN BASINS

Away from the margins of continents the structure of the ocean floor is quite different. Here the seafloor is a blanket of sediments up to 5 kilometers (3 miles) thick overlying basaltic rocks. Deep-ocean basins constitute more than half of Earth's surface.

The deep-ocean floor consists mainly of oceanic ridge systems and the adjacent sediment-covered plains. Deep basins may be rimmed by trenches or by masses of sediment. Flat expanses are interrupted by islands, hills, active and extinct volcanoes, and active zones of seafloor spreading. The sediments on the deep floor reflect the history of the surrounding continents, the biological productivity of the overlying water, and the ages of the basins themselves.

Oceanic Ridges

If the ocean evaporated, the oceanic ridges would be Earth's most remarkable and obvious feature. An **oceanic ridge** is a mountainous chain of young basaltic rock at the active spreading center of an ocean. Stretching 65,000 kilometers (40,000 miles), more than 1½ times Earth's circumference, oceanic ridges girdle the globe like seams surrounding a softball (**Figure 4.21**). The rugged ridges, which often are devoid of sediment, rise about 2 kilometers

(1.25 miles) above the seafloor. In places they project above the surface to form islands such as Iceland (**Box 4.2**), the Azores, and Easter Island. Oceanic ridges and their associated structures account for 22% of the world's solid surface area (all the land above sea level accounts for 29%). Although these features are often called mid-ocean ridges, less than 60% of their length actually exists along the centers of ocean basins.

As we saw in our discussion of plate tectonics, the rift zones associated with oceanic ridges are sources of new ocean floor where lithospheric plates diverge. The oceanic ridges are widest where they are most active. The youngest rock is located at the active ridge center, and rock becomes older with distance from the center. As lithosphere cools, it shrinks and subsides. Slowly spreading ridges have a steeper profile than rapidly spreading ones because slowly diverging seafloor cools and shrinks closer to the spreading center. **Figure 4.22a,** a bathymetric map of the North Atlantic, clearly shows the great extent of the Mid-Atlantic Ridge, a typical oceanic ridge. **Figure 4.22b,** a multibeam image, provides a detailed look at the young central rift. A side view is provided in **Figure 4.22c.**

As can be seen in Figure 4.22a, the Mid-Atlantic Ridge is offset at more or less regular intervals by transform faults. A **fault** is a fracture in the lithosphere along which movement

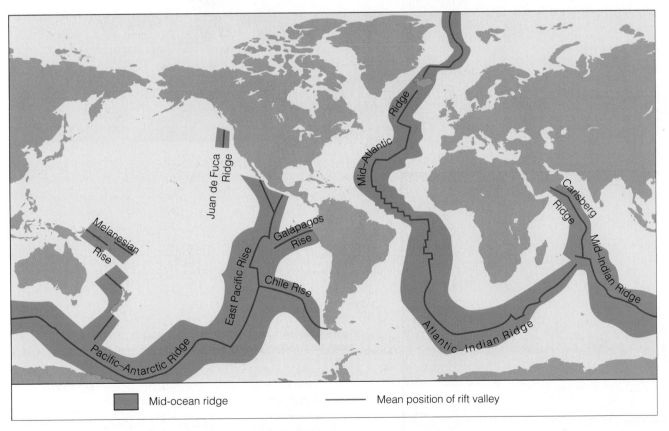

Figure 4.21 The oceanic ridge system stretches some 65,000 kilometers (40,000 miles) around Earth. If the ocean evaporated, the ridge system would be Earth's most remarkable and obvious feature.

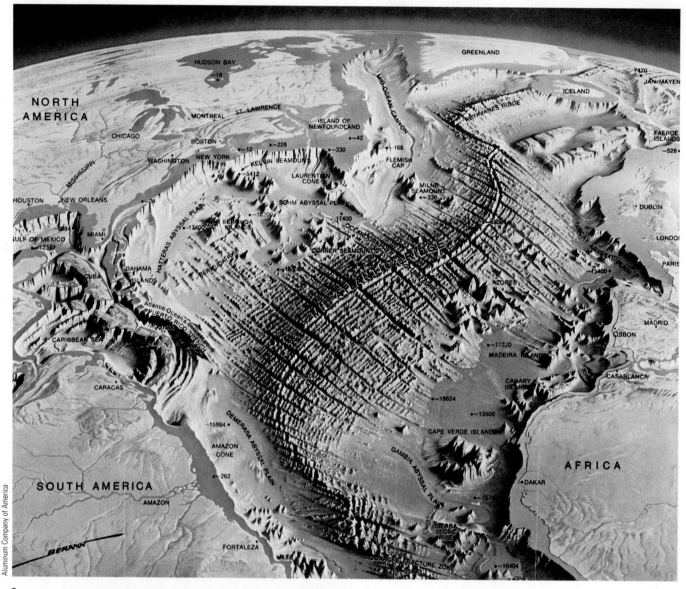

a

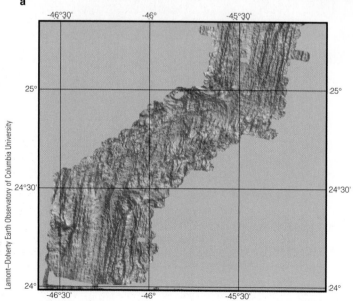

b

Figure 4.22 (**a**) A hand-drawn map of a portion of the Atlantic Ocean floor showing some major oceanic features: mid-ocean ridge, transform faults, fracture zones, submarine canyons, seamounts, continental rises, trenches, and abyssal plains. Depths are in feet. The map is vertically exaggerated. (**b**) The fine structure of the central portion of the Mid-Atlantic Ridge between Florida and western Africa. The depressed central valley (the spreading center, shown in blue) is clearly visible in this computer-generated multibeam image.

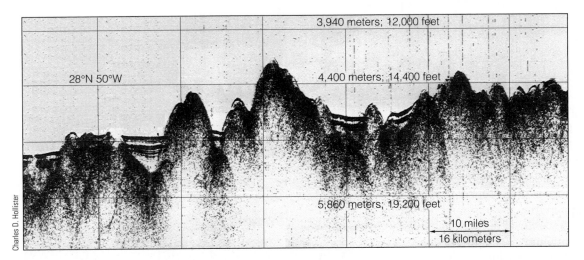

Charles D. Hollister

c

Figure 4.22 continued

Figure 4.22 continued
(c) As this seismic profile shows, the rugged relief of the North Atlantic's mid-ocean ridge is gradually being buried by slowly accumulating sediments.

has occurred, and **transform faults** are fractures along which lithospheric plates slide horizontally (**Figure 4.23**). As you saw in Figure 3.23, when segments of a ridge system are offset, the fault connecting the axis of the ridge is a transform fault. Shallow earthquakes are common along transform faults. Since the ocean floor cannot expand evenly on the surface of a sphere, plate divergence on the spherical Earth can only be irregular and asymmetrical, and transform faults and **fracture zones** result.

Transform faults are the active part of fracture zones. Extending outward from the ridge axis, fracture zones are seismically inactive areas that show evidence of past transform

fault activity. While segments of a lithospheric plate on either side of a transform fault move in *opposite* directions from each other, the plate segments adjacent to the outward segments of a fracture zone move in the *same* direction, as Figure 4.23 shows.

Hydrothermal Vents

Some of the most exciting features of the ocean basins are the **hydrothermal vents.** Hot springs were discovered on oceanic ridges in 1977 by Robert Ballard and J. F. Grassle of the Woods Hole Oceanographic Institution. Diving in *Alvin*

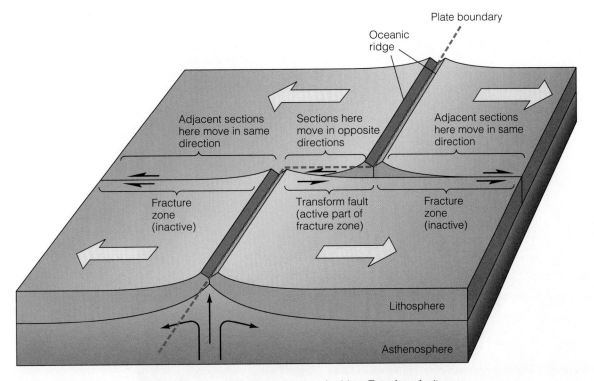

Figure 4.23 Transform faults and fracture zones along an oceanic ridge. Transform faults are fractures along which lithospheric plates slide horizontally past one another. Transform faults are the active part of fracture zones. (For a larger-scale look at this process, please see Figure 3.23.)

The 253,000 inhabitants of the island country of Iceland live on a section of the Mid-Atlantic Ridge that rises above sea level just south of the Arctic Circle. Maintaining civilization on the exposed seafloor at an active spreading center is an interesting experience. There are a few benefits: Irrigation water is piped from hot springs to protect the crops from summer frosts; steam is tapped from underground reservoirs to generate electrical power to all residents in their homes; and nearly every community is supplied with hot-water swimming pools and fresh vegetables, grown year-round in steam-heated greenhouses. But there are two important drawbacks: frequent earthquakes and vigorous volcanoes. It is estimated that the volcanoes of Iceland have produced one-third of the total lava that has flowed onto dry land since the year 1500. With more hot springs and vapor vents than any other country in the world, Iceland is the most geologically active place on Earth.

Iceland sits atop a vast mantle plume. Probably originating near the mantle–core boundary, the uppermost part of the rising cylindrical plume has been traced by seismic tomography (**Figure a**). The plume has a diameter of about 300 kilometers (190 miles). Iceland itself is broken by parallel fissures stretching north and south along a central rift zone (**Figure b**); they are similar to fissures on the adjacent submerged ridge. As along the rest of the ridge, tectonic forces are stretching Iceland in the east-west direction, and

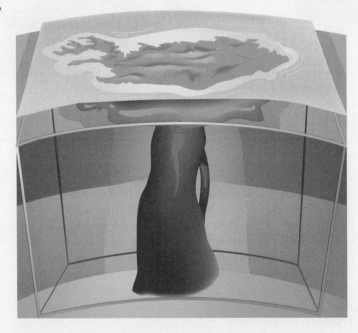

a A mantle plume beneath Iceland. Low seismic waves reveal a cylindrical plume 300 kilometers (190 miles) in diameter. The plume probably originates near the core–mantle boundary, and it powers the intense geological activity in the Mid-Atlantic Ridge in and around Iceland.

b Thingvellir Graben, a rift valley in Iceland. The valley is actually the central part of the Mid-Atlantic Ridge, exposed above sea level. The arrows show the direction of spreading on each side of the rift.

Dr. Jian Lin, WHOI

magma is rising from the plume to fill the fissures and become new seafloor. The lava venting from Iceland's volcanoes is very similar to lava rising into nearby submerged segments of the Mid-Atlantic Ridge, but the cooling lava doesn't assume the overstuffed pillow shape of submerged lava—for two reasons. First, lava in contact with cold seawater cools and solidifies more rapidly than lava in contact with air; second, the water pressure at the ocean bottom keeps most of the volatile gases in solution during the cooling process, thus maintaining the extruded bunlike shape. Instead, the Icelandic lava blasts from the crevices, explosively releases its gases, and spreads over the desolate countryside.

The central rift zone of Iceland is almost completely covered by lava deposited since the last ice age—that is, within the last 20,000 years. Its forbidding central desert— the Odádahraun, the "Desert of Evil Deeds"—is one of the most lifeless places in the world, so alien-seeming that it was used as a training ground for the American astronauts who walked on the moon. It was here in 1783 that a series of fissures more than 24 kilometers (15 miles) long yawned open and began disgorging lava and corrosive gases. Over the next year 585 square kilometers (226 square miles) of central Iceland was covered by new rock. Sulfurous vapors and acidic rains poisoned the crops and killed most of the livestock on the island. One-fifth of the 49,000 inhabitants died of famine that winter.

The section of the Mid-Atlantic Ridge of which Iceland is a part spreads about 15 centimeters (6 inches) per year. Icelanders must contend with considerable geological inconvenience (and, as in 1938 and 1996, occasional bouts of terror), but at least their country is growing bigger by the day.

4-15

at 3 kilometers (1.9 miles) near the Galápagos Islands along the East Pacific Rise (an oceanic ridge), they came across rocky chimneys up to 20 meters (66 feet) high, from which dark, mineral-laden water was blasting at 350°C (660°F) (**Figure 4.24**). Only the great pressure at this depth prevented the escaping water from flashing to steam. These *black smokers*, as they were quickly nicknamed, fascinate marine geologists. It is believed that water descends through fissures and cracks in the ridge floor until it comes into contact with very hot rocks associated with active seafloor spreading. There the superheated, chemically active water dissolves minerals and gases and escapes upward through the vents by convection (**Figure 4.25**).

Since that first discovery, vents have been found on the Mid-Atlantic Ridge east of Florida, in the Sea of Cortez east and south of Baja California, and on the Juan de Fuca Ridge off the coast of Washington and Oregon. Scientists now believe that hydrothermal vents may be very common on oceanic ridges, especially in zones of rapid seafloor spreading. In July 1990 vents were discovered in fresh water, at the bottom of Lake Baikal in southern Siberia. This discovery

WHOI

Figure 4.24 A black smoker discovered at a depth of about 2,800 meters (9,200 feet) along the East Pacific Rise.

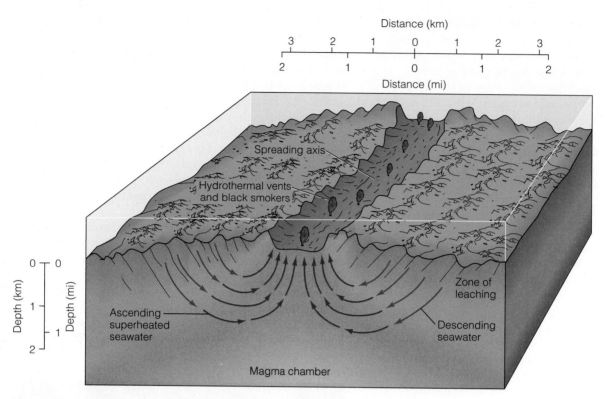

Figure 4.25 A cross section of the central part of a mid-ocean ridge—similar to that shown in Figure 4.22b—showing the origin of hydrothermal vents. Cool water (blue arrows) is heated as it descends toward the hot magma chamber, leaching sulfur, iron, copper, zinc, and other materials from the surrounding rocks. The heated water (red arrows) returning to the surface carries these elements upward, discharging them at the hydrothermal springs on the seafloor. The areas around the vents support unique communities of organisms (see Chapter 16).

suggests that the world's oldest and deepest lake may someday become part of the ocean as Asia slowly breaks apart.

Not all vents form chimneys of mineral deposits—some are simply cracks in the seabed, or porous mounds, or broad segments of ocean ridge floor through which warm, mineral-laden water percolates upward. Cooler vents result when hot, rising water mixes with cold bottom water before reaching the surface. Water temperature in the vicinity of most hydrothermal vents averages 8–16°C (46–61°F), much warmer than usual for ocean bottom water, which has an average temperature of 3–4°C (37–39°F). We will study the unusual communities of animals that populate the vents in Chapter 16.

A volume of water equal to the volume of the world ocean is thought to circulate through the hot oceanic crust at spreading centers about every 10 million years! The water coming from the vents and seeping from the floor is more acidic, is enriched with metals, and has higher concentrations of dissolved gas than seawater does. The heat and chemicals issuing from these structures may play important roles in the chemical composition of seawater and the atmosphere, and in the formation of mineral deposits.

Abyssal Plains and Abyssal Hills

A quarter of Earth's surface consists of abyssal plains and abyssal hills. *Abyssal* is an adjective derived from a Greek word meaning "without bottom." While this is obviously not literally true, you can appreciate how the term came into use following the *Challenger* Expedition's laborious soundings of these extremely deep areas!

Abyssal plains are flat, featureless expanses of sediment-covered ocean floor found on the periphery of all oceans. They are most common in the Atlantic, less so in the Indian Ocean, and relatively rare in the active Pacific, where peripheral trenches trap most of the sediments flowing from the continents. They lie between the continental margins and the oceanic ridges about 3,700 to 5,500 meters (12,000 to 18,000 feet) below the surface (see again Figure 4.22a). The Canary Abyssal Plain, a huge plain west of the Canary Islands in the North Atlantic, has an area of about 900,000 square kilometers (350,000 square miles).

Abyssal plains are extraordinarily flat. A 1947 survey by the Woods Hole Oceanographic Institution ship *Atlantis* found that a large Atlantic abyssal plain varies no more than a

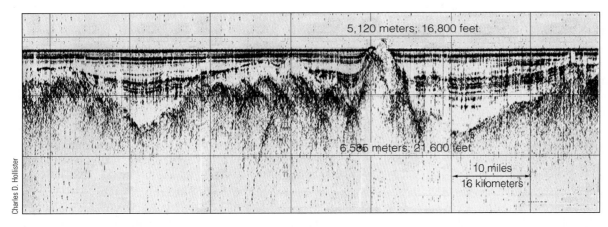

5,120 meters; 16,800 feet

6,585 meters; 21,600 feet

10 miles
16 kilometers

Charles D. Hollister

Figure 4.26 The deep, smooth sediments of the Atlantic's Northern Madeira Abyssal Plain bury 100-million-year-old mountains.

few meters in depth over its entire area. Such flatness is caused by the smoothing effect of the layers of sediment, which often exceed 1,000 meters (3,300 feet) in thickness. Most of the sediment that forms the abyssal plains appears to be of terrestrial or shallow-water origin, not derived from biological activity in the ocean above. Some of it may have been transported to the plains by winds or turbidity currents. These deep sediment layers mask irregularities in the underlying ocean crust, but a powerful type of echo sounder can "see" through this sediment to reveal the complex topography of the basaltic basin floor below (**Figure 4.26**). The broad basaltic shoulders of the Mid-Atlantic Ridge extend beneath this cloak of sediment almost as far as the bordering continental slopes.

Abyssal plain sediments may not be thick enough to cover the underlying basaltic floor near the edges bordering the oceanic ridges. Here the plains are punctuated by **abyssal hills**—small, sediment-covered extinct volcanoes or intrusions of once-molten rock, usually less than 200 meters (650 feet) high (one of which is seen extending above the flat surface in Figure 4.26). These abundant features are associated with seafloor spreading; they form when newly formed crust moves away from the center of a ridge, stretches, and cracks. Some blocks of the crust drop to form valleys, and others remain higher as hills. Lava erupting from the ridge flows along the fractures, coating the hills. This helps explain why abyssal hills occur in lines parallel to the flanks of the nearby oceanic ridge and why they occur most abundantly in places where the rate of seafloor spreading is fastest. Abyssal plains and hills account for nearly all of the area of deep-ocean floor that is not part of the oceanic ridge system.

Seamounts and Guyots

4-17

The ocean floor is dotted with thousands of volcanic projections that do not rise above the surface of the sea. These projections are called **seamounts.** Seamounts are circular or elliptical, more than 1 kilometer (0.6 mile) in height, with relatively steep slopes of 20° to 25°. (Abyssal hills, in contrast, are much more abundant, less than a kilometer high, and not as steep.) Seamounts may be found alone or in groups of from 10 to 100. Though many form at hot spots (see Figure 3.29), most are thought to be submerged inactive volcanoes that formed at spreading centers (see **Figure 4.27**). Movement of the lithosphere away from spreading centers has carried them outward and downward to their present positions. As many as 10,000 seamounts are thought to exist in the Pacific, about half the world total.

Guyots are flat-topped seamounts that once were tall enough to approach or penetrate the sea surface. Generally they are confined to the west-central Pacific. The flat top

Figure 4.27 An undersea volcano on the Mid-Atlantic Ridge is "photographed" by a side-scan sonar device towed 3,000 meters (10,000 feet) below the surface

Deborah Smith, WHOI

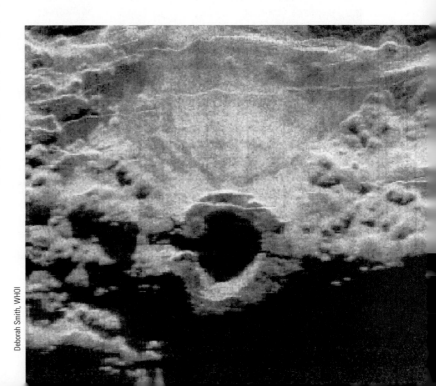

suggests that they were eroded by wave action when they were near sea level. Their plateaulike tops eventually sank too deep for wave erosion to continue wearing them down. Like the more abundant seamounts, most guyots were formed near spreading centers and transported outward and downward as the seafloor moved away from a spreading center and cooled (see again Figure 3.31).

Trenches and Island Arcs

A **trench** is an arc-shaped depression in the deep-ocean floor. These flat-bottomed creases in the seafloor occur where a converging oceanic plate is subducted. The water temperature at the bottom of a trench is slightly cooler than the near-freezing temperatures of the adjacent flat ocean floor, reflecting the fact that trenches are underlain by old, relatively cold ocean crust sinking into the upper mantle. Trenches (and their associated island arcs topped by erupting volcanoes) are among the most active geological features on Earth. Great earthquakes and tsunami (huge waves we will discuss in Chapter 10) often originate in them. **Figure 4.28** shows the distribution of the ocean's major trenches. Not surprisingly, most are around the edges of the active Pacific.

Trenches are the deepest places in Earth's crust, 3 to 6 kilometers (1.9 to 3.7 miles) deeper than the adjacent basin floor. The ocean's greatest depth is the Mariana Trench of the western Pacific, where *Trieste* set the world's deep-diving record (described in the chapter opener). Near the place where *Trieste* came to rest, the ocean bottom is 11,022 meters (36,163 feet) below sea level, 20% deeper than Mt. Everest is high (see **Figure 4.29**). The Mariana Trench is about 70 kilometers (44 miles) wide and 2,550 kilometers (1,600 miles) long, typical dimensions for these structures.

Trenches are curving chains of V-shaped indentations. The trenches are curved because of the geometry of plate interactions on a sphere. The convex sides of these curves generally face the open ocean (see again Figure 4.28). The trench walls on the island side of the depressions are steeper than those on the seaward side, indicating the direction of plate subduction. The sides of trenches become steeper with depth, normally reaching angles of about 10° to 16° before flattening to a floor underlain by thick sediment. (Parts of the concave wall of the Kermadec–Tonga Trench are the world's steepest at 45°.) No continental rise occurs along coasts with trenches because the sediment that would form the rise ends up at the bottom of the trench.

Island arcs, curving chains of volcanic islands and seamounts, are almost always found paralleling the concave edges of trenches. As you may remember from Chapter 3, trenches and island arcs are formed by tectonic and volcanic activity associated with subduction. The descending lithospheric plate contains some materials that melt as the plate sinks into the mantle. These materials rise to the surface as magmas and lavas that form the chain of islands behind the trench. The Aleutian Islands, most Caribbean islands, and the Marianas Islands are island arcs. (See Figure 3.21 for a review of the processes involved in their construction.)

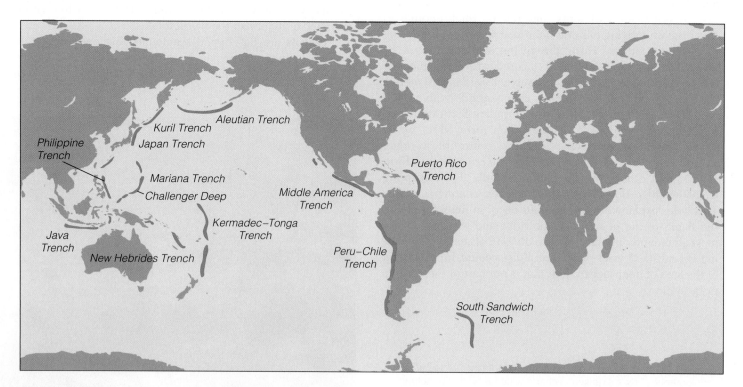

Figure 4.28 Oceanic trenches of the world.

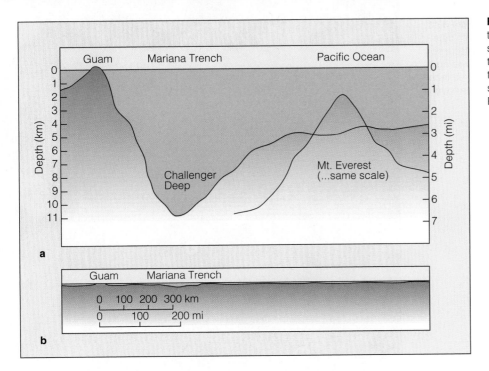

Figure 4.29 The Mariana Trench. (**a**) Comparing the Challenger Deep and Mt. Everest at the same scale shows that the deepest part of the Mariana Trench is about 20% deeper than the mountain is high. (**b**) The Mariana Trench shown without vertical exaggeration. (Source: Reprinted by permission of Chet Raymo.)

THE GRAND TOUR

Researchers at the National Oceanic and Atmospheric Administration and the National Geophysical Data Center have generated a map of the world ocean floor based on satellite observations of the shape of the sea surface (**Figure 4.30**). The graphic shows all the features discussed in this chapter. These features—and a basic understanding of the geological reasons for their existence—will help you recall the dramatic nature and history of the seafloor that we have discussed in the past two chapters.

QUESTIONS FROM STUDENTS

1. **You mentioned that Walsh and Piccard saw animals at a depth of 11,022 meters (6.85 miles). The pressure there is more than 8 tons per square inch. How could any living thing withstand such crushing pressure?**

It seems impossible until you think about your own situation. Humans live at the bottom of a surprisingly heavy "ocean" of air: A column of air 1 inch square extending from sea level to the top of the atmosphere weighs about 6.7 kilograms (14.7 pounds). This pressure presents no problem to us because we have the same pressure inside of us pushing out. The animals at great depths also have the same pressure inside as outside; that is, they live in perfect hydrostatic balance. They notice the weight of water above them no more than you notice the weight of air above you.

2. **Why is Iceland one of only a few places in the world where oceanic crust is found above sea level?**

Oceanic crust is thin and dense. Because of isostasy, the ocean floor lies at a lower elevation than the thicker, less dense, higher continents. Water filled the lower elevations first, submerging nearly all the basaltic basement we now call ocean floor. In areas of rapid seafloor spreading, or at hot spots, peaks are occasionally pushed toward the ocean surface by the large volume of mantle material rising in plumes from below. Large quantities of erupted magma (lava) then build the crests above sea level to form islands. The Azores are another place on the Mid-Atlantic Ridge where this is happening.

3. **Turbidity currents seem important in forming canyons and distributing deep sediments over abyssal plains. Has anybody ever seen a turbidity current in action?**

Yes, surprisingly. In the late 1940s the Dutch geologist Philip Kuenen produced turbidity currents in his laboratory by pouring muddy water into a trough with a sloping bottom. His observations confirmed nineteenth-century reports that the muddy Rhône River continued to flow in a dense stream along the bottom of Lake Geneva. In the 1960s Robert Dill and Francis Shepard viewed sandfalls in southern California's Scripps Canyon, and French researchers have recently photographed these currents in the Mediterranean.

Figure 4.30 A technological *tour de force,* derived from data provided to the National Geophysical Data Center from satellites and shipborne sensors, shows all the features discussed in this chapter. These features—and a basic understanding of the geological reasons for their existence—will help you recall the dramatic nature and history of the seafloor that we have discussed in the past two chapters. Key to features:

(1) Aleutian Trench
(2) Hawaiian Islands
(3) Juan de Fuca Ridge
(4) Clipperton Fracture Zone
(5) Peru–Chile Trench
(6) East Pacific Rise
(7) Mariana Trench, Challenger Deep
(8) Mid-Atlantic Ridge
(9) South Sandwich Trench
(10) Red Sea
(11) Easter Island
(12) Galápagos Rift
(13) Iceland
(14) Gulf of Aden
(15) Pacific–Antarctic Ridge
(16) Azores
(17) Tristan de Cunha
(18) Kermadec Trench
(19) Tonga Trench
(20) Hatteras Abyssal Plain
(21) Grand Banks
(22) Mid-Indian Ridge
(23) Atlantic–Indian Ridge
(24) Maldive Islands
(25) Eltanin Fracture Zone
(26) Emperor Seamounts
(27) Java Trench
(28) Great Barrier Reef
(29) Kuril Trench
(30) Japan Trench

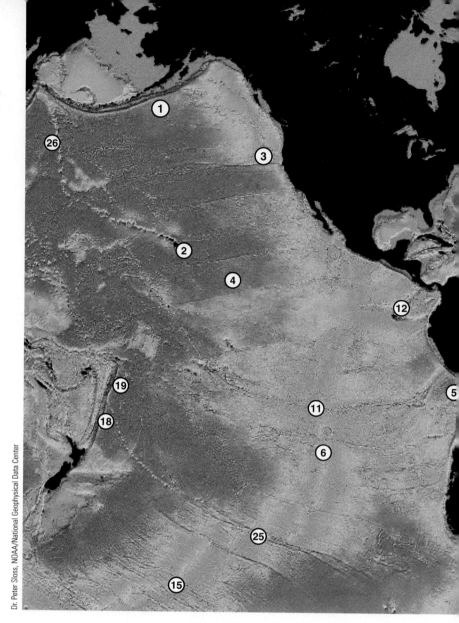

Dr. Peter Sloss, NOAA/National Geophysical Data Center

4. I might be interested in exploring some of the continental shelf using scuba. How does scuba work?

Scuba is an acronym for self-contained underwater breathing apparatus. Scuba divers uses a regulator to valve air from the high-pressure tank on their backs into their lungs. At sea level the regulator puts just enough air into the lungs to balance the pressure of air above, so the diver can take a breath. At 30 meters (100 feet) depth, the regulator must provide a total of four atmospheres of pressure to allow the same breath—one atmosphere to compensate for the weight of the air, plus three atmospheres to compensate for the weight of the water above. If the diver were suddenly transported to sea level, the imbalance would result in a dangerous—and messy—incident. That's why divers ascend slowly, breathe regularly, and never skimp on regulator quality.

Scuba divers breathe compressed air, a mixture of nitrogen and oxygen. Pure oxygen is not used because it has dangerous side effects at high pressures. The practical limit for normal scuba operation is approximately 125 feet. Deeper dives require mixtures of helium and oxygen to prevent the toxic effects of high-pressure nitrogen on the nervous system. If a way could be found to perfuse all of a diver's air-filled spaces with fluid (an oxygen-carrying, low-viscosity silicon compound has been proposed), humans could possibly dive to depths of thousands of feet on helium–oxygen mixtures, requiring only protection from the cold.

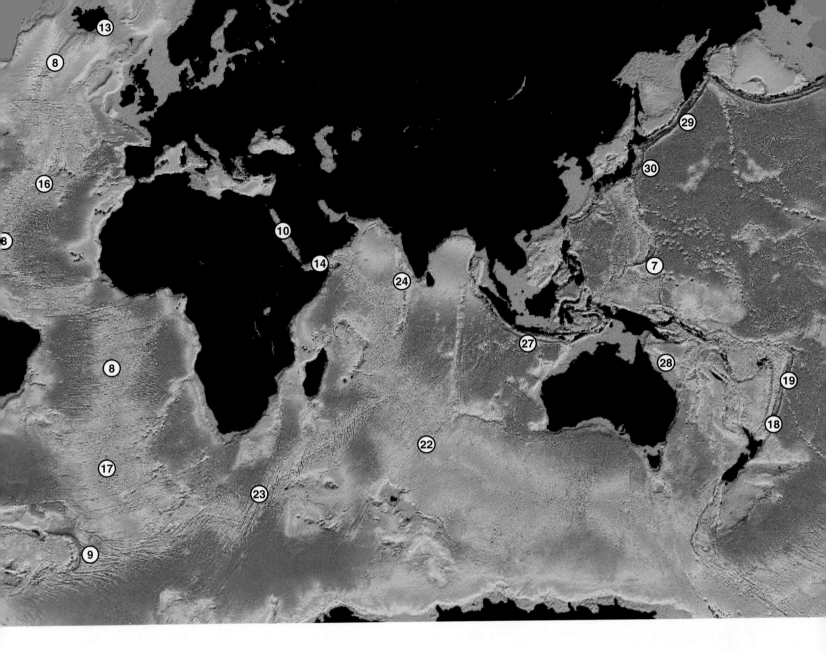

- Seafloor features result from a combination of tectonic activity and the processes of erosion and deposition. New bathymetric devices being deployed to study these features include side-scan sonar and satellites using sensitive radar for altimetry.

- Near shore, the features of the ocean floor are similar to those of the adjacent continent because they share the same granitic basement. The transition to basalt marks the true edge of the continent and divides ocean floors into two major provinces. The submerged outer edge of a continent is called the continental margin. The deep-sea floor beyond the continental margin is properly called the ocean basin.

- Features of the continental margins include continental shelves, continental slopes, submarine canyons, and continental rises.

- Features of the deep-ocean basins include oceanic ridges, hydrothermal vents, abyssal plains and hills, seamounts, guyots, trenches, and island arcs.

The Web site for this book contains helpful study aids. Log on to:

www.brookscole.com/product/053437557xs

and click on the Chapter-by-Chapter area. Choose Chapter 4 and select a resource:

- **Flash Cards** allows you to test your mastery of the Terms and Concepts to Remember for this chapter.
- **Tutorial Quizzes** provides a multiple-choice practice quiz.
- **Student Guide to InfoTrac College Edition** will lead you to Critical Thinking Projects that use InfoTrac College Edition as a research tool.

TERMS AND CONCEPTS TO REMEMBER

abyssal hill	hydrothermal vent
abyssal plain	ice age
active margin	island arc
bathymetry	ocean basin
bathyscaphe	oceanic ridge
continental margin	passive margin
continental rise	seamount
continental shelf	shelf break
continental slope	submarine canyon
epicenter	telepresence
fault	transform fault
fracture zone	trench
guyot	turbidity current

STUDY QUESTIONS

Review Questions

1. How can satellites be used to measure ocean surface height? Why does the surface of the ocean "bunch up" over submerged mountains and ridges?

2. How does a multibeam sonar system work?

3. Imagine that you could walk from the shore to the deep-sea bed off the east coast of the United States. What features would you encounter?

4. What happens to continental shelves during ice ages?

5. Which is greater: the height of Mt. Everest or the depth of the Mariana Trench?

6. What do we think causes submarine canyons?

7. Why are abyssal plains relatively rare in the Pacific?

8. Why are trenches and island arcs curved? Is the descent to the bottom steeper on the convex side of the arc or on the concave side? Why the difference? Why do you think most trenches are in the western Pacific? (Hint: Check the position and action of the East Pacific Rise.)

9. Distinguish among abyssal hills, seamounts, guyots, and island arcs.

10. What is an oceanic ridge? Are they always literally in mid-ocean? How are oceanic ridges and trenches related?

Critical Thinking Questions

1. Why did people think an ocean was deepest at its center? What changed their minds?

2. What do the facts that (a) granite underlies the edges of continents and (b) basalt underlies deep-ocean basins suggest? (Hint: Consider thicknesses and densities.)

3. The terms *leading* and *trailing* are also used to describe continental margins. How do you suppose these words relate to *passive* and *active,* or *Atlantic-type* and *Pacific-type,* used in the text?

4. What forces control the shape of a continental shelf? A continental slope? A continental rise?

5. Answer this question if you have already read Chapter 3: Your time machine has been programmed to deliver you to Frankfurt, Germany, on a chilly evening in January 1912 to hear Wegener's lectures on continental drift. What two illustrations from this chapter would you take with you to cheer him up after the lecture? Why did you select those particular illustrations?

6. **InfoTrac College Edition Project** In the twentieth century, ships carrying sensors of various kinds were the dominant means of exploring the oceans, including the ocean floor. In recent years, satellite data have also contributed much to our knowledge. What new opportunities for exploration of the ocean floor do recent technological improvements provide? What strategies are scientists proposing to observe undersea processes and collect data? Do some research using InfoTrac College Edition to find out about proposals for future undersea investigations. Write a brief essay on your findings, and describe some of the types of observations that could not be made using traditional research tools.

Earth Systems Today CD Question

1. Go to "Plate Tectonics," then "Expedition." Can you find the boundaries? Using your understanding of the influence of tectonic activity on the location and characteristics of a coast, can you predict coastal features after surveying and exploring the coast in question?

RESOURCES FOR FURTHER READING AND RESEARCH

The Web site for this book contains many ideas for further reading and research. Log on to:

 www.brookscole.com/program/053437557xs

and click on the Chapter-by-Chapter area. Choose Chapter 4 and select a resource:

- **References** lists the major books and articles consulted in writing this chapter, along with comments from the author about their content and reading level.
- **Hypercontents** takes you to an extensive list of sites with news, research, and images related to individual sections of the chapter.

- In **Student Guide to InfoTrac College Edition** scroll down to Suggested Readings from InfoTrac College Edition for brief descriptions of the articles listed and search hints for finding them in InfoTrac College Edition.
- **Regional InfoTrac College Edition** articles are organized into East Coast, West Coast, and Gulf Coast regions, allowing you to study oceanography on a more local level. You can also access Regional InfoTrac College Edition at www.localocean.com.

 For additional readings, go to InfoTrac College Edition, your online research library, at:

http://infotrac.thomsonlearning.com/

Sediments 5

THE MEMORY OF THE OCEAN

In the early 1960s, researchers aboard the American research ship *Chain* discovered a layer of sediments of unknown composition beneath the Mediterranean Sea. For the next ten years scientists used continuous seismic profilers (powerful echo sounders) to trace this mysterious layer, which appeared everywhere they probed this inland sea. Unlike most marine sediments, this material reflected back virtually all of the acoustical energy sent toward it. The M-reflector, as it came to be called, covered the basement of the Mediterranean like a thick blanket of snow in a mountain valley.

On 24 August 1970 geologists aboard the deep-sea drilling ship *Glomar Challenger* inspected the first samples from the layer. They had expected to

Did huge waterfalls along the Gibraltar precipice refill the Mediterranean Sea?

Ron Miller

see finely compacted particles of silts and clays; instead they found pea-size gypsum (calcium sulfate) gravel, which forms when seawater evaporates. Gypsum occurs in muddy sediments on arid coasts, and on land at the sites of dry lakes and ponds. It had never been reported on the deep-sea floor and does not occur in gravel deposits on the surrounding continents. What could account for its presence?

The mystery deepened as more samples were brought to the surface. One core contained a different form of calcium sulfate, along with other salts commonly found in evaporating seawater. (This core looked like marble and was quickly nicknamed the "Pillar of Atlantis.") Another core showed evidence of algal mats, which form only in water less than 10 meters (33 feet) deep. All these puzzling samples were taken about 200 meters (650 feet) beneath the seabed in 2,000 meters (6,600 feet) of water.

From the evidence contained in these sediment samples, two scientists aboard *Glomar Challenger*, William B. F. Ryan of the Lamont–Doherty Earth Observatory and Kenneth J. Hsü of the Swiss Federal Institute of Technology, pieced together a controversial new history of the Mediterranean Sea. About 6 million years ago, they said, the shallow Strait of Gibraltar—the narrow passage between the Mediterranean and the Atlantic—was blocked by an uplift of the land—a dam, in effect. Water could no longer flow into the Mediterranean, so the sea east of Gibraltar gradually evaporated, leaving behind a desert of gypsum-rich salts—the future M-reflector. Half a million years later, however, either the strait eroded or subsided, or the level of the Atlantic rose—and the ocean flooded back into the basin. The types of fossil microorganisms in the drill cores suggest that the rate of refilling was very rapid. In little more than a century, waters equal to a thousand Niagaras falling across the Gibraltar precipice refilled the Mediterranean Sea!

At least that's how Ryan and Hsü reconstructed the tale of the drill cores. But some scientists who have studied the same evidence doubt that the Mediterranean was ever completely dry. They suggest that most of the floor was studded by alkaline lakes or that perhaps the entire bottom was covered by very shallow, very salty water. The M-reflector could still have formed in this scenario.

Sediments have many stories to tell, some of them still open to interpretation. In a sense, sediments are the memory of the ocean because they provide a record—if we can read it—of the recent history of the ocean, and therefore of the planet itself.

5-1

WHAT TO WATCH FOR IN CHAPTER 5

A glance at the first few photographs in this chapter will show you the true face of the ocean floor. The basalt and lava we have been discussing are nearly always hidden—covered by dust and gravel, silt and mud. Sediment includes particles from land, from biological activity in the ocean, from chemical processes within water, and even from space. Analysis of this sedimentary material can tell us the recent history of an ocean basin, and sometimes the recent history of Earth.

Notice that the blanket of marine sediment is thickest at the continental margins and thinnest over the active oceanic ridges. You'll find that the ocean's sedimentary "memory" is not long. Your understanding of tectonics and basin ages should suggest why this is so. As you read, think about movement and transport: How did this sediment arrive? Where did it come from? How can we decipher the story it has to tell?

You probably have more daily contact with marine sediment than you think. Components of the building materials for roads and structures, toothpaste, paint, and swimming pool filters come directly from sediments. About a third of the oil and natural gas we use is extracted from marine sedimentary deposits.

Sediment is particles of organic or inorganic matter that accumulate in a loose, unconsolidated form. The particles originate from the weathering and erosion of rocks, from the activity of living organisms, from volcanic eruptions, from chemical processes within the water itself, and even from space. Most of the ocean floor is being slowly dusted by a continuing rain of sediments. Accumulation rates on the deep-sea floor vary from a few centimeters per year to the thickness of a dime every thousand years.

Marine sediments occur in a broad range of sizes and types. Beach sand is sediment; so are the mud of a quiet bay and the mix of silt and tiny shells found on the continental margins. Less familiar sediments are the fine clays of the deep-ocean floor, the biologically derived oozes of abyssal plains, and the nodules and coatings that form around hard objects on the seafloor. The origin of these materials—and the distribution and sizes of the particles—depends on a combination of physical and biological processes.

WHAT SEDIMENTS LOOK LIKE

What do sediments look like? That depends on where you look. **Figure 5.1** shows the Mid-Atlantic Ridge, 48°N, at a depth of 2,629 meters (8,626 feet). Sponges and some relatives of coral grow from young rocky outcrops only lightly powdered with sediment. Contrast that rough ridge with the smooth seafloor shown in **Figure 5.2.** The sediment there is about 35 meters (115 feet) thick and marked by the tracks of brittle stars. These widely distributed organisms feed on surface bacteria and fallen particles of organic sediment.

Figure 5.1 Sediment near the crest of the Mid-Atlantic Ridge. The rock outcrops are dusted only lightly with sediment; crinoids and gorgonians (relatives of coral) attach to the rocks.

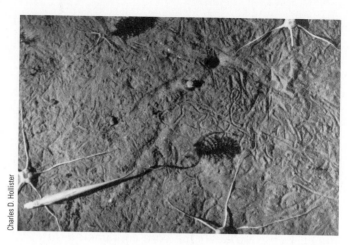

Figure 5.2 Brittle stars and their tracks on the continental slope off New England. The depth is 1,476 meters (4,842 feet).

The surface of the sediment is not always smooth. Where bottom currents are swift and persistent, they can cause ripples like those on a streambed (**Figure 5.3**).

The extraordinary thickness of some layers of marine sediment can be seen in **Figure 5.4,** a seismic profile of the eastern edge of a seamount in the North Atlantic's Sohm Abyssal Plain south of Nova Scotia. The sediment at the eastern boundary of this profile covers the oceanic crust to a depth of more than 1.8 kilometers (1.1 miles).

The colors of marine sediments are often quite striking. Sediments of biological origin are white or cream-colored, with deposits high in silica tending toward gray. Though traditionally termed "red clays" from the rusting (oxidation) of iron within the sediments to form iron oxide, deep-sea clays can range from tan to a chocolate brown color. Other clays are shades of green or tan. Nodular sediments are a dark sooty brown or black. Some nearshore sediments contain decomposing organic material and smell of hydrogen sulfide, but most are odorless.

A very few areas of the seabed are altogether free of overlying sediments. The water over these areas is not completely sediment-free, but for some reason sediment does not collect on the bottom. Strong currents may scour the sediments away, the seafloor may be too young in these areas for sediments to have had time to accumulate, or hot water percolating upward through a porous seafloor may dissolve the material as fast as it settles.

CLASSIFYING SEDIMENT BY PARTICLE SIZE

Particle size is frequently used to classify sediments. The scheme shown in **Table 5.1** was first devised in 1898 and with refinements has been used by geologists, soil scientists, and oceanographers ever since. In this classification the coarsest particles are boulders, which are more than 256 millimeters (about 10 inches) in diameter. Although

Figure 5.3 Ripples on the sediment beneath the swift Antarctic Circumpolar Current in the northern Drake Passage. The depth here is 4,010 meters (13,153 feet).

Figure 5.4 The deep sediments of the Sohm Abyssal Plain in the North Atlantic south of Nova Scotia have buried the base of this seamount. This seismic profile shows the depth of the sediments above the geologic base of the seamount to be more than 1.8 kilometers (1.1 miles). Note the scour moat—depression along the boundary of seamount and sediment—caused by a persistent deep boundary current in the area. The vertical exaggeration in this figure is about 12:1.

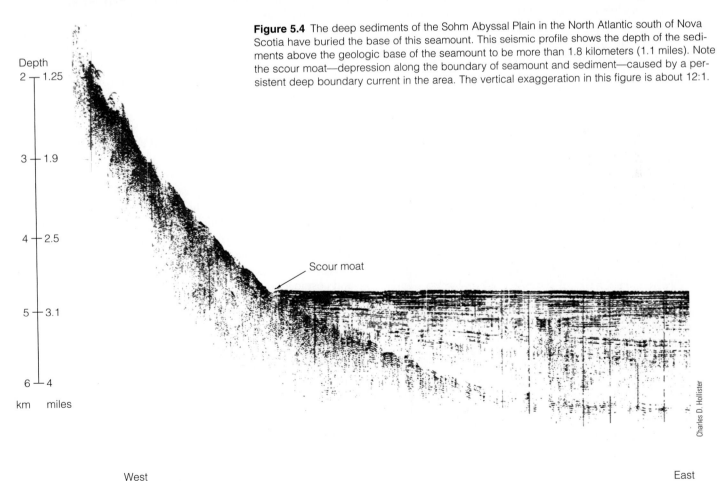

Depth

km	miles
2	1.25
3	1.9
4	2.5
5	3.1
6	4

Scour moat

West

East

Table 5.1	Particle Sizes and Settling Rates in Sediment		
Type of Particle	**Diameter**	**Settling Velocity**	**Time to Settle 4 km (2.5 mi)**
Boulder	>256 mm (10 in.)	—	—
Cobble	64–256 mm (>2 1/2 in.)	—	—
Pebble	4–64 mm (1/6–2 1/2 in.)	—	—
Granule	2–4 mm (1/12–1/6 in.)	—	—
Sand	0.062–2 mm	2.5 cm/sec (1 in./sec)	1.8 days
Silt	0.004–0.062 mm	0.025 cm/sec (1/100 in./sec)	6 months
Clay	<0.004 mm	0.00025 cm/sec	50 years[a]

[a]Though the theoretical settling time for individual clay particles is usually very long, under certain conditions clay particles in the ocean can interact chemically with seawater, clump together, and fall at a faster rate. Small biogenous particles are often compressed by organisms into fecal pellets that can fall more rapidly than would otherwise be possible. A fecal pellet is shown in Figure 5.11.

boulders, cobbles, and pebbles occur in the ocean, most marine sediments are made of finer particles: **sand, silt,** and **clay.**

Generally, the smaller the particle, the more easily it can be transported by streams, waves, and currents. As sediment is transported it tends to be sorted by size; coarser grains, which are moved only by turbulent flow, tend not to travel as far as finer grains, which are more readily moved. The clays, particles less than 0.004 millimeter in diameter, can remain suspended for very long periods and may be transported great distances by ocean currents before they are deposited. As indicated in **Figure 5.5,** the cohesiveness of smaller particles can be as important as grain size in determining whether they will be eroded and transported. Once in suspension, the finest clays may circulate in the ocean for decades.

A layer of sediment can contain particles of similar size, or it can be a mixture of different-size particles. Sediments composed of particles of one size are said to be **well-sorted sediments.** Sediments with a mixture of sizes are **poorly sorted sediments.** Sorting is a function of the energy of the environment—the exposure of that area to the action of waves, tides, and currents. Well-sorted sediments occur in environments where energy fluctuates within narrow limits. Sediments of the calm deep-ocean floor are typically well sorted (see again Figure 5.2). Poorly sorted sediments form in environments where energy fluctuates over a wide spectrum. The mix of rubble at the base of a rapidly eroding shore cliff is a good example of poorly sorted sediment. Sediments that have been transported by turbidity currents (which can transport a wide range of grain sizes) tend to be poorly sorted.

CLASSIFYING SEDIMENTS BY SOURCE

Another way to classify marine sediments is by their origin. Such a scheme was first proposed in 1891 by Sir John Murray and A. F. Renard after a thorough study of sediments collected during the *Challenger* Expedition. A modern modification of their organization is shown in **Table 5.2.** This scheme separates sediments into four categories by source: terrigenous, biogenous, hydrogenous (also called authigenic), and cosmogenous.

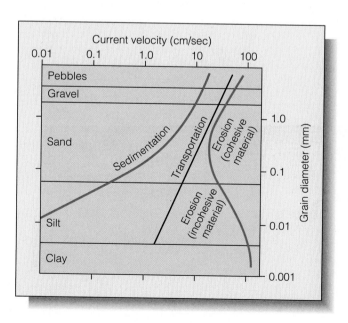

Figure 5.5 The velocities of currents required for erosion, transportation, and deposition of sediment particles of different sizes. The green curve indicates that rapid currents are needed to erode clay, fine silt, and coarse sand. Weaker currents are sufficient to move fine or medium sand off the seafloor. Note that the *cohesiveness* of smaller particles can be more important than grain size in determining whether they will be eroded and transported. Mud, a mixture of clay and silt, is "stickier" than sand and resists erosion. As the black line indicates, less sticky materials would behave in flowing water in the near-linear manner we might expect.

Sediment Type	Source	Examples	Distribution	Percent of All Ocean Floor Area Covered
Terrigenous	Erosion of land, volcanic eruptions, blown dust	Quartz sand, clays, estuarine mud	Dominant on continental margins, abyssal plains, polar ocean floors	~45%
Biogenous	Organic; accumulation of hard parts of some marine organisms	Calcareous and siliceous oozes	Dominant on deep-ocean floor (siliceous ooze below about 5 km)	~55%
Hydrogenous (authigenic)	Precipitation of dissolved minerals from water, often by bacteria	Manganese nodules, phosphorite deposits	Present with other, more dominant sediments	<1%
Cosmogenous	Dust from space, meteorite debris	Tektite spheres, glassy nodules	Mixed in very small proportion with more dominant sediments	0%

Sources: Kennett, 1982; Weihaupt, 1979; Sverdrup, Johnson, and Fleming, 1942.

Terrigenous Sediments

Terrigenous (*terra* = Earth, *generare* = to produce) **sediments** are the most abundant. As the name implies, terrigenous sediment originates on the continents or on islands near them. The rocks of Earth's crust are made up of **minerals,** inorganic crystalline materials with specific chemical compositions. The texture of igneous rocks—rocks that crystallize from magma—is determined by how rapidly they cool. Igneous rocks that cool rapidly, such as the basalt that forms the ocean floor at spreading centers or pours from volcanic vents on land, solidify so quickly that obvious crystals do not have a chance to form. Igneous rocks that cool very slowly, as would occur in masses rising from a subducted lithospheric plate, solidify very slowly and form large crystals. Nearly all terrigenous sediments are derived directly or indirectly from these crystals.

The most familiar continental igneous rock is granite, the source of quartz and clay, the two most common components of terrigenous marine sediments. Quartz, an important mineral in granite, is hard, relatively insoluble, very durable, and can withstand lengthy weathering and transport. Quartz sands washed from the adjacent land are important components of the sediments along continental margins. Feldspar, another important mineral in granite, ultimately combines with carbonic acid (a mild acid that forms when carbon dioxide dissolves in water) and seawater to form clay. These tiny particles, which are the chief component of soils, are carried to the ocean by wind, rivers, or streams. Because of their small size, they are easily transported across the continental shelf to settle slowly to the deep-ocean floor. Although estimates vary, it appears that about 15 billion metric tons (16.5 billion tons) of terrigenous sediments is transported in rivers to the sea each year, with an additional 100 million metric tons transported annually from land to ocean as fine airborne dust and volcanic ash (see **Figure 5.6**).

Biogenous Sediments

Biogenous (*bio* = life, *generare* = to produce) **sediments** are the next most abundant marine sediment. The siliceous (silicon-containing) and calcareous (calcium carbonate–containing) compounds that make up these sediments of biological origin were originally brought to the ocean in solution by rivers or dissolved in the ocean at mid-ocean ridges. The siliceous and calcareous materials were then extracted from the seawater by the normal activity of tiny plants and animals to build protective shells and skeletons. Some of this sediment derives from larger mollusk shells or from stationary colonial animals such as corals, but most of the organisms that produce biogenous sediments drift free in the water as plankton. After the death of their owners, the hard structures fall slowly to the bottom and accumulate in layers. Biogenous sediments are most abundant where ample nutrients encourage high biological productivity, usually near continental margins. Over millions of years, organic molecules within these sediments can form oil and natural gas (see Chapter 17 for details).

Note in Table 5.2 that biogenous sediments cover a larger percentage of the *area* of the ocean floor than terrigenous sediments, but the terrigenous sediments dominate in total *volume.*

Hydrogenous Sediments

Hydrogenous (*hydro* = water, *generare* = to produce) **sediments** are minerals that have precipitated directly from seawater. The sources of the dissolved minerals include submerged rock and sediment, leaching of the fresh crust at oceanic ridges, material issuing from hydrothermal vents, and substances flowing to the ocean in river runoff. As we shall see, the most prominent hydrogenous sediments are manganese nodules, which litter some deep seabeds, and phosphorite nodules, seen along some continental margins. Hydrogenous sediments are also

a

USGS

Department of Geology, University of Delaware

b

NASA

c

Figure 5.6 Sources of terrigenous sediments. (**a**) Rivers are the main source of terrigenous sediments. This photo, taken from space, shows sediment entering the Gulf of Mexico from the Mississippi River. (**b**) The wind may transport ash from a volcanic eruption for hundreds of kilometers and deposit it in the ocean. This ash cloud was caused by the summer 1991 eruption of Mt. Pinatubo in the Philippines. (**c**) Wind also transports particles. A huge dust cloud streams over the Atlantic from Africa's Sahara Desert on 26 February 2000. The particles will fall to the ocean surface and descend slowly to the ocean floor to end up as terrigenous sediments.

called **authigenic** (*authis* = in place, on the spot) **sediments** because they were formed in the place they now occupy. Though they usually accumulate very slowly, rapid deposition of hydrogenous sediments is possible—in a rapidly drying lake, for example.

Cosmogenous Sediments

Cosmogenous (*cosmos* = universe, *generare* = to produce) **sediments,** which are of extraterrestrial origin, are the least abundant. These sediments are typically greatly diluted by other sediment components and rarely constitute more than a few parts per million of the total sediment in any layer. Scientists believe that cosmogenous sediments come from two major sources—interplanetary dust that falls constantly into the top of the atmosphere, and rare impacts by large asteroids and comets.

Interplanetary dust consists of silt- and sand-size micrometeoroids that come from asteroids and comets or from collisions between asteroids. The silt-size particles settle gently to Earth's surface, but larger, faster-moving dust is heated by friction with the atmosphere and melts, sometimes glowing as the meteors we see in a dark night sky. Though much of this material is vaporized, some may persist in the form of iron-rich cosmic spherules. Most of these dissolve in seawater before reaching the ocean floor. About 15,000 to 30,000 metric tons (16,500 to 33,000 tons) of interplanetary dust enters Earth's atmosphere every year.

The highest concentrations of cosmogenous sediments occur when large volumes of extraterrestrial matter arrive all at once. Fortunately, this happens only rarely, when Earth is hit by a large asteroid or comet. Very few examples of this are known, but most geologists now believe that 65 million years ago an asteroid 10 kilometers (6 miles) in diameter struck Earth on what is now the northern coast of Yucatán (see Box 13.1). The debris ejected from this impact was blown into space around Earth. Much of it fell back and was deposited in a layer that has now been found in more than a hundred places around Earth (mostly in marine sediments). Cosmogenous components may comprise between 10% and 20% of these extraordinary sediments!

Occasionally cosmogenous sediment includes translucent oblong particles of glass known as **microtektites** (**Figure 5.7**). Microtektites are thought to form from the violent impact of large meteors or small asteroids on the crust of Earth. The impact melts some of the crustal material and splashes it into space; the material melts again as it rushes through the atmosphere, producing the various raindrop shapes shown in the photo. Microtektites do not dissolve easily and usually reach the ocean floor. Most are less than 1.5 millimeter (1/16 inch) long.

Sediment Mixtures

Sediments on the ocean floor only rarely come from a single source; most sediment deposits are a mixture of biogenous and terrigenous particles, with an occasional hydrogenous or cosmogenous supplement. The pattern and composition of

sediment layers on the seabed are of great interest to researchers studying conditions in the overlying ocean. Different marine environments have characteristic sediments, and these sediments preserve a record of past and present conditions within those environments.

THE DISTRIBUTION OF MARINE SEDIMENTS

The sediments on the continental margin are generally different in quantity, character, and composition from those on the deeper basin floors. Continental shelf sediments—called **neritic** (*neritos* = of the coast) **sediments**—consist primarily of terrigenous material. Deep-ocean floors are covered by finer sediments than those of the continental margins, and a greater proportion of deep-sea sediment is of biogenous origin. Sediments of the slope, rise, and deep-ocean floor that originate in the ocean are called **pelagic** (*pelagios* = of the sea). **sediments.** The distribution and average thickness of the marine sediments in each oceanic region are shown in **Table 5.3**. Note that 72% of all marine sediment is associated with continental slopes and rises.

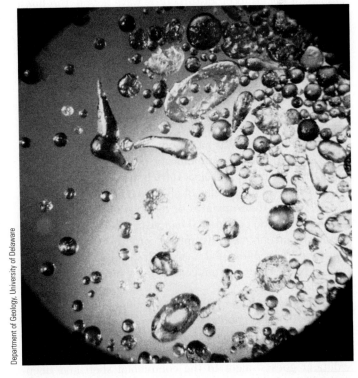

Department of Geology, University of Delaware

Figure 5.7 Microtektites, very rare particles that began a long journey when a large body impacted Earth and ejected material from Earth's crust. Some of this material traveled through space, re-entered Earth's atmosphere, melted, and took on a rounded or teardrop shape. These specimens of sculptured glass range from 0.2 to 0.8 millimeter in length. Glassy dust much finer in size, as well as nut-size chunks, has also fallen on Earth.

Table 5.3	The Distribution and Average Thickness of Marine Sediments		
Region	Percent of Ocean Area	Percent of Total Volume of Marine Sediments	Average Thickness
Continental shelves	9%	15%	2.5 km (1.6 mi)
Continental slopes	6%	41%	9 km (5.6 mi)
Continental rises	6%	31%	8 km (5 mi)
Deep-ocean floor	78%	13%	0.6 km (0.4 mi)

Sources: Emery in Kennett, *Marine Geology*, 1982 (Table 11.1); Weihaupt, *Exploration of the Oceans*, 1979; Sverdrup, Johnson, and Fleming, *The Oceans: Their Physics, Chemistry and General Biology*, 1942.

THE SEDIMENTS OF CONTINENTAL MARGINS

The bulk of terrigenous sediment is eroded and carried to streams, where it is transported to the ocean. Currents distribute sand and larger particles along the coast, while wave action carries the silts and clays to deeper water. When the water is too deep to be disturbed by wave action, the finest sediment may come to rest or continue to be transported by the turbulence of deep currents toward the deeper ocean floor. Ideally, these processes produce an orderly sorting of particles by size from relatively large grains near the coast to relatively small grains near the shelf break.

There are exceptions, however. Shelf deposits are subject to further modification and erosion as sea level fluctuates: Larger particles may be moved toward the shelf edge when sea level is low, as it was in periods of extensive glaciation—ice ages. Poorly sorted sediments are also found as glacial deposits. In polar regions glaciers and ice shelves give rise to icebergs. These carry particles of all sizes, and when they melt they distribute their mixtures of rocks, gravel, sand, and silt onto high-latitude continental margins and deep-ocean floor. Turbidity currents also disrupt the orderly sorting of sediments on the continental margin by transporting coarse-grained particles away from coastal areas and onto the deep-ocean floor.

Ice ages have other effects on sediment deposition. Note in Figure 4.14 that continental shelves are almost completely exposed by the lowered sea level during times of extensive glaciation. Rivers carry their sediment right to the shelf edge. Sediment goes straight to the continental shelf and deep-sea bed.

Between ice ages, when the shelves are covered with water, the rate of sediment deposition on continental shelves is variable, but it is almost always greater than the rate of sediment deposition in the deep ocean. Near the mouths of large rivers, 1 meter (about 3 feet) of sediment may accumulate every thousand years. Along the east coast of the United States, however, many large rivers terminate in estuaries, which trap most of the sediment brought to them. The continental shelf of eastern North America is therefore covered mainly by sediments laid down during the last period of glaciation, when sea level was lower.

In addition to terrigenous material, the continental margins almost always contain biogenous sediments. Biological productivity in coastal waters is often quite high, and the skeletal remains of creatures living on the bottom or in the water above mix with the terrigenous sediments and dilute them.

Sediments can build to impressive thickness on continental shelves. In some cases, shelf sediments undergo **lithification:** They are converted into sedimentary rock by pressure or by cementation. If these lithified sediments are thrust above sea level by tectonic forces, they can form mountains or plateaus. The top of Mount Everest, the world's tallest peak, is a shallow-water biogenic marine limestone (a calcareous rock). Much of the Colorado Plateau, with its many stacked layers, was formed by sedimentary deposition and lithification beneath a shallow continental sea beginning about 570 million years ago. The Colorado River has cut and exposed the uplifted beds to form the Grand Canyon. Hikers walking from the canyon rim down to the river pass through spectacular examples of continental-shelf sedimentary deposits. Their journey takes them deep into an old ocean floor!

THE SEDIMENTS OF DEEP-OCEAN BASINS

Sediment thickness is highly variable. When averaged, the Atlantic Ocean bottom is covered by sediments to a thickness of about 1 kilometer (3,300 feet), while the Pacific floor has an average sediment thickness of less than 0.5 kilometer (1,650 feet). There are three reasons for this difference. First, the Atlantic Ocean is smaller in area. Second, the Atlantic is fed by a greater number of rivers laden with sediment. Third, in the Pacific Ocean many oceanic trenches trap sediments moving toward basin centers. Beyond this, the composition and thickness of pelagic sediments also vary with location, being thickest on the abyssal plains and thinnest (or absent) on the oceanic ridges.

Turbidites

Dilute mixtures of sediment and water periodically rush down the continental slope in turbidity currents, the erosive force of which is thought to help cut submarine canyons (see again Figure 4.18). These underwater avalanches of thick muddy fluid can reach the continental rise and often continue moving onto an adjacent abyssal plain before eventually coming to rest. The resulting deposits are called **turbidites;** they are graded layers of terrigenous sand interbedded with the finer sediments typical of the deep-sea floor.

Clays

About 38% of the deep-sea bed is covered by clays and other fine terrigenous particles. As we have seen, the finest terrigenous sediments are easily transported by wind and water currents. Microscopic waterborne particles and tiny bits of wind-borne dust and volcanic ash settle slowly to the deep-ocean floor, forming fine brown, olive-colored, or reddish clays. As Table 5.1 shows, the velocity of particle settling is related to particle size, and clay particles usually fall very slowly indeed. Terrigenous sediment accumulation on the deep-ocean floor is typically about 2 millimeters (⅛ inch) every thousand years.

Oozes

Seafloor samples taken farther from land usually show a greater proportion of biogenous sediments than those obtained near the continental margins. This is not because biological productivity is higher farther from land (the opposite is usually true), but because there is less terrigenous material far from shore and thus the deposits contain a greater proportion of biogenous material.

Deep-ocean sediment containing at least 30% biogenous material is called an **ooze** (surely one of the most descriptive terms in the marine sciences). Oozes are named after the dominant remanent organism constituting them. The organisms contributing their remains to deep-sea oozes are small, single-celled, drifting, plantlike organisms and the single-celled animals that feed on them. The hard shells and skeletal remains of these creatures are of relatively dense glasslike silica or calcium carbonate (limy) substances. When these organisms die, their shells settle slowly toward the bottom, mingle with fine-grained terrigenous silts and clays, and accumulate as ooze. The silica-rich residues give rise to **siliceous ooze,** the calcium-containing material to **calcareous ooze.**

Oozes accumulate slowly, at a rate of about 1 to 6 centimeters (½ to 2½ inches) per thousand years. But they collect almost ten times more quickly than deep-ocean terrigenous clays. The accumulation of any ooze therefore depends on a delicate balance between the abundance of organisms at the surface, the rate at which they dissolve once they reach the bottom, and the rate of accumulation of terrigenous sediment.

Calcareous ooze forms mainly from shells of the amoeba-like **foraminifera** (**Figure 5.8a** and **b**), small drifting mollusks called **pteropods,** and tiny algae known as **coccolithophores** (**Figure 5.8c**). Although these creatures live in nearly all surface ocean water, calcareous ooze does not accumulate everywhere on the ocean floor because the shells are

a

b

c

Figure 5.8 Organisms that contribute to calcareous ooze. (**a**) A living foraminiferan, an amoeba-like organism. The shell of this beautiful foram, genus *Hastigerina,* is surrounded by a bubble-like capsule. It is one of the largest of the planktonic species with spines, reaching nearly 5 centimeters (2 inches) in length. (**b**) A much smaller foraminiferan, the snail-like planktonic *Rosalina.* (**c**) Coccoliths, individual plates of coccolithophores, a form of planktonic algae. Because of their tendency to dissolve, calcareous oozes very rarely occur at bottom depths below 4,500 meters (14,800 feet).

Howard Spero

Wim van Egmond

Roger Witmer

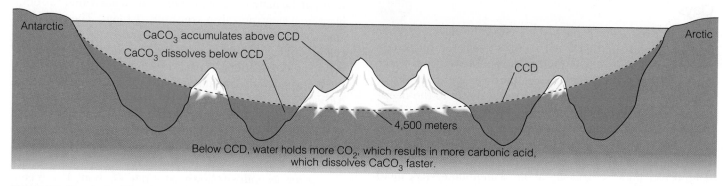

Antarctic

CaCO₃ accumulates above CCD

CaCO₃ dissolves below CCD

Arctic

CCD

4,500 meters

Below CCD, water holds more CO_2, which results in more carbonic acid, which dissolves $CaCO_3$ faster.

Figure 5.9 The dashed line shows the calcium carbonate ($CaCO_3$) compensation depth (CCD). At this depth the rate at which calcareous sediments accumulate equals the rate at which those sediments dissolve.

dissolved by seawater. At great depths seawater contains more CO_2 and becomes slightly acid.[1] As you will see in Figure 7.10, this acidity, combined with the increased solubility of calcium carbonate in cold water under pressure, dissolves the shells. At a certain depth, the **calcium carbonate compensation depth** (**CCD**), the rate at which calcareous sediments are supplied to the seabed equals the rate at which those sediments dissolve. Below this depth, the tiny skeletons of calcium carbonate dissolve on the seafloor, so no calcareous oozes form. Calcareous sediment dominates on the deep-sea floor at depths less than about 4,500 meters (14,800 feet), the usual calcium carbonate compensation depth. Sometimes a line analogous to a snow line on a terrestrial mountain can be seen on undersea peaks: Above the line the white sprinkling of calcareous ooze is visible; below it the "snow" is absent (see **Figure 5.9**). About 48% of the surface of deep-ocean basins is covered by calcareous oozes.

Siliceous (silicon-containing) ooze predominates at greater depths and in colder polar regions. Siliceous ooze is formed from the hard parts of another amoeba-like animal, the beautiful glassy **radiolarian** (**Figure 5.10a**), and from single-celled algae called **diatoms** (**Figure 5.10b**). After a radiolarian or diatom dies, its shell will also dissolve back into the seawater, but this dissolution occurs *much* more slowly than the dissolution of calcium carbonate. Slow dissolution, combined with very high diatom productivity in some surface waters, leads to the buildup of siliceous ooze. Diatom ooze is most common in the Antarctic because strong ocean currents and seasonal upwelling in this area support large populations of diatoms. Radiolarian oozes occur in equatorial regions, most notably in the zone of equatorial upwelling west of South America (as will be seen in Figure 5.13). About 14% of the surface of the deep-ocean floor is covered by siliceous oozes.

---

[1] A discussion of acid–base balance may be found in Chapter 7 on pages 179–182.

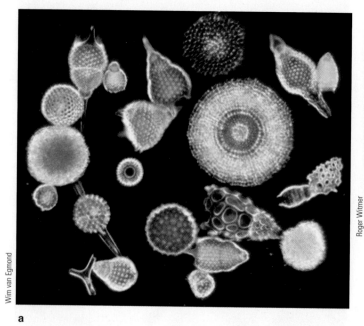

Wim van Egmond

a

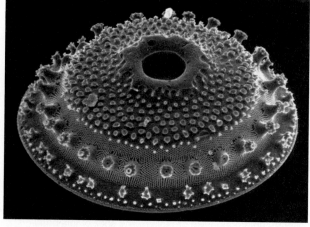

Roger Witmer

b

Figure 5.10 Micrographs of siliceous oozes, which are most common at great depths. (**a**) Shells of radiolarians, amoeba-like organisms. Radiolarian oozes are found primarily in the equatorial regions. (**b**) A test (shell) of a diatom, a single-celled alga. Diatom oozes are most common at high latitudes.

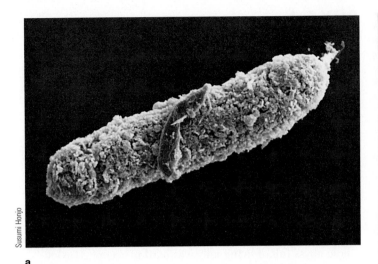

a

Figure 5.11 A fecal pellet of a small planktonic animal. (**a**) The compressed pellet is about 80 μm in length. (**b**) Enlargement of the pellet's surface, magnified about 2,000 times. The pellet consists of the indigestible remains of small microscopic plantlike organisms, mostly coccolithophores. Unaided by pellet packing, these remains might take months to reach the seabed, but compressed in this way, they can be added to the ooze in perhaps two weeks.

b

The very small particles that make up most of these deep-ocean sediments would need between 20 and 50 years to sink to the bottom. By that time they would have drifted a great lateral distance from their original surface position. But researchers have noted that the composition of deep-bottom sediments is usually similar to particle composition in the water directly above. How could such tiny particles fall quickly enough to avoid great horizontal displacement? The answer appears to involve their compression into fecal pellets. While still quite small, the fecal pellets of small animals are much larger than the tiny individual skeletons of diatoms, foraminifera, and other plantlike organisms that they have consumed (**Figure 5.11**), so they fall much faster, reaching the deep-ocean floor in about two weeks.

Some deep-sea oozes have been uplifted by geologic processes and are now visible on land. The calcareous chalk White Cliffs of Dover in eastern England are partially lithified deposits composed largely of foraminifera and coccolithophores. Fine-grained siliceous deposits called *diatomaceous earth* are mined from other deposits. This fossil material is a valued component in flat paints, pool and spa filters, and mildly abrasive car and tooth polishes.

Hydrogenous Materials

Hydrogenous sediments also accumulate on deep-sea floors. They are associated with terrigenous or biogenous sediments and very rarely form sediments by themselves. Most hydrogenous sediments originate from chemical reactions that occur on particles of the dominant sediment.

The most famous hydrogenous sediments are manganese **nodules,** which were discovered by the hard-working crew of

HMS *Challenger.* The nodules consist primarily of manganese and iron oxides but also contain small amounts of cobalt, nickel, chromium, copper, molybdenum, and zinc. They form in ways not fully understood by marine chemists, "growing" at an average rate of 1 to 10 millimeters (0.04 to 0.4 inch) per *million* years, one of the slowest chemical reactions in nature. Though most are irregular lumps the size of a potato, some nodules exceed 1 meter (3.3 feet) in diameter. Manganese nodules often form around nuclei such as sharks' teeth, bits of bone, microscopic algae, animal skeletons, and tiny crystals—as the cross section of a manganese nodule in **Figure 5.12a** shows. Bacterial activity may play a role in the development of a nodule. Between 20% and 50% of the Pacific Ocean floor may be strewn with nodules (**Figure 5.12b**).

Why don't these heavy lumps disappear beneath the constant rain of accumulating sediment? Possibly the continuous churning of the underlying sediment by creatures living there keeps the dense lumps on the surface, or perhaps slow currents in areas of nodule accumulation waft particulate sediments away.

HMS *Challenger* scientists also discovered nodules of phosphorite. The first of these irregular brown lumps was taken from the continental rise off South Africa, and phosphorite nodule fields have since been found on shallow bank tops, outer continental shelves, and the upper parts of continental slope areas off California, Argentina, and Japan. Phosphorus is an important ingredient in fertilizer, and the nodules may someday be collected as a source of agricultural phosphates. Like manganese nodules, phosphorite nodules are found only in areas with low rates of sediment accumulation.

Ooze may be one of the most graphic words in science, but most oozes aren't slimy or sticky or gelatinous, and they don't slither across the seabed. Despite their name, oozes are submerged deposits containing the hard parts of once-living organisms that collectively look and act like fine sand or silt.

But there was once an ooze that lived up to the name. This mythical muck generated quite a controversy and changed the history of marine science even though it never really existed!

The story begins in 1857, when officers aboard HMS *Cyclops* ran a line of soundings from Ireland to Newfoundland and collected small samples of bottom sediment at the same time. The trapped sediments were preserved in alcohol and stored in bottles. Eleven years later, Thomas Henry Huxley, the preeminent British biologist, studied the samples. In every sample he analyzed, Huxley found a grayish-white, filmy ooze coating the solid mud. When stirred, the goo broke up into long strings; one scientist remarked that the stuff looked like an egg white stirred in water. The first chemical analysis indicated calcareous granules suspended in a proteinaceous base. After watching bits of the slime beneath his microscope and seeing the granules slowly shift about, Huxley decided "that the granule-heaps and the transparent gelatinous matter in which they are embedded represent masses of protoplasm." He quickly named the ooze **bathybius** (*bathy* = deep, *bios* = life) and leaped to the conclusion that the deep ocean was full of the glop. He first thought it was part of the body of a giant abyssal organism but later decided it must be a remanent of the "primeval

living slime" from which all life may have arisen. He studied bathybius, lectured about it, and wrote articles on its properties for scholarly publications.

Many of Huxley's fellow scientists thought he was mistaken about the nature of bathybius. They had been convinced only recently that animal life was possible in the cold, dark depths past about 300 fathoms (550 meters, 1,800 feet), when submarine telegraph cables encrusted with organisms were raised from the ocean bottom. Even so, a whole abyss full of living jelly was hard to accept. Huxley countered with the opinion that the protoplasmic beast could live on organic matter raining down from above. Besides, Huxley was a friend and supporter of Charles Darwin, and this supposed remanent of the primordial ooze seemed to fit neatly into the developing evolutionary scheme.

Huxley's attentions had stirred up scientific and popular interest in bathybius. This publicity, combined with a growing interest in submerged fossils, chalk deposits, and the newly formulated theory of evolution, was one factor that contributed to the rise of oceanography and the commissioning of the *Challenger* Expedition.

Challenger scientists were on the lookout for bathybius but didn't find any. What they did find was that when too much alcohol was added to some sediment samples, a gelatinous calcium sulfate compound was precipitated from seawater, forming the famous slime. Unpreserved samples showed no evidence of bathybius. Bathybius had never existed—Huxley had been studying an artifact of specimen preservation!

At least Huxley was graceful in his concession. In an open letter written in 1875 to the scientific journal *Nature,* Huxley wrote: "Professor Wyville Thomson . . . informs me that the best efforts of the *Challenger's* staff have failed to discover bathybius in a fresh state, and that it is seriously suspected that the thing to which I gave that name is little more than sulphate of lime. I am mainly responsible for the mistake. . . ." But the public's appetite for the weird (combined with Huxley's reputation as a first-rate scientist) kept bathybius alive in popular magazines and Sunday supplement articles until the early 1930s.

The story of bathybius demonstrates how important it is for marine scientists to go to sea to obtain and preserve their own samples. There's no substitute for immediate observation and analysis.

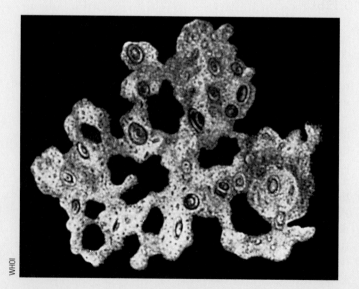

The supposed living slime—primordial ooze—"discovered" by Thomas Henry Huxley in 1868. He believed that this "animal jelly" carpeted the deep floor of the ocean.

a

b

Figure 5.12 Manganese nodules. (**a**) A cross section cut through a manganese nodule, showing the concentric layers of manganese and iron oxides. This nodule is about 11 centimeters (4½ inches) long, a typical size. (**b**) Lemon-size manganese nodules littering the abyssal Pacific.

For now, the low market value of the minerals in manganese and phosphorite nodules makes them too expensive to recover. As techniques for deep-sea mining become more advanced and raw material prices grow higher, however, the nodules' concentration of valuable materials will almost certainly be exploited.

Powdery deposits of metal sulfides have been found in the vicinity of hydrothermal vents at oceanic ridges. Hot, metal-rich brines blasting from the vents meet cold water, cool rapidly, and lose the heavy metal sulfides by precipitation. Iron sulfides and manganese precipitates fall in thick blankets around the vents. The cobalt crusts of rift zones also seem to be associated with this phenomenon. These areas may one day be mined for their metal content.

Evaporites

Evaporites are an important group of hydrogenous deposits that include many salts important to humanity. These salts precipitate as water evaporates from isolated arms of the ocean or from landlocked seas or lakes. For thousands of years people have collected sea salts from evaporating pools or deposited beds. Evaporites are forming today in the Gulf of California, the Red Sea, and the Persian Gulf. The first evaporites to precipitate as water's salinity increases are the carbonates, such as calcium carbonate (from which limestone is formed). Calcium sulfate, which gives rise to gypsum, is next. (The M-reflector at the bottom of the Mediterranean consists of a layer of gypsum-rich evaporites.) Crystals of sodium chloride (table salt) will form if evaporation continues.

Not all hydrogenous calcium carbonate deposits are caused by evaporation, however. A small decrease in the acidity of seawater, or an increase in its temperature, can cause calcium carbonate to precipitate from water of normal salinity. In shallow areas of high biological productivity where sunlight heats the water, microscopic plants use up dissolved carbon dioxide, making seawater slightly less acidic (see Figure

7.9). Molecules of calcium carbonate then may precipitate around shell fragments or other particles. These white, rounded grains are called *ooliths* (*oon* = egg) because they resemble fish eggs. **Oolite sands**—comprised of ooliths—are abundant in many warm, shallow waters such as those of the Bahama Banks.

Sediments: An Overview

Figure 5.13 is a simplified look at the distribution of marine sediments. Notice especially the lack of radiolarian deposits in much of the deep North Pacific; the strand of siliceous oozes extending west from equatorial South America; and the broad expanses of the Atlantic, South Pacific, and Indian Ocean floors covered by calcareous oozes. The broad, deep, relatively old Pacific contains extensive clay deposits, most delivered in the form of airborne dust. As you might expect, the poorly sorted glacial deposits are found only at high latitudes.

This map summarizes more than a century of effort by marine scientists. Studies of sediments will continue because of their importance to natural resource development and because of the details of Earth's history that remain locked beneath their muddy surfaces.

Studying Sediments

Deep-water cameras have enabled researchers to photograph bottom sediments. The first of these cameras was simply lowered on a cable and triggered by a trip wire. Other, more elaborate cameras have been taken to the seafloor on towed sleds or deep submersibles.

Actual samples usually provide more information than photographs do. HMS *Challenger* scientists used weighted, wax-tipped poles and other tools attached to long lines to obtain

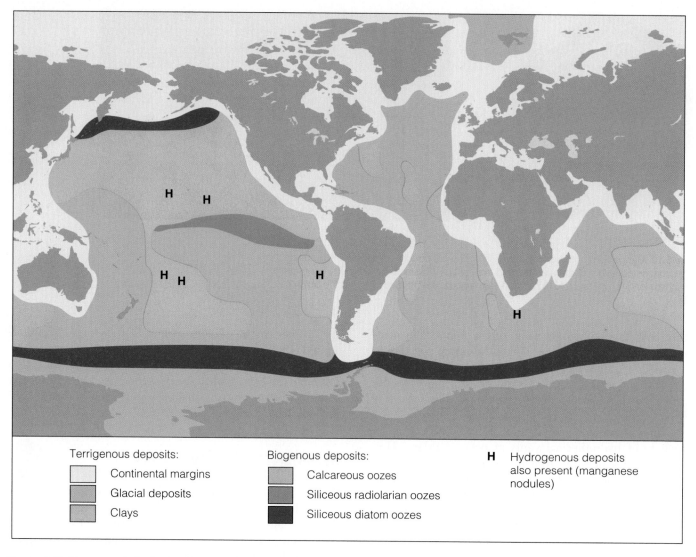

Terrigenous deposits:
- ☐ Continental margins
- ☐ Glacial deposits
- ☐ Clays

Biogenous deposits:
- ☐ Calcareous oozes
- ☐ Siliceous radiolarian oozes
- ☐ Siliceous diatom oozes

H Hydrogenous deposits also present (manganese nodules)

Figure 5.13 The general distribution of sediments on the ocean floor. Note the dominance of diatom oozes at high latitudes.

samples, but today's oceanographers have more sophisticated equipment. Shallow samples may be taken using a **clamshell sampler** (named because of its method of operation, not its target; see **Figure 5.14**). Deeper samples are taken by a **piston corer** (**Figure 5.15**), a device capable of punching through as much as 25 meters of sediment and returning an intact plug of material. Using a rotary drilling technique similar to that used to drill for oil, the drilling ship *Glomar Challenger* returned much longer core segments, some more than 1,100 meters (3,600 feet) in length! As we saw in Chapter 2, *Glomar Challenger* drilled more than a thousand holes for the Deep Sea Drilling Project, a program begun in 1968 with funds from the National Science Foundation. These cores are stored in core libraries, a very valuable scientific resource (**Figure 5.16**). Analysis of sediments and fossils from the Deep Sea Drilling Project cores helped verify the theory of plate tectonics. It has also shed light on the evolution of life forms and helped researchers decipher the history of changes in Earth's climate over the last 100,000 years. A newer and larger ship, *JOIDES Resolution* (**Figure**

5.17), is carrying this work forward as part of the Ocean Drilling Program, an international research consortium (**Box 5.2**).

Powerful new continuous seismic profilers have also been used to determine the thickness and structure of layers of sediment on the continental shelf and slope and to assist in the search for oil and natural gas (see **Figure 5.18**). Recent improvements in computerized image processing of the echoes returning from the seabed now permit detailed analysis of these deeper layers.

SEDIMENTS AS HISTORICAL RECORDS

In 1899 the British geologist W. J. Sollas theorized that deep-sea deposits could reveal much of the planet's history. In the era before plate tectonics theory this certainly seemed reasonable—the deep-ocean bottom was thought to be a calm,

a

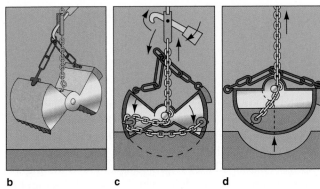

b   c   d

Figure 5.14 (**a**) A clamshell sampler (**b**) before sampling, (**c**) during sampling, and (**d**) after the sample has been taken. Note that the sample is relatively undisturbed. (Source for b–d: From *Exploration of the Oceans* by John G. Weihaupt. Copyright © 1979 John G. Weihaupt.)

a

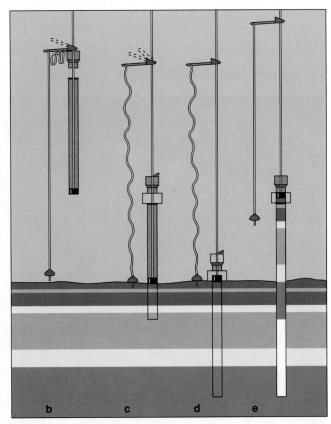

b   c   d   e

Figure 5.15 (**a**) A piston corer. (**b**) The corer is allowed to fall to the bottom. (**c**) The corer reaches the bottom and continues, forcing a sample partway into the cylinder. (**d**) Tension on the cable draws a small piston within the corer toward the top of the cylinder, and the pressure of the surrounding water forces the corer deeper into the sediment. (**e**) The corer and sample being hauled in.

The Ocean Drilling Program (ODP), currently the largest and most successful multinational Earth science research project, is the direct successor to the Deep Sea Drilling Project (DSDP), which began in 1968. The two drilling ships commissioned for the projects pioneered drilling technology to retrieve cores, long cylinders of sediment and rock extracted from beneath the seafloor. The cores arrive on the deck of *JOIDES Resolution*—the ship currently in active service—in 9.5-meter (30-foot) sections encased in plastic sheaths. Immediately after retrieval, a core is marked to indicate its original location on the seafloor, coded to distinguish top from bottom, measured, and cut into sections for study and storage. Each section is slit lengthwise; one half is stored for the ODP archives, and the other is taken to the first of many shipboard laboratories for study.

Paleontologists then examine the sediments at the bottom of the core to determine the age of the oldest material. Other researchers measure its density, strength, molecular composition, radioactivity, and ability to conduct heat. Magnetic specialists read paleomagnetic data from the core to determine the ages of the rock fragments and the latitude at which they probably formed. Sensors are also lowered into the hole from which the core was removed, to gather additional information on the physical and chemical properties of the site.

ODP scientists have recently recovered fragments of the oldest remaining seafloor. The sample is about 175 million years old, a relic of the Middle Jurassic period, when the continents were massed in one huge cluster. From this and other cores they have learned about cycles of global climate change, information that will be useful in evaluating the present potential for global warming. They have also discovered how fluids move through the lithosphere, found evidence of an ice-free Antarctica, and noted the influence of plate tectonics on worldwide weather and current patterns. From trapped pollen grains, they have been able to tell what land plants were thriving on Earth at the time the core sediments were laid down. As analysis technology evolves, stored cores are restudied and new information obtained.

The size and shape of a typical deep-core section may be seen in **Figure a.** This core was taken 250 meters (820

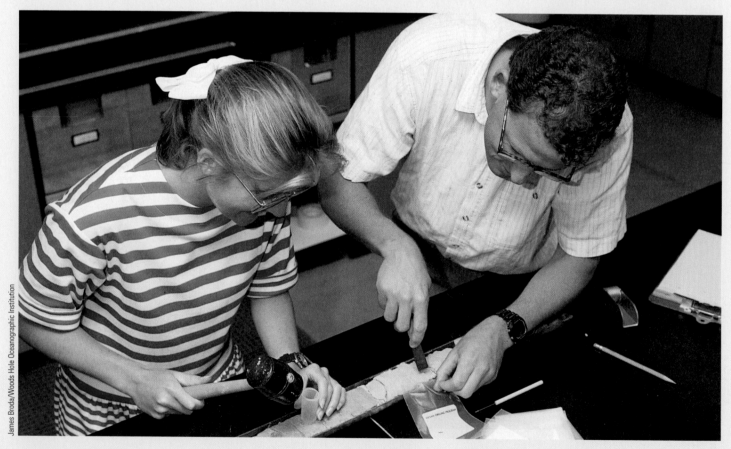

James Broda/Woods Hole Oceanographic Institution

a Researchers examine a deep core taken off the northwestern coast of Australia.

feet) beneath the seafloor off the northwestern coast of Australia by *JOIDES Resolution* on 22 July 1988. The ocean in the area was around 2,000 meters (6,500 feet) deep.

The material in the core progresses from sandstone (near the bottom of the core, at the right side of the photo) to fine claystone (near the top). Foraminifera and microfossils are abundant in the sandstone, and small squidlike fossils, easily visible to the unaided eye, are present in the claystone. The sediments date from the Late Jurassic period, about 140 million years ago. The progression of slowly deposited sediments and fossils suggests that the seabed in this area has slowly subsided, possibly because of tectonic effects.

Details in a core from DSDP Hole 480, leg 64, are shown in **Figure b.** This core was obtained by the now-retired drilling ship *Glomar Challenger* on 1 January 1979 near Guaymas in the Gulf of California. The gulf is an area of active seafloor spreading (see Figure 3.24), and researchers were interested in sampling the bottom in this place where basaltic magma is intruding into soft, wet, young sediments. The sediments here are very deep and have been accumulating at the extremely rapid rate of about 1,200 meters (4,000 feet) per million years.

Glomar Challenger arrived in the area on 29 December 1978 and used a conventional core barrel to penetrate 444 meters (1,456 feet) into diatomaceous ooze and mudstones. Nearly 273 meters (900 feet) of core was recovered from the hole drilled before Hole 480. Because of the disturbance caused by the conventional coring process, researchers were unable to determine the details of fine structure tantalizingly seen in a few lengths of that core.

On 31 December, *Glomar Challenger* moved to a new location 7 kilometers (4.4 miles) northwest. Hole 480, source of the core segment pictured here, was begun later that day. Unlike the previous hole, Hole 480 was drilled using a newly designed hydraulic corer developed by three DSDP engineers. The new device allowed 80% recovery of a core containing essentially undisturbed laminated diatomaceous ooze and muds.

As can be seen in Figure b, the core consists of thin alternating brown and gray bands of sediments. These are believed to be annual couplets, formed about 2 million years ago in response to two seasonal events that still occur in the gulf: the winter rains that introduce terrigenous clays into the region, and diatom blooms produced by seasonal upwelling and northwest winds. Preservation of these fine details depends upon a low level of free oxygen at the seafloor. Burrowing animals would normally churn these fine layers into mush, but in a low-oxygen environment these animals are absent, so the thin alternating sheets of clay and diatom tests (shells) are preserved.

Note that parts of the core have already been removed for study. Trenches across the width of the core mark places where particular sets of layers have been removed for isotope analysis. A larger sample for paleomagnetic study was taken from the square depression in the sample. Considering the vast expense and skill necessary to retrieve and analyze them, these small, gray, gritty bits of sediment are probably more costly than their weight in fine diamonds!

5-22

James Broda/Woods Hole Oceanographic Institution

b This core from the Gulf of California shows thin alternating bands of clay and ooze. Voids in the core show where samples were removed for study.

Figure 5.16 Sediment cores in storage. Cores are sectioned longitudinally, placed in trays, and stored in hermetically sealed cold rooms. The Gulf Coast Repository of the Ocean Drilling Program, located at Texas A&M University (pictured here), stores about 75,000 sections taken from more than 80 kilometers (50 miles) of cores recovered from the Pacific and Indian Oceans. Smaller core libraries are maintained at the Scripps Institution in California (Pacific and Indian Oceans) and at the Lamont–Doherty Earth Observatory in New York State (Atlantic Ocean).

Figure 5.17 *JOIDES Resolution,* the deep-sea drilling ship operated by the Joint Oceanographic Institutions for Deep Earth Sampling. The vessel is 124 meters (407 feet) long, with a displacement of over 16,000 tons. The rig can drill to a depth of 9,150 meters (30,000 feet) below sea level.

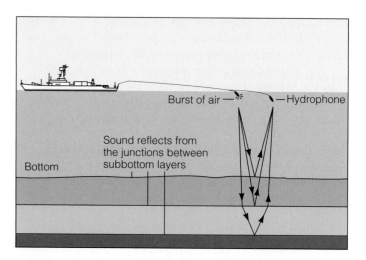

Figure 5.18 A typical method of continuous seismic profiling. A moving ship trails a sound transmitter and receiver at a distance sufficient to minimize interference from the ship's noise. Bubbles and turbulence from a burst of compressed air act as a sound source. The sound is reflected from sediment layers beneath the surface and is detected by a sensitive hydrophone for analysis.

changeless place where an unbroken accumulation of sediment could be probed to discover the entire history of the ocean. Unfortunately for this promising idea, difficulties began to crop up almost immediately. For one thing, the sediments should have been much *thicker* than early probes indicated. If Earth's ocean is truly older than a few hundred thousand years, and if life has existed within it for most of that time, the sediment layer should be thicker than was observed. Another difficulty lay in the uneven distribution of sediments. Sollas thought that the center of an ocean basin should contain the thickest layers of sediment, yet the ridged mid-Atlantic bottom was nearly naked. There didn't seem to be any difference in the nature of the overlying seawater that could account for the variations in thickness and composition of the sediments across the bottom of the Atlantic. Oozes were especially puzzling: The organisms that form ooze grow well at the surface of the mid-Atlantic, yet the mid-Atlantic floor seemed to bear little ooze.

Today we know the tectonic reasons for these discrepancies, but turn-of-the-century geologists were understandably confused. Because the deep-sea sediment record is ultimately destroyed in the subduction process, the ocean's sedimentary "memory" does not start with the ocean's formation as originally reasoned by early marine scientists. But modern studies of deep-sea sediments using seafloor samples, cores obtained by deep drilling, and continuous seismic profiling have demonstrated that these deposits contain a remarkable record of relatively recent ocean history. However, these same data have also shown that the record is not uninterrupted as originally assumed by early workers. In fact, some of the gaps in the deep-sea deposits represent erosional events and constitute evidence of changes in deep-sea circulation, and hence are valuable in their own right. The analysis of layered sedimentary

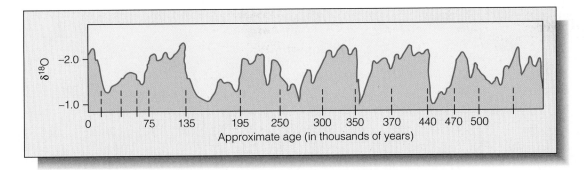

Figure 5.19 Interpreting changes in climate from analysis of a deep-sea sediment core. Scientists know that oxygen atoms can have different numbers of neutrons in their nuclei—they are known as *isotopes* of oxygen. The ratios of two stable (nonradioactive) oxygen isotopes found in biogenous sediments, ^{16}O and ^{18}O, can be used to determine the relative extent of water frozen into ice sheets, thus providing a chronology of periods of extensive glaciation, or ice ages.

Although the two isotopes of oxygen have identical chemical properties, the lighter (and much more common) ^{16}O has a higher vapor pressure. During evaporation, higher concentrations of this lighter isotope leave the ocean in the form of water vapor, while more of the heavier ^{18}O isotope remains behind. The water vapor precipitates as rain and snow. Ice sheets lock up some of the ^{16}O-enriched fresh water, preventing its recirculation to the ocean. Marine organisms (such as foraminifera) then incorporate the heavier isotopes into their bodies in greater proportion. When they die, their skeletons fall to the seabed as biogenous sediment. When the glaciers melt and the fresh water returns to the ocean, the original ratio of ^{16}O to ^{18}O is restored, and foraminifera will build skeletons with a different oxygen isotope ratio. Researchers use sensitive mass spectrometers to determine the ratio of ^{16}O to ^{18}O in foraminifera and infer sea-level height from the data.

In the curve seen here, the $\delta^{18}O$ notation defines the deviation in parts per thousand (o/oo) of the isotopic ratio of the samples analyzed from that of an arbitrary standard. Ratios around -1.0 o/oo indicate times of relatively high oceanic ^{18}O concentrations, and thus of extensive glaciation. Ratios greater than -2.0 o/oo—which we see today—suggest low oceanic ^{18}O concentrations and warmer conditions. Compare this curve to that of Figure 4.14. (*Sources:* Shackleton and Opdyke, 1973; Kennett, 1982; Carew and Mylroie, 1995.)

deposits in the ocean (or on land) represents the discipline of **stratigraphy** (*stratum* = layer, *graph* = a drawing). Deep-sea stratigraphy utilizes variations in the composition of rocks, microfossils, depositional patterns, geochemical character, and physical character (density and such) to trace or correlate distinctive sedimentary layers from place to place, establish the age of the deposits, and interpret changes in ocean and atmospheric circulation, productivity, and other aspects of past ocean behavior. In turn, these sorts of studies and the advent of deep-sea drilling have given rise to the emerging science of **paleoceanography** (*palaios* = ancient), the study of the ocean's past.

Early attempts to interpret ocean and climate history from evidence in deep-sea sediments occurred in the 1930s through 1950s as cores became available. These initial studies relied primarily on down-core variations in the abundance and distribution of glacial marine sediments, carbonate and siliceous oozes, and temperature-sensitive microfossils. Modern paleoceanographic studies continue to utilize these same features but with much greater understanding of their significance and the aid of seismic imaging of the deposits over large areas. In addition, newer and more precise methods of dating deep-sea sediments have allowed events to be placed in a proper time context. Finally, the appearance of instru-

ments capable of analyzing very small variations in the relative abundances of the stable isotopes of oxygen preserved within the carbonate shells of microfossils found in deep-sea sediments has allowed scientists to interpret changes in the temperature of surface and deep water over time (**Figure 5.19**). These same data are also used to estimate variations in the volume of ice stored in continental ice sheets. Other geochemical evidence contained in the shells of marine microfossils, including variations in carbon isotopes and trace metals such as cadmium, provides insights into ancient patterns of ocean circulation, productivity of the marine biosphere, and ancient upwelling. These sorts of data have already provided quantitative records of the glacial-interglacial climatic cycles of the past 2 million years. Future drilling and analysis of deep-sea sediments are poised to extend our paleoceanographic perspective much farther back in time.

Figure 5.20 shows the age of the Pacific Ocean floor using data obtained largely from analyses of the overlying sediment. Note that sediments get older with increasing distance from the East Pacific Rise spreading center.

Global analysis of marine sediments in the modern basins can shed light on unexpected details of the last 180 million years of Earth's history. One of the most significant events is the

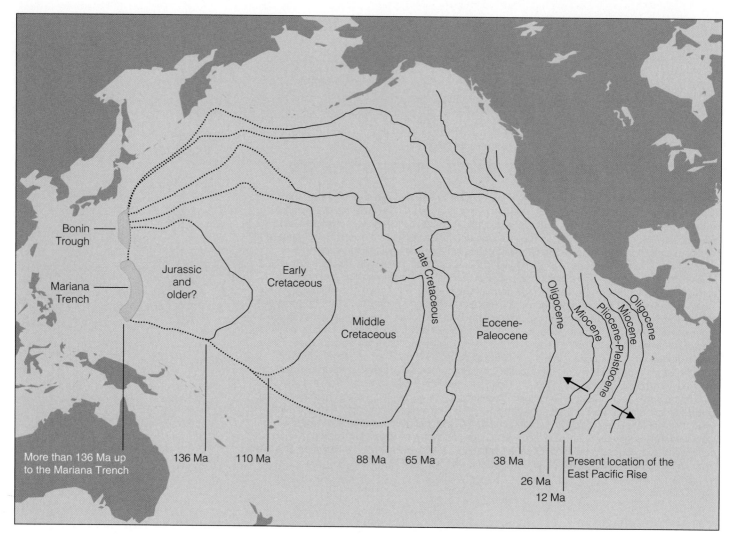

Figure 5.20 The age of portions of the Pacific Ocean floor, based on core samples of sediments just above the basalt seabed, in millions of years ago (Ma = mega-annums). The youngest sediments are found near the East Pacific Rise, and the oldest are found close to the eastern side of the troughs and trenches. Contrast this figure with Figure 3.27. (Source: A. G. Fischer, *Geological History of the Western North Pacific,* vol. 168, p. 1210, June 5, 1970. © 1970 AAAS. Reprinted by permission.)

unexplained extinction of up to 52% of known marine animal species (and the dinosaurs) at the end of the Cretaceous period 65 million years ago. Many researchers now believe that a sudden and violent impact of one or more asteroids or comets caused this catastrophe. The clouds of dust and ash thrown into the atmosphere by any of these events would have drastically reduced incident sunlight and greatly affected the lives of organisms and the photosynthetic base of ecosystems. Oceanographers are currently searching for evidence of the cause of the Cretaceous extinctions in layers of deep sediment. (For more on this fascinating topic, please see Box 13.1, page 327.)

Earth might not be the only planet where marine sediments have left historical records. As you read in Chapter 1, Mars probably had an ocean between 3.2 and 1.2 billion years

ago. In May 1998 *Mars Global Surveyor* photographed sediments that look suspiciously marine near the edge of an ancient bay (**Figure 5.21**). One can only wonder what stories they will tell.

THE ECONOMIC IMPORTANCE OF MARINE SEDIMENTS

Study of sediments has brought practical benefits. By 2005 an estimated 36% of the world's crude oil and 28% of its natural gas will be extracted from the sedimentary deposits of continental shelves and continental rises. Offshore hydrocarbons currently

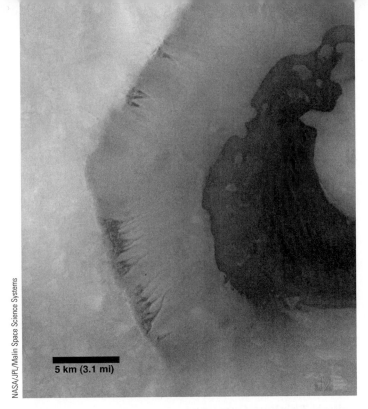

NASA/JPL/Malin Space Science Systems

Figure 5.21 Marine sediments on Mars? Dark windblown dunes of sandy sediments rest on the floor of an ancient bay in the southern Martian lowlands. At some time in the past, water seeped out of layers within the cliffs (to the left) and flowed downhill, flooding part of the basin. The zones of contrast between the dark floor material and the lighter material of the slopes suggest the formation of bays and peninsulas. The appearance of dunes within the crater may be coincidental, or the sand may have been generated by wind and wave action. The lack of superimposed fresh impact craters suggests this process may have been active relatively recently. (Photograph by *Mars Global Surveyor,* May 1998.)

generate annual revenues in excess of $125 billion. Deposits within the sediments of continental margins account for about one-third of the world's estimated oil and gas reserves.

In addition to oil and gas, in 1998 sand and gravel valued at more than $480 million were taken from the ocean. This is about 1% of world needs. Commercial mining of manganese nodules has also been considered. In addition to manganese, these nodules contain substantial amounts of iron and other industrially important chemical elements. The high iron content of these nodules has prompted a proposal to rename them *ferromanganese* nodules. We will investigate these resources in more detail in Chapter 17.

1. The question of sediment age seems to occupy much of sedimentologists' time. Why?

The dating of sediments has been a central problem in marine science for many years. In 1957, during the International Geophysical Year, sedimentologists designed a coordinated effort to determine sediment age, which included plans for the *Glomar Explorer* and *Glomar Challenger* drilling surveys. Their primary interest was to seek evidence of the hypothesis of the then-new idea of seafloor spreading. Cores returned by the Deep Sea Drilling Project in 1968 enabled researchers including J. Tuzo Wilson, Harry Hess, and Maurice Ewing to put the evidence together. Much of the proof for plate tectonics rests on the interpretation of sediment cores.

2. Where are sediments thickest?

Sediment is thickest close to eroding land and beneath biologically productive neritic waters, and thinnest over the fast-spreading oceanic ridges of the eastern South Pacific. The thickest accumulation of sediments may be found along and beneath the continental margins (especially on continental rises). Some are typically more than 1,500 meters (5,000 feet) thick. Remember, much of the rocky material of the Grand Canyon was once marine sediment atop an ancient seabed. The Grand Canyon is nearly 2 kilometers (1¼ miles) deep, and the uppermost layer of sedimentary rock has already been eroded completely away!

3. What's the relationship between deep-sea animals and the sediments on which they live?

Though microscopic bacteria and benthic foraminifera may be very abundant on the seabed, visible life is not abundant on the bottom of the deep ocean. There are no plants at great depths because there is no light, but animals do live there. Some, like the brittle stars in Figure 5.2, move slowly along the surface searching for bits of organic matter to eat. Others burrow through the muck in search of food particles. Worms eat quantities of sediment to extract any nutrients that may be present and then deposit strings of fecal material as they move forward. The deeps are uninviting places, but life is tenacious and survives even in this hostile environment.

- Sediment is particles of organic or inorganic matter that accumulate in a loose, unconsolidated form. Sediment may be classified by grain size or by the origin of the majority of its particles.

- Sediment deposited on a quiet seabed can provide a sequential record of events in the water column above. Sediments can act only as the *recent* memory of the ocean—marine sediments are not of great age because they are recycled into Earth at subduction zones.

- The most abundant marine sediments are broadly classified as terrigenous and biogenous. Terrigenous sediments are of geological origin and arise on the continents or on islands near them. They are the more abundant. Biogenous sediments are of biological origin and may be calcareous or siliceous. Hydrogenous sediments are formed directly from seawater. Of less importance are cosmogenous sediments that have come from space.

- Though there are exceptions, the sediments of continental margins tend to be mostly terrigenous, while the generally finer sediments of the deep-ocean floor contain a larger proportion of biogenous material.

- Deep-sea oozes—forms of biogenous sediment—contain the remains of some of the ocean's most abundant and important organisms.

- Sediments are economically important. By 2005 an estimated 36% of the world's crude oil will be extracted from marine sedimentary deposits.

INTERNET STUDY RESOURCES

The Web site for this book contains helpful study aids. Log on to:

www.brookscole.com/product/053437557xs

and click on the Chapter-by-Chapter area. Choose Chapter 5 and select a resource:

- **Flash Cards** allows you to test your mastery of the Terms and Concepts to Remember for this chapter.
- **Tutorial Quizzes** provides a multiple-choice practice quiz.
- **Student Guide to InfoTrac College Edition** will lead you to Critical Thinking Projects that use InfoTrac College Edition as a research tool.

TERMS AND CONCEPTS TO REMEMBER

authigenic sediment	lithification
bathybius	microtektite
biogenous sediment	mineral
calcareous ooze	neritic sediment
calcium carbonate compensation depth (CCD)	nodule
	oolite sands
clamshell sampler	ooze
clay	paleoceanography
coccolithophore	pelagic sediment
cosmogenous sediment	piston corer
diatom	poorly sorted sediments
evaporite	pteropod
foraminiferan	radiolarian
hydrogenous sediment	sand

sediment	terrigenous sediment
siliceous ooze	turbidite
silt	well-sorted sediments
stratigraphy	

STUDY QUESTIONS

Review Questions

1. In what ways are sediments classified?

2. List the four types of marine sediments. Explain the origin of each.

3. How are neritic sediments generally different from pelagic ones?

4. Where are sediments thickest? Are there any areas of the ocean floor free of sediments?

5. Are sediments commercially important? In what ways?

Critical Thinking Questions

1. Is the thickness of ooze always an accurate indication of the biological productivity of surface water in a given area? (Hint: See next question.)

2. What is the calcium carbonate compensation depth? Is there a compensation depth for the siliceous components of once-living things?

3. What sediments accumulate most rapidly? What sediments accumulate least rapidly?

4. Can marine sediments tell us about the history of the ocean from the time of its origin? Why or why not?

5. How do paleoceanographers infer water temperatures, and therefore terrestrial climate, from sediment samples?

6. **InfoTrac College Edition Project** The continental slopes and ocean bottom are thought to contain oil and gas reserves, as well as other mineral reserves. But many oppose extraction of these minerals because of the dangers of contamination. Can these reserves be tapped safely? Would you support or oppose these efforts? Why? Use information from articles found on InfoTrac College Edition to support your answers.

Earth Systems Today CD Questions

1. Go to "Rocks and the Rock Cycle," then "Rock Cycle." How do sediments fit into the rock cycle? From the animation, can you tell the origin and fate of sediments?

Water and Ocean Structure

6

ICEBERGS

Water occurs in three states: liquid, solid, and gas. Oceanographers are most familiar with water's liquid form, but about 6% of the world ocean is covered by ice. A small fraction of this ice is contained in the fantastic shapes of icebergs.

Icebergs in the southern ocean originate as huge ice sheets attached to the Antarctic continent; lengths of up to 8 kilometers (5 miles) are not unusual, with flat tops rising 45 meters (150 feet) above sea level. In 1927 a frozen section of 26,000 square kilometers (10,000 square miles)—about eight times the size of Rhode Island—broke from the Antarctic shore and

Penguins hitch a ride on an iceberg off the Antarctic coast. About 6% of the world ocean is covered by ice.

floated north along the coast of Argentina! The giant iceberg carried an overheated load of Adélie penguins. Favorable currents on the west coast of South America have driven large icebergs (and itinerant penguins) as far north as the tropics.

Pinnacled icebergs, more characteristic of the northern polar ocean, usually begin life as glaciers in western Greenland. These slowly flowing ice rivers form at the edge of the great Greenland ice sheet, squeeze between the sharp-crested ridges of the mountains, and eventually reach the sea. Rising and falling tides at the end of the deep fjords break off huge tongues of ice, which drift from shore as icebergs. The great weight of ice in a glacier compacts these glistening masses and makes them very dense; only about one-seventh of their bulk floats above sea level.

The jagged underwater edge of a pinnacled iceberg broke the hull of the "unsinkable" White Star liner RMS *Titanic* on the night of 14 April 1912; 1,517 people lost their lives. The iceberg had reached the Atlantic shipping lanes after separating from its glacier, being swept down the west coast of Greenland, and then moving south toward the Gulf Stream and North Atlantic Current.

Despite the danger, icebergs are often extraordinarily beautiful:

> On the afternoon of the day we knew the storm had passed, I stood on the starboard side of the bridge at a window, with the heavy protective glass lowered. I rested my forearms on the sill, feeling the warmth of the bridge heaters around my legs and a slipstream of cool air past my face. The first icebergs we had seen, just north of the Strait of Belle Isle, listing and guttered by the ocean, seemed immensely sad, exhausted by some unknown calamity. We sailed past them.
>
> I occasionally drew back from the starboard window to make a sketch, or to bring the binoculars up to my eyes. I marveled as much at the behavior of light around the icebergs as I did at their austere, implacable progress through the water. They took their color from the sun, and from the clouds and the water. But they also took

their dimensions from the light: the stronger and more direct it was, the greater the contrast upon the surface of the ice, of the ice itself with the sea. And the more finely etched were the dull surfaces of their walls. The bluer the sky, the brighter their outline against it. . . .

Where the walls entered the water, the surf pounded them, creating caverns, grottoes, and ice bridges, strengthening an impression of sea cliffs. At the waterline the ice gleamed aquamarine against its own gray-white walls above. Where meltwater had filled cracks or made ponds, the pools and veins were milk-blue, or shaded to brighter marine blues, depending on the thickness of the ice. If the iceberg had recently fractured, its new face glistened greenish blue—the greens in the older, weathered faces were grayer. In twilight the ice took on the colors of the sun: rose, reddish yellows, watered purples, soft pinks. The ice both reflected the light and trapped it within its crystalline corners and edges, where it intensified. . . . How utterly still, unorthodox, and wondrous they seem.

6-1

Source: Barry Lopez, *Arctic Dreams*. New York: Scribner's, 1986.

WHAT TO WATCH FOR IN CHAPTER 6

Water is so familiar and abundant that we don't always appreciate its unusual characteristics. If liquid methane, or alcohol, or any other flowing substance dominated the surface of Earth, physical conditions here would be strikingly different. Life as we know it would be impossible.

Watch for the characteristics that make water unusual—the molecule's polarity and the bonds that hold it together, the large amount of heat needed to change its temperature, and the heat needed to change its physical state. And, no, heat and temperature are not the same thing.

Two big lessons follow. One is the influence of water on global temperatures. Water has an important thermostatic balancing effect. On an oceanless Earth it would be much colder in winter and much hotter in summer. The other lesson is the influence of density on ocean structure. You'll see that the ocean's structure and large-scale movement depend on changes in density, with density dependent on temperature and salinity.

The chapter ends with an explanation of light and sound in the ocean. Why is the ocean blue? Why do noises sound different—and travel farther—in water? Here are the answers.

In Chapter 7, we'll continue our examination by looking at the many dissolved substances in seawater, and we'll see how those substances alter seawater's physical properties.

ater is a compound. **Compounds** are substances that contain two or more different elements in a fixed proportion. An **element** is a substance composed of identical particles, called **atoms,** that cannot be broken into simpler substances by chemical means. Water's familiar chemical formula, H_2O, shows that two atoms of hydrogen (H) are present for each atom of oxygen (O). Some other common compounds are carbon dioxide (CO_2), rust (Fe_2O_3)[1], and simple sugar ($C_6H_{12}O_6$).

THE WATER MOLECULE

A **molecule** is a group of atoms held together by chemical bonds. **Chemical bonds,** the energy relationships between atoms that hold them together, are formed when **electrons**—tiny negatively charged particles found toward the outside of an atom—are shared between atoms or moved from one atom to another. A water molecule forms when electrons are shared between two hydrogen atoms and one oxygen atom. The bonds formed by shared pairs of electrons are known as **covalent bonds.** Covalent bonds hold together many familiar molecules, including CO_2, CH_4 (methane gas), and O_2 (atmospheric oxygen). Because of the way a water molecule's oxygen electrons are distributed, the overall geometry of the molecule is a bent or angular shape. The angle formed by the two hydrogen atoms and the central oxygen atom is about 105°. The formation of a water molecule is depicted in **Figure 6.1.**

---

[1] Rust is iron oxide. Fe, the chemical symbol for iron, is derived from *ferrum,* the Latin name for iron.

The angular shape of the water molecule makes it electrically asymmetrical, or a **polar molecule**. Each water molecule can be thought of as having a positive (+) end and a negative (−) end. Positively charged particles at the center of the hydrogen atoms—called **protons**—are left partially exposed when the negatively charged electrons bond more closely to oxygen. The polar water molecule acts something like a magnet; its positive end attracts particles having a negative charge, and its negative end attracts particles having a positive charge. When water comes into contact with compounds whose elements are held together by the attraction of opposite electrical charges (most salts, for example), the polar water molecule will separate that compound's component elements from each other. This explains why water can dissolve so many other compounds so easily.

The polar nature of water also permits it to attract other water molecules. When a hydrogen atom (the positive end) in one water molecule is attracted to the oxygen atom (the negative end) of an adjacent water molecule, a **hydrogen bond** forms. The water molecules are bonded together by electrostatic forces. The resulting loosely held webwork of water molecules is shown in **Figure 6.2.** Hydrogen bonds greatly influence the properties of water by enabling individual water molecules to stick to each other, a property called **cohesion.** Cohesion gives water an unusually high surface tension, which results in a surface "skin" capable of supporting needles, razor blades, and even walking insects. **Adhesion,** the tendency of water to stick to other materials, allows water to adhere to solids—that is, to make them wet. Cohesion and adhesion are the causes of capillary action, the tendency of water to spread through a towel when one corner is dipped in water.

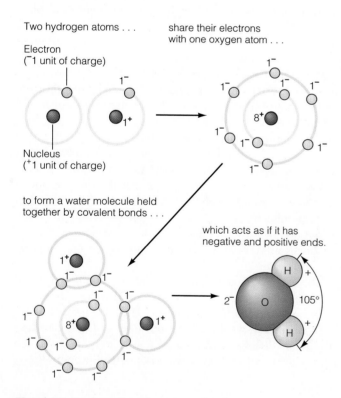

Figure 6.1 The formation of a water molecule.

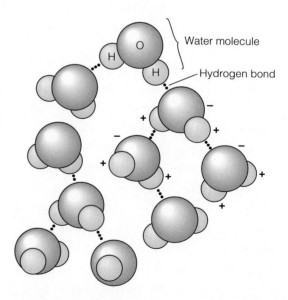

Figure 6.2 Hydrogen bonds in liquid water. The attractions between adjacent polar water molecules form a webwork of hydrogen bonds. These bonds are responsible for cohesion and adhesion, the properties of water that cause surface tension and wetting. Hydrogen bonds among water molecules also make it difficult for individual molecules to escape from the surface.

Hydrogen bonds are also what gives pure water its pale blue hue. When water molecules vibrate, adjacent molecules tug and push against their hydrogen-bonded neighbors. This action absorbs a small amount of red light, leaving proportionally more blue light to scatter back to our eyes. The same blue color is seen in ice formations.

If hydrogen bonds did not hold water molecules together, water would fly apart to form a gas. Hydrogen sulfide (H_2S) is chemically similar to water but lacks water's ability to form networks of hydrogen bonds. H_2S is therefore a gas rather than a liquid at normal temperatures and pressures.

WATER AND HEAT

Perhaps the most important physical properties of water are related to its behavior as it absorbs or loses heat. Water's unusual thermal characteristics prevent wide temperature variation from day to night and from winter to summer, permit vast amounts of heat to flow from equatorial to polar regions, and power Earth's great storms, wind waves, and ocean currents.

Heat and Temperature

Heat and temperature are not the same thing. **Heat** is energy produced by the random vibration of atoms or molecules. On the average, water molecules in hot water vibrate more rapidly than water molecules in cold water. *Heat* is a measure of how many molecules are vibrating and how rapidly they are vibrating. *Temperature* records only how rapidly the molecules of a substance are vibrating. **Temperature** is an object's response to an input (or removal) of heat. The amount of heat required to bring a substance to a certain temperature varies with the nature of that substance.

An example will help. Which has a higher temperature— a candle flame or a bathtub of hot water? The flame. Which contains more heat? The tub. The molecules in the flame vibrate very rapidly, but there are relatively few of them. The molecules of water in the tub vibrate more slowly, but there are a great many of them, so the total amount of heat energy in the tub is greater.

Temperature is measured in **degrees.** One degree Celsius (°C) = 1.8 degrees Fahrenheit (°F). Though we are more familiar with the older Fahrenheit scale, Celsius degrees are more useful in science because they are based on two of pure water's most significant properties: its freezing point (0°C) and its boiling point (100°C).

Heat Capacity

Heat capacity is a measure of the heat required to raise the temperature of 1 gram (0.035 ounce) of a substance by 1°C (1.8°F). Different substances have different heat capacities. *Not all substances respond to identical inputs of heat by rising in temperature the same number of degrees* (see **Table 6.1**). Heat capacity is measured in calories per gram. A **calo-**

rie is the amount of heat required to raise the temperature of 1 gram (0.035 ounce) of pure water by 1°C.[2]

Because of the great strength and large number of the hydrogen bonds between water molecules, more heat energy must be added to speed up molecular movement and raise water's temperature than would be necessary in a substance held together by weaker bonds. Water's heat capacity is therefore among the highest of all known substances. This means that *water can absorb (or release) large amounts of heat while changing relatively little in temperature.*

Anyone who waits by a stove for water to boil knows much about water's heat capacity—it seems to take a very long time to warm water for soup or coffee. By contrast, ethyl alcohol has a much lower heat capacity. If both liquids absorb heat from identical stove burners at the same rate, pure ethyl alcohol (the active ingredient in alcoholic beverages) will rise in temperature about three times as fast as an equal mass of water. Beach sand has an even lower heat capacity: A gram of sand requires as little as 0.2 calorie to rise 1°C (1.8°F). So on sunny days beaches can get too hot to stand on with bare feet, while the water remains pleasantly cool.

As we will soon see, the concept of heat capacity is very important in oceanography. But for now, remember this: Water has an extraordinarily high heat capacity—it resists changing *temperature* when *heat* is added or removed.

Water Temperature and Density

The uniqueness of water becomes even more apparent when we consider the effect of a temperature change on water's *density* (its mass per unit of volume). You may recall from Chapter 3 that

---

[2] A nutritional calorie, the unit we see on cereal boxes, equals 1,000 of these calories. A gram is about ten drops of seawater.

Table 6.1	Heat Capacity of Common Substances
Substance	**Heat Capacity**[a] **in calories/gram/°C**
Silver	0.06
Granite	0.20
Aluminum	0.22
Alcohol (ethyl)	0.30
Gasoline	0.50
Acetone	0.51
Pure water	**1.00**
Ammonia (liquid)	1.13

[a] Heat capacity is a measure of the heat required to raise the temperature of 1 gram (0.035 ounce) of a substance by 1°C (1.8°F). Different substances have different heat capacities. *Not all substances respond to identical inputs of heat by rising in temperature the same number of degrees.* Notice how little heat is required to raise the temperature of 1 gram of silver 1 degree.

Because of the great strength and large number of the hydrogen bonds between water molecules, water can gain or lose large amounts of *heat* with very little change in *temperature.* This *thermal inertia* moderates temperatures worldwide. Of all common substances, only liquid ammonia has a higher heat capacity than liquid water.

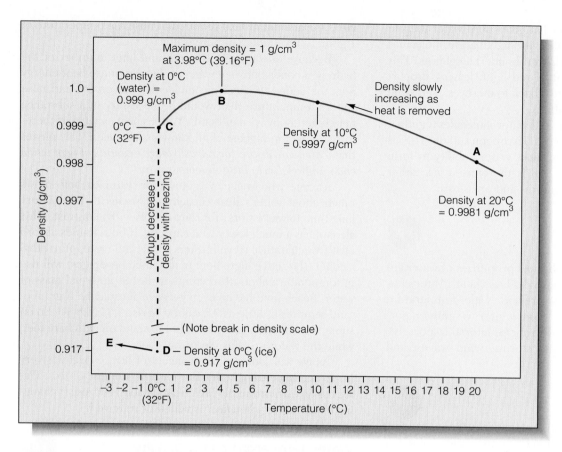

Figure 6.3 The relationship of density to temperature for pure water. Note that points C and D both represent 0°C (32°F) but different densities, and thus different states of water. Ice floats because the density of ice is less than the density of liquid water.

the density of pure water is 1 gram per cubic centimeter (1 g/cm³), and granite rock is heavier, with a density of about 2.7 g/cm³. Air is lighter, with a density of about 0.0012 g/cm³. Most substances become denser (weigh more per unit of volume) as they get colder. Pure water generally becomes denser as heat is removed and temperature falls, but water's density behaves unexpectedly as the temperature approaches the freezing point.

A **density curve** shows the relationship between the temperature or salinity of a substance and its density. Most substances become progressively denser as they cool—their temperature–density relationships are linear (that is, appear as a straight line on graphs). But **Figure 6.3** shows the unusual temperature–density relationship of pure water. Imagine heat being removed from some water placed in a freezer. Initially, the water is at room temperature, point A on the graph. As expected, the density of water increases as its temperature drops along the line from point A toward point B. Approaching point B the density increase slows, reaching a maximum at point B of 1 g/cm³ at 3.98°C (39.2°F). As cooling continues, the water begins to set up a more rigid framework of hydrogen bonds, causing the liquid to expand slightly. Water thus becomes slightly less dense as cooling continues, until point C (0°C, 32°F) is reached. At point C the water begins to freeze—to change state by crystallizing into ice.

State is an expression of the internal form of a substance (**Figure 6.4**). Changes in state are accompanied by either an input or an output of energy. Water exists on Earth in three physical states: liquid, gas (water vapor), and solid (ice). If the freezer continues to remove heat from the water at point C, the water will change from liquid to solid state. Through this transition from water to ice—from point C to point D—the density of the water *decreases* abruptly. Ice is therefore lighter than an equal volume of water. Ice increases in density as it gets colder than 0°C, but no matter how cold it gets, ice never reaches the density of liquid water. Being less dense than water, ice "freezes over" as a floating layer instead of "freezing under" like the solid forms of virtually all other liquids.

As we'll see in a moment, the implications of water's high heat capacity and the ability of ice to float are vital in maintaining Earth's moderate surface temperature. First, we'll look at the transition from point C to point D in Figure 6.3.

Freezing Water

During the transition from liquid to solid state at the **freezing point,** the bond angle between the oxygen and hydrogen atoms in water expands from about 105° to slightly more than 109°. This change allows the hydrogen bonds in ice to form a

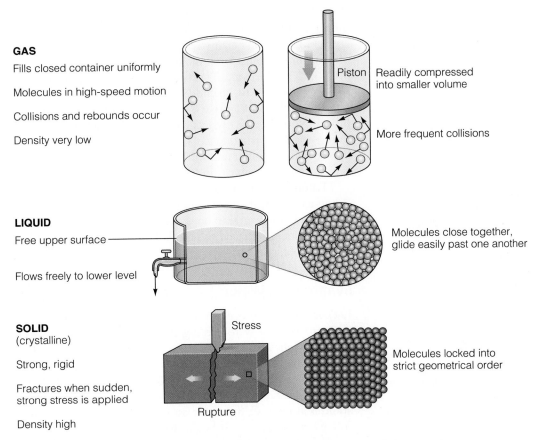

GAS

Fills closed container uniformly

Molecules in high-speed motion

Collisions and rebounds occur

Density very low

Piston — Readily compressed into smaller volume

More frequent collisions

LIQUID

Free upper surface

Flows freely to lower level

Molecules close together, glide easily past one another

SOLID
(crystalline)

Strong, rigid

Fractures when sudden, strong stress is applied

Density high

Stress

Rupture

Molecules locked into strict geometrical order

Figure 6.4 The three common states of matter—solid, liquid, and gas—apply to pure substances (elements and compounds) and to mixtures. A *gas* is a substance that can expand to fill any empty container. Atoms or molecules of gas are in high-speed motion and move in random directions. A *liquid* is a substance that flows freely in response to unbalanced forces but has a free upper surface in a container it does not fill. Atoms or molecules of a liquid move freely past one another as individuals or small groups, and liquids compress only slightly under pressure. Gases and liquids are classed as *fluids* because both substances flow easily. A *solid* is a substance that resists changes of shape or volume. A solid can typically withstand stresses without yielding permanently. When a solid does break, it usually breaks suddenly. On Earth, water can occur in all three states: gas, liquid, and solid. (Source: From Strahler, A. N. *Physical Geology,* published by Addison Wesley Longman, Inc. © 1981 by Arthur N. Strahler. Reprinted by permission of Pearson Education, Inc.)

crystal lattice (see **Figure 6.5**). The space taken by 27 water molecules in the liquid state could be occupied by only 24 water molecules in the solid lattice, so water has to expand about 9% as the crystal forms. Because the molecules are packed less efficiently, ice is less dense than liquid water—and so it floats. A cubic centimeter of ice at 0°C (32°F) has a mass of only 0.917 gram, but a cubic centimeter of liquid water at 0°C has a mass of 0.999 gram.

The ice lattice is not a flat sheet but a three-dimensional network. The water molecules in the ice lattice are still linked by hydrogen bonds, but they now form regular hexagons. This hexagonal pattern explains the lovely six-sided symmetry of snowflakes and other ice crystals.

The transition from liquid water to ice crystal (point C to point D in Figure 6.3) requires continued removal of heat energy—the change in state does not occur instantly throughout the mass when the cooling water reaches 0°C (32°F). Again, consider water in a freezer. **Figure 6.6,** a plot of heat re-

moval versus temperature, details the water's progress to ice. As in Figure 6.3, point A represents 20°C (68°F) water just placed into the freezer. The removal of heat does not stop when the water reaches point C, *but the decline in temperature stops.* Even though heat continues to be removed, the water will not get colder until all of it has changed state from liquid (water) to solid (ice). Heat may therefore be removed from water when it is changing state (that is, when it is freezing) without the water dropping in temperature. Indeed, the continued removal of heat is what makes the change in state possible. Heat is released as bonds form to make ice, and that heat must be removed to allow more ice to form.

The removal of heat from point A to point C in Figures 6.3 and 6.6 produces a *measurable* lowering of temperature detectable by a thermometer. Removing just 1 calorie of heat from 1 gram of liquid water causes its temperature to drop 1°C. This detectable decrease in heat is called **sensible heat** loss. But the loss of heat as water freezes between points

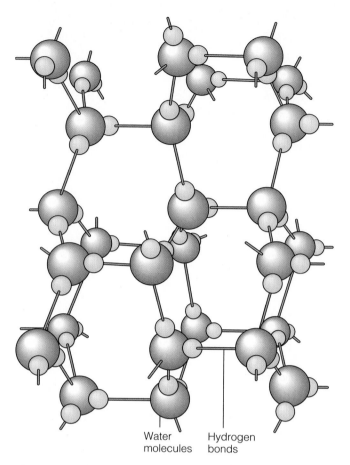

Figure 6.5 The lattice structure of an ice crystal, showing its hexagonal arrangement at the molecular level. The space taken by 24 water molecules in the solid lattice could be occupied by 27 water molecules in the liquid state, so water expands about 9% as the crystal forms. Because molecules of liquid water are packed less efficiently, ice is less dense than liquid water and will float.

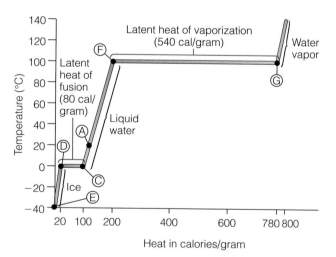

Figure 6.6 A graph of temperature versus heat as water freezes, melts, or vaporizes. The horizontal line between points C and D represents the latent heat of fusion, when heat is being added or removed but temperature is not changing. The horizontal line between points G and F represents the latent heat of vaporization, when heat is being added or removed but temperature is not changing. (Note that points A–E on this graph are the same as those in Figure 6.3.)

C and D is not measurable (that is, not sensible) by a thermometer. Removing 1 calorie of heat from freezing water at 0°C (32°F) won't change its temperature at all; 80 calories of heat energy must be removed per gram of pure water at 0°C (32°F) to form ice. This heat is called the **latent heat of fusion** (*latere* = to be hidden). The straight line between points C and D in Figure 6.6 represents water's latent heat of fusion. No more ice crystals can form when all the water in the freezer has turned to ice. If the removal of heat continues, the ice will get colder and will soon reach the temperature inside the freezer, point E in Figures 6.3 and 6.6.

Latent heat of fusion is also a factor during thawing. When ice melts, it *absorbs* large quantities of heat (the same 80 calories per gram) but does not change in temperature until all the ice has turned to liquid. This is why ice is so effective in cooling drinks.

Evaporating Water

Let's reverse the process now, and warm the ice. Imagine (as in Figure 6.6) the water resting at −40°C (−40°F), point E at the lower left. Add heat and the ice warms toward point D. It begins to melt. The horizontal line between point D and point C represents the latent heat of fusion—heat is absorbed, but temperature does not change as the ice melts. When all the liquid is at point C, the water warms past our original point A and arrives at point F. It begins to boil—it *vaporizes*.

When water vaporizes (or evaporates), individual water molecules diffuse into the air (**Figure 6.7**). Since each water molecule is hydrogen-bonded to adjacent molecules, heat energy is required to break those bonds and allow the molecule to fly away from the surface. Evaporation cools a moist surface because departing molecules of water vapor carry this energy away with them. (This is how perspiring cools us when we're hot. The heat energy required to evaporate water from our skin is taken away from our bodies, thus cooling us.)

Hydrogen bonds are numerous, and the amount of energy required to break them—known as the **latent heat of vaporization**—is very high. The long horizontal line between points F and G in Figure 6.6 represents the latent heat of vaporization. Even though more heat is applied, the water cannot get warmer until all of it has vaporized. At 540 calories per gram at 20°C (68°F), water has the highest latent heat of vaporization of any known substance. As before, the term *latent* applies to heat input that does not cause a temperature change but does produce a change of state—in this case from liquid to gas.

About 1 meter (3.3 feet) of water evaporates each year from the surface of the ocean, a volume of water equivalent to 334,000 cubic kilometers (80,000 cubic miles). The great quantities of solar energy that cause this evaporation are carried from the ocean by the escaping water vapor. When a gram of water vapor condenses back into liquid water, the same 540 calories are again available to do work. As we shall see, winds, storms, ocean currents, and wind waves are all powered by that heat.

Figure 6.7 Water vapor is invisible, but as water vapor evaporates from the sea surface and rises into cool air it can condense into the tiny droplets that form clouds and fog.

Why the big difference between water's latent heat of *fusion* (80 calories per gram) and its latent heat of *vaporization* (540 calories per gram)? Only a small percentage of hydrogen bonds is broken when ice melts, but *all* must be broken during evaporation. Breaking these bonds requires additional energy in proportion to their number.

A summary of water's unusual properties is provided in **Figure 6.8** and **Table 6.2.**

GLOBAL THERMOSTATIC EFFECTS   6-9

The **thermostatic properties** (*therme* = heat, *stasis* = standing still) of water are those properties that act to moderate changes in temperature. Water temperature rises as the sun's energy is absorbed and changed to heat, but, as we've

seen, water has a very high heat capacity, so its temperature will not rise very much even if a large quantity of heat is added. This tendency of a substance to resist change in temperature with the gain or loss of heat energy is called **thermal inertia.**

Remember the hot sand and cool water on a hot summer afternoon? Think about Earth as a whole. The highest temperatures on land, in the North African desert, exceed 50°C (122°F); the lowest, on the Antarctic continent, drop below −90°C (−129°F). That's a difference of 140°C (or 250°F). On the ocean surface, however, the range is from −2°C (29°F) where sea ice is forming to about 32°C (90°F) in the tropics—a difference of only 34°C (61°F). Consisting of water, the ocean rises very little in temperature as it absorbs heat. The ocean's thermal inertia is much greater than the land's.

A practical example of thermal inertia can be seen in **Figure 6.9.** San Francisco, California, and Norfolk, Virginia, are

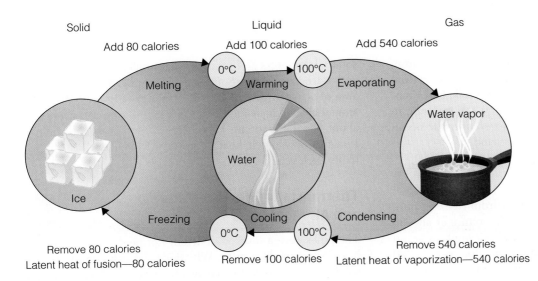

Solid

Add 80 calories

Liquid

Add 100 calories

Gas

Add 540 calories

0°C

100°C

Melting

Warming

Evaporating

Water vapor

Water

Ice

Freezing

Cooling

Condensing

0°C

100°C

Remove 80 calories
Latent heat of fusion—80 calories

Remove 100 calories

Remove 540 calories
Latent heat of vaporization—540 calories

Figure 6.8 We must add 80 calories of heat energy to change 1 gram of ice to liquid water. After it is melted, about 1 calorie of heat is needed to raise each gram of water by 1 degree. But 540 calories must be added to each gram of water to vaporize it—to boil it away. This process is reversed for condensation and freezing.

Table 6.2 Properties of Water

Property	Remarks	Importance to the Ocean Environment
Physical state	Only substance occurring naturally in all three phases as solid, liquid, and gas (vapor) on Earth's surface	Transfer of heat between ocean and atmosphere by phase change
Dissolving ability	Dissolves more substances in greater quantities than any other common liquid	Important in chemical, physical, and biological processes
Density: mass per unit volume	Density determined by (1) temperature, (2) salinity, and (3) pressure, in that order of importance. The temperature of maximum density for pure water is 4°C. For seawater, the freezing point decreases with increasing salinity.	Controls oceanic vertical circulation, aids in heat distribution, and allows seasonal stratification
Surface tension	Highest of all common liquids	Controls drop formation in rain and clouds; important in cell physiology
Conduction of heat	Highest of all common liquids	Important on the small scale, especially on cellular level
Heat capacity: quantity of heat required to raise the temperature of 1 g of a substance 1°C	Highest of all common solids and liquids	Prevents extreme range in Earth's temperatures; great heat moderator
Latent heat of fusion: quantity of heat gained or lost per unit mass by a substance changing from a solid to a liquid or a liquid to a solid phase without an accompanying rise in temperature	Highest of all common liquids and most solids (80 cal/g)	Thermostatic heat regulating effect due to the release of heat on freezing and absorption on melting
Latent heat of vaporization: quantity of heat gained or lost per unit mass by a substance changing from a liquid to a gas or a gas to a liquid phase without an increase in temperature	Highest of all common substances (540 cal/g)	Immense importance: a major factor in the transfer of heat in and between ocean and atmosphere, driving weather and climate
Refractive index	Increases with increasing salinity and decreases with increasing temperature	Objects appear closer than in air
Transparency	Relatively great for visible light; absorption high for infrared and ultraviolet	Important in photosynthesis
Sound transmission	Good compared with other fluids	Allows sonar and precision depth recorders to rapidly determine water depth, detect subsurface features and animals; sounds can be heard great distances underwater
Compressibility	Only slight	Density changes only slightly with pressure/depth
Boiling and melting points	Unusually high	Allows water to exist as a liquid on most of Earth

Sources: Sverdrup, et al., 1942; Ingmanson and Wallace, 1995.

on the same line of latitude—each is the same distance from the equator. As Chapter 8 will show, wind tends to flow from west to east at this latitude. Compared to Norfolk, San Francisco is warmer in the winter and cooler in the summer, in part because air in San Francisco has moved over the ocean while air in Norfolk has approached over land.

Thermostatic Effects of Ice

As you know, removing a calorie of heat from freezing pure water at 0°C (32°F) won't change its temperature at all—80 calories of heat energy must be removed per gram of liquid water to form ice. More than 18,000 cubic kilometers (4,300 cubic miles) of polar ice, covering as many as 20 million square kilometers (7.7 million square miles) of surface, thaws and refreezes in the Southern Hemisphere each year—an area of ocean larger than South America (**Figure 6.10**)! The annual change in sea ice cover is less in the Arctic, averaging about 5 million square kilometers (2 million square miles). Imagine how much heat is stored and released as this ice melts each year.

Thermostatic Effects of Water and Air Movement

The poles have a marked deficiency of heat, and the equator has a pronounced surplus. Why don't the polar oceans freeze solid and the equatorial oceans boil away? The reason is that

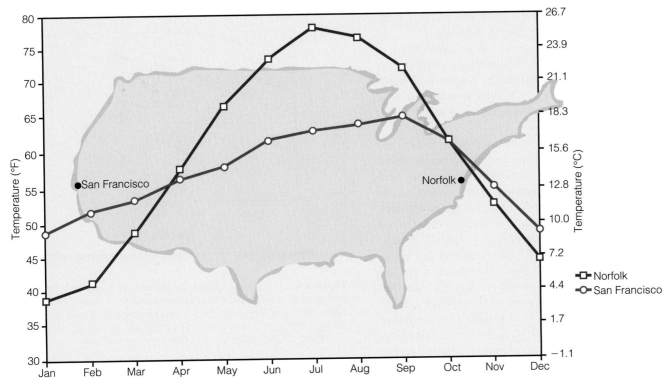

Figure 6.9 San Francisco, California, and Norfolk, Virginia, are on the same line of latitude. Wind tends to flow from west to east at this latitude. Compared to Norfolk, San Francisco is warmer in the winter and cooler in the summer, in part because air in San Francisco has moved over the ocean while air in Norfolk has approached over land. Water doesn't warm as much as land in the summer, nor cool as much in the winter—a demonstration of thermal inertia.

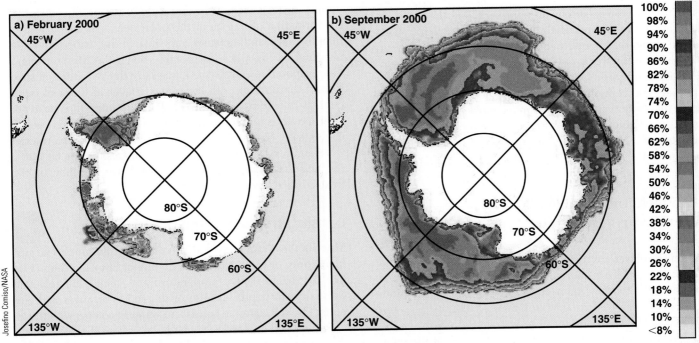

Josefino Comiso/NASA

Figure 6.10 About 20 million square kilometers (7.7 million square miles) of ocean surface thaws and refreezes in the Southern Hemisphere each year—an area of ocean larger than South America! The autumn cooling of the atmosphere is delayed because heat energy is released as masses of water turn to ice. Heat is absorbed during ice melt in the spring. Seasonal extremes are moderated by the absorption and release of heat energy as ice thaws and refreezes. (Remember, the seasons are reversed in the southern hemisphere.) The scale shows the precent of ocean surface completely covered by ice.

currents in the atmosphere and ocean are moving huge amounts of heat from the tropics toward the poles.

Water's high heat capacity makes it an ideal fluid to equalize the polar–tropical heat imbalance. Ocean currents and atmospheric weather result from the response of water and air to unequal solar heating. Although weather and currents are discussed in more detail in Chapters 8 and 9, here's a brief overview.

Ocean currents carry heat from the tropics (where incoming energy exceeds outgoing) to the polar regions (where outgoing energy exceeds incoming). The amount of heat transferred in this way is astonishing. For example, "outbound" water in the warm Gulf Stream (a large northward-flowing ocean current just offshore of the eastern United States) is about 10°C (18°F) warmer than "inbound" returning water, meaning that about 10 million calories are transported per cubic meter. Since the flow rate of the Gulf Stream is about 55 million cubic meters per second, some 550 trillion calories are being transported northward in the western North Atlantic each *second!*[3] Nearly half of these calories reach the high latitudes above 40°N. As we will see in Chapter 9, this warmth has a dramatic moderating influence on the winter climate of northwestern Europe.

As impressive as the figures are for ocean currents, the amount of heat transported by water vapor in the atmosphere is even greater. About half of the solar energy entering water results in evaporation. The solar energy required for this evaporation is later surrendered during condensation and cloud formation (and rain), but usually at a distance from where the initial evaporation occurred. So the ocean surface near Cuba may be cooled by evaporation today, the water vapor may then be moved north by winds, and eastern Canada may be warmed by condensation of the same water in a rainstorm later in the week.

Both atmosphere and ocean transfer heat by movement, but water's exceptionally high latent heat of vaporization means that water vapor transfers much more heat (per unit of mass) than liquid water. Masses of moving air account for about two-thirds of the poleward transfer of heat; ocean currents move the other third.

TEMPERATURE, SALINITY, AND WATER DENSITY

The density of water is mainly a function of its temperature and salinity. We will learn more about the specifics of salinity in Chapter 7, but for now you need to know that the solids (often called salts) dissolved in seawater raise its density between 1.020 and 1.030 g/cm³. A liter of seawater weighs between 2% and 3% more than a liter of pure water (1 g/cm³) at the same temperature. *Cold, salty water is denser than warm, less salty water.* Seawater's density increases with increasing salinity, increasing pressure, and decreasing temperature. **Figure 6.11**

---

[3] If you can't wait for a look, see Figure 9.11

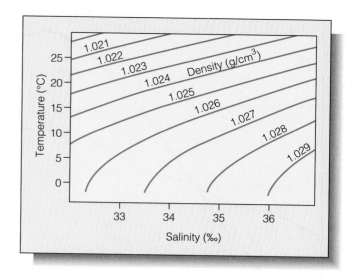

Figure 6.11 The complex relationship between temperature, salinity, and density of seawater. Note that two samples of water can have the *same* density at *different* combinations of temperature and salinity. (Source: From G. P. Kuiper, ed., *The Earth as a Planet*, © 1954 The University of Chicago Press. Reprinted by permission.)

shows the relationship between temperature, salinity, and density. Notice that two samples of water can have the *same* density at *different* combinations of temperature and salinity.

Ocean water tends to form into stable layers with the heaviest water at the bottom—a form of density stratification. Curiously, even the deepest of these layers originates at the ocean's surface. Very cold and salty water produced during the formation of sea ice at the polar ocean surface is denser than the surrounding water and sinks to the seabed. In a few marginal basins, the most notable being the Mediterranean Sea, evaporation can also produce salty, dense water that sinks toward the bottom until it reaches a layer of water of equal density. In contrast, warm fresh water entering the ocean at a river mouth is much less dense and floats for miles above the cooler, salty layers below. Early explorers of South America were amazed to find a surface layer of fresh water far out to sea, and they discovered the Amazon River by following this fresh surface layer to its source.

Density Structure of the Ocean

Much of the ocean is divided into three density zones. The **surface zone,** or **mixed layer,** is the upper layer of ocean (**Figure 6.12a**), in which temperature and salinity are relatively constant with depth because of the action of waves and currents. The surface zone consists of water in contact with the atmosphere and exposed to sunlight; it contains the ocean's least dense water, only about 2% of total ocean volume. The surface zone (or mixed layer) typically extends to a depth of about 150 meters (500 feet), but depending on local conditions, it may reach a depth of 1,000 meters (3,300 feet) or be absent entirely.

THE PYCNOCLINE The **pycnocline** (*pyknos* = strong, *clinare* = slope, to lean) is a zone in which density increases

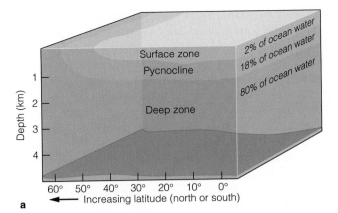

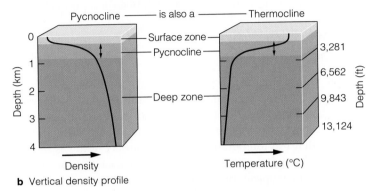

b Vertical density profile

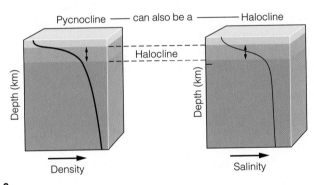

c

Figure 6.12 Density stratification in the ocean. (**a**) In most of the ocean, a surface zone (or mixed layer) of relatively warm, low-density water overlies a layer called the *pycnocline*. Density increases rapidly with depth in the pycnocline. Below the pycnocline lies the deep zone of cold, dense water—about 80% of total ocean volume. (**b**) The rapid density increase in the pycnocline is mainly due to a decrease in temperature with depth in this area—the *thermocline*. (**c**) In some regions, especially in shallow water near rivers, a pycnocline may develop in which the density increase with depth is due to vertical variations in salinity. In this case, the pycnocline is a *halocline*.

with increasing depth. This zone isolates surface water from the denser layer below. The pycnocline contains about 18% of all ocean water. The **deep zone** lies below the pycnocline, at depths below about 1,000 meters (3,300 feet) in mid-latitudes (40°S to 40°N). There is little additional change in water density with increasing depth through this zone. This deep zone contains about 80% of all ocean water.

THE THERMOCLINE The pycnocline's rapid density increase with depth is due mainly to a decrease in water temperature.

Figure 6.12b shows the general relationship of temperature with depth in the open sea. The surface zone is well mixed, with little decrease in temperature with depth. In the next layer, temperature drops rapidly with depth; beneath it lies the deep zone of cold, stable water. The middle layer, the zone in which temperature changes rapidly with depth, is called the **thermocline** (*therm* = heat).

Thermoclines are not identical in form for all areas or latitudes. Surface temperature is proportional to available sunlight. More solar energy is available in the tropics than in the polar regions, so the water there is warmer. The ocean's sunlit upper layer is also thicker in the tropics, both because the solar angle there is more nearly vertical and because water in the open tropical ocean contains fewer suspended particles (and is therefore clearer than water in open temperate or polar regions). Because the ocean is heated to a greater depth, the tropical thermocline is deeper than thermoclines at higher latitudes. It is also much more pronounced—the transition to the colder, denser water below is more abrupt in the tropics than at high latitudes.

Polar waters, which receive relatively little solar warmth, are not stratified by temperature and generally lack a thermocline because surface water in the polar regions is nearly as cold as water at great depths.

Figure 6.13 contrasts polar, tropical, and temperate thermal profiles, showing that the thermocline is primarily a mid- and low-latitude phenomenon. Thermocline depth and intensity also vary with season, local conditions (storms, for example), currents, and many other factors.

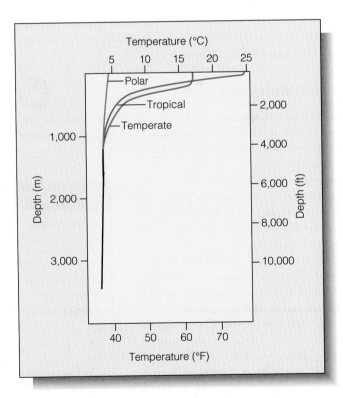

Figure 6.13 Typical temperature profiles at polar, tropical, and middle (temperate) latitudes. Note that polar waters lack a thermocline.

Below the thermocline water is very cold, ranging in temperature from −1°C to 3°C (30.5°F to 37.5°F). Because this deep and cold layer contains the bulk of ocean water, the average temperature of the world ocean is a chilly 3.9°C (39°F).

THE HALOCLINE Low salinity can also contribute to the pycnocline, especially in cool regions where precipitation is high, or along coasts where freshwater runoff mixes with surface water. Wherever precipitation exceeds evaporation, salinity will be low. These differences produce the **halocline** (*halos* = salt), a zone of rapid salinity increase with depth (**Figure 6.12c**). The halocline often coincides with the thermocline, and the combination produces a pronounced pycnocline.

Density Stratification and Water Masses

The layers we have been discussing are distinct water masses. A **water mass** is a body of water with characteristic temperature and salinity, and therefore density. Layering by density traps dense water masses at great depths where they are not exposed to daily heating and cooling, to surface circulation driven by winds and storms, or to light. The pycnocline effectively isolates 80% of the world ocean's water from the 20% involved in surface circulation.

Dense water masses form near polar continental shelves (as cold water freezes and excludes salt) or in enclosed areas such as the Mediterranean Sea (where evaporation exceeds precipitation and river input, raising salinity). These heavy water masses sink, sometimes overlapping one another and often retaining their identity for long periods. Separate water masses below the pycnocline tend not to merge because little energy is available for mixing in these quiet depths. (Density differences in water masses power the deep-ocean currents, as will be discussed in Chapter 9.)

Vertical Water Movement

The vertical movement of large volumes of water from the surface to great depths (and vice versa) is possible only where surface water density is similar to deep water density. The great difference in temperature, and therefore density, between surface water and deep water in the tropics makes the water column very stable and prevents an exchange of surface and deep water. This stability is maintained despite the fact that the surface of the tropical ocean is in constant horizontal motion, churned by tropical cyclones and stirred by currents.

Vertical movement of water in the northern polar ocean is also limited. There, however, the stratification is caused largely by a salinity difference between surface water and water at great depths. The surface of the Arctic Ocean receives a large volume of fresh-water runoff from Siberian and Canadian rivers. Continental masses block the formation of large currents, and the land-locked northern ocean communicates sluggishly with other ocean areas, so the surface water tends not to mix with deeper water or to flow to lower latitudes.

In contrast, the southern polar ocean is only weakly stratified. The cold temperature of southern ocean surface water closely matches that of deep water, so no thermocline divides surface water from deep water (see again Figure 6.13). The absence of confining continental margins and mixing at the boundaries of the Antarctic Circumpolar Current minimize salinity differences. Turbulence and weak stratification encourage a huge volume of deep-water upwelling, which contributes to high surface nutrient levels and high biological productivity.

An Overview of Ocean Surface Conditions

Table 6.3 summarizes a few important properties of seawater at polar, temperate, and tropical latitudes. The temperature data show the effects of solar radiation at various latitudes, along with the ratio of evaporation to precipitation. Notice that the temperature of ocean surface water is more variable through the year in the temperate zone than in either the polar or tropical areas, but that temperate zone salinity stays relatively constant. Notice also that evaporation generally exceeds precipitation in the tropics, but that precipitation dominates in temperate and polar zones.

Ocean surface temperature is highest in the Pacific north and east of Borneo, where westward-driving currents funnel seawater warmed by long exposure to tropical sunlight. The highest open-ocean surface temperatures range up to 32°C (90°F). Surface salinity is especially high in the warm central North and South Atlantic, where evaporation rates are high and surface water is isolated by the currents flowing around the ocean's periphery. Some of this information is summarized graphically in **Figures 6.14** and **6.15,** which show typical ocean surface temperatures and salinities worldwide.

Table 6.3 Some Characteristics of the World Ocean Surface, by Latitude			
Characteristic	**Tropical Oceanic Waters**	**Temperate Oceanic Waters**	**Polar Oceanic Waters**
Winter temperature	20–25°C (68–77°F)	5–20°C (41–68°F)	About −2°C (28°F)
Annual variation of temperature	Less than 5°C (9°F)	About 10°C (18°F)	Less than 5°C (9°F)
Average salinity	35‰–37‰	About 35‰	28‰–32‰[a]
Annual variation of air temperature	Less than 5°C (9°F)	About 10°C (18°F)	Up to 40°C (72°F)
Precipitation–evaporation balance	E exceeds P	P exceeds E	P exceeds E

Source: H. Charnock, 1971.
[a] Polar surface water is a bit less saline because the very dense, cold, salty water left behind after the formation of sea ice sinks rapidly toward the seabed.

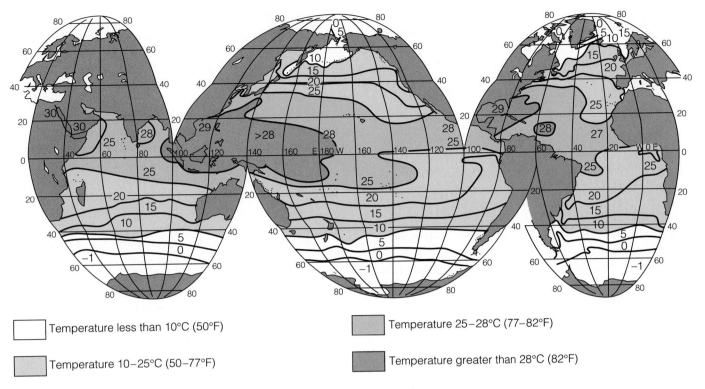

Temperature less than 10°C (50°F)

Temperature 10–25°C (50–77°F)

Temperature 25–28°C (77–82°F)

Temperature greater than 28°C (82°F)

Figure 6.14 Sea surface temperatures in degrees Celsius during Northern Hemisphere summer. (Another way to view surface temperature is shown in Figure 9.8a.)

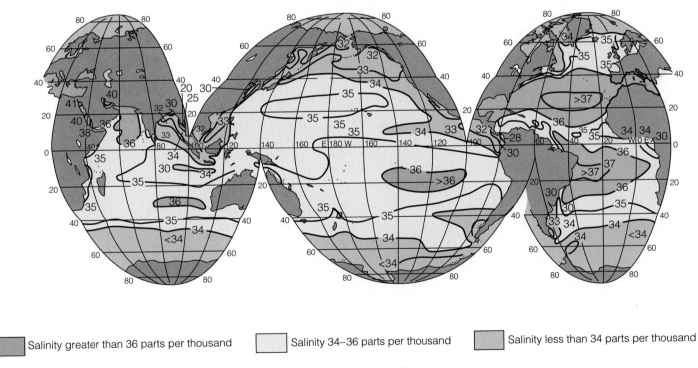

Salinity greater than 36 parts per thousand

Salinity 34–36 parts per thousand

Salinity less than 34 parts per thousand

Figure 6.15 Sea surface salinities in parts per thousand (‰) during Northern Hemisphere summer.

Remember, though, that variations in temperature, salinity, and density between high and low latitudes are usually confined to the uppermost 2,000 meters (6,500 feet) of water. Below this depth conditions are similar at all latitudes. The deep ocean is nearly as cold, salty, and dense in the equatorial Pacific as it is off the Siberian coast.

REFRACTION, LIGHT, AND SOUND

The ring of light sometimes seen around the moon and the safe concealment of a submarine may not seem related, but both events depend on **refraction,** the bending of waves. Light and sound are wave phenomena. When a light wave or a sound wave leaves a medium of one density—such as air—and enters a medium of a different density—such as water—at an angle other than 90°, it is bent from its original path. The reason for this bending is that light or sound waves travel at different speeds in different media.

The situation is analogous to a line of people marching along a desert highway with their arms linked over each other's shoulders. The marchers can walk faster if they stay on the pavement than if they walk in the sand next to the highway. Their speed on the pavement, then, is greater than their speed in the sand. As long as they stay on the pavement, they won't change direction. But if their marching angle gradually takes them off the edge (into a medium in which their speed is *lower*), the people who reach the sand first will suddenly slow down, and the line will pivot quickly off the highway. They have been refracted. Their progress is depicted in **Figure 6.16a**. Note that the transition from one medium to another must occur at an angle other than 90° for refraction to occur—our marchers will not change direction if they march straight off the asphalt into the sand. (They will still slow down, however; see **Figure 6.16b**.)

The refraction of light waves by water happens in much the same way. The speed of light in water is only about three-quarters its speed in air, so water effectively refracts light. (Glass bends light even more.) The degree to which light is refracted from one medium to another is expressed as a ratio called the **refractive index.** The higher the refractive index, the greater the bending of waves between media. The refrac-

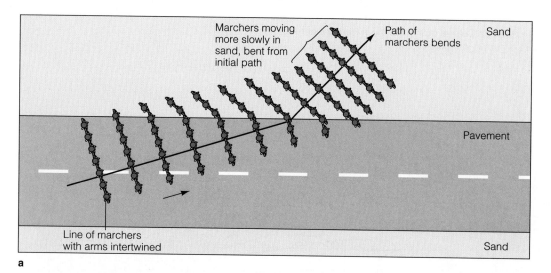

a

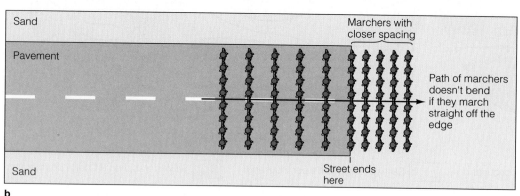

b

Figure 6.16 An analogy for refraction. The ranks of marchers represent light or sound waves; the pavement and sand represent different media. (**a**) If the marchers head off the pavement at an angle other than 90°, their path will bend (refract) as they hit the sand because some will be walking more slowly than others. (**b**) If they march straight off the pavement, the ranks will slow down but not bend as they hit the sand.

Figure 6.17 A broken rainbow, a beautiful demonstration of the difference between the refractive index of fresh water and that of seawater. Broken rainbows occur when an atmospheric rainbow (caused by raindrops of pure water) meets a surf-droplet rainbow. The rainbow angle for rainwater is 0.8° larger than for seawater.

Gunther Konnen, by permission Bob Greenler

tive index of water increases with increasing salinity. A beautiful example of the difference between the refractive index of fresh water and that of seawater is seen in **Figure 6.17.**

Examples of the refraction of light by water are all around you. A pencil sticking out of a glass of water looks bent because of refraction, and the submerged steps of a swimming pool ladder look closer than they are because of refraction. Divers exaggerate the size of the fish that got away because refraction can magnify objects.

LIGHT IN THE OCEAN

Light is electromagnetic radiation propagated as small, nearly massless particles. Light has a dual nature—it behaves like a wave and like a stream of particles. The wavelength of light determines its color: Shorter wavelengths are bluer, longer wavelengths are redder. The visible spectrum—the wavelengths that human eyes can detect—is only a small part of the entire electromagnetic spectrum. Except for very long radio waves, water rapidly absorbs nearly all electromagnetic radiation. Only blue and green wavelengths pass through water in any appreciable quantity or distance.

Scattering and Absorption

Sunlight has a difficult time reaching and penetrating the ocean. Once past the sea surface, light is rapidly weakened by scattering and absorption. **Scattering** occurs as light is bounced between air or water molecules, dust particles, water droplets, or other objects before being absorbed. The greater density of water (along with the greater number of suspended and dissolved particles) makes scattering more prevalent in water than in air. The **absorption** of light is governed by the structure of the water molecules it happens to strike. When light is absorbed, molecules vibrate and the light's electromagnetic energy is converted to heat.

Even the clearest seawater is not perfectly transparent. If it were, the sun's rays would illuminate the greatest depths of the ocean, and seaweed forests would fill its warmed basins. The thin film of lighted water at the top of the surface zone is called the **photic zone** (*photo* = light). In the very clearest tropical waters the photic zone may extend to a depth of 600 meters (2,000 feet), but a more typical value for the open ocean is 100 meters (330 feet). (In contrast, light may penetrate the coastal waters in which we typically swim only to about 40 meters, or 130 feet.) All the production of food by photosynthetic marine organisms occurs in this thin, warm surface zone. Here, water is heated by the sun, heat is transferred from the ocean into the atmosphere and space, and gases are exchanged with the atmosphere. The thermostatic effects we've discussed function largely within this zone. Most of the ocean's life is found here. The photic zone may be extraordinarily thin, but it is also extraordinarily important.

The ocean below the photic zone lies in blackness. Except for light generated by living organisms, the region is perpetually dark. This dark water beneath the photic zone is called the **aphotic zone** (a = without, photo = light).

Water Colors

The light energy of some colors is converted into heat nearer to the surface than the light energy of other colors. **Figure 6.18** shows this differential absorption by color. The top 1 meter (3.3 feet) of the ocean absorbs nearly all the infrared radiation reaching the ocean surface, significantly contributing to surface warming. The top 1 meter also absorbs 71% of red light. The dimming light becomes bluer with depth because the red, yellow, and orange wavelengths are being absorbed. By 300 meters (1,000 feet) even the blue light has been converted into heat.

From above, clear ocean water looks blue because blue light can travel through water far enough to be scattered back

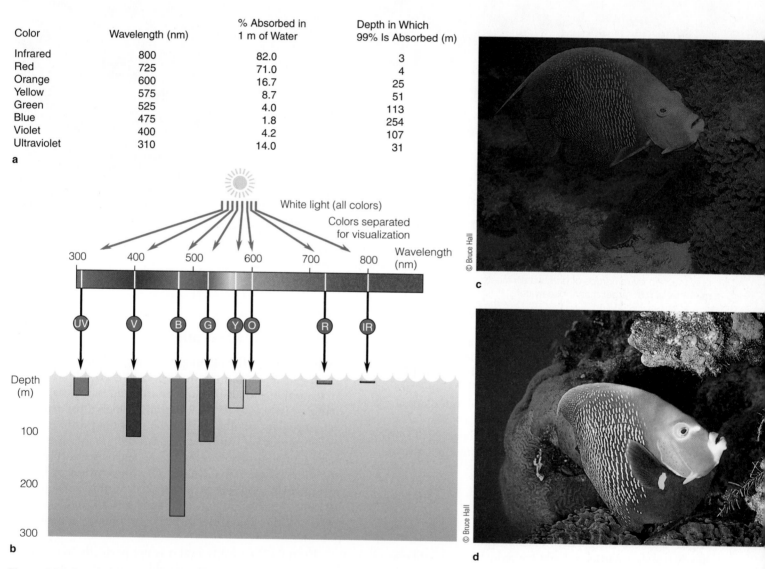

Color	Wavelength (nm)	% Absorbed in 1 m of Water	Depth in Which 99% Is Absorbed (m)
Infrared	800	82.0	3
Red	725	71.0	4
Orange	600	16.7	25
Yellow	575	8.7	51
Green	525	4.0	113
Blue	475	1.8	254
Violet	400	4.2	107
Ultraviolet	310	14.0	31

Figure 6.18 Only a thin film of seawater is illuminated by the sun. Except for light generated by living organisms, most of the ocean lies in complete blackness. Absorption of light of different wavelengths (colors) by seawater: (**a**) The table shows the percentage of light absorbed in the uppermost meter of the ocean, and the depths at which only 1% of the light of each wavelength remains. (**b**) The bars show the depth of penetration of 1% of the light of each wavelength (as in the last column of the table). (**c**) A fish photographed in normal oceanic light—the color blue predominates. (**d**) The same fish photographed with a strobe light. The flash contains all colors, and the distance from the strobe to the fish and back to the camera is not far enough to absorb all the red light. The fish shows bright, warm colors otherwise invisible.

through the surface to our eyes. Divers near the surface see an even brighter blue color. Because nearly all red light is converted to heat in the first few meters of ocean water, red objects a short distance beneath the surface look gray. If you were a diver working at a depth of 10 meters (33 feet) and you cut your hand, you would see gray blood rather than red, because there is not enough red light at that depth to reflect from blood's red pigment and stimulate your eye. The underwater pictures of red organisms you've seen are possible only because the diver has brought along a source of white light (which contains all colors). This is shown in Figure **6.18c** and **d**.

Suspended particles scatter some colors of light and absorb others. Some sediments reflect yellow light, giving the ocean a yellow cast. The Red Sea gets its name from its abundance of cyanobacteria, small plantlike organisms that contain a reddish pigment.

SOUND IN THE OCEAN

Sound is a form of energy transmitted by rapid pressure changes in an elastic medium. Sound energy decreases as it travels through seawater because of spreading, scattering, and absorption. Energy loss due to spreading is proportional to the square of the distance from the source. Scattering occurs as sound bounces off bubbles, suspended particles, organisms, the surface, the bottom, or other objects. Eventually sound is absorbed and converted by molecules into a very small amount of heat. Absorption of sound is proportional to the square of the frequency of the sound—higher frequencies are absorbed sooner. Sound waves can travel for much greater distances through water than light waves can before being absorbed. Because sound travels through water so efficiently, many marine animals use sound rather than light to "see" in the ocean.

The speed of sound in seawater is about 1,500 meters per second (3,345 miles per hour), almost five times the speed of sound in air. The speed of sound in seawater increases as temperature and pressure increase. Sound travels faster at the warm ocean surface than it does in deeper, cooler water. Its speed decreases with depth, eventually reaching a minimum at about 1,000 meters (3,300 feet). Below that depth, however, the effect of increasing pressure offsets the effect of decreasing temperature, so speed increases again. Near the bottom of an ocean basin the speed of sound may actually be higher than at the surface. Though important in the behavior of oceanic sound, these variations amount to only 2% or 3% of the average speed of sound in seawater. The relationship between depth and sound speed is shown in **Figure 6.19**.

Sofar Layers and Shadow Zones

The depth at which the speed of sound reaches its *minimum* varies with conditions, but it is usually located near 1,200 meters (3,900 feet) in the North Atlantic or about 600 meters (2,000 feet) in the North Pacific. Although its speed is relatively slow, the transmission of sound in this minimum-velocity layer is very efficient because refraction tends to

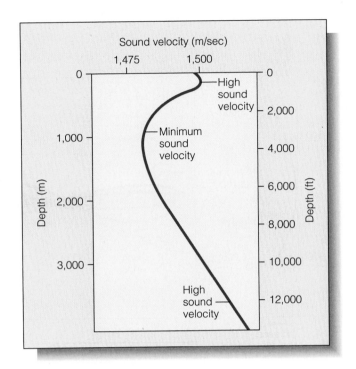

Figure 6.19 The relationship between water depth and sound velocity.

cause sound energy to remain within the layer. The outer edges of sound waves escaping from this layer will enter water in which the speed of sound is higher, then speed up and cause the wave to pivot back into the minimum-velocity layer, as shown in **Figure 6.20.** Upward-traveling sound waves that are generated within the minimum-velocity layer will tend to be refracted downward, and downward-traveling sound waves will tend to be refracted upward. In short, sound waves bend *toward* layers of lower sound velocity and so tend to stay within the zone. Therefore, loud noises made at this depth can be heard for thousands of kilometers.

Navy depth charges detonated in the minimum-velocity layer in the Pacific have been heard 3,680 kilometers (2,290 miles) from the explosion. In a recent test, sound generated by a U.S. Navy ship in the Indian Ocean was heard at the Oregon Coast! (See **Box 6.1**.) In the early 1960s, the U.S. Navy experimented with the use of sound transmission in the minimum-velocity layer as a lifesaving tool. Survivors in a life raft would drop a small charge into the water that was set to explode at the proper depth. A number of widely spaced listening stations ashore would compare the differences in the arrival times of the signal and then compute the position of the raft. The project—which has since been abandoned in favor of radio beacons—was called **sofar** (for *so*und *f*ixing *a*nd *r*anging). The minimum-velocity layer has come to be known as the **sofar layer.**

Sound travels slowly in the sofar layer, but it moves rapidly near the bottom of the well-mixed surface layer. Temperature and salinity conditions are homogeneous there, so they do not produce any refraction. Pressure still increases with depth, however, causing a thin high-velocity layer at around 80 meters (260 feet), just above the pycnocline.

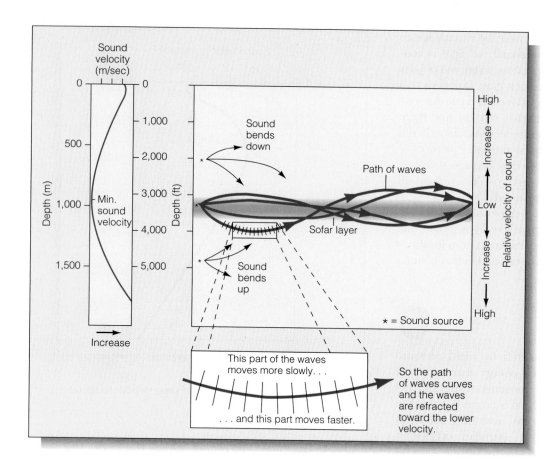

Some ships project pulses of sound into the water to search for animals, submarines, or hazards to navigation. Depending on the angle at which the sound waves arrive at the high-velocity layer, they will sometimes split and refract to the surface or bend into the depths. An object beyond the area of divergence may be undetectable; it would be within a **shadow zone,** a region into which very little sound energy penetrates. Shadow zones are of particular interest to submariners and to the ship captains who use sound to hunt them. As depicted in **Figure 6.21,** a smart submarine captain can use the shadow zone to hide from a pursuer.

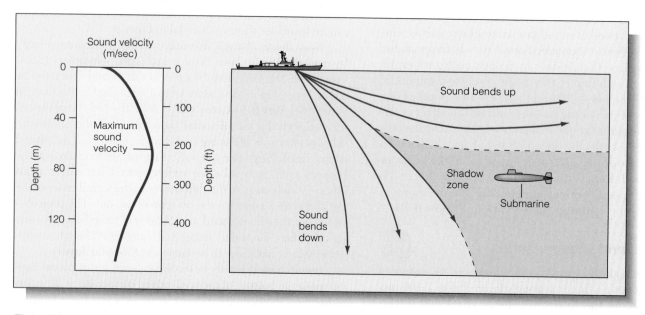

Figure 6.21 The shadow zone. The thin, high-sound-velocity layer, which forms at a depth of about 80 meters (260 feet), deflects sound. The shadow zone creates a good place for submarines to hide from sonar. (Note the difference in depth scales in Figures 6.19 and 6.20.)

Is Climate Change Sofar Away?   BOX 6.1

Early in 1991 underwater sounds generated by a ship near Heard Island in the southern Indian Ocean were detected off the west coast of the United States, 17,960 kilometers (11,160 miles) away. The foghornlike sounds were produced by powerful underwater loudspeakers suspended 200 meters (660 feet) beneath a research vessel. The ship's five transmitters emitted a booming, low-frequency hum in all directions for an hour and then were silent for two hours. The pattern was repeated through most of five days.

Some of the sound was trapped in the Indian Ocean's minimum-velocity (sofar) layer, in which it traveled both into the Atlantic and around Australia into the Pacific. The sound was detected by a ship equipped with sensitive hydrophones lowered to the sofar depth off the California coast, near Point Conception. It took about 3½ hours for the sound to travel that far, nearly halfway around the world. Similarly equipped ships and offshore stations detected the sound at other locations (**Figure a**).

The experiment was part of a plan to measure ocean temperature within a few thousandths of a degree. Since sound travels more rapidly through warmer water, water temperature differences can be calculated by recording the time taken for sound to reach receivers around the world. The temperature of the ocean is an important measure of potential global warming, and the sound-transmission experiment collected baseline data for a study designed to help settle the argument over whether carbon dioxide and other greenhouse gases are warming the planet artificially. Computer models of Earth's climate suggest that greenhouse gases could cause the ocean to warm by about 0.005°C (0.009°F) per year at a depth of 1,000 meters (3,300 feet). Such a subtle temperature change would be very difficult to detect by conventional monitoring, but over a decade this warming would cut a few seconds from the travel time of a sound signal. (More information on the possibilities and consequences of global warming may be found in Chapter 18.)

In early 1993 nearly 100 researchers at 13 institutions began a larger five-year $40 million experiment. They constructed and tested sound sources capable of projecting a 75-cycle-per-second signal with 260 watts of acoustic power. To address concerns that the low-frequency sound would in some way harm marine organisms, especially whales and other marine mammals that depend on sound for communication and feeding, the scientists carefully analyzed mammal

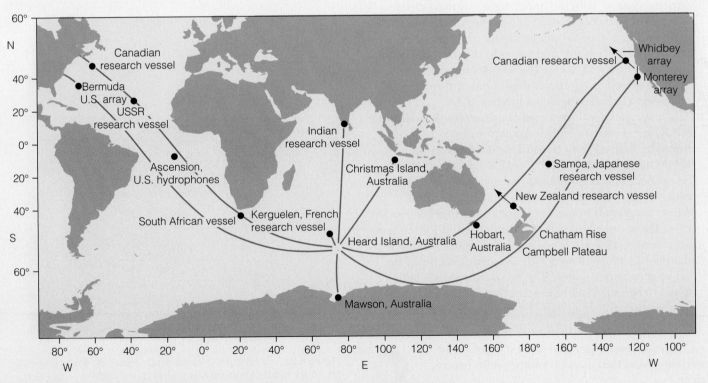

a Paths taken by sound in the Heard Island experiment. The sound sources were suspended from the R/V *Cory Chouest,* located about 50 kilometers (31 miles) southeast of Heard Island. Black circles indicate receiver sites. Ray paths from the source to receivers are along approximate great circle routes.

distribution and behavior around transmitters being tested. No significant evidence of disruption was found. Environmental impact statements have been prepared for two transmitter sites: off the coast of Kaua'i, Hawaii, and on the Pioneer Seamount about 80 kilometers (50 miles) off the central California coast.

The Acoustic Thermography of Ocean Climate (ATOC) program began transmitting acoustic signals in late 1995 from the Pioneer Seamount. The signals were recorded on horizontal receiving arrays at 11 U.S. Navy SOSUS stations in the northeast Pacific and on two vertical receiving arrays, one near the island of Hawaii and the other near Kiritimati (Christmas) Island, just north of the equator in the central Pacific. They were also recorded by a single receiver off New Zealand.

By January 1998, 15 months of data had been collected by some of these receivers. Researchers are now retooling transmitters, receivers, and computer programs before beginning a new series of observations. The key results obtained to date:

- Acoustic travel times can be measured with a precision of about 20 to 30 milliseconds at 3,000- to 5,000-kilometer (1,860- to 3,100-mile) ranges. (For comparison the total travel time at 5,000 kilometers or 3,100 miles is nearly an hour.) This measurement is more precise than originally thought possible.
- The observed travel-time changes can be clearly related to known ocean processes. One of the significant sources of lower-frequency variability in ocean temperature is the seasonal change, with the upper ocean warming during summer and cooling during winter. The ATOC data show corresponding seasonal changes in travel times, as expected, particularly for acoustic paths that travel north of the Subarctic Front, where the seasonal temperature changes extend to significant depths rather than being confined to a shallow seasonal thermocline.
- The range and depth-averaged temperatures derived from ATOC are consistent with and complementary to related estimates derived from measurements of sea-surface height made by the *TOPEX/Poseidon* satellite altimeter. Sea-surface height is related to ocean temperature because of thermal expansion. One primary goal of the ATOC program is to use the acoustic data to refine models of the ocean circulation. Consistent results for the seasonal heat storage are found when the acoustic and altimetric data are combined with a computer model of the ocean general circulation. The two data types are both found to be important in constraining the model, with the combination providing more information than either data type alone.

As experiments proceed, the extent of ocean warming—and thus global warming—may become clear.

6-23

Sonar

6-24

Crews aboard surface ships and submarines employ **active sonar** (*s*ound *n*avigation *a*nd *r*anging), the projection and return through water of short pulses (*pings*) of high-frequency sound to search for objects in the ocean (**Figure 6.22**). In a modern system, electrical current is passed through crystals, which respond by producing powerful sound pulses pitched above the limit of human hearing. (Though this high-frequency sound is absorbed rapidly in the ocean, its use greatly improves the resolution of images.) Some of the sound from the transmitter bounces off any object larger than the wavelength of sound employed and returns to a microphone-like sensor. Signal processors then amplify the echo and reduce the frequency of the sound to within the range of human hearing. An experienced sonar operator can tell the direction of the contact, its size and heading, and even something about its composition (whale or submarine or school of fish) by analyzing the characteristics of the returned ping.

Side-scan sonar is a type of active sonar. Operating with as many as 60 transmitter/receivers tuned to high sound frequencies, side-scan systems towed in the quiet water beneath a ship are sometimes capable of near-photographic resolution (see **Figure 6.23**). Side-scan systems are used for geological investigations, archaeological studies, and the location of downed ships and airplanes.

For deeper soundings, or to "see" into sediment layers below the surface, geologists use *seismic reflection profilers* employing powerful electrical sparks, explosives, or com-

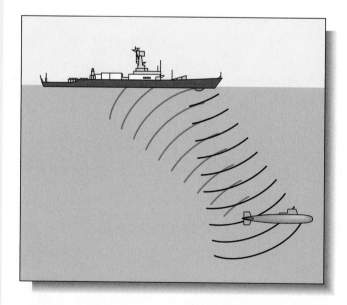

Figure 6.22 The principle of active sonar. Pulses of high-frequency sound are radiated from the sonar array of the sending vessel. Some of the energy of this ping reflects from the submerged submarine and returns to the sending vessel. The echo is analyzed to plot the position of the submarine.

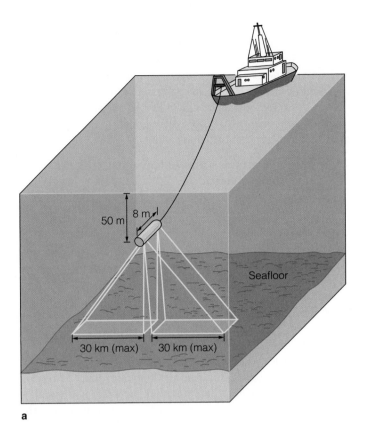

a

b

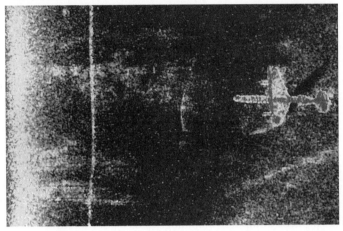

c

Figure 6.23 Side-scan sonar. (**a**) Side-scan sonar in action. Sound pulses leave the submerged towed array (**b**), bounce off the bottom, and return to the device. One of the two slits through which sound pulses depart and return is visible on its side. Computers process the impulses into images: (**c**) A side-scan sonar record showing a downed World War II–era PB4Y-2 Privateer at a depth of 50 meters (165 feet). The artificial colors enhance target detail and clarity.

pressed air to generate a very energetic low-frequency sound pulse. Again, the round-trip travel time of the sound waves is crucial. The low-frequency sound cannot resolve great detail, but the echo can usually provide an image of the sedimentary layers beneath the surface (see Figures 5.4 and 5.18). Low-frequency sound also has the advantage of efficient travel with less absorption.

The first human use of sea sound was passive: Mariners listened through their hulls to the whistles and clicks of whales and other animals. When submarine warfare emerged during World War I, the British invented a simple underwater listening device to detect noises made by enemy submarines. Operators would listen through a sensitive directional microphone for the telltale sounds of a propeller, a torpedo, or even a dropped wrench or slammed hatch cover. The first systems were primitive, but **passive sonar,** as these listening-only devices are now called, is currently undergoing a renaissance. Modern passive devices are much more sophisticated and sensitive than their World War I predecessors. Passive sonar confers the benefit of surprise—unlike active sonar, in which a listener can hear the loud ping long before an operator aboard the sending vessel can

hear his faint echo. Usually, it's safer just to listen. The combination of computerized signal processing, microphones towed at a distance from the listener's noisy ship, and better knowledge of ocean physics will improve the usefulness of both kinds of sonar.

Humans are not the only organisms to use sonar. Whales and other marine mammals use clicks and whistles to find food and avoid obstacles. We'll discuss this form of active sonar in Chapter 15.

QUESTIONS FROM STUDENTS

1. **If water has the highest latent heat of vaporization, why does a drop of alcohol make my hand feel colder when it evaporates than a drop of water does?**

Heat versus temperature again. The alcohol makes your hand *colder*, but it evaporates much *more quickly*. The water stays around longer (takes longer to evaporate), and although it may not cause your skin to become as *cold*, it will remove more *heat*.

2 **In physics, I learned that water's latent heat of evaporation is 585 calories per gram. In this chapter you say it's 540 calories per gram. What's up?**

Actually, I wrote that water's latent heat of *vaporization* is 540 calories per gram. The latent heat of vaporization is the amount of heat that must be added to 1 gram of a substance, at its boiling point, to change it completely from a liquid to a gas. But water doesn't have to be at its boiling point to vaporize. When water molecules evaporate at temperatures *below* the boiling point (as from your skin during a workout), more heat must be supplied to evaporate each molecule than is needed to vaporize it *at* the boiling point. The latent heat of vaporization increases with decreasing temperature—at 20°C (68°F), the latent heat of vaporization is 585 calories per gram. This is sometimes called the latent of heat of *evaporation*.

3. **Liquid methane appears to be the only other substance found in quantity in a liquid state in the solar system. If our ocean were comprised of liquid methane rather than liquid water, how would conditions here be different?**

Liquid methane freezes at −183°C (−272°F) and boils at −162°C (−234°F). Methane has a much lower heat capacity than liquid water because it is not dipolar—it does not form a great bonded mass as water does—and consequently does not require a large energy input to break the maze of hydrogen bonds to release molecules. Because of methane's low specific heat, the difference between Earth's polar and tropical daytime temperatures would be drastically greater unless the circulation rate of liquid and vapor currents accelerated to keep pace. Computer modeling yields a very unattractive vision of a planet with a boiling equatorial ocean, crushing atmospheric pressures from the greater amount of vapor in the air, torrential polar methane rainfall, cataclysmic cyclonic storms, and average wind speeds at mid-latitudes of hundreds of kilometers per hour. We miss all this excitement because our ocean is made of water.

4. **Input of solar radiation in the Northern Hemisphere peaks on 22 June and reaches a minimum on 22 December, because of Earth's orbital tilt (more about this in Chapter 8). Why, then, do our warmest days occur in August or September and our coldest days in January or February?**

Because of thermal inertia. There is a lag between maximum sunlight and maximum warmth because of water's great heat capacity. The sun must shine on this watery planet for many weeks to raise the summer hemisphere's temperature. Of course water also retains heat well, so the coldest days in the winter hemisphere come well after the darkest ones.

5. **What about pressure? Is seawater compressible? Does compression have anything to do with water density?**

Seawater is nearly incompressible—it doesn't squeeze very much as pressure builds. Still, there is some effect. Pressure in the deepest ocean trench is equal to the weight of 1,100 atmospheres (about 8½ tons per square inch), and a cubic centimeter of water lowered to that level will lose about 2% of its volume. The average pressure acting on the world ocean's volume compresses water enough to lower sea level by about 37 meters (121 feet). If gravity (and therefore pressure) were released, sea level would rise that much. Still, the effect of pressure on density is very small in comparison to the effects of temperature and salinity.

6. **Why don't sounds seem to travel easily from air to ocean, or vice versa?**

Sound waves can make the transition from one medium to another with little energy loss only when the speeds of sound waves in the two different media are similar. The speeds of sound in water and air, however, are too different for an efficient transition to be made. Too great a contrast in speed produces reflection (not refraction) of the sound waves at the junction. This is why you can't hear people shouting from the edge of the pool while you're underwater, even though the weak sound of a submerged pebble clicking against the side is very clear and sharp.

If you place a solid medium in which the speed of sound is intermediate between air and water, the sound can move across one junction and then the other, for a more efficient total transition. Wood works well for this, which explains why ocean noises are easy to hear in wooden boats. Even if the speed of sound in the intermediate medium is higher than in water (as in steel, for instance), some sound will be audible simply because the hard surface provides a good radiating surface for noises coming from the water.

7. **I can't tell where sound is coming from when I'm underwater. It seems to be coming from inside my head. Why?**

Our normal sensation of stereo hearing depends in part on the difference in the arrival times of sound from one ear to the other. The speed of sound in water, however, is more than four times greater than its speed in air. Our brains are unable to sense arrival-time differences from sounds originating underwater. You'd need to be more than four times as sensitive—or have a head more than four times as large—to hear stereophonically underwater.

8. **Why don't divers see shadows even when the sun is shining directly on the ocean at their location?**

The greater density of water (along with the greater number of suspended and dissolved particles) makes scattering more prevalent in water than in air. A few meters below the surface, light in the ocean seems to come from all directions (rather than primarily from one direction).

KEY CONCEPTS TO REVIEW

• Water is a polar chemical compound composed of two hydrogen atoms and one oxygen atom. Its remarkable thermal properties are dependent on the large number and relatively great strength of hydrogen bonds between water molecules.

- Heat and temperature are not the same thing. Heat is energy produced by the random vibration of atoms or molecules. Heat is a measure of how many molecules are vibrating and how rapidly they are vibrating. Temperature records only how rapidly the molecules of a substance are vibrating. Temperature is an object's response to an input (or a removal) of heat.

- The thermal properties of water are responsible for the mild physical conditions at Earth's surface, a global thermostatic effect. Heat is carried from equatorial regions to polar regions by seawater currents and atmospheric water vapor.

- Water density is greatly influenced by changes in temperature and salinity. Water masses are usually layered by density, with the densest water (coldest and saltiest) on or near the ocean floor. Differences in water-mass density power deep-ocean circulation.

- Light and sound are affected by the physical properties of water, with refraction and absorption effects playing important roles.

INTERNET STUDY RESOURCES

The Web site for this book contains helpful study aids. Log on to:

www.brookscole.com/product/053437557xs

and click on the Chapter-by-Chapter area. Choose Chapter 6 and select a resource:

- **Flash Cards** allows you to test your mastery of the Terms and Concepts to Remember for this chapter.
- **Tutorial Quizzes** provides a multiple-choice practice quiz.
- **Student Guide to InfoTrac College Edition** will lead you to Critical Thinking Projects that use InfoTrac College Edition as a research tool.

TERMS AND CONCEPTS TO REMEMBER

absorption	electron
active sonar	element
adhesion	freezing point
aphotic zone	halocline
atom	heat
calorie	heat capacity
chemical bond	hydrogen bond
cohesion	latent heat of fusion
compound	latent heat of vaporization
covalent bond	light
deep zone	mixed layer
degree	molecule
density curve	passive sonar

photic zone	sofar
polar molecule	sofar layer
proton	sound
pycnocline	state
refraction	surface zone
refractive index	temperature
scattering	thermal inertia
sensible heat	thermocline
shadow zone	thermostatic property
side-scan sonar	water mass

STUDY QUESTIONS

Review Questions

1. Why is water a polar molecule? What properties of water derive from its polar nature?

2. How is heat different from temperature?

3. What is the difference between sensible and nonsensible heat? What is meant by *latent* heat?

4. Why does ice float? Why is this fact important to thermal conditions on Earth?

5. How is heat transported from tropical regions to polar regions?

6. What factors affect the density of water? Why does cold air or water tend to sink? What role does salinity play?

7. How is the ocean stratified by density? What physical factors are involved? What names are given to the ocean's density zones?

8. What factors influence the intensity and color of light in the sea? What factors affect the depth of the photic zone? Could there be a photocline in the ocean?

9. Which moves faster through the ocean—light or sound?

10. How is the pycnocline related to the thermocline and halocline?

Critical Thinking Questions

1. What is the latent heat of vaporization? Of fusion? Which requires more heat to transform water's physical state? Why?

2. How does water's high latent heat influence the ocean? Leaving aside its effect on beach parties, how do you think conditions on Earth would differ if our ocean consisted of ethyl alcohol?

3. How do the seasonal freezing and thawing of the polar ocean areas influence global temperatures?

4. How does refraction permit sound to be transmitted in the ocean for thousands of miles?

5. What factors influence the density of seawater?

6. **InfoTrac College Edition Project**   Sound travels in water both faster and for longer distances than it does in air. Light, by contrast, does not travel as far, and shorter wavelengths, such as red, do not penetrate water more than a few meters. What do these differences suggest for the use of sound rather than light for underwater life and exploration? Use InfoTrac College Edition to back up your answer.

**RESOURCES FOR FURTHER READING
AND RESEARCH**

The Web site for this book contains many ideas for further reading and research. Log on to:

www.brookscole.com/product/053437557xs

and click on the Chapter-by-Chapter area. Choose Chapter 6 and select a resource:

- **References** lists the major books and articles consulted in writing this chapter, along with comments from the author about their content and reading level.

- **Hypercontents** takes you to an extensive list of sites with news, research, and images related to individual sections of the chapter.
- In **Student Guide to InfoTrac College Edition** scroll down to Suggested Readings from InfoTrac College Edition for brief descriptions of the articles listed and search hints for finding them in InfoTrac College Edition.
- **Regional InfoTrac College Edition** articles are organized into East Coast, West Coast, and Gulf Coast regions, allowing you to study oceanography on a more local level. You can also access Regional InfoTrac College Edition at www.localocean.com.

 For additional readings, go to InfoTrac College Edition, your online research library, at:

http://infotrac.thomsonlearning.com/

Seawater Chemistry 7

COOKING UP THE RECIPE FOR WATER

In the fifth century B.C., the Greek philosopher Empedocles believed that all matter was composed of four primary substances in various combinations: earth, air, fire, and water. Divide anything into small enough bits, he wrote, and you would end up with one or more of these four elements, none of which could be further divided. A century later the philosopher Aristotle, considered by many to be the father of modern science, endorsed this concept and gave it credence.

By the 1700s, the four-element theory was being seriously challenged by chemists making many discoveries that could not adequately be explained using Empedocles' theory. A test that would unequivocally disprove the four-element theory would be to separate one of the four elements into smaller components, and water became the target. On 24 June 1783, the French chemist Antoine-Laurent Lavoisier discovered that water could be subdivided, a landmark event in the history of chemistry.

Lavoisier burned two gases in a special vessel. One gas was produced by dripping acid onto iron or copper. This clear, odorless gas was called "inflammable air" because it burned when ignited. The other gas—also colorless and odorless—was produced by heating a red precipitate of mercury. This second gas was known as the "acid maker" because of its tendency to combine with other substances to make mild acids. During the combustion of these gases, droplets of liquid condensed on the cool interior of an attached flask. Tests showed this liquid to be "as pure as distilled water." Lavoisier argued that

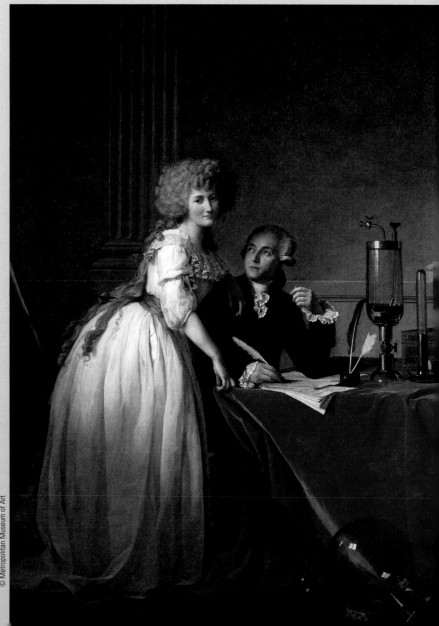

Monsieur and Madame Antoine-Laurent Lavoisier, in a painting by Jacques-Louis David, 1788. M. Lavoisier, a chemist, is credited with the "discovery" of water in 1783. Though a pioneering chemist and one of the founders of the metric system of measurements, Lavoisier could not escape the terrors of the French Revolution. He was sentenced to death by a revolutionary tribunal and guillotined on 8 May 1794.

© Metropolitan Museum of Art

in chemistry, as in mathematics, the whole is equal to the sum of its parts. "Since only water and nothing else was formed . . . the water was the sum of the [masses] of the two gases from which it was produced."

In short, water was not the irreducible element it had always been thought to be. It was a combination—a compound—of the acid maker (*oxys* = acid, *gen* = making) and the newly named "begetter of water" (*hydro* = water, *gen* = making). Oxygen and hydrogen were found to be simpler substances than water. Elements were clearly more numerous than Empedocles had thought, and water more complex.

We know today that water *is*, in a way, more than the sum of its parts. The parts that Lavoisier described are simple enough: hydrogen, the most abundant atom in the universe, and oxygen, an element that accounts for nearly half the mass of Earth's crust. But their combination—water itself—behaves in complex and remarkable ways.

7-1

WHAT TO WATCH FOR IN CHAPTER 7

In Chapter 1 you read that Earth's surface waters are thought to have escaped from the crust and mantle through the process of outgassing. Outgassing of other substances, and the water's ability to dissolve crustal material, have added solids and gases to the ocean. In Chapter 6 you learned about some of water's physical properties and the implications of these properties for the ocean and Earth as a whole. In this chapter we will investigate the chemistry of seawater with its many dissolved substances.

Notice how easily water can dissolve substances—a legacy of its polar construction. We informally call some of these substances "salts," but salts don't actually exist in seawater. Ions do. Watch for them. What are the most abundant ions? How do they cycle in the ocean? Do their proportions change as salinity varies?

Gases also dissolve in seawater. Marine organisms "breathe" dissolved oxygen, and the ocean is an important reservoir of carbon dioxide, a gas critical to photosynthesis and climate control. The chapter ends with a brief discussion of acid–base balance (pH), a concept we will explore again in our upcoming look at marine biology.

When you've finished reading, consider Chapters 6 and 7 together. Your appreciation of the uniqueness of water will grow as your oceanographic perspective expands.

As we have learned, water is a polar molecule (look again at Figures 6.1 and 6.2). Because of the way a water molecule's oxygen electrons are distributed, the overall geometry of the molecule is a bent or angular shape. Each water molecule can be thought of as having a positive (+) end and a negative (−) end. The polar water molecule acts something like a magnet; its positive end attracts particles having a negative charge, and its negative end attracts particles having a positive charge. When water comes into contact with compounds whose elements are held together by the attraction of opposite electrical charges (most salts, for example), the polar water molecule will separate that compound's component elements from each other.

No wonder, then, that seawater and most other liquids in nature are water solutions. A **solution** is made of two components: The **solvent,** usually a liquid, is always the more abundant constituent; the **solute,** often a dissolved solid or gas, is less abundant. In a true solution (sugar in well-stirred coffee, for example), the molecules of the solute are homogeneously dispersed among the molecules of solvent; that is, the solution has uniform properties throughout. In a **mixture,** different substances are closely intermingled but retain separate identities. The properties of a mixture are heterogeneous; they may vary from place to place within the mixture. Think of noodle soup as a mixture of noodles and liquid.

THE DISSOLVING POWER OF WATER

7-2

Water's dissolving power is due to the polar nature of the water molecule. Consider how water dissolves sodium chloride (NaCl), the most common salt.[1] An **ion** is an atom (or small group of atoms) that has an unbalanced electrical charge because it has gained or lost one or more electrons. Unlike the electron *sharing* found in covalently bonded molecules, in NaCl the sodium atoms have *lost* electrons, and the chlorine atoms have *gained* them. The resulting ions are linked by the mutual attraction of their opposite electrical charges. The ions of sodium and chloride in NaCl are said to be held together by **ionic bonds** (**Figure 7.1**), electrostatic attraction that exists between ions that have opposite charge. When NaCl dissolves in water (**Figure 7.2**), the polarity of water reduces the electrostatic attraction (ionic bonding) between the sodium ion (Na^+) and the chloride ion (Cl^-). This causes the sodium ion to separate from the chloride ion. The ions move away from the salt crystal, permitting water to attack the next layer of NaCl. *Note that NaCl does not exist as "salt" in seawater*—its components are separated when salt crystals dissolve in water, but they are joined when crystals re-form as water evaporates.

Some other salts form ions in which groups of atoms bond together and carry an electrical charge. The sulfate ion (SO_4^{2-}), for example, carries two extra electrons. The charge on the overall ion is therefore −2.

---

[1] Na, the chemical symbol for sodium, is derived from *natrium*, the Latin name for sodium.

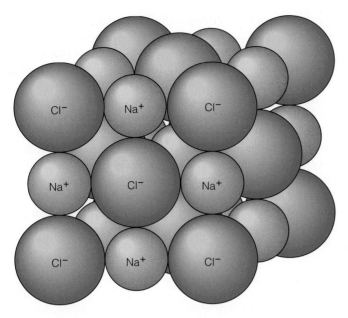

Figure 7.1 Salt in the solid state. In the compound sodium chloride (NaCl), ionic bonds hold the sodium ions (Na$^+$) and the chloride ions (Cl$^-$) together.

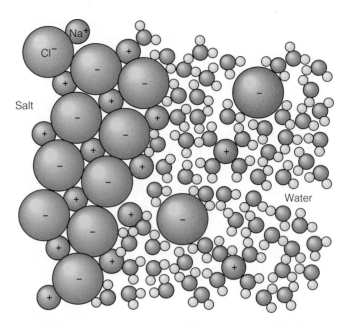

Figure 7.2 Salt in solution. When a salt such as NaCl is placed in water (left side of figure), the positively charged hydrogen end of the polar water molecule is attracted to the negatively charged Cl$^-$ ion, and the negatively charged oxygen end is attracted to the positively charged Na$^+$ ion. The ions are surrounded by water molecules that are attracted to them, and they become solute ions in the solvent (right side of figure).

By contrast, oil doesn't dissolve in water even if the two are very thoroughly shaken together. When oil is dispersed in water it forms a mixture, because molecules of oil are nonpolar in character. This means that oil has no positive or negative charges to attract the polar water molecule. In a way this is fortunate—living tissues would readily dissolve in water if the oils within their membranes didn't blunt water's powerful attack.

As we'll see in a moment, seawater is a complex solution of ions and nonionic solutes. These particles can move through still water by **diffusion,** the random movement of particles through a solution. If you were to drop a large crystal of NaCl into a container of pure water, for example, the water would quickly begin to dissolve the salt. At first, the concentration of Na$^+$ and Cl$^-$ ions would be greater next to the crystal than a short distance away from it; but eventually, the crystal would dissolve completely, and the ions would diffuse evenly through the entire water volume, forming a homogeneous solution. Atmospheric gases dissolve in the ocean surface and diffuse easily through water. Dissolved substances tend to diffuse from regions of high concentration of those substances to regions of low concentration.

When no more of a substance will dissolve in water, the water is said to be *saturated* with that substance. At **saturation** the rate at which molecules of the solute are being dissolved equals the rate at which they are **precipitating** (reforming into crystals) at another location in the solution.

SEAWATER

About 97.2% of the 1,370 million cubic kilometers (329 million cubic miles) of Earth's surface water is marine. By mass, seawater is about 96.5% water and 3.5% dissolved substances,

most of which form salts. The world ocean contains some 5,000 trillion kilograms (5.5 trillion tons) of salt. If the ocean's water evaporated completely, leaving its salts behind, the dried residue could cover the entire planet with an even layer 45 meters (150 feet) thick.

Salinity: Dissolved Solids and Water Together

The total quantity (or concentration) of dissolved inorganic solids in water is its **salinity.** The ocean's salinity varies from about 3.3% to 3.7% by mass, depending on such factors as evaporation, precipitation, and freshwater runoff from the continents, but the average salinity is usually given as 3.5%. Most of the dissolved solids in seawater are salts that have been separated into ions. Sodium and chloride are the most abundant.

The many ions present in seawater react with each other (and with water molecules) in complex ways to modify the physical properties of pure water. Consider the following:

- The heat capacity of water decreases with increasing salinity; that is, less heat is necessary to raise the temperature of seawater by 1° than is required to raise the temperature of fresh water by the same amount.
- Dissolved salts disrupt the webwork of hydrogen bonding in water. As salinity increases, the freezing point of water becomes lower; the salts act as a sort of antifreeze. Sea ice therefore forms at a lower temperature than ice in freshwater lakes.

- Because dissolved salts tend to attract water molecules, seawater evaporates more slowly than fresh water. Swimmers usually notice that fresh water evaporates quickly and completely from their skin, but seawater lingers.
- Osmotic pressure, the pressure exerted on a biological membrane when the salinity of the environment is different from that within the cells, rises with increasing salinity. This is a key factor in transmitting water into and out of cells.

These four properties, which vary with the quantity of solutes dissolved in the water, are called water's **colligative properties** (*colligatus* = to bind together). Because colligative properties are the properties of *solutions*, the more concentrated (saline) water is, the more important these properties become. (Because it is not a solution, pure water has no colligative properties.)

The Components of Salinity

About 3.5% of seawater consists of dissolved substances. Boiling away 100 kilograms of seawater theoretically produces a residue with a mass of 3.5 kilograms. Variations of 0.1% are significant, and oceanographers prefer to use parts-per-thousand notation (‰) rather than percent (%, parts per hundred) in discussing these materials.[2] The seven ions listed below oxygen and hydrogen in **Table 7.1** make

---

[2] Note that 3.5% = 35‰. If you begin with 1,000 kilograms of seawater, you would expect 35 kilograms of residue.

up more than 99% of this residual material. When seawater evaporates, its ionic components combine in many different ways to form table salt ($NaCl$), epsom salts ($MgSO_4$), and other mineral salts. **Figure 7.3** shows the proportions of ions in seawater.

Seawater also contains minor constituents. The ocean is sort of an "Earth tea"—every element present in the crust and atmosphere is also present in the ocean, though sometimes in extremely small amounts. Only 14 elements have concentrations in seawater larger than 1 part per million (ppm). Elements present in amounts less than 0.001‰ (1 part per million) are known as **trace elements.** These are more easily given in parts per billion (ppb). **Table 7.2** lists some minor and trace elements, many of which are crucial to life processes.

The Source of the Ocean's Salts

Remembering the effectiveness of water as a solvent, you might think that the ocean's saltiness has resulted from the ability of rain, groundwater, or crashing surf to dissolve crustal rock. Much of the sea's dissolved material originated in that way, but is crustal rock the source of all the ocean's solutes? An easy way to find out would be to investigate the composition of salts in river water and compare these figures to those of the ocean as a whole. If crustal rock is the only source, the salts in the ocean should be like those of concentrated river water. But they are not. River water is usually a dilute solution of calcium and bicarbonate ions, while the principal ions in seawater are chloride and sodium. The magnesium content of seawater would also be higher if seawater were simply concentrated river water. The proportions

Table 7.1	Major Constituents of Seawater at 34.4‰ Salinity		
Constituent	Concentration in Parts per Thousand (‰), or Grams per Kilogram (g/kg)	Percent by Mass	Total Percent
Water Itself			
Oxygen	857.8	85.8%	96.5%
Hydrogen	107.2	10.7%	
The Most Abundant Ions			
Chloride (Cl^-)	18.980	1.9%	
Sodium (Na^+)	10.556	1.1%	
Sulfate (SO_4^{2-})	2.649	0.3%	
Magnesium (Mg^{2+})	1.272	0.1%	3.4%
Calcium (Ca^{2+})	0.400	0.04%	
Potassium (K^+)	0.380	0.04%	
Bicarbonate (HCO_3)	0.140	0.01%	
Total	**999.377 g/kg**	**99.99%**	

Source: Adapted from Walton-Smith, 1994.

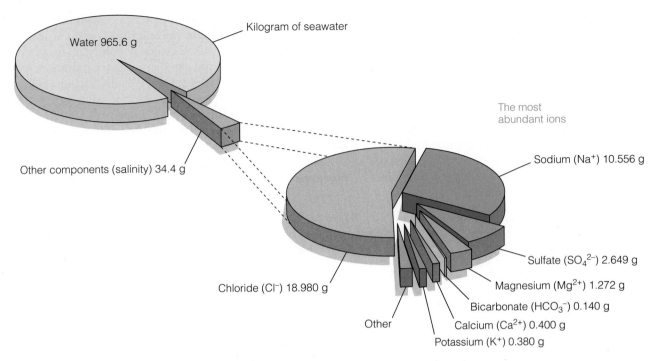

Figure 7.3 A diagrammatic representation of the most abundant components of a kilogram of seawater at 35‰ salinity. Note that the specific ions are represented in grams per kilogram, equivalent to parts per thousand (‰).

Constituent	Concentration in Parts per Thousand (‰), or g/kg		Concentration in Parts per Million, or mg/kg (mg/L)	Concentration in Parts per Billion, or mg/1,000 kg
Table 7.2 Minor and Trace Elements in Seawater at 35‰ Salinity				
Minor Elements				
Bromine (Br)	0.065	=	65	
Strontium (Sr)			8	
Boron (B)			4	
Silicon (Si)			3	
Fluorine (F)			1	
Important Trace Elements				
Nitrogen (N)[a]			0.28 = 280	
Lithium (Li)				125
Iodine (I)				60
Phosphorus (P)				30
Zinc (Zn)				10
Iron (Fe)				6
Aluminum (Al)				2
Manganese (Mn)				2
Lead (Pb)				0.04
Mercury (Hg)				0.03
Gold (Au)				0.000004

Source: Adapted from Kennish, 1994.
[a]Refers to nitrogen available as a nutrient, not as dissolved gas.

of salts in isolated salty inland lakes, such as Utah's Great Salt Lake or the Dead Sea, are much different from the proportions of salts in the ocean. Thus, weathering and erosion of crustal rocks cannot be the only source of sea salts.

The components of ocean water whose proportions are *not* accounted for by the weathering of surface rocks are called **excess volatiles.** To find the source of these excess volatiles, we must look to Earth's deeper layers. The upper mantle appears to contain more of the substances found in seawater (including the water itself) than are found in surface rocks, and their proportions are about the same as found in the ocean. As you read in Chapter 3, convection currents slowly churn Earth's mantle, causing the movement of tectonic plates. This activity allows some deeply trapped volatile substances to escape to the exterior, outgassing through volcanoes and rift vents. These excess volatiles include carbon dioxide, chlorine, sulfur, hydrogen, fluorine, nitrogen, and, of course, water vapor. This material, along with residue from surface weathering, accounts for the chemical constituents of today's ocean.

Some of the ocean's solutes are hybrids of the two processes of weathering and outgassing (see **Figure 7.4**).

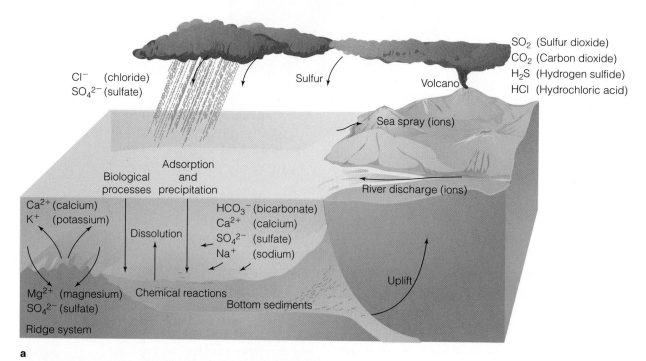

a

b

Figure 7.4 Processes that regulate the major constituents in seawater. (**a**) Ions are *added* to seawater by rivers running off crustal rocks, volcanic activity, groundwater, hydrothermal vents and cold springs, and the decay of once-living organisms. Ions are *removed* from the ocean by chemical entrapment as water percolates through the mid-ocean ridge systems, or by spray, uptake by living organisms, or incorporation into sediments (and ultimately by subduction). (**b**) An active hydrothermal vent spouts superheated seawater. Recent research at a spreading center east of the Galápagos Islands suggests that mid-ocean rifts may play a role in reducing the magnesium content of seawater and increasing its calcium content. The water that circulates through new ocean floor at these sites is apparently stripped of magnesium and a few other elements. The magnesium seems to be incorporated into mineral deposits, but calcium is added to the seawater as hot water dissolves adjacent rocks. All the water in the ocean is thought to cycle through the seabed at rift zones every 1 to 10 million years.

Table salt, or sodium chloride, is an example. The sodium ions come from the weathering of crustal rocks, while the chlorine ions come from the mantle by way of volcanic vents and outgassing from mid-ocean rifts. As for the lower-than-expected quantity of magnesium and sulfate ions in the ocean, recent research at a spreading center east of the Galápagos Islands suggests that mid-ocean rifts may play a role in reducing the magnesium and sulfate content and increasing the calcium and potassium content of seawater. The water that circulates through new ocean floor at these sites is apparently stripped of magnesium and a few other elements. The magnesium seems to be incorporated into mineral deposits, but calcium is added to the seawater as hot water dissolves adjacent rocks.

The Principle of Constant Proportions

In 1865, the chemist Georg Forchhammer noted that although the total *amount* of dissolved solids (salinity) might vary between samples, the *ratio* of major salts in samples of seawater from many locations was constant. In other words, the percentage of various salts in seawater is the same in samples from many places, regardless of how salty the water is. This constant ratio is known as **Forchhammer's principle,** or the **principle of constant proportions.** Forchhammer was also the first to observe that seawater contains fewer silica and calcium ions than concentrated river water, and the first to realize that removal of these compounds by marine animals and plants to form shells and other hard parts might account for some of the difference. The English chemist William Dittmar, working with HMS *Challenger* samples ten years after they had been collected, confirmed Forchhammer's principle of constant proportions. Building on Forchhammer's and Dittmar's work, and taking advantage of improved analytical techniques, researchers have established a reliable way to determine salinity.

Determining Salinity

Water's salinity by mass would seem an easy property to measure. Why not simply evaporate a known weight of seawater and weigh the residue? This simple method yields imprecise results because some salts will not release all the molecules of water associated with them. If these salts are heated to drive off the water, other salts (carbonates, for example) will decompose to form gases and solid compounds not originally present in the water sample.

Because of seawater's constancy of composition, we need to measure the concentration of only one major constituent to find the salinity of a water sample. The chloride ion is abundant, comparatively easy to measure, and always accounts for the same proportion of dissolved solids (55.04%), so marine chemists devised the concept of chlorinity to simplify the measurement of salinity. **Chlorinity** is a measure of the total mass of halogen ions in seawater. (The halogens are a chemical family that includes fluorine, chlorine, bromine, and io-

dine.) By international agreement, the following formula is used to determine salinity:

$$\text{Salinity in \textperthousand} = 1.80655 \times \text{Chlorinity in \textperthousand}$$

Chlorinity is about 19.2‰, so salinity is around 34.7‰.

Seawater samples can be obtained by methods ranging from tossing a clean bucket over the side of the ship to sophisticated tube-and-pump systems. Typically water samples are collected using a string of sampling bottles similar to those perfected early in this century by the Norwegian scientist and explorer Fridtjof Nansen. A **Nansen bottle** (and its modern equivalent, a **Niskin bottle**) is open at both ends (**Figure 7.5**). A series of bottles is lowered on a wire to known depths. Each bottle is then triggered to close by a brass weight (appropriately called a "messenger") sent sliding down the line. Sometimes electrical or acoustic impulses are used to signal the bottles to close. The bottles are hauled to the surface and their contents analyzed.

Until recently, marine chemists used a delicate chemical procedure involving a silver nitrate solution to measure the chlorinity of seawater. Conversion to salinity was made by a set of mathematical tables. The procedure was calibrated against a standard sample of seawater of precisely known chlorinity. Today's marine scientists usually use an electronic device called a **salinometer** that measures the electrical conductivity of seawater (see **Figure 7.6**). Conductivity varies with the concentration and mobility of ions present and with water temperature. Circuits in the salinometer adjust for water temperature, convert conductivity to salinity, and then display salinity. Salinometers are calibrated against a sample of known conductivity and salinity. The best salinometers can determine salinity to an accuracy of 0.001%. Some salinometers are designed for remote sensing—the electronics stay aboard ship while the sensor coil is lowered over the side. A quick way to measure salinity in the field is with a **refractometer** (**Figure 7.7**), a compact optical device that compares the degree to which light is bent by a sample of seawater to the degree of bending for water of known salinity. A refractometer is not as accurate as a good salinometer.

Chemical Equilibrium and Residence Times

If outgassing and the chemical weathering of rock are constant processes, shouldn't the ocean become progressively saltier with age? Landlocked seas and some lakes usually become saltier as they grow older, but the ocean does not. The ocean appears to be in **chemical equilibrium;** that is, the proportion *and amounts* of dissolved salts per unit volume of ocean are nearly constant. Evidently, whatever goes in must come out somewhere else.

Geologists in the 1950s developed the concept of a "steady state ocean." The idea suggests that ions are added to the ocean at the same rate as they are being removed. This theory helps explain why the ocean is not growing saltier. The idea was quantified by T.F.W. Barth, who in 1952 devised the

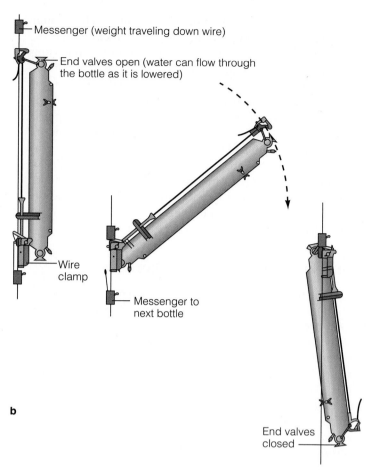

Messenger (weight traveling down wire)

End valves open (water can flow through the bottle as it is lowered)

Wire clamp

Messenger to next bottle

End valves closed

a

b

Figure 7.5 Sampling bottles. (**a**) A Nansen bottle being prepared for deployment. (**b**) When the messenger (a small weight) slides down the line, it triggers the bottle to flip upside down, closing the end valves at the same time. Nansen bottles are being replaced by sampling bottles that are able to hold greater volumes of water and are more resistant to additional triggering, such as the 10-liter Niskin bottles shown in (**c**). Each Niskin bottle is remotely triggered by a signal from the research ship when the array reaches a predetermined depth.

concept of **residence time,** the average length of time an element spends in the ocean. Residence time for a particular element may be calculated by this equation:

$$\text{Residence time} = \frac{\text{Amount of element in the ocean}}{\text{The rate at which the element is added to (or removed from) the ocean}}$$

Additions of salts from the mantle or from the weathering of rock are balanced by subtractions of minerals being bound into sediments. Dissolved salts precipitate out of the water and the hard parts of living organisms containing silicon and calcium carbonate drift slowly down to the seabed. Some of these sediments are removed from the ocean and drawn into the mantle at subduction zones by the cycling of crustal plates. Input from runoff and outgassing equals outfall (binding into sediments) for each dissolved component.

The residence time of an element depends on its chemical activity. Atoms (or ions) of some elements, such as aluminum and iron, remain in seawater for a relatively short time before becoming incorporated in sediments; others, such as

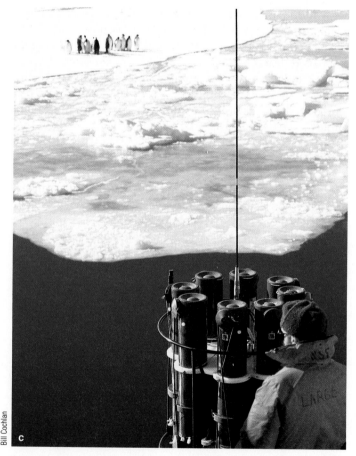

c

chloride, sodium, and magnesium, remain in water for millions of years. The approximate residence times for the major constituents of seawater are shown in **Table 7.3.** Even seawater itself has a residence time (see **Box 7.1**).

Mixing Time 7-10

If constituent minerals remain in ocean water longer than the ocean's mixing time, they will become evenly distributed through the ocean. Because of the vigorous activity of currents, the **mixing time** of the ocean is thought to be on the order of 1,600 years—thus, the ocean has been mixed hundreds of thousands of times during its long history. The relatively long residence times of seawater's major constituents assure thorough mixing, which is the basis of Forchhammer's principle of constant proportions.

Conservative and Nonconservative Constituents 7-11

As we have seen, the major constituent salts maintain constant ratios in seawater, but the quantity and proportions of minor and trace elements can change because living things take them up or excrete them or local geological events add or remove them.

Figure 7.6 This portable salinometer reads temperature, pH, and dissolved oxygen as well as conductivity. Designed to be lowered at the end of a line over the side of a research vessel, the self-powered device contains a small pump that passes water over sensors. This model takes two readings every second and operates down to about 200 meters (660 feet). Data may be retrieved by a cable connected to the research vessel or stored in the salinometer's memory to be retrieved when it is brought back aboard ship. A microprocessor contained within the white tube converts conductivity to salinity and, when connected to a portable computer, displays the results of all readings as graphs.

Figure 7.7 A refractometer in operation. The observer places a drop of water on a glass plate on top of the device and then looks into the refractometer to determine the position of a line against a scale. The degree to which the line has been refracted (bent) depends on the salinity of the water sample.

Ocean water has a residence time. Scientists estimate that about 334,000 cubic kilometers (80,000 cubic miles) of pure water evaporates from the ocean each year. Assuming a total ocean volume of 1,370 million cubic kilometers (329 million cubic miles), we can calculate that about 0.024% of the water in the ocean vaporizes each year. Thus, if it were not replenished by precipitation and runoff from land, the ocean would evaporate completely in about 4,100 years.

For a complete picture, more than oceanic water should be considered. Earth's total surface-water volume—including the ocean, lakes, rivers, groundwater, ice caps, and so on—is about 1,399 million cubic kilometers (336 million cu-

bic miles). The total yearly evaporation/precipitation estimate is 396,000 cubic kilometers (95,000 cubic miles) of pure water. From these figures we again calculate a recycling time of about 4,100 years. With the exception of water subducted into the mantle along with sediments at subduction zones, then, water molecules have a residence time in the ocean of about 4,100 years.

Earth's ocean is more than 4 billion years old. Thus, on average, individual water molecules have evaporated from and returned to the ocean more than a million times since the world ocean formed in its basins!

7-12

Those constituents of seawater that occur in constant proportion or change very slowly through time are **conservative constituents.** Conservative elements have long residence times. Not surprisingly, these are the most abundant dissolved constituents of seawater—the ones that constitute the bulk of the ocean's dissolved material, listed at the top of Table 7.3. The inert gases dissolved in the ocean (and the water of the ocean itself) are also conservative constituents.

Nonconservative constituents are those substances dissolved in seawater that are tied to biological or seasonal cycles or to very short geological cycles. They have short resi-

dence times. Biologically important nonconservative constituents include dissolved oxygen produced by plants, carbon dioxide produced by animals, silica and calcium compounds needed for plant and animal shells, and the nitrates and phosphates needed for production of protein and other biochemicals. Aluminum, with a residence time of only 600 years, is rapidly removed by absorption onto clay sediment particles and other objects, so it is a nonconservative element. Aluminum is very rare in seawater (10 parts per billion). Many trace elements are in great biological demand or tend to form insoluble compounds in water. (Oceanic residence times calculated to be less than 1,600 years—the ocean's mixing time—are mainly a function of strong local demand variations. The definition of residence time assumes the existence of steady-state conditions and a well-mixed ocean, so a residence time less than the ocean's mixing time is mainly an indication of an element's reactivity.)

DISSOLVED GASES

Most gases in the air readily dissolve in seawater at the ocean's surface. Plants and animals living in the ocean require these dissolved gases to survive—no marine animal has the ability to break down water molecules to obtain oxygen directly, and no marine plant can manufacture enough carbon dioxide to support its own metabolism. In order of their relative abundance, the major gases found in seawater are nitrogen, oxygen, and carbon dioxide (see **Table 7.4**). The proportions of dissolved gases in the ocean are very different from the proportions of the same gases in the atmosphere, because of differences in their solubility in water and in air.

Unlike solids, gases dissolve most readily in *cold* water. A cubic meter of chilly polar water usually contains a greater

Table 7.3	Approximate Residence Times for Constituents of Seawater
Constituent	**Residence Time (years)**
Chloride (Cl⁻)	100,000,000
Sodium (Na⁺)	68,000,000
Magnesium (Mg²⁺)	13,000,000
Potassium (K⁺)	12,000,000
Sulfate (SO₄²⁻)	11,000,000
Calcium (Ca²⁺)	1,000,000
Carbonate (CO₃²⁻)	110,000
Silicon (Si)	20,000
Water (H₂O)	4,100
Manganese (Mn)	1,300
Aluminum (Al)	600
Iron (Fe)	200

Sources: Data from Broecker and Peng, 1982; Bruland, 1983; Riley and Skirrow, 1975.

Table 7.4 Major Gases in the Atmosphere and Ocean			
Gas	Percent of Gas in Atmosphere, by Volume	Percent of Dissolved Gas in Seawater, by Volume	Concentration in Seawater in Parts per Million, by Mass
Nitrogen (N_2)	78.08%	48%	10–18 ppm
Oxygen (O_2)	20.95%	36%	0–13 ppm
Carbon dioxide (CO_2)[a]	0.035%	15%	64–107 ppm

Sources: Data from Weihaupt, 1979; Hill, 1963.
[a]Also present in seawater as carbonic acid, carbonate ions, and bicarbonate ions.

volume of dissolved gases than a cubic meter of warm tropical water. Dissolved nitrogen gas is a conservative component of seawater, while biologically active oxygen and carbon dioxide are nonconservative components.

Nitrogen

About 48% of the dissolved gas in seawater is nitrogen. (In contrast, the atmosphere is slightly more than 78% nitrogen by volume.) The upper layers of ocean water are usually saturated with nitrogen; that is, additional nitrogen will not dissolve. Living organisms require nitrogen to build proteins and other important biochemicals, but they cannot use the free nitrogen in the atmosphere and ocean directly. It must first be "fixed" into usable chemical forms by specialized organisms. Though some species of bottom-dwelling bacteria can manufacture usable nitrates from the nitrogen dissolved in seawater, most nitrogen compounds needed by living organisms must be recycled among the organisms themselves.

Oxygen

About 36% of the gas dissolved in the ocean is oxygen, but there is about a hundred times more gaseous oxygen in Earth's atmosphere than is dissolved in the whole ocean. An average of 6 milligrams of oxygen is dissolved in each liter of seawater (that is, 6 parts of oxygen per million parts of seawater, by mass). Yet this small amount of oxygen is a vital resource for animals that extract oxygen with gills. The sources of the ocean's dissolved oxygen are the photosynthetic activity of plants and plantlike organisms and the diffusion of oxygen from the atmosphere.

Carbon Dioxide

The amount of carbon dioxide (CO_2) in the atmosphere is very small (0.03%) because CO_2 is in great demand by photosynthetic plants as a source of carbon for growth. Carbon dioxide is very soluble in water though; the proportion of dissolved CO_2 in water is about 15% of all dissolved gases. Because CO_2 combines chemically with water to form a weak acid (carbonic acid, H_2CO_3), water can hold perhaps a thousand times more carbon dioxide than either nitrogen or oxygen at saturation. Carbon dioxide is quickly used by marine

photosynthesis, however, so dissolved quantities of CO_2 are almost always much less than this theoretical maximum. Even so, at the present time there is about 60 times as much CO_2 dissolved in the ocean as in the atmosphere.

CO_2 moves quickly from atmosphere to ocean, but more slowly from ocean to atmosphere. This occurs because some dissolved CO_2 forms carbonate ions, which combine with calcium ions in seawater to form the calcium carbonate ($CaCO_3$) used by many marine organisms to build shells and skeletons. When these organisms die, their coverings and bones sink to form sediments that may, in time, become limestone rock. Most of Earth's surface carbon—10,000 times that in the total mass of all life on Earth—is stored in sediments. This carbon slowly reenters the cycle as sediments dissolve and re-form CO_2 that can enter the atmosphere. Geological uplift can also bring sediments to the surface, exposing the carbonate rock to chemical attack by oxygen and conversion to CO_2. Acidic rain falling on exposed limestone rock also releases some CO_2 back into the atmosphere. (This cycle can be seen in Figure 7.10.)

Figure 7.8 illustrates how carbon dioxide and oxygen concentrations vary with depth. Carbon dioxide concentrations increase with depth, but oxygen concentrations usually decrease through the middle depths and then rise again toward the bottom. High concentrations of oxygen at the surface are usually by-products of photosynthesis in the ocean's brightly lit upper layer. Since plants and plantlike organisms require carbon dioxide for photosynthesis, surface CO_2 concentrations tend to be low. A decrease in oxygen below the sunlit upper layer is usually due to the respiration of bacteria and marine animals, which leads to higher concentrations of carbon dioxide. Oxygen levels are slightly higher in deeper water because fewer animals are present to take up oxygen reaching these depths and because oxygen-rich polar water is the greatest source of deep water.

ACID–BASE BALANCE

Water can separate to form hydrogen ions (H^+) and hydroxide ions (OH^-). These two ions are present in equal concentrations in pure water. An imbalance in the proportions of ions produces an acidic or a basic solution. An **acid** is a substance that *releases* a hydrogen ion in solution; a **base** is a substance that *combines with* a hydrogen ion in solution.

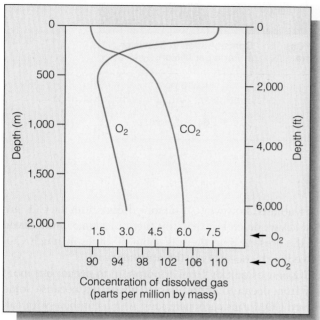

Figure 7.8 How concentrations of oxygen and carbon dioxide vary with depth. Oxygen is abundant near the surface because of the activity of marine photosynthesizers. Oxygen concentration decreases below the sunlit layer because of the respiration of marine animals and bacteria and the oxygen consumed by the decay of tiny dead organisms slowly sinking through the area. In contrast, plants and plantlike organisms use carbon dioxide during photosynthesis, so surface levels of CO_2 are low. Photosynthesis cannot take place in the dark, so CO_2 given off by animals and bacteria tends to build up at depths below the sunlit layer. CO_2 also increases with depth because its solubility increases as pressure increases and temperature decreases.

A solution containing a base is also called an **alkaline** solution.

Acidity or alkalinity, determined by the concentration of hydrogen ions in a solution, is measured in terms of the **pH scale**. An excess of hydrogen ions (H^+) in a solution makes that solution acidic. An excess of hydroxide ions (OH^-) makes a solution alkaline. **Figure 7.9** shows a pH scale and the pH of a few familiar solutions. The scale is logarithmic, which means a change of 1 pH unit represents a ten-fold change in hydrogen ion concentration. Thus, a modern nonphosphate detergent is a thousand times more alkaline than seawater, and black coffee is a hundred times more acidic than pure water. Pure water, which is neutral (neither acidic nor alkaline), has a pH of 7; lower numbers indicate greater acidity (more H^+ ions), and higher numbers indicate greater alkalinity (fewer H^+ ions).

Seawater is slightly alkaline; its average pH is about 8.0. This seems odd because of the large amount of CO_2 dissolved in the ocean. If dissolved CO_2 combines with water to form carbonic acid, why is the ocean mildly alkaline and not slightly acidic? When dissolved in water, CO_2 is actually present in several different forms. Carbonic acid (H_2CO_3) is only one of these. In water solutions, some carbonic acid breaks down to produce the hydrogen ion (H^+), the bicarbonate ion (HCO_3^-), and the carbonate ion (CO_3^{2-}). This sequence is shown in **Figure 7.10.** If acid is added to seawater, and the pH of

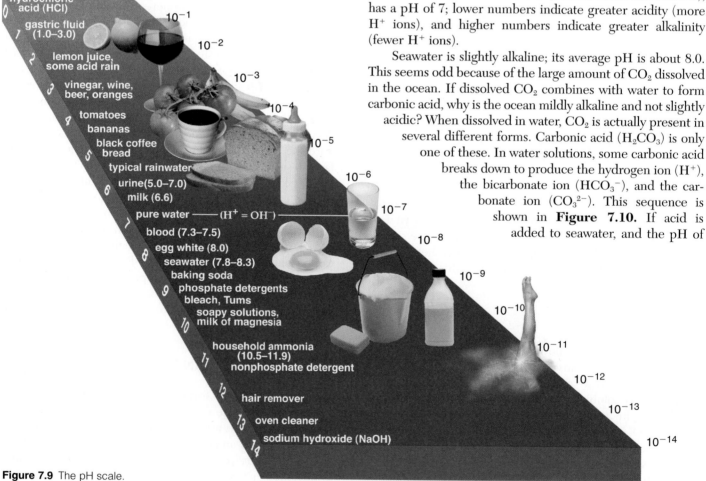

Figure 7.9 The pH scale.

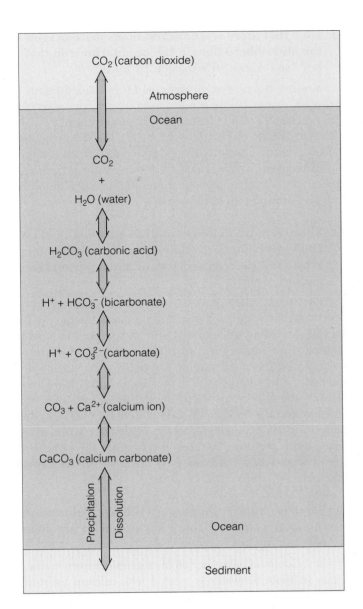

CO₂ (carbon dioxide)

Atmosphere

Ocean

CO_2

+

H_2O (water)

H_2CO_3 (carbonic acid)

$H^+ + HCO_3^-$ (bicarbonate)

$H^+ + CO_3^{2-}$ (carbonate)

$CO_3 + Ca^{2+}$ (calcium ion)

$CaCO_3$ (calcium carbonate)

Precipitation | Dissolution

Ocean

Sediment

Figure 7.10 Carbon dioxide (CO_2) combines readily with seawater to form carbonic acid (H_2CO_3). Carbonic acid can then lose an H^+ ion to become bicarbonate (HCO_3^-), or two H^+ ions to become carbonate (CO_3^{2-}). Some bicarbonate dissociates to form carbonate ions, which combine with calcium ions in seawater to form calcium carbonate, used by some organisms to form hard shells and skeletons. When their builders die, these structures may fall to the seabed as sediments, eventually to be redissolved. As the double-headed arrows indicate, all these reactions may move in either direction.

the ocean drops (becomes more acidic), the reaction in Figure 7.10 proceeds *upward,* raising the pH by removing H^+ ions. If an alkaline solution is added, and the pH of the ocean rises (becomes more alkaline), the reaction proceeds *downward,* lowering the pH by releasing H^+ ions. This behavior acts to **buffer** the water, preventing broad swings of pH when acids or bases are introduced. Rivers and lakes lack buffering, making them more susceptible to pH swings when acids or bases are added at waste outfalls.

Though seawater remains slightly alkaline, it is subject to some variation. In areas of rapid photosynthesis, for example, pH will rise because CO_2 is used by plants and plantlike organisms for the production of sugars. The reactions of Figure 7.10 go upward, removing free H^+ ions. Because temperatures are generally warmer at the surface, less CO_2 can dissolve in the first place. Thus, surface pH in warm productive water is usually around 8.5. **Figure 7.11** shows the relative abundance of carbonic acid, bicarbonate, and carbonate in seawater as a function of pH. Seawater has an average pH of about 8, and the bicarbonate ion is most prevalent. As the graph shows, carbonic acid dominates at low pH, and the carbonate ion dominates at higher pH.

At middle depths and in deep water, more CO_2 may be present. Its source is the respiration of animals and bacteria and the decay of the remains of tiny organisms falling from

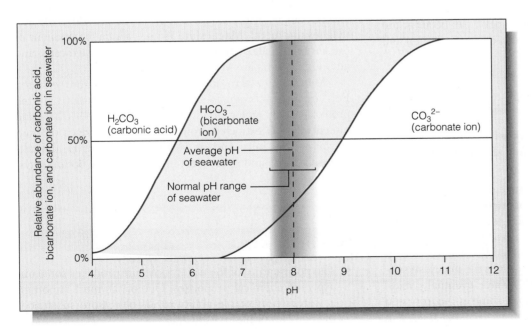

Figure 7.11 The relative abundance of carbonic acid, bicarbonate, and carbonate in seawater is a function of pH. Seawater has an average pH of about 8, at which level the bicarbonate ion is most prevalent. As the graph shows, carbonic acid dominates at low pH, and the carbonate ion dominates at higher pH.

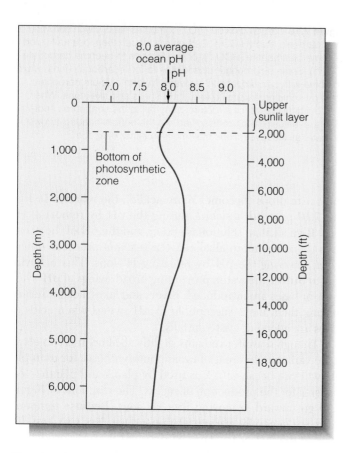

8.0 average
ocean pH

Figure 7.12 The variation in pH with depth.

the sunlit layer above. With cold temperatures, high pressure, and no photosynthetic plants to remove it, this CO_2 will lower the pH of seawater, making it more acidic with depth (see **Figure 7.12**). Thus, deep, cold seawater below 4,500 meters (15,000 feet) has a pH of around 7.5. This lower pH can dissolve calcium-containing marine sediments. (Recall from Chapter 5 that sediments containing calcium carbonate are rarely found in deep water.) A drop to pH 7 can occur at the deep-ocean floor when bottom bacteria consume oxygen and produce hydrogen sulfide.

QUESTIONS FROM STUDENTS

1. If outgassing continues today, why aren't the oceans getting bigger? By now shouldn't they have covered the continents?

Well, yes, outgassing does continue, but the water is being drawn from the mantle. Some marine geologists believe that the mantle shrinks a little, making the ocean basins a bit deeper, which accommodates the ocean's greater volume. Also, geological processes have increased the amount of continental crust, and water is drawn back into the mantle at subduction zones. Though some of this water eventually reappears at the surface in volcanic vapor, some is incorporated back into mantle material. It probably just about balances out.

2. I see that there is a hundred times more oxygen in the atmosphere than in the ocean. How can that be? Isn't water 86% oxygen by mass?

Yes, but the oxygen of water (H_2O) is bonded tightly to hydrogen atoms. Unlike atmospheric oxygen, the oxygen in water molecules cannot be used by organisms for respiration. Fish can't extract this oxygen with their gills. Marine animals must depend on dissolved oxygen, paired molecules of oxygen (O_2) present in the water and free to move through the gill membranes. Compared to the atmosphere, very little of this free oxygen is present in the ocean.

3. There is more than $3.1 million worth of gold (at 2000 prices) in each cubic kilometer of seawater. Why don't we remove some of it and pay off the national debt?

Easy to say, difficult to do. For one thing, extracting the gold would require sophisticated chemical treatment. But for the sake of argument, let's assume you perfect a magic method of precipitating gold from seawater with a simple wave of your hand. All you have to do is pump water out of the ocean into a large holding tank to do your trick. Unfortunately the energy cost of lifting the water just a few millimeters into the tank would eat up all the profits, and then some. German economists made a study after World War I to see if this method could be used to retire their war debt. It was shown to be fruitless. So much for get-rich-quick schemes.

4. Some essential resources are relatively abundant in ocean water. Are there any prospects for chemical "mining" of seawater?

Yes. Commercial extraction of magnesium from seawater began in 1940. Seawater is treated with calcium hydroxide and hydrochloric acid to form magnesium chloride. This substance is dried and electrolytically separated into chlorine gas (which can be used to make more hydrochloric acid) and magnesium metal. Magnesium is an essential component of the aluminum alloys from which aircraft, beverage cans, and automobile parts are manufactured.

Many nonmetal resources are also obtained from seawater. For example, bromine—an element important in the production of motor fuels, photographic film, dyes, and insecticides—is extracted from concentrated brines or from crystallized sea salts. And, as we'll see in Chapter 17, the salts themselves are among the most important of the ocean's physical resources.

KEY CONCEPTS TO REVIEW

• The polar nature of the water molecule produces some unexpected chemical properties. One of the most important is water's remarkable ability to dissolve more substances than any other natural solvent.

- The quantity of dissolved inorganic solids in water is its salinity. The proportion of ions in seawater is not the same as the proportion in concentrated river water, which indicates that ongoing geological and chemical processes affect the ocean's salinity.

- The most abundant ions dissolved in seawater are chloride, sodium, and sulfate.

- Though most solids and gases are soluble in water, the ocean is in chemical equilibrium, and neither the proportion nor the amount of most dissolved substances changes significantly through time.

- Gases dissolve in water in proportions that vary with their physical properties. Nitrogen is the most abundant dissolved gas in seawater; oxygen is the second most abundant. Carbon dioxide is the most soluble gas and is one of many substances that affect the ocean's pH balance.

- Seawater acts as a buffer, preventing broad swings of pH when acids or bases are introduced.

INTERNET STUDY RESOURCES

The Web site for this book contains helpful study aids. Log on to:

www.brookscole.com/product/053437557xs

and click on the Chapter-by-Chapter area. Choose Chapter 7 and select a resource:

- **Flash Cards** allows you to test your mastery of the Terms and Concepts to Remember for this chapter.
- **Tutorial Quizzes** provides a multiple-choice practice quiz.
- **Student Guide to InfoTrac College Edition** will lead you to Critical Thinking Projects that use InfoTrac College Edition as a research tool.

TERMS AND CONCEPTS TO REMEMBER

acid	ion
alkaline	ionic bond
base	mixing time
buffer	mixture
chemical equilibrium	Nansen bottle
chlorinity	Niskin bottle
colligative properties	nonconservative constituent
conservative constituent	pH scale
diffusion	precipitate
excess volatiles	principle of constant
Forchhammer's principle	proportions

refractometer	solute
residence time	solution
salinity	solvent
salinometer	trace element
saturation	

STUDY QUESTIONS

Review Questions

1. Why is water a polar molecule? What properties of water derive from its polar nature?

2. How does a crystal of common salt dissolve in water?

3. How does a solution differ from a mixture?

4. What are some of seawater's colligative properties? Does pure water have colligative properties?

5. Other than hydrogen and oxygen, what are the most abundant elements in seawater?

6. How is salinity determined?

7. What factors affect seawater's pH? How does the pH of seawater change with depth? Why?

Critical Thinking Questions

1. Where did the water of the ocean come from?

2. How are modern methods of determining salinity dependent on the principle of constant proportions?.

3. What was the earthly origin of the sodium and chloride ions in common table salt?

4. Technically, there are no "salts" in seawater. How can that be?

5. How are seawater's conservative constituents different from its nonconservative constituents? Give an example of each.

6. Which dissolved gas is present in the ocean in much greater proportion than in the atmosphere? Why the disparity?

7. **InfoTrac College Edition Project** Water is considered a "universal solvent" because a great number of chemical compounds can be dissolved in it. However, insoluble compounds that end up at the bottom of the ocean may stay there a very long time, unavailable to life forms on the surface. By contrast, too much of a dissolved substance may be toxic to marine life. Do some research using InfoTrac College Edition and write an essay exploring some of the consequences of soluble and insoluble nutrients.

 **Systems Today CD Question**

1. Go to "Atoms and Crystals," then "Atomic Behavior." Although the example deals with magma rather than water, note the presence of covalent bonds (shown as translucent rods in the silica tetrahedron) and Brownian motion. Both are present in water molecules in liquids.

RESOURCES FOR FURTHER READING AND RESEARCH

The Web site for this book contains many ideas for further reading and research. Log on to:

www.brookscole.com/product/053437557xs

and click on the Chapter-by-Chapter area. Choose Chapter 7 and select a resource:

- **References** lists the major books and articles consulted in writing this chapter, along with comments from the author about their content and reading level.

- **Hypercontents** takes you to an extensive list of sites with news, research, and images related to individual sections of the chapter.
- In **Student Guide to InfoTrac College Edition** scroll down to Suggested Readings from InfoTrac College Edition for brief descriptions of the articles listed and search hints for finding them in InfoTrac College Edition.
- **Regional InfoTrac College Edition** articles are organized into East Coast, West Coast, and Gulf Coast regions, allowing you to study oceanography on a more local level. You can also access Regional InfoTrac College Edition at www.localocean.com.

 For additional readings, go to InfoTrac College Edition, your online research library, at:

http://infotrac.thomsonlearning.com/

Atmospheric Circulation

8

RIDING THE WINDS

In March 1999 Brian Jones and Bertrand Piccard rode the winds around the world. Housed in a cylindrical capsule beneath a balloon 55 meters (180 feet) high, the adventurers were blown eastward in fast-moving ribbons of air—jet streams—that form between warm and cold air masses (**Figure a**). Their flight illustrates the general west-to-east flow of the winds at mid-latitudes—a flow caused by Earth's uneven solar heating and daily rotation. This circulation of air is important in creating climate and weather, driving ocean currents, and steering storms.

Piccard, a Swiss psychiatrist, is the son of Jacques Piccard, inventor of the bathyscaphe and builder of *Trieste* (see Chapter 4 opener). Jones is a former British Royal Air Force pilot. Their journey in *Breitling Orbiter 3* began in a quiet valley between snow-capped peaks in Chateau-d'Oex, Switzerland, where compartments in the balloon were filled with helium and hot air (**Figure b**). It ended near the Egyptian oasis town of Mût 19 days and 21 hours later. They traveled about 45,800 kilometers (28,500 miles), setting records for distance and duration of flight.

In March 1999, Bertrand Piccard and Brian Jones became the first people to fly around the world in a balloon. The balloonists took advantage of jet streams (**a**), swiftly flowing currents of air that move in a west-to-east direction. (**b**) *Breitling Orbiter 3* floats near its launch point in Switzerland. The flight lasted 19 days, 21 hours.

AP/Wide World Photos

b

The flight was not without its difficulties. Balloons necessarily go wherever the wind is blowing, but by adjusting altitude the pilots could move between layers of air going in different directions. Sometimes they found themselves in winds headed the wrong way, or becalmed and going nowhere. Cruising at an average height of 6,400 meters (21,000 feet) meant they needed supplemental oxygen. Failure of a cabin heating system subjected them to 1.7°C (33°F) temperatures for the last half of the trip. And they ventured outside the balloon once so Piccard could chip away at heavy ice that had formed on the cables and capsule. While waiting the six hours for Egyptian helicopter crews to find and retrieve them after landing, explorers Piccard and Jones must have amused themselves by reflecting on the world-spanning power of the winds and, of course, contemplating the $1 million prize offered by the Anheuser-Busch corporation for the first nonstop circumnavigation of the globe by balloon.

8-1

WHAT TO WATCH FOR IN CHAPTER 8

Water, gases, and energy are shared between the atmosphere and the ocean. The two bodies are in continuous contact, and conditions in one are certain to influence conditions in the other. Be alert for examples of this influence. What happens to air when it is heated? How does excess heat near the equator affect the flow of winds over Earth? How do storms and air currents prevent the overheated tropical ocean from boiling away? As always, water's unique thermal properties are involved.

The sun's rays make the equatorial regions of Earth warmer than the poles. The atmosphere responds to this uneven solar heating by flowing in three great circulating cells over each hemisphere. The flow of air within these cells is influenced by the rotation of Earth, which causes moving air (or any moving mass) in the Northern Hemisphere to curve to the right of its initial path, and in the Southern Hemisphere to curve to the left. The apparent curvature of path is known as the Coriolis effect, which is the mysterious "force" that people believe causes toilets to flush clockwise in one hemisphere, counterclockwise in the other. Does it really happen? Read on.

Weather and storms are part of atmospheric circulation. Large storms are spinning areas of unstable air between or within air masses. Look for similarities and differences in the two kinds of large storms—extratropical cyclones (frontal storms) and tropical cyclones (hurricanes).

In this chapter we investigate the movement of air, and in Chapter 9 the movement of water.

COMPOSITION AND PROPERTIES OF THE ATMOSPHERE

8-2

The lower atmosphere is a nearly homogeneous mixture of gases, most plentifully nitrogen (78.1%) and oxygen (20.9%). Other elements and compounds, as listed in **Table 8.1,** make up less than 1% of its composition.

Air is never completely dry; **water vapor,** the gaseous form of water, can occupy as much as 4% of its volume. Sometimes liquid droplets of water are visible as clouds or fog, but more often the water is simply there—invisible in vapor form, having entered the atmosphere from the ground, plants, and the sea surface. The residence time of water vapor in the lower atmosphere is about ten days. Water leaves the atmosphere by condensing into dew, rain, or snow.

Air has mass. A 1-square-centimeter (0.16-square-inch) column of air, extending from sea level to the top of the atmosphere, weighs about 1.04 kilograms (2.3 pounds). A 1-square-foot column of air the same height weighs more than a ton.

The temperature and water content of air greatly influence its density. Because the molecular movement associated with heat causes a mass of warm air to occupy more space than an equal mass of cold air, warm air is less dense than cold air. But contrary to what we might guess, humid air is *less* dense than dry air at the same temperature—because molecules of water vapor weigh less than the nitrogen and oxygen molecules that the water vapor displaces.

Near Earth's surface, air is packed densely by its own mass. Air lifted from near sea level to a higher altitude is subjected to less pressure and will expand. As anyone knows who has felt the cool air rushing from an open tire valve, air becomes *cooler* when it *expands.* The opposite effect is also familiar: Air *compressed* in a tire pump becomes *warmer.* Air descending from high altitude toward sea level warms as it is compressed by the higher atmospheric pressure near Earth's surface.

Warm air can hold more water vapor than cold air. Water vapor in rising, expanding, cooling air will often condense into clouds (aggregates of tiny droplets) because the cooler air can no longer hold as much water vapor. If rising and cooling continue, the droplets may coalesce into raindrops or snowflakes. The atmosphere will then lose water as **precipitation,** liquid

Table 8.1 Average Composition of Dry Air, by Volume	
Component	**Percentage**
Nitrogen	78.084%
Oxygen	20.946%
Argon	0.934%
Carbon dioxide	0.033%
Neon, helium, methane, and other elements and compounds	0.003%
	100.000%

or solid water that falls from the air to Earth's surface. These rising-expanding-cooling and falling-compressing-heating relationships are important in understanding atmospheric circulation, weather, and climate (see **Figure 8.1**).

ATMOSPHERIC CIRCULATION

As we'll see, atmospheric circulation is powered by sunlight. Only about 1 part in 2.2 billion of the sun's radiant energy is intercepted by Earth, but that amount averages 7 million calories per square meter per day at the top of the atmosphere, or, for Earth as a whole, an impressive 17 trillion kilowatts (23 trillion horsepower)! As can be seen in **Figure 8.2,** on a global basis about 51% of the incoming energy is absorbed by Earth's land and water surface. How much light penetrates the ocean depends greatly on the angle at which it approaches, the sea state (surface turbulence), the presence of an ice covering or light-colored foam, and other factors.

The 51% of solar energy striking land and sea is converted to heat and then transferred into the atmosphere by conduction, radiation, and evaporation. The atmosphere, like the land and ocean, eventually radiates this heat back into

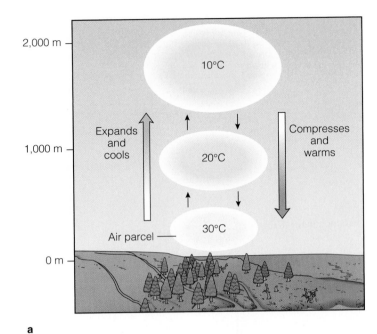

a

Figure 8.1 (**a**) Ascending air cools as it expands. Cooler air can hold less water in solution, so water vapor condenses into tiny droplets—clouds. Descending air warms as it compresses—the droplets (clouds) evaporate. (**b**) Warm, moist air evaporating from the ocean expands as it cools and rises, leading to the formation of clouds.

b

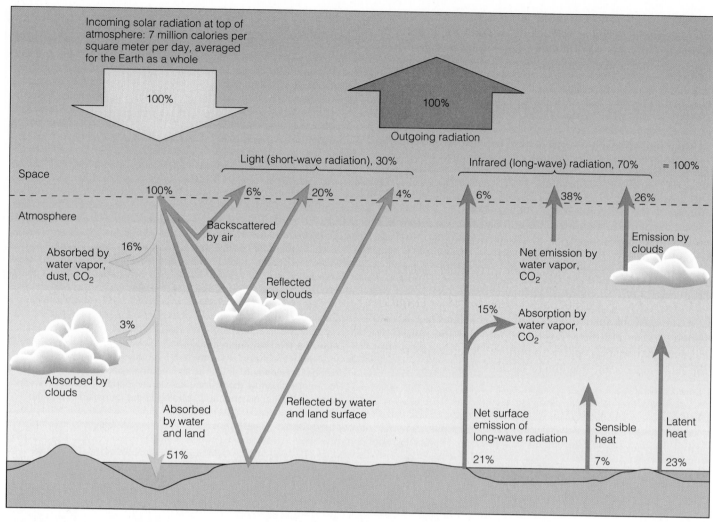

Figure 8.2 An estimate of the heat budget for Earth. On an average day, about half of the solar energy arriving at the upper atmosphere is absorbed at Earth's surface. Light (short-wave) energy absorbed at the surface is converted into heat. Heat leaves Earth as infrared (long-wave) radiation. Since input equals output over long periods of time, the heat budget is balanced.

space in the form of infrared radiation. The heat input and outflow "account" for Earth can be thought of as a **heat budget.** As in your personal financial budget, income must eventually equal outgo. Over long periods of time the total *incoming* heat (plus that from earthly sources) equals the total *outgoing* heat, so Earth is in **thermal equilibrium;** it is growing neither significantly warmer nor colder.[1]

Uneven Solar Heating and Latitude

While the heat budget for Earth *as a whole* is in balance, the heat budget for its *different latitudes* is not. As can be seen in **Figure 8.3,** sunlight striking polar latitudes spreads over

a greater area, filters through more atmosphere, and approaches the surface at a low angle, favoring reflection. Polar regions receive no sunlight at all during the depths of local winter. In contrast, at the tropical latitudes the high solar angle distributes the same amount of sunlight over a much smaller area. The more nearly vertical angle at which the sunlight approaches means that it passes through less atmosphere, minimizing reflection. Tropical regions thus receive significantly more solar energy than the polar regions, and mid-latitude areas receive more heat in summer than in winter.

Figure 8.4 shows heat received versus heat reradiated at different latitudes. Near the equator the amount of solar energy received by Earth greatly exceeds the amount of heat radiated into space. In the polar regions the opposite is true.

So why don't the polar ocean areas freeze solid and the equatorial oceans boil away? The water itself is moving huge amounts of heat between tropics and poles, and wa-

---

[1]Changes in heat balance do occur over short periods of geological time. Increasing amounts of carbon dioxide and methane in Earth's atmosphere may contribute to an increase in surface temperature called the *greenhouse effect.* More on this subject can be found in Chapter 18.

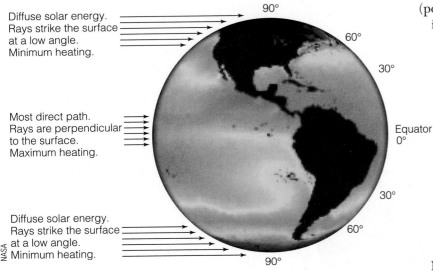

Diffuse solar energy. Rays strike the surface at a low angle. Minimum heating.

Most direct path. Rays are perpendicular to the surface. Maximum heating.

Diffuse solar energy. Rays strike the surface at a low angle. Minimum heating.

NASA

90°
60°
30°
Equator 0°
30°
60°
90°

Figure 8.3 How solar energy input varies with latitude. Equal amounts of sunlight are spread over a greater surface area near the poles than in the tropics. Ice near the poles reflects much of the energy that reaches the surface there. Solar energy approaching the polar regions must also pass through more of the atmosphere because of its low angle of approach. (Source: From Hamblin & Christiansen, *Earth's Dynamic Systems*, 9/e, p. 214. Reprinted by permission of Pearson Education, Inc., Upper Saddle River, NJ 07458.)

ter's thermal properties make it an ideal fluid to equalize the polar–tropical heat imbalance. Water's heat capacity moves heat poleward in ocean currents, but water's exceptionally high latent heat of vaporization (540 calories per gram) means that water vapor transfers much more heat

(per unit of mass) than liquid water. Masses of moving air account for about two-thirds of the poleward transfer of heat; ocean currents move the other third.

Uneven Solar Heating and the Seasons

At mid-latitudes the Northern Hemisphere receives about three times as much solar energy per day in June as it does in December. This difference is due to the 23½° tilt of Earth's rotational axis relative to the plane of its orbit around the sun (**Figure 8.5**). As Earth revolves around the sun, the constant tilt of its rotational axis causes the Northern Hemisphere to lean *toward* the sun in June but *away* from it in December. The sun therefore appears higher in the sky in summer but lower in winter. The inclination of Earth's axis also causes days to become longer as summer approaches but shorter with the coming of winter. Longer days mean more time for the sun to warm Earth's surface. The tilt of Earth is what causes the seasons.

Uneven Solar Heating and Atmospheric Circulation

The concentration of solar energy near the equator affects the atmosphere. We know that warm air rises and cool air sinks. Think of air circulation in a room with a hot radiator opposite

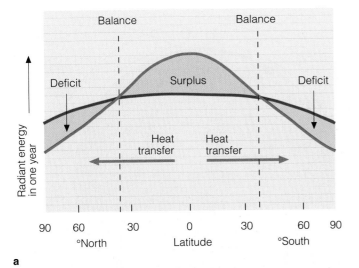

a

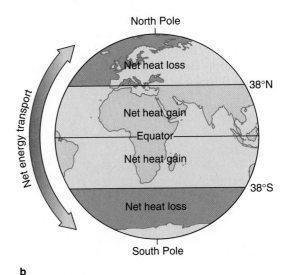

b

Figure 8.4 Areas of heat gain and loss on Earth's surface. (**a**) The average annual incoming solar radiation (red line) *absorbed* by Earth is shown along with the average annual infrared radiation (blue line) *emitted* by Earth. Note that polar latitudes lose more heat to space than they gain, and tropical latitudes gain more heat than they lose. Only at about 38° N and S latitudes does the amount of radiation received equal the amount lost. Since the area of heat gained (orange area) equals the area of heat lost (blue areas), Earth's total heat budget is balanced. (**b**) The ocean does not boil away near the equator or freeze solid near the poles because heat is transferred by winds and ocean currents from equatorial to polar regions.

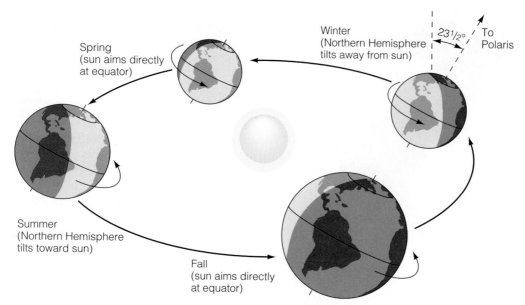

Spring
(sun aims directly
at equator)

Winter
(Northern Hemisphere
tilts away from sun)

23½° To
Polaris

Summer
(Northern Hemisphere
tilts toward sun)

Fall
(sun aims directly
at equator)

Figure 8.5 The seasons (shown here for the Northern Hemisphere) are caused by variations in the amount of incoming solar energy as Earth makes its annual rotation around the sun on an axis tilted by 23½°. During the Northern Hemisphere winter, the Southern Hemisphere is tilted toward the sun, and the Northern Hemisphere receives less light and heat. During the Northern Hemisphere summer, the situation is reversed.

a cold, closed window (**Figure 8.6**). Air warms, expands, becomes less dense, and *rises* over the radiator. Air cools, contracts, becomes denser, and *falls* near the cold glass window. The circular current of air in the room, a **convection current,** is caused by the difference in temperature between the ends of the room.

A similar process occurs over the surface of Earth. As we have seen, surface temperatures are higher at the equator than at the poles, and air can gain heat from warm surroundings. Since air is free to move over Earth's surface, it would be reasonable to assume that an air circulation pattern like the one shown in **Figure 8.7** would develop. In this ideal model, air heated in the tropics would expand and become less dense, rise to high altitude, turn poleward, and "pile up" as it converged near the poles. The air would then cool and contract by radiating heat into space, sink to the surface, and turn

equatorward, flowing along the surface back to the tropics to complete the circuit.

But this is *not* what happens. Global circulation of air is governed by two factors: uneven solar heating *and* the rotation of Earth. The eastward rotation of Earth on its axis deflects the moving air or water (or any moving object having mass) away from its initial course. This deflection is called the **Coriolis effect** in honor of **Gaspard Gustave de Coriolis,** the French scientist who worked out its mathe-

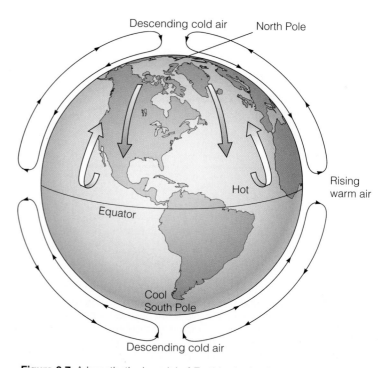

Descending cold air   North Pole

Rising warm air

Hot

Equator

Cool
South Pole

Descending cold air

Figure 8.7 A hypothetical model of Earth's air circulation if uneven solar heating were the only factor to be considered. The thickness of the atmosphere is greatly exaggerated in this drawing.

Warm air rising

Cold window (closed)

Hot radiator

Cool air falling

Figure 8.6 A convection current forms in a room when air flows from a hot radiator to a cold, closed window and back. (For a practical oceanic application of this principle, look ahead to Figure 8.16.)

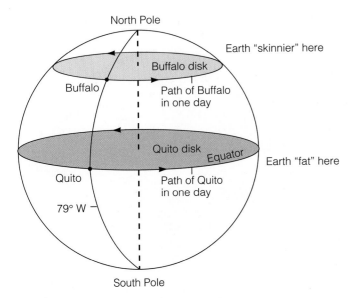

North Pole

Earth "skinnier" here

Buffalo disk

Buffalo

Path of Buffalo
in one day

Quito disk

Equator

Earth "fat" here

Quito

Path of Quito
in one day

79° W

South Pole

Figure 8.8 Sketch of the thought experiment in the text, showing that Buffalo travels a shorter path on the rotating Earth than does Quito.

matics in 1835. An understanding of the Coriolis effect is important to an understanding of atmospheric and oceanic circulation.

The Coriolis Effect

To an earthbound observer, any object moving freely across the globe appears to curve slightly from its initial path. In the Northern Hemisphere this curve is to the right, or *clockwise*, from the expected path; in the Southern Hemisphere it is to the left, or *counterclockwise*. To earthbound observers the deflection is very real; it isn't caused by some mysterious force, and it isn't an optical illusion or some other trick caused by the shape of the globe itself. *The observed deflection is caused by the observer's moving frame of reference on the spinning Earth.*

The influence of this deflection can be illustrated by performing a mental experiment involving concrete objects—in this case, cities and cannonballs—and then applying the principle to atmospheric circulation. Let's pick as examples for our experiment the equatorial city of Quito, capital of Ecuador, and Buffalo, New York. Both cities are on almost the same line of longitude (79°W), so Buffalo is almost exactly north of Quito (as **Figure 8.8** shows). Like everything else attached to the rotating Earth, both cities make one trip around the world each 24 hours. Through one day, the north-south relationship of the two cities never changes—Quito is *always* due south of Buffalo.

A complete trip around the world is 360°, so each city moves eastward at an angular rate of 15° per hour (360°/24 hours = 15°/hour). Yet, even though their angular rates are the same, the two cities move eastward at different speeds. Quito is on the equator, the "fattest" part of Earth. Buffalo is farther north at a "skinnier" part. Imagine both cities isolated on flat disks, and imagine Earth's sphere made up of a great number of these disks strung together on a rod connecting the North Pole to the South Pole. From Figure 8.8, you can see that Buffalo's disk has a smaller circumference than Quito's. Thus, Buffalo doesn't have as far to go in one day as Quito. Buffalo must move eastward *more slowly* than Quito to maintain its position due north of Quito. (**Figure 8.9** puts another spin on this idea.)

Look at Earth from above the North Pole in **Figure 8.10.** The Quito disk and the Buffalo disk must turn through 15° of longitude each hour (or Earth would rip itself apart), but the city on the equator must move faster to the east to turn its 15° each hour because its "slice of the pie" is larger. Buffalo must move at 1,260 kilometers (783 miles) per hour to go around the world in one day, while Quito must move at 1,668 kilometers (1,036 miles) per hour to do the same.

Now imagine a massive object moving between the two cities. A cannonball shot north from Quito toward Buffalo would carry Quito's eastward component as it goes; that is, regardless of its northward speed, the cannonball is also moving *east* at 1,668 kilometers (1,036 miles) per hour. The fact of

Calvin and Hobbes by Bill Watterson

Figure 8.9 Same idea, different approach. (Calvin and Hobbes. © 1990 Universal Press Syndicate. Reprinted by permission of Universal Press Syndicate.)

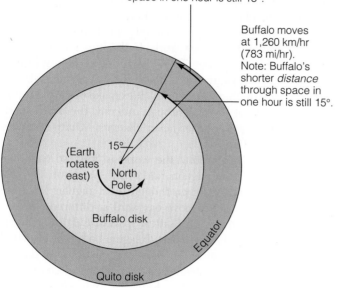

Quito moves at 1,668 km/hr (1,036 mi/hr). Note: Quito's longer *distance* through space in one hour is still 15°.

Buffalo moves at 1,260 km/hr (783 mi/hr). Note: Buffalo's shorter *distance* through space in one hour is still 15°.

Figure 8.10 A continuation of the thought experiment. A look at Earth from above the North Pole shows that Buffalo and Quito move at different velocities.

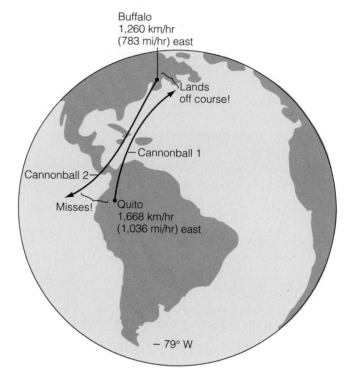

Figure 8.11 The final step in the experiment. As observed from space, cannonball 1 (shot northward) and cannonball 2 (shot southward) move as we might expect; that is, they travel straight away from the cannons and fall to Earth. Observed from the ground, however, cannonball 1 veers slightly east and cannonball 2 veers slightly west of the intended target. The effect depends on the observer's frame of reference.

being fired northward by the cannon does not change its eastward movement in the least. As the cannonball streaks north, an odd thing happens. The cannonball veers from its northward path, angling slightly to the right (east) (**Figure 8.11**). Actually, this first cannonball is moving just as an observer from space would expect it to, but to those of us on the ground the cannonball "gets ahead of Earth." As cannonball 1 moves north, the ground beneath it is no longer moving eastward at 1,668 kilometers per hour. During the ball's time of flight, Buffalo (on its smaller disk) *has not moved eastward enough to be where the ball will hit.* If the time of flight for the cannonball is one hour, a city 408 kilometers east of Buffalo (1,668 [Quito's speed] − 1,260 [Buffalo's speed] = 408) will have an unexpected surprise. Albany may be in for some excitement!

If a cannonball were fired south from Buffalo toward Quito, the situation would be reversed. This second cannonball has an eastward component of 1,260 kilometers (783 miles) per hour even while it sits in the muzzle. Once fired and moving southward, cannonball 2 travels over portions of Earth that are moving ever faster in an eastward direction. The ball again appears to veer off course to the right (see Figure 8.11 again), falling into the Pacific to the west of Ecuador. Don't be deceived by the word *appears* in the last sentence. The cannonballs actually do veer to the right, or *clockwise*. Only to an observer in space would they appear to go straight, and points on Earth would appear to move out from underneath them.

The Coriolis effect is a real effect dependent on our rotating frame of reference. Part of that frame of reference involves the direction from which you view the problem. Thus, in Figure 8.11 cannonball 2 looks to you as if it is veering left, but to the citizens of Buffalo facing south to watch the cannonball disappear it curves to the right (west). Coriolis deflection works *counterclockwise* in the Southern Hemisphere because the frame of reference there is reversed. Also, except at the equator (where the Coriolis effect is nonexistent), the Coriolis effect influences the path of objects moving from east to west, or west to east.

A quick way to remember the Coriolis effect is to remember that in the Northern Hemisphere moving objects veer off course clockwise, to the right, and in the Southern Hemisphere moving objects veer off course counterclockwise, to the left.

Because the Coriolis effect influences any object with mass—*as long as that object is moving*—it plays a large role in the movements of air and water on Earth. The Coriolis effect is most apparent in mid-latitude situations involving the almost frictionless flow of fluids: between layers of water in the ocean and in the circuits of winds. Does the Coriolis effect influence the directions of cars and airplanes? Yes, but in these cases friction (of tires on pavement, of wings on air) is much

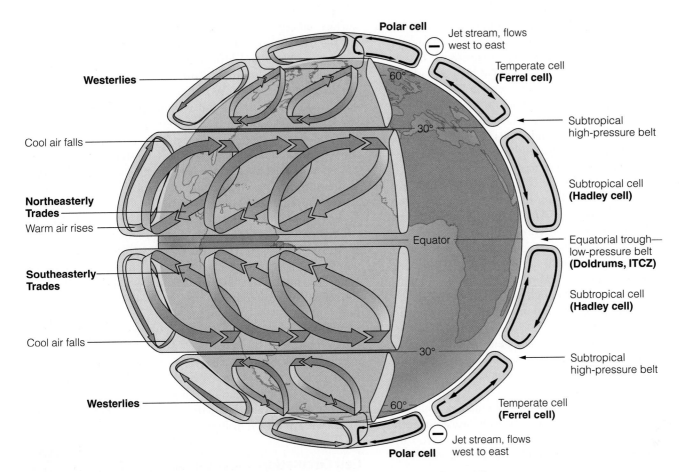

Polar cell

Jet stream, flows
west to east

Westerlies

60°

Temperate cell
(Ferrel cell)

Cool air falls

30°

Subtropical
high-pressure belt

**Northeasterly
Trades**

Warm air rises

Subtropical cell
(Hadley cell)

Equator

Equatorial trough—
low-pressure belt
(Doldrums, ITCZ)

**Southeasterly
Trades**

Subtropical cell
(Hadley cell)

Cool air falls

30°

Subtropical
high-pressure belt

Westerlies

60°

Temperate cell
(Ferrel cell)

Jet stream, flows
west to east

Polar cell

Figure 8.12 Global air circulation as described in the six-cell circulation model. As in Figure 8.7,
air rises at the equator and falls at the poles. But instead of one great circuit in each hemisphere
from equator to pole, there are three in each hemisphere. Note the influence of the Coriolis effect
on wind direction. The circulation shown here is ideal—that is, a *long-term average* of wind flow.
Contrast this view with Figure 8.14, a snapshot of a moment in 1996.

greater than the influence of the Coriolis effect, so the deflection is not observable. (For a mathematical treatment of the Coriolis effect, see Appendix V.)

The Coriolis Effect and Atmospheric Circulation Cells

We can now modify our original model of atmospheric circulation (Figure 8.7) into the more correct representation provided in **Figure 8.12.** Yes, air does warm, expand, and rise at the equator; and air does cool, contract, and fall at the poles. But instead of continuing all the way from equator to pole in a continuous loop in each hemisphere, air rising from the equatorial region moves poleward and is gradually deflected eastward; that is, it turns to the *right* in the Northern Hemisphere and to the *left* in the Southern Hemisphere. This eastward deflection is caused by the Coriolis effect. (Note that the Coriolis effect does not *cause* the wind; it only *influences* the wind's direction.)

As air rises at the equator, it loses moisture by precipitation (rainfall) caused by expansion and cooling. This drier air now grows denser in the upper atmosphere as it radiates

heat to space and cools. When it has traveled about a third of the way from the equator to the pole—that is, to about 30°N and 30°S latitude—the air becomes dense enough to fall back toward the surface. Most of the descending air turns back toward the equator when it reaches the surface. In the Northern Hemisphere the Coriolis effect again deflects this surface air to the right, and the air blows across the ocean or land from the northeast. (The Northeasterly Trades are shown in Figure 8.12.) Though it has been heated by compression during its descent, this air is generally still colder than the surface over which it flows. The air soon warms as it moves equatorward, however, evaporating surface water and becoming humid. The warm, moist, less dense air then begins to rise as it approaches the equator and completes the circuit.

Such a large circuit of air is called an **atmospheric circulation cell.** A pair of these tropical cells exists, one on each side of the equator. They are known as **Hadley cells** in honor of George Hadley, the London lawyer and philosopher who worked out an overall scheme of wind circulation in 1735. Look for them in Figure 8.12.

A more complex pair of circulation cells operates at mid-latitudes in each hemisphere. Some of the air descending at 30° latitude turns poleward rather than equatorward. Before this air descends to the surface it is joined at high altitude by air returning from the north. As can be seen in Figure 8.12, a loop of air forms between 30° and about 50°–60° of latitude. As before, the air is driven by uneven heating and influenced by the Coriolis effect. Surface wind in this circuit is again deflected to the right, this time flowing from the west to complete the circuit. (The Westerlies are shown in Figure 8.12.) The mid-latitude circulation cells of each hemisphere are named **Ferrel cells** after William Ferrel, the American who discovered their inner workings in the mid-nineteenth century. They, too, can be seen in Figure 8.12.

Meanwhile, air that has grown cold over the poles begins blowing toward the equator at the surface, turning to the west as it does so. At between 50° and 60° latitude in each hemisphere, this air has taken up enough heat and moisture to ascend. However, this polar air is denser than the air in the adjacent Ferrel cell and does not mix easily with it. The unstable zone between these two cells generates most mid-latitude weather. At high altitude the ascending air from 50°–60° latitude turns poleward to complete a third circuit. These are the **polar cells.**

Thus, three large atmospheric circulation cells—a Hadley cell, a Ferrel cell, and a polar cell—exist in each hemisphere. Air circulation within each cell is powered by uneven solar heating and influenced by the Coriolis effect.

WIND PATTERNS

The model of atmospheric circulation described above has many interesting features. At the bands between circulation cells, the air is moving *vertically,* and surface winds are weak and erratic. Such conditions exist at the equator (where air rises and atmospheric pressure is generally low) and at 30° latitude in each hemisphere (where air falls and atmospheric pressure is generally high). Places within these circulation cells where air moves rapidly *horizontally* across the surface from zones of high pressure to zones of low pressure are characterized by strong, dependable winds.

Sailors have a special term for the calm equatorial areas where the surface winds of the two Hadley cells converge—the equatorial low called the **doldrums.** The word has come to be associated with a gloomy, listless mood, perhaps reflecting the sultry air and variable breezes found there. Scientists who study the atmosphere call this area the **intertropical convergence zone,** or **ITCZ,** to reflect the influence of wind convergence on conditions near the equator. Strong heating in the ITCZ causes surface air to expand and rise. The humid, rising, expanding air loses moisture as rain, some of which contributes to the success of tropical rain forests.

Sinking air, in contrast, is generally arid. The great deserts of both hemispheres, dry bands centered around 30°, mark the intersection of the Hadley and Ferrel cells. Because evaporation is higher than precipitation in these areas, ocean

surface salinity tends to be highest at these latitudes (as can be seen in Figure 6.15). At sea, these areas of high atmospheric pressure and little surface wind are called the subtropical high, or **horse latitudes.** Spanish ships laden with supplies for the New World were often becalmed there, sometimes for weeks on end. When the mariners ran out of water and feed for their livestock, they were forced to throw the dead horses over the side.

Of much more interest to sailing masters were the bands of dependable surface winds *between* the zones of ascending and descending air. Most constant of these are the persistent **trade winds,** or easterlies, centered at about 15°N and 15°S latitude. The trade winds are the surface winds of the Hadley cells as they move from the horse latitudes to the doldrums. In the Northern Hemisphere they are the northeast trade winds; the southeast trade winds are the Southern Hemisphere counterpart.[2] The **westerlies,** surface winds of the Ferrel cells centered at about 45°N and 45°S latitude, flow between the horse latitudes and the boundaries of the polar cells in each hemisphere. Thus, the westerlies approach from the southwest in the Northern Hemisphere and from the northwest in the Southern Hemisphere. Sailors outbound from Europe to the New World learned to drop south to catch the trade winds and to return home by a more northerly route to take advantage of the westerlies. Trade winds and westerlies are shown in Figure 8.12.

Cell Circulation: Ideal Versus Actual

The six-cell model of atmospheric circulation (three cells in each hemisphere) shown in Figure 8.12 represents an *average* of air flow through many years over the planet as a whole. Though the model is accurate in a general sense, local details of cell circulation vary because surface conditions are different at different longitudes. The ocean's thermostatic effect is the major factor reducing irregularities in cell circulation over water.

For example, the equator-to-pole patterns of airflow within circulation cells along the 20°E line of longitude, a meridian crossing Africa and Europe, are much more complex than the patterns of flow along 170°W longitude, which is almost exclusively over the ocean. The ITCZ is also much narrower and more consistent over the ocean than over land. Since the Northern Hemisphere contains much less ocean surface than the Southern Hemisphere and land masses have a lower specific heat than the ocean, seasonal differences in temperature and cell circulation are more extreme in the north. Also, cell circulation is much more symmetrical in the Southern Hemisphere.

Another consequence of the markedly different proportions of land to ocean surface in the two hemispheres is the position of the ITCZ. The convergence zone does not coincide with the **geographical equator** (0° latitude). Instead it lies at

---

[2]Winds are named by the direction *from* which they blow. A west wind blows *from* the west toward the east; a northeast wind blows *from* the northeast toward the southwest.

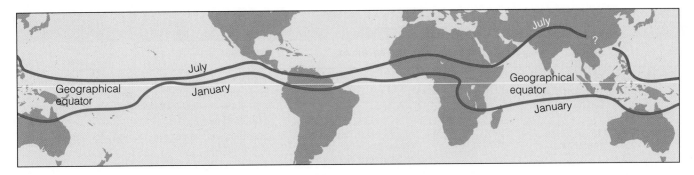

Figure 8.13 Seasonal changes in the position of the intertropical convergence zone (ITCZ). The zone reaches its most northerly location in July and its most southerly location in January. Because of the thermostatic effect of water, the seasonal north-south movement is generally less over the ocean than over land. (Source: *Meteorology*, 5/e by Moran and Morgan. © 1997 Prentice-Hall, Inc.)

the **meteorological equator** (or *thermal equator*), an irregular imaginary line of thermal equilibrium between the hemispheres, situated about 5° north of the geographic equator. The positions of the meteorological equator and the ITCZ generally coincide and change with the seasons, moving slightly farther north in the northern summer and returning toward the equator in the northern winter (**Figure 8.13**). Atmospheric and oceanic circulation in the two hemispheres is approximately symmetrical about the meteorological equator, not the geographical equator. Thus, the doldrums, trade winds, horse latitudes, and westerlies shift north in the northern summer, and south in the northern winter.

There are also east-west variations in the expected patterns of

the circulation cells. In the northern winter, air above the chilled continental masses of North America and Siberia becomes very cold and dense. This air sinks and forms zones of high atmospheric pressure over the continents. Air over the relatively warmer waters near the Aleutians and Iceland rises and forms zones of low atmospheric pressure. Air flows from the high-pressure zones toward low-pressure zones, modifying the flow of air within the cells. In summer the situation is reversed: Low pressure forms over the heated land masses, and higher pressure forms over the cooler ocean. These effects are most pronounced in the middle latitudes of the Northern Hemisphere, where land and water are present in near-equal amounts.

Figure 8.14 is a depiction of winds over the Pacific on two days in September 1996. As you can see, the patterns can depart from what we would expect in the six-cell

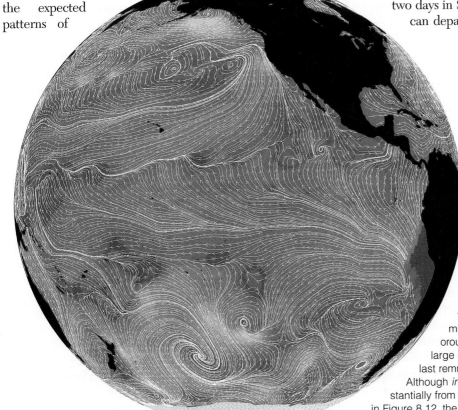

Figure 8.14 Winds over the Pacific Ocean on 20 and 21 September 1996. Wind speed increases as colors change from blue-purple to yellow-orange, with the strongest winds at 20 meters per second (45 miles per hour). Wind direction is shown by the small white arrows. The measurements were made with a NASA radar scatterometer aboard Japan's Advanced Earth Orbiting Satellite, launched 16 August 1996. Note the Hawaiian islands in the midst of the persistent northeast trade winds, the vigorous westerlies driving toward western Canada, a large extratropical cyclone east of New Zealand, and the last remnants of a tropical cyclone off the coast of Japan. Although *instantaneous* views such as this one depart substantially from wind flow predicted in the six-cell model developed in Figure 8.12, the *average* wind flow over many years looks remarkably like what we would expect from the model.

Region	Name	Pressure	Surface Winds	Weather
Equator (0°)	Doldrums (ITCZ) (equatorial low)	Low	Light, variable winds	Cloudiness, abundant precipitation in all seasons; breeding ground for hurricanes. Relatively low sea surface salinity because of rainfall (see Figure 17.4)
0°–30°N and S	Trade winds (easterlies)	—	Northeast in Northern Hemisphere, southeast in Southern Hemisphere	Summer wet, winter dry; pathway for tropical disturbances
30°N and S	Horse latitudes (subtropical high)	High	Light, variable winds	Little cloudiness; dry in all seasons. Relatively high sea surface salinity because of evaporation
30°–60°N and S	Prevailing westerlies	—	Southwest in Northern Hemisphere, northwest In Southern Hemisphere	Winter wet, summer dry; pathway for subtropical high and low pressure
60°N and S	Polar front	Low	Variable	Stormy, cloudy weather zone; ample precipitation in all seasons
60°–90°N and S	Polar easterlies	—	Northeast in Northern Hemisphere, southeast in Southern Hemisphere	Cold polar air with very low temperatures
90°N and S	Poles	High	Southerly in Northern Hemisphere, northerly in Southern Hemisphere	Cold, dry air; sparse precipitation in all seasons

Table 8.2   Major Wind and Pressure Systems and Related Weather

Note: Compare to Figure 8.12. (*Source:* From *Earth in Crisis: An Introduction to Earth Sciences,* 2/e, Thomas L. Burrus, Herbert J. Spiegel, 1980. C.V. Mosby Co. Reprinted by permission of Thomas L Burrus.)

model of Figure 8.12. Most of the difference is caused by the geographical distribution of land masses—and the different responses of land and ocean to solar heating—and by chaotic flow. But, as noted earlier, over long periods of time (many years), *average* flow looks remarkably like what we would expect.

The major surface wind and pressure systems of the world, and their prevailing weather conditions, are summarized in **Table 8.2.** These great wind patterns are responsible for about two-thirds of the heat transfer from the tropics to the polar regions. (As noted earlier, ocean currents account for the other third.)

Monsoons

A **monsoon** is a pattern of wind circulation that changes with the season. (The word *monsoon* is derived from *mausim,* the Arabic word for season.) Areas subject to monsoons generally have wet summers and dry winters.

Monsoons are linked to the different specific heats of land and water and to the annual north-south movement of the ITCZ. In the spring, land heats more rapidly than the adjacent ocean. Air above the land becomes warmer and thus rises. Relatively cool air flows from over the ocean to the land to take its place. Continued heating causes this humid air to rise, condense, and form clouds and rain. In autumn the land cools more rapidly than the adjacent ocean. Air cools and sinks over the land, and dry surface winds move seaward. The intensity and location of monsoon activity depend on the position of the ITCZ. Note in **Figure 8.15** that the monsoons

follow the ITCZ south in the Northern Hemisphere's winter, and north in its summer.

In Africa and Asia, more than 2 billion people depend on summer monsoon rains for drinking water and agriculture. The most intense summer monsoons occur in Asia. The great landmass of Asia draws vast quantities of warm, moist air from the Indian Ocean (see again Figure 8.15). Southerly winds drive this moisture toward Asia, where it rises and condenses to produce a months-long deluge. Much smaller monsoons occur in North America as warming and rising air over the South and West draws humid air and thunderstorms from the Gulf of Mexico.

Sea Breezes and Land Breezes

Land breezes and sea breezes are small, daily mini-monsoons. Morning sunlight falls on land and adjacent sea, warming both. The temperature of the water doesn't rise as much as the temperature of the land, however. The warmer inland rocks transfer heat to the air, which expands and rises, creating a zone of low atmospheric pressure over the land. Cooler air from over the sea then moves toward land; this is the **sea breeze** (see **Figure 8.16a**). The situation reverses after sunset, with land losing heat to space and falling rapidly in temperature. After a while, the air over the still-warm ocean will be warmer than the air over the cooling land. This air will then rise, and the breeze direction will reverse, becoming a **land breeze** (see **Figure 8.16b**). Land breezes and sea breezes are common and welcome occurrences in coastal areas.

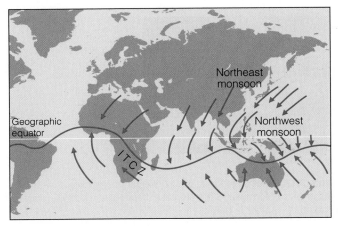

a January

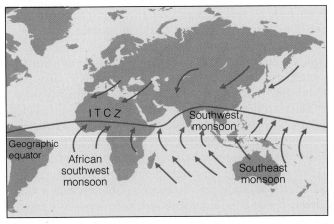

b July

Figure 8.15 Monsoon patterns. During the monsoon circulations of January and July, surface winds are deflected to the right in the Northern Hemisphere and to the left in the Southern Hemisphere. (Source: Reprinted by permission of Alan D. Iselin.)

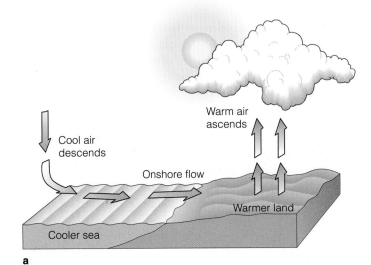

a

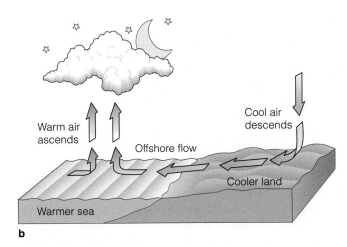

b

Figure 8.16 The flow of air in coastal regions during stable weather conditions. (**a**) In the afternoon, the land is warmer than the ocean surface, and warm air rising from the land is replaced by an onshore sea breeze. (**b**) At night, as the land cools, the air over the ocean is now warmer than air over the land. The ocean air rises and air flows offshore to replace it, generating an offshore flow (a land breeze).

STORMS

Storms are regional atmospheric disturbances characterized by strong winds often accompanied by precipitation. Few natural events underscore human insignificance like a great storm. When powered by stored sunlight, the combination of atmosphere and ocean can do fearful damage.

In Bangladesh on 13 November 1970, a tropical cyclone (a hurricane) with wind speeds of more than 200 kilometers (125 miles) per hour roared up the mouths of the Ganges River, carrying with it masses of seawater up to 12 meters (40 feet) high. Water and wind clawed at the aggregation of small islands, most just above sea level, that make up this impoverished country. In only 20 minutes at least 300,000 lives were lost, and estimates ranged to 1 million dead! Property damage was absolute. Photographs taken soon after the storm showed a horizon-to-horizon morass of flooded, deep-gashed ground tortured by furious winds. There was almost no trace of human inhabitants, farms, domestic animals, or villages. An-

other great storm, which struck in May 1991, killed another 200,000 people. The economy of the shattered country may never fully recover.

A much different type of storm hammered the U.S. East Coast in March 1993. Mountainous snows from New York to North Carolina (1.3 meters, or 50 inches, at Mount Mitchell), winds of 175 kilometers (109 miles) per hour in Florida, and record cold in Alabama (−17°C, or 2°F, in Birmingham) were elements of a four-day storm that spread chaos from Canada to Cuba. At least 238 people died on land; another 48 were lost at sea. At one point more than 100,000 people were trapped in offices, factories, vehicles, and homes; 1.5 million were without electricity. Economic damage exceeded $1 billion.

These two great storms exemplify *tropical cyclones* and *extratropical cyclones* at their worst. As the name implies, tropical cyclones like the hurricane that struck Bangladesh are primarily a tropical phenomenon. Extratropical cyclones—the winter weather disturbances with which residents of the U.S. eastern seaboard and other mid-latitude dwellers are most familiar—are found mainly in the Ferrel cells of each hemisphere. (Note that the prefix *extra-* means "outside" or "beyond," so *extratropical* refers to the location of the storm, not its intensity.)

Both kinds of storms are **cyclones,** huge rotating masses of low-pressure air in which winds converge and ascend. The word *cyclone,* derived from the Greek noun *kyklon* (meaning "an object moving in a circle"), underscores the spinning nature of these disturbances. (Don't confuse a cyclone with a **tornado,** a much smaller funnel of fast-spinning wind associated with severe thunderstorms.)

Air Masses

Cyclonic storms form between or within air masses. An **air mass** is a large body of air with nearly uniform temperature, humidity, and therefore density throughout. Air pausing over water or land will tend to take on the characteristics of the surface below. Cold, dry land causes the mass of air above to become chilly and dry. Air above a warm ocean surface will become hot and humid. Cold, dry air masses are dense and form zones of high atmospheric pressure. Warm, humid air masses are less dense and form zones of lower atmospheric pressure.

Air masses can move within or between circulation cells. Density differences, however, will prevent the air masses from mixing when they approach one another. Energy is required to mix air masses. Since that energy is not always available, a dense air mass may slide beneath a lighter air mass, lifting the lighter one and causing its air to expand and cool. Water vapor in the rising air may condense. All of these effects contribute to turbulence at the boundaries of the air masses.

The boundary between air masses of different density is called a **front.** The term was coined by **Vilhelm Bjerknes** (**Figure 8.17**), a pioneering Norwegian meteorologist who saw a similarity between the zone where air masses meet and the violent battle fronts of World War I.

Extratropical cyclones form at a front between *two* air masses. Tropical cyclones form from disturbances within *one* warm and humid air mass.

Extratropical Cyclones

Extratropical cyclones form at the boundary between each hemisphere's polar cell and its Ferrel cell—the **polar front.** These great storms occur mainly in the winter hemisphere when temperature and density differences across the polar front are most pronounced. Remember, the cold wind poleward of the front is generally moving from the east; the warmer air equatorward of the front is generally moving from the west (see again Figure 8.12). The smooth flow of winds past each other at the front may be interrupted by zones of alternating high and low atmospheric pressure that bend the front into a series of waves. Because of the difference in wind

Figure 8.17 Vilhelm Bjerknes, the Norwegian meteorologist who, with his son Jacob, formulated the air-mass theories on which our understanding of weather is based.

direction in the air masses north and south of the polar front, the wave shape will enlarge, and a twist will form along the front. The different densities of the air masses prevent easy mixing, so the cold, dense air mass will slide beneath the warmer, lighter one. Formation of this twist in the Northern Hemisphere, as seen from above, is shown in **Figure 8.18.** The twisting mass of air becomes an extratropical cyclone.

The twist that generates an extratropical cyclone circulates counterclockwise in the Northern Hemisphere, seemingly in opposition to the Coriolis effect. The reasons for this paradox become clear, however, when we consider the wind directions and the nature of the interruption of airflow between the cells. (In fact, the counterclockwise motion of the cyclone *is* Coriolis-driven because the large-scale airflow pattern at the edges of the cells is generated in part by the Coriolis effect.) Wind speed increases as the storm "wraps up" in much the same way that a spinning skater increases rotation speed by pulling in her arms close to her body. Air rushing toward the center of the spinning storm rises to form a low-pressure zone at the center. Extratropical cyclones are embedded in the westerly winds and thus move eastward. They are typically 1,000 to 2,500 kilometers (620 to 1,600 miles) in diameter and last from two to five days. **Figure 8.19** provides a beautiful example.

Precipitation can begin as the circular flow develops. Figure 8.18b shows why. Precipitation is caused by the lifting and consequent expansion and cooling of the mass of mid-latitude air involved in the twist. As it rises and cools, this air can no longer hold all of its water vapor, so clouds and rain result. When *cold* air advances and does the lifting (as on the left side of Figure 8.18b), a *cold front* occurs. A *warm front* happens when *warm* air is blown on top of the retreating edge of cold air (as on the right side of Figure 8.18b). The wind and precipitation associated with these fronts are sometimes referred to as **frontal storms.** They are the principal cause of weather in the mid-latitude regions, where most of the world's people live.

North America's most violent extratropical cyclones are the **nor'easters** (**northeasters**) that sweep the eastern seaboard in winter. The name indicates the direction from which the storm's most powerful winds approach. About 30 times a year, nor'easters moving along the mid-Atlantic and New England coasts generate wind and waves with enough force to erode beaches and offshore barrier islands, disrupt

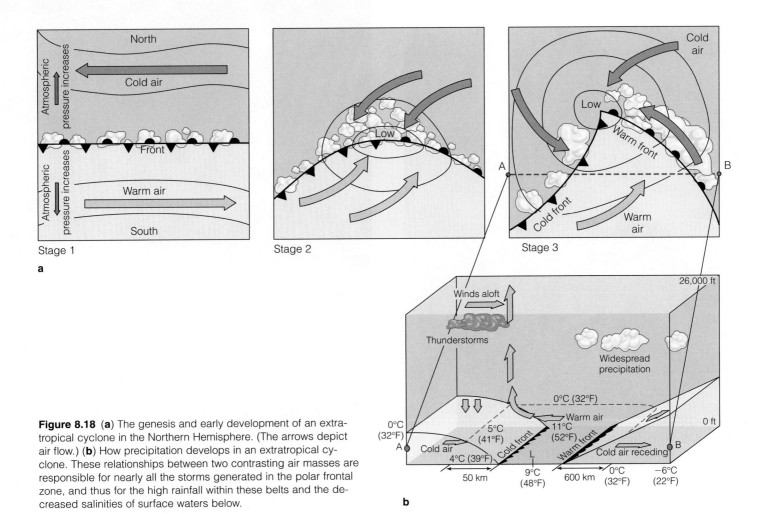

Stage 1

Stage 2

Stage 3

a

b

Figure 8.18 (**a**) The genesis and early development of an extratropical cyclone in the Northern Hemisphere. (The arrows depict air flow.) (**b**) How precipitation develops in an extratropical cyclone. These relationships between two contrasting air masses are responsible for nearly all the storms generated in the polar frontal zone, and thus for the high rainfall within these belts and the decreased salinities of surface waters below.

Figure 8.19 A well-developed extratropical cyclone over the northeastern Pacific on 27 October 2000. Looking like a huge comma, the cloud-dense cold front extends southward and westward from the storm's center. Spotty cumulus clouds and thunderstorms have formed in the cold, unstable air behind the front. This picture was taken by *GOES-10* in visible light.

© Bettman Archive/CORBIS

Figure 8.20 Storm damage from a particularly severe nor'easter, the Ash Wednesday (7 March) storm of 1962. Massive damage such as this, on Fire Island, New York, is most common for buildings near the shoreline.

communication and shipping schedules, damage shore and harbor installations, and break power lines. About every hundred years a nor'easter devastates coastal settlements. In spite of a long history of destruction, people continue to build on unstable exposed coasts (see **Figure 8.20**).

Tropical Cyclones

Tropical cyclones are great masses of warm, humid, rotating air. They occur in all tropical oceans except the equatorial South Atlantic. Large tropical cyclones are called **hurricanes** (*Huracan* = the god of the wind of the Caribbean Taino people) in the North Atlantic and eastern Pacific, *typhoons* (*Tai-fung* = Chinese for great wind) in the western Pacific, *tropical cyclones* in the Indian Ocean, and *willi-willis* in the waters near Australia. To qualify formally as a hurricane or typhoon, the tropical cyclone must have winds of at least 119 kilometers (74 miles) per hour. About a hundred tropical cyclones grow to hurricane status each year. A very few of these develop into superstorms, with winds near the core that exceed 250 kilometers (155 miles) per hour! (To imagine what winds in such a storm might feel like, picture yourself clinging to the wing of a twin-engine private airplane in flight!) Tropical cyclones containing winds less than hurricane force are called *tropical storms* and *tropical depressions*.

NASA

Figure 8.21 Hurricane Florence over the North Atlantic, photographed from the space shuttle *Discovery* in November 1994. The cloud-free eye of the storm is clearly visible. Note the spiral bands of cloud extending outward from the eye.

From above, tropical cyclones appear as circular spirals, as **Figure 8.21** shows. They may be 1,000 kilometers (620 miles) in diameter and 15 kilometers (9.3 miles, 50,000 feet) high. The calm center, or *eye*, of the storm—a zone some 13 to 16 kilometers (8 to 10 miles) in diameter—is sometimes surrounded by clouds so high and dense that the daytime sky above looks dark. Farther out, churned by furious winds, the rainband clouds condense huge amounts of water vapor into rain. A tropical cyclone is diagrammed in **Figure 8.22**.

Unlike extratropical cyclones, these greatest of storms form within *one* warm, humid air mass between 10° and 25° latitude in both hemispheres (**Figure 8.23**). (Though air conditions would be favorable, the Coriolis effect closer to the equator is too weak to initiate rotary motion.) The origins of tropical cyclones are not well understood. A tropical cyclone usually develops from a small tropical depression. Tropical depressions form in easterly waves, areas of lower pressure

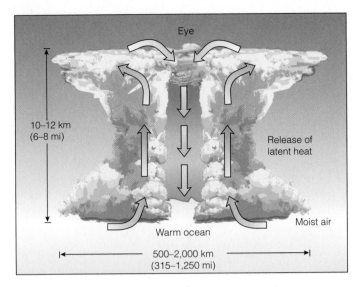

Figure 8.22 The internal structure of a hurricane. The vertical dimension is greatly exaggerated in this drawing.

within the easterly trade winds that are thought to originate over a large, warm landmass. When air containing the disturbance is heated over tropical water with a temperature of about 26°C (79°F) or higher, circular winds begin to blow in the vicinity of the wave, and some of the warm humid air is forced upward. Condensation begins, and the storm takes shape. Under ideal conditions the embryo storm reaches hurricane status—that is, with wind speeds in excess of 119 kilo-

meters (74 miles) per hour—in two to three days. The centers of most tropical cyclones move westward and poleward at 5 to 40 kilometers (3 to 25 miles) per hour. Typical tracks of these storms are shown in **Figure 8.24.**

Although its birth process is somewhat mysterious, the source of the storm's power is well understood. Its strength comes from the same seemingly innocuous process that warms a chilled soft-drink can when water from the atmosphere condenses on its surface. As you may recall from Chapter 6, it takes quite a bit of energy to evaporate water into the atmosphere—water's latent heat of vaporization is very high (540 calories per gram). That heat energy is released when the water vapor recondenses as liquid. It tends to warm your drink very quickly, and the more humid the air, the faster the condensing and warming. Think of the situation with a hair dryer. The heat energy generated by the dryer causes water to evaporate rapidly from your hair. When that water recondenses to liquid (on a nearby can of soda, for instance), the original heat used for vaporization is released. The cycle of evaporation and condensation has carried heat from the hair dryer to your soda. In tropical cyclones the condensation energy generates air movement (wind), not more heat. Fortunately only 2% to 4% of this energy of condensation is converted into motional energy!

A tropical cyclone is an ideal machine for "cashing in" water vapor's latent heat of vaporization. Warm, humid air forms in great quantity only over a warm ocean. As already noted, tropical cyclones originate in ocean areas having surface temperatures in excess of 26°C (79°F) (see Figure 8.23). As hot, humid tropical air rises and expands, it cools and is unable to retain the

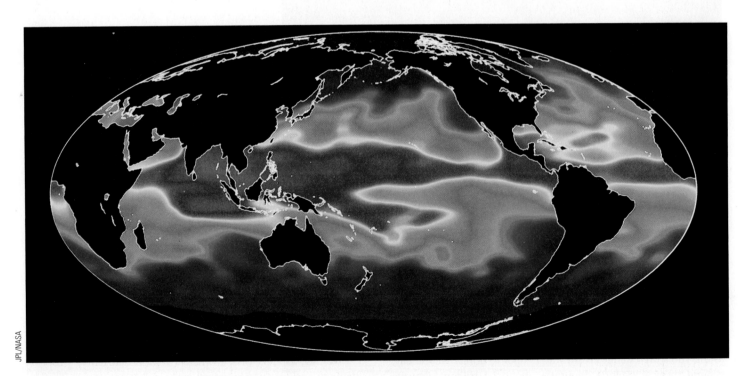

Figure 8.23 Tropical cyclones can develop in the hot, humid air over a sea surface exceeding 26°C (79°F). The base map in this diagram is derived from satellite data showing water vapor in the atmosphere in October 1992. Red zones indicate high humidity and warm temperatures.

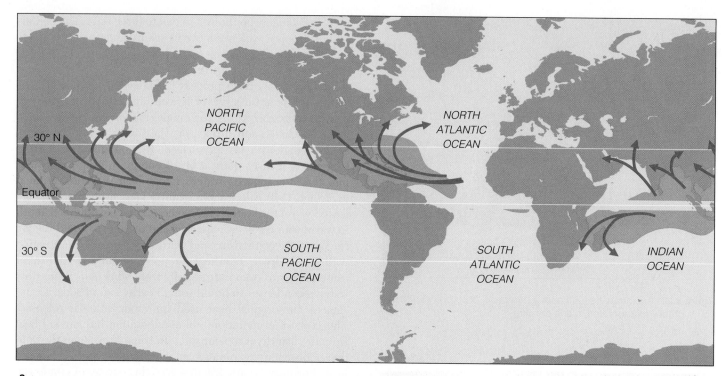

a

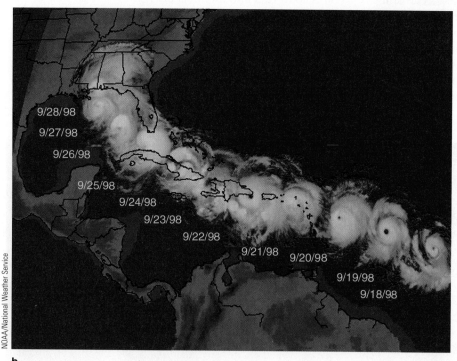

b

9/28/98
9/27/98
9/26/98
9/25/98
9/24/98
9/23/98
9/22/98
9/21/98 9/20/98
9/19/98
9/18/98

NOAA/National Weather Service

Figure 8.24 (**a**) The breeding grounds of tropical cyclones are shown as orange shaded areas. The storms follow curving paths: First they move westward with the trade winds. Then they either die over land or turn eastward until they lose power over the cooler ocean of mid-latitudes. Cyclones are not spawned over the South Atlantic or the southeast Pacific, because their waters are too chilly, nor in the still air—the doldrums—within 5° of the equator. (**b**) A composite of infrared satellite images of Hurricane Georges from 18 September to 28 September 1998. Its westward and poleward trek across the Caribbean and into the United States is clearly shown.

moisture it held when warm. Rainfall begins. The rainfall rate in some parts of the storm routinely exceeds 2.5 centimeters (1 inch) per hour, and 20 billion metric tons of water can fall from a large tropical cyclone in a day! Tremendous energy is released as this moisture changes from water vapor to liquid. In one day, a large tropical cyclone generates about 2.4 trillion kilowatt-hours of power, equivalent to the electrical energy needs of the entire United States for a year! Thus, solar energy ultimately

powers the storm in a cycle of heat absorption, evaporation, condensation, and conversion of heat energy to kinetic energy. This energy is available as long as the storm stays over warm water and has a ready source of hot, humid air.

Three aspects of a tropical cyclone can cause property damage and loss of life: wind, rain, and storm surge. The destructive force of winds of 250 kilometers (155 miles) per hour—or more—is self-evident (and is described in **Box 8.1**).

The most lethal Atlantic hurricane of modern times began as a chain of rolling thunderclouds drifting westward across the coast of central Africa. By the morning of 22 October 1998, the clouds had become organized into a tropical depression. Earth's rotation had given the clouds a gentle twist, and warm oceanic air rose into their core. Like water spinning down a bathtub drain, humid air was sucked into the growing vortex. The moisture condensed to form rain, liberating heat energy that caused the storm to grow into the thirteenth hurricane of the season: Mitch.

Three days passed. Mitch tracked uncertainly toward the west, paralleling the north coast of Honduras. On two occasions, shearing winds threatened to tear the storm apart; each time it survived and strengthened. Then, on the morning of 26 October, Mitch metamorphosed into a towering black mountain of wind and rain, a rare category-5 storm (**Figure a**). Its spreading violence sucked the sea surface into a 6-meter (20-foot) dome tens of kilometers across. The wind speed near its center rose to 290 kilometers (180 miles) per hour, driving ocean waves 15 meters (50 feet) high against the coast. In the predawn hours of 28 October, the leading edge of the storm touched the shore. During the next three days Mitch stalked Honduras and Nicaragua—a tearing, clawing force that entered every window and door, that tore the roofs off shacks and children from their parents' grasp, that hurled cars and structures about like toy blocks. The division between ocean and land blurred beneath an onrushing wall of windblown seawater. At the height of the storm the hot, wet air glowed yellow-green and was ear-poppingly thin.

As **Figure b** shows, devastation was all but absolute. More than 9,000 people died, 5,500 were missing, and 570,000 were homeless. The steep terrain of Honduras and Nicaragua was covered with poorly consolidated volcanic soil, and mudflows and landslides buried at least ten communities. The banana crop, economic mainstay of the region, was all but wiped out, the area's fragile infrastructure decimated. Fifty years of progress disappeared in fifty hours.

The damage done by Hurricane Mitch in Nicaragua alone has been estimated at $1.36 billion, or 67% of that country's gross domestic product (GDP). If a natural disaster in the United States caused damage equivalent to 67% of our GDP, the cost would be a staggering $4.3 trillion, equivalent to 143 hurricane landfalls the magnitude of Andrew (Florida, 1992) or 108 Northridge (California, 1994) earthquakes. Taken as a whole, Mitch was one of the Western Hemisphere's greatest twentieth-century natural disasters, a sobering demonstration of uneven solar heating and the power of water's latent heat of vaporization.

8-17

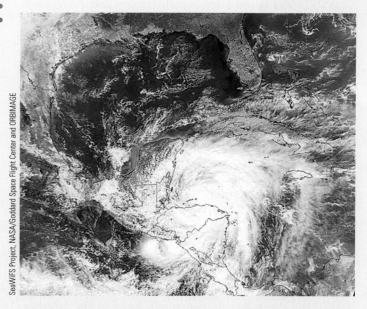

a Hurricane Mitch rakes the Central American Atlantic coast, 28 October 1998.

b Residents of the Honduran capital of Tegucigalpa dig themselves out after floods caused by Hurricane Mitch. Officials say the hurricane damage set their country's infrastructure back 50 years.

Rapid rainfall can cause severe flooding when the storm moves onto land. But the greatest danger lies in **storm surge,** a mass of water driven by the storm. The low atmospheric pressure at the storm's center produces a dome of seawater that can reach a height of 1 meter (3.3 feet) in the open sea. The water height increases when waves and strong hurricane winds ramp the water mass ashore. If a high tide coincides with the arrival of all this water at a coast or if the coastline converges (as is the case at the mouths of the Ganges in Bangladesh), rapid and catastrophic flooding will occur. Storm surges of up to 12 meters (40 feet) were reported at Bangladesh in 1970. Much of the $3 billion in damage done by Hurricane Andrew to south Florida in August 1992 was caused by a 4-meter (13-foot) storm surge arriving at high tide. You'll learn more about storm surges in Chapter 11's discussion of large waves.

You may have noticed that tropical cyclones turn counterclockwise in the Northern Hemisphere and clockwise in the Southern Hemisphere. Does this mean that the Coriolis effect does not apply to tropical cyclones? No. Their spin is caused by the Coriolis deflection of winds approaching the center of a low-pressure area from great distances. In the Northern Hemisphere, there is rightward deflection of the *approaching air.* The edge spin given by this approaching air causes the storm to spin counterclockwise in the Northern Hemisphere.

Tropical cyclones last from three hours to three weeks; most have lives of five to ten days. They eventually run down when they move over land or over water too cool to supply the humid air that sustains them. The friction of a land encounter rapidly drains a tropical cyclone of its energy, and a position above ocean water cooler than 24°C (75°F) is a sure harbinger of the storm's demise. When deprived of energy, the storm "unwinds" and becomes a mass of unstable humid air pouring rain, lightning, and even tornadoes from its clouds. Tropical cyclones can be dangerous to the end: Torrential rain streaming from the remnants of Hurricane Agnes in 1976 caused more than $2 billion in damage, mostly to Pennsylvania. Chesapeake and Delaware Bays were flooded with fresh water and sediments, destroying much of the shellfish industry there.

Tropical cyclones are nature's escape valves, flinging solar energy poleward from the tropics. They are beautiful, dangerous examples of the energy represented by water's latent heat of vaporization.

QUESTIONS FROM STUDENTS

1. Earth's orbit brings it closer to the sun in the Northern Hemisphere's winter than in its summer. Yet it's warmer in summer. Why?

Earth's orbit around the sun is elliptical, not circular. The whole Earth receives about 7% more solar energy through the half of the orbit during which we are closer to the sun than through the other half. The time of greater energy input comes during our winter, but the entire Northern Hemisphere is tilted toward the sun during the summer, which results in much more light reaching it in the summer. Three times more energy enters the Northern Hemisphere each day at midsummer than at midwinter.

2. How did the trade winds get their name? Is their name a reminder of the assistance they provided to shipboard traders anxious to sell their wares in distant corners of the world?

The trade winds are not named after their contribution to commerce in the days of sail. This use of the word *trade* derives from an earlier English meaning equivalent to our adverbs *steadily* or *constantly.* These persistent winds were said to "blow trade."

3. If the Coriolis effect acts on all moving objects, why doesn't it pull my car to the right? And does the Coriolis effect really make explorers wander to the left in the snows of Antarctica, or tree trunks grow in rightward spirals in Canadian forests, or water swirl clockwise down a toilet in Springfield?

The Coriolis effect depends on the speed, mass, and latitude of the moving object. Your car's motion is affected. When you drive along the road at 112 kilometers (70 miles) per hour in an average-size car, the Coriolis "force" would pull your car to the right about 460 meters (1,500 feet) for every 160 kilometers (100 miles) you travel *if* it were not for the friction between your tires and the road surface.

Small, lightweight objects moving slowly are subject to many forces and conditions (such as wind currents, natural variations in basin shape, and friction) that overwhelm Coriolis acceleration. For example, think of how *very* small the difference in eastward speed of the northern edge of a toilet is in comparison to the southern edge. Any small irregularity in the toilet's shape will be hundreds of times more important than the Coriolis effect in determining whether water will exit in a rightward spin or leftward spin! Explorers and trees aren't massive enough and don't move quickly enough to be affected by the Coriolis effect.

But if the moving object is at mid-latitude, is heavy, and is moving quickly, it *will* be deflected to the right of its intended path. Jet airliners, trains, artillery shells, and weather systems are all noticeably influenced by the Coriolis effect.

4. Does the ocean affect weather at the centers of continents?

Absolutely. In a sense, *all* large-scale weather on Earth is oceanically controlled. The ocean acts as a solar collector and heat sink, storing and releasing heat. Most great storms (tropical and extratropical cyclones alike) form over the ocean and then sweep over land.

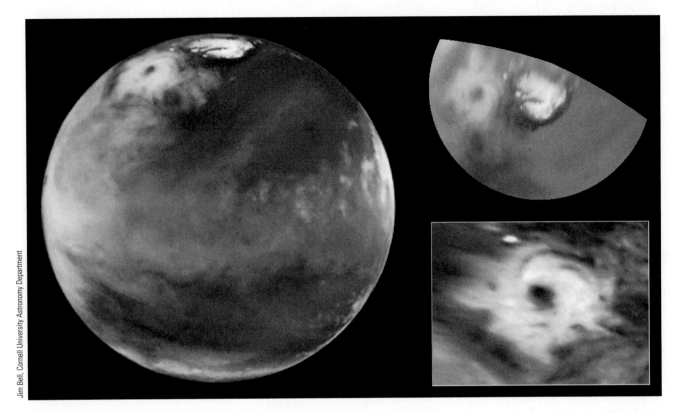

Figure 8.25 A vast extratropical cyclone on Mars. The storm is more than 1,600 kilometers (1,000 miles) across.

5. On average, the meteorological equator lies about 5° north of the geographical equator. Why the displacement?

The meteorological equator lies in the Northern Hemisphere because, overall, that hemisphere is slightly warmer than the Southern Hemisphere. At least three factors are responsible for this. First, the unbroken ice covering of Antarctica reflects more light back into space than the surface of the Arctic Ocean, which contains occasional light-absorbing patches of open water. Second, there is more land surface in the tropical latitudes of the Northern Hemisphere than in the Southern Hemisphere. Since land warms more than water with the same input of solar energy, tropical latitudes are warmer in the Northern Hemisphere. And third, ocean currents transport more warm water toward the north than toward the south.

6. Are any of the results of large-scale atmospheric circulation apparent to the casual observer?

Yes. The view of atmospheric circulation developed in this chapter explains some phenomena you may have experienced. For instance, flying from Los Angeles to New York takes about 40 minutes less than flying from New York to Los Angeles because of westerly head winds (indicated in Figure 8.12).

Because of these same prevailing westerlies, most storms travel over the United States from west to east. Weather prediction is based on observations and samples taken from the air masses as they move. Forecasting is often easier in the East and Midwest than in the West because more data are available from an air mass when it's over land.

7. Has anything similar to Earth's weather patterns been seen on other planets?

Yes, indeed. Saturn, Jupiter, and Neptune have tremendous cyclonic storms. A few on Jupiter are large enough to be seen from Earth through small telescopes. Tracks of tornadoes have recently been identified on Mars, and a huge cyclonic storm was photographed there in 1999 (**Figure 8.25**). Venus has huge cloud banks suggestive of polar fronts and extratropical cyclones. The Coriolis effect and uneven solar heating are common to all planets with atmospheres in this solar system, so we shouldn't be surprised by similarities.

KEY CONCEPTS TO REVIEW

- Different amounts of solar energy are absorbed at different latitudes. The tropics are warmer than the polar regions because of this difference.

- The interaction of ocean and atmosphere moderates surface temperatures, shapes Earth's weather and climate, and creates most of the sea's waves and currents.

- The atmosphere responds to uneven solar heating by flowing in three great circulating cells over each hemisphere. The flow of air within these cells is influenced by the rotation of Earth.

- To observers on the surface, Earth's rotation causes moving air (or any moving mass) in the Northern Hemisphere to curve to the right of its initial path, and in the Southern Hemisphere to curve to the left. The apparent curvature of path is known as the Coriolis effect.

- The Coriolis effect does not initiate the movement of air or water but, once moving, does influence the direction of air and water movement.

- Circulation of air is responsible for about two-thirds of the heat transfer from tropical to polar regions. (Ocean currents account for the other third.)

- Large storms are spinning areas of unstable air that develop between or within air masses. Extratropical cyclones originate at the boundary between air masses. Tropical cyclones, the most powerful of Earth's atmospheric storms, occur within a single humid air mass.

INTERNET STUDY RESOURCES

The Web site for this book contains helpful study aids. Log on to:

 www.brookscole.com/product/053437557xs

and click on the Chapter-by-Chapter area. Choose Chapter 8 and select a resource:

- **Flash Cards** allows you to test your mastery of the Terms and Concepts to Remember for this chapter.
- **Tutorial Quizzes** provides a multiple-choice practice quiz.
- **Student Guide to InfoTrac College Edition** will lead you to Critical Thinking Projects that use InfoTrac College Edition as a research tool.

TERMS AND CONCEPTS TO REMEMBER

air mass	extratropical cyclone
atmospheric circulation cell	Ferrel cell
	front
Bjerknes, Vilhelm	frontal storm
convection current	geographical equator
Coriolis, Gaspard Gustave de	Hadley cell
Coriolis effect	heat budget
cyclone	horse latitudes
doldrums	hurricane

intertropical convergence zone (ITCZ)	sea breeze
	storm
land breeze	storm surge
meteorological equator	thermal equilibrium
monsoon	tornado
nor'easter (northeaster)	trade winds
polar cell	tropical cyclone
polar front	water vapor
precipitation	westerlies

STUDY QUESTIONS

Review Questions

1. What is the composition of air?

2. Can more water vapor be held in warm air or cool air?

3. Which is denser at the same temperature and pressure: humid air or dry air?

4. What happens when air containing water vapor rises?

5. What causes precipitation (rain and snow)?

6. How many atmospheric circulation cells exist in each hemisphere?

7. Where does air rise, and fall, in each hemisphere? How does this movement affect weather?

8. How do the two kinds of large storms differ? How are they similar? What causes an extratropical cyclone? What happens in one?

9. What triggers a tropical cyclone? From what is its great power derived? What causes the greatest loss of life and property when a tropical cyclone reaches land?

10. How does weather differ from climate? Which is easier to predict? Why?

Critical Thinking Questions

1. What factors contribute to the uneven heating of Earth by the sun?

2. How does the atmosphere respond to uneven solar heating? How does the rotation of Earth affect the resultant circulation?

3. Why doesn't the ocean boil away at the equator and freeze solid near the poles?

4. Describe the atmospheric circulation cells in the Northern Hemisphere. At what latitudes does air move vertically? Horizontally? What are the trade winds? The westerlies? Where are deserts located? Why? What do you think ocean surface salinity is like in these desert bands?

5. How are the geographic equator, meteorological equator, and ITCZ related? What happens at the ITCZ?

6. What is a monsoon? How is monsoon circulation affected by the position of the ITCZ?

7. If the Coriolis effect causes the clockwise deflection of moving objects in the Northern Hemisphere, why does air rotate counterclockwise around zones of low pressure in that hemisphere?

8. **InfoTrac College Edition Project** People have always been attracted to the seashore as a place to live. Suppose a friend of yours has just told you of a great opportunity to purchase beachfront property in an area on the Atlantic coast that is known to have hurricane warnings every year. Your friend asks you what you think of the idea. Do some research using InfoTrac College Edition and compose an answer for your friend.

 Systems Today CD Questions

1. Go to the "Weather and Climate," then to "Weather Forecasting." Note the many parameters from which to select. For our purposes, temperature, humidity, wind speed, and satellite pictures will be most useful. Review the general flow pattern of Earth's surface winds (see Figure 8.12). Now predict wind direction over the continental United States. Toggle the wind vector and jet stream overlays and see if your predictions hold up. How does the wind flow affect temperature and humidity?

2. Go to "Weather and Climate" and connect to the Internet through **Earth Systems Today** to see today's satellite images and forecasts for your zip code.

 **RESOURCES FOR FURTHER READING AND RESEARCH**

The Web site for this book contains many ideas for further reading and research. Log on to:

www.brookscole.com/product/053437557xs

and click on the Chapter-by-Chapter area. Choose Chapter 8 and select a resource:

- **References** lists the major books and articles consulted in writing this chapter, along with comments from the author about their content and reading level.
- **Hypercontents** takes you to an extensive list of sites with news, research, and images related to individual sections of the chapter.
- In **Student Guide to InfoTrac College Edition** scroll down to Suggested Readings from InfoTrac College Edition for brief descriptions of the articles listed and search hints for finding them in InfoTrac College Edition.
- **Regional InfoTrac College Edition** articles are organized into East Coast, West Coast, and Gulf Coast regions, allowing you to study oceanography on a more local level. You can also access Regional InfoTrac College Edition at www.localocean.com.

For additional readings, go to InfoTrac College Edition, your online research library, at:

http://infotrac.thomsonlearning.com/

Ocean Circulation 9

GOING WITH THE FLOW

The first navigators who sailed past Gibraltar into the Atlantic noticed a persistent southward flow of water, which could drive them off course. Pytheas of Massalia, a Greek ship's captain who explored the northeastern Atlantic in the fourth century B.C., was the first observer to record this slow, continuous movement and to estimate its speed. Though his principal work, *On the Ocean,* is lost, we know of his discoveries from the writings of the historians of his time. Pytheas imagined the flow was part of an immense river too wide to sail across. Greek traders plying the eastern Atlantic coast later used the term *okeanos* (oceanus), meaning "great river," to describe it.

Today, we know that the river idea is not far off the mark: The world ocean contains moving masses of water called **currents.** Ocean currents transport water, heat, and nutrients to every corner of the sea in a manner resembling the circulatory system of a living organism. The volume of water moving in any one of the ocean's largest surface currents is about 50 times the combined flow of all Earth's freshwater rivers. The term *current* is usually reserved for water flowing horizontally (that is, parallel to the ocean's surface), but masses of wa-

Gibraltar from space; the Mediterranean lies beyond. Spain and Portugal are on the left, Africa on the right. Note the thinness of Earth's atmosphere.

ter can also move vertically. Downward movement transports gases to the deep ocean for the maintenance of life there, and upward movement brings fallen nutrients back toward the surface.

The southward-flowing "river" that Pytheas discovered is now called the Canary Current after a cluster of islands lying in its midst. The Canary Current forms as surface water sweeps eastward and southward along the northern boundary of the Atlantic. Though not particularly fast or deep as currents go, the Canary Current moves about 16 million cubic meters (21 million cubic yards) of water per second along the African coast between the Canary and Cape Verde Islands. Had Pytheas gone with the flow, his ship would probably have turned westward with the water and headed toward North America. His chroniclers might have had a much different tale to tell had he persisted all the way around the North Atlantic and back to Gibraltar—along a path his *oceanus* would lead him.

9-1

. . . .

WHAT TO WATCH FOR IN CHAPTER 9

Ocean water circulates in currents that transfer heat from tropical regions to polar regions, influence weather and climate, distribute nutrients, and scatter organisms. There are different kinds of currents above and below the pycnocline, the zone in which density increases with depth.

Surface currents driven by the wind affect the world ocean in and above the pycnocline. As you'll see, some surface currents are rapid and riverlike, with well-defined boundaries; others are slow and diffuse. But why should surface currents be different from one another? Could the Coriolis effect play a role? Of course!

Circulation of the 90% of ocean water beneath the pycnocline is driven by the force of gravity, as denser water sinks and less dense water rises. Since density is largely a function of temperature and salinity, the movement of deep water because of density differences is called *thermohaline circulation.* Currents near the seafloor move as slow streams in a few places, but the greatest volumes of deep water flow through the ocean at an almost imperceptible pace. The whole ocean slowly falls, rises, and creeps from place to place.

As you read, notice the interplay between wind and water, heat and cold, salinity and density. Although El Niño and La Niña grab all the headlines, you'll find that ocean currents can influence your daily life in surprising ways.

. . . .

SURFACE CURRENTS

About 10% of the water in the world ocean is involved in **surface currents,** water flowing horizontally in the uppermost 400 meters (1,300 feet) of the ocean's surface, driven mainly by wind friction. Most surface currents move water above the pycnocline, the zone of rapid density change with depth.

The primary force responsible for surface currents is wind. As you read in Chapter 8, surface winds form global patterns within latitude bands (see Figure 8.12 and **Figure 9.1**). Most of Earth's surface wind energy is concentrated in each hemisphere's trade winds (easterlies) and westerlies. Waves on the sea surface transfer some of the energy from the moving air to the water by friction. This tug of wind on the ocean surface begins a mass flow of water. The water flowing beneath the wind forms a surface current.

The moving water will "pile up" in the direction the wind is blowing. Water pressure will be higher on the "piled up" side, and the force of gravity will act to pull the water down the slope—against the *pressure gradient*—in the direction from which it came. But the Coriolis effect intervenes. Because of the Coriolis effect (discussed in Chapter 8), Northern Hemisphere surface currents flow to the *right* of the wind direction.[1] Southern Hemisphere currents flow to the *left*. Continents and basin topography often block continuous flow and help to deflect the moving water into a circular pattern. This flow around the periphery of an ocean basin is called a **gyre** (*gyros* = a circle). Two gyres are shown in **Figure 9.2**.

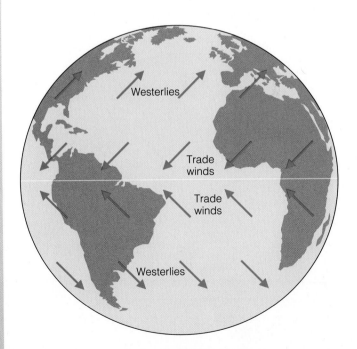

Figure 9.1 Winds, driven by uneven solar heating and Earth's spin, drive the movement of the ocean's surface currents. The prime movers are the powerful westerlies and the persistent trade winds (easterlies).

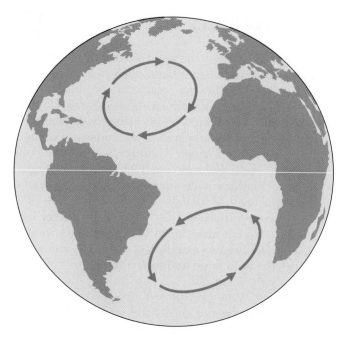

Figure 9.2 A combination of four forces—surface winds, the sun's heat, the Coriolis effect, and gravity—circulates the ocean surface clockwise in the Northern Hemisphere and counterclockwise in the Southern Hemisphere, forming gyres.

Flow Within a Gyre

Figure 9.3 shows the North Atlantic gyre in more detail. Though it flows continuously without obvious places where one current ceases and another begins, oceanographers subdivide the North Atlantic gyre into four interconnected currents because each has distinct flow characteristics and tem-

peratures. (Gyres in other ocean basins are similarly divided.) Notice that the east-west currents in the North Atlantic gyre flow to the right of the driving winds; once initiated, water flow in these currents continues in a roughly east-west direction. Where their flow is blocked by continents, the currents turn clockwise to complete the circuit.

Why does water flow around the *periphery* of the ocean basin instead of spiraling to the center? After all, the Coriolis effect influences any moving mass *as long as it moves*, so water in a gyre might be expected to curve to the center of the North Atlantic and stop. To understand this aspect of current movement, imagine the forces acting on the surface water at 45°N latitude (point A in **Figure 9.4**). Here the westerlies blow from the southwest, so initially the water will move toward the northeast. The rightward Coriolis deflection then causes the water to flow almost due east. A particle at 15°N latitude (point B) responds to the push of the trade winds from the northeast, however, and with Coriolis deflection it will flow almost due west.

When driven by the wind, the topmost layer of ocean water in the Northern Hemisphere flows at about 45° to the right of the wind direction, a flow consistent with the arrows leading away from points A and B in Figure 9.4. But what about the water in the next layer down? It can't "feel" the wind at the surface; it "feels" only the movement of the water immediately above. This deeper layer of water moves *at an angle to the right* of the overlying water. The same thing happens in the layer below that, and the next layer, and so on, to a depth of about 100 meters (330 feet) at mid-latitudes. Each layer slides horizontally over the one beneath it like cards in a deck, with each lower card moving at an angle slightly to the right of the one above. Because of frictional losses, each lower layer also moves more slowly than the layer above. The re-

Figure 9.3 The North Atlantic gyre, a series of four interconnected currents with different flow characteristics and temperatures.

Figure 9.4 Surface water blown by the winds at point A will veer to the right of its initial path and continue eastward. Water at point B veers right and continues westward.

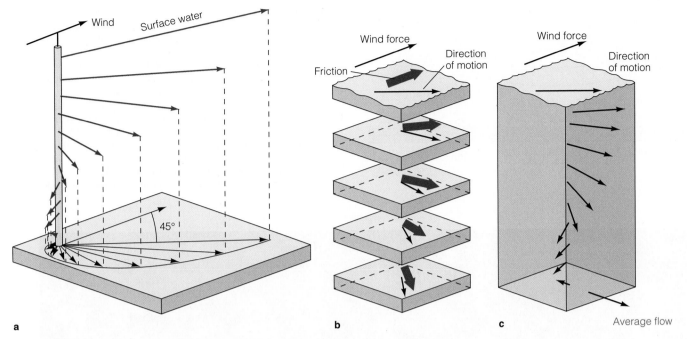

a   b   c

Figure 9.5 Ekman spiral and the mechanism by which it operates. (**a**) The Ekman spiral model. (**b**) A body of water can be thought of as a set of layers. The top layer is driven forward by the wind, and each layer below is moved by friction. Each succeeding layer moves with a slower speed and at an angle to the layer immediately above it—to the right in the Northern Hemisphere, to the left in the Southern Hemisphere—until water motion becomes negligible. (**c**) Though the direction of movement varies for each layer in the stack, the theoretical net flow of water in the Northern Hemisphere is 90° to the right of the prevailing wind force. The length of the arrows is proportional to the speed of the current in each layer. (Source for 9.5a: From *Laboratory Exercises in Oceanography*, 4/e, by Pipkin, Gorslin, Casey, and Hammond. © 1987 by W. H. Freeman and Company. Source for 9.5b: NASA.)

sulting situation, portrayed in **Figure 9.5,** is known as **Ekman spiral** after the Swedish oceanographer who worked out the mathematics involved. The term is somewhat misleading; the water itself does not spiral downward in a whirlpool-like motion. Rather, the spiral is a way of conceptualizing the horizontal movements in a layered water column, with each layer moving in a slightly different horizontal direction. An unexpected result of Ekman spiral is that at some depth (known as the friction depth) water will be flowing in the opposite direction from the surface current!

The *net* motion of the water down to about 100 meters, after allowance for the summed effects of Ekman spiral (the sum of all the arrows indicating water direction in the affected layers), is known as **Ekman transport.** In theory, the direction of Ekman transport is 90° to the *right* of wind direction in the Northern Hemisphere, and 90° to the *left* in the Southern Hemisphere.

Armed with this information, we can look in more detail at the area around point B in Figure 9.4, which is enlarged in **Figure 9.6.** In nature, Ekman transport in gyres is less than 90°; in most cases the deflection barely reaches 45°. This deviation from theory occurs because of an interaction between the Coriolis effect and the pressure gradient. Some flowing Atlantic water has turned to the right to form a hill of water—it followed the rightward dotted-line arrow in Figure 9.6. Why does the water now go straight west from point B with-

out turning? Because, as **Figure 9.7a** shows, to turn further *right* the water would have to move uphill against the pressure gradient (and in defiance of gravity), but to turn *left* in response to the pressure gradient would defy the Coriolis

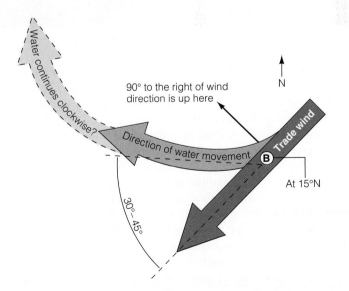

Figure 9.6 The movement of water away from point B in Figure 9.4 is influenced by the rightward tendency of the Coriolis effect and the gravity-powered movement of water down the pressure gradient.

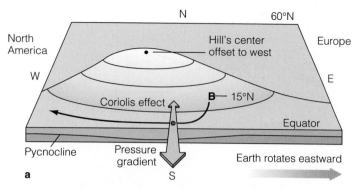

a

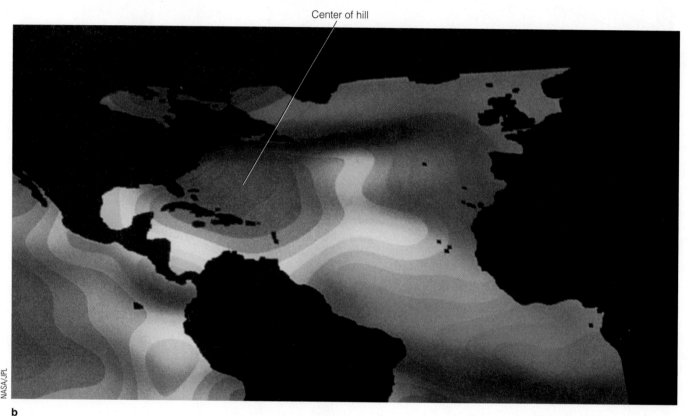

Center of hill

NASA/JPL

b

Figure 9.7 (**a**) The surface of the North Atlantic is raised through wind motion and Ekman transport to form a low hill. The eastward rotation of Earth offsets the center of the hill to the west. Water from point B (see also Figures 9.4 and 9.6) turns westward and flows along the side of this hill. The westward-moving water is balanced between the Coriolis effect (which would turn the water to the right) and flow down the pressure gradient, driven by gravity (which would turn it to the left). Thus, water in a gyre moves along the outside edge of an ocean basin. (**b**) The average height of the surface of the North Atlantic is shown in color in this image derived from data taken in 1992 by the *TOPEX/Poseidon* satellite. Red indicates the highest surface, green and blue the lowest. Note that the measured position of the hill is offset to the west as seen in (a). The gradually sloping hill is only 2 meters (6.5 feet) high and would not be apparent to anyone traveling from coast to coast.

effect. So the water continues westward and then clockwise around the whole North Atlantic gyre, dynamically balanced between the downhill urge of the pressure gradient and the uphill tendency of Coriolis deflection.

Yes, there really *is* a hill near the middle of the North Atlantic, centered in the area of the Sargasso Sea (see **Figure 9.7b**). This hill is formed of surface water gathered at the ocean's center of circulation. It is not a steep mountain of water—its maximum height is an unspectacular 2 meters (6.5 feet)—but rather a gradual rise and fall from coastline to open ocean and back to opposite coastline. Its slope is so gradual you wouldn't notice it on a transatlantic crossing.

The hill is maintained by wind energy. If the winds did not continuously inject new energy into currents, friction within the fluid mass and with the surrounding ocean basins would slow the flowing water, gradually converting its motion into heat. The balance of wind energy and fric-

tion, and of the Coriolis effect and the pressure gradient (through the effect of gravity), propels the currents of the gyre and holds them along the outside edges of the ocean basin.

Geostrophic Gyres

Gyres in balance between the pressure gradient and the Coriolis effect are called **geostrophic** gyres (*Geos* = Earth, *strophe* = turning), and their currents are *geostrophic currents*. Because of the patterns of driving winds and the present positions of continents, the geostrophic gyres are largely independent of each other in each hemisphere.

There are six great current circuits in the world ocean, two in the Northern Hemisphere and four in the Southern Hemisphere. They are shown in **Figure 9.8a** and **b.** Five are geostrophic gyres: the North Atlantic gyre, the South Atlantic

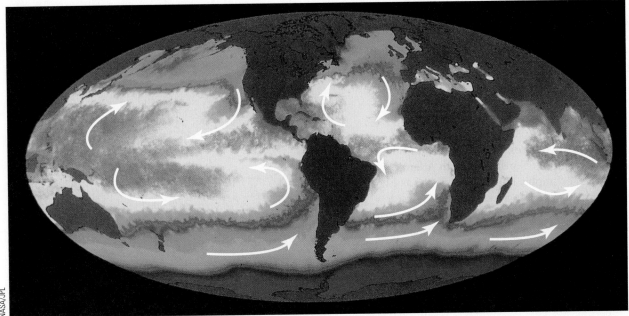

NASA/JPL

a

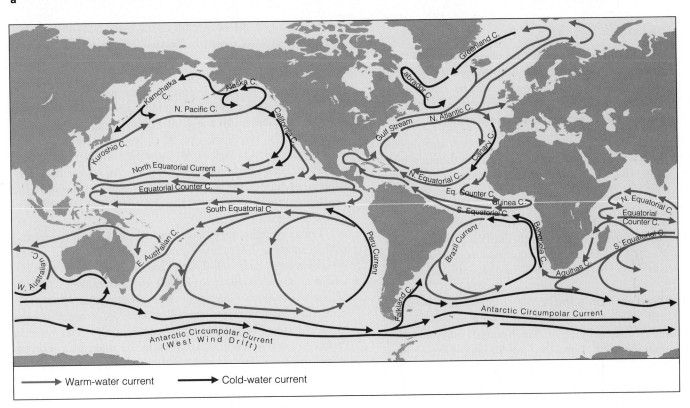

⟶ Warm-water current   ⟶ Cold-water current

b

Figure 9.8 Two ways of viewing the major surface currents of the world ocean. (**a**) An illustration of sea surface temperature showing the general direction and pattern of surface current flow. Sea surface temperatures were measured by a radiometer aboard *NOAA-7* in July 1984. The warmest temperatures, shown in red, are 25–28°C (77–82°F). Yellow represents 20–25°C (68–77°F); green, 15–20°C (59–68°F); blue, 0–15 °C (32–59°F). The purple color around Antarctica and west of Greenland indicates water below 0°C, the freezing point of fresh water. Note the distortion of the temperature patterns we might expect from the effects of solar heating alone—the patterns twist clockwise in the Northern Hemisphere, counterclockwise in the Southern Hemisphere. (**b**) A chart showing the names and usual direction of the world ocean's major surface currents. The powerful western boundary currents flow along the western boundaries of ocean basins in *both* hemispheres.

gyre, the North Pacific gyre, the South Pacific gyre, and the Indian Ocean gyre. Though it is a closed circuit, the sixth and largest current is technically not a gyre because it does not flow around the periphery of an ocean basin. The **West Wind Drift,** or **Antarctic Circumpolar Current,** as this exception is called, flows endlessly eastward around Antarctica, driven by powerful, nearly ceaseless westerly winds. This greatest of all surface ocean currents is never deflected by a continent.

We might expect the two gyres in the North and South Pacific (and the two gyres in the North and South Atlantic) to converge exactly at the geographic equator. However, as Figure 9.8b shows, the junction of equatorial currents lies a few degrees north of the geographical equator, at the meteorological equator. As noted in Chapter 8, the meteorological equator and the intertropical convergence zone (the band at which the trade winds converge) are displaced 5°–8° northward primarily because of the heat accumulated in the Northern Hemisphere's greater tropical land surface area. Ocean circulation, like atmospheric circulation, is balanced around the meteorological equator.

Currents Within Gyres

Because of the different factors that drive and shape them, the currents constituting geostrophic gyres have different characteristics. Geostrophic currents may be classified by their position within the gyre as western boundary currents, eastern boundary currents, or transverse currents.

WESTERN BOUNDARY CURRENTS The fastest and deepest geostrophic currents are found at the western boundaries of ocean basins (that is, off the *east* coast of continents). These narrow, fast, deep currents move warm water poleward in each of the gyres. There are five large **western boundary currents:** the Gulf Stream (in the North Atlantic), the Japan or Kuroshio Current (in the North Pacific), the Brazil Current (in the South Atlantic), the Agulhas Current (in the Indian Ocean), and the East Australian Current (in the South Pacific).

The **Gulf Stream** is the largest of the western boundary currents. Studies of the Gulf Stream, such as those undertaken by the drift submarine *Ben Franklin* (see **Box 9.1**), have revealed that off Miami the Gulf Stream moves at an average speed of 2 meters per second (5 miles per hour) to a depth of over 450 meters (1,500 feet). Water in the Gulf Stream can move more than 160 kilometers (100 miles) in a day. Its average width is about 70 kilometers (43 miles).

The volume of water transported in western boundary currents is extraordinary. The unit used to express volume transport in ocean currents is the **sverdrup (sv),** named in honor of Harald Sverdrup, one of this century's pioneering oceanographers. A sverdrup equals 1 million cubic meters per second.[1] The Gulf Stream flow is at least 55 sv (55 million me-

---

[1] One million cubic meters is about half the volume of the Louisiana Superdome.

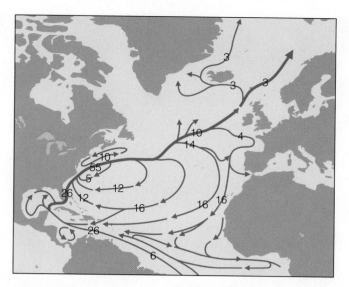

Figure 9.9 The general surface circulation of the North Atlantic. The numbers indicate flow rates in sverdrups (1 sv = 1 million cubic meters of water per second).

ters per second), about 300 times the usual flow of the Amazon, the greatest of rivers. In **Figure 9.9** the surface currents of the North Atlantic gyre are shown with their volume transport (in sverdrups) indicated.

Water in a current, especially a western boundary current, can move for surprisingly long distances within well-defined boundaries. In the Gulf Stream the current-as-river analogy can be startlingly apt: The western edge of the current is often clearly visible. Water within the current is usually warm, clear, and blue, often depleted of nutrients and incapable of supporting much life. By contrast, water over the continental slope adjacent to the current is often cold, green, and teeming with life.

Long, straight edges are the exception rather than the rule in western boundary currents, however. Unlike rivers, ocean currents lack well-defined banks, and friction with adjacent water can cause a current to form waves along its edges. Western boundary currents meander as they flow poleward. The looping meanders will sometimes connect to form turbulent rings, or **eddies,** that trap cold or warm water in their centers and then separate from the main flow. For example, *cold-core eddies* form in the Gulf Stream as it meanders eastward upon leaving the coast of North America off Cape Hatteras (see **Figure 9.10**). *Warm-core eddies* can form north of the Gulf Stream when the warm current loops into the cold water lying to the north. When the loops are cut off, they become freestanding spinning masses of water. Warm-core eddies rotate clockwise, and cold-core eddies rotate counterclockwise.

The slowly rotating eddies move away from the current and are distributed across the North Atlantic. Some may be 1,000 kilometers (620 miles) in diameter and retain their identity for more than three years. In mid-latitudes as much as one-fourth of the surface of the North Atlantic may consist of old, slow-moving, cold-core eddy remnants! Both cold and

In July 1969 six researchers sealed themselves into a specially designed submarine for a 2,640-kilometer (1,650-mile) drift within the Gulf Stream, the great ocean current that flows northward off the east coast of the United States. Their vehicle, named the *Ben Franklin* in honor of the man who first took a scientific interest in the current, was designed by Dr. Jacques Piccard, the builder of the bathyscaphe *Trieste* and father of balloonist Bertrand Piccard.

The 15-meter (50-foot) *Ben Franklin* was towed into the Gulf Stream off West Palm Beach, Florida, and positioned at a depth of 150 meters (500 feet). The month-long program of observations was designed to track the Gulf Stream and measure physical conditions within it. The scientists aboard were in regular contact with two surface ships that followed the drifting submarine and monitored its exact position. They kept a nearly continuous record of water temperature, salinity, and density.

The team made surprising discoveries right from the start. First, the submarine drifted more rapidly than expected. The speed of the surface current often exceeds 8 kilometers (about 5 miles) per hour, but oceanographers believed the current would move more slowly at greater depths. It didn't; the surface ships drifted at about the same rate as the submarine. And the drift was not always smooth. Underwater hills looming from the seabed, many of them previously uncharted, disrupted the flow of the submerged current and caused the submarine to rise and fall alarmingly.

Gulf Stream eddies also caused problems. Thirteen days into the mission, a huge eddy spun the submarine 80 kilometers (50 miles) out of the Gulf Stream core, forcing it to surface in order to be towed back into the center of the stream. These great eddies were found to reach much deeper than previously believed (we now know that they can reach all the way to the ocean bottom).

Another surprise was the general lack of marine life. The crew did sight several species of sharks, a huge jellyfish with tentacles 10 centimeters (4 inches) thick, and an aggressive broadbilled swordfish that jousted with their vessel before swimming away unharmed; but this was less life than they had expected. The biology program did make some strides, however. Tape recordings of underwater sounds of biological origin were studied on board, and some of these recordings were later used in the first detailed analyses of whale vocalizations. Dr. Piccard photographed some unusual bioluminescent organisms, a few of which were new to science. The crew also completed medical and psychological investigations on each other and analyzed the performance of the vehicle's life-support systems. These studies were used in later submarine and spacecraft design.

When the *Ben Franklin* finally bobbed to the surface 510 kilometers (317 miles) south of Halifax, Nova Scotia, researchers were confident of their biggest finding: They hadn't known as much about the Gulf Stream as they thought they did.

9-6

The submarine *Ben Franklin* enters New York harbor in August 1969 after drifting for 31 days within the Gulf Stream.

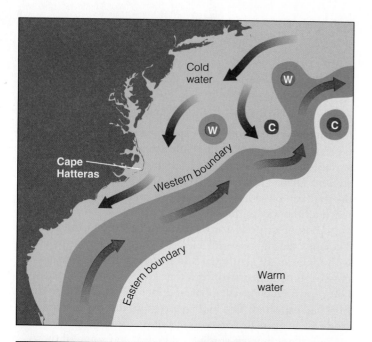

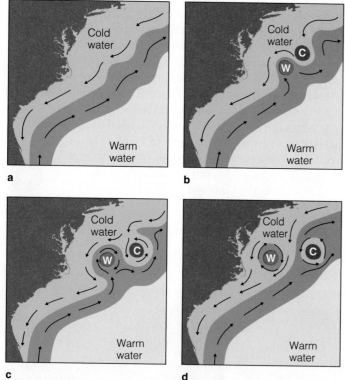

Figure 9.10 Eddy formation. The western boundary of the Gulf Stream is usually distinct, marked by abrupt changes in water temperature, speed, and direction. Meanders (eddies) form at this boundary as the Gulf Stream leaves the U.S. coast at Cape Hatteras (**a**). The meanders can pinch off (**b**), and eventually become isolated cells of warm water between the Gulf Stream and the coast (**c**). Likewise, cold cells can pinch off and become entrained in the Gulf Stream itself (**d**). (C = cold water, W = warm water; blue = cold, red = warm). Figure 9.11 shows the Gulf Stream from space with meanders and eddies clearly visible.

warm eddies are visible in the satellite image of **Figure 9.11.** Recent research suggests that their influence reaches to the seafloor. Warm- and cold-core eddies may be responsible for the slowly moving *abyssal storms* that leave often-observed ripple marks in deep sediments.

EASTERN BOUNDARY CURRENTS There are also five **eastern boundary currents** found at the eastern edge of ocean basins (that is, off the *west* coast of continents): the Canary Current (in the North Atlantic), the Benguela Current (in the South Atlantic), the California Current (in the North Pacific), the West Australian Current (in the Indian Ocean), and the Peru or Humboldt Current (in the South Pacific).

Eastern boundary currents are the opposite of their western boundary counterparts in nearly every way: They carry cold water equatorward; they are shallow and broad, sometimes more than 1,000 kilometers (620 miles) across; their boundaries are not well defined; and eddies tend not to form. Their total flow is less than that of their western counterparts. The Canary Current in the North Atlantic carries only 16 sv of water at about 2 kilometers (1.2 miles) per hour. The current is so shallow and broad that sailors may not notice it. Contrast the flow rates of the North Atlantic's western and eastern boundary currents in Figure 9.9. **Table 9.1** summarizes the major differences between eastern and western boundary currents in the Northern Hemisphere.

WESTWARD INTENSIFICATION Why should western boundary currents be concentrated and eastern boundary currents be diffuse? One reason is the converging flow of the trade winds on either side of the equator. Water moved by the trades approaches the meteorological equator and is shepherded west, where it piles up at the western edge of the basin before turning swiftly poleward. This concentration of water produces the poleward-moving western boundary currents. This poleward-moving water is traveling fast, and because the Coriolis effect increases with velocity, the pressure gradient must also be stronger on the steeper (western) side of the hill described in Figure 9.7. The heightened rightward tendency of the Coriolis effect is then balanced by the stronger, steeper pressure gradient, resulting in even more rapid poleward flow. (In contrast, the westerly winds of each hemisphere do not converge, and water driven by them is not swept along a line of convergence. Coriolis deflection can therefore move some of the eastward-moving water equatorward before the basin's eastern boundary is reached.)

A second reason is the rotation of Earth itself. The hill of Figure 9.7 is offset to the west because of Earth's eastward rotation, so water must squeeze closer to the ocean basin's western edge to pass around the hill at the western boundary. The combined effect on current flow is known as **westward intensification (Figure 9.12)**, a phenomenon clearly visible in Figures 9.7 and 9.9.

TRANSVERSE CURRENTS As we have seen, most of the power for ocean currents is derived from the trade winds at the fringes of the tropics and from the mid-latitude westerlies. The stress of winds on the ocean in these bands gives rise

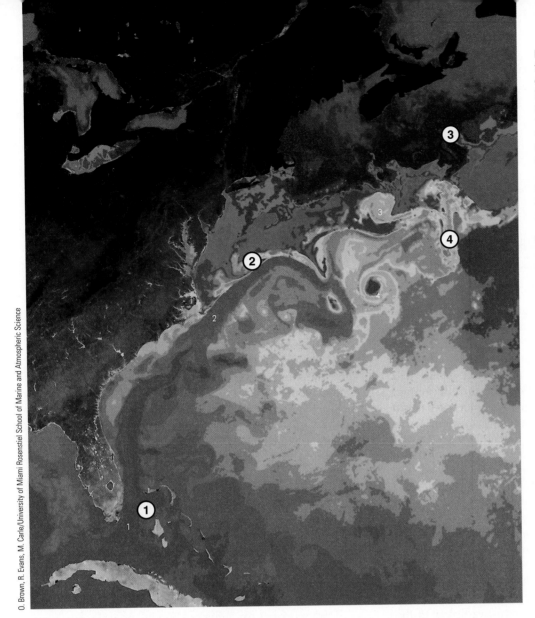

Figure 9.11 The Gulf Stream viewed from space. This image is a composite of temperature data returned from NOAA polar orbiting meteorological satellites during the first week of April 1984. The composite image is printed with an artificial color scale: Reds and oranges are a warm 24–28°C (76–82°F); yellows and greens are 17–23°C (63–74°F); blues are 10–16°C (50–61°F); and purples are a cold 2–9°C (36–48°F). The Gulf Stream appears like a red (warm) river as it moves from the southern tip of Florida ① north along the east coast. Moving offshore at Cape Hatteras ②, it begins to meander, with some meanders pinching off to form warm-core ③ and cold-core ④ eddies. As it moves northeastward, the water cools dramatically, releasing heat to the atmosphere and mixing with the cooler surrounding waters. By the time it reaches the middle of the North Atlantic, it has cooled so much that its surface temperature can no longer be distinguished from that of the surrounding waters.

Table 9.1 Boundary Currents in the Northern Hemisphere

Type of Current (example)	General Features	Speed	Transport (millions of cubic meters per second	Special Features
Western Boundary Currents Gulf Stream, Kuroshio (Japan) Current	**Warm** Narrow, <100 km. Deep—substantial transport to depths of 2 km	Swift, hundreds of kilometers per day.	Large, usually 50 sv or greater.	Sharp boundary with coastal circulation system; little or no coastal upwelling; waters tend to be depleted in nutrients, unproductive; waters derived from trade wind belts.
Eastern Boundary Currents California Current. Canary Current	**Cold** Broad, ~1,000 km. Shallow, <500 m.	Slow, tens of kilometers per day.	Small, typically 10–15 sv.	Diffuse boundaries separating from coastal currents; coastal upwelling common; waters derived from mid-latitudes.

Source: From M. Grant Gross, *Oceanography: A View of the Earth,* 5/e, © 1990, p. 173. Prentice-Hall. Reprinted by permission.

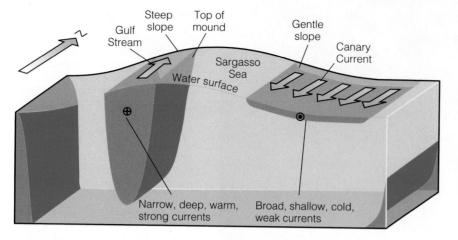

Figure 9.12 Geostrophic flow in the North Atlantic. The Gulf Stream, a western boundary current, must squeeze between the continental shelf and the steeper (western) side of the hill described in Figure 9.7. The Gulf Stream is narrow and deep and carries warm water northward. The Canary Current moves at a much more leisurely pace because it is farther away from the hill on a gentle slope. It can spread out across a larger area of ocean surface as it moves cold water south. Remember, although the gradually sloping hill is only 2 meters (6.5 feet) high and would not be apparent to anyone traveling from coast to coast, it is large enough to steer currents in the North Atlantic gyre. This figure is vertically exaggerated.

to the **transverse currents**—currents that flow from east to west and west to east, thus linking the eastern and western boundary currents.

The trade wind–driven North and South Equatorial Currents in the Atlantic and Pacific are moderately shallow and broad, but each transports about 30 sv westward. Because of the thrust of the trades, Atlantic water at Panama is usually 20 centimeters (8 inches) higher, on average, than water across the isthmus in the Pacific. The Pacific's greater expanse of water at the equator and stronger trade winds develop more powerful westward-flowing equatorial currents, and the height differential between the western and eastern Pacific is thought to approach 1 meter (3.3 feet)!

Westerly winds drive the eastward-flowing transverse currents of the mid-latitudes. Because they are not shepherded by the trade winds, eastward-flowing currents are wider and flow more slowly than their equatorial counterparts. The North Pacific and North Atlantic Currents are Northern Hemisphere examples.

As can be seen in Figure 9.8, the westward flow of the transverse currents near the equator proceeds unimpeded for great distances, but the eastward flow of transverse currents at middle and high latitudes in the northern ocean basins is interrupted by continents and island arcs. In the far south, however, eastward flow is almost completely free. Intense westerly winds over the southern ocean drive the greatest of all ocean currents, the unobstructed West Wind Drift (or Antarctic Circumpolar Current). This current carries more water than any other—at least 100 sv west-to-east in the Drake Passage between the tip of South America and the adjacent Palmer Peninsula of Antarctica.

Countercurrents and Undercurrents

Equatorial currents are typically accompanied by **countercurrents** flowing on the surface in the opposite direction from the main current. As you may remember, at the meteorological equator air is rising and the trade winds do not blow across the boundary. Without persistent winds to drive water to the west, some backward flow of water occurs at the meteorological equator. Some water flows away from the main equatorial currents (which flow westward a bit north and south of the meteorological equator) to return on the surface to the east. Look for this in the Pacific in Figure 9.8b.

Countercurrents can also exist *beneath* surface currents. These are sometimes referred to as **undercurrents.** The first undercurrent was discovered in 1951 in the central Pacific by Townsend Cromwell, a researcher employed by the U.S. Fish and Wildlife Service to investigate deep long-line fishing techniques pioneered by the Japanese. The Pacific Equatorial Undercurrent, or Cromwell Current, flows beneath the North Equatorial Current with an average velocity of 5 kilometers (3 miles) per hour at a depth of 100 to 200 meters (330 to 660 feet). It is about 300 kilometers (190 miles) wide and carries a volume equivalent to about half the Gulf Stream. It has been traced for over 14,000 kilometers (8,700 miles), from New Guinea to Ecuador. Undercurrents have since been found under most major currents. They can be very large—their volumes sometimes approach the volume of the current above.

Undercurrents can influence conditions at the ocean surface. The ocean around the Galápagos Islands is cold at the surface, even though the islands straddle the equator. For many years the cold was blamed on Antarctic water driven north in the Peru (Humboldt) Current, but recent research has shown that this surface water turns west before reaching the islands. The chill is actually caused by the upwelling of the Pacific Equatorial Undercurrent (the Cromwell Current) near the eastern end of its travel. Once at the surface, the cool water moves toward the west as part of the South Equatorial Current. The Galápagos Islands lie in its path.

Oceanographers are uncertain about what causes undercurrents, but near the equator (where the Coriolis effect is negligible) some water is not deflected north or south when it reaches the west side of the basin. Rather, it stays on the equator and underflows the surface current back to the east. This eastward-flowing current is kept on course by the Coriolis effect. If the current veers north from the equatorial region, the Coriolis effect bends it to the right, or south, back to its original course. Should the current wander too far south, the Coriolis effect in the Southern Hemisphere deflects the water to the left (north).

Exceptional Surface Currents

Some currents don't behave in ways you might first expect. Local or seasonal variations in wind and geography sometimes intrude on the flow patterns we have discussed. Two of these exceptions are noted here.

MONSOON CURRENTS From November to March the northeast trade winds propel water in the usual way, and normal surface circulation results. In the summer months, however, the intertropical convergence zone (ITCZ) moves north. Winds over the Indian Ocean change direction to blow from the south and southwest (see again Figure 8.15). Water in the relatively small northern Indian Ocean begins to flow clockwise, and the normally westward-flowing North Equatorial Current south of India reverses flow to become the eastward-moving Southwest Monsoon Current. (Due to its temporary nature, this current is not shown in Figure 9.8b.)

HIGH-LATITUDE CURRENTS Another exception to expected surface flow is seen in the many lesser current circuits at high northern latitudes in Figure 9.8b. Some of these—such as the Greenland and Labrador Currents in the North Atlantic and the Kamchatka and Alaska Currents in the North Pacific—move in apparent contradiction to the influence of the Coriolis effect, gravity, and friction. These currents form when larger currents are split and deflected by collision with a continent. The polar easterly winds at these latitudes keep the water moving westward. The size of these high-latitude currents on the map is exaggerated by the distortion inherent in the Mercator projection (see Appendix IV). The direction and rate of flow of these smaller currents are controlled primarily by the tendency of water to flow around obstacles rather than by the direct action of winds or Coriolis deflection. Technically these are not geostrophic currents, but their importance is often great. The Greenland and Labrador Currents, for example, transport icebergs to the Atlantic shipping lanes and deliver nutrients to the rich fishing grounds off Newfoundland.

Gyres: A Final Word

Although we have stressed individual currents in our discussion, remember that gyres consist of currents that blend into one another. Flow is continuous without obvious places where one current ceases and another begins. The *balance* of wind energy, friction, the Coriolis effect, and the pressure gradient propels gyres and holds them along the outside of ocean basins.

EFFECTS OF SURFACE CURRENTS ON CLIMATE

Along with the winds, surface currents distribute tropical heat worldwide. Warm water flows to higher latitudes, transfers heat to the air and cools, moves back to low latitudes, absorbs heat again, and repeats the cycle. The greatest amount of heat transfer occurs at mid-latitudes, where about 10 million bil-lion calories of heat are transferred *each second*—more than a million times the power consumed by all the world's human population in the same length of time! This combination of water flow and heat transfer from and to water influences climate and weather in several ways.

Commercial airliners flying to Europe from U.S. West Coast cities fly over central Ontario. In winter and spring months, the ground is hidden beneath masses of ice and snow, but passengers can often make out the frozen surface of James Bay. When those airliners land in London or Edinburgh, their passengers find a much milder climate than the barren whiteness of central Canada. Yet London's latitude of 51°N is the same as that of the southern tip of James Bay. Edinburgh lies at 55°N, the same latitude as Ontario's Polar Bear Provincial Park, a sanctuary for migrating polar bears. Why the difference in climate? The predominant direction of air flow is over the water eastward toward the British Isles. Even in winter, Edinburgh and London are bathed in air only recently in contact with the relatively warm North Atlantic Current. Scotland and England therefore have a maritime climate, and their only polar bears are found in zoos. These cities are warmed in part by the energy of tropical sunlight transported to their high latitudes by the Gulf Stream (see again Figure 9.8b).

At lower latitudes on an ocean's eastern boundary the situation is often reversed. Mark Twain is supposed to have said that the coldest winter he ever spent was a summer in San Francisco. Summer months in that West Coast city are cool, foggy, and mild, while Washington, D.C., on nearly the same line of latitude (but on the western boundary of an ocean basin), is known for its August heat and humidity. Why the difference? Look at Figure 9.8b and follow the currents responsible. The California Current, carrying cold water from the north, comes close to the coast at San Francisco. As shown in **Figure 9.13,** air normally flows clockwise in summer around an offshore zone of high atmospheric pressure. Wind approaching the California coast loses heat to the cold sea and comes ashore to chill San Francisco. Summer air often flows around a similar high off the East Coast (the

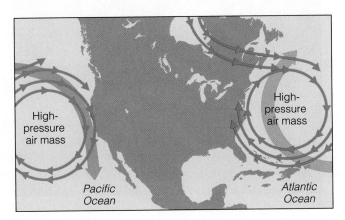

Figure 9.13 General summer air circulation patterns of the east and west coasts of the United States. Warm ocean currents are shown in red, cold currents in blue. Air is chilled as it approaches the west coast and warmed as it approaches the east coast.

Bermuda High). Winds approaching Washington, D.C., therefore blow from the south and east. Heat and moisture from the Gulf Stream contribute to the capital's oppressive summers. (In winter, on the other hand, Washington, D.C., is colder than San Francisco because westerly winds approaching Washington are chilled by the cold continent they cross.)

UPWELLING AND DOWNWELLING

The wind-driven *horizontal* movement of water can sometimes induce *vertical* movement in the surface water. This movement is called **wind-induced vertical circulation.** Upward movement of water is known as **upwelling;** the process brings deep, cold, usually nutrient-laden water toward the surface. Downward movement is called **downwelling.**

Equatorial Upwelling

Because the meteorological equator usually lies about 5° north of the geographic equator, the South Equatorial Currents of the Atlantic and Pacific straddle the geographic equator. Though the Coriolis effect is weak near the geographic equator (and absent *at* the geographic equator), water moving in the currents on either side of the geographic equator is deflected slightly poleward and replaced by deeper water (**Figure 9.14**). Thus, **equatorial upwelling** occurs in these

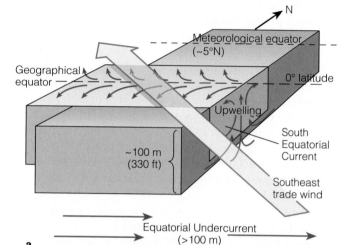

Figure 9.14 Equatorial upwelling. (**a**) The South Equatorial Current, especially in the Pacific, straddles the geographic equator (see again Figure 9.8b). Water north of the equator veers to the right (northward), and water to the south veers to the left (southward). Surface water therefore diverges, causing upwelling. Most of the upwelled water comes from the area above the equatorial undercurrent, at depths of 100 meters (330 feet) or less. (**b**) The phenomenon of equatorial upwelling is worldwide, but most pronounced in the Pacific.

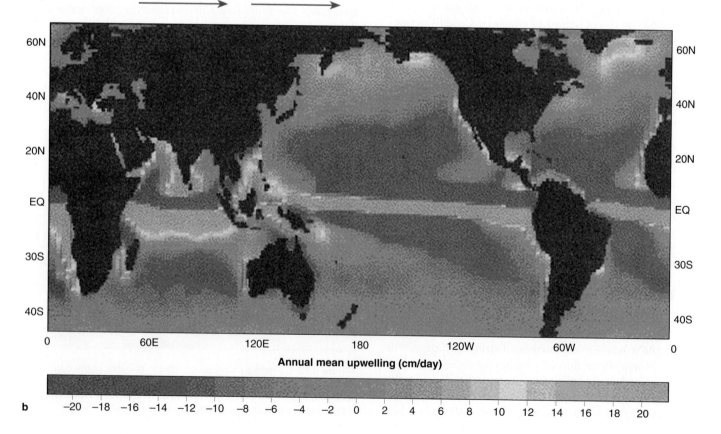

Annual mean upwelling (cm/day)

westward-flowing equatorial surface currents. Upwelling is an important process because this water from within and below the pycnocline is often rich in the nutrients needed by marine organisms for growth. The long, thin band of upwelling and biological productivity extending along the equator westward from South America is clearly visible in Figures 9.17b and 9.20. The layers of ooze on the equatorial Pacific seabed (Figure 5.13) are testimony to the biological productivity of surface water there. By contrast, generally poor conditions for growth prevail in most of the open tropical ocean—because strong layering isolates deep, nutrient-rich water from the sunlit ocean surface.

Coastal Upwelling

Wind blowing parallel to shore or offshore can cause **coastal upwelling.** The friction of wind blowing along the ocean surface causes the water to begin moving, the Coriolis effect deflects it to the right (in the Northern Hemisphere), and the resultant Ekman transport moves it offshore. As shown in **Figure 9.15a,** coastal upwelling occurs when this surface wa-

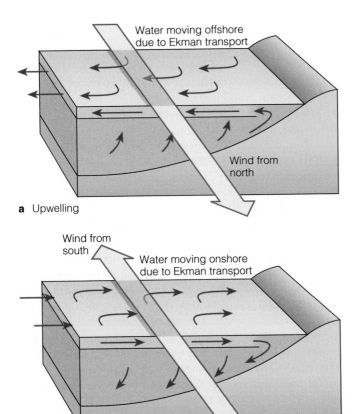

a Upwelling

b Downwelling

Figure 9.15 Coastal upwelling and downwelling. (**a**) In the Northern Hemisphere, coastal upwelling can be caused by winds from the north blowing along the west coast of a continent. Water moved offshore by Ekman transport is replaced by cold, deep, nutrient-laden water. (**b**) A prolonged southerly wind along a Northern Hemisphere west coast can result in downwelling.

ter is replaced by water rising along the shore. Again, because the new surface water is often rich in nutrients, prolonged wind can result in increased biological productivity. Coastal upwelling along the coast of South America is visible in Figure 9.17b.

Upwelling can also influence weather. Wind blowing from the north along the California coast causes offshore movement of surface water and subsequent coastal upwelling. The overlying air becomes chilled, contributing to San Francisco's famous fog banks and cool summers. Wind-induced upwelling is also common in the Peru Current, along the west coast of Antarctica's Palmer Peninsula, in parts of the Mediterranean, and near some large Pacific islands.

Downwelling

Water driven toward a coastline will be forced downward, returning seaward along the continental shelf. This downwelling (**Figure 9.15b**) helps supply the deeper ocean with dissolved gases and nutrients, and it assists in the distribution of living organisms. Unlike upwelling, downwelling has no direct effect on the climate or productivity of the adjacent coast.

Langmuir Circulation

Winds that blow steadily across the ocean, and the small waves that such winds generate, can induce long sets of counter-rotating cells in the surface water. These slowly twisting cells, or vortices, align in the direction of the wind (**Figure 9.16**). It takes about an hour for a particle in a cell to complete one revolution. Streaks of foam, seaweed, or debris, known as *windrows*, collect in areas where adjacent vortices converge, while regions of divergence remain relatively clear. Observed by mariners for hundreds of years, these windrow lines (and the underlying vortices) were first explained in 1938 by Irving Langmuir, who observed them while crossing the calm Sargasso Sea in the center of the North Atlantic gyre. Named in his honor, **Langmuir circulation** rarely disturbs the ocean below a depth of about 20 meters (66 feet). Unlike the equatorial and coastal upwellings mentioned above, Langmuir circulation operates within the surface layer and thus does not lift nutrients trapped within or below the pycnocline.

EL NIÑO, LA NIÑA

Surface winds across most of the tropical Pacific normally move from east to west (review Figure 8.14). The trade winds blow from the normally high-pressure area over the eastern Pacific (near Central and South America) to the normally stable low-pressure area over the western Pacific (north of Australia). However, for reasons that are still unclear, these pressure areas change places at irregular intervals of roughly three to eight years: High pressure builds in the western Pacific, and low pressure dominates the eastern Pacific. Winds across the tropical Pacific then reverse direction and blow from west

Mobil Oil Corporation

a

Figure 9.16 Winds that blow steadily across the ocean, and the small waves that such winds generate, can induce long sets of counter-rotating cells in the surface water. These shallow, slowly twisting cells of surface water are known as Langmuir circulation in honor of the researcher who explained their motion.

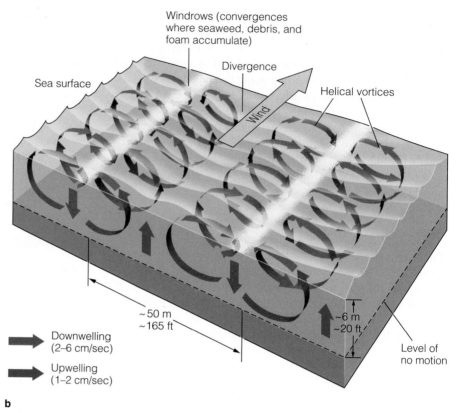

Windrows (convergences where seaweed, debris, and foam accumulate)

Divergence

Sea surface

Wind

Helical vortices

~50 m
~165 ft

~6 m
~20 ft

Level of no motion

→ Downwelling (2–6 cm/sec)

→ Upwelling (1–2 cm/sec)

b

to east—the trade winds weaken or reverse. This change in atmospheric pressure (and thus in wind direction) is called the **Southern Oscillation.** Ten of these attention-getting oscillations have occurred since 1950.

The trade winds normally drag huge quantities of water westward along the ocean's surface on each side of the equator, but as the winds weaken these equatorial currents crawl to a stop. Warm water that has accumulated at the western side of the Pacific—the warmest water in the world ocean—can then build to the east along the equator toward the coast of Central and South America. The eastward-moving warm water usually arrives near the South American coast around Christmastime. In the 1890s it was reported that Peruvian fishermen were using the expression *Corriente del Niño* ("current of the Christ child") to describe the flow; hence the current's name, **El Niño.** The phenomena of the Southern Oscillation and El Niño are coupled, so the terms are often combined to form the acronym **ENSO,** for El Niño/Southern Oscillation. An ENSO event typically lasts about a year, but some have persisted for more than three years. The effects are felt not only in the Pacific; all ocean areas at trade-wind latitudes in both hemispheres can be affected.

Normally, a current of cold water, rich in upwelled nutrients, flows north and west away from the South American continent (see **Figure 9.17a** and **b**). When the propelling trade winds falter during an ENSO event, warm equatorial water that would normally flow westward in the equatorial Pacific backs up to flow east (see **Figure 9.17c** and **d**). The normal northward flow of the cold Peru Current is interrupted or overridden by the warm water. Upwelling within

the nutrient-laden Peru Current is responsible for the great biological productivity of the ocean off the coasts of Peru and Chile. Although upwelling may continue during an ENSO event, the source of the upwelled water is nutrient-depleted water in the thickened surface layer approaching from the west. When the Peru Current slows and its upwelled water lacks nutrients, fish and seabirds dependent on the abundant life it normally contains die or migrate elsewhere. Peruvian fishermen are never cheered by this Christmas gift!

During major ENSO events, sea level rises in the eastern Pacific, sometimes by as much as 20 centimeters (8 inches) in the Galápagos. Water temperature also increases by up to 7°C (13°F). The warmer water causes more evaporation, and the area of low atmospheric pressure over the eastern Pacific intensifies. Humid air rising in this zone, centered some 2,000 kilometers (1,200 miles) west of Peru, causes high precipitation in normally dry areas. The increased evaporation intensifies coastal storms, and rainfall inland may be much higher than normal. Marine and terrestrial habitats and organisms can be affected by these changes.

The two most severe ENSO events of this century occurred in 1982–83 and 1997–98 (**Figure 9.18**). In both cases, effects associated with El Niño were spectacular over much of the Pacific and some parts of the Atlantic and Indian Oceans. In February 1998, 40 people were killed and 10,000 buildings damaged by a "wall" of tornadoes advancing over the southeastern United States. This record-breaking tornado event was spawned by the collision of warm, moist air that had lingered over the warm Pacific and a polar front that dropped from the north. In the eastern Pacific, heavy rains through the 1997–98

Figure 9.17 A non–El Niño year: (**a**) Normally the air and the surface water flow westward, the thermocline rises, and upwelling of cold water occurs along the west coast of Central and South America. (**b**) This map from satellite data shows the temperature of the equatorial Pacific on 31 May 1988. The warmest water is indicated by dark red, and progressively cooler water by yellow and green. Note the coastal upwelling along the coast at the lower right of the map and the tongue of recently upwelled water extending westward along the equator from the South American coast.

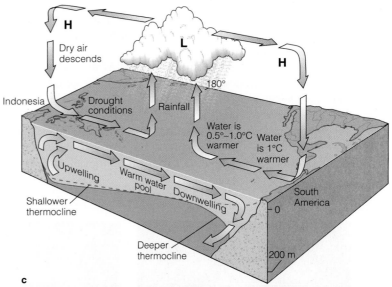

c

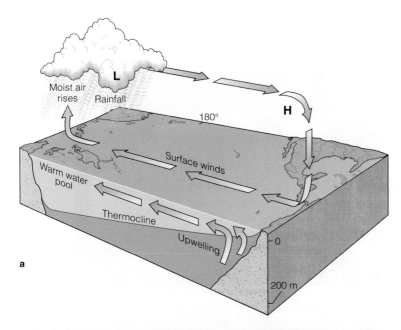

a

b

d

An El Niño year: (**c**) When the Southern Oscillation develops, the trade winds diminish and then reverse, leading to an eastward movement of warm water along the equator. The surface waters of the central and eastern Pacific become warmer, and storms over land may increase. (**d**) Sea-surface temperatures on 13 May 1992, a time of El Niño conditions. The thermocline was deeper than normal, and equatorial upwelling was suppressed. Note the absence of coastal upwelling along the coast and the lack of a tongue of recently upwelled water extending westward along the equator.

winter in Peru left at least 250,000 people homeless, destroyed 16,000 dwellings, and closed every port in the country for at least one month. Hawaii, however, was left with record drought, and some parts of southwestern Africa and Papua New Guinea received so little rain that crops failed completely and whole villages were abandoned due to starvation. Most of the United States escaped serious consequences—indeed, the midwestern states, Pacific Northwest, and eastern seaboard enjoyed a relatively mild fall, winter, and spring. But California's trials were widely reported—rainfall in most of the state exceeded twice normal amounts, and landslides, avalanches, and other weather-related disasters crowded the evening news. Conditions did not return to near-normal until the late spring of 1998. Estimates of worldwide 1997–98 ENSO-related damage exceed 23,000 deaths and $33 billion.

Normal circulation sometimes returns with surprising vigor, producing strong currents, powerful upwelling, and chilly and stormy conditions along the South American coast. These contrasting colder-than-normal events are given a contrasting name: **La Niña** (the girl). As conditions to the east

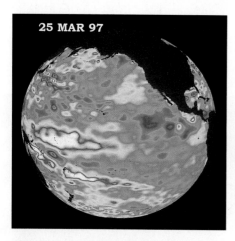

a March 1997. The slackening of the trade winds and westerly wind bursts allow warm water to move away from its usual location in the western Pacific Ocean. Red and white colors indicate sea level above average height.

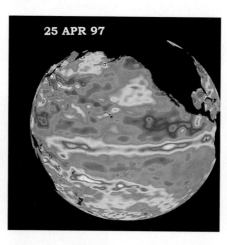

b April 1997. About a month after it began to move, the leading edge of the warm water reaches South America.

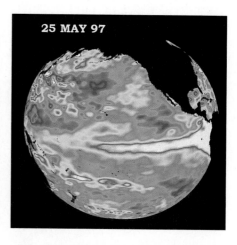

c May 1997. Warm water piles up against the South American continent. The white area of sea level is 13 to 30 centimeters (5 to 12 inches) above normal height, and 1.6–3°C (3–5°F) warmer.

d October 1997. By October, sea level is up to 30 centimeters (12 inches) lower than normal near Australia. The bulge of warm water has spread northward along the coast of North America from the equator to Alaska. Fisheries in Peru are severely affected—the warm water prevents upwelling of cold, nutrient-rich water necessary for the support of large fish populations.

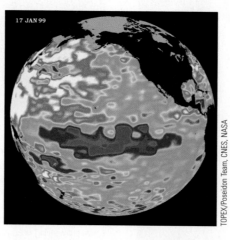

e Normal circulation sometimes returns with surprising vigor after an El Niño event, producing strong currents, powerful upwelling, and chilly and stormy conditions along the South American coast. This image was prepared from data for 17 January 1999. Note the mass of cold surface water and relatively low sea level (purple). Such cold water tends to deflect winds around it, changing the course of weather systems locally and the nature of weather patterns globally.

TOPEX/Poseidon Team, CNES, NASA

Figure 9.18 Development of the 1997–98 El Niño, observed by the *TOPEX/Poseidon* satellite.

cool off, the ocean to the west (north of Australia) warms rapidly. The renewed thrust of the trades piles this water upon itself, depressing the upper curve of the thermocline to more than 100 meters (328 feet). In contrast, the thermocline at this time in the eastern equatorial Pacific rests at about 25 meters (82 feet). A vigorous La Niña followed the 1997–98 El Niño and persisted for nearly a year (**Figure 9.18e**).

Studies of the ocean and atmosphere in 1982–83 and 1997–98 have given researchers new insight into the behavior and effects of the Southern Oscillation. Some researchers believe that the 1982–83 event was triggered by the violent 1982 eruption of El Chichón, a Mexican volcano, which injected huge quantities of obscuring dust and sulfur-rich gases into the atmosphere. No similar trigger occurred before the 1997–98 ENSO, however, and researchers have recently suggested that the Southern Oscillation may be triggered by pulses of heat entering the Pacific at crustal spreading zones near the equator. Some of this heat is moved by currents to the ocean surface. The warming of surface water may cause air above the ocean to warm, atmospheric pressure to fall, and the east-to-west trade winds to slacken. Though the exact cause or causes of the Southern Oscillation are not yet understood, subtle changes in the atmosphere permit meteorologists to predict a severe El Niño nearly a year in advance of its most serious effects.

THERMOHALINE CIRCULATION

The surface currents we have discussed affect the uppermost layer of the world ocean (about 10% of its volume), but horizontal and vertical currents also exist below the pycnocline in the ocean's deeper waters. The slow circulation of water at great depths is driven by density differences rather than by wind energy. Because density is largely a function of water temperature and salinity, the movement of water due to differences in density is called **thermohaline circulation** (*therme* = heat, *halos* = salt). Virtually the entire ocean is involved in slow thermohaline circulation, a process responsible for most of the vertical movement of ocean water and the circulation of the global ocean as a whole.

Water Masses

As you may recall from Chapter 6, the ocean is density stratified, with the densest water near the seafloor and the least dense near the surface. Each water mass has specific temperature and salinity characteristics. Density stratification is most pronounced at temperate and tropical latitudes because the temperature difference between surface water and deep water is greater there than near the poles.

The water masses possess distinct, identifiable properties. Like air masses, water masses don't often mix easily when they meet due to their differing densities; instead, they usually flow above or below each other. Water masses can be remarkably persistent and will retain their identity for great distances and long periods of time. Oceanographers name water masses according to their relative position.

In temperate and tropical latitudes, there are five common water masses:

- *Surface water,* to a depth of about 200 meters (660 feet)
- *Central water,* to the bottom of the main thermocline (which varies with latitude)
- *Intermediate water,* to about 1,500 meters (5,000 feet)
- *Deep water,* water below intermediate water but not in contact with the bottom, to a depth of about 4,000 meters (13,000 feet)
- *Bottom water,* water in contact with the seafloor

Surface currents move in the relatively warm upper environment of surface and central water. The boundary between central water and intermediate water is the most abrupt and pronounced.

No matter at what depth they are located, the characteristics of each water mass are usually determined by conditions of heating, cooling, evaporation, and dilution that occurred at the ocean surface when the mass was formed. The densest (and deepest) masses were formed by surface conditions that caused water to become very cold and salty. Water masses near the surface can be warmer and less saline; they may have formed in warm areas where precipitation exceeded evaporation. Water masses at intermediate depths are intermediate in density.

In spite of this differentiation, the relatively cold water masses lying beneath the thermocline exhibit smaller variations in salinity and temperature than the water in the currents that move across the ocean's surface.

The Temperature–Salinity Diagram

Perhaps the best way to visualize ocean layering is with a **temperature–salinity (T-S) diagram** like the one in **Figure 9.19.** The S-shaped curve through the center of the figure shows the temperature and salinity of water at each depth indicated. Note that many *combinations* of temperature and salinity can yield the *same* density, and that the density of the water tends to increase with depth. The shape of the S-curve is governed by the position and nature of water masses.

Formation and Downwelling of Deep Water

Antarctic Bottom Water, the most distinctive of all deep-water masses, is characterized by a salinity of 34.65‰, a temperature of −0.5°C (30°F), and a density of 1.0279 grams per cubic centimeter. This water is noted for its extreme density (the densest in the world ocean), for the great amount of it produced near Antarctic coasts, and for its ability to migrate north along the seafloor.

Most Antarctic Bottom Water forms in the Weddell Sea during winter. Sea ice can incorporate only about 15% of seawater's salt, and the salt remaining in the unfrozen water beneath the ice forms a frigid brine. Between 20 and 50 million

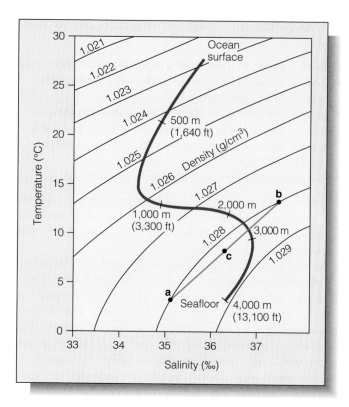

Figure 9.19 A general temperature–salinity (T-S) diagram, with density curves added. The S curve tracks the combination of temperature and salinity (and therefore density) with increasing depth. Note that the top of the S curve represents the temperature and salinity (and therefore density) of water at the ocean surface; the bottom of the curve represents the density of water at the ocean floor. Intermediate depths are shown on the curve.

Points a and b are on the same gently curving *isopycnal*—line of equal density. Note that seawater at points a and b has *different* temperatures and salinities, but the *same* density. If these two masses of water were to merge, their combined density would be greater than either of their separate densities—a situation represented at point c. This process is known as *caballing* and can lead to the formation of deep water (see Figure 9.25).

cubic meters of this brine forms every second! The water's great density causes it to sink toward the continental shelf where it mixes with nearly equal parts of water from the southern Antarctic Circumpolar Current.

The mixture settles along the edge of Antarctica's continental shelf, descends along the slope, and spreads along the deep-sea bed, creeping north in slow sheets. Antarctic Bottom Water flows many times more slowly than the water in surface currents: In the Pacific it may take a thousand years to reach the equator. Six hundred years later it may be as far away as the Aleutian Islands at 50°N! Antarctic Bottom Water also flows into the Atlantic Ocean basin, where it flows north at a faster rate than in the Pacific. Antarctic Bottom Water has been identified as high as 40° *north* latitude on the Atlantic floor, a journey that has taken some 750 years.

Some dense bottom water also forms in the northern polar ocean, but the topography of the Arctic Ocean basin prevents most of it from escaping, except in the deep channels formed in the submarine ridges separating Scotland, Iceland,

and Greenland. These channels allow the cold, dense water formed in the Arctic to flow into the North Atlantic, forming North Atlantic Deep Water (look ahead to Figure 9.25). Very little deep Arctic Ocean water enters the Pacific.

Other distinct deep-water masses exist. Their positions in relation to each other are always determined by their relative densities. North Atlantic Deep Water forms when the relatively warm and salty North Atlantic Ocean cools as cold winds from northern Canada sweep over it. Exposed to the chilled air, water at the latitude of Iceland releases heat, cools from 10°C to 2°C (50°F to 36°F), and sinks. (Transferred to the air, this bonus heat helps to moderate European winters.) As can be seen in **Figure 9.20,** similar water forms at the Antarctic Circumpolar Current in the South Atlantic. Pacific Deep Water also forms in the Antarctic Circumpolar Current and along the east coast of the Kamchatka Peninsula. Atlantic or Pacific Deep Water is less dense than Antarctic Bottom Water and so floats above it, out of contact with the ocean floor.

A deep-water mass also forms in the enclosed Mediterranean Sea, where surface water is made more saline by the excess of evaporation over freshwater input. About 300,000 cubic kilometers (72,000 cubic miles) more water evaporates annually from the Mediterranean than is replaced by river runoff or precipitation. In the cool winter months, Mediterranean water with a salinity of about 38‰ flows past the lip of Gibraltar and spreads into the Atlantic as Mediterranean Deep Water (**Figure 9.21**). Mediterranean Deep Water underlies much of the central water mass in the Atlantic, and some of this water can be traced as far south as the basins of the Antarctic. Though saltier than Antarctic Bottom Water or Atlantic Deep Water, Mediterranean Deep Water is considerably warmer and therefore not as dense. It will lie atop the layers of denser water at high southern latitudes.

Researchers can determine the age of deep water by analyzing its dissolved oxygen content. Ocean water picks up free oxygen only at or near the surface by contact with the atmosphere or through the action of photosynthetic plants. After leaving the surface, the water gradually loses its oxygen through the respiration of organisms or by chemical reactions with sediments, rocks, or dissolved components. The dissolved oxygen content of a water mass is therefore a rough index of its age—the length of time since the water left the surface. Researchers have also found that water masses slowly mix with the surrounding water as they flow away from their sources, losing their individual identity as they grow older.

Thermohaline Circulation Patterns

The great quantities of dense water sinking at ocean basin edges must be offset by equal quantities of water rising elsewhere. **Figure 9.22** shows an idealized model of thermohaline flow. Note that water sinks relatively rapidly in a small area where the ocean is very cold and rises much more gradually across a very large area in the warmer temperate and tropical zones. It then slowly returns poleward near the surface to repeat the cycle. The continual diffuse upwelling of

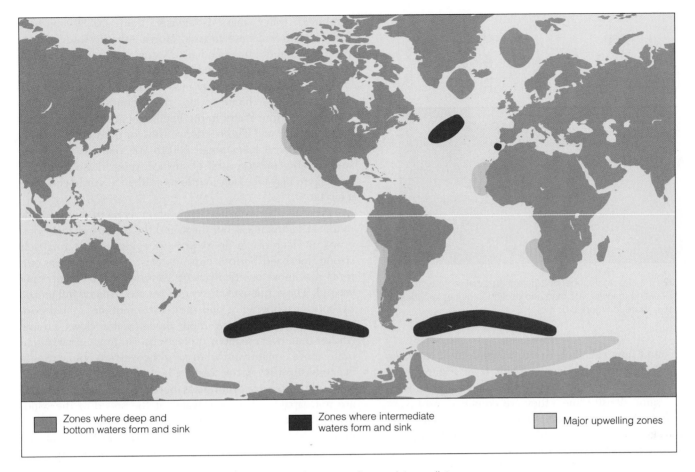

Zones where deep and bottom waters form and sink

Zones where intermediate waters form and sink

Major upwelling zones

Figure 9.20 Places where most surface water sinks and where most deep or intermediate water rises.

deep water maintains the existence of the permanent thermocline found everywhere at low and mid-latitudes. This slow upward movement is estimated to be about 1 centimeter (½ inch) per day over most of the ocean. If this rise were to stop, downward movement of heat would cause the thermocline to descend and would reduce its steepness. In a sense, the thermocline is "held up" by the continual slow upward movement of water.

Most features of this ideal circulation pattern exist in nature. **Figure 9.23** shows deep circulation in the Atlantic. The water masses, each of distinct density and sandwiched in layers, are slowly propelled by gravity. Masses butt against one another in **convergence zones,** and the heavier water can slide beneath the lighter water (**Figure 9.24**). In some cases, however, two distinct water masses with the same *density* but with different *temperatures and salinities* will combine at a convergence zone to produce a new water mass of greater density. This mixing-and-sinking process is called **caballing.**

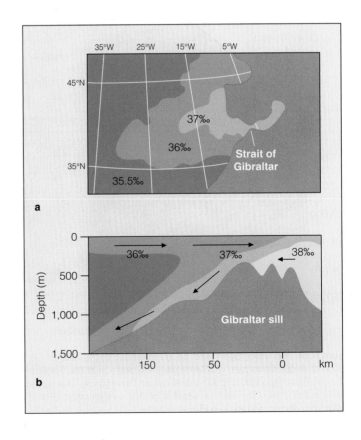

Figure 9.21 Saline flow over the sill of the Mediterranean Sea at Gibraltar. (**a**) A map showing the Strait of Gibraltar; salinity was measured at 1,000 meters. (**b**) A vertical section at the strait, showing the movement of water masses of different salinity.

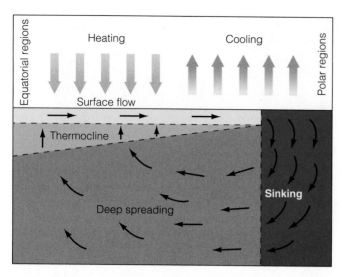

Figure 9.22 The classical model of a pure thermohaline circulation, caused by heating in lower latitudes and cooling in higher latitudes. Compare this figure to Figure 8.6.

To understand how caballing works, imagine the two water masses indicated by point a and point b in Figure 9.19. Though both have different temperatures and salinities, they lie on the same density line—they are equally dense. Now imagine the water at these points mixing together. The density of the new water mass could be represented by a point halfway along a straight line connecting them—point c. The combined water mass is denser than 1.0280—perhaps 1.0283—and will tend to sink. North Atlantic Intermediate Water, Antarctic Intermediate Water, and some Antarctic Bottom Water are produced by caballing.

Hundreds of years may pass before water masses complete a circuit or blend to lose their identities. Remember, Antarctic Bottom Water in the Pacific retains its character for up to 1,600 years! The residence time of most deep water is less, however; it takes about 200 to 300 years to rise to the surface. (By contrast, a bit of surface water in the North Atlantic gyre may take only a little more than a year to complete a circuit.)

Not all thermohaline circulation is so sedate. Ripple marks in sediments, scour lines, and the erosion of rocky outcrops on deep-ocean floors provide evidence that relatively strong, localized bottom currents exist. Some of these currents may move as rapidly as 60 centimeters (24 inches) per second. These relatively fast currents are strongly influenced by bottom topography, and they are sometimes called **contour currents** because their dense water flows around (rather than over) seafloor projections. Bottom currents generally move equatorward at or near the western boundaries of ocean basins (below the western boundary surface currents). The subtle density differences between deep-water masses are not capable of moving water at the speed of the wind-driven surface currents. Water in some bottom currents may move only 1 to 2 meters (3 to 7 feet) per day. Even at that slow speed the Coriolis effect modifies their pattern of flow.

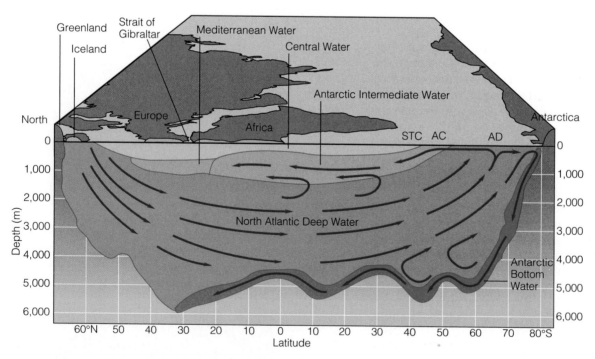

Figure 9.23 The water layers and deep circulation of the Atlantic Ocean. Arrows indicate the direction of water movement. STC indicates the position of the Subtropical Convergence, AC the Antarctic Convergence, and AD the Antarctic Divergence. The surface layer is too thin to show clearly in this scale. The vertical scale is greatly exaggerated. (For a view of the convergence zones from above, please see Figure 9.24.)

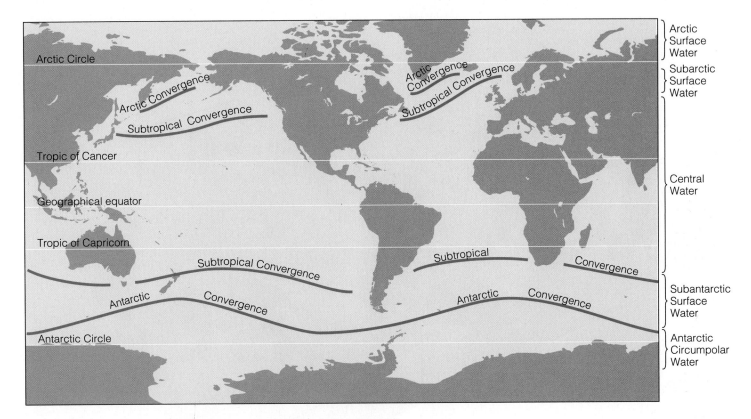

Figure 9.24 Convergence zones, areas where surface water masses meet. The most pronounced convergence zone is the *Antarctic Convergence,* which encircles the continent of Antarctica between about 50°S and 60°S. South of the Antarctic Convergence is cold *Antarctic Circumpolar Water,* in which flows the Antarctic Circumpolar Current, the greatest of all ocean currents. The *Subtropical Convergence,* between about 40°S and 50°S, marks the northern boundary of *Subantarctic Surface Water.* The position of the Subtropical Convergence varies with the time of year, the position of currents, and other factors. The surface water equatorward of the Subtropical Convergence—appropriately known as *Central Water*—is the most abundant, warmest, and saltiest of the three surface water masses.

The organization of surface water masses in the Northern Hemisphere is less orderly because of the different configuration of continents there. As in the south, *Central Water* is most abundant, warmest, and saltiest. The transition from Central Water to *Subarctic Surface Water* occurs at about 45°N in the Pacific and at about 60°N in the Atlantic, at the Northern Hemisphere's Subtropical Convergence. The transition to Arctic Water is a subtle one, and the *Arctic Convergence* is often poorly defined. A circumpolar current like that of the southern polar ocean does not form in the north because continents interrupt its flow.

Thermohaline Flow and Surface Flow: The Global Heat Connection

As we have seen, swift and narrow currents along the western margins of ocean basins carry warm tropical surface waters toward the poles. In a few places, as shown in Figure 9.20, the water loses heat to the atmosphere and sinks to become deep water and bottom water. This sinking is most pronounced in the North Atlantic. The cold, dense water moves at great depths toward the Southern Hemisphere and eventually wells up into the surface layers of the Indian and Pacific Oceans. Almost a thousand years are required for this water to make a complete circuit.

The transport of tropical water to the polar regions is part of a global conveyor belt for heat. A simplified outline of the global circuit, the result of three decades of concentrated effort to understand deep circulation, is shown in **Figure 9.25.** This slow circulation straddles the hemispheres and is superimposed upon the more rapid flow of water in surface gyres. Recent analysis of this global circuit suggests that some of the heat warming the European coast enters the ocean in the vicinity of Indonesia and Australia, travels to the Indian Ocean, and enters the Gulf Stream by way of the Agulhas Current rounding the southern tip of Africa. The surface water that leaves the Pacific is driven in part by excess rainfall and river runoff throughout the Pacific basin. The slow, steady, three-dimensional flow of water in the conveyor belt distributes dissolved gases and solids, mixes nutrients, and transports the juvenile stages of organisms between ocean basins.

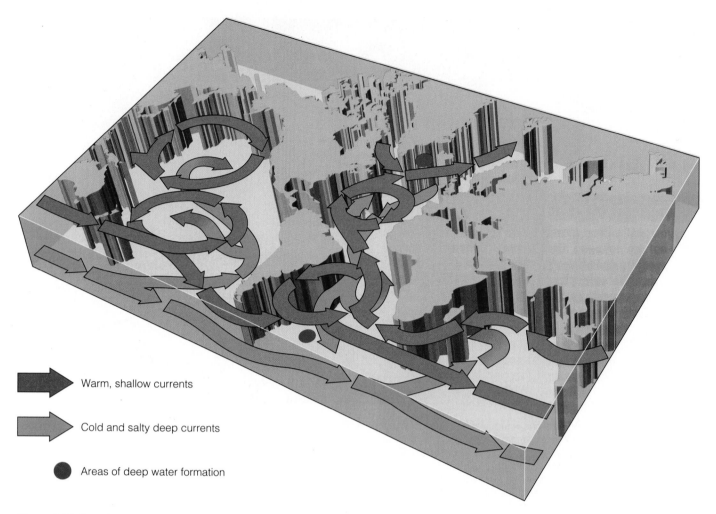

Warm, shallow currents

Cold and salty deep currents

 Areas of deep water formation

Figure 9.25 The global pattern of deep circulation resembles a vast "conveyor belt" that carries surface water to the depths and back again. Begin with the formation of North Atlantic Deep Water north of Iceland. This water mass flows south through the Atlantic, then flows over (and mixes with) deep water formed near Antarctica. The combined mass circumnavigates Antarctica and then moves north into the Indian and Pacific Ocean basins. Diffuse upwelling in all of the ocean returns some of this water to the surface. Water in the conveyor gradually warms and mixes upward to be returned to the North Atlantic by surface circulation. The whole slow-moving system is important in transporting water and heat.

STUDYING CURRENTS

Traditional methods of measuring currents divide into two broad categories: the float method and the flow method. The **float method** depends on the movement of a drift bottle or other free-floating object. In the **flow method,** the current is measured as it flows past a fixed object.

Surface currents can be traced with drift bottles or drift cards. These tools are especially useful in determining coastal circulation, but they provide no information on the path the drift bottle or card may have taken between its release and collection points. Researchers wishing to know the precise track taken by a drifting object can deploy more elaborate drift devices, such as the drogue in **Figure 9.26a** that can be tracked continuously by radio direction finders or radar. Surface currents can also be tracked by noting the difference between the daily expected and observed positions of ships at sea.

Bottle, card, or drogue studies are almost always carefully planned and executed, but not all surface drift releases have been intentional. In May 1990, a violent storm struck the containership *Hansa Carrier* enroute between Korea and Seattle. Twenty-one boxcar-size cargo containers were lost overboard, among them containers holding 30,910 pairs of Nike athletic shoes. About six months later, shoes from the broken containers began washing up on beaches from British Columbia to Oregon. Because the shoes were not tied together in pairs, beachcombers placed advertisements in local newspapers and held swap meets to exchange the shoes (which were in excellent condition despite having been exposed to the ocean). Oceanographers noticed the ads and asked the media to request that individuals let them know where and when they found shoes. By knowing the place where the shoes were lost and the places where they were found, researchers have been able to refine their computer

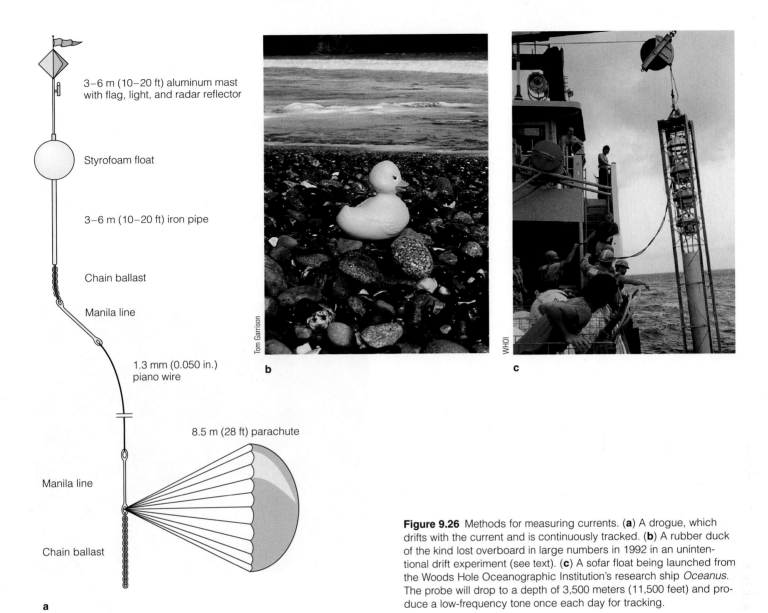

3–6 m (10–20 ft) aluminum mast with flag, light, and radar reflector

Styrofoam float

3–6 m (10–20 ft) iron pipe

Chain ballast

Manila line

1.3 mm (0.050 in.) piano wire

8.5 m (28 ft) parachute

Manila line

Chain ballast

a

b

c

Figure 9.26 Methods for measuring currents. (**a**) A drogue, which drifts with the current and is continuously tracked. (**b**) A rubber duck of the kind lost overboard in large numbers in 1992 in an unintentional drift experiment (see text). (**c**) A sofar float being launched from the Woods Hole Oceanographic Institution's research ship *Oceanus*. The probe will drop to a depth of 3,500 meters (11,500 feet) and produce a low-frequency tone once each day for tracking.

models of the North Pacific gyre. Some of the shoes have completed a full circuit of the North Pacific. A similar spill occurred on 10 January 1992, when a storm-beset freighter lost a container filled with 29,000 rubber ducks, turtles, and other bathtub toys in the North Pacific (**Figure 9.26b**). At last report, 400 of the toys had been recovered from 500 miles of Alaskan shoreline.

Deeper currents can also be surveyed by free-floating devices. The Swallow float (named after its developer) is used to detect the drift of intermediate water masses. Adjusted to descend to a specific density, the Swallow float emits sonar "pings" as it drifts so that a tracking vessel can follow the movement of the water mass in which it is embedded. More sophisticated devices developed in the 1970s and 1980s depend on sofar channels (see Figure 6.21) to transmit sound. Low-frequency tones are broadcast from autonomous submerged probes to moored listening stations (see **Figure 9.26c**). These sofar probes have accumulated more than 240 float-years of data in the North Atlantic, at depths of from 700

to 2,000 meters (2,300 to 6,600 feet). One of these drifting probes has been sending data for nine years!

Bottom currents can also be investigated. The Woodhead seabed drifter (**Figure 9.26d**), a type of drift card for bottom currents, looks like a small weighted parasol. Released in groups, a few of these objects may eventually wash on shore to be returned by interested beachcombers.

Current meters, or flow meters, measure the speed and direction of a current from a fixed position. Most flow meters, such as the Ekman type shown in **Figure 9.26e,** use rotating vanes to measure current speed and a recording compass to measure direction. Bottom-water movements are usually too slow to be measured by flow meters.

Advances in electronics and computer design have made possible several new methods of study. One new device pioneered by the U.S. Office of Naval Research measures current speed by sensing the electromagnetic force generated by seawater as it moves in Earth's magnetic field. Buoys equipped with these sensors can record current

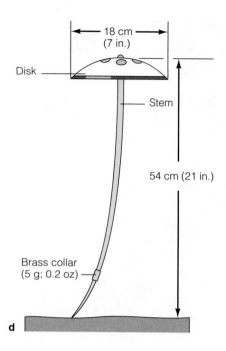

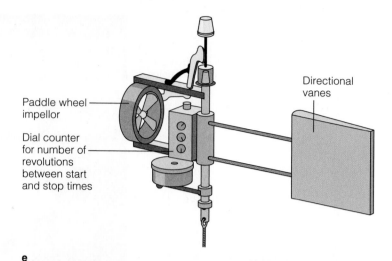

Figure 9.26 Methods for measuring currents. (**d**) A Woodhead seabed drifter, for charting bottom currents. (**e**) An Ekman flow meter, which measures current speed and direction at a fixed position. (**f**) A Slocum glider—a probe that uses energy from gravity, buoyancy, heat, and batteries to power long-range exploration of water masses.

Webb Research

speed and direction without dependence on delicate moving parts. Another device (**Figure 9.26f**) glides smoothly up and down through the water column powered by gravity and buoyancy. Energy to pump ballast overboard, allowing the little glider to rise, is provided by a simple heat engine powered by the difference in temperature between ocean surface and great depths. To fall again, the glider pumps seawater aboard. Named "Slocums" after Joshua Slocum, the first person to circumnavigate the globe alone, these gliders could operate in fleets to map the ocean's thermohaline depth profiles for years, transmitting data to satellites on their occasional visits to the surface.

Yet another method, developed for study of thermohaline circulation, senses the presence in seawater of chemical tracers—artificial substances with known histories of production or release. Because they dissolve easily in seawater, chlorofluorocarbons (CFCs) can be used as such tracers. They are completely man-made and were first produced in the 1930s for use as refrigerants, aerosol propellants, and blowing agents for foam. Once in the ocean, they spread into the in-

terior like a dye, following oceanic circulation. The speed of deep currents has been measured by careful analysis of their CFC content. **Box 9.2** discusses the process.

Satellites can sense sea-surface temperature and topography, ocean color, and even chlorophyll content (an indication of biological productivity and therefore of nutrients), giving fresh insight into ocean circulation.[3] So far, satellites can observe only the ocean surface, but satellite-borne lasers may someday probe the ocean to greater depths.

Acoustical tomography (*tomos* = to cut or slice, *graph* = to write) uses pulses of low-frequency sound to sense variations in water temperature, salinity, and movement beneath the surface. Experiments begun in 1981 have used multiple transmitters and receivers to investigate the formation and movement of Gulf Stream eddies and deep-ocean circulation at a depth of 760 meters (2,500 feet). Preliminary results suggest that a great deal of the ocean's energy of

---

[3] For temperature examples, see Figures 9.8a , 9.11, and 9.18. For an example of chlorophyll content, look ahead to Figure 14.16.

An Environmentally Friendly Use of CFCs

BOX 9.2

by Denise Smythe-Wright, Southampton Oceanography Centre, U.K.

What would you answer if after a hard week taking minutes at an international meeting, with a party for delegates the night before, your senior colleague asked, "How are you physically?" When it happened to me, I certainly didn't realise he was enquiring about the depth of my training in

Aboard the RRS *Discovery*, Denise Smythe-Wright prepares to take seawater samples. The samples will be collected at known depths in ultraclean Niskin bottles and analyzed for CFCs. By measuring the concentration of CFCs in seawater, she is able to estimate the last time a parcel of water was in contact with the atmosphere. This information provides insight into deep ocean circulation.

D. Smythe-Wright

physics. When I explained that although majoring in chemistry I had graduated in physics and physical oceanography, his reply was, "Want another job?" This is how I became a tracer chemist.

Chemical tracers are substances that have entered the environment with known histories of production or release. Chlorofluorocarbons (CFCs) can be used as such tracers. They are completely man-made and were first produced in the 1930s for use as refrigerants, aerosol propellants, and blowing agents for foam. From 1930 to the mid-1970s their atmospheric concentrations grew exponentially until many of them were banned because of the damage they cause to the ozone layer. CFCs are soluble in seawater and the increasing atmospheric concentration has been mirrored in surface seawater. Once in the ocean, they spread into the interior like a dye, following oceanic circulation.

CFC-11 and CFC-12 were the first CFCs to be produced but there are now a number of such compounds. Between the late 1940s and the late 1970s the ratio of CFC-11 to CFC-12 in the atmosphere gradually increased. By measuring the concentration of these two CFCs in seawater we can compare the ratio with the known changes in the atmospheric ratio and so estimate the last time a parcel of water was in contact with the atmosphere. It seems rather ironic that we are able to use substances that have been banned because they cause environmental problems to look at ocean circulation in relation to changes that affect Earth's climate.

It was great to be offered a new job, but my senior colleague didn't tell me that there was very little money for the specialised equipment required to make CFC measurements in seawater. Measuring CFCs is not easy because the source is the atmosphere, so any contact with air will cause the concentration in a seawater sample to increase. Analysis of samples has to be done immediately at sea. Therefore I needed portable equipment that was completely gas tight from the atmosphere, and large-volume water sample bottles to collect the water samples from oceanic depths with minimum risk of contamination. Not even the tiniest pinhead of an air bubble can be allowed to get into the syringes used to collect the CFC samples from the 10-litre Niskin bottles used for deep sea collection. One good thing about CFCs is that they are easily detected by electron capture gas chromatography, so the first thing I did was to purchase one of these instruments. But I also needed a purge and trap system to extract and preconcentrate the CFCs from the seawater sample. Since no such equipment was commercially available, I built it myself with the help of my husband (an electronics engineer), mainly at home on the kitchen table. I don't think we actually used the kitchen table for its real purpose for over three months.

At last it was all complete and I took it off to sea. Like all new equipment it had teething problems but by the time of my third cruise I was in business. Since then the basic equipment has been rebuilt and upgraded as technology has advanced, and I now have a team of chemists who help measure a suite of CFCs and other halogenated compounds rather than just CFC-11 and CFC-12. With these we are able to date seawater from the 1920s to the present day.

Using the CFCs I have been able to show that water from the Brazil Current makes its way in eddies across the South Atlantic to the African coast. This was an important finding since up to then it was thought that all eddies off the coast of Africa originated in the Agulhas Current, which brings warm saline Indian Ocean water into the Atlantic. My work has also shown that the abyssal circulation in the Argentine Basin is more complex than the simple western boundary theory of classical oceanography. It is now clear that Antarctic Bottom Water entering the Argentine Basin from the south splits into two flows. One turns west and hugs the continental slope of South America as it moves north into the Brazil Basin. The other turns northeast and makes a complete counterclockwise circulation around the basin, with the consequence that much of the water never leaves the basin.

Recently, I have turned my attention to the North Atlantic and have been looking at the spread of seawater from the Labrador Sea to the coast of the British Isles. Results from three oceanographic cruises to the North Atlantic over the last six years suggest that the spread of seawater is variable depending on the depth to which the source water convects in the Labrador Sea in the winter.

Measuring CFCs in seawater is not easy, requiring long working hours at sea, but the results are exciting and I don't regret for one minute my quick positive response to the question, "Want another job?" after the party the night before.

9-24

motion is involved in relatively small fluctuating flows analogous to atmospheric weather systems. Unlike weather systems—which generally extend across 1,000 kilometers (620 miles) and last less than a week—this midscale oceanic turbulence typically extends only 100 kilometers (62 miles) but may persist for more than a hundred days.

Researchers are also using acoustical tomography to conduct a long-term study of the formation of North Atlantic Deep Water south of Greenland. The technique may revolutionize our understanding of deep-sea circulation.

Currents are the very heart of physical oceanography. Their global effects, great masses of water, complex flow, and possible influence on human migrations make their study of particular importance.

QUESTIONS FROM STUDENTS

1. **If the Gulf Stream warms Britain during the winter and keeps Baltic ports free of ice, why doesn't it moderate New England winters? After all, Boston is closer to the warm core of the Gulf Stream than London is.**

Yes, but remember the direction of prevailing winds in winter. Winter winds at Boston's latitude are generally from the west, so any warmth is simply blown out to sea. On the other side of the Atlantic the same winds blow toward London. It does get cold in London, but winters in London are generally much milder than those in Boston.

2. **Could currents be used as a source of electrical power? With the Gulf Stream so close to Florida, it seems that some way could be devised to take advantage of all that water flow to turn a turbine.**

It's been considered. The total energy of the Gulf Stream flowing off Miami has been estimated at 25,000 megawatts! A Woods Hole Oceanographic Institution team has proposed a honeycomb-like array of turbines for the layer between 30 and 130 meters (100 and 430 feet) across 20 kilometers (13 miles) of the current. They estimate a power output of around 1,000 megawatts, equal to the generation potential of two large nuclear power plants. Engineering difficulties would be prohibitive, however.

3. **Are there any *non*geostrophic currents? Are any currents not noticeably influenced by gravity, the Coriolis effect, uneven solar heating, planetary winds, and so forth?**

Yes. Besides the high-latitude currents mentioned in the text, there are small-scale currents that are not noticeably affected. Currents of fresh water from river mouths, rip currents in surf, and tidal currents in small harbors are much more affected by basin and bottom topography than by the Coriolis effect and gravity.

4. **Why are western boundary currents strong in *both* hemispheres? I thought things went the other way**

(counterclockwise) in the Southern Hemisphere. Shouldn't the *eastern* boundary currents be stronger down there?

Western boundary currents are strong in part because the "Coriolis hill" is offset to the west, forcing water to move in a relatively narrow path along the ocean's western boundary. The bulge is offset to the west because Earth turns toward the east—and remember, Earth turns eastward in *both* hemispheres! Western boundary currents are therefore strong in both hemispheres.

5. Which takes the lead in producing the hill in the middle of a geostrophic gyre—the pressure gradient or the Coriolis effect?

The question reminds me of the "chicken-and-egg" question—which came first? In ocean currents, the pressure gradient and the Coriolis effect act together, in balance, to form both the hill and the circular flow around its crest. Imagine the situation some 150 million years ago when the Atlantic was first forming—Pangaea was splitting and the rift began to fill with water. Driven by winds, a small amount of water would have turned right to begin forming the hill. A pressure gradient was immediately formed, and water would have been forced back downhill by gravity. On its way back down, that water would achieve a balance with the Coriolis effect and come to a "compromise" position of clockwise flow around the apex.

6. A north wind comes from the north, but a north current is going north. Why the difference?

Traditions die hard, it seems. For thousands of years winds have been named by where they come *from*. A north wind comes from the north, and a west wind comes from the west. Currents, though, are named by where they are *going*. A southern current is headed south; a western current is moving west. An exception is the Antarctic Circumpolar Current, or West Wind Drift, which moves eastward. This current, however, is named after the wind that drives it, the powerful polar westerlies.

It may also be a matter of perspective. Ancient peoples took shelter *from* winds (and referred to the wind's place of origin), but early oceanic travelers were aware of where the currents were carrying them *to*.

KEY CONCEPTS TO REVIEW

- Ocean water circulates in currents caused mainly by wind friction at the surface and by differences in water mass density beneath the surface zone.

- Surface currents affect the uppermost 10% of the world ocean, mostly the area above the pycnocline. Some surface currents are rapid and riverlike, with well-defined boundaries; others are slow and diffuse. The largest surface currents are organized into huge circuits known as gyres.

- Upwelling and downwelling describe the vertical movement of water masses. Upwelling is often due to the divergence of surface currents; downwelling is often caused by surface current convergence or an increase in the density of surface water.

- El Niño, an anomaly in surface circulation, occurs when the trade winds falter, allowing warm water to build eastward across the Pacific at the equator.

- Circulation of the 90% of ocean water beneath the surface zone is driven by the force of gravity, as denser water sinks and less dense water rises. Since density is largely a function of temperature and salinity, the movement of deep water due to density differences is called thermohaline circulation.

- The Coriolis effect modifies the courses of currents, with currents turning clockwise in the Northern Hemisphere and counterclockwise in the Southern Hemisphere. The Coriolis effect is largely responsible for the phenomenon of westward intensification in both hemispheres.

INTERNET STUDY RESOURCES

The Web site for this book contains helpful study aids. Log on to:

www.brookscole.com/product/053437557xs

and click on the Chapter-by-Chapter area. Choose Chapter 9 and select a resource:

- **Flash Cards** allows you to test your mastery of the Terms and Concepts to Remember for this chapter.
- **Tutorial Quizzes** provides a multiple-choice practice quiz.
- **Student Guide to InfoTrac College Edition** will lead you to Critical Thinking Projects that use InfoTrac College Edition as a research tool.

TERMS AND CONCEPTS TO REMEMBER

acoustical tomography	eddy
Antarctic Bottom Water	Ekman spiral
Antarctic Circumpolar Current (West Wind Drift)	Ekman transport
	El Niño
	ENSO
caballing	equatorial upwelling
coastal upwelling	float method
contour current	flow method
convergence zone	geostrophic
countercurrent	Gulf Stream
current	gyre
downwelling	La Niña
eastern boundary current	Langmuir circulation

Southern Oscillation
surface current
sverdrup (sv)
temperature–salinity (T-S)
  diagram
thermohaline circulation
transverse current

undercurrent
upwelling
western boundary current
westward intensification
wind-induced vertical
  circulation

STUDY QUESTIONS

Review Questions

1. What forces are responsible for the *movement* of ocean water in currents?

2. What forces and factors influence the *direction* and *nature* of ocean currents?

3. What is a gyre? How many large gyres exist in the world ocean? Where are they located?

4. What are water masses? Where are distinct water masses formed? What determines their relative position in the ocean?

5. What drives the vertical movement of ocean water? What is the general pattern of thermohaline circulation?

6. What methods are used to study ocean currents?

Critical Thinking Questions

1. Why does water tend to flow around the periphery of an ocean basin? Why are western boundary currents the fastest ocean currents? How do they differ from eastern boundary currents?

2. What are countercurrents? Undercurrents? How might El Niño be related to these currents?

3. What is the role of ocean currents in the transport of heat? How can ocean currents affect climate? Contrast the climate of a mid-latitude coastal city at a western ocean boundary with that of a mid-latitude coastal city at an eastern ocean boundary.

4. Can you think of ways ocean currents have (or *might* have) influenced history?

5. What holds up the thermocline? Wouldn't water slowly warm in an even gradient all the way to the bottom? (Hint: See Figure 9.22.)

6. **InfoTrac College Edition Project** The El Niño Southern Oscillation has been going through its cycles for hundreds of thousands of years. Is El Niño getting worse, or was 1997–98 just a bad year? Do some research using InfoTrac College Edition to gather information about El Niño, and use your findings to back up your view about El Niño's severity and predictability.

 **Earth Systems Today CD Questions**

1. Go to "Waves, Tides, and Currents," then to "World Currents." Toggle on the prevailing winds. Now predict the direction of the resulting surface currents. (Don't forget to include Coriolis effect.) Now toggle the currents on, animate them, and check your predictions. Did you predict which currents were hot and which were cold?

2. Go to "Waves, Tides, and Currents," then to "Dissect a Current." Before beginning the animation, review the information on Ekman spiral in the text. Now predict the resultant direction of flow of all the water in the surface current based only on the red wind arrow at the surface. Run the animation and see if you were successful.

3. Go to "Weather and Climate," then to "El Niño." Watch the sea surface temperatures from 1982 to 1990 in the animation. Note the back-and-forth oscillation of warm and cold water in the Pacific. Noting the pattern, could you predict ocean temperatures in the central eastern Pacific through 1992? How do you think warm water affects upwelling along the western coast of South America?

RESOURCES FOR FURTHER READING AND RESEARCH

The Web site for this book contains many ideas for further reading and research. Log on to:

www.brookscole.com/product/053437557xs

and click on the Chapter-by-Chapter area. Choose Chapter 9 and select a resource:

- **References** lists the major books and articles consulted in writing this chapter, along with comments from the author about their content and reading level.
- **Hypercontents** takes you to an extensive list of sites with news, research, and images related to individual sections of the chapter.
- In **Student Guide to InfoTrac College Edition** scroll down to Suggested Readings from InfoTrac College Edition for brief descriptions of the articles listed and search hints for finding them in InfoTrac College Edition.
- **Regional InfoTrac College Edition** articles are organized into East Coast, West Coast, and Gulf Coast regions, allowing you to study oceanography on a more local level. You can also access Regional InfoTrac College Edition at www.localocean.com.

 For additional readings, go to InfoTrac College Edition, your online research library, at:

http://infotrac.thomsonlearning.com/

Wave Dynamics and Wind Waves

SURFING

More than 2 million Americans surf regularly. Rushing down the face of a growing, breaking wave is exhilarating! People willingly endure cold, boredom, and some danger for a ride lasting only a few seconds. If you're a good swimmer, give it a try. You will need to paddle your board or swim vigorously to match your speed to that of the advancing wave crest. As the wave rises to break, your forward speed (and sense of timing) will place you on the leading edge of the crest, accelerating downward and forward. The technique takes time to master, but the feeling is worth the effort!

Although riding a surfboard can be great fun, I have long believed that the most rewarding surfing is body surfing. The buoyant forces board surfing depends on are diminished in body surfing, but forces within the wave propel the

Body surfing. Many surfers prefer body surfing to board surfing because of the body surfer's proximity to the water and the sensation of great speed and acceleration.

© Ron Romanosky

body surfer with greater acceleration. A body surfer's intimate contact with the water and close proximity to the wave surface greatly increase the sensation of movement.

The ultimate trick is to surf the wave completely submerged, as a dolphin does. You push powerfully off the bottom (or swim rapidly toward the surface) as the wave crest approaches, inserting yourself into the freshly breaking wave from below. The rapid flow of water within the toppling wave will ripple your skin, and the sudden acceleration can thrust you from the face of the wave like a wet bar of soap shooting from a fist. Altogether a *wonderful* oceanic experience!

WHAT TO WATCH FOR IN CHAPTER 10

Waves are not what they seem. Waves transmit energy, not water mass, across the ocean's surface. Beware of the wave illusion—*water* isn't moving at the speed of the wave, but *energy* is. What generates ocean waves? From personal experience we think first of the wind, but can you think of other forces that could be responsible?

Note as you read that the speed of ocean waves usually depends on their wavelength, with long waves moving fastest. Arranged from short to long wavelengths (and therefore from slowest to fastest), capillary waves are generated by very small disturbances, wind waves by wind, seiches by the rocking of water in enclosed spaces, tsunami by seismic and volcanic activity or other sudden displacements, and tides by gravitational attraction. You will find that the behavior of a wave depends largely on the relationship between a wave's size and the depth of water through which it is moving. Waves can refract, reflect, break, and interfere with one another.

In this chapter you will read about the basic properties of all ocean waves and the occasionally enormous "small" waves generated by wind. Chapter 11 will extend this information to great waves of longer wavelengths.

To most people an ocean wave in deep water appears to be a massive moving object—a ridge of water traveling across the sea surface. An ocean wave is one of several kinds of **waves,** all of which are disturbances caused by the movement of energy from a source through some medium (solid, liquid, or gas). As the energy of the disturbance travels, the medium through which it passes moves in specific ways. Sometimes this movement is visible to us as crests or ridges in the medium. The traveling crests produce the appearance of movement we see in a wave. In an ocean wave, a ribbon of *energy* is moving at the speed of the wave, but *water* is not. In a sense, the wave is an illusion.

OCEAN WAVES

Picture a resting sea gull as it bobs on the wavy ocean surface far from shore. The gull moves in *circles*—up and forward as the tops of the waves move to his position, down and backward as the tops move past. Each circle is equal in diameter to the wave's height. As can be seen in **Figure 10.1,** energy in waves flows past the resting bird, but the gull and its patch of water move only a very short distance forward in each up-and-forward, down-and-backward wave cycle. The water on which the bird rests does not move continuously across the sea surface as the wave illusion suggests.[1]

The transfer of energy from water particle to water particle in these circular paths (or **orbits**) transmits the wave energy across the ocean surface and causes the waveform to move. This kind of wave is known as an **orbital wave**—a wave in which particles of the medium (water) move in closed circles as the wave passes. Orbital ocean waves occur at the boundary between two fluid media (air and water) and between layers of water of different densities. Because the waveform moves forward, these waves are a type of **progressive wave.**

The progressive wave that moved the gull was probably caused by wind. Other forces can generate much greater progressive waves in which water molecules move through much larger circular or elliptical orbits. Some of these waves are so large that they do not appear to us as waves at all, but rather as the slow sloshing of water in a harbor or bay, as dangerous flooding surges of water, or as rhythmic and predictable ocean tides.

Ocean waves have distinct parts. The **wave crest** is the highest part of the wave above average water level; the **wave trough** is the valley between wave crests below average water level. **Wave height** is the vertical distance between a wave crest and the adjacent trough, while **wavelength** is the horizontal distance between two successive crests (or troughs). The relationship between these parts is shown in **Figure**

---

[1] To clarify the important idea of wave-as-illusion, imagine yourself at a sports stadium where spectators are doing "the wave." Your role in wave propagation is simple—you stand up and sit down in precise synchronization with your neighbors. Though you move only a few feet vertically, the wave you were a part of circles the arena at high speed. You and all the other participants stay in place, but the wave moves faster than anyone can run.

Figure 10.1 A floating sea gull demonstrates that waveforms travel but the *water* itself does not. In this sequence, a wave moves from left to right as the gull (and the water in which it is resting) revolves in an imaginary circle, moving slightly to the left up the front of an approaching wave, then rising to the crest, then sliding to the right down the back of the wave.

10.2. The time it takes for a wave to move a distance of one wavelength is known as the **wave period. Wave frequency** is the number of waves passing a fixed point per second.

The circular motion of water particles at the surface of a wave continues underwater. As **Figure 10.3** shows, the diameter of the orbits through which water particles move diminishes rapidly with depth. For all practical purposes wave motion is negligible below a depth of half the wavelength, where the circles are only $\frac{1}{23}$ the diameter of those at the surface. This means

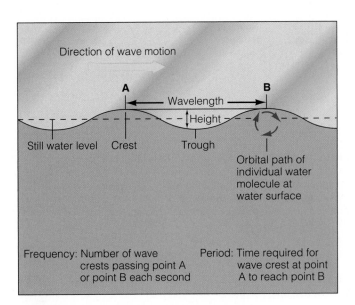

Figure 10.2 The anatomy of a progressive wave.

divers in 20 meters of water would not notice the passage of a wind wave with a 30-meter wavelength, and might barely notice the wave if they were at a depth of 15 meters. Since most ocean waves have moderate wavelengths, the circular disturbance of the ocean that propagates these waves affects only the uppermost layer of water. *Note that the movement of water in circles doesn't resemble interlocking mechanical gears. Instead, there is a coordinated, uniform circular movement of water molecules in one direction as the waves pass.*

Figure 10.3 indicates that water molecules in the crest of a passing wave move in the same direction as the wave, but molecules in the trough move in the opposite direction. If particle speed decreases with depth, molecules in the top half of the orbit will move farther forward in the direction the wave is moving than molecules in the bottom half of the orbit will move backward (**Figure 10.4**). This flow, known as **Stokes drift,** is important in driving the ocean surface currents you studied in Chapter 9.

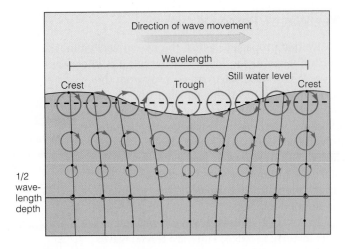

Figure 10.3 The orbital motion of water particles in a wave, which extends to a depth of about half the wavelength.

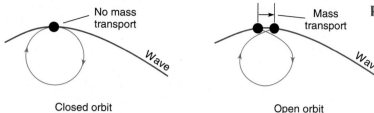

Figure 10.4 The orbits in Figures 10.1 and 10.3 are not completely closed. Instead, each particle moves a very small distance forward (exaggerated in this figure) during each orbit. There is a small net mass transport of water in the direction of the wave. This mass transport contributes to the movement of the wind-driven surface currents discussed in Chapter 9.

CLASSIFYING WAVES

Ocean waves are classified by the *disturbing force* that creates them, the extent to which the disturbing force *continues* to influence the waves once they are formed, the *restoring force* that tries to flatten them, and their *wavelength*. (Wave height is not often used for classification because it varies greatly depending on water depth, interference between waves, and other factors.)

Disturbing Force

Energy that causes ocean waves to form is called a **disturbing force.** Wind blowing across the ocean surface provides the disturbing force for *wind waves.* Arrival of a storm surge or seismic sea wave in an enclosed harbor or bay, or a sudden change in atmospheric pressure, is the disturbing force for the resonant rocking of water known as a *seiche.* Landslides, volcanic eruptions, and faulting of the seafloor associated with earthquakes are the disturbing forces for *tsunami.* The disturbing forces for *tides* are changes in the magnitude and direction of gravitational forces among Earth, the moon, and the sun, combined with Earth's rotation.

Free Waves, Forced Waves

A wave that propagates across the sea surface without the further influence of the force that formed it is known as a **free wave.** When wind waves move away from the storm that created them or when the storm ceases, they continue without the injection of additional wind energy. Likewise, the tsunami we will encounter in the next chapter—waves caused by submerged landslides or earthquakes—continue to move across the ocean surface long after the landslide or earthquake has stopped moving.

In contrast, a **forced wave** is maintained by its disturbing force. Tides are forced waves dependent on the gravitational attraction of the moon and the sun.

Restoring Force

Restoring force is the dominant force trying to return the water surface to flatness after a wave has formed in it. If the restoring force of a wave were quickly and fully successful, a disturbed sea surface would immediately become smooth, and the energy of the embryo wave would be dissipated as heat. But that isn't what happens. Waves continue after they form because the restoring force overcompensates and causes oscillation. The situation is analogous to a weight bobbing at the bottom of a very flexible spring, constantly moving up and down past its normal resting point.

The dominant restoring force for very small water waves—those with wavelengths less than 1.73 centimeters (0.68 inch)—is cohesion, the property that allows individual water molecules to stick to each other by means of hydrogen bonds (see Figure 6.2). These **capillary waves** are transmitted across a puddle because cohesion—the force that makes tea creep up on the sides of a teacup—tugs the tiny wave troughs and crests toward flatness.

Capillary waves are the first waves to form when the wind blows. These small ripples are important in transferring energy from air to water to drive ocean currents (as we saw in Chapter 9) but are of little consequence in the overall picture of ocean waves because they are tiny and carry very little energy.

All waves with wavelengths greater than 1.73 centimeters depend primarily on gravity to provide the restoring force. Like the spring weight moving up and down, gravity pulls the crests downward, but the momentum of the water causes the crests to overshoot and become troughs. The repetitive nature of this movement gives rise to the circular orbits of individual water molecules in an ocean wave. These larger waves are called **gravity waves.** Since the circular motion of water molecules in a wave is nearly friction-free, gravity waves can travel across thousands of miles of ocean surface without disappearing, eventually to break on a distant shore.

Wavelength

Wavelength is the horizontal measure of wave size. **Table 10.1** lists the causes and typical wavelengths of wind waves, seiches, seismic sea waves, and tides. **Figure 10.5** shows the relationship between disturbing and restoring forces, the period, and the relative amount of energy present in the ocean's

Table 10.1	Wavelengths and Disturbing Forces of Important Ocean Waves	
Wave Type	**Typical Wavelength**	**Disturbing Force**
Wind wave	60–150 m (200–500 ft)	Wind over ocean
Seiche	Large, variable; a function of basin size	Change in atmospheric pressure, storm surge, tsunami
Seismic sea wave (tsunami)	200 km (125 mi)	Faulting of seafloor, volcanic eruption, landslide
Tide	½ circumference of Earth	Gravitational attraction, rotation of Earth

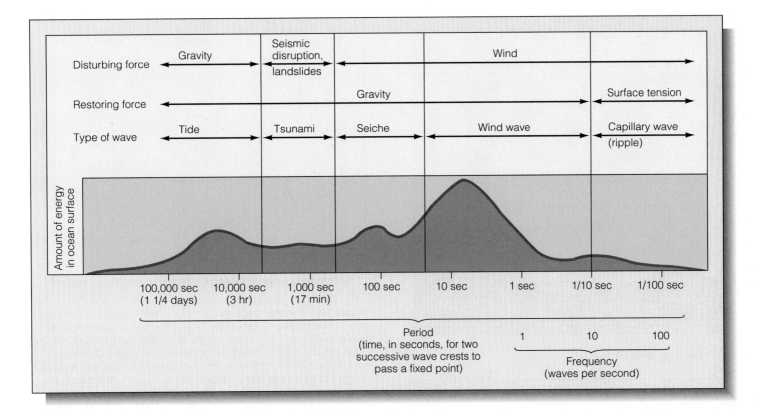

Disturbing force	← Gravity →	Seismic disruption, landslides	←		Wind			→
Restoring force	←		Gravity				Surface tension	
Type of wave	← Tide →	← Tsunami →	← Seiche →	←	Wind wave	→	← Capillary wave (ripple) →	

Amount of energy in ocean surface

| 100,000 sec (1 1/4 days) | 10,000 sec (3 hr) | 1,000 sec (17 min) | 100 sec | 10 sec | 1 sec | 1/10 sec | 1/100 sec |

Period
(time, in seconds, for two successive wave crests to pass a fixed point)

1   10   100

Frequency
(waves per second)

Figure 10.5 Wave energy in the ocean as a function of the wave period. As the graph shows, most wave energy is typically concentrated in wind waves. However, large tsunami, rare events in the ocean, can transmit more energy than all wind waves for a brief time.

surface for each wave type. Note that more energy is stored in wind waves than in any other wave type.

DEEP-WATER WAVES, SHALLOW-WATER WAVES

Most characteristics of ocean waves depend on the relationship between their wavelength and water depth. Wavelength determines the *size* of the orbits of water molecules within a wave, but water depth determines the *shape* of the orbits. The paths of water molecules in a wind wave are circular only when the wave is traveling in deep water. A wave cannot "feel" the bottom when it moves through water deeper than half its wavelength because too little wave energy is contained in the small circles below that depth. Waves moving through water deeper than half their wavelength are known as **deep-water waves.** A wave has no way of "knowing" how deep the water is, only that it is in water deeper than about half its wavelength. For example, a wind wave with a 20-meter wavelength will act as a deep-water wave if it is passing through water more than 10 meters deep (**Figure 10.6a**).

The situation is different for wind-generated waves close to shore. The orbits of water molecules in waves moving through shallow water are flattened by the proximity of the bottom. Water just above the seafloor cannot move in a circular path, only forward and backward. Waves in water shallower than ¹⁄₂₀ their

original wavelength are known as **shallow-water waves (Figure 10.6b)**. A wave with a 20-meter wavelength will act as a shallow-water wave if the water is less than 1 meter deep.

Transitional waves travel through water deeper than 1/20 their original wavelength but shallower than half their original wavelength. In our example, this would be water between 1 and 10 meters (3 and 30 feet) deep. The orbital motion of water molecules in a transitional wave is shown in **Figure 10.6c.**

Of the four wave types listed in Table 10.1, only wind waves can ever be deep-water waves. To understand why, remember that most of the ocean floor is deeper than 125 meters (400 feet), half the wavelength of very large wind waves. The wavelengths of the larger waves are much longer; the wavelength of seismic sea waves usually exceeds 100 kilometers (62 miles). No part of the ocean is 50 kilometers (31 miles) deep, so seiches, seismic sea waves, and tides are always in water that—to them—is shallow or transitional in depth. Their huge orbital circles flatten against a distant bottom always less than half a wavelength away.

In general, the longer the wavelength, the faster the wave energy will move through the water. For *deep-water* waves this relationship is shown in the formula

$$C = \frac{L}{T}$$

in which C represents speed (celerity), L is wavelength, and T is time, or period (in seconds).

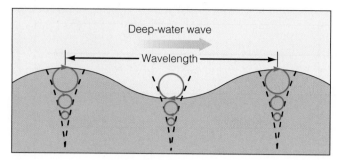

a Depth $\geq \frac{1}{2}$ wavelength

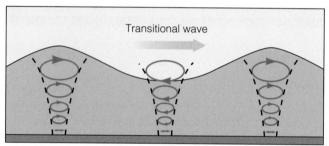

b Depth $\leq \frac{1}{20}$ wavelength

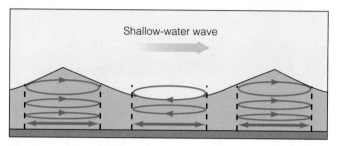

c $\frac{1}{20}$ wavelength $\leq$ depth $\leq \frac{1}{2}$ wavelength

Figure 10.6 Progressive waves: (**a**) a deep-water wave; (**b**) a shallow-water wave; (**c**) a transitional wave. These diagrams are not to scale.

The speed of all ocean waves is controlled by gravity, wavelength, and water depth. The speed of a deep-water wave may also be approximated by

$$C = \frac{\sqrt{gL}}{2\pi}$$

where g is the acceleration due to gravity. Because g (9.8 meters per second per second) and π (3.142) are constants,

$$C = 1.25\sqrt{L}$$

when C is measured in meters per second and L in meters. Note in both instances that wave speed is proportional to wavelength.

Now look at **Figure 10.7**, a diagram showing the relationship between wavelengths of deep-water waves and their

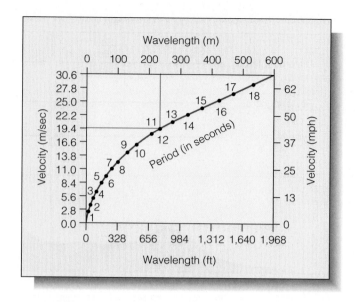

Figure 10.7 The theoretical relationship between speed, wavelength, and period in deep-water waves. Speed is equal to wavelength divided by period. If one characteristic of a wave can be measured, the other two can be calculated. The easiest to measure exactly is period. In the example shown in red, the speed of a wave with a wavelength of 233 meters and a period of 12 seconds is 19.4 meters per second.

speed and period. For the example indicated by the red lines, the speed of a wave with a wavelength of 233 meters and a period of 12 seconds is

$$C = \frac{L}{T} = \frac{233 \text{ meters}}{12 \text{ sec}} = \frac{19.4 \text{ meters}}{\text{sec}}$$

Wavelength is difficult to determine at sea, but period is comparatively easy to find—for example, by an observer timing the movement of waves past the bow of a stopped ship. If period (T) is known, speed (S) can be calculated from the relation

$$C \text{ (in meters/sec)} = \frac{gT}{2\pi}$$
$$= \frac{9.8\text{m/sec}^2 \times T \text{ (in seconds)}}{2 \times (3.14)} = 1.56T$$

where g is acceleration due to gravity.

The speed of *shallow-water* waves is described by a different equation that may be written as

$$C = \sqrt{gd} \quad \text{or} \quad C = 3.1\sqrt{d}$$

where C is speed (in meters per second), g is the acceleration due to gravity (an average of 9.8 meters per second per second), and d is the depth of water in meters. The *period* of a wave remains unchanged regardless of the depth of water through which it is moving, but as deep-water waves enter the shallows and feel bottom, their speed is reduced and their crests "bunch up," so their wavelength shortens.

The following lists describe two very different kinds of ocean waves (a comparison of apples and oranges), but notice the general relationship between wavelength and wave speed: *The longer the wavelength, the greater the speed.*

Wind Waves (Deep-Water Waves):

- Period to about 20 *seconds*
- Wavelength to perhaps 600 meters (2,000 feet) in extreme cases
- Speed to perhaps 112 kilometers (70 miles) per hour in extreme cases

Seismic Sea Waves (Shallow-Water Waves):

- Period to perhaps 20 *minutes*
- Wavelength typically 200 kilometers (125 miles)
- Speeds of 760 kilometers (470 miles) per hour

Remember that it is *energy*—not the water mass itself—that is moving through the water at the astonishing speed of 760 kilometers (470 miles) per hour in seismic sea waves (the speed of a jet airliner)!

We will discuss seiches, seismic sea waves, and tides in Chapter 11. The rest of this chapter deals primarily with wind waves.

WIND WAVES

Wind waves are gravity waves formed by the transfer of wind energy into water. Most wind waves are less than 3 meters (10 feet) high. Wavelengths of 60 to 150 meters (200 to 500 feet) are most common in the open ocean.

Wind waves grow from capillary waves, tiny waves nearly always present on the ocean. Capillary waves form as wind friction stretches the water surface and surface tension tries to restore it to smoothness. Capillary waves interrupt the smooth sea surface and deflect surface wind upward and slow it, which causes some of the wind's energy to be transferred into the water to drive the capillary wave crest forward. The wind may eddy briefly behind the tiny crest, creating a slight partial vacuum there. Atmospheric pressure pushes the trailing crest forward (downwind) toward the trough, adding still more energy to the water surface. The process is shown in **Figure 10.8.** The increasing energy in the water surface expands the circular orbits of water particles in the direction of the wind, enlarging the small wave's size. The capillary wave becomes a wind wave when its wavelength exceeds 1.73 centimeters (0.68 inch), which is the wavelength at which gravity supersedes capillary action as the dominant restoring force.

If the wind wave remains in water deeper than half its wavelength and the wind continues to blow, the wave becomes larger. Its crest is thrust higher into faster wind, extracting even more energy from the moving air. The circular orbits of water particles within the wave grow larger with more energy input; height, wavelength, and period increase proportionally. The irregular peaked waves in the area of wind wave formation are called **sea;** the chaotic surface is formed by simultaneous wind waves of many wavelengths,

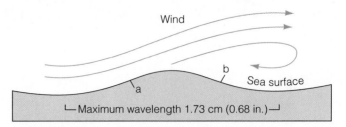

Figure 10.8 Wind forces acting on a capillary wave. Capillary waves interrupt the smooth sea surface, deflect surface wind upward, and slow it, which causes some of the wind's energy to be transferred into the water to drive the capillary wave crest forward (point a). The wind may eddy briefly downwind of the tiny crest, creating a slight partial vacuum there. Atmospheric pressure pushes the trailing crest forward (downwind) toward the trough (point b), adding still more energy to the water surface. The increasing energy in the water surface expands the circular orbits of water particles in the direction of the wind, enlarging the small wave's size. The capillary wave becomes a wind wave when its wavelength exceeds 1.73 centimeters (0.68 inch), the wavelength at which gravity supersedes capillary action as the dominant restoring force.

periods, and heights. When the wind slows or ceases, as it does away from a storm, the wave crests become rounded and regular.

Swell Formation and Dispersion

Because waves with the longest wavelengths move fastest, they leave the area of wave formation sooner.[2] They outrun their smaller relatives. Mature waves from a storm sort themselves into groups that have similar wavelengths and speeds. The process of wave separation, or **dispersion,** produces the familiar smooth undulation of the ocean surface called **swell** (**Figure 10.9**). Swell often move thousands of kilometers from a storm to a shore, announcing the storm's impending arrival. Contrary to what you might expect, observers far from the storm would first encounter large, quick-moving waves with long wavelength, then middle-size waves, and then slow, small ones. Because the circular movement of water particles in deep-water waves is virtually friction-free, the waves will continue until they break upon a shore. There they release their absorbed wind energy as random movement, heat, and sound.

Progressing groups of swell with the same origin and wavelength are called **wave trains.** The leading waves in the wave train are drained of energy because they must begin the circular movement of the undisturbed water into which they are intruding. These leading waves gradually disappear, but after the wave train has passed some energy remains behind in the circles to form new waves. New waves thus form behind as the leading waves disappear at the front of the wave train. This process is shown in **Figure 10.10.**

---

[2] Remember, speed is directly proportional to wavelength in deep-water waves: $C = 1.25\sqrt{L}$.

Tom Garrison

Figure 10.9 Swell—mature, regular wind waves sorted by dispersion—off the Oregon coast. The small waves superimposed on the large swell are the result of local wind conditions.

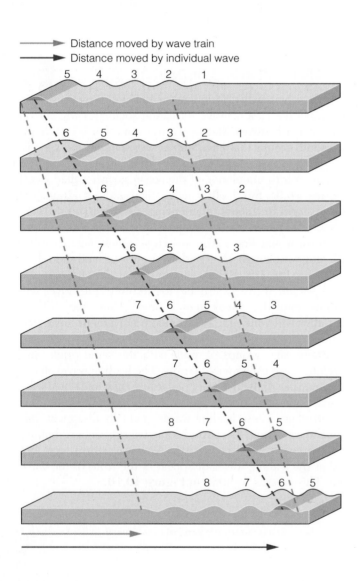

→ Distance moved by wave train
→ Distance moved by individual wave

The implications of this detail are surprising. Though each wave moves forward with a speed proportional to its wavelength in deep water, the wave train itself moves forward at only *half* that speed. Groups of waves, therefore, move ahead at half the speed of individual waves within the group. The half-speed advance of the wave train is called **group velocity** and is the speed with which wave energy advances. Note that individual waves do not persist in the ocean. Individual waves last only as long as they take to pass through the group. As deep-water waves move into shallow water, the speed of individual waves within the group slows until wave speed equals group velocity.

Factors Affecting Wind Wave Development

Three factors affect the growth of wind waves. Wind must be moving faster than the wave crests for energy transfer to continue from air to sea, so the mean wind speed, or **wind strength,** is clearly important to wind wave development. A second factor is the length of time the wind blows, or **wind duration;** high winds that blow for only a short time will not generate large waves. The third factor is the uninterrupted distance over which the wind blows without significant change in direction—the **fetch** (**Figure 10.11**).

Figure 10.10 Waves travel in groups called wave trains. As the leading wave of the group travels forward, it transfers half of its energy forward to initiate motion in the undisturbed surface ahead. The other half is transferred to the wave behind to maintain wave motion. The leading wave in the wave train continuously disappears, while a new wave is continuously formed at the back of the train. Follow wave number 5 in this diagram. The wave train travels at *half* the speed of any individual wave, a speed known as group velocity.

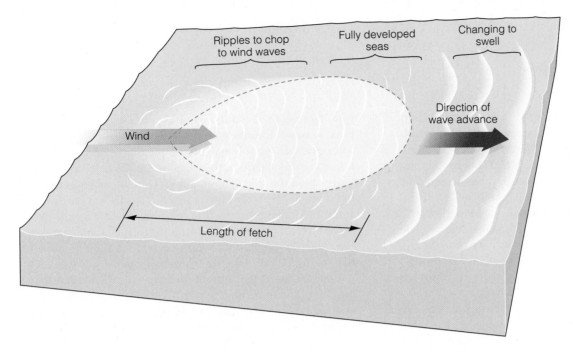

Figure 10.11 The fetch, the uninterrupted distance over which the wind blows without significant change in direction. Wave size increases with increased wind speed, duration, and fetch. A strong wind must blow continuously for three days for the largest waves to fully develop.

A strong wind must blow continuously for three days for the largest waves to fully develop. A **fully developed sea** is the maximum wave size theoretically possible for a wind of a specific strength, duration, and fetch. Longer exposure to wind at that speed will not increase the size of the waves because energy is lost due to the breaking of wave tops and the formation of whitecaps. The first three columns of **Table 10.2** give examples of the combinations of wind speed, fetch, and duration required to form waves of the fully developed sea described in the remaining columns of the table. Note that waves in a fully developed sea are small if wind speed is low. However, if the wind speed is 74 kilometers (46 miles) per hour through 1,313 kilometers (816 miles) for 42 hours—conditions not wholly unrealistic in the Pacific—waves that average 8.5 meters (28 feet) high can result. The highest 10% of the waves in this sea will exceed 17.2 meters (56.6 feet) in height! The combination of factors required to produce truly great seas doesn't occur very often. In the example given at the bottom of Table 10.2, it seems unlikely that a 92-kilometer- (58-mile-) per-hour wind would blow steadily in one direction for 69 hours over 2,627 kilometers (1,633 miles). Of course, as Lieutenant Marggraff's experience in 1933 suggests (see **Box 10.1**), the ocean is full of surprises!

Table 10.2 Conditions Necessary for a Fully Developed Sea at Given Wind Speeds, and the Parameters of the Resulting Waves					
Wind Conditions			**Wave Size**		
Wind Speed in One Direction	**Fetch**	**Wind Duration**	**Average Height**	**Average Wavelength**	**Average Period**
19 km/hr (12 mi/hr)	19 km (12 mi)	2 hr	0.27 m (0.9 ft)	8.5 m (28 ft)	3.0 sec
37 km/hr (23 mi/hr)	139 km (86 mi)	10 hr	1.5 m (4.9 ft)	33.8 m (111 ft)	5.7 sec
56 km/hr (35 mi/hr)	518 km (322 mi)	23 hr	4.1 m (13.6 ft)	76.5 m (251 ft)	8.6 sec
74 km/hr (46 mi/hr)	1,313 km (816 mi)	42 hr	8.5 m (27.9 ft)	136 m (446 ft)	11.4 sec
92 km/hr (58 mi/hr)	2,627 km (1,633 mi)	69 hr	14.8 m (48.7 ft)	212.2 m (696 ft)	14.3 sec

Source: Data from U.S. Army Corps of Engineers Coastal Research Center, Richmond, Virginia.

Think of the ocean and you think of waves. On the night of 7 February 1933, Lieutenant Frederick Marggraff's world consisted of little *but* waves. Mr. Marggraff was standing a predawn watch on the bridge of the USS *Ramapo*, a large U.S. Navy tanker steaming from Manila to San Diego. February 7 was the third full day of a furious storm, and the tanker's crew had become professionally and emotionally interested in the intense wind and mountainous seas sweeping past their ship. A few of the officers may have thought about the so-called 60-foot law, a paragraph in a Navy Hydrographic Office bulletin implying that all sightings of waves over 60 feet high (about 18 meters) were exaggerations. Indeed, despite persistent reports to the contrary, few theorists believed that wind-driven ocean waves exceeded 18 meters in height. Lieutenant Marggraff was about to prove them very wrong.

February is the stormiest month in the North Pacific, and *Ramapo* had encountered the largest and most severe storm of the year, a storm made more intense by the coalescence of three low-pressure centers. For days a steady wind had blown at 107 kilometers (67 miles) per hour, and gusts to 126 kilometers (78 miles) per hour often lashed the decks. But the wind blew persistently from one direction, and though the monstrous waves it generated dwarfed the tanker, they were surprisingly orderly in form.

"The conditions for observing the seas from the ship were ideal," wrote *Ramapo*'s executive officer. "We were running directly down the wind with the sea. There were no cross seas and therefore no peaks along wave crests. There was practically no rolling, and the pitching motion was easy because of the fact that the sides of the waves were much longer than the ship. The moon was out astern and facilitated observations during the night. The sky was partly cloudy." At about three in the morning, Mr. Marggraff observed a train of giant waves looming in the moonlight. As the trough of the first wave approached the ship, he noted that its distant crest was on a level with the crow's nest on the mainmast. At that instant the ship's stern sank into the bottom of the onrushing trough. The next two waves were about the same size. Not surprisingly, such immense waves made an indelible impression on all who witnessed them.

How big were these waves? When the executive officer had some time to spare, he did some calculations. **Figure a** illustrates the attitude of the ship when the largest waves were measured. The height of the waves was determined by using a set of the ship's plans, a calculation of the height of the observer above the sea surface, the draft of the ship, and a sight to the horizon. The largest wave for which a dependable observation had been made was 34 meters (112 feet) high, still a record! After the executive officer published his findings, the paragraph on the 60-foot law was quietly dropped from the next edition of the Hydrographic Office bulletin.

10-12

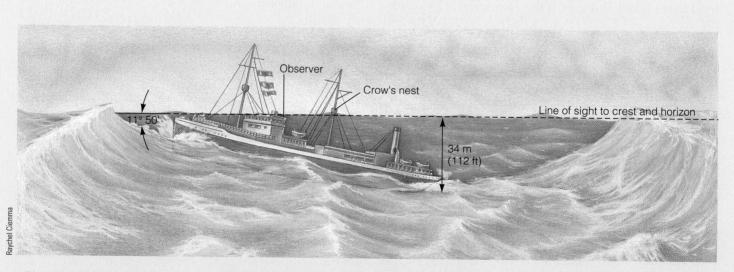

a How the great wave observed from the USS Ramapo was measured. An officer on the bridge was looking toward the stern and saw the crow's nest in his line of sight to the crest of the wave, which had just come in line with the horizon. Wave height was calculated based on the geometry of the situation.

The greatest potential for large waves occurs beneath the strong and nearly continuous winds of the Antarctic Circumpolar Current surrounding Antarctica. The early nineteenth-century French explorer of the South Seas Jules Dumont d'Urville encountered a wave train with heights estimated "in excess" of 30 meters (100 feet) in Antarctic waters. In 1916 Ernest Shackleton contended with occasional waves of similar size in the Antarctic Circumpolar Current during a heroic voyage to remote South Georgia Island in an open boat. Satellite observations (like those of **Figure 10.12**) have shown that wave heights to 11 meters (36 feet) are fairly common in the Antarctic Circumpolar Current.

In zones of high winds a less than fully developed sea can also attract attention. Though wind speed within cyclonic storms is often very strong, the circular motion of air doesn't allow long fetches, and fully developed seas rarely occur beneath them. Officers standing deck watches during storms rarely quibble with theoretical maximum height versus observed height, however. Wind waves can be overwhelming even if they are not fully developed (**Figure 10.13**).

Wind Wave Height

Wave height is not directly represented in the deep-water wave formula $C = L/T$. During their formation, moderately sized wind waves in the open ocean exhibit a maximum 1:7 ratio of wave height to wavelength (see **Figure 10.14**); this ratio is the **wave steepness.** Waves 7 meters long will not be more than 1 meter high, and waves with a 70-meter wavelength will not exceed 10 meters in height. The angle at their crest will not exceed 120°. A peaked appearance usually indicates the continuing injection of wind energy. If a wave gets any higher than the 1:7 ratio for its wavelength it will break, and excess energy from the wind will be dissipated as turbulence—hence the *white-caps* or *combers* associated with a fully developed sea.

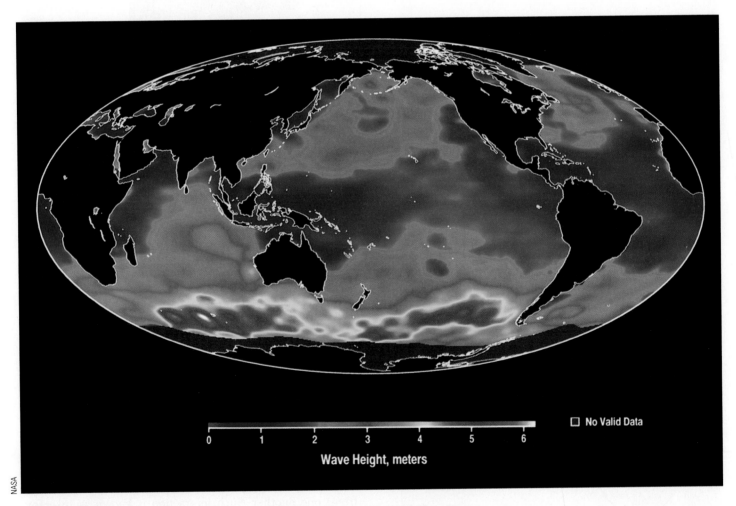

Figure 10.12 Global wave height acquired by a radar altimeter aboard the *TOPEX/Poseidon* satellite in October 1992. In this image, the highest waves occur in the southern ocean, where waves were measured at more than 6 meters (19.8 feet) high (represented in white). The lowest waves (indicated by dark blue) are found in the tropical and subtropical ocean, where wind speed is lowest.

Figure 10.13 A NOAA research ship rides large waves in a storm off North Carolina.

As wind waves mature into swell, their wavelengths increase somewhat and the ratio of wave height to wavelength decreases: The waves become less steep. Swell continue in this rounded-top shape until they break on shore.

Wavelength Records

How big can wind waves get? This is an interesting question, and one not dispassionately answered. Calculating the wavelength from wave period (T in $C = L/T$), Bigelow and Edmondson (1953) reported that mariners have sighted waves with wavelengths of 451 meters (1,480 feet) off the west coast of Ireland, 583 meters (1,913 feet) off the Cape of Good Hope, and 829 meters (2,720 feet) in the equatorial Atlantic. (That last monster would have had a period of 25 sec-

onds!) The nineteenth-century French admiral J. Mottez reported a wave with a wavelength of 790 meters (2,600 feet), a period of 23 seconds, and a speed of 123 kilometers (76 miles) per hour in the equatorial Atlantic west of Africa. The wavelength of the *Ramapo* wave (discussed in Box 10.1) was calculated at 360 meters (1,180 feet), so these sightings are perhaps exaggerated. I can attest to the fact that smaller waves can also add a great deal of excitement to a deck officer's day (see **Figure 10.15**)!

INTERFERENCE AND ROGUE WAVES

The real situation of wind waves at sea is not as simple as has been suggested here. The ideal vision of one set of waves moving in one direction at one speed across an otherwise smooth surface is almost never observed in the ocean.

Independent wave trains exist simultaneously in the ocean most of the time. Because long waves outrun shorter ones, wind waves from different storm systems can overtake and interfere with one another. One wave doesn't crawl over the others when they meet—instead they add to (or subtract from) one another. Such interaction is known as **interference**. In **Figure 10.16a,** a wave of one wavelength is represented as a green line, a slightly longer wavelength wave as a blue line. In the sea surface where these waves coincide (shown in **Figure 10.16b**), you can see the alternation be-

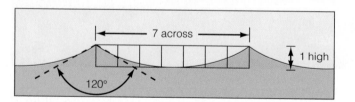

Figure 10.14 A wind wave of moderate size shown during the time of its formation. The ratio of wave height to wavelength, called *wave steepness*, is 1:7; the crest angle does not exceed 120°.

Figure 10.15 A U.S. Navy destroyer battles large waves generated by a storm in the Pacific. Despite the battering, the ship is holding a steady enough course to take on fuel and stores from the aircraft carrier barely visible on the right.

tween addition (large crests and troughs) and subtraction (almost no waves at all). The cancellation effect of subtraction is called **destructive interference,** not because of harm to lives or property but because wave interference destroys or cancels the waves involved. **Constructive interference** is the additive formation of large crests or deep troughs, the size of which exceeds the size of each of the participating waves. The huge waves observed by officers aboard *Ramapo* were surely the result of constructive interference.

You have probably noticed that the surf along a coast seems to rise to a few big waves, diminish, and then build again. Surfers wait for the big "sets" to arrive, ride toward the shore, and then use the relatively calm interval to swim out into position for the next big set of waves. Constructive and destructive interferences explain this behavior, called **surf beat.** Constructive interference between waves of different wavelengths creates the sought-after big waves; destructive interference diminishes the waves and makes it easier to swim back out. The characteristics of surf beat explain why in some instances every ninth wave might be quite large. As wavelengths and interference change, though, every seventh wave might be large, or every fifth or twelfth. Contrary to folklore there is no set ratio.

Interference can have sudden unpleasant consequences on the open sea. In or near a large storm, wind waves at many wavelengths and heights may approach a single spot from different directions. If such a rare confluence of crests occurred at your position, a huge wave crest would suddenly erupt from a moderate sea to threaten your ship. The freak wave—called a **rogue wave**—would be much larger than any noticed before or after and would be higher than the theoretical

maximum wave capable of being sustained in a fully developed sea. In such conditions one wave in about 1,175 is more than three times average height, and one in every 300,000 is more than four times average height! **Box 10.2** describes some encounters between ocean liners and rogue waves.

Currents can contribute to the formation of a kind of rogue wave that does not depend on interference. This wave is formed by the interaction of a wind wave and a swift surface current. The southeastern tip of Africa seems to breed a dis-

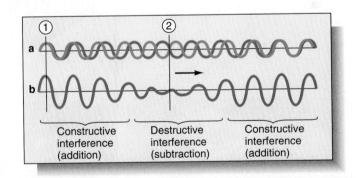

Figure 10.16 Constructive and destructive interference. (**a**) Two overlapping waves of different wavelength are shown in blue and green. Note that the wave shown in blue has a slightly longer wavelength. If both are present in the ocean at the same time, they will interfere with each other to form the composite wave shown in (**b**). At the position of line 1, the two waves in (a) will constructively interfere to form very large crests and troughs in (b). At the position of line 2, the two waves will destructively interfere, and crests and troughs in (b) will be very small.

In 1942 the Cunard liner RMS *Queen Mary,* then the largest ship in the world at 310 meters (1,020 feet) and 81,000 tons, encountered a rogue wave while transporting 15,000 American soldiers to Glasgow. About 1,100 kilometers (700 miles) off the coast of Scotland, the mountainous wave struck the ship broadside and heeled her past 45°. An eyewitness reported that the *Queen* ". . . listed until her upper decks were awash and those who had sailed in her since she first took to the sea were convinced she would never right herself." She did, but as the captain calmly noted, "She took rather a long time to do so."

Earlier, smaller ocean liners suffered even more severely. An encounter with a rogue wave nearly capsized another Cunard liner, RMS *Carmania,* in 1912. Already reeling from the effects of a huge North Atlantic extratropical cyclone, *Carmania* was struck broadside by a towering rogue wave:

> The ship went through a series of violent 50° rolls that carried kitchen ranges away and sent them grinding back and forth across the tiles. Deck chairs, in the words of one observer, "floated through the air like pieces of paper." Some cabins were made untenable as beds, only recently introduced as a luxury, battered their way through wooden partitions. Passengers thus dispossessed finally lay flat in the ship's square, clutching the base of the pillars.

Modern ships are not immune to the effects of rogue waves. The Italian Lines' beautiful flagship *Michelangelo* was severely damaged by a rogue wave in a 1966 storm only 1,500 kilometers (900 miles) out of New York. Inch-thick glass was shattered on the bridge 24 meters (80 feet) above the waterline, steel railings on the upper decks were ripped away, and the bow was flattened. Many passengers were injured; three died. Even the contemporary Cunard liner *Queen Elizabeth 2* has suffered from rogue wave encounters. On 16 September 1996, 14-meter (45-foot) seas and 192-kilometer- (120-mile-) per-hour winds from fading Hurricane Luis battered the ship en route to New York from Southampton. Shortly after midnight, while 1,200 passengers huddled below deck, a 29-meter (95-foot) rogue wave formed ahead of the ship. "It looked as if we were going straight into the white cliffs of Dover," said Captain Ronald Warwick. "It broke with tremendous force and washed over the ship. A tremendous shudder went through the vessel. . . ." The last liner specifically designed for North Atlantic service, *QE2* sustained no serious damage, and there were no injuries. She was eight hours late arriving in New York. "We *do* prefer to be on time," Captain Warwick noted.

10-16

Source: John Maxtone-Graham, *The Only Way to Cross* (New York: Macmillan, 1972), 297–8.

W. C. Stillwell

A rare photo of a rogue wave, taken in 1935 from the liner SS *Washington,* eastbound across the North Atlantic. Rogue waves occur when two or more wave crests add together to form a short-lived high wave.

Figure 10.17 *Neptune's Horse* by Walter Crane, a graphic representation of the power of breaking waves. The energy of a wave is proportional to the square of its height. Researchers report that each linear meter of a wave 2 meters (6.5 feet) above average sea level represents an energy flow of about 25 kilowatts (34 horsepower), enough to light 250 100-watt light bulbs; a wave twice as high would contain four times as much energy. A single wave 1.2 meters (3.5 feet) high striking the west coast of the United States may release as much as 50 million horsepower. (Reprinted by permission of Art Resource.)

proportionately large number of these giant waves. The Agulhas Current, a strong western boundary current, flows close to shore there, moving in the opposite direction to wind waves generated by high-latitude southwesterly gales. When large waves suddenly hit the 5-kilometer- (3-mile-) per-hour current, they can "trip" on it, rise suddenly to double their original height, and break precipitously. Mathematical modeling suggests a swift current or large fields of random eddies can act like a lens to refract and focus wave trains at a point. Waves at that location would have a steep forward face preceded by a deep trough. Mariners who have experienced these kinds of rogue waves describe them as "holes in the sea." Such waves have broken tankers in half; many lives have been lost.

Figure 10.17 gives graphic presentation of the power of breaking waves.

WIND WAVES
APPROACHING SHORE

10-17

Most wind waves eventually find their way to a shore and break, dissipating all their order and energy. The process begins with the transition of our now familiar deep-water wave to a shallow-water wave in water less than half a wavelength deep. **Figure 10.18** outlines the events leading to the break:

1. The wave train moves toward shore. When the water is less than half the wavelength in depth, the wave "feels" bottom.
2. The circular motion of water molecules in the wave is interrupted. Circles near the bottom flatten to ellipses. The wave's energy must now be packed into less water depth, so the wave crests become peaked rather than rounded.
3. Interaction with the bottom slows the wave. Waves behind it continue toward shore at the original rate. Wavelength therefore decreases but period remains unchanged.
4. The wave becomes too high for its wavelength, approaching the critical 1:7 ratio.
5. As the water becomes even shallower, the part of the wave below average sea level slows because of the restricting effect of the ocean floor on wave motion. When the wave was in deep water, molecules at the top of the crest were supported by the molecules ahead (thus transferring energy forward). This is now impossible because the *water* is moving faster than the *wave*. As the crest moves ahead of its supporting base, the wave breaks. The break occurs at about a 3:4 ratio of wave height to water depth. (That is, a 3-meter wave will break in 4 meters of water.) The turbulent mass of agitated water rushing shoreward during and after the break is known as **surf**. The **surf zone** is the region between the breaking waves and the shore.

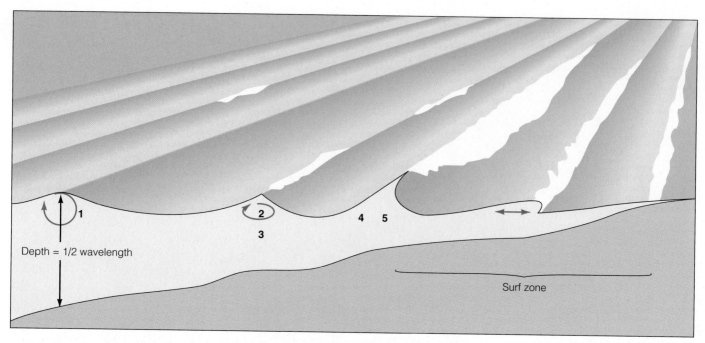

Figure 10.18 How a wave train breaks against the shore. (1) The swell "feels" bottom when the water is shallower than half the wavelength. (2) The wave crests become peaked because the wave's energy is packed into less water depth. (3) Constraint of circular wave motion by interaction with the ocean floor slows the wave, while waves behind it maintain their original speed. Therefore, wavelength shortens but period remains unchanged. (4) The wave approaches the critical 1:7 ratio of wave height to wavelength. (5) The wave breaks when the ratio of wave height to water depth is about 3:4. The movement of water particles is shown in red. Note the transition from a deep-water wave to a shallow-water wave.

Waves break against the shore in different ways, depending in part on the slope of the bottom. The break can be violent and toppling, leaving an air-filled channel (or *tube*) between the falling crest and the foot of the wave. These **plunging waves** (**Figure 10.19a**) form when waves approach a steeply sloping bottom (as in this chapter's opener). A more gradually sloping bottom generates a milder **spilling wave** as the crest slides down the face of the wave (**Figure 10.19b**).

Slope alone does not determine the position and nature of the breaking wave. The contour and composition of the

a & b © Dennis Junor, Creation Captured

a

b

Figure 10.19 Types of breaking waves. (**a**) Plunging waves form when the bottom slopes steeply toward shore. (**b**) Spilling waves form when the bottom slopes gradually.

bottom can also be important. Gradually shoaling bottoms can sap waves of their strength due to prolonged interaction against the bottom of the lowest elliptical water orbits. Energy may be lost even more rapidly if the bottom is covered with loose gravel or irregular growths of coral. Masses of moving seaweed or jostling chunks of sea ice can also extract energy from a wave. In a few rare cases the shore is configured in such a way that waves don't break at all—the waves have lost virtually all their energy by the time their remnants arrive at the beach.

Wave Refraction

What happens when a wave line approaches the shore at an angle, as it almost always does (**Figure 10.20**)? The line does not break simultaneously because different parts of it are in different depths of water. The part of the wave line in shallow water slows down, but the attached segment still in deeper water continues at original speed, so the wave line bends (or refracts). The bend can be as much as 90° from the original direction of the wave train. This slowing and bending of

waves in shallow water is called **wave refraction**.[3] The refracted waves break in a line almost parallel to the shore.

Wave refraction can produce some odd surf patterns. A swell with a wavelength of 600 meters "feels" bottom at 300 meters (1,000 feet), while a swell with a wavelength of 200 meters "feels" bottom at 100 meters (330 feet). You may remember that the average depth at the outer continental shelves is only about 150 meters (500 feet). As these two wave trains approach shore, the longer swell will be slowed by the shelf for perhaps 60 kilometers (37 miles) before the shorter swell is slowed by the bottom. With one wave bending and the other heading in its original direction, the potential for complex interference is great. Waves may break for a time at one spot, cease, and then begin breaking at another spot a few hundred meters down the beach. The ocean surface might have a checkerboard appearance from the constructive and destructive interference of these crests (see **Figure 10.21**).

Wave Diffraction

Wave diffraction is the propagation of a wave around an obstacle. Unlike wave refraction (which depends on a wave's response to a change in speed), wave diffraction depends on the interruption of the wave train by an obstacle to provide a new point of departure for the waves. This is illustrated by the gap in the breakwater shown in **Figure 10.22**. Wave crests excite the water in the gap. Water moving in the gap generates smaller waves in the harbor, which radiate from the gap and disturb the quiet water. The diffracted waves are much

---

[3] To review the principle of refraction, please see Chapter 6, page 158.

Figure 10.20 Wave refraction. (**a**) Diagram showing the elements that produce refraction. (**b**) Wave refraction around Maili Point, Oahu, Hawaii. Note how the wave crests bend almost 90° as they move around the point.

Figure 10.21 A checkerboard sea surface beyond the surf zone off La Jolla, California—the result of wave interference.

Figure 10.22 Diffraction of waves by a breakwater gap at Morro Bay, California.

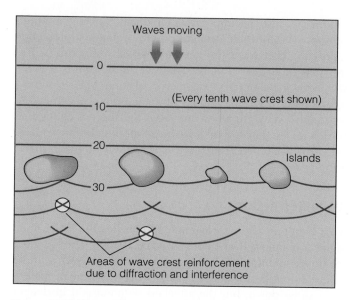

Figure 10.23 Wave diffraction past an island chain. Polynesian navigators used diffraction patterns to sense the presence of islands out of sight over the horizon (see Figure 2.4).

smaller than those in the open ocean, but boats in the harbor might still be jostled.

A more complex case of diffraction is shown in **Figure 10.23,** in which waves are interrupted by a chain of islands. Some of the wave energy "squeezes" through the spaces between the islands. These spaces act as new sources of waves. The waves spread into the space past the islands, creating areas of reinforcement and cancellation. Polynesian navigators learned early in their training to sense the disturbances in wave patterns caused by diffraction. The best among them could feel the presence of islands beyond the horizon by sensing subtle irregularities in the rocking of their canoe.

Wave Reflection

10-20

The waves discussed so far in this chapter—the familiar ocean waves in which the disturbance travels in one direction along the surface of the transmission medium—are known as progressive waves. A vertical barrier such as a seawall, large ship hull, or smooth jetty will reflect progressive waves with little loss of energy. If the waves approach the obstruction straight on, the reflected waves will move away from the obstruction in the direction from which they came. This **wave reflection** will cause interference in the form of vertical oscillations called **standing waves.** As their name suggests, standing waves do not progress but appear as alternating crests and troughs at a fixed position. **Figure 10.24** shows how a standing wave oscillates in a motion that resembles water sloshing back and forth in a half-filled bathtub. Because of constructive interference between crests (and troughs), these waves can be dangerous to boats or swimmers near the obstruction. As you will see in the next chap-

ter, standing waves are important in the physics of tsunami and tides.

Waves approaching the obstruction at an angle can also reflect; in this case the sea surface near the reflector will not form standing waves, but complicated sea-surface motions can develop. Watch for these effects the next time you visit a solid breakwater or a steep beach.

INTERNAL WAVES

Progressive waves can occur at the junction between air and water, as we have seen, or they can form at the boundary between water layers of different densities. These subsurface waves are called **internal waves.** As is the case with ocean waves at the air–ocean interface, internal waves possess troughs, crests, wavelength, and period. "Desktop-ocean" devices, which are sold at some gift shops, demonstrate internal waves. These sealed bottles contain nonmixing liquids of contrasting colors and slightly different densities. When you tilt the bottle, a very slow internal wave forms at the junction of the liquids, moves slowly to the low end, and breaks.

Normal ocean waves move rapidly because the difference in density between air and ocean is relatively great. Internal waves usually move very slowly because the density difference between the joined media is very small. Internal waves occur in the ocean at the base of the pycnocline, especially at the bottom edge of a steep thermocline (**Figure 10.25a**). The wave height of internal waves may be greater than 30 meters (100 feet), causing the thermocline to undulate slowly through a considerable depth. Their wavelength often exceeds 0.8 kilometer (0.5 mile); periods are typically

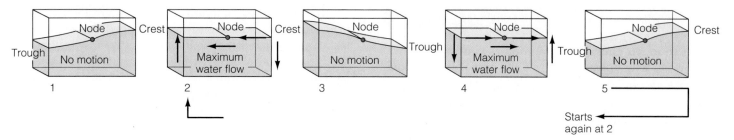

Figure 10.24 A standing wave in a basin. The wave oscillates about a node, a location at which there is no vertical movement. The rocking movement of water generates alternating crests and troughs at fixed positions a distance from the node. As their name suggests, standing waves do not progress, and there is no net movement of water in them. (See also Figure 11.3.)

5 to 8 minutes. Oceanographers are not certain how these large, slow-motion waves are generated, but wind energy, tidal energy, or ocean currents may be responsible. Surface manifestations of internal waves have been photographed from space (**Figure 10.25b**).

Are internal waves important? They may mix nutrients into surface water and trigger plankton blooms. They can also affect submarines and oil platforms. In 1963 the nuclear-powered USS *Thresher*, a fast attack submarine, was lost with all hands off the coast of Massachusetts. Running at high speed, *Thresher* may have encountered an internal wave and been forced beyond her test depth. In 1980 an oil production platform was slowly rotated nearly 90° from its original orientation by a series of internal waves. Also, the slow-motion breaking of internal waves against a shore may occasionally exaggerate tidal height.

Figure 10.25 Internal waves can form between masses of water with different densities, especially at the base of the pycnocline. (**a**) The crest of an internal wave. (**b**) Internal waves diffracting around the Seychelles Islands in the Indian Ocean, photographed by the crew of the space shuttle *Columbia* in January 1990. The waves are visible because their crests have altered the reflectivity of the ocean surface. Patterns of constructive and destructive interference can be seen.

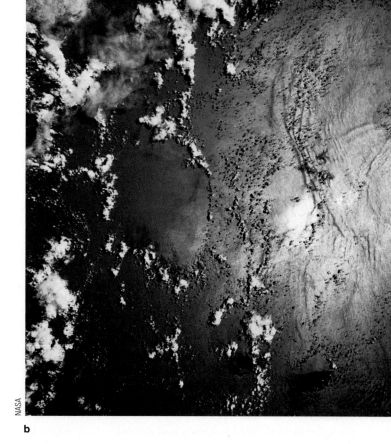

a

b

WAVE DYNAMICS AND WIND WAVES • • • • 255

1. If so much energy is expended as wind waves break, why doesn't water in the surf zone get hot?

The amount of energy moving through the ocean in progressive waves is impressive. The energy of a wave is proportional to the square of its height. Inman and Brush (1973) report that each linear meter of a wave 2 meters (6.6 feet) above average sea level represents an energy flow of about 25 kilowatts (34 horsepower), enough to light 250 100-watt light bulbs; a wave twice as high would contain four times as much energy. A single wave 1.2 meters (4 feet) high striking the west coast of the United States may release as much as 50 million horsepower!

The energy of wind waves is dissipated mostly as heat in the surf, but since water has such a high heat capacity (discussed in Chapter 6), the injection of heat into the surf zone doesn't significantly increase water temperature. The surf zone is also an area of vigorous mixing, so any heat released is quickly distributed through a large volume of water.

A number of schemes have been proposed to take advantage of this energy before it is dissipated. Among the most promising is a design proposed by Kvaerner Industrier A/S, Oslo, Norway (see Figure 17.5).

2. What about wind waves and the Coriolis effect? Do wind waves turn right in the Northern Hemisphere as they move across the ocean surface?

The Coriolis effect has no influence on waves with periods less than about 5 minutes. Large shallow-water waves such as tsunami, seiches, and tides involve the mass movement of water and are influenced by Earth's rotation. Wind waves, however, with a period rarely exceeding 20 seconds, are not.

3. I love to surf. Where should I plan to live?

The Pacific has the largest potential fetch distances, so the best chances for large wind waves are there. The biggest wind waves are made in the Antarctic Circumpolar Current, but temperatures there are much too cold for comfortable surfing. Besides, the sea state in areas of continuous high wind is chaotic, and surfing requires orderly waves. It is best to let the wave trains sort out. Hawaii is in the middle of the Pacific, so wind waves from polar storms in either hemisphere strike its shores only after much sorting by wavelength. Order is assured. The water is warm, too. I vote for Hawaii.

4. Could a method be invented to surf internal waves?

Think of the possibilities: a wave 30 meters (100 feet) high moving in slow motion with a period of 5 to 8 minutes. It would be a calm and stately ride, and the density differences would require a very delicate balancing act at the interface between the two water densities. Still, with the right equipment to adjust buoyancy and provide air, it might be done.

Indeed, surfing an internal wave may already have been inadvertently and disastrously accomplished. As noted in the text, the slow undulation of an internal wave may have led to the loss in 1963 of USS *Thresher*. An equipment failure at high speed coupled with a surprise wave encounter may have driven the new submarine past its designed pressure limit, causing the violent implosion that cost 129 lives.

- Waves transmit energy, not water mass, across the ocean's surface.

- Waves are classified by the disturbing force that creates them, the extent to which the disturbing force continues to influence the waves once they are formed, the restoring force that tries to flatten them, and their wavelength.

- The speed of ocean waves usually depends on their wavelength, with long waves moving fastest.

- In order of wavelength from short to long (and therefore from slowest to fastest), ocean waves are generated by very small disturbances (capillary waves), wind (wind waves), rocking of water in enclosed spaces (seiches), seismic and volcanic activity or other sudden displacements (tsunami), and gravitational attraction (tides).

- The behavior of a wave depends largely on the relation between the wave's size and the depth of water through which it is moving. Deep-water waves move through water deeper than half their wavelength. Only wind waves can be deep-water waves.

- Wind waves form when energy is transferred from wind into water. Wave size depends on wind strength and duration.

- When waves approach shore, movement of water in orderly orbits is interrupted, causing the wave to break.

- Waves can change direction by refraction and diffraction, interfere with one another, and reflect from solid objects.

The Web site for this book contains helpful study aids. Log on to:

www.brookscole.com/053437557xs

and click on the Chapter-by-Chapter area. Choose Chapter 10 and select a resource:

- **Flash Cards** allows you to test your mastery of the Terms and Concepts to Remember for this chapter.
- **Tutorial Quizzes** provides a multiple-choice practice quiz.

TERMS AND CONCEPTS TO REMEMBER

$C = L/T$
$C = \sqrt{gd}$
capillary waves
constructive interference
deep-water wave
destructive interference
dispersion
disturbing force
fetch
forced wave
free wave
fully developed sea
gravity wave
group velocity
interference
internal wave
orbit
orbital wave
plunging wave
progressive wave
restoring force
rogue wave
sea
shallow-water wave

spilling wave
standing wave
Stokes drift
surf
surf beat
surf zone
swell
transitional wave
wave
wave crest
wave diffraction
wave frequency
wave height
wave period
wave reflection
wave refraction
wave steepness
wave train
wave trough
wavelength
wind duration
wind strength
wind wave

STUDY QUESTIONS

Review Questions

1. Draw an ocean wave and label its parts. Show the orbits of water particles. Include a definition of *wave period*. How would you measure frequency?

2. Make a list of ocean waves, arranged by disturbing force and wavelength. What is restoring force? How is a capillary wave different from a gravity wave?

3. What factors influence the growth of a wind wave? What is a fully developed sea? Where would we regularly expect to find the largest waves? Are waves in fully developed seas always huge?

4. What happens when a wind wave breaks? What factors affect the break? How are plunging waves different from spilling waves?

5. Describe *wave reflection, wave refraction,* and *wave diffraction.*

6. What are internal waves? Are internal waves larger or smaller than most wind waves? Slower or faster?

Critical Thinking Questions

1. How do particles move in an ocean wave? How is that movement similar to or different from the movement of particles in a wave in a spring or a rope? How does this movement relate to a *stadium wave*—a waveform made by sports fans in a circular arena?

2. What is the general relationship between wavelength and wave speed? How does water movement in a wave change with depth? Can deep-water waves and shallow-water waves exist at the same point offshore (that is, in the same depth of water)?

3. How can a rogue wave be larger than the theoretical maximum height of waves in a fully developed sea?

4. How is a progressive wave different from a standing wave? Must standing waves be orbital waves only, or can standing waves also form in shaken ropes or pushed-and-pulled springs?

5. How can large waves generated by a distant storm arrive at a shore first, to be followed later by small waves?

6. **InfoTrac College Edition Project** Breaking waves and the currents they create near the shore can be dangerous. For some, they can also be helpful and even a source of enjoyment. Do some research on rip currents, surf, and breaking waves using InfoTrac College Edition. What do swimmers, divers, surfers, and others need to watch for?

Earth Systems Today CD Questions

1. Go to "Waves, Tides, and Currents," then to "Wave Properties" and review your understanding of terms used to describe progressive waves.

2. Go to "Waves, Tides, and Currents," then to "Wave Progression." Note in the animation the circular movement of water molecules in waves, and how the movement is affected by shallow water. This is a particularly effective demonstration of a difficult-to-understand concept.

3. Go to "Waves, Tides, and Currents," then to "Wind-Wave Relationships." Perform some experiments to determine the factors that influence a fully developed sea. What is the best possible combination to generate truly large wind waves?

RESOURCES FOR FURTHER READING AND RESEARCH

The Web site for this book contains many ideas for further reading and research. Log on to:

www.brookscole.com/product/053437557xs

and click on the Chapter-by-Chapter area. Choose Chapter 10 and select a resource.

- **References** lists the major books and articles consulted in writing this chapter, along with comments from the author about their content and reading level.
- **Hypercontents** takes you to an extensive list of sites with news, research, and images related to individual sections of the chapter.
- In **Student Guide to InfoTrac College Edition** scroll down to Suggested Readings from InfoTrac College Edition for brief descriptions of the articles listed and search hints for finding them in InfoTrac College Edition.

- **Regional InfoTrac College Edition** articles are organized into East Coast, West Coast, and Gulf Coast regions, allowing you to study oceanography on a more local level. You can also access Regional InfoTrac College Edition at www.localocean.com.

 For additional readings, go to InfoTrac College Edition, your online research library, at:

http://infotrac.thomsonlearning.com/

Tsunami, Seiches, and Tides

11

THE GODS OF LITUYA BAY

On 1 April 1946 a fracture along the Aleutian Trench generated a seismic sea wave—a form of tsunami—that quickly engulfed the Scotch Cap lighthouse on Unimak Island in Alaska's Aleutian Islands. The lighthouse is depicted here moments before being destroyed; all five coastguardsmen operating the lighthouse died. The inset shows the site a few days later. The remnants of the lighthouse's foundation are 15 meters (49 feet) above sea level, and rubble has been carried to the upper plateau 36 meters (118 feet) above sea level. The slopes are heavily washed.

An artist's conception of Scotch Cap lighthouse moments before its destruction in 1946 by a seismic sea wave. The wreckage of the lighthouse, located at Unimak Pass in Alaska's Aleutian Islands, is shown in the inset.

D. Millsap

USGS

Though undeniably huge, this wave was far from a record breaker. One of the largest ocean waves ever observed on Earth formed on 9 July 1958, and four of the six eyewitnesses lived to tell about it.

The great wave formed in Lituya Bay, a narrow glacier-fed inlet on the northeast shore of the Gulf of Alaska. The area was explored in 1788 by the French explorer Jean-François La Pérouse, but the only notable oceanic feature mentioned in his account was the strong tidal current encountered at the bay's narrow entrance. Had he inquired, the local inhabitants could have told him of the awesome magic contained within the bay: Waves ". . . as big as gods."

Three pleasure boats were anchored in the bay on the calm evening the great wave appeared. At 10:16 P.M. the crew of the U.S. Geologic Survey ship *Stephen R. Capps,* lying at anchor in another bay about 100 kilometers (62 miles) to the east, felt a strong earthquake along the Fairweather Fault, a fault that angles west along the steep mountains rimming the innermost reaches of Lituya Bay. The earthquake dislodged 61 million cubic meters (80 million cubic yards) of rock, mud, and glacial ice, which plummeted into upper Lituya Bay. Water displaced by this enormous mass of debris generated a huge surge that raced to the bay's opposite wall, where it climbed 530 meters (1,740 feet) above sea level. (The Sears Tower in Chicago, the world's second-tallest inhabited structure, is 87 meters, or 286 feet, shorter than the height of this "splash"!) The displacement also generated a gravity wave that roared out into the bay at a speed of between 155 and 210 kilometers (96 and 130 miles) per hour. This wave was at least 49 meters (160 feet) high, roughly the height of a 15-story building.

Mr. Howard Ulrich, aboard the small boat *Edrie,* was probably the first to see the wave in the gathering darkness. Ulrich and his seven-year-old son had been awakened by a deafening roar and had gone on deck to investigate. Ulrich did not comprehend the wave at first;

it looked like a low cloud or fogbank extending from shore to shore. The monster began to break as it rounded the north side of an island in the center of the bay, but on the south side, where his boat rested at anchor, the wave had not started to curl. Ulrich was unable to free the anchor, but the line snapped as the boat rose to meet the wave. *Edrie* accelerated upward to the crest, was carried forward over a submerged peninsula, and was deposited into a different section of the now wildly churning bay. Amazingly, the two were able to leave the bay under their boat's own power the next day.

Mr. and Mrs. William Swanson, aboard their fishing trawler *Badger* anchored on the north side of the bay, were not so fortunate. The Swansons were awakened by a violent vibration. At first, William Swanson saw nothing unusual; then a blue wall moved past the northern end of the island toward their boat, traversing the distance in about 90 seconds. The trawler's anchor chain parted, and she ascended the wave bow-first at a terrifying angle. Then the top of the wave broke, engulfing the boat. *Badger* shuddered and then rushed stern-first down the face of the shattered wave, wind howling through her rigging. She was carried at a height of 25 meters (80 feet) over the tops of the highest trees, over the bay mouth bar, and out to sea. The Swansons' boat was fatally broken. A deserted skiff was floating nearby, and they clambered aboard as *Badger* sank. They were rescued by a fishing boat two hours later.

The two-person crew of the third boat was unluckiest of all. They perished within the wave. William Swanson last saw the *Sunmore* in the trough of the wave moving toward a rocky cliff. No sign of crew or boat was ever found.

The 49-meter (161-foot) height of the wave was established by survey crews analyzing the line of tree devastation around the periphery of the bay. It gives one pause to learn that even this wave was comparatively small. Though they have not been directly observed, much larger waves have left trails of smashed timber even far-ther inland. Evidence of a 61-meter (200-foot) wave was reported in 1899, an 1853 wave reached 120 meters (390 feet) in height, and on 27 October 1936 a land-slide-generated wave rose to 150 meters (490 feet) in Lituya Bay!

The "gods" of Lituya, known as tsunami, are mem-bers of the class of long waves. Although they have a dif-ferent origin from wind waves, the wavelength, wave height, and wave period of long waves are measured in the same way. And as was the case for wind waves, en-ergy in most large waves passes through the ocean by the orbital movement of water particles.

Listed in order of wavelength, from shortest to longest, the waves described in this chapter are storm surges, seiches, tsunami, and tides. Even the smallest of these waves is much longer in wavelength than the largest wind waves. Because of their very long wavelengths, each wave in this chapter is classified as a shallow-water wave. As we shall see, all of these waves move through the ocean very rapidly.

WHAT TO WATCH FOR IN CHAPTER 11

We now turn our attention to immense waves. The waves dis-cussed in this chapter are caused by the low atmospheric pressure and wind associated with large cyclonic storms (storm surges), the rocking of water in confined harbors and bays (seiches), the sudden displacement of the water surface by phenomena such as earthquakes and landslides (tsunami), and gravitational attraction (tides). How do these waves com-pare to the wind waves discussed in the previous chapter? How do the generating forces differ?

Notice that, unlike wind waves, these great waves have such long wavelengths that they are always in shallow water (that is, water that is shallower than half their wavelength). It seems odd at first to imagine waves that moving across the deepest ocean basins are always in shallow water, but re-member that wave behavior depends in part on the ratio of wavelength to water depth. Because of their long wave-lengths, these long waves travel at high speeds. Some of the waves can be destructive, but their ability to cause damage is fortunately not proportional to their wavelength.

And. yes, here you will find impressive information on the ocean's largest waves.

A WORD ABOUT "TIDAL WAVES"

The term *tidal wave* is often misused to describe some of the waves discussed in this chapter. The popular media and gen-eral public tend to label *any* unusually large wave a "tidal wave" regardless of its origin. Press accounts of storms at sea usually list a rogue wave as a tidal wave, and very large sets of wind waves are called tidal waves by some yacht owners or surfers. The sea waves associated with earthquakes are almost always called tidal waves in media damage reports; certainly the Lituya Bay waves have been misnamed this way. The waves caused by the approach of a tropical cyclone to land may also incorrectly be termed tidal waves. The term *tidal wave* is *not* synonymous with *large wave,* however. As we will see, the only true tidal waves are relatively harmless waves as-sociated with the tides themselves.

STORM SURGES

The abrupt bulge of water driven ashore by a tropical cyclone (hurricane) or frontal storm is called a **storm surge.** The low atmospheric pressure associated with a great storm will draw the ocean surface into a broad dome as much as 1 meter (more than 3 feet) higher than average sea level. This dome of water accompanies the storm to shore, becoming much higher as the water gets shallower at the coast. There the wa-ter ramps ashore, driven forward by large, storm-generated wind waves (**Figure 11.1**). The crest of a storm surge can temporarily add up to 7.5 meters (25 feet) to coastal sea level. Water can reach even greater heights when the surge is fun-neled into a confined bay or estuary.

A storm surge is a short-lived phenomenon. Technically it is not a wave because it is only a crest, so wavelength and period cannot be assigned to it. Water in a storm surge does not come ashore as a single breaking wave but rushes inland in what looks like a sudden, very high, windblown tide. In-deed, storm surges are sometimes called **storm tides.** The wrong combination of low atmospheric pressure, strong on-shore winds, high tide, and bottom contour can be very dangerous—especially if estuaries in the area have been swollen by heavy rainfall preceding the storm.

Storm surges have had catastrophic consequences. The frightful tropical storm of November 1970 in Bangladesh (described in Chapter 8) generated a storm surge up to 12 meters (40 feet) high that caused the death of more than 300,000 people. Storm surges associated with extratropical cyclones (frontal storms) can also do tremendous damage. On 1 February 1953, a storm surge and high tide arrived si-multaneously at the Dutch coast. Wind waves breached the dikes and flooded the low country, covering more than 3,200 square kilometers (800,000 acres) and drowning 1,783 peo-ple. The Dutch anticipate that this coincidence of events will occur only about once in 400 years. The dikes have been rebuilt higher and the land reclaimed from the North Sea at the cost of billions of dollars. On the opposite side of the North Sea, Londoners have spent $1.3 billion on a flood

Figure 11.1 Low atmospheric pressure and high winds in a cyclonic storm can produce a storm surge up to 9 meters (30 feet) high. In 1992, storm surge waves up to 4.5 meters (15 feet) high attacked homes on Peggotty Beach, Scituate, Massachusetts.

AP/Wide World Photo

defense system at the mouth of the Thames River. The centerpiece of the project is an immense barrier against storm surges (see **Figure 11.2**). Experts expect the barrier to prevent a devastating flood on an average of once every 50 years.

The U.S. coast is also at risk. In 1900, a storm surge topped the Galveston, Texas, seawall, swept into the city, and killed more than 6,000 area residents. (In contrast, no lives were lost during a similar Texas storm in 1961 because coastal barriers had been constructed and advance warning permitted preparation and evacuation.) Hurricane Andrew's 1992 assault on the Florida coast was made even more lethal by its storm surge. Anyone living in a low-lying coastal area frequented by violent storms should be aware of the potential danger of storm surge and should heed calls for evacuation if conditions warrant.

a & b: Thames Barrier Visitors Centre

a

Figure 11.2 The Thames tidal barrier near London. The funnel shape of the Thames estuary concentrates the surge as it nears London. The barrier sections can be raised (**a**) to prevent flooding upstream. Each barrier section is controlled by hydraulic arms within towers. The great size of the $1.3 billion project can be seen in (**b**).

b

SEICHES

When disturbed, water confined to a small space (such as a bucket, a bathtub, or a bay) will slosh back and forth at a specific resonant frequency. The frequency changes with different amounts of water or with different sizes or shapes of containers. If you carry a shallow tray of water from one place to another, you're careful not to move the tray at its resonant frequency to avoid a spill. Most of the water's random motion quickly settles down after you place the tray on a tabletop, but the water in the tray may rock gently for some seconds at this one resonant frequency. That rocking is a **seiche** (pronounced *saysh*).

The seiche phenomenon was first studied in Switzerland's Lake Geneva by eighteenth-century researchers curious about why the water level at each end of the long, narrow lake rises and falls at regular intervals after windstorms. They found that constant breezes tend to push water into the downwind end of the lake. When the wind stops, the water is released to rock slowly back and forth at the lake's resonant frequency, completing a crest-trough-crest cycle in a little more than an hour. At the ends of the lake the water rises and falls nearly a meter (more than 3 feet); at the center it moves back and forth without changing height (**Figure 11.3**). This kind of wave is called a **standing wave** because it oscillates vertically with no forward movement (see again Figure 10.24). The point (or line) of no vertical wave action in a standing wave—the place in the lake where the water moves only back and forth—is called a

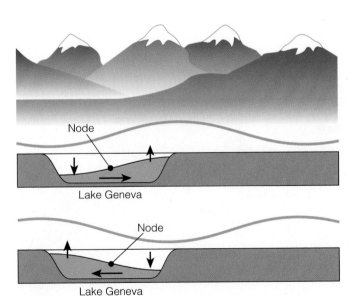

Figure 11.3 Seiche in Lake Geneva. After a windstorm, the lake sloshes back and forth, forming a standing wave, or seiche, with a wavelength of 220 kilometers (132 miles), which is twice the length of the lake. The period of the seiche is about 72 minutes, and its height is about 0.5 meter (1.5 feet). At the node, water moves sideways and does not rise or fall. (See also Figure 10.24.) The blue line represents the hypothetical whole wave of which the seiche is a part.

node. In Lake Geneva the wavelength of the seiche is twice the length of the lake itself; the node lies at the center. The lake acts like a large version of the tray of water in the previous example.

Energy can be added rhythmically to a seiche. The process is analogous to adding energy to a child in a swing. Insert a little energy at the right time in each cycle, and the child will swing higher and higher. (As our daughter discovered at an early age, the same thing applies to water in a bathtub. By moving her body in rhythm with the growing wave, she quickly learned how to slop most of the bathwater onto the floor.)

This kind of rhythmic oscillation in a seiche can contribute to lakeside property damage during storms. Lake Michigan began to rock because of successive storms during the first week of February 1987. On 8 February the water level at the south end of the lake was swinging 1 meter (more than 3 feet) above normal. North-northwest winds from a storm driving down the long axis of the lake generated wind waves as high as 3 to 6 meters (10 to 20 feet) above this bulge. The waves caused severe flooding and erosion in Chicago.

This kind of activity can also occur in confined areas of ocean ranging in size from bays and harbors to entire ocean basins. As we have seen, a drop in atmospheric pressure beneath a storm can draw water to a slightly higher level. When the storm moves inland the water is released, and the oscillation begins. The wavelength of the oscillation is a multiple of the length of the ocean basin, often thousands of miles. Tides and surf beat can also cause seiches in bays or harbors, as can internal waves moving toward shore. Seiche periods may range from a few minutes to more than a day. In large areas the wavelength of these standing waves may rival that of tsunami.

Damage from seiches along most ocean coasts is rare. The wavelength may be tremendous, but seiche wave height in the open ocean rarely exceeds a few inches. Larger seiches can occur in harbors—coastal seiches at Nagasaki, on the southern coast of Japan, occasionally reach 3 meters (10 feet). Seiches may disturb shipping schedules by interfering with the predicted arrival times of tides, or they may cause currents in harbors, which could snap mooring lines. In rare instances, seiches contribute to the peril of people on shore, but usually because the seiche is coincidentally accompanied by wind waves of considerable height or by a tsunami.

TSUNAMI AND SEISMIC SEA WAVES

Long-wavelength, shallow-water progressive waves caused by the rapid displacement of ocean water are called **tsunami,** a descriptive Japanese term combining *tsu* (harbor) with *nami* (wave). The word is both singular and plural. Tsunami caused by the sudden vertical movement of earth along faults (the same forces that cause earthquakes) are properly called **seismic sea waves.** Tsunami can also be caused by landslides, icebergs falling from glaciers, volcanic eruptions, and other direct displacements of the water surface. Note that all seismic sea waves are tsunami, but not all tsunami are seismic sea waves.

Origins

"Small" tsunami are caused by the displacement of surface water. Although less energy is released by landslides than by most seismic fractures, the resulting sea waves are still very destructive for people or structures near their point of origin. This is especially true if the wave is formed within a confined area, as we saw in the Lituya Bay example at the beginning of this chapter.

Seismic sea waves originate on the ocean floor when underwater landslides or earth movement along faults displaces ocean water. **Figure 11.4** shows a debris trail off the Oregon coast, the remains of a slump some 6 kilometers (3.7 miles) across. Probably triggered by an earthquake, the landslide displaced water and probably caused a tsunami. (Faults recently discovered in the Hawaiian Islands and off the coast of southern Virginia and North Carolina suggest that similar events could happen in those areas.) **Figure 11.5** shows the birth of a seismic sea wave in the Aleutians. Rupture along a submerged fault lifts the sea surface above. Gravity pulls the crest downward, but the momentum of the water causes the crest to overshoot and become a trough. The oscillating ocean surface generates progressive waves that radiate from the epicenter in all directions (see **Figure 11.6**). Waves would also form if the fault movement were downward. In that case a depression in the water surface would propagate outward as a trough. The trough would be followed by smaller crests and troughs caused by surface oscillation.

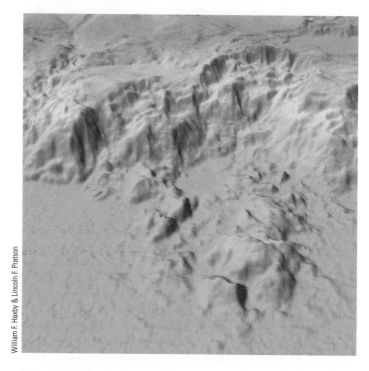

William F. Haxby & Lincoln F. Pratson

Figure 11.4 Failure of the edge of the continental shelf off the Oregon coast produced an underwater landslide that probably caused a large tsunami sometime around the end of the last ice age. The pocket from which the debris originated is about 6 kilometers (3.7 miles) wide. (See Figure 4.15 for a larger view of this area.) Faults recently discovered in the Hawaiian Islands and off the coast of southern Virginia and North Carolina suggest that similar events could happen in those areas.

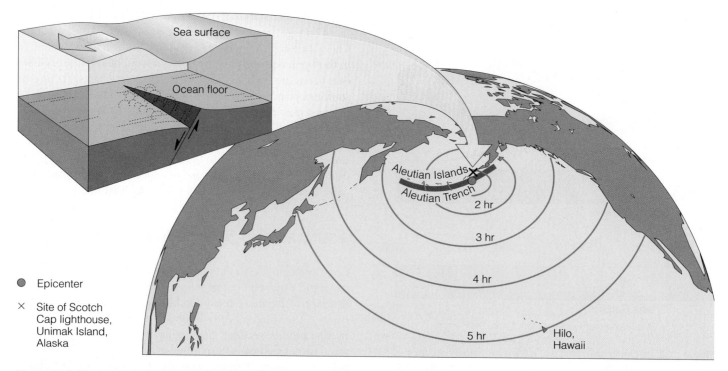

● Epicenter

✕ Site of Scotch Cap lighthouse, Unimak Island, Alaska

Figure 11.5 The tsunami of 1 April 1946 began when a rupture along a submerged fault lifted the sea surface above. The resulting seismic sea wave quickly engulfed the Scotch Cap lighthouse (see chapter opener). The wave moved outward at a speed of about 212 meters per second (472 miles per hour) and took only about five hours to travel to the Hawaiian Islands (see Figure 11.8).

a 2.5 hours after

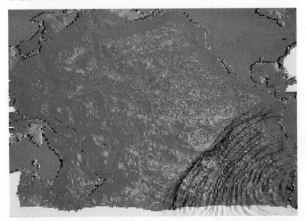

b 5 hours after

c 12.5 hours after

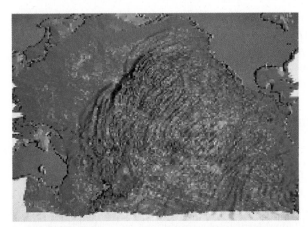

d 17.5 hours after

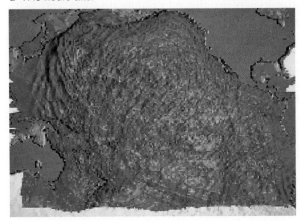

e 22.5 hours after

Figure 11.6 A computer simulation of the movement of a 1960 tsunami that originated in western Chile and sent destructive waves to Japan. The images represent the successive positions of the waves (**a**) 2.5 hours after the earthquake, (**b**) 5 hours after, (**c**) 12.5 hours after, (**d**) 17.5 hours after, and (**e**) reaching Japan after 22.5 hours of travel. (Source: Research by Philip L. F. Liv, Seung Nam Seo, and Sung Bum Yoon, Civil and Environmental Engineering, Cornell University. Visualization by Catherine Devine, Cornell Theory Center. Used by permission.)

It seems strange to refer to tsunami—with wavelengths of up to 200 kilometers (125 miles)—as shallow-water waves. Yet half their wavelength would be 100 kilometers (62 miles), and even the deepest ocean trenches do not exceed 11 kilometers (7 miles) in depth. These immense waves therefore never find themselves in water deeper than half their wavelength. Like any shallow-water wave, seismic sea waves are affected by the contour of the bottom and are commonly refracted, sometimes in unexpected ways.

Speed

The speed (C) of a tsunami is calculated by the same formula as that for the speed of a shallow-water wave:

$$C = \sqrt{gd}$$

The acceleration due to gravity (g) is 9.8 meters (32.2 feet) per second per second, and a typical Pacific abyssal depth (d)

is 4,600 meters (15,000 feet), so solving for C shows that the wave would move at 212 meters per second (472 miles per hour). At this speed a seismic sea wave will take only about five hours to travel from Alaska's seismically active Aleutians to the Hawaiian Islands.

Encountering Tsunami

We are familiar with the steepness of a wind wave and the short period of a few seconds between its crests. Tsunami are much different. Once a tsunami is generated, its steepness (ratio of height to wavelength) is extremely low. This lack of steepness, combined with the wave's very long period (5 to 20 minutes), enables it to pass unnoticed beneath ships at sea. A ship on the open ocean that encounters a tsunami with a 16-minute period would rise slowly and imperceptibly for about 8 minutes, to a crest only 0.3 to 0.6 meter (1 or 2 feet) above average sea level. It would then ease into the following trough 8 minutes later. With all the wind waves around, such a movement would not be noticed.

As the tsunami crest approaches shore, however, the situation changes rapidly and often dramatically. The period of the wave remains constant, velocity drops, and wave height greatly increases. As the crest arrived at the coast, observers would see water surge ashore in the manner of a very high, very fast tide (**Figure 11.7**). In confined coastal waters relatively close to their point of origin, tsunami can reach a height of perhaps 30 meters (100 feet). The wave is a fast, onrushing flood of water, not the huge, plunging breaker of popular folklore.

The wave energy spreads through an enlarging circumference as a tsunami expands from its point of origin. People on shore near the generating shock have reason to be concerned because the energy will not have dissipated very much. The seismic sea wave that destroyed the Scotch Cap lighthouse on Unimak Island (the wave shown on this chapter's opening page) reached the Hawaiian Islands about five hours later. By this time the wave circumference was enormous and its energy more dispersed. Even so, successive waves surged onto Hawaiian beaches at 15-minute intervals for more than two hours. One 9-meter (30-foot) wave struck the town of Hilo, and water rose to a height of 17 meters (56 feet) in an exposed valley near Polulu! At least 150 people were killed in Hawaii that morning, and property damage was in excess of $25 million. A photograph taken in Hilo that morning is shown in **Figure 11.8.**

V-shaped Hilo Bay is an especially dangerous place during a tsunami because its funnel-like shore outline concentrates the energy of the waves. Fourteen years after the 1946 disaster, a seismic disturbance in the subduction zone off western South America generated a tsunami that drowned 61 people in Hilo. Wave-cut scars on cliffs north of the town suggest that visits by large tsunami are not rare occurrences.

Note that the destruction in Hawaii was caused not by one wave (as at Scotch Cap lighthouse), but by a series of waves following one another at regular intervals. Some energy from the main tsunami wave was distributed into smaller

a

b

c

Photos a–c: USGS

Figure 11.7 Sequential photos showing the arrival of a tsunami at Laie Point, Oahu, Hawaii, on 9 March 1957. An 8.3 earthquake had occurred in the Aleutian Islands about five hours earlier. Note that the wave resembles a fast, onrushing flood of water, not the huge, plunging breaker of movies and folklore.

Figure 11.8 A tsunami in progress. The largest wave of the 1 April 1946 tsunami rushes ashore in Hilo. Terrified residents run for their lives; more than 150 died. V-shaped Hilo Bay is an especially dangerous place during a tsunami because its funnel-like outline concentrates the energy of the waves. Fourteen years after the 1946 disaster, a seismic disturbance in the subduction zone off western South America generated a tsunami that drowned 61 people in Hilo. Wave-cut scars on cliffs north of the town suggest that visits by large tsunami are not rare occurrences.

waves ahead of or behind the main wave as it moved. If the source of energy responsible for a tsunami is very far away, sea level at shore will rise and fall as these waves arrive. The interval between crests (the wave period) is usually about 15 minutes. Coastal residents far from a tsunami's origin can be lulled into thinking the waves are over; they return to the coastline only to be injured or killed by the next crest. This behavior has contributed to loss of life in Hawaii.

Tsunami in History

Destructive tsunami strike somewhere in the world an average of once each year. An earthquake along the Peru–Chile Trench on 22 May 1960 killed more than 4,000 people, and tsunami reaching Japan, 14,500 kilometers (9,000 miles) away, killed 180 people and caused $50 million in structural damage. Los Angeles and San Diego harbors were badly disrupted by seiches excited by the tsunami. In 1992 a tsunami struck the coast of Nicaragua and killed 170 people; 13,000 Nicaraguans were left homeless. In 1993 an earthquake in the Sea of Japan generated a tsunami that washed over areas 31 meters (101 feet) above sea level and killed 239 people (**Figure 11.9**). Sometimes even a relatively minor wave can cause considerable loss of life and property. In 1998 a wave 15 meters (49 feet) high destroyed remote villages in New Guinea, killing more than 2,200 people. Some recent lethal tsunami are indicated in **Figure 11.10.**

Half the time a tsunami's trough will arrive first. This occurred with catastrophic results in the Tagus estuary at Lisbon, Portugal, on 1 November 1755. Many curious people were attracted to the shore when sea level dropped several meters over about 15 minutes. People ventured out to pick up fish, inspect grounded vessels, and look around. Thousands were killed when the first large crest arrived, and would-be rescuers were drowned by later crests, some of which exceeded 10 meters (33 feet) in height. In all, more than 50,000 people in and around Lisbon died from the effects of the earthquake and tsunami.

There have been even greater tsunami disasters. In 1703 at Awa, Japan, 100,000 people died. On 27 August 1883 the enormous volcanic explosion of Krakatoa in Indonesia gener-

ated 35-meter (115-foot) waves that destroyed 163 villages and killed more than 36,000 people. Evidence of even larger waves has been discovered. Researchers have found signs of a huge wave, perhaps as high as 91 meters (300 feet), that crashed against the Texas coast 66 million years ago. It may have been caused by a comet or asteroid striking the Gulf of Mexico near Yucatán (see Box 13.1). The wave scoured the floor of the gulf; picked up sand, gravel, and sharks' teeth; and deposited the material in what is now central Texas.

The Tsunami Warning Network

Since 1948, an international tsunami warning network has been in operation around the seismically active Pacific to alert coastal residents of possible danger. Warnings must be issued rapidly because of the speed of these waves. Telephone books in coastal Hawaiian towns contain maps and evacuation instructions for use when the warning siren sounds.

The tsunami warning system was responsible for averting the loss of many lives after the great 27 March 1964 earthquake in Alaska (discussed in Chapter 3). A 3.7-meter (12-foot) wave, probably the fourth crest to reach the coast, swept into Crescent City, California, about six hours after the quake. Though more than 300 buildings were destroyed or damaged, five gasoline storage tanks set ablaze, and 27 blocks of the city demolished, there were relatively few casualties.

It has been over 30 years since a tsunami caused substantial damage in the United States. Some public safety experts suggest we have become complacent about the risks associated with these destructive waves. As can be seen in the photos in **Figure 11.11,** some coastal communities remind residents and visitors of the danger.

TIDES AND THE FORCES THAT GENERATE THEM

Tides are periodic, short-term changes in the height of the ocean surface at a particular place caused by a combination of the gravitational force of the moon and the sun and the motion

Figure 11.9 Massive property damage in the fishing village of Aonae on the island of Okushiri, Japan, was caused by a tsunami generated by the Hokkaido Nansei-Oki earthquake in 1993. Maximum wave height was 31 meters (101 feet); 239 people died. This event has become the best-documented tsunami disaster in history.

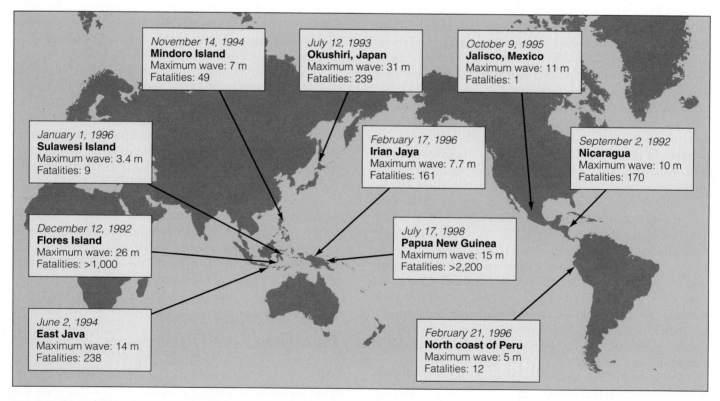

November 14, 1994
Mindoro Island
Maximum wave: 7 m
Fatalities: 49

July 12, 1993
Okushiri, Japan
Maximum wave: 31 m
Fatalities: 239

October 9, 1995
Jalisco, Mexico
Maximum wave: 11 m
Fatalities: 1

January 1, 1996
Sulawesi Island
Maximum wave: 3.4 m
Fatalities: 9

February 17, 1996
Irian Jaya
Maximum wave: 7.7 m
Fatalities: 161

September 2, 1992
Nicaragua
Maximum wave: 10 m
Fatalities: 170

December 12, 1992
Flores Island
Maximum wave: 26 m
Fatalities: >1,000

July 17, 1998
Papua New Guinea
Maximum wave: 15 m
Fatalities: >2,200

June 2, 1994
East Java
Maximum wave: 14 m
Fatalities: 238

February 21, 1996
North coast of Peru
Maximum wave: 5 m
Fatalities: 12

Figure 11.10 Ten destructive tsunami have claimed more than 4,000 lives since 1990.

a

b

Figure 11.11 (a) Tsunami hazard warning sign in a town in central coastal Oregon. The configuration of the coast and the slope and instability of the bottom in this area make tsunami particularly dangerous (see Figure 11.4). **(b)** An unfortunate effect of the tsunami warning network. Sightseers flocked to Oahu's Makapu'u Beach in Hawaii, awaiting a tsunami generated by a 6.5 earthquake centered in the Aleutian Trench on 7 May 1986.

of Earth (**Figure 11.12**). With a wavelength that can equal half of Earth's circumference, tides are the longest of all waves. Unlike the other waves we have met, these huge shallow-water waves are never free of the forces that cause them and thus are called *forced* waves; (After they are formed, wind waves, seiches, and tsunami are *free* waves; that is, they are no longer being acted upon by the force that created them and do not require a maintaining force to keep them in motion.)

The Greek navigator and explorer Pytheas first wrote of the connection between the position of the moon and the height of a tide around 300 B.C., but full understanding of tides had to await Isaac Newton's analysis of gravitation.

Among many other things, Isaac Newton's brilliant 1687 book *Philosophiae Naturalis Principia Mathematica* (*Mathematical Principles of Natural Philosophy*) described the motions of planets, moons, and all other bodies in gravitational fields. A central finding: The pull of gravity between two bodies is proportional to the masses of the bodies but inversely proportional to the square of the distance between them. This means that heavy bodies attract each other more strongly than light bodies do, and gravitational attraction weakens as the distance grows larger. This may be expressed mathematically as

$$F = G\left(\frac{m_1 m_2}{r^2}\right)$$

where F is the gravitational attraction, G is the universal gravitational constant, m_1 and m_2 are the masses of the two bodies, and r is the distance between their centers.

While the main cause of tides is the gravity of the moon and the sun acting on the ocean, the forces that actually generate the tides vary inversely with the *cube* of the distance from Earth's center to the center of the tide-generating object (the moon or the sun). Distance is thus even more important in this relationship, which may be expressed as

$$T = G\left(\frac{m_1 m_2}{r^3}\right)$$

where T is the tide-generating force, G is the universal gravitational constant, m_1 and m_2 are the masses of the two bodies, and r is the distance between their centers.[1] The sun is about 27 million times more massive than the moon, but the sun is about 387 times farther away than the moon, so the sun's influence on the tides is only about half that of the moon's.

As we will see, Newton's gravitational model of tides—the *equilibrium theory*—deals primarily with the position and attraction of Earth, the moon, and the sun and does not allow for the influence of ocean depth or the positions of continental land masses on tides. The equilibrium theory would accurately describe tides on a planet uniformly covered by water. A modification proposed by Pierre-Simon Laplace about a century later—the *dynamic theory*—takes into account the speed of the long-wavelength tide wave in relatively shallow water, the presence of interfering continents, and the circular movement or rhythmic back-and-forth rocking of water in ocean basins. We will explore the idealized situation of the equilibrium theory before moving to the real-world dynamic view.

---

[1] For a mathematical treatment of the tide-generating force, please see Appendix VII on page 509.

Figure 11.12 Tides in the eastern Bay of Fundy on the Atlantic coast of Canada. The tidal range is nearly 15 meters (50 feet). At the peak of the flood, water rises 1 meter (3.3 feet) in 23 minutes. These photos were taken near St. John, New Brunswick (see Figure 11.3b). Other places in the world have large tides: the Port of Bristol in England; the Sea of Okhotsk, northeast of Japan; Turnagain Arm in Alaska; the Bay of Saint-Malo in France; the Feuilles River in Ungava Bay, Quebec. All have tidal ranges of about 10 meters (33 feet). However, not as much water is moved at these locations as in the Bay of Fundy, where the tidal contribution equals the total daily discharge of all the world's rivers—about 100 cubic kilometers (24 cubic miles) of water.

THE EQUILIBRIUM THEORY OF TIDES

The **equilibrium theory of tides** explains many characteristics of ocean tides by examining the balance and effects of the forces that enable a planet to stay in a stable orbit around the sun, or the moon to orbit Earth. The equilibrium theory assumes the ocean conforms instantly to the forces affecting the position of its surface; that is, the ocean surface is presumed always to be in equilibrium (balance) with the forces acting on it.

The Moon and Tractive Forces

We begin our examination of these forces by looking at the moon's effect on the ocean surface. Gravity tends to pull Earth and the moon toward each other, but inertia—the tendency of moving objects to continue in a straight line—keeps them apart. (In this context, we sometimes call inertia *centrifugal "force,"* the "force" that keeps water against the bottom of a bucket when you swing it overhead in a vertical circle.) Earth and the moon don't smash into each other (or fly apart) because they are in a stable orbit; their mutual gravitational attraction is exactly offset by their inertia. Contrary to what you might think, the moon does not revolve around the center of Earth. Rather, the Earth–moon *system* revolves once a month (every 27.3 days) around the system's center of mass. Because Earth's mass is 81 times that of the moon, this common center of mass is located not in space, but 1,700 kilo-

meters (1,060 miles) *inside* Earth. This center of mass is shown in **Figure 11.13.**

Note in Figure 11.13a that the center of Earth is beginning to describe a circle around the center of mass. You can see that circle completed at the center of Earth in **Figure 11.14.** Because Earth and the moon revolve as a system, all points on and in Earth move in similar circles of the same radius; the circles described by four points on Earth are shown in Figure 11.14.

Now imagine the forces working on those points. In **Figure 11.15** we see the outward-flinging force of inertia. All the outward-directed force arrows are equal and parallel to one another. In **Figure 11.16** we see the inpulling force of gravity, directed toward the center of the moon. Unlike the centrifugal arrows of Figure 11.15, the arrows of Figure 11.16 are *not* parallel to one another. The inward pull of gravity and the outward-moving tendency of inertia (centrifugal "force") don't always act in exactly the same balanced way on each particle of Earth and the moon. In **Figure 11.17,** notice that points 1 and 2 are closer to the moon, so gravitational attraction at those points slightly exceeds the outward-moving tendency of inertia. Water there tends to be attracted toward the moon and so is pulled along the ocean surface toward a spot beneath the moon. At points 3 and 4, slightly farther from the moon, inertia exceeds gravitational attraction. Water at those points tends to be flung away from the moon and so moves along the ocean surface toward a spot opposite the moon.

The unbalanced forces doing the pulling and flinging are known as *tractive forces.* Together, the tractive forces cause two small bulges in the ocean, one in the direction of the

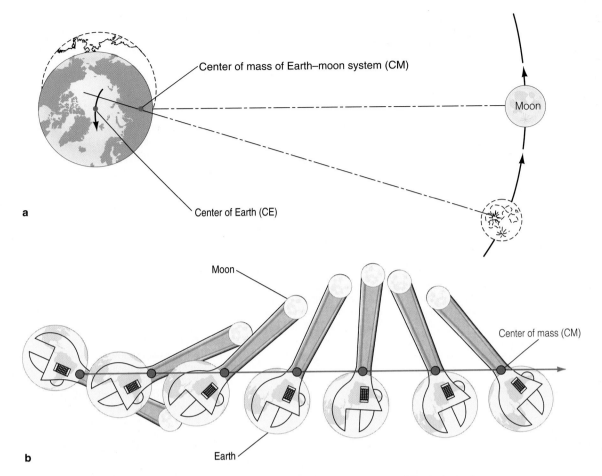

a

b

Figure 11.13 (**a**) The Earth–moon system does not revolve around the center of Earth, but rather around its common center of mass (CM). The moon revolves around Earth once a month, and through each month the center of Earth describes a circle in space around the center of mass. (The drawing is not to scale.) (**b**) The rotation of a thrown wrench is analogous to the motion of the Earth–moon system. Imagine that Earth's mass is located at the heavy head of the wrench and the moon is located at the end of the handle to visualize the motion of the system through space.

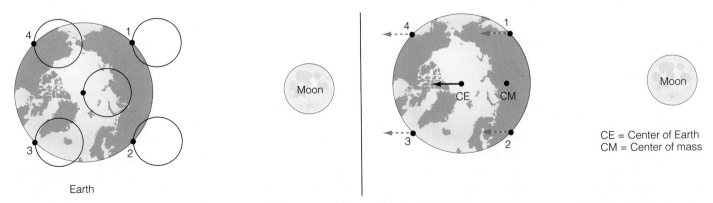

Figure 11.14 Because Earth and the moon revolve as a system, all points on and in Earth move monthly in a circle identical in size to the circle described by Earth's center. The movement of Earth's center and four points on Earth's surface are shown here.

Figure 11.15 The outward-flinging force of inertia (often called *centrifugal "force"*) at five places on and in Earth. Note that all the outward-directed force arrows are *equal* and *parallel* to one another.

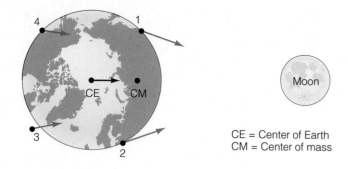

CE = Center of Earth
CM = Center of mass

Figure 11.16 The inpulling force of gravity, directed toward the center of the moon. Unlike the centrifugal arrows of Figure 11.15, these arrows are *not* equal and parallel to one another.

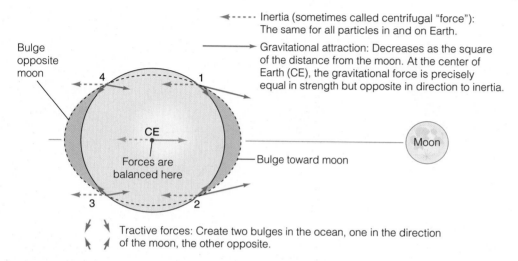

◄----- Inertia (sometimes called centrifugal "force"): The same for all particles in and on Earth.

───► Gravitational attraction: Decreases as the square of the distance from the moon. At the center of Earth (CE), the gravitational force is precisely equal in strength but opposite in direction to inertia.

Bulge opposite moon

CE

Forces are balanced here

Bulge toward moon

Moon

Tractive forces: Create two bulges in the ocean, one in the direction of the moon, the other opposite.

The two forces that can move the ocean are balanced only at the center of Earth (point CE). Elsewhere the net imbalance is a small force that causes ocean water to converge into two equal "bulges," as shown.

Figure 11.17 The actions of gravity and inertia on particles at five different points on Earth. At points 1 and 2, the gravitational attraction of the moon slightly exceeds the outward-moving tendency of inertia, and the imbalance of forces causes water to move along Earth's surface to converge at a point *toward* the moon. At points 3 and 4, inertia exceeds gravitational force, so water moves along Earth's surface to converge at a point *opposite* the moon. Forces are balanced only at the center of Earth (point CE).

moon, the other in the opposite direction. Note that there is no point on Earth's surface where the force of the moon's gravity exactly equals the outward-moving tendency of inertia. Only at point CE—the center of Earth—are the inward pull of gravity and the outward-moving tendency of inertia exactly equal and opposite. The solid Earth cannot move much in response to these forces, but the fluid atmosphere and ocean can. We don't notice the changes in the height of the atmosphere, but changes in water level are visible to coastal observers. In **Figure 11.18,** tractive forces pull water toward a point beneath the moon and to a point opposite the moon.

How do these bulges cause the rhythmic rise and fall of the tides? In the idealized equilibrium model we are discussing, the bulges tend to stay aligned with the moon as Earth spins around its axis. **Figure 11.19** shows the situation in Figure 11.18 as it would look from above the North Pole. As Earth turns eastward, an island on the equator is seen to move in and out of these bulges through one rotation (one day). The bulges are the crests of the planet-size waves that cause **high tides. Low tides** correspond to the troughs, the area between bulges. Starting at 0000 (midnight) we see the

island in shallow water at low tide. Around six hours later, at 0613 (6:13 A.M.), the island is submerged in the lunar bulge at high tide. At 1226 (about noon) the island is within the tide wave trough at low tide. At 1838 (6:38 P.M.) the island is again submerged, this time in the opposite crest caused by inertia. About an hour after midnight (0050) on the next day the island is back in shallow water, where it began.

The wave crests and troughs that cause high and low tides are actually very small: A 2-meter (7-foot) rise or fall in sea level is insignificant in comparison to the ocean's great size. Earth rotates beneath the bulges (tide wave crests) at about 1,600 kilometers (1,000 miles) per hour at the equator. The bulges appear to move across the ocean surface at this speed in an attempt to keep up with the moon. Theoretically, the wavelength of these tide waves is as long as 20,000 kilometers (12,500 miles)! The bulges tend to stay aligned with the moon as Earth spins around its axis. *The key to understanding the equilibrium theory of tides is to see Earth turning beneath these bulges.*

There are complications, of course. For example, the **lunar tides** (tides caused by gravitational and inertial interaction

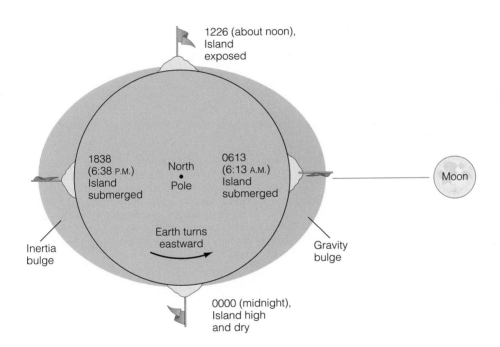

Water bulge resulting from inertia (centrifugal "force")

N

Water bulge resulting from gravitational attraction

S

Moon

Figure 11.18 The formation of tidal bulges at points toward and away from the moon. (This and the similar diagrams that follow are not drawn to scale.)

1226 (about noon), Island exposed

1838 (6:38 P.M.) Island submerged

North Pole

0613 (6:13 A.M.) Island submerged

Earth turns eastward

Inertia bulge

Gravity bulge

Moon

0000 (midnight), Island high and dry

Figure 11.19 How Earth's rotation beneath the tidal bulges produces high and low tides. Notice that the tidal cycle is 24 hours 50 minutes long because the moon rises 50 minutes later each day.

of the moon and Earth) complete their cycle in a tidal day (also called a lunar day). A complete tidal day is 24 hours 50 minutes long because the moon, which exerts the greatest tidal influence, rises 50 minutes later each day (**Figure 11.20**). Thus, the highest tide also arrives 50 minutes later each day.

Another complication arises from the fact that the moon does not stay right over the equator; each month, it moves from a position as high as $28\frac{1}{2}°$ above Earth's equator to $28\frac{1}{2}°$ below.[2] When the moon is above the equator, the bulges are offset accordingly (as in **Figure 11.21**). When the moon is

$28\frac{1}{2}°$ north of the equator, an island north of the equator will pass through the bulge on one side of Earth but miss the bulge on the other side. During one day the island passes through a very high tide, a low tide, a lower high tide, and another low tide. This is shown in **Figure 11.22.**

The Sun's Role

The sun's gravity also attracts particles on Earth. Remember, closeness counts for much in determining the strength of gravitational attraction. As we saw earlier, the sun is about 27 million times more massive than the moon, but the sun is about 387 times farther away than the moon, so the sun's

---

[2] If Earth, the moon, and the sun were all moving in the same plane, lunar and solar eclipses would happen every two weeks.

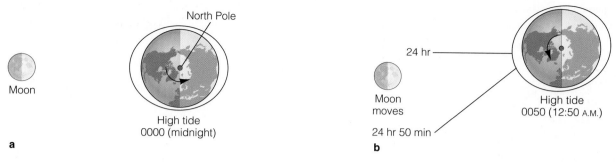

a

Moon

North Pole

High tide
0000 (midnight)

24 hr

Moon
moves

24 hr 50 min

High tide
0050 (12:50 A.M.)

b

Figure 11.20 (**a**) A high tide at midnight. (**b**) The high tide one day later. Earth rotates once a day, and the Earth–moon system revolves around its center of mass once a month. Both turn in the same direction (eastward). During one complete Earth rotation (one day), the moon moves eastward 12.2°. So Earth must rotate for an additional 50 minutes for the moon to reach the same place in the sky on consecutive days. Because Earth must turn an additional 50 minutes for the same tidal alignment, lunar tides usually arrive 50 minutes later each day.

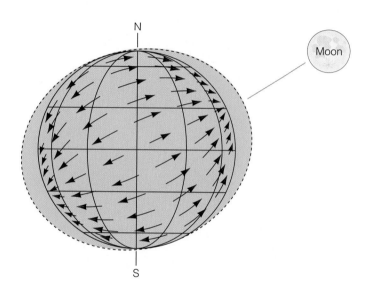

Moon

Figure 11.21 The bulges follow the moon. When the moon's position is north of the equator, the bulge toward the moon is also located north of the equator. The opposite inertia bulge is correspondingly below the equator. (Compare with Figure 11.18.) At intervals of about 18½ years the plane of the moon's orbit is inclined 28½° relative to the plane of Earth's equator, so the moon can sometimes appear directly overhead as far north as Florida or southern Texas. Lunar tides are extreme at such times.

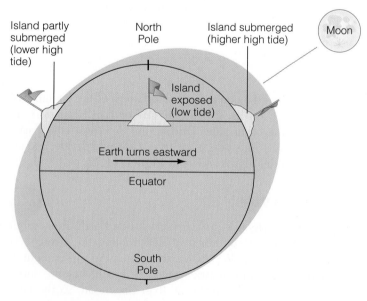

Island partly submerged (lower high tide)

North Pole

Island submerged (higher high tide)

Moon

Island exposed (low tide)

Earth turns eastward

Equator

South Pole

Figure 11.22 How the changing position of the moon relative to Earth's equator produces higher and lower high tides. Sometimes the moon is below the equator, and sometimes it is above.

influence on the tides is only about half that of the moon's. The sun's tractive forces develop in the same way as the moon's, and the smaller solar bulges tend to follow the sun through the day. These are the **solar tides,** caused by the gravitational and inertial interaction of the sun and Earth.

Like the moon, the sun also appears to move above and below the equator ($23\frac{1}{2}°$ north to $23\frac{1}{2}°$ south, as you may recall from Chapter 8), so the position of the solar bulges varies like that of the lunar bulges. Earth revolves around the sun only once a year, however, so the position of the solar bulges above or below the equator changes much more slowly than does the position of the lunar bulges. (Figure 8.4, used to explain the cause of the seasons, shows this well.)

Sun and Moon Together

The ocean responds simultaneously to inertia and to the gravitational force of both the sun and the moon. If Earth, the moon, and the sun are all in a line (as shown in **Figure 11.23a**), the lunar and solar tides will be additive, resulting in higher high tides and lower low tides. But if the moon, Earth, and the sun form a right angle (as shown in **Figure 11.23b**), the solar tide will tend to diminish the lunar tide. Because the moon's contribution is more than twice that of the sun, the solar tide will not completely cancel the lunar tide.

The large tides caused by the linear alignment of the sun, Earth, and the moon are called **spring tides** (*springen* = to move quickly). During spring tides, high tides are very high, and low tides very low. These tides occur at two-week intervals corresponding to the new and full moons. (Please note that spring tides don't happen only in the spring of the year.) **Neap tides** (*naepa* = hardly disturbed) occur when the moon, Earth, and the sun form a right angle. During neap tides, high tides are not very high, and low tides not very low. Neap tides also occur at two-week intervals, with the neap tide arriving a week after the spring tide. **Figure 11.24** plots tides at two coastal sites through spring and neap cycles.

Because their orbits are ellipses, not perfect circles, the moon and the sun are closer to Earth at certain times than at others. The difference between **apogee** (the moon's most distant point) and **perigee** (its closest) is 30,600 kilometers (19,015 miles). Because the tidal force is inversely proportional to the cube of the distance between the bodies, the closer moon raises a noticeably higher tidal crest. The difference between **perihelion** (Earth's closest approach to the sun) and **aphelion** (its greatest distance) is 3.7 million kilometers (2.3 million miles). If the moon and the sun are over nearly the same latitude and if Earth is also close to the sun, extreme spring tides will result.

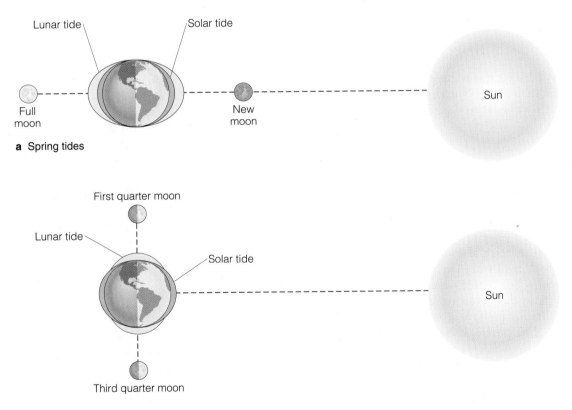

Figure 11.23 Relative positions of the sun, the moon, and Earth during spring and neap tides. (**a**) At the new and full moons, the solar and lunar tides reinforce each other, making spring tides, the highest high and lowest low tides. (**b**) At the first and third quarter moons, the sun, Earth, and the moon form a right angle, creating neap tides, the lowest high and highest low tides.

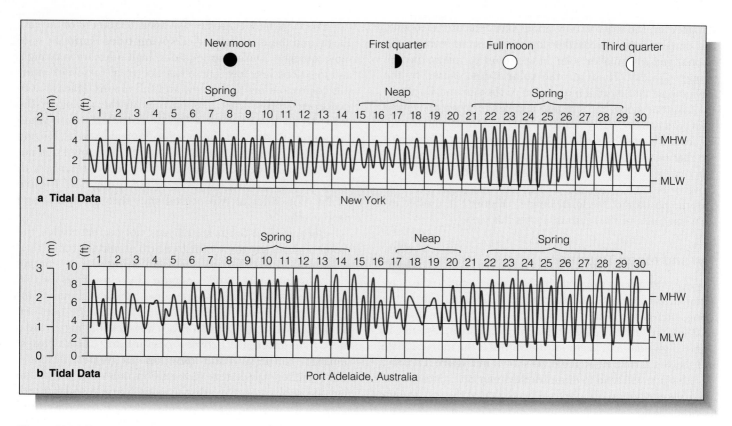

Figure 11.24 Tidal records for a typical month at (**a**) New York and (**b**) Port Adelaide, Australia. Note the relationship of spring and neap tides to the phases of the moon. MHW = mean high water; MLW = mean low water.

Interestingly, spring tides will have greater ranges in the Northern Hemisphere winter than in the Northern Hemisphere summer because Earth is closest to the sun during the northern winter.

THE DYNAMIC THEORY OF THE TIDES

Newton knew his explanation was incomplete. For one thing, the maximum theoretical range of a lunar tidal bulge is only 55 centimeters (about 22 inches), and that of a solar tide only 24 centimeters (about 10 inches), both considerably smaller than the 2-meter (7-foot) average tidal range we observe in the world ocean. This is because the ocean surface never comes completely to the equilibrium position at any instant. The moon and the sun change their positions so rapidly that the water cannot keep up.

The **dynamic theory** of the tides, first proposed in 1775 by Laplace, added a fundamental understanding of the problems of fluid motion to Newton's breakthrough in celestial mechanics. The dynamic theory explains the differences between predictions based on Newton's model and the observed behaviors of tides.

Remember that tides are a form of *wave*. The crests of these waves—the tidal bulges—are separated by a distance of half of Earth's circumference (see again Figure 11.19). In the equilibrium model, the crests would remain stationary, pointing steadily toward (or away from) the moon (or the sun) as Earth turned beneath them. They would appear to move across the idealized water-covered Earth at a speed of about 1,600 kilometers (1,000 miles) per hour. But how deep would the ocean have to be to allow these waves to move freely? For a tidal crest (or tide wave) to move at 1,600 kilometers per hour, the ocean would have to be 22 kilometers (13.7 miles) deep. As you may recall, the average depth of the ocean is only 3.8 kilometers (2.4 miles). So tidal crests (tidal bulges) move as forced waves, and their velocity is determined by ocean depth.

Tidal Patterns, Amphidromic Points

This behavior of tide waves as *shallow-water waves* is only one variation from the ideal that the dynamic theory explains. The continents also get in the way. As Earth turns, land masses obstruct the tidal crests, diverting, slowing, and otherwise complicating their movements. This interference produces different patterns in the arrival of tidal crests at different places. Imagine, for example, a continent directly facing the moon. There would be no oceanic bulge, and the shores of the continent would experience high tide. A few hours later the moon would be over the ocean. When the continent is not aligned with the moon but the ocean is, the tidal bulge would re-form and the continent's edges would experience low tide.

The shape of the basin itself has a strong influence on the patterns and heights of tides. As we have seen, water in large basins can rock rhythmically back and forth in seiches. Though they are small, tidal crests can stimulate this resonant oscillation, and the configuration of coasts around a basin can alter its rhythm.

For these and other reasons, some coastlines experience **semidiurnal** (twice daily) **tides**: two high tides and two low tides of nearly equal level each lunar day. Others have **diurnal** (daily) **tides**: one high and one low. The tidal pattern is called a **mixed tide** if successive high tides or low tides are of significantly different heights through the cycle.[3] **Figure 11.25** gives an example of each tidal pattern. At Cape Cod two tidal crests arrive per lunar day, a semidiurnal pattern (Figure 11.25a). The natural tendency of water in an enclosed ocean basin to rock at a specific frequency modifies the pattern in the Gulf of Mexico, so New Orleans sees one crest per lunar day, a diurnal pattern (Figure 11.25b). The Pacific coast of the United States has a mixed tidal pattern: often a *higher* high tide, followed by a *lower* low tide, a *lower* high tide, and a *higher* low tide each lunar day. (That's not as confusing as it first sounds—see Figure 11.25c.)

The Pacific Ocean has a unique pattern of diurnal, semidiurnal, and mixed tides. As **Figure 11.26** shows, it has the most complex of all tidal patterns. The east coast of Australia, all of New Zealand, and much of the west coasts of Central and South America have a semidiurnal tidal pattern. The Aleutians have diurnal tides. The Pacific coasts of North America and some of South America have mixed tides. *Why the differences?*

Remember the surface of Lake Geneva in our discussion of seiches? The water level at the center of the lake remains at the same height while water at the ends rises and falls (see again Figure 11.3). The long axis of the lake stretches east and west. Because of the Coriolis effect, water moving east at the center

of the lake is deflected slightly to the right (to the south). If the lake were larger and the flow of water greater (and thus the Coriolis effect stronger), water would hug the southern shore as it traveled eastward. When the water began to rock the other way—back to the west—the Coriolis effect would move the water to the right toward (and along) the northern shore. Note that the overall movement of water would be counterclockwise.

Water moving in a tide wave tends to stay to the right of an ocean basin for the same reason. As water moves north in a Northern Hemisphere ocean, it moves toward the eastern boundary of the basin; as it moves south, it moves toward the western boundary. A wave crest moving counterclockwise will develop around a node if this motion continues to be stimulated by tidal forces. This rotary motion is shown in **Figure 11.27**.

The node (or nodes) near the center of an ocean basin is called an **amphidromic point** (*amphi* = around + *dromas* = running). An amphidromic point is a no-tide point in the ocean around which the tidal crest rotates through one tidal cycle. Because of the shape and placement of land masses around ocean basins, the tidal crests and troughs cancel each other at these points. The crests sweep around amphidromic points like wheel spokes from a rotating hub, radiating crests toward distant shores. Tide waves are influenced by the Coriolis effect because a large volume of water moves with the wave; they move counterclockwise around the amphidromic point in the Northern Hemisphere, and clockwise in the Southern Hemisphere. The height of the tides increases with distance from an amphidromic point.

About a dozen amphidromic points exist in the world ocean; **Figure 11.28** shows their location. Notice the complexity of the Pacific, which contains five. It's no wonder that the arrival of tidal crests at the Pacific's edge produces such a complex mixture of tide patterns, depending on shoreline location.

---

[3] Mixed tides are also called *mixed semidiurnal*.

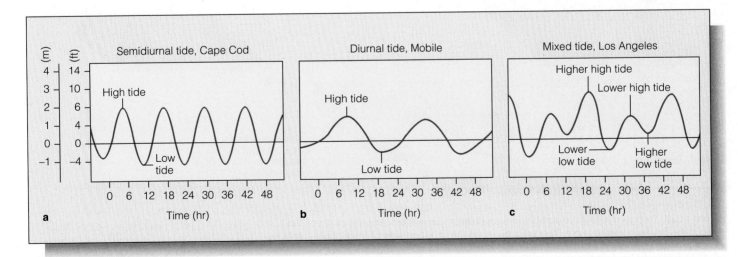

Figure 11.25 Tide curves for the three common types of tides. (**a**) A semidiurnal tide pattern at Cape Cod, Massachusetts. (**b**) A diurnal tide pattern at Mobile, Alabama. (**c**) A mixed tide pattern at Los Angeles, California.

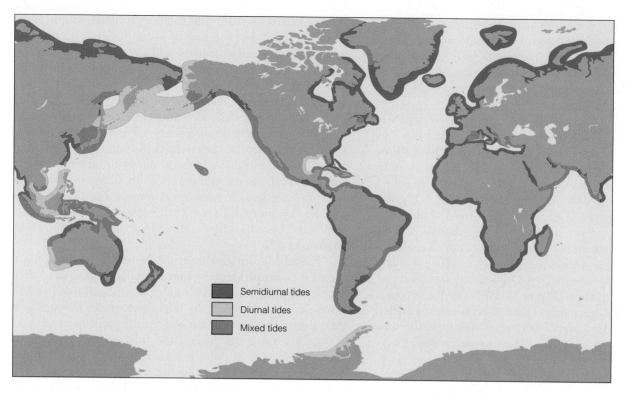

Figure 11.26 The worldwide geographic distribution of the three tidal patterns. Most of the world's ocean coasts have semidiurnal tides.

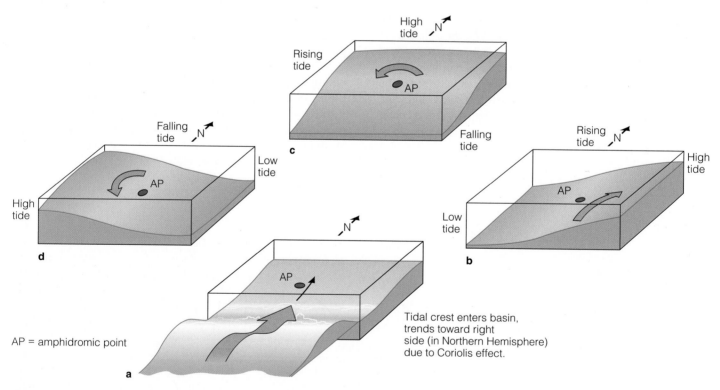

AP = amphidromic point

Tidal crest enters basin, trends toward right side (in Northern Hemisphere) due to Coriolis effect.

Figure 11.27 The development of amphidromic circulation. (**a**) A tidal crest enters an ocean basin in the Northern Hemisphere. The wave trends to the right because of the Coriolis effect (**b**), causing a high tide on the basin's eastern shore. Unable to continue turning to the right because of the interference of the shore, the crest moves northward, following the shoreline (**c**) and causing a high tide on the basin's northern shore. The wave continues its progress around the basin in a counterclockwise direction (**d**), forming a high tide on the western shore and completing the circuit. The point around which the crest moves is an amphidromic point (AP).

Figure 11.28 Amphidromic points in the world ocean. Tidal ranges generally increase with increasing distance from amphidromic points. Lines radiating from the points indicate tide waves moving around these points, counterclockwise in the Northern Hemisphere and clockwise in the Southern Hemisphere. The numbers represent the positions of a hypothetical tidal crest in passing hours.

Tidal Datum

The reference level to which tidal height is compared is called **tidal datum.** The tidal datum is the zero point (0.0) seen in tide graphs such as Figures 11.24 and 11.25. This reference plane is not always set at **mean sea level,** the height of the ocean surface averaged over a few years' time. On coasts with mixed tides, the zero tide level is the average level of the lower of the two daily low tides (*mean lower low water*, or MLLW). On coasts with diurnal and semidiurnal tides, the zero tide level is the average level of all low tides (*mean low water*, or MLW).

Tides in Confined Basins

The **tidal range** (high-water to low-water height difference) varies with basin configuration. In small areas such as lakes, the tidal range is small. In larger enclosed areas such as the Baltic or Mediterranean Sea, tidal range is also moderate. The tidal range is not the same over a whole ocean basin; it varies from the coasts to the centers of oceans. The largest tidal ranges occur at the edges of the largest ocean basins, especially in bays or inlets that concentrate tidal energy because of their shape.

If a basin is wide and symmetrical, like the Gulf of St. Lawrence in eastern Canada (**Figure 11.29**), a miniature amphidromic system develops that resembles the large systems of the open ocean. If the basin is narrow and restricted,

the tidal crest cannot rotate around an amphidromic point and simply moves into and out of the bay (**Figure 11.30**). Extreme tides occur in places where arriving tidal crests stimulate natural oscillation periods of around 12 or 24 hours. In rare cases, water in the bay naturally resonates (seiches) at the same frequency as the lunar tide (12 hours 25 minutes). This rhythmic sloshing results in extreme tides. In the eastern reaches of the Bay of Fundy near Moncton, New Brunswick (Canada), the tidal range is especially wide: up to 15 meters (50 feet) from highs to lows (see **Figure 11.31**). The northern reaches of the Sea of Cortez east of Baja California have a tidal range of about 9 meters (30 feet). Tide waves sweeping toward the narrow southern end of the North Sea can build to great heights along the southeast coast of England and the north coast of France.

A **tidal bore** (*bara* = wave) will form in some inlets (and their associated rivers) exposed to great tidal fluctuation. Here, at last, is a true **tidal wave**—a steep wave moving upstream generated by the action of the tidal crest in the enclosed area of a river mouth (**Figure 11.32**). The confining river mouth forces the tide wave to move toward land at a speed that exceeds the theoretical shallow-water wave speed for that depth. The forced wave then breaks, forming a spilling wave front that moves upriver. Though most are less than 1 meter (3 feet) high, bores on China's Qiantang River may be up to 8 meters (26 feet) high and move at 11 meters

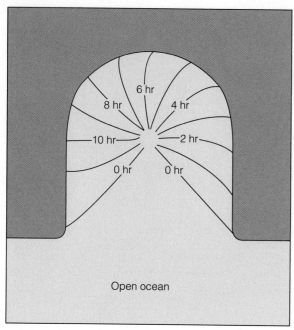

a Broad basin

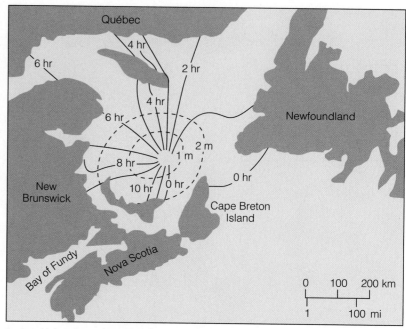

b Amphidromic system: Gulf of St. Lawrence. Dashed lines show tide height

Figure 11.29 Tides in broad confined basins. (**a**) An imaginary amphidromic system in a broad, shallow basin. The numbers indicate the hourly positions of tidal crests as a cycle progresses. (**b**) The amphidromic system for the Gulf of St. Lawrence between New Brunswick and Newfoundland, southeastern Canada. Compare the situation in this limited basin with the tides surrounding the amphidromic point in the open ocean south of Greenland in Figure 11.28.

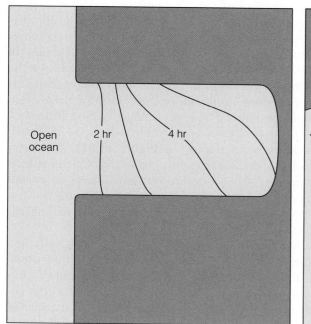

a Narrow basin

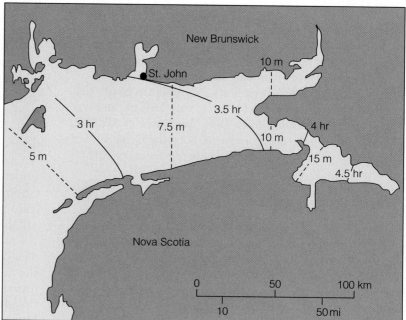

b Tidal crests: Bay of Fundy

Figure 11.30 Tides in narrow restricted basins. (**a**) True amphidromic systems do not develop in narrow basins because space for rotation is not available. (**b**) Tides in the Bay of Fundy, Nova Scotia, are extreme because water in the bay naturally resonates (seiches) at the same frequency as the lunar tide. (See also Figure 11.12.)

a

b

Figure 11.31 Tides in the eastern Bay of Fundy on the Atlantic coast of Canada. (**a**) Tidal range is near 15 meters (50 feet). (**b**) At the peak of the flood, water rises 1 meter (3.3 feet) in 23 minutes.

per second (25 miles per hour). Their potential danger is lessened by their predictability. Accurately predicting the arrival of tidal bores is essential to safe navigation. In addition to the Bay of Fundy, tidal bores are common in the Amazon, the Ganges Delta, and England's Severn River. Some rivers in southern China may have three or four simultaneous tidal bores at different places along the river's length.

Figure 11.32 A tidal bore on the Severn River in southwestern England. A tidal crest sweeping up the funnel-shaped Bristol channel from the Atlantic Ocean encounters the increasingly shallow and narrow river passage, and it rushes forward to cause the bore. Like all tidal bores, the Severn bore reaches its greatest height during the days surrounding spring tides. The waves that form the bore in this example are nearly 1 meter (3.3 feet) high and are traveling at about 5 meters per second (11 miles per hour). Note that the smooth waves are followed by turbulent water also moving rapidly upriver. The tidal bore on the Severn River is large enough for surfing, and accomplished board surfers can ride upstream for many miles.

Tidal Currents

The rise or fall in sea level as a tidal crest approaches and passes will cause a **tidal current** of water to flow into or out of bays and harbors. Water rushing into an enclosed area because of the rise in sea level as a tidal crest approaches is called a **flood current.** Water rushing out because of the fall in sea level as the tidal trough approaches is called an **ebb current.** (The terms *ebb tide* and *flood tide* have no technical meaning.) Tidal currents reach maximum velocity midway between high tide and low tide. **Slack water,** a time of no currents, occurs at high and low tide when the current changes direction.

Anyone who has stood at the narrow mouth of a large bay or harbor cannot help but be impressed with the speed and volume of the tidal current that occurs between tidal extremes. Midway between high and low spring tides, the ebb current rushing from San Francisco Bay strikes the base of the south tower of the Golden Gate Bridge with such force that a bow wave is formed, giving the convincing illusion that the bridge itself is moving rapidly. Tidal currents at the Golden Gate can reach 3 meters per second (about 7 miles per hour) because of the volume of enclosed water and the narrowness of the channel through which it must escape. Navigators must know the times of tidal currents to safely negotiate any harbor entrance or other narrow strait—in some places, this knowledge may save their lives (see **Box 11.1**).

Tidal currents become more complex in the open sea. One's position relative to an amphidromic point, the shape of the basin, and the magnitude of gravitational forces and inertia must all be considered in order to calculate the speed and direction of tidal currents over a deep bottom. The velocity of tidal currents is less in the open sea because the water is not confined, as it is in a harbor. Open-sea tidal currents have been measured at a few centimeters per second, and their velocity tends to decrease with depth.

According to Homer's account in *The Odyssey*, across Ulysses' homeward path lay a lethal whirlpool named Charybdis:

> We sailed up the narrow strait. On one side was shining Charybdis, who made her terrible ebb and flow of the sea's water. When she vomited it up, like a caldron over a strong fire, the whole sea would boil up in turbulence, and the foam flying spattered the pinnacles of the rocks in either direction; but when in turn again she sucked down the sea's salt water, the turbulence showed all the inner sea, and the rock around it groaned terribly, and the ground showed at the sea's bottom, black with sand. My crew turned sallow with fright, staring into this abyss from which we expected our immediate death.

For millennia, the prospect of navigating past Charybdis terrified sailors in the Strait of Messina, which separates Italy and Sicily. Its activity was greatly reduced in 1908 when the great Calabria–Messina earthquake and tsunami altered the geology of the area.

Edgar Allan Poe wrote about a whirlpool called the Maelstrom, which lies among the southern Lofoten Islands off Norway's west coast. The rapid spinning of water in the Maelstrom raises its outer edge and depresses the central core. Here is Poe's description, from his 1841 story "A Descent into the Maelstrom":

> The edge of the whirl was represented by a broad belt of gleaming spray; but no particle of this slipped into the mouth of the terrific funnel, whose interior, as far as the eye could fathom it, was a smooth, shining, and jet-black wall of water, inclining to the horizontal at an angle of some forty-five degrees, speeding dizzily round and round with swaying and sweltering motion, and sending forth to the winds an appalling voice, half shriek, half roar, such as not even the mighty cataract of Niagara ever lifts up in its agony to Heaven.

Arthur Rackham

a Maelstrom: fiction. A representation by the early-twentieth-century English illustrator Arthur Rackham.

BODØ Arrangement, Norway

b Maelstrom: fact. This maelstrom, the world's strongest, is located in the strait between the Norwegian islands of Moskenes and Moskenesøya.

Do these whirlpools genuinely deserve the dread they inspired, or is their reputation largely a figment of artistic imagination (**Figure a**)?

Whirlpools arise in shallow, restricted straits connecting two large bodies of deep water. They are caused by the tides. If the large bodies of water move to different tidal cycles, high tide in one will not occur at the same time as high tide in the other. Turbulence develops when one mass of water tries to pass the other during changes in tide. Under ideal conditions, bottom contours and the rush of opposing tidal currents can cause seawater in the confined area to spin vigorously. The larger the opposing bodies of water and the smaller the passage through which the confined water must pass, the greater the vortex caused by the tidal currents.

Maelstrøm is a Norwegian word derived from the Dutch *malen* (to grind in a circle, as a millstone grinds grain) and *strøm* (stream). Today's maelstrom is a heaving mass of active tidal water connecting Vest Fjord and the Norwegian Sea between the islands of Moskenesøya and Moskenes. Look for the area on a chart at 67°48'N, 12°50'E, about 165 kilometers (100 miles) north of the Arctic Circle. Currents in the passage can reach a speed of 13 kilometers (8 miles) per hour when the tides change, and strong local winds make the passage even more dangerous. The shallow "saddle" separating the large bodies of water to the east and west is only about 20 meters (70 feet) deep, and water rushing over its rocky bottom is driven to spin.

But is the chaos as horrifying as Homer, Poe, Jules Verne, and others have suggested? Well, no. Fishing boats in this rich fishing region carefully avoid the area during times of maximum tidal difference, and no sailor likes to risk his or her life in turbulent confined waters during times of high winds. But, as can be seen in **Figure b,** the sucking, roaring, cavernous maw of the maelstrom is more fiction than fact.

Tidal Friction

The daily rise and fall of the tides consumes a very large amount of energy, and this energy is ultimately dissipated as heat. Most of this energy comes directly from the rotation of Earth itself, and tidal friction is gradually slowing Earth's rotation by a few hundredths of a second per century. Even such a small change has long-term planetary effects, however. Geologists studying the daily growth rings of fossil corals and clams estimate that the length of the day has grown longer, so the number of days in a year has become less as planetary rotation has slowed. Evidence suggests that 350 million years ago a year contained between 400 and 410 days, with each day being about 22 hours long; 280 million years ago there were about three hundred ninety 22½-hour days in a year.

Tidal friction affects other bodies. Tidal forces have locked the rotation of the moon to that of Earth. As a result, the same side of the moon is always facing Earth, and a day on the moon is a month long.

Predicting Tides

There are at least 140 tide-generating and tide-altering forces and factors in addition to the ones previously discussed. About seven of the most important of these must be considered if we wish to predict tides mathematically. The interactions of all the forces and factors are so complex that if a previously unknown continent were discovered on Earth the coastal tide times and ranges on its shores could not be accurately predicted. It is the study of past records that enables tide tables to be projected into the future. Experience permits prediction of tidal height to an accuracy of within about 3 centimeters (1.2 inches) for years in advance.

Even so, extraneous factors can affect the estimates. For example, arrival of a storm surge will greatly affect the height or timing of a tide—as will gentle, atmospherically induced seiching of the basin or excitement of large-scale resonances by a tsunami. Even a strong, steady wind onshore or offshore will affect tidal height and the arrival time of the crest. Weather-related alterations are sometimes called **meteorological tides** after their origin.

TIDES AND MARINE ORGANISMS

Not surprisingly, as pervasive a phenomenon as the tides has a significant influence on coastal marine life. Organisms living between the high-tide and low-tide marks experience very different conditions from those residing below the low-tide line. Within the intertidal zone itself, organisms are exposed to varying amounts of emergence and submergence. Because some organisms can tolerate many hours of exposure while others are able to tolerate only a very few hours per week or month, the animals and plants sort themselves into three or more horizontal bands, or subzones, within the intertidal zone. Each distinct zone is an aggregation of animals and plants best adapted to the conditions within that particular narrow habitat. The zones are often strikingly different in appearance, even to a person unfamiliar with shoreline characteristics. (This zonation is clearly evident in the rocky shore of Figure 16.7b.)

Less obvious are periodic visitors that time their arrivals and departures to the rise and fall of the tide. At low tide, the tiny diatom (one-celled alga) *Hantzschia* migrates upward through wet sand to photosynthesize at the sunlit surface. At the first hint of the return of the tide, these plantlike organisms descend to a relatively safe depth in the sand, protected from wave action. Fiddler crabs (genus *Uca*) return to their burrows at high tide to avoid marine predators but emerge at low tide to look for any bits of food the ocean might have deposited at their doorsteps. Filter-feeding animals such as sand crabs (*Emerita*) and bean clams (*Donax*) migrate up and down the beach to stay in the surf zone, where the chaos can provide both food and protection (see Figure 16.9a).

Among the most famous tide-driven visitors are the grunion (*Leuresthes*), small fish named after the Spanish word *gruñion* ("grunter"), a reference to the squeaking noise they sometimes make during spawning. From late February through early September, these small fish (which reach a length of 15 centimeters, or 6 inches) swim ashore at night in large numbers just after the highest spring tides, deposit and fertilize their eggs below the sand surface, and return to the sea (**Figure 11.33**). Safe from marine predators, the eggs develop and are ready to hatch nine days after spawning. After a few more days, the spring tides return, soak and erode the beach, and stimulate the eggs to hatch. No one is certain how grunion

time their reproductive behavior so precisely to the tidal cycle, but some research suggests that they sense very small changes in hydrostatic pressure caused by tidal change or visually time their spawning to three or four nights following each full and new moon. Not surprisingly, these accommodating little fish were an important food for Native Americans.

POWER FROM THE TIDES

Humans have found ways to use the tides. Ships sail to sea and return to port with the tides. Intentional grounding of a ship with the fall of a tide can provide a convenient, if temporary, drydock. To these traditional uses has been added a potential alternative to our growing dependence on fossil fuels: taking advantage of trapped high-tide water to generate electricity.

Tidal power is the only marine energy source that has been successfully exploited on a large scale. The first major tidal power station was opened in 1966 in France on the estuary of the river Rance (see **Figure 11.34**), where tidal range reaches a maximum of 13.4 meters (44 feet). Built at a cost of $75 million, this 850-meter- (2,800-foot-) long dam contains 24 turbo-alternators capable of generating 544 million kilowatt-hours of electricity annually. At high tide, seawater flows from the ocean through the generators into the estuary. At low tide the seawater and river water from the estuary flow out through the same generators. Power is generated in both directions. A smaller but similar installation is

Cabrillo Marine Aquarium

Figure 11.33 Students enjoying a grunion run. During spring and summer months these small fish (genus *Leuresthes*) swim ashore at night in large numbers just after the highest spring tides, deposit and fertilize their eggs below the sand surface, and return to the sea. Nearly two weeks later, when spring tides return, the eggs hatch. No one is certain how grunion time their reproductive behavior so precisely to the tidal cycle. The grunion are found only along the Pacific coast of North America and in the Gulf of California. Unlike the Pacific coast species, Gulf of California grunion spawn during daylight.

Figure 11.34 Tidal power installation at the Rance estuary in western France.

generating power on the Annapolis River in Nova Scotia. A much larger generating facility has been proposed for Passamaquoddy Bay, a part of the Bay of Fundy between Maine and New Brunswick, Canada.

Tidal power has many advantages: Operating costs are low, the source of power is free, and no carbon dioxide or other pollutants are added to the atmosphere. But even if tidal power stations were built at every appropriate site worldwide, the power generated would amount to less than 1% of current world needs. And, of course, this method of power generation is not free of trade-offs. The dam and electrical generators can be damaged by storms, and the large, finely made metal valves and vanes at the heart of the plant are easily corroded by seawater. Computer simulations have suggested that installing a dam would change the resonance modes of a bay or estuary—and therefore the height of the tide wave. Studies also suggest that sensitive planktonic and benthic marine life would be disrupted and even that increased tidal friction would cause a tiny decrease in the rate of Earth's rotation.

QUESTIONS FROM STUDENTS

1. One of my friends has discovered I'm studying waves. He demands to know about the world's largest wave. What's the record, anyway?

If your friend classifies the largest wave by wavelength, the answer is easy: Tides take the prize, with a wavelength that is ideally half of Earth's circumference. Tides also win the speed race: Under ideal circumstances, their crests can move at speeds up to 1,600 kilometers (1,000 miles) per hour.

If your friend classifies the largest wave by height, things become more complicated. As you read in Chapter 10, the crew of *Ramapo* measured a wave 34 meters (112 feet) high in 1933. In general, the highest wind waves probably develop in the Antarctic Circumpolar Current, the nasty area of windy ocean ringing Antarctica. Multiple storms there, some with huge fetch distances, can cause very large waves and opportunity for constructive interference between them. In 1916 during an agonizing small-boat trip from an island on which their expedition was marooned, Frederic Worsley, captain of Ernest Shackleton's ship *Endurance*, encountered occasional tremendous waves greater than any he had encountered in a lifetime at sea. He estimated a few to be more than 30 meters (100 feet) high!

Some tsunami may be higher, however. As we have seen, landslides in Lituya Bay, Alaska, have generated swash waves in the 50-meter (160-foot) range. And a wave of about 530 meters (1,740 feet) formed briefly as water surged up the opposite side of the bay on 9 July 1958. (It was more a titanic splash than a classic tsunami; one writer has likened the effect to dropping a sack of concrete into a half-filled bathtub.) The greatest height recently recorded for a tsunami in the ocean was 84 meters (278 feet) in 1971 off Ishigaki, Japan. But don't forget the 91-meter (300-foot) tsunami thought to have occurred 66 million years ago on the coast of what is now Texas.

2. Are there tides in the solid Earth?

Yes. Even Earth isn't stiff enough to resist the tidal pulls of the moon and sun. Bulges occur in solid Earth just as they appear in the ocean or the atmosphere. The crests (bulges) of Earth are, of course, much smaller; 25 to 30 centimeters (10 to 12 inches) is about average. They pass unnoticed beneath us twice a day. On the other hand, tidal variability in the height of the atmosphere has been measured in miles.

3. In my newspaper one of today's low tides is listed as −1.0, 1445. What does that mean?

In the United States, it means that the water will be 1.0 feet below tidal datum (that is, below mean lower low water, MLLW—the long-term average position of the lower of the daily low tides) at 2:45 P.M. local time. This might be a good afternoon to spend at the shore digging for clams, because a tide that low would expose intertidal organisms only rarely seen above water. See Figure 16.7 for more information on the relationship between tidal height and exposure.

4. Are there tides in a glass of water?

Yes. They're too small to detect, but equilibrium tides do exist in very small bodies of water. Each molecule of water in the glass responds to the same planetary forces that affect molecules in the ocean.

5. Will people in California be in great danger from a seismic sea wave when a major earthquake occurs on the San Andreas Fault?

Probably not, for two reasons. First, San Andreas quakes are usually lateral-displacement quakes: The ground moves suddenly sideways rather than up or down. Tsunami are usually generated by vertical movement. Second, the motion along the San Andreas Fault will be parallel to the coast, not at right angles to it. If the ground movement were toward the coast (as could happen north of San Francisco), the "shove" might result in a wave. There may be some slopping about at the coastline during a large earthquake, but probably not a classic tsunami.

6. You mentioned that the process of extracting power from the tides slows Earth's rotation. What would happen if all the world's electrical power needs could somehow be met by tidal generating plants?

If we assume it could actually be done, and if we assume a constant energy demand of today's requirement of 2×10^{20} joules per year (somewhat unrealistic in light of the recent strong rise in electrical power generation), the length of a day would increase by an additional second in roughly a million and a half years. Clearly the "cashing in" of Earth's rotational kinetic energy to supply energy needs is not of great concern. A much more serious hazard would arise in the disruption of ocean currents, biological cycles, and esthetic values.

KEY CONCEPTS TO REVIEW

- The great waves described in this chapter have such long wavelengths that they are always in "shallow water" (in water that is shallower than half their wavelength). In wavelength order from shortest to longest, they are seiches, tsunami, and tides. (Technically, a storm surge is not a wave because it has only a crest.)

- Because wave speed is proportional to wavelength, these great waves move rapidly through the ocean. Tides move fastest of all—up to 1,600 kilometers (1,000 miles) per hour.

- A seiche is a pendulum-like rocking of water in a basin. Its resonant frequency depends on the dimensions of the basin. Seiches take the form of standing waves.

- Tsunami are caused by displacement of water by the forces that cause earthquakes, by landslides, by eruptions, or by asteroid impacts. Typically they resemble a fast-onrushing tide, not a breaking wave. They are not dangerous in the open sea.

- Tides are caused by the gravitational attraction of the sun and the moon, by inertia, and by basin resonances. The equilibrium theory of tides explains tides by analyzing the gravity and inertia of heavenly bodies, while the dynamic theory of tides extends this understanding to problems of fluid motion in enclosed basins.

- Tidal patterns—the heights and arrival times of tidal crests—at a continent's edge may be semidiurnal (twice daily), diurnal (daily), or mixed depending on local combinations of gravity, inertia, and basin shape.

INTERNET STUDY RESOURCES

The Web site for this book contains helpful study aids. Log on to:

www.brookscole.com/product/053437557xs

and click on the Chapter-by-Chapter area. Choose Chapter 11 and select a resource:

- **Flash Cards** allows you to test your mastery of the Terms and Concepts to Remember for this chapter.
- **Tutorial Quizzes** provides a multiple-choice practice quiz.
- **Student Guide to InfoTrac College Edition** will lead you to Critical Thinking Projects that use InfoTrac College Edition as a research tool.

TERMS AND CONCEPTS TO REMEMBER

amphidromic point	neap tide
aphelion	node
apogee	perigee
diurnal tide	perihelion
dynamic theory	seiche
ebb current	seismic sea wave
equilibrium theory of tides	semidiurnal tide
flood current	slack water
high tide	solar tide
low tide	spring tide
lunar tide	standing wave
mean sea level	storm surge
meteorological tide	storm tide
mixed tide	tidal bore

tidal current
tidal datum
tidal range

tidal wave
tide
tsunami

STUDY QUESTIONS

Review Questions

1. What combination of events makes a storm surge most destructive? Are all storms approaching a coast associated with a storm surge? Is a storm surge a progressive wave?

2. What factors affect the wavelength and period of a seiche? Do seiches themselves often cause damage?

3. What causes tsunami? Are all seismic sea waves tsunami? Are all tsunami seismic sea waves? How fast do tsunami travel? Do they move in the same way or at the same speed in a confining bay as they would in the open ocean?

4. What causes the rise and fall of the tides? What celestial bodies are most important in determining tides?

5. What is a high tide? A low tide? A spring tide? A neap tide?

Critical Thinking Questions

1. Though they move through all the ocean, the waves described in this chapter are referred to as shallow-water waves. How can that be?

2. Are tsunami ever dangerous if encountered in the open sea? What happens when they reach shore?

3. What are the most important factors influencing the heights and times of tides? What tidal patterns are observed? Are there tides in the open ocean? If so, how do they behave?

4. How does the latitude of a coastal city affect the tides there—or does it?

5. From what you learned about tides in this chapter, where would you locate a plant that generated electricity from tidal power? What would be some advantages and disadvantages of using tides as an energy source?

6. **InfoTrac College Edition Project** High tides inundate land at the shore, often with heavy wave action, and low tides expose land. The intertidal zone, between high tide and low tide, is an area of rich biodiversity. What adaptations would you expect plants and animals living in the intertidal zone to have? Do some research using InfoTrac College Edition to find examples for your answer.

 Earth Systems Today CD Questions

1. Go to " Earhquakes and Tsunami," and then to "Tsunami." Note the three choices: Prediction, Most Destructive, and Papau Simulation. Play each animation noting the time of propagation and the height of the waves near shore.

2. Go to "Earthquakes and Tsunami," and then to "Seismic Case Study, Alaska, 1964." Study each section of the case study, paying particular attention to the video and the speed of wave propagation through the whole Pacific ocean. Did you see any references to 1964 in your research of question #1? Try your hand at finding the epicenter in the interactive part of this case study.

3. Go to "Waves, Tides, and Currents," and to "High/Low Tides." Drag the moon to different elevations to see its effect on ocean tides.

4. Go to "Waves, Tides, and Currents," and to "Tidal Forces: Center of Mass." There are two interactive opportunities and one demonstration here. In "Center of Mass," you will see that the center of mass of the Earth–moon system is beneath the surface of Earth. In "Calculate Forces" and "Apply Forces," you can move the moon to find out how its influence changes with its proximity.

5. Go to "Waves, Tides, and Currrents," then to "Tidal Shifts: Earth–Moon." Note the effect of the moon's position above or below Earth's equator on the types and heights of ocean tides. Now add the influence of sun by going to "Earth–Moon–Sun." Which has greater influence, moon or sun?

RESOURCES FOR FURTHER READING AND RESEARCH

The Web site for this book contains many ideas for further reading and research. Log on to:

www.brookscole.com/product/053437557xs

and click on the Chapter-by-Chapter area. Choose Chapter 11 and select a resource:

- **References** lists the major books and articles consulted in writing this chapter, along with comments from the author about their content and reading level.
- **Hypercontents** takes you to an extensive list of sites with news, research, and images related to individual sections of the chapter.
- In **Student Guide to InfoTrac College Edition** scroll down to Suggested Readings from InfoTrac College Edition for brief descriptions of the articles listed and search hints for finding them in InfoTrac College Edition.
- **Regional InfoTrac College Edition** articles are organized into East Coast, West Coast, and Gulf Coast regions, allowing you to study oceanography on a more local level. You can also access Regional InfoTrac College Edition at www.localocean.com.

For additional readings, go to InfoTrac College Edition, your online research library, at:

http://infotrac.thomsonlearning.com/

VERY CAREFULLY!

Coasts are dynamic, changeable places. Few coasts are more susceptible to change than the wave-swept outer banks of North Carolina. Like many of the world's large barrier islands, the outer banks probably began to form during the last major rise in sea level, about 6,000 years ago. Coastal plains near the edge of the continental shelf were fronted by lines of sand dunes. Rising sea level caused the ocean to break through the dunes and form a lagoon—a long, shallow body of seawater isolated from the ocean. The high lines of coastal dunes became islands. As sea level continued to rise, wave action caused the islands and lagoons to migrate landward. One island supports the tallest and most famous lighthouse in North America, a heroic structure guarding a dangerous stretch of coast deservedly called the "graveyard of the Atlantic."

When it was built in 1870, Cape Hatteras lighthouse was more than 460 meters (1,500 feet) from the ocean. Sea level rise and frequent Atlantic storms caused Hatteras Island to retreat westward, and by 1997 the structure was only 37 meters (120 feet) from the waves. Groins built to protect the lighthouse had actually accelerated the rate of shoreline retreat immediately south of the structure. This famous landmark's destruction was probably only one hurricane away (**Figure a**).

In 1987 the National Research Council initiated studies to determine how to save the lighthouse. After many other options were considered, the National Park Service decided to sever the lighthouse

a Cape Hatteras light in peril—the coast is only 37 meters (120 feet) away.

© Billy E. Barnes/Stock, Boston

from its foundation and move all 4,381 metric tons (9.7 million pounds) of it to a new location 884 meters (2,900 feet) inland. Twelve million dollars was granted by Congress; work began in late 1998.

How does one move a 60-meter (198-foot) brick lighthouse? *Very* carefully! First, the structure was sliced from its foundation with a diamond-encrusted cable. Workers chipped away at the foundation as they cut through the bottom of the lighthouse, replacing granite blocks with huge lifting jacks. They then slid seven steel beams between the jacks and in a week's time gingerly lifted the lighthouse far enough to insert steel roller dollies into the gap. Meanwhile, trucks had dumped more than 10,000 tons of crushed stone to form a smooth path to the new location. Steel tracks were laid, and five hydraulic rams delicately shoved the structure toward its new home 5 feet at a time (**Figure b**). Computers and laser beams monitored deviation from the vertical.

A new foundation was built, and 23 days after the horizontal move began the lighthouse was safely lowered onto the rebuilt foundation. The lamp was relit in May 1999, ready to warn ships of the deadly nearness of shore for another 130 years.

AP/Wide World Photos

b Cape Hatteras lighthouse during the move.

WHAT TO WATCH FOR IN CHAPTER 12

Our personal experience with the ocean usually begins at the coast. Have you ever wondered why a coast is in a particular location, or why it is shaped as you see it? These temporary, often beautiful junctions of land and sea are subject to rearrangement by waves and tides, gradual changes in sea level, biological processes, and tectonic activity.

One of the more useful ways to classify coasts is based on the degree to which a coast has been modified by marine processes. As you'll learn, *primary coasts* are relatively young coasts in which terrestrial influences dominate, while *second-ary coasts* have been significantly changed by wave action and the growth of marine organisms or other marine processes. Beaches, changeable places you probably find quite pleasant, are secondary coasts. Some long coasts combine primary and secondary features.

Coasts are active areas where marine, terrestrial, atmospheric, and human factors converge. No single one of these factors dominates for long. We enjoy visiting and living near coasts, but human interference in coastal processes does not always preserve or "improve" a stretch of rocky coast or a beach. The interference may have the opposite effect, and coastal residents do not always learn by example.

COAST AND SHORE

Coastal areas join land and sea. The place where ocean meets land is usually called the **shore,** while **coast** refers to the larger zone affected by the processes occurring at this boundary. A sandy beach might form the shore in an area, but the coast (or coastal zone) includes the marshes, sand dunes, and cliffs just inland of the beach, as well as the sandbars and troughs immediately offshore. The world ocean is bounded by about 440,000 kilometers (273,000 miles) of shore.

Because of its proximity to both ocean and land, a coast is subject to natural events and processes common to both realms. A coast is an active place. Here is the battleground on which wind waves break and expend their energy. Tides sweep water on and off the rim of land, rivers drop most of their terrigenous sediments at the coasts, and ocean storms batter the continents. The *location* of a coast depends primarily on global tectonic activity and the volume of water in the ocean. The *shape* of a coast is a product of many processes: uplift and subsidence, the wearing down of land by erosion, and the redistribution of material by sediment transport and deposition.

CLASSIFYING COASTS

A good classification scheme allows us to group things into categories determined by common origins or the influence of common processes. The process of categorizing things often makes them easier to comprehend. Coasts have been hard to classify because, as we will see, many factors combine to shape them.

As we saw in Chapter 3, no area of geology has been left undisturbed by the revelations of plate tectonics. In the 1960s geologists began to classify coasts according to tectonic position. *Active* coasts, near the leading edge of moving continental plates, were found to be fundamentally different from the more *passive* coasts near trailing edges. The shapes, composition, and ages of coasts are better understood by taking plate movements into account. But the slow forces of plate movement are often obscured by the more rapid action of waves, the erosion of land, and the transport of sediments.

Another important factor that must be considered in the classification of coasts is long-term change in sea level. Five factors can cause sea level to change. Three of these factors are responsible for **eustatic change**—variations in sea level that can be measured all over the world ocean:

- The amount of water in the world ocean can vary. Sea level is lower during periods of global glaciation (ice ages) because there is less water in the ocean. It is higher during warm periods, when the glaciers are smaller. Periods of abundant volcanic outgassing or an increase in the number of icy comets striking Earth can also add water to the ocean, which raises sea level.

- The volume of the ocean's "container" can vary. As noted in Chapter 4, high rates of seafloor spreading are associated with the expansion in volume of the mid-ocean ridges. This displaces the ocean's water, which climbs higher on the edges of the continents. Sediments shed by the continents during periods of rapid erosion can also decrease the volume of ocean basins and raise sea level.

- The water itself may occupy more or less volume as its temperature varies. During times of global warming, seawater expands and occupies more volume, raising sea level.

Of course the continents rarely stay still as sea level rises and falls. Local changes are bound to occur, and two other factors produce variations in *local* sea level:

- Tectonic motions and isostatic adjustment can change the height and shape of the coast. Coasts can experience uplift as lithospheric plates converge, or can be weighted down by masses of ice during a period of widespread glaciation. The continents slowly rise when the ice melts.

- Wind and currents, seiches, storm surges, an El Niño or La Niña event, and other effects of water in motion can force water against the shore or draw it away.

Sea level has been at its current elevation (give or take 0.5 meter, or 1.5 feet) for only about 2,500 years. Over the past 2 million years, worldwide sea level has varied from about 6 meters (20 feet) above to about 125 meters (410 feet) below its present position. The recent low point occurred about 18,000 years ago at the height of the most recent glaciation (see Figure 4.14). Indeed, sea level has been at the modern "high" only rarely in the past 2 million years—the predominant state for Earth is a much lower eustatic sea level position. It is important to realize that coastlines have not yet come into equilibrium with modern sea level.

Changes in sea level produce major differences in the position and nature of coastlines, especially in areas where the edge of the continent slopes gradually or where the coast is rising or sinking. **Figure 12.1** shows an estimate of previous shore positions along the southeastern coast of the United States in the geologically recent past and a prediction for the distant future should the present warming trend cause more of the polar ice to melt.

What coastal classification scheme best combines these elements? One of the most useful was popularized by Francis Shepard and is used by many geological oceanographers: *primary* and *secondary coasts.* If a coast is essentially in almost the same condition as it was when sea level stabilized after the last ice age, Shepard called it a **primary coast.** Primary coasts are young coasts in which terrestrial influences (that is, processes that occur at the boundary between land and air) dominate. If the coast has been significantly changed by wave action and other marine processes since sea level stabilized, he termed it a **secondary coast.** Secondary coasts are usually older than primary coasts—they have been exposed to marine

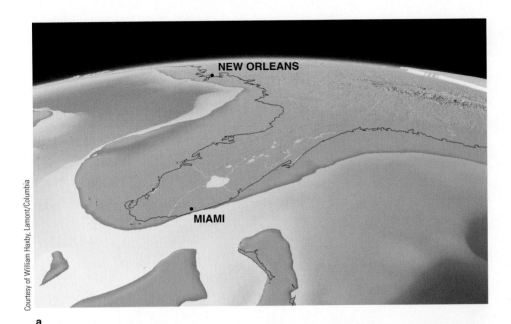

a

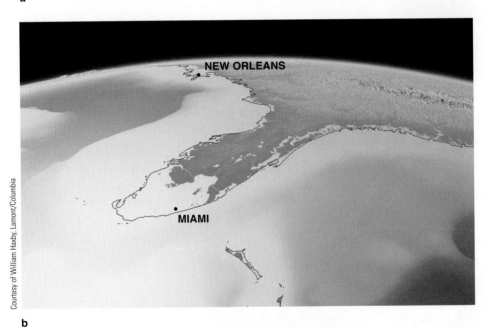

b

Figure 12.1 (**a**) The southeastern coast of the United States looked much different 18,000 years ago, during the last ice age. Because of lower sea level, the position of the gently sloping eastern coast has been as much as 200 kilometers (125 miles) seaward from the present shoreline, leaving much of the continental shelf exposed. (**b**) In the distant future, if the ocean were to expand and the polar ice caps were to melt because of global warming, sea level could rise perhaps 60 meters (200 feet), driving the coast inland as much as 250 kilometers (160 miles).

action for a longer time. Secondary coasts retain little (if any) evidence of the nonmarine processes that produced them. Some coasts show both characteristics: A single, long shoreline can consist of both primary and secondary coasts.

PRIMARY COASTS

Primary coasts are often rough and irregular. The ocean has not had time to modify the terrestrial features provided by changes in sea level, the scouring of glaciers, deposition of sediment at the mouths of rivers, volcanic eruptions, or the movement of earth along faults.

Land Erosion Coasts

When sea level was lower during the last glaciations, rivers cut across the land and eroded sediment to form coastal river valleys. When higher sea level returned, the valleys were flooded, or *drowned*, with seawater. Chesapeake Bay (**Figure 12.2**) and the Hudson River valley are examples of drowned river mouths. Coasts dominated by these sunken river valleys are sometimes called ria coasts, from the Spanish word for river.

Glaciers sometimes form in river valleys when rivers cut through the edges of continents at high latitudes. Deep, narrow bays known as **fjords** are often formed by tectonic forces and later modified by glaciers eroding valleys into deep, U-shaped troughs. Fjords (**Figure 12.3**) are found in British

Figure 12.2 A false-color photograph of Chesapeake Bay taken from space. The complex bay is an example of a drowned river valley.

Figure 12.3 Milford Sound, a deep fjord on the western coast of New Zealand's South Island.

Columbia, Greenland, Alaska, Norway, New Zealand, and other cold, mountainous places.

Coasts Built Out by Land Processes

In a few places, sediments washing off the land have built out the coasts extensively. The shoreline in such places is much different from its configuration at the end of the last ice age. The most important of these coastal features are **deltas.** This term, first used by Herodotus in the fifth century B.C. to describe the triangular mass of sediment at the terminus of the Nile, is derived from the triangular shape of the capital Greek letter delta (Δ).

Deltas do not form at the mouth of every sediment-laden river. A broad continental shelf must be present to provide a platform on which sediment can accumulate. In addition (as befits a primary coast), marine processes must not dominate: Tidal range should be low, and waves and currents generally mild. There are no large deltas along the Atlantic coast of the United States because sediments arriving at the coast are deposited in the sunken river mouths or dispersed by tides and currents. Also, there are no large deltas along the western margins of North and South America because these coasts are converging margins, where an oceanic plate is being subducted and the continental shelf is very narrow; sediment that would form a delta is swept down the continental slope or dispersed along the coast by waves. Deltas are most common on the low-energy shores of enclosed seas (where the tidal range is not extreme) and along the tectonically stable trailing edges of some continents. The largest deltas are those of the Gulf of Mexico (the Mississippi, **Figure 12.4**), the Mediterranean Sea (the Nile, **Figure 12.5,** and the Rhône, Po, and Ebro), the Ganges–Brahmaputra river system in the Bay of Bengal, and the huge deltas formed by the rivers of China that empty into the South China Sea.

The shape of a delta represents a balance between the accumulation of sediments and their removal by the ocean. For a delta to maintain its size or grow, the river must carry enough sediment to keep marine processes in check. The combined effects of waves, tides, and river flow determine the shape of a delta. In 1975 William Galloway, a geologist at the University of Texas, classified deltas by the relative influence of those three factors. (See **Figure 12.6.** Note that any delta strongly affected by only one of the three factors is placed at an apex of the triangle. A delta with a balance of forces is placed near the middle of the triangle.) *River-dominated deltas* are fed by a strong flow of fresh water and continental sediments, and they form in protected marginal seas. They terminate in a well-developed set of *distributaries*—the split ends of the river—in a characteristic bird's-foot shape (as shown in the Mississippi, Figure 12.4). In *tide-dominated deltas,* freshwater discharge is overpowered by tidal currents that mold sediments into long

USGS

Figure 12.4 River deltas form at places where sediment-laden rivers enter enclosed or semienclosed seas, where wave energy is limited. The bird's-foot shape of the Mississippi Delta is seen clearly in this photograph. Submerged overlapping lobes of the Mississippi Delta extend completely across a wide continental shelf, and sediment from the river is now being deposited directly on the continental slope. Lobed and bird's-foot deltas form where deposition overwhelms the processes of coastal erosion and sediment transportation. The sediment-laden water looks brown or tan in this photograph taken from low orbit.

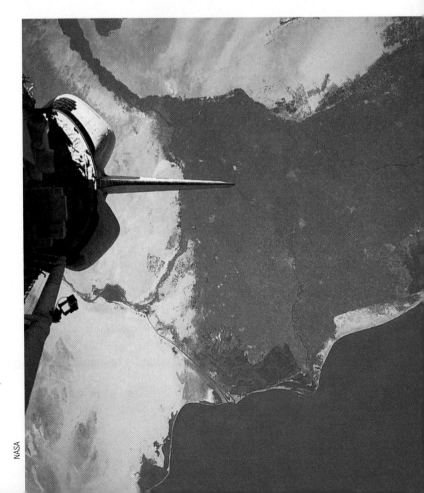

NASA

Figure 12.5 The Nile Delta, which has been smoothed into a more rounded shape by postdepositional coastal erosion and sediment transportation. (The tail of the space shuttle from which this photo was taken can be seen in the upper left of the photo.)

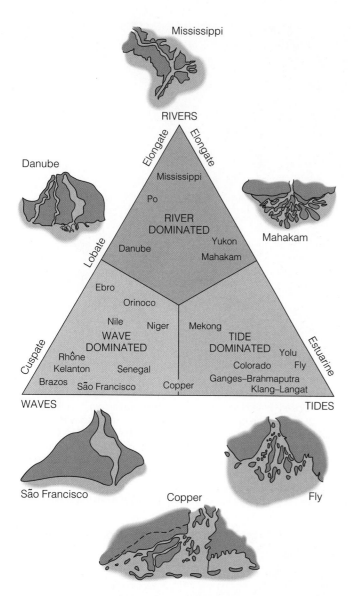

Figure 12.6 The Galloway classification of deltas. Each delta reflects the balance between the rate of sediment delivery at the river mouth and the mechanisms of its dispersal. This triangular diagram classifies river deltas according to the influence of the three major factors affecting their development: the river, waves, and tides. (Source: Used by permission of Dr. William Galloway, University of Texas, Austin and the Houston Geological Survey.)

islands parallel to the river flow and perpendicular to the trend of the coast. The largest tide-dominated delta has formed at the mouths of the Ganges–Brahmaputra river system on the Bay of Bengal. *Wave-dominated deltas* are generally smaller than either tide- or river-dominated deltas and have a smooth shoreline punctuated by beaches and sand dunes. Instead of a bird's-foot pattern of distributaries, a wave-dominated delta has one primary exit channel.

Deltas are not the only types of coasts built out by the land. The glaciers that covered the poleward parts of the continents during the ice age deposited great quantities of sediments and rocks near their outer margins. When the glaciers retreated, they left streamlined hills known as **drumlins** and **moraines**—hills and ridges of sediments—some of which

still stand above sea level. Part of Long Island, New York, is a glacial moraine, as are the oval-shaped hills of Boston and sections of the Puget Sound coast at Seattle. Perhaps the most famous glacial moraine in the United States is the area around Cape Cod, Massachusetts. The cores of the islands of Martha's Vineyard and Nantucket represent the farthest advance of the glaciers around 18,000 years ago, and the spine of the cape itself is a remanent of material dropped by the glacier as it stabilized and then retreated northward when the climate warmed (see **Figure 12.7**).

Volcanic Coasts

As we saw in Chapter 4, most islands that rise from the deep ocean are of volcanic origin. If the volcanism has been recent, the coasts of a volcanic island will consist of lobed lava flows extending seaward, common features in the Hawaiian Islands. Concave shorelines (**Figure 12.8**) result when a small coastal volcano explodes or collapses and seawater fills the crater.

Coasts Shaped by Earth Movements

Coasts can coincide with places where Earth's crust is being warped or faulted. When the seabed on the seaward side of a coastal fault moves *downward*, a steep escarpment that continues to a greater depth than a wave-cut cliff can result. When the landward side of the fault moves *upward*, the part of the

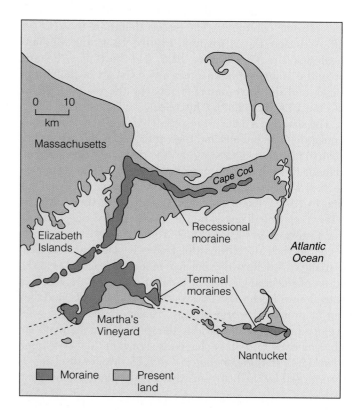

Figure 12.7 Most of the islands of Martha's Vineyard and Nantucket were formed from debris deposited or shaped by an advancing glacier. The spine of Cape Cod is built of material dropped as the glaciers stabilized and then retreated.

Figure 12.8 Concave shapes may also occur on a volcanic coast. The explosion and collapse of a volcano close to shore caused this indentation on the coast of Isla Encantada, Gulf of California, Mexico.

Figure 12.9 This former seafloor at Prince William Sound, Alaska, was raised 8 meters (26 feet) above sea level by tectonic uplift during the great earthquake of 27 March 1964. The exposed surface, which slopes gently from the base of the sea cliffs to the water, is about 400 meters (¼ mile) wide. The light-colored coating on the rocks consists mainly of the dried remains of small marine organisms. The photo was taken at about 0.0 tide, 30 May 1964.

coast previously submerged during high tides can be left high and dry. The 1964 Alaska earthquake (described in Chapter 3) caused parts of the Prince William Sound shore to rise more than 3.5 meters (12 feet). As can be seen in **Figure 12.9,** in some places a band of the old sea-cut bench nearly half a kilometer (¼ mile) wide was exposed, even at high tide.

Fault coasts sometimes occur along transform faults. The Pacific coast of North America provides two striking examples: the Gulf of California (located along the San Andreas Fault between Baja California and the mainland of Mexico) and the area around Point Reyes, north of San Francisco. Baja California was once a part of the North American continent, but movement at the boundary between the Pacific and North American Plates has torn the finger of land on the west side of the fault from the land mass to the east. The young, narrow, straight gulf formed as seawater intruded. As can be seen in **Figure 12.10,** a photo of Tomales Bay north of San Francisco, fault coasts can be startlingly straight.

SECONDARY COASTS

Secondary coasts are those coasts that have been significantly changed by wave action and other marine processes after sea level has stabilized. Land erosion and marine erosion both work to change a primary coast to a secondary coast. Secondary coasts are shaped and attacked from the land by stream erosion, the abrasion of wind-driven particles, the alternate freezing and thawing of water in rock cracks, the probing of plant roots, glacial activity, rainfall, dissolution by acids from soil, and slumping. (As any mountaineer will confirm, these factors are not limited to coasts.)

Attack from the sea is by waves and currents. The continual onslaught of waves does most of the work, with currents distributing the results of the waves' labor. Some indication of the violence of this activity may be inferred from **Figure 12.11.**

Large storm surf routinely generates pressures of 27,000 kilograms per square meter (25 tons per square yard). The crashing waves push air and water into tiny rock crevices. The repeated buildup and release of pressure within these crevices can weaken and fracture the rock. But it is not the hydraulic pressure of moving water alone that abrades the coasts. Tiny pieces of sand, bits of gravel, or stones hurled by waves toward the shore are even more effective at eroding the shore. **Dissolution,** the dissolving of minerals in the rocks by water, contributes to the erosion of easily soluble coastal rocks such as limestone. Even the digging and scraping activities of marine organisms have an effect.

The rate at which a shore erodes depends on the hardness and resistance of the rock, the violence of the wave shock to which it is exposed, and the local range of tides. Hard rock resists wear. Coasts made of granite or basalt may retreat an insignificant amount over a human lifetime; the granite coast of Maine erodes only a few centimeters per decade. Coasts of soft sandstone or other weak (or soluble) materials, however, may disappear at a rate of a few meters per year. In an extreme case, soft shore cliffs facing the North Sea in East Suffolk, England, eroded more than 11 meters (36 feet) during two hours of a severe storm!

Figure 12.10 A characteristically straight fault coast at Tomales Bay, California. Point Reyes is visible to the left (west); the city of San Francisco is out of view to the south. The San Andreas Fault trace disappears below sea level in the bay (arrow); the straight sides of the bay closely parallel the submerged fault.

Figure 12.11 Waves from a large Pacific storm strike the north coast of the Hawaiian island of Oahu. Shore erosion may be rapid in areas of pounding surf.

Marine erosion is usually most rapid on **high-energy coasts,** areas frequently battered by large waves. High-energy coasts are most common adjacent to stormy ocean areas of great fetch and along the eastern edges of continents exposed to tropical storms. The coasts of Maine and British Columbia and the southern tips of South America and South Africa are typical high-energy coasts. **Low-energy coasts** are only infrequently attacked by large waves. Because of their generally protected location in the Gulf of Mexico, the U.S. Gulf states share a low-energy coast—at least between hurricanes!

Waves can affect the coast only where they strike, so erosion is concentrated near average sea level. A shore with little tidal variation can erode quickly because the wave action is concentrated near one level for longer times. Low-energy coasts protected by offshore islands usually erode slowly, as do areas below the low-tide line. Some erosion does occur below the surface because of the orbital motion of water in waves, but even the largest waves have little erosive effect at depths greater than about 15 meters (50 feet) below average sea level. Cliffs above shore are subject to pounding either directly from waves or by rocks hurled by waves. In one case, windows of a Scottish lighthouse 100 meters (330 feet) above sea level were broken by wave-thrown rocks during an Atlantic winter storm.

Some Features of Secondary Coasts

Erosive forces can produce a wave-cut shore showing some or all of the features illustrated in **Figure 12.12.** Note the complex, small-scale irregularities of this rocky erosional coastline. **Sea cliffs** slope abruptly from land into the ocean, their steepness usually resulting from the collapse of undercut

notches. The position of the sea cliffs marks the shoreward limit of marine erosion on a coast. The parade of waves cuts **sea caves** into the cliffs at local zones of weakness in the rocks. Most sea caves are accessible only at low tide. A blowhole can form if erosion follows a zone of weakness upward to the top of the cliff. When the tide is at just the right height, spray can blast from the fissure as waves crash into the cliff. Offshore features of rocky coasts can include natural arches, rock stacks, and a smooth, nearly level **wave-cut platform** just offshore, which marks the submerged limit of rapid marine erosion. Much of the debris removed from cliffs during the formation of these structures is deposited in the quieter water farther offshore, but some can rest at the bottom of the cliffs as exposed beaches. Indeed, as we shall soon see, broad beaches are often features of secondary coasts.

Shore Straightening

The first effect of marine erosion on a newly exposed coast is to intensify the irregularity of the coastline. This happens because coastal rocks are usually not uniform in composition over long horizontal distances. Some hard rocks will resist erosion well, while softer rocks on the same coast may disappear almost overnight. (This explains the uneven character of stacks, arches, and sea cliffs.)

Eventually, however, coastal erosion tends to produce a smooth shoreline. Because of wave refraction (see Figure 10.20), wave energy is focused onto headlands and away from bays (**Figure 12.13**). Sediment eroded from the headlands tends to collect as beaches in the relatively calm bays. As erosion continues, the deposits may eventually protect the base

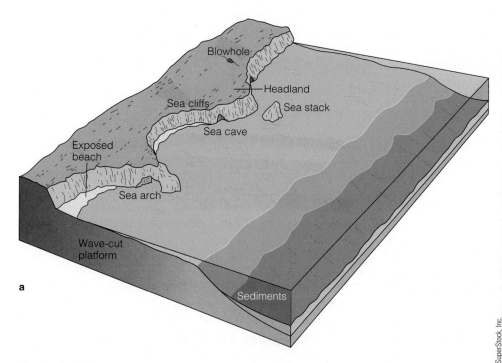

Figure 12.12 (**a**) Features of a secondary coast at low tide. (**b**) Sea stacks off the Australian coast.

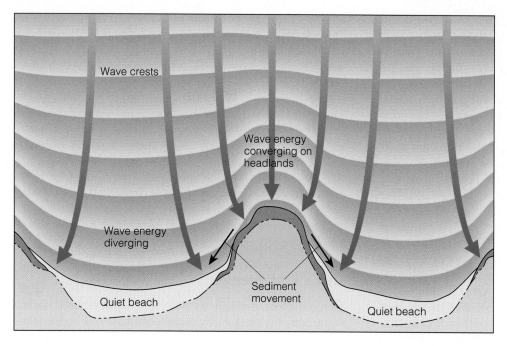

Figure 12.13 Wave energy converges on headlands and diverges in the adjoining bays. The accumulation of sediment derived from the headland in the tranquil bays eventually smooths the contours of the shore. See Figure 10.20 for a discussion of the process of wave refraction responsible for this pattern.

of the shore cliffs from the waves. Coastal irregularities are thus smoothed with the passage of time. As you might expect, straightening occurs most rapidly on high-energy coasts.

The Accumulation of Beaches

If the submerged slope of the seafloor is steep, eroded sediments will soon drain to deeper waters. If the slope is not too steep, sediments will be transported along the coast by wave and current action. The movement of sediment (usually sand) along the coast, driven by wave action, is referred to as **longshore drift.** Longshore drift occurs in two ways: the wave-driven movement of sand along the exposed beach, and the current-driven movement of sand in the surf zone just offshore (**Figure 12.14**).

As we have seen, wind waves arriving from distant storms do not always approach the coast straight on. Most wind waves approach at an angle and then refract in shallow water to break almost parallel to the shore. Refraction is usually incomplete, however, and some angle remains when the waves break. If sediments have accumulated to form a beach, water from the breaking wave will rush up the beach at a slight angle but return to the ocean by running straight downhill under the influence of gravity. The sand grains disturbed by the wave will follow the water's path, moving up the beach at an angle but retreating down the beach straight down the slope. Net transport of the grains is longshore, parallel to the coast, away from the direction of the approaching waves. Net sand flow along the Pacific and Atlantic coasts of the United States is usually to the south, because the waves that drive the transport system usually approach from the north, where storms most commonly occur.

Sediments are also transported in the surf zone in a **longshore current.** The waves breaking at a slight angle distrib-

ute a portion of their energy away from their direction of approach. This energy propels a narrow current in which sediment already suspended by wave action can be transported downcoast. The speed of the longshore current sometimes approaches 4 kilometers (about 2 ½ miles) per hour.

Sand moving in the wash of waves along the beach and sediments propelled in the longshore current just offshore are often joined by much greater loads of sediments brought to the coast by local rivers. Net southward transport of all this material along the central California coast exceeds 230,000 cubic meters (300,000 cubic yards) per year. Typical figures for the Atlantic coast of the United States are about two-thirds of this value.

BEACHES

The most familiar feature of a secondary coast is the beach. A **beach** is a zone of unconsolidated (loose) particles that covers part or all of a shore. The landward limit of a beach may be vegetation, a sea cliff, relatively permanent sand dunes, or construction such as a seawall. The seaward limit occurs where sediment movement onshore and offshore ceases—a depth of about 10 meters (33 feet) at low tide. The continental United States has 17,672 kilometers (10,983 miles) of beaches, about 30% of the total shoreline.

Beaches result when sediment, usually sand, is transported to places suitable for deposition. Such places include the calm spots between headlands, shores sheltered by offshore islands, and regions with moderate surf or broad stretches of high-energy coast. Sometimes the sediment is transported a very short distance—particles may simply fall from the cliff above and accumulate at the shoreline—but

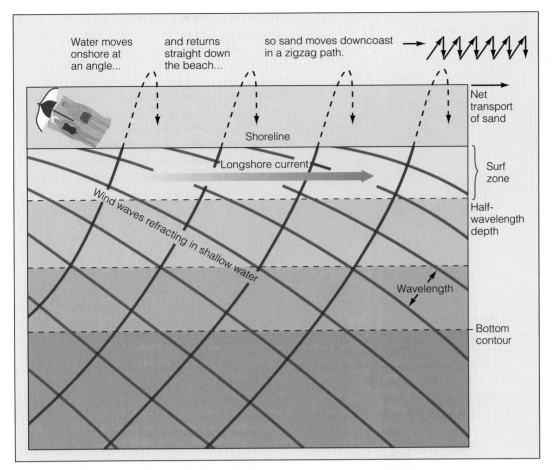

Figure 12.14 How wind-wave refraction and longshore current result in longshore drift of beach sand.

In the figure: "Water moves onshore at an angle...", "and returns straight down the beach...", "so sand moves downcoast in a zigzag path.", "Net transport of sand", "Shoreline", "Longshore current", "Surf zone", "Wind waves refracting in shallow water", "Half-wavelength depth", "Wavelength", "Bottom contour"

more often the sediment on a beach has been moved over long distances to its present location.

Wherever they are found, beaches are in a constant state of change. They may be thought of as rivers of sand—zones of continuous sediment transport.

The Composition and Slope of Beaches   12-14

The material of beaches can range from boulders through cobbles, pebbles, and gravel to very fine silt. The rare black sand beaches of Hawaii are made of finely fragmented lava. Some beaches consist of shells and shell debris or of fragments of coral. Unfortunately, some also include large quantities of human junk—glass or plastic beaches are not unknown. Cobble beaches can be very steep (occasionally with slopes in excess of 20°), but wide beaches of fine sand are sometimes nearly as flat as a parking lot.

In general, the flatter the beach, the finer the material from which it is made (**Table 12.1**). The relation between particle size and beach slope depends on wave energy, particle shape, and the porosity of the packed sediments. Water from waves washing onto a beach—the **swash**—carries particles onshore, increasing the beach's slope. If water returning to the ocean—the **backwash**—carries back the *same* amount of material as it delivered, the beach slope will be in equilibrium, so the beach will not become larger or steeper.

On fine-grain beaches, the ability of small, sharp-edged particles to interlock discourages water from percolating down into the beach itself, so water from waves runs quickly back down the beach, carrying surface particles toward the ocean. This process results in a very gradual slope. Broad, flat beaches also have a large area on which to dissipate wave energy, and

Table 12.1	The Relationship Between the Particle Size of Beach Material and the Average Slope of the Beach	
Type of Beach Material	**Size (mm)**	**Average Slope of Beach**
Very fine sand	0.0625–0.125	1°
Fine sand	0.125–0.25	3°
Medium sand	0.25–0.50	5°
Coarse sand	0.50–1.0	7°
Very coarse sand	1–2	9°
Granules	2–4	11°
Pebbles	4–64	17°
Cobbles	64–256	24°

Source: Shepard, 1973.

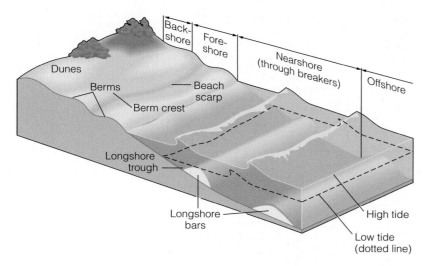

Figure 12.15 A typical beach profile.

they can provide a calm environment for the settling of fine sediment particles. In contrast, coarse particles (gravel, pebbles) do not fit together well and readily allow water to drain between them. Onrushing water disappears *into* a beach made of coarse particles, so little water is left to rush down the slope, thereby minimizing the transport of sediments back to the ocean. Thus, larger particles tend to build up at the back of the beach where they are thrown by large waves, increasing its steepness.

Beach Shape

Figure 12.15 shows a profile, or cross section, of a beach affected by small-to-moderate wave and tidal action. The scale is exaggerated vertically to show detail. The key fea-

ture of any beach is the **berm** (or berms), an accumulation of sediment that runs parallel to shore and marks the normal limit of sand deposition by wave action. The peaked top of the highest berm, called the **berm crest,** is usually the highest point on a beach. It corresponds to the shoreward limit of wave action during the most recent high tides. Inland of the berm crest, extending to the farthest point where beach sand has been deposited, is the **backshore.** The backshore is the relatively inactive portion of the beach, which may include windblown dunes and grasses. The **foreshore,** seaward of the berm crest, is the active zone of the beach, washed by waves during the daily rise and fall of the tides. It extends from the base of the berm—where a **beach scarp** (a vertical wall of variable height) is often carved by wave action at high tide—to the low-tide mark, where the offshore zone begins. Below the low-tide mark, wave action, turbulent backwash, and longshore currents excavate a **longshore trough** parallel to shore. Irregular **longshore bars** (submerged or exposed accumulations of sand) complete the seaward profile.

This beach profile is only temporary, generated by the interplay of sediments, waves, and tides. Great storm waves can rearrange a beach in a day, transporting thousands of tons of sediment from the beach to hidden sandbars offshore. Indeed, most temperate-climate beaches undergo a seasonal transformation. Beaches are cut to a lower level in winter than in summer because higher waves accompany winter storms. Changes from summer to winter on a southern California beach are shown in **Figure 12.16.**

a   b

Figure 12.16 The movement of sand on Carlsbad State Beach, California. (**a**) The beach in the summer of 1978. (**b**) The same beach in the spring of 1983 after a winter of severe storms. The sand has moved offshore, and coastal structures have been damaged.

Minor Beach Features

Small features are among the things that make beaches such interesting places. *Ripples* in the sand (**Figure 12.17a**) are caused by rushing currents; similar structures can be seen in dry river bottoms or streambeds. *Rills* (**Figure 12.17b**) are small, branching surface depressions that channel water back to the ocean from a saturated beach during a falling tide. Diamond-shaped *backwash marks* (**Figure 12.17c**) form when projecting shells, pebbles, or animals interrupt the backwash along the low-tide terrace and cause uneven deposition or erosion of very fine sediment of a contrasting color. Another result of uneven deposition can be seen in beach layering. A vertical slice into a calm beach will often reveal layer upon layer of sed-

iments. The sediments are often separated by their density and may be different colors and textures (**Figure 12.17d**).

On a somewhat larger scale, the beach shoreline is sometimes scalloped by *cusps* (*cuspis* = point) (**Figure 12.18**). These repeating points, interrupted by miniature bays, are evenly spaced every several meters. Cusps tend to form at neap-tide periods and are destroyed during the greater range of spring tides. Cusp size appears to be correlated with wave energy. Their origin is not well understood, but they are probably caused by interference between approaching waves. Offshore sandbars and troughs are also more pronounced during periods of moderate tides. The variability of cusps and sandbars reflects the transitory nature of beach processes.

a

b

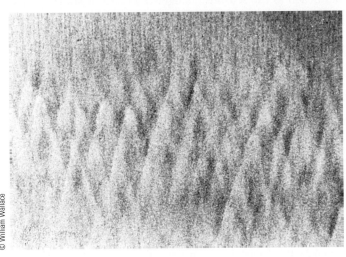

c

d

Figure 12.17 Minor features of beach sand. (**a**) Ripple marks. (**b**) The streamlike pattern of rill marks. (**c**) Backwash marks. Each V-shaped feature is caused by a small irregularity in the sand; each is about 15 centimeters (6 inches) long. (**d**) Layers of sediments are clearly seen in this freshly eroded beach face. The alternating layers were deposited by small waves during many ebb-and-flood tide cycles. Larger, storm-driven waves have exposed the layers to view. The keys provide scale.

Figure 12.18 Beach cusps at Point Reyes National Seashore, northern California, formed in a gently sloping fine-sand beach.

Figure 12.19 A rip current. Rip currents are created when waves pile water inside the surf zone. The excess water then returns seaward as a swift rip current, ending in turbulence outside the surf. The rip current can slice through sandbars present offshore, and the narrow opening can focus the current into an even faster moving mass.

Rip Currents

You may have noticed currents moving seaward through the surf on high-energy beaches (**Figure 12.19**). These small currents are caused by the movement offshore of a large amount of water at one time. Because they're not caused by large-scale gravitational and inertial forces, *riptide* isn't a good term for them; they're properly called **rip currents.**

Rip currents form when a group of incoming waves piles an excess of water on the landward side of the surf zone faster than the longshore current can carry it. The water breaks through the wave line in a few places and flows rapidly through the surf back to sea. Rip currents are often made visible by the muddy color of their suspended sediments contrasting with the cleaner water just offshore. A strong swimmer can use the rip current to his or her advantage, hitching a ride seaward through the churning surf to where the body surfing is best. To escape this narrow ocean-going band, however, a swimmer is advised to slowly swim parallel to shore and then return to shallow water. The higher the surf, the greater the probability of rip currents.

Rip currents are sometimes called *undertows,* a word as deceptive and inaccurate as *riptide.* There aren't any small-scale, nearshore features that suck swimmers beneath the surface; even the legendary whirlpools would have trouble accomplishing that task (see, for example, Box 11.1).

Coastal Cells

As we have seen, most new sand on a coast is brought in by rivers. The sand is moved parallel to the beach by longshore drift, and it is moved onshore and offshore at right angles to the beach as the seasons change. If a beach is stable in size, neither growing nor shrinking, the amount of new sand entering must be balanced by the amount of old sand being removed. Sand that drifts below the reach of wave action is lost from the coast and may migrate farther out on the continental shelf. Some sand is driven by longshore currents into the nearshore heads of submarine canyons. Sand moving away from shore in these canyons sometimes forms impressive sandfalls (see Figure 4.19 for an example) and is lost from the beaches above. The bulk of this material is transported by gravity down the axis of the canyon and ultimately deposited on a submarine fan at the base of the slope.

The natural sector of a coastline in which sand *input* and sand *outflow* are balanced may be thought of as a **coastal cell.** The main features of such a cell are illustrated in **Figure 12.20a.** Coastal cells are usually bounded by submarine canyons that conduct sediments to the deep sea. Their size varies greatly. They are often very large along the relatively smooth, tectonically passive trailing edges of continents; coastal cells along the southeastern coast of the United States, for example, are hundreds of kilometers long. On the active leading edge of the continent, they are smaller. Four coastal cells exist in the 360 kilometers (225 miles) between southern California's Point Conception and the Mexican border. Each terminates in a submarine canyon at the downcoast end (**Figure 12.20b**).

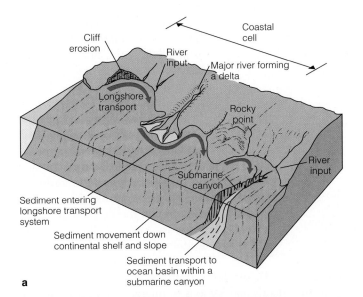

a

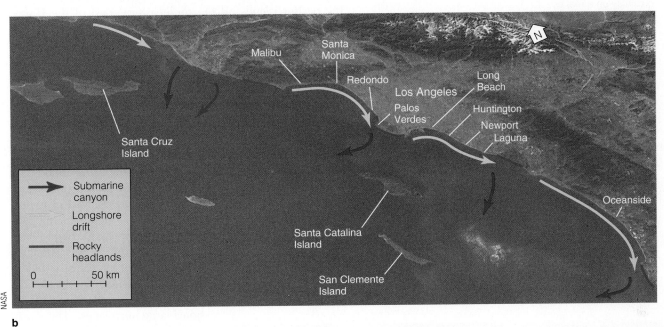

b

Figure 12.20 Coastal sediment transport cells. **(a)** The general features of coastal cells, in which sand is introduced by rivers, transported southward by the longshore drift, and trapped within the nearshore heads of submarine canyons. **(b)** A series of coastal cells in southern California. The yellow arrows show sand flowing in rivers toward the beaches; the red arrows show sand flowing into submarine canyons.

LARGE-SCALE FEATURES OF SECONDARY COASTS

We have already noted the presence of small longshore bars off beaches (see Figure 12.15), but secondary coasts, especially the coasts along a subsiding irregular continental margin, also exhibit much larger features. Some of these features are illustrated in **Figure 12.21.**

Sand Spits and Bay Mouth Bars

Sand spits are among the most common features of secondary coasts. A sand spit forms where the longshore current slows as it clears a headland and approaches a quiet bay. The slower current in the mouth of the bay is unable to carry as much sediment, so sand and gravel are deposited in a line downcurrent of the head-land (**Figure 12.22**). As can be seen in Figures 12.21 and 12.22, sand spits often have a curl at the tip. This is caused by the current-generating waves being refracted around the tip of the spit.

A **bay mouth bar** forms when a sand spit closes off a bay by attaching to a headland adjacent to the bay. The bay mouth bar protects the bay from waves and turbulence and encourages the accumulation of sediments there. An inlet—a passage to the ocean—may be cut through a bay mouth bar by tidal action, water flowing from a river emptying into the bay, or heavy rains. A bay mouth bar is shown in **Figure 12.23.**

Barrier Islands and Sea Islands

Secondary coasts can also develop narrow, exposed sandbars that are parallel to but separated from land. These are known as **barrier islands** (see **Figure 12.24**). About 13% of the world's coasts are fringed with barrier islands.

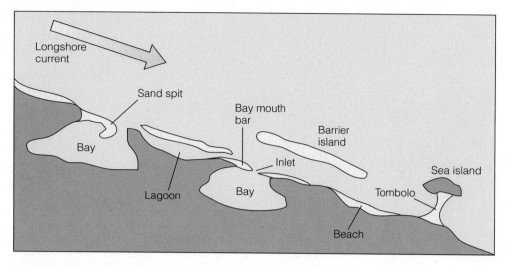

Figure 12.21 A composite diagram of the large-scale features of an imaginary secondary coast. Not all these features would be found in such close proximity on a real coast.

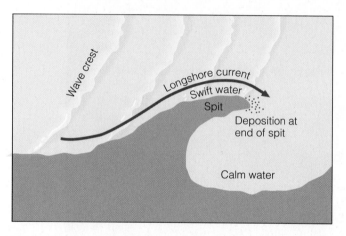

Figure 12.22 A sand spit forms where the longshore current slows as it clears a headland and approaches a quiet bay.

Figure 12.23 A bay mouth bar. The inlet is now closed, but increased river flow (from inland rainfall) or large waves combined with very high and low tides could break through the bar.

Barrier islands can form when sediments accumulate on submerged rises paralleling the shoreline. Some islands off the Mississippi–Alabama coast developed in this way. Larger barrier islands are thought to form in a different way, however. Near the end of the last major rise in sea level, about 6,000 years ago, coastal plains near the edge of the continental shelf were fronted by lines of sand dunes. Rising sea level caused the ocean to break through the dunes and form a **lagoon**—a long, shallow body of seawater isolated from the ocean. The high lines of coastal dunes became islands. As sea level continued to rise, wave action caused the islands and lagoons to migrate landward. Most of the barrier islands off the southeastern coast of the United States originated in this way. They are still migrating slowly landward as sea level continues to rise.

There are 295 barrier islands along the Atlantic and Gulf coasts of the United States, with a combined length of 2,591 kilometers (1,610 miles). Every year severe storms generate waves intense enough to erode barrier island beaches. The largest of these storms can generate waves that overwash the low islands. Runoff from rivers swollen by rains, coupled with water driven by wind waves and storm surge, can rapidly flood a lagoon and cut new inlets through barrier islands.

Despite these dangers, about 70 of the barrier islands have been commercially developed, and millions of people live on them. The most famous barrier islands include Atlantic City, New Jersey; Ocean City, Maryland; Miami Beach and Palm Beach, Florida; and Galveston, Texas. Roughly once every hundred years a winter storm has catastrophic effects on populated areas of Atlantic barrier islands, and the southeastern Atlantic and Gulf coasts must contend with occasional large hurricanes.[1] The rise in sea level and continuing subsidence of these passive coasts (combined with changes caused by commercial development and the ongoing rise in sea level) will undoubtedly cost lives and destroy property. **Figure 12.25** suggests the extent of the threat.

Unlike barrier islands, **sea islands** contain a firm central core that was part of the mainland when sea level was lower. The rising ocean separated these high points from land, and sedimentary processes

---

[1]Galveston, Texas, was the site of the worst natural disaster in U.S. history. On 8 September 1900 more than 6,000 people died when a hurricane-driven storm surge swept the barrier island.

NASA

Figure 12.24 Padre Island and Laguna Madre, barrier islands off the Texas Gulf coast. (This photo, taken from space, is on a much larger scale than Figure 12.23.)

© Gerald G. Kuhn

a

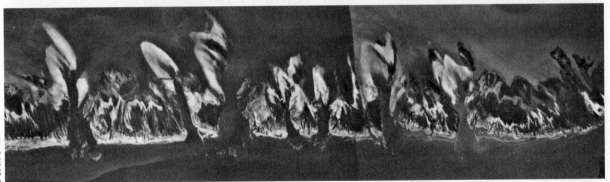

© Gerald G. Kuhn

b

© W. Demetrakas, O.C. Camera

c

Figure 12.25 Barrier island modification, actual and potential. (**a**) The beach extending along the Matagorda Peninsula (Texas) barrier in September 1960. (**b**) The same area six days after the passage of Hurricane Carla in September 1961. The beach and island have been breached, and washover deltas are clearly seen. (**c**) Ocean City, Maryland, a developed barrier island. Host to 8 million visitors a year, this city (and others similarly situated) has no effective protection against flooding and damage from severe storms.

surrounded them with beaches. Hilton Head, South Carolina, and Cumberland Island, Georgia, are sea islands. If the island is close to shore, a bridge of sediments called a **tombolo** may accumulate to connect the island to the mainland. Tombolos can also connect offshore rocky outcrops or volcanoes to the mainland. A sea island and a tombolo are shown in Figure 12.21.

COASTS FORMED BY BIOLOGICAL ACTIVITY

Coasts can be extensively modified by the activities of animals and plants. As we will see, some kinds of marine algae and plants can build coasts, but the most dramatic biological modifications occur in the tropics, where coral polyps form reefs around volcanic islands or along the margin of a continent.

Coral Reefs

The greatest of all reefs is the Australian Great Barrier Reef (**Figure 12.26**), which begins in the Torres Strait separating New Guinea and Australia and runs down the northeastern coast of Australia for 2,500 kilometers (1,560 miles). The reef is not a single object, but a composite of more than 3,000 individual coral reefs covering 350,000 square kilometers (135,000 square miles)—collectively the largest structure made by living organisms on Earth.

Coral animals (which you will learn more about in Chapter 15) are related to the familiar sea anemones found along U.S. coasts. Reef-building coral polyps secrete a cup-shaped calcium carbonate skeleton, which remains behind after the animal dies; the accumulation of skeletons gradually forms the reef. These corals grow best in brightly lighted water about 5 to 10 meters (16 to 33 feet) deep, and in ideal conditions they grow at a rate of about 1 centimeter (½ inch) per year.

Coral Reef Types

In 1842 Charles Darwin classified tropical reef structures into three types: fringing reefs, barrier reefs, and atolls. We still use his classification today.

FRINGING REEFS As their name implies, **fringing reefs** cling to the margin of land. As can be seen in **Figure 12.27a**, a fringing reef connects to shore near the water surface. Fringing reefs form in areas of low rainfall runoff primarily on the leeward (downwind) side of tropical islands. The greatest concentration of living material will be at the reef's seaward edge, where plankton and clear water of normal salinity are dependably available. Most new islands anywhere in the tropics have fringing reefs as their first reef form. Permanent fringing reefs are common in the Hawaiian Islands and in similar areas near the boundaries of the tropics.

BARRIER REEFS **Barrier reefs** are separated from land by a lagoon (**Figure 12.27b**). They tend to occur at lower latitudes than fringing reefs and can form around islands or in

Great Barrier Reef Marine Park Authority, Queensland, Australia

Figure 12.26 A small section of the Great Barrier Reef, Queensland, Australia. This coast has been extensively modified by biological activity.

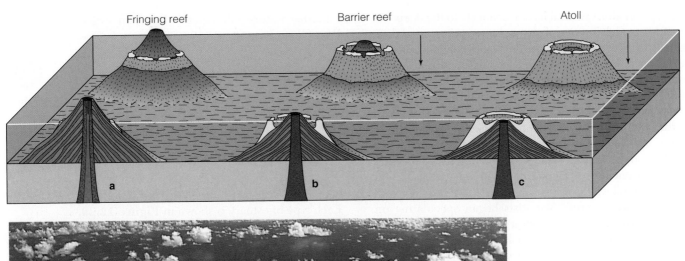

Fringing reef   Barrier reef   Atoll

Figure 12.27 The development of an atoll. (**a**) A fringing reef forms around an island in the trop-ics. (**b**) The island sinks as the oceanic plate on which it rides moves away from a spreading center. In this case, the island does not sink at a rate faster than coral organisms can build up-ward. (**c**) The island eventually disappears beneath the surface, but the coral remains at the sur-face as an atoll. (**d**) The typical ring shape of an atoll.

lines parallel to continental shores. The outer edge—the bar-rier—is raised because the seaward part of the reef is sup-plied with more food and is able to grow more rapidly than the shore side. The lagoon may be from a few meters to 60 meters (200 feet) deep, and it may separate the barrier from shore by only tens of meters or as much as 300 kilometers (190 miles). Coral grows slowly within the lagoon because fewer nutrients are available and sediments and fresh water run off from shore. As you would expect, conditions and species within the lagoon are much different from those of the wave-swept barrier. The calm lagoon is often littered with eroded coral debris moved from the barrier by storms.

The Great Barrier Reef of Australia is the largest biologi-cal construction on the planet. It isn't a single reef but a con-glomeration of thousands of interlinked segments. The seg-ments present a steep outer wall to the prevailing currents and

trade winds. At a growth rate of 1 centimeter (½ inch) per year, the structure is obviously of great age and astonishing volume. The huge reef is younger and thinner at its southern end; the slow northward movement of the Indian–Australian lithospheric plate in which Australia is embedded is thought responsible for this. The variety of organisms within the Aus-tralian Barrier Reef staggers the imagination.

ATOLLS An **atoll** (**Figure 12.27 c and d**) is a ring-shaped island of coral reefs and coral debris enclosing, or almost en-closing, a shallow lagoon from which no land protrudes. Coral debris may be driven onto the reef by waves and wind to form an emergent arc on which coconut palms and other land plants take root. These plants stabilize the sand and lead to colonization by birds and other species. Here is the tropical island of the travel posters.

Though an atoll's central lagoon connects to the deep water outside through a series of channels or grooves, coral does not usually thrive there because the water may become too fresh during rains or too hot, and feeding opportunities for the coral in the enclosed lagoon are limited.

Some atolls are isolated, but most occur in loose groups in shallow continental-shelf areas or in the deep open ocean. More than 300 atolls exist—most in the Pacific. They range in size from a few kilometers in diameter to Kwajalein in the Marshall Islands, whose slender 280-kilometer (176-mile) ring of coral encloses a lagoon of 2,850 square kilometers (1,100 square miles).

How do atolls form? Scientists began speculating on the cause of their ring shape soon after the first scientific voyages published their reports. Charles Darwin imagined a volcanic island growing from the sea, accumulating a skirt of coral around its shore, and then slowly subsiding at a rate equal to the growth rate of the coral. The central volcanic island would eventually sink from view, but the coral could grow continuously atop skeletons of past generations to maintain a living presence near the surface. Figure 12.27a–c shows the progression. Note that the island begins with a fringing reef, passes through a barrier-reef stage as it sinks, and eventually becomes an atoll as the peak disappears beneath the ocean surface. Should the island subside faster than about 1 centimeter (½ inch) per year (the growth rate of coral), all traces of both the island and the reefs would disappear. (The submerged island might become a guyot.)

This theory seemed reasonable, but Darwin couldn't explain what would cause volcanic islands to subside because he didn't know about plate tectonics. Now we know that volcanoes can form near spreading centers, ride outward and downward from their birthplaces, cease to be active as they leave their source of mantle heat, and sink as they are carried into deeper water—just slowly enough to permit coral growth to continue as they go.

Mangrove Coasts

Other coasts have been formed by mangroves, trees that can grow in salt water. The coast of southwestern Florida has been extended and shaped by the activity of mangroves (**Figure 12.28**), whose root systems trap and hold sediments around the plant. The root complex forms an impenetrable barrier and safe haven for organisms around the base of the trees. You will learn more about mangroves in the section on marine plants in Chapter 14.

ESTUARIES

An **estuary** (*æstus* = tide) is a body of water partially surrounded by land, where fresh water from a river mixes with ocean water. Estuaries are areas of remarkable biological productivity and diversity. The coasts of the United States contain about 15,150 square kilometers (5,850 square miles) of estuarine waters. Chesapeake Bay, San Francisco Bay, and Puget Sound are all estuaries.

Classification of Estuaries

Estuaries are classified into four types depending on their origins:

- Drowned river mouths
- Fjords
- Bar-built
- Tectonic

Each type of estuary is shown in **Figure 12.29.**

Estuaries formed at drowned river mouths are common throughout the world, particularly along the Atlantic coast of the United States. Remember, sea level has risen about 125 meters (410 feet) since the end of the last major period of glaciation some 18,000 years ago, and this has resulted in the incursion of seawater into river mouths. The mouths of the York, James, and Susquehanna Rivers and Chesapeake Bay are examples of this type of estuary.

As Figure 12.3 suggests, fjords are steep, glacially eroded U-shaped troughs. They are often about 300 to 400 meters (1,000 to 1,300 feet) deep but typically terminate in a shallow lip, or sill, of terminal glacial deposits. In fjords with shallow sills, little vertical mixing occurs below the sill depth, and the bottom waters can become stagnant (look ahead to Figure 12.30d). In fjords with deeper sills, the bottom waters mix slowly with adjacent oceanic waters. While fjords are common in Norway, Greenland, New Zealand, Alaska, and

Figure 12.28 A mangrove coast in Florida. Mangrove trees trap sediments, building and stabilizing the coast.

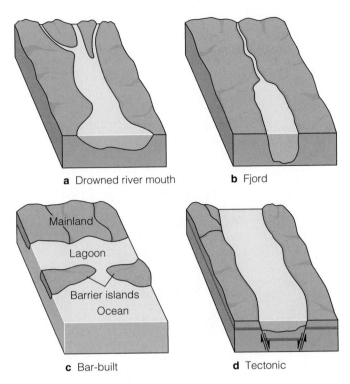

a Drowned river mouth   **b** Fjord

c Bar-built   **d** Tectonic

Mainland
Lagoon
Barrier islands
Ocean

Figure 12.29 Estuaries classified by their origins. (**a**) Drowned river mouths: the mouths of the James, York, and Susquehanna Rivers, and Chesapeake Bay. (**b**) Fjords: Milford Sound (see Figure 12.3), the Strait of Juan de Fuca north of Washington state. (**c**) Bar-built: Albemarle and Pamlico Sounds in North Carolina. (**d**) Tectonic: San Francisco Bay, Tomales Bay (see Figure 12.10).

western Canada, they are not common in the lower 48 states. The Strait of Juan de Fuca in Washington is a unique example.

Bar-built estuaries form when a barrier island or a barrier spit is built parallel to the coast above sea level. Since these estuaries are shallow and usually have only a small inlet connecting them to the ocean, tidal action is limited. Waters in bar-built estuaries are mixed mainly by the wind. Albemarle and Pamlico Sounds in North Carolina and Chincoteague Bay in Maryland are bar-built estuaries.

Estuaries produced by tectonic processes are coastal indentations formed by faulting and local subsidence. Fresh water and seawater both flow into the depression, and an estuary results. San Francisco Bay is, in part, a tectonic estuary.

Characteristics of Estuaries

`12-28`

Three factors determine the characteristics of estuaries: the shape of the estuary, the volume of river flow at the head of the estuary, and the range of tides at the estuary's mouth. The mingling of waters of different densities, the rise and fall of the tide, and the variations in river flow—along with the actions of wind, ice, and the Coriolis effect—guarantee that patterns of water circulation in an estuary will be complex.

Estuaries are categorized by their circulation patterns. The simplest circulation patterns are found in **salt wedge estuaries,** which form where a rapidly flowing large river enters

the ocean in an area where tidal range is low or moderate (**Figure 12.30a**). The exiting fresh water holds back a wedge of intruding seawater. Note that density differences cause fresh water to flow over salt water. The seawater wedge retreats seaward at times of low tide or strong river flow, and it returns landward as the tide rises or when river flow diminishes. Some seawater from the wedge joins the seaward-flowing fresh water at the steeply sloped upper boundary of the wedge, and new seawater from the ocean replaces it. Nutrients and sediments from the ocean can enter the estuary in this way. Examples of salt wedge estuaries are the mouths of the Hudson and Mississippi Rivers.

A different pattern occurs where the river flows more slowly and tidal range is moderate to high. As their name

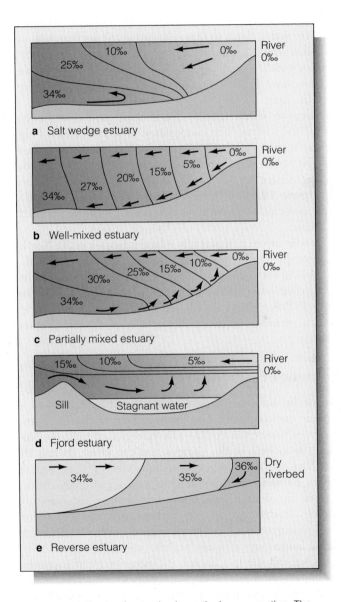

a Salt wedge estuary

b Well-mixed estuary

c Partially mixed estuary

d Fjord estuary

e Reverse estuary

Figure 12.30 Types of estuaries in vertical cross section. The salinity values show the amount of mixing between fresh water (0‰) and seawater (34‰) in the various types. (**a**) Salt wedge estuary. (**b**) Well-mixed estuary. (**c**) Partially mixed estuary. (**d**) Fjord estuary. (**e**) Reverse estuary, in which evaporation plays a major role.

implies, **well-mixed estuaries** contain differing mixtures of fresh water and salt water through most of their length. Tidal turbulence stirs the waters together as river runoff pushes the mixtures to sea. A well-mixed estuary is illustrated in **Figure 12.30b.** The mouth of the Columbia River is an example.

Deeper estuaries exposed to similar tidal conditions but greater river flow become **partially mixed estuaries.** Partially mixed estuaries share some of the properties of salt wedge and well-mixed estuaries. Note in **Figure 12.30c** the influx of seawater beneath a surface layer of fresh water flowing seaward; mixing occurs along the junction. Energy for mixing comes from both tidal turbulence and river flow. England's Thames River, San Francisco Bay, and Chesapeake Bay are examples.

Fjord estuaries form where glaciers have gouged steep U-shaped valleys below sea level. Typically, fjord estuaries have small surface areas, high river input, and little tidal mixing. River water tends to flow seaward at the surface with little contact with the seawater below (**Figure 12.30d**). In fjord estuaries with steep sills, a layer of stagnant water—cold water containing little oxygen and few nutrients—can form above the floor.

Reverse estuaries can form along arid coasts when rivers cease to flow. The evaporation of seawater in the uppermost reaches of these estuaries will cause water to flow from the ocean into the estuary, producing a gradient of *increasing* salinity from the ocean to the estuary's upper reaches (see **Figure 12.30e**). Reverse estuaries are common on the Pacific coast of Mexico's Baja Peninsula and along the U.S. Gulf coast.

In well-mixed and partially mixed estuaries in the Northern Hemisphere, the incoming seawater will press against the right side of the estuary because of the Coriolis effect. Outflowing river water will also trend to the right of its direction of travel. This rightward drift can be seen in the contour of lines representing surface salinity in Chesapeake Bay (**Figure 12.31**).

The Value of Estuaries

Some of the oldest continuous civilizations have flourished in estuarine environments. The lower regions of the Tigris and Euphrates Rivers, the Po River Delta region of Italy, the Nile Delta, the mouths of the Ganges, and the lower Hwang Ho valley have supported dense human habitation for thousands of years. Estuaries continue to be irresistibly attractive to developers. In areas of high population density, estuaries are routinely dredged to provide harbors, marinas, and recreational resources and filled to make space for homes and agricultural land.

As we will see in Chapter 16, estuaries often support a tremendous number of living organisms. The easy availability of nutrients and sunlight, protection from wave shock, and the presence of many habitats permit the growth of many species and individuals. Estuaries are frequently nurseries for marine animals; several species of perch, anchovy, and Pacific

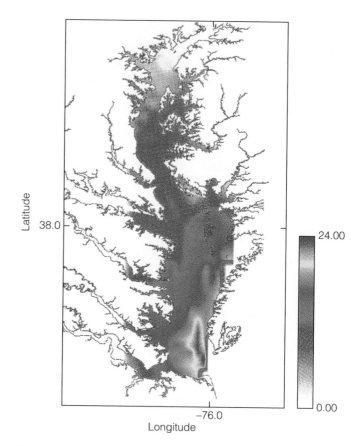

Figure 12.31 Range of salinity during May in Chesapeake Bay, an example of a partially mixed estuary. The colors indicate salinity in parts per thousand. The typical distribution of surface salinity in the estuary ranges from 28‰ at the mouth to 1‰ near the upper reaches. The Coriolis effect forces the inflowing salt water against the right (eastern) bank. (Notice how the 20‰ contour lines trend toward the right bank.) Compare this diagram to the photograph in Figure 12.2. (Source: Used by permission of Glen Wheless.)

herring take advantage of the abundant food in estuaries during their first weeks of life. Unfortunately for their inhabitants, the high demand for development is incompatible with a healthy estuarine ecosystem.

Estuaries have also become the most polluted of all marine environments. Some plants growing in shallow temperate estuaries have the ability to "scrub" polluted water—to remove inorganic nitrogen compounds and metals from seawater polluted by sources on land. Such plants use electrostatic attraction and sticky surface layers to flocculate clay-size particles from the water and deposit them on their surfaces. When the tide falls, this material is deposited at the base of the plant, helping to protect it from erosion. Bacteria in the mud can decompose the nitrogen compounds and bind the metals, making the water cleaner. Ironically, development is destroying the very ecosystems capable of helping to clean the water entering them. More than half the nation's estuaries and other wetlands have been lost. Of the original 870,000 square kilometers (215 million acres) of wetlands that once existed in the lower 48 states, only about 360,000 square kilometers (90 million acres) remains.

Lagoons and Wetlands

The term *lagoon* is used for the region between a barrier island or spit and the mainland, but the term also refers to a shallow estuary essentially isolated from the ocean. Laguna Madre (Figure 12.24), one of the best known, extends along the Texas coast from Corpus Christi to the mouth of the Rio Grande. Most of the lagoon is less than 1 meter (3.3 feet) deep. It is protected from the ocean by Padre Island, a low offshore island about 160 kilometers (100 miles) long. Laguna Madre is hypersaline; high evaporation keeps the salinity above 50‰ for most of the year. Because tidal range in the Gulf of Mexico is so small and the inlets at the ends of the barrier island are constricted, there is very little exchange of water between the lagoon and the ocean.

Several terms are used interchangeably (and often incorrectly) with the term *estuary*. The term *wetland* refers to land on which water dominates the soil development and consequently the types of plant and animal life that live in the soil and on its surface.

A *slough* is a shallow estuary in which large areas of the bottom are exposed during low tide. The term is often used to refer to parts of the Great Lakes and the Everglades. A *salt marsh* is a shallow tidal estuary inhabited by plants that can withstand only limited tidal submergence. A *swamp*, like a salt marsh, is a lowland area saturated with water, but here the water is fresh; trees dominate the vegetation in swamps.

CHARACTERISTICS OF U.S. COASTS

Plate tectonic forces have had immense influence on the margins of continents, and the edges of the United States are no exception. The results of plate movement on the Pacific coast differ greatly from those on the Atlantic and Gulf coasts, primarily because the Pacific coast is near an active plate margin while the Atlantic and Gulf coasts are not.

The Pacific Coast

The Pacific coast is an actively rising margin on which volcanoes, earthquakes, and other indications of recent tectonic activity are easily observed. Pacific coast beaches are typically interrupted by jagged rocky headlands, volcanic intrusions, or the effects of submarine canyons. Wave-cut terraces (**Figure 12.32**) are found as much as 400 meters (1,300 feet) above sea level, evidence that tectonic uplift has exceeded the general rise in sea level through the past million years.

Most of the sediments on the Pacific coast originated from the erosion of relatively young granitic or volcanic rocks of nearby mountains. The particles of quartz and feldspar that constitute most of the sand were transported to the shore by flowing rivers. The volume of sedimentary material transported to Pacific coast beaches from inland areas greatly exceeds the amount originating at the coastal cliffs. Deltas tend not to form at Pacific coast river mouths because the continental shelf is narrow, river flow is generally low (except for the Columbia River), and beaches are usually high in wave energy. The predominant direction of longshore drift is to the south because northern storms provide most of the wave energy.

The Atlantic Coast

The Atlantic coast is a passive margin, tectonically calm and subsiding because of its trailing position on the North American Plate. Subsidence along the coast has been

© John S. Shelton

Figure 12.32 Wave-cut terraces on San Clemente Island, off the coast of southern California. The highest terraces are about 400 meters (1,300 feet) above sea level. The island has been periodically uplifted by tectonic forces.

considerable—3,000 meters (10,000 feet) over the last 150 million years. A deep layer of sediment has built up offshore, material that helped produce today's barrier islands. Relatively recent subsidence has been more important in shaping the present coast, however. Except for the coast of Maine (which is still in isostatic rebound after the recent departure of the glaciers), coastal sinking and rising sea level have combined to submerge some parts of the Atlantic coast at a rate of about 0.3 meter (1 foot) per century. This process has formed the huge flooded valleys of Chesapeake and Delaware Bays, the landward-migrating barrier islands, and the shrinking lowlands of Florida and Georgia.

Rocks to the north (in Maine, for example) are among the hardest and most resistant to erosion of any on the continent, so beaches are uncommon in Maine. But from New Jersey southward the rocks are more easily fragmented and weathered, and beaches are much more common. As on the Pacific coast, sediments are transported coastward by rivers from eroding inland mountains, but the transported material is trapped in estuaries and therefore plays a less important role on beaches. Eastern beaches are typically formed of sediments from shores eroding nearby or from the shoreward movement of offshore deposits laid down when the sea level was lower. The amount of sand in an area thus depends in part on the resistance or susceptibility of nearby shores to erosion. Sand moves generally south on these beaches just as it does on the Pacific coast, but the volume of moving sand is less here.

As we have seen, glaciers have also contributed to the shaping of the northern part of the Atlantic coast: Large portions of Long Island and all of Cape Cod are remnants of debris deposited by glaciers.

The Gulf Coast

12-34

The Gulf coast experiences a smaller tidal range and—hurricanes excepted—a smaller average wave size than either the Pacific or Atlantic coast. Reduced longshore drift and an absence of interrupting submarine canyons allow the great volume of accumulated sediments from the Mississippi and other rivers to form large deltas, barrier islands, and a long raised "super berm" that prevents the ocean from inundating much of this sinking coast.

These are fortunate conditions because the rate of subsidence in the Gulf coast is greater than that for most of the Atlantic coast. Subsidence here is not the result of tectonic activity but rather is due to sediment compaction, dewatering, and the removal of oil and natural gas. Sediment starvation and dredging have exacerbated the situation around some large cities. At Galveston, Texas, for example, sea level is nearly 64 centimeters (25 inches) higher than it was a century ago, and parts of New Orleans are now 2 meters (6.6 feet) below sea level. As we have seen, hurricanes at such places can have tragic results. The protective natural berm can easily be breached, and floodwaters can surge far inland.

HUMAN INTERFERENCE IN COASTAL PROCESSES

12-35

Beaches exist in a tenuous balance between accumulation and destruction. Human activity can tip the balance one way or the other (see **Box 12.1**). For example, consider the rocky **breakwater** shown in **Figure 12.33**. The breakwater inter-

a

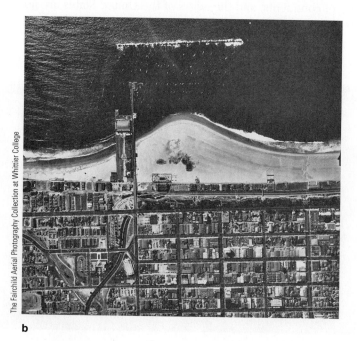

b

Figure 12.33 Growth of a beach protected by a breakwater: Santa Monica, California. (**a**) The shoreline as it appeared in 1931. (**b**) The same shoreline in 1949 after the breakwater was built. The boat anchorage formed by the breakwater is filling with sand deposited by disruption of the longshore current.

Nowhere is *direct* human coastal intervention more obvious than in the Netherlands, or Holland. As its name suggests (*Holland* = hollow land), this is a low-lying country—more than half its landmass lies below sea level. Some 7,800 square kilometers (3,000 square miles) of this land has been wrested from the ocean by an extensive program of water management, drainage, and reclamation that began in the twelfth century. The greatest reclamation project reached a climax in 1932 with the completion of a barrier dam to bisect the Zuider Zee, originally an estuary of the Rhine River. On the protected inner side of the huge dam, engineers constructed a series of dikes and began pumping out water to expose the fertile sediments. By the early 1980s about 2,020 square kilometers (500,000 acres) more land was available to the Dutch for homes and farms.

Interference with coastal processes on such a vast scale is not without hazards, however. More than 1 million citizens still remember the disaster of 1 February 1953. In the middle of the night, a combination of storm waves and a high tide drove the North Sea over the rim of dikes sur-rounding the province of Zeeland (*Zee* = sea). Floodwaters reached 64 kilometers (40 miles) inland; 2,000 square kilometers (495,000 acres) of farmland was inundated. More than 100,000 were left homeless; 1,800 people died.

In response, the Dutch launched the Delta Plan, a massive project that shortened the coastline by about 700 kilometers (435 miles), developed an extensive system of dikes, and built dams, bridges, locks, and a major canal. The land lost in 1953—and more—has been reclaimed from the sea.

But the diligent Dutch (and their pumps) cannot rest. *Indirect* human intervention is also at work. Researchers now predict a significant rise in sea level through the next century. Global warming (see Chapter 18) may melt land ice at a faster-than-anticipated rate, and a warming ocean would expand. Estimates of sea-level rise over the next hundred years range from 20 centimeters (8 inches) to almost 1 meter (3.3 feet). Netherlanders are now planning a costly program to accommodate a 60-centimeter (24-inch) rise in sea level.

12-36

At high tide, the level of the North Sea lies between 4 and 6 meters (13 and 20 feet) above the streets of the Dutch town of Westkapelle.

Peter Van Elk

rupts the progress of waves toward the beach, weakening the longshore current and allowing sand to accumulate there. Without dredging, the beach will eventually reach the breakwater and fill the small-boat anchorage the breakwater was built to provide. This is a minor example of human alteration of a beach, yet it serves to introduce the growing problem of human influences on coastal processes.

We often divert or dam rivers, build harbors, and develop property with surprisingly little understanding of the impact our actions will have on the adjacent coast. Our role then becomes that of powerless observers. Residents of eroding coasts can only accept the inevitable loss of their property to the attack of natural forces, but residents of coasts in which deposition exceeds erosion are sometimes presented with alternatives. The choices are almost never simple. For example, should rivers be dammed to control devastating floods? If the dams are built, they will trap sediments on their way from mountains to coast. Beaches within the coastal cell fed by the dammed river will shrink because the sand on which they depend (to replenish losses at the shore) is blocked. Alarmed coastal residents will then take steps to hang onto whatever sand remains. They may try to trap "their" beaches by erecting **groins,** short extensions of rock or other material placed at right angles to longshore drift to stop the longshore transport of sediments. This temporary expedient usually accelerates erosion downcoast (**Figure 12.34a**). Diminished beaches then expose shore cliffs to accelerated erosion. Wind-wave energy that would have harmlessly churned sand grains now speeds the destruction of natural and artificial structures. Seawalls don't help, either (**Figure 12.34b**). They increase beach erosion by deflecting wave energy onto the sand. Churning by this increased energy eventually undermines the seawall, causing it to collapse. The importation of sand trapped behind dams (or from other sources) is also only a temporary—and very expensive—expedient (**Figure 12.34c**). Scenes like the one shown in **Figure 12.35** will be more common as beaches shrink.

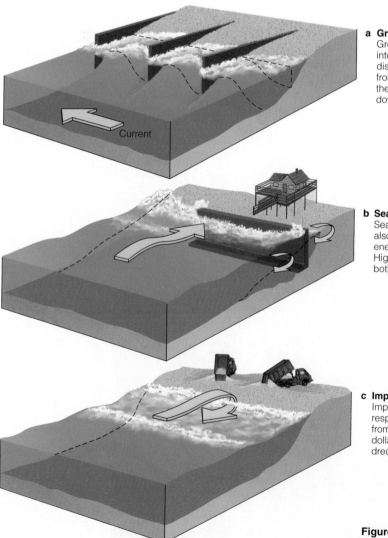

a Groin
Groins are structures that extend from the beach into the water. They help counter erosion by dissipating wave energy and by trapping sand from the current. Groins accumulate sand on their updrift side, but erosion is worse on the downdrift side, which is deprived of sand.

b Seawall
Seawalls protect property temporarily, but they also increase beach erosion by deflecting wave energy onto the sand in front of and beside them. High waves can wash over seawalls and destroy both the seawalls and the protected property.

c Importing sand
Importing sand to a beach is considered the best response to erosion. The new sand is often dredged from offshore and can cost tens of millions of dollars. Because it is often finer than beach sand, dredged sand erodes more quickly.

Figure 12.34 Some measures taken to slow beach erosion, and why these measures are largely ineffective.

Figure 12.35 A resident of Rodanthe, North Carolina, rides his bicycle past a house undercut by Hurricane Dennis in September 1999. The extent of beach erosion is dramatically clear.

AP/Wide World Photos

What are the implications of these unlooked-for sand movements? Douglas Inman, director of the Center for Coastal Studies at Scripps Institution of Oceanography, feels that *at least 20% of the beach-bounded coastline of the United States is in danger of serious or catastrophic alteration.* On the Pacific coast a 30-year period of relatively mild weather may be ending. During this time, people felt it was safe to build close to the shore. Increased dam building and breakwater, jetty, and groin construction have made southern California's beaches more vulnerable. In the 1997–98 El Niño, coastal California alone suffered losses exceeding $750 million. The barrier islands of the Atlantic and Gulf coasts are at least as vulnerable. **Figure 12.36** summarizes the progress of shore erosion by region.

Shores that look permanent through the short perspective of a human lifetime are in fact among the most temporary of all marine structures. Let's enjoy them in their present stages.

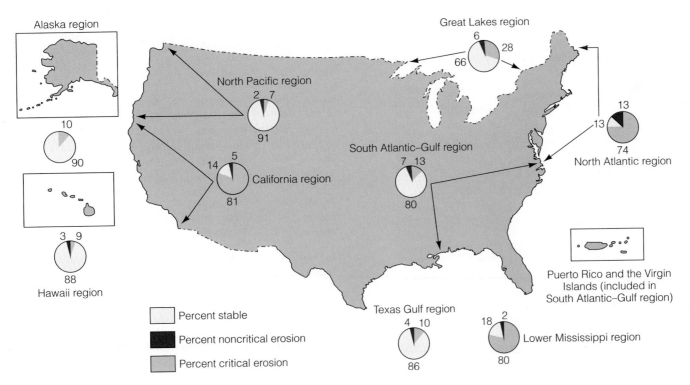

Figure 12.36 Shore erosion by region. Critical erosion is potentially threatening to property.

1. My foot tends to sink whenever I stand on the beach and let water from a wave run over it. The sand moves away from the edges of my foot, and I sink in. Why?

This is a good example of water's ability to carry more sediment as its speed of flow increases. Your foot interrupts the flow of water up or down the beach after a wave breaks, and the water must speed up to get around your foot. Fast-running water moves sand more effectively than slow-running water, so the sand immediately next to your foot is removed. This process, termed *scouring*, becomes a problem when structures are placed in shallow water.

2. How can any beach accumulate material when it slopes *seaward*? Why doesn't all the sediment making up the beach just flow downhill into the ocean and disappear?

Wave action and beach permeability combine to build beaches. Small and moderate waves tend to predominate in places where there are beaches. These waves keep sand on or near the shore because the orbital motion of water within them before they break moves sand more easily *toward* shore than *away* (see **Figure 12.37**). After the waves break, a certain amount of the water that rushes onto the beach soaks into the sand. That much less water is therefore available to push sand grains downhill and out to sea. Coarse-grain beaches are proportionally steeper because a higher percentage of the uprushing water percolates into the beach, so less water is left to rush down the slope and "unpack" the beach.

3. What are those little white pellets I find on the beach along the high-tide line? Surfers call them "nurdles."

Those ubiquitous, insidious particles are the raw material for molded plastic goods. They are transported from producers to fabricators in containers loaded onto container ships. The pellets escape if a container is mishandled, breaks open during a storm, or is lost overboard. Virtually indestructible and able to float, these small "nurdles" float with the winds and currents until they encounter a shore. One researcher has calculated that just 25 containers would carry enough plastic pellets to spread 100,000 "nurdles" per mile along all the seashores of the world!

4. I hope someday to live near the ocean. What should I look for in buying property there?

Firm ground! Coastal Maine would be an ideal bet. The dense metamorphic rock of much of the Maine shore is stable (within the human time frame) and hard—ideal footings for a house. Make sure the site is far enough inland to avoid high surf, storm surge, and tides. If the winters in Maine don't appeal to you, coastal Florida might make a good choice—if your children aren't hoping to inherit the property. If you insist on building on a barrier island, make sure your home is on the mainland side of the southern end! Parts of the Pacific coast are all right, but local variability on that active margin makes some knowledge of the geological history of the area very valuable. For example, some parts of the San Diego shoreline are eroding at a rate of about 3 meters (10 feet) per year—hardly a solid investment.

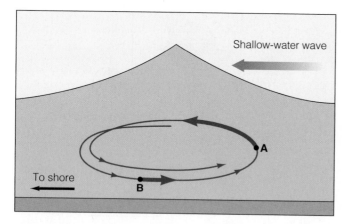

Figure 12.37 In small or moderate waves, a sand grain at point A moves easily *toward* shore, free of most bottom friction. Movement *offshore* by the sand grain at point B is impeded by friction with the bottom. New flow may be reversed from onshore to offshore by high waves because of strong backwash.

KEY CONCEPTS TO REVIEW

- Coasts are temporary structures, often subject to rapid change.

- The *location* of a coast depends primarily on global tectonic activity and the volume of water in the ocean. The *shape* of a coast is a product of many processes: uplift and subsidence, the wearing down of land by erosion, and the redistribution of material by sediment transport and deposition.

- Coasts are classified as primary coasts (on which terrestrial influences dominate) or secondary coasts (on which marine influences dominate).

- Secondary coasts often support beaches, accumulations of loose particles. Generally, the finer the particles on the beach, the flatter its slope. Beaches change shape and volume as a function of wave energy and the balance of sediment input and removal.

- Coral reefs are a form of secondary coast. They represent the largest biological constructions on the planet.

- Estuaries are among the most complex and biologically productive coasts. They are often under pressure for development into harbors and recreational resources.
- Human interference in coastal processes rarely increases the long-term stability of a coast.

INTERNET STUDY RESOURCES

The Web site for this book contains helpful study aids. Log on to:

www.brookscole.com/053437557xs

and click on the Chapter-by-Chapter area. Choose Chapter 12 and select a resource:

- **Flash Cards** allows you to test your mastery of the Terms and Concepts to Remember for this chapter.
- **Tutorial Quizzes** provides a multiple-choice practice quiz.
- **Student Guide to InfoTrac College Edition** will lead you to Critical Thinking Projects that use InfoTrac College Edition as a research tool.

TERMS AND CONCEPTS TO REMEMBER

atoll	high-energy coast
backshore	lagoon
backwash	longshore bar
barrier island	longshore current
barrier reef	longshore drift
bay mouth bar	longshore trough
beach	low-energy coast
beach scarp	moraine
berm	partially mixed estuary
berm crest	primary coast
breakwater	reverse estuary
coast	rip current
coastal cell	salt wedge estuary
delta	sand spit
dissolution	sea cave
drumlin	sea cliff
estuary	sea island
eustatic change	secondary coast
fjord	shore
fjord estuary	swash
foreshore	tombolo
fringing reef	wave-cut platform
groin	well-mixed estuary

STUDY QUESTIONS

Review Questions

1. How is a primary coast different from a secondary coast?

2. What features would you expect to see along a primary coast? A secondary coast? What determines how long the features will last?

3. What are some of the features of a sandy beach? Are they temporary or permanent? Is there a relationship between wave energy on a coast and the size (or slope or grain size) of beaches found there?

4. How are deltas classified? Why are there deltas at the mouths of the Mississippi and Nile Rivers, but not at the mouth of the Columbia River?

5. What is a coastal cell? Where does sand in a coastal cell come from? Where does it go?

6. How are estuaries classified? Upon what does the classification depend? Why are estuaries important?

Critical Thinking Questions

1. In what other ways can coasts be classified? What are the advantages of classification? Are there disadvantages?

2. What two processes contribute to longshore drift? What powers longshore drift? What is the predominant direction of drift on U.S. coasts? Why?

3. What physical conditions might limit or encourage the development of a coral reef? An atoll?

4. Compare and contrast the U.S. Pacific, Atlantic, and Gulf coasts.

5. How do human activities interfere with coastal processes? What steps can be taken to minimize loss of life and property along U.S. coasts?

6. **InfoTrac College Edition Project** Do increased human use and occupation of coastal areas always result in erosion and pollution? Why or why not? What steps could be taken to prevent degradation of coastal areas? Use InfoTrac College Edition to answer this question.

Earth Systems Today CD Question

1. Go to "Volcanism" and then to "Discover Volcanic Landforms." In "Discovering Volcanic Landforms," estimate the ways volcanism can alter the shapes of coasts. Think specifically about the differences between coasts on the leading edges of tectonic plates versus those on the trailing edges.

The Web site for this book contains many ideas for further reading and research. Log on to:

www.brookscole.com/053437557xs

and click on the Chapter-by-Chapter area. Choose Chapter 12 and select a resource:

- **References** lists the major books and articles consulted in writing this chapter, along with comments from the author about their content and reading level.
- **Hypercontents** takes you to an extensive list of sites with news, research, and images related to individual sections of the chapter.

- In **Student Guide to InfoTrac College Edition** scroll down to Suggested Readings from InfoTrac College Edition for brief descriptions of the articles listed and search hints for finding them in InfoTrac College Edition.
- **Regional InfoTrac College Edition** articles are organized into East Coast, West Coast, and Gulf Coast regions, allowing you to study oceanography on a more local level. You can also access Regional InfoTrac College Edition at www.localocean.com.

For additional readings, go to InfoTrac College Edition, your online research library, at:

http://infotrac.thomsonlearning.com/

Life in the Ocean 13

THE BREATH OF THE PLANET

Most biologists agree that life on Earth arose in the ocean (see Chapter 1). The chemical reactions of this planet's living things proceed only in water that is slightly saline, and life, whether terrestrial or aquatic, depends on substances dissolved in water. All life on Earth reflects its marine origin.

Life and the ocean and the atmosphere influence one another through a large and complex set of interactions. In recent years, two contrasting views of these interactions have been put forward. One view holds that life has prospered in spite of many obstacles because it adapts to conditions as it finds them. The competing view, suggested in 1979 by James Lovelock, a British chemist and inventor, holds that all life on Earth actually defines and exerts control over the conditions necessary for its survival. Lovelock's original hypothesis envisions the surface of Earth as a single huge organism intentionally creating an optimum environment for itself. It is called the Gaia hypothesis (pronounced *guy*'a), named for *Gaia*, the early Greek Earth goddess. A less dramatic version of this hypothesis, accepted today by many scientists, acknowledges that life influences the environment but does so without intentional control.

What evidence do we have that life influences the environment? Consider a couple of the many examples involving atmospheric gases. About 19% of the solar radiation reaching Earth from the sun is absorbed by carbon dioxide (CO_2) and water vapor in the atmosphere. This is a near-optimal amount of

A coral reef in the Red Sea.

© Peter Scoones/Planet Earth Pictures

energy to maintain atmospheric temperature at levels ideal for life: Too little heat absorbance and the ocean would freeze, too much and it would boil. As we will see shortly, plants consume CO_2 as they grow, and an excess of CO_2 in the atmosphere would lead to runaway atmospheric warming. However, plants and other photosynthetic producers consume CO_2 as they grow, and excess CO_2 would stimulate the growth of producers; this would in turn cause the CO_2 level to fall, cooling the atmosphere. Too cold an atmosphere would cause the growth of producers to slow, allowing CO_2 levels to rebuild until temperature was again optimal. Thus, the CO_2–producer–energy system acts as a global thermostat.

Dimethyl sulfide (DMS), a chemical used by small marine plants to balance water pressure within their cells, may be another atmospheric temperature regulator. More marine plants would mean more DMS. Increased quantities of DMS in the ocean would lead to more gaseous DMS in the atmosphere over the ocean. Marine air is usually so clean that the cloud cover over the ocean is limited by the supply of condensation nuclei needed to form clouds, but DMS (and its oxidation by-products) can act as condensation nuclei and help to form dense—and highly reflective—cloud layers. Thus, more plant activity again would cause Earth to cool and would decrease the solar energy available for photosynthesis. Less plant activity would cause Earth to warm, encouraging plant growth. Similar feedback loops involving marine plants have been proposed as the mechanisms that modulate seawater salinity, the quantities of atmospheric oxygen and nitrogen, and the ocean's acid–base balance.

Are these delicate interactions proof of *intentional* control? Most researchers say no. But the fact of the interactions is clear: Life, atmosphere, ocean, and land do profoundly affect one another. Rifting lithospheric plates and volcanic eruptions spewing carbon dioxide, sulfurous gases, water vapor, and other substances are thus more than geological phenomena. These and other physical processes affect life in hundreds of ways. Geology and biology are different faces of the same coin.

13-1

WHAT TO WATCH FOR IN CHAPTER 13

We now turn our attention to life in the sea, introducing the biological players on the ocean's physical-chemical-geological stage. It is important to understand that the atoms in living things are no different from the atoms in nonliving things; in fact, they move between the living and nonliving realms in biogeochemical cycles. Also, the energy that powers living things is the same energy found in inanimate objects. So how is it possible to distinguish between life and nonlife? The definition of life in this chapter highlights the highly organized nature of living material and the complex ways that living things manipulate matter and energy.

Life on Earth is notable for both unity and diversity: *diversity* because there are at least 5 million different species (kinds) of living organisms on Earth; *unity* because each species shares the same underlying mechanisms for capturing and storing energy, manufacturing proteins, and transmitting information between generations. That last point is especially important: In a sense, all life on Earth is fundamentally the same; it's just packaged in different ways. All Earth's life forms are related; all have apparently evolved from a single instance of origin.

How does the ocean fit into this? In a sense all life on Earth is marine because of its watery nature and origin. In part at least, life in the ocean is so successful because of the ocean's benign physical characteristics.

This chapter covers general concepts. It is a prelude to our discussion of the ocean's producers, consumers, and decomposers and the communities they form.

BEING ALIVE

It would seem easy to differentiate between the living and nonliving components of the marine environment, but we cannot always see the difference, as **Figure 13.1** points out. The same atoms move continuously in and out of living and nonliving systems. With your last breath you exhaled millions of carbon atoms that entered your body as food in your last meal. Before the day is done, some of those atoms may be incorporated into a nearby houseplant. The carbon atoms have moved from life to nonlife and back again, all in the space of an afternoon. There is nothing special about the atoms or energy of life, nothing to distinguish them from their nonliving counterparts. The free exchange of identical components between life and nonlife complicates any attempts at formal definition. Yet we all make this distinction intuitively by observing the organization of matter in living things, and by considering the ways living things manipulate energy.

Matter

All Earth's organisms are composed of about 23 of the 107 known chemical elements. Four of these—carbon, hydrogen, oxygen, and nitrogen—make up 99% of the mass of all living things, with nine additional elements comprising nearly all the remainder (see **Table 13.1**).

These elements (atoms) combine in living things to form classes of biological chemicals common to all life. The main categories of these chemicals are probably familiar to you: carbohydrates, lipids (fats, waxes, oils), proteins, and nucleic acids (such as DNA, the primary molecule of heredity). Despite the astonishing variety of life forms on Earth—more

Table 13.1	The Chemical Elements Used by Organisms in Their Metabolism	
Element	**Symbol**	**Representative Use**
Major Components[a]		
Carbon	C	All organic molecules
Oxygen	O	Almost all organic molecules
Hydrogen	H	All organic molecules
Nitrogen	N	Proteins, nucleic acids
Macronutrients[b]		
Sodium	Na	Body fluid, osmotic regulation
Magnesium	Mg	Osmotic balance; in chlorophyll
Phosphorus	P	Nucleic acids, teeth/bone/shell
Sulfur	S	Proteins, cell division
Chlorine	Cl	Nerve discharge, osmotic balance, ATP formation
Potassium	K	Nerve discharge, osmotic balance, enzyme activation
Calcium	Ca	Shell, bone, coral, teeth
Iodine	I	Thyroid hormone
Silicon	Si	Rigid parts
Micronutrients[c]		
Iron	Fe	Electron transport, N assimilation
Copper	Cu	Electron transport
Zinc	Zn	Nucleic acid replication and transcription

Source: Adapted from Milne, 1995, p. 281.
[a]Major components are elements that make up 100,000 or more of every million atoms in an organism.
[b]Macronutrients constitute between 1,000 and 100,000 of every million atoms.
[c]Micronutrients constitute fewer than 1,000 of every million atoms.

than 5 million species are suspected—biologists have come to appreciate the central message of biology as *unity*, not diversity. That is, the extraordinary sameness of these biological chemicals as they are found in the structure of all living things suggests that all life is related. (Indeed, as we saw in Chapter 1, all life on Earth may have evolved from a common ancestor.) The complex organization of atoms into biological chemicals and the organization of these biochemicals into repeating structures such as cells (**Figure 13.2**) suggest a definition: If we look closely at a candidate and find functioning cells and the ability to transmit this intricate organization to a new generation, the candidate is probably alive.

Energy

Living matter cannot function without **energy,** the capacity to do work. Living organisms cannot create new energy, but they can transform one kind of energy to a different kind. A

© Linda E. Tway

Figure 13.1 Living or nonliving? The brightly colored "rock" in this picture is a colony of small marine plants. The distinction between life and nonlife lies not in composition or outward appearance, but in the ability to manipulate energy.

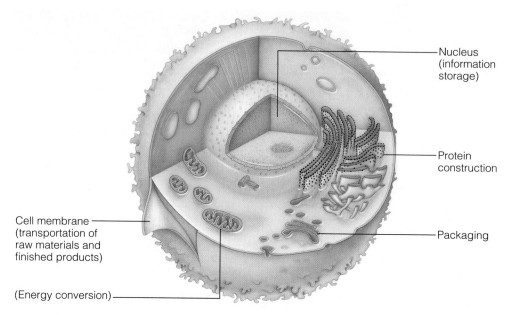

Nucleus
(information
storage)

Protein
construction

Cell membrane
(transportation of
raw materials and
finished products)

Packaging

(Energy conversion)

Figure 13.2 A cell and a few of its functions. A living cell is a complex, self-contained factory that absorbs raw materials from its surroundings and uses them to maintain itself and manufacture finished products for the use of the organism as a whole. If we look closely at an object and find functioning cells and the ability to transmit this intricate organization to a new generation, the object is probably alive.

plant can transform light energy into chemical energy; an animal can transform chemical energy into energy of movement by the muscles, can transform energy of movement into heat, and so on. In one way or another, all life activity is involved, directly or indirectly, in energy transformation and transfer. Energy must therefore be central to anyone's definition of life.

Living organisms are complex assemblies; they need energy to build that complexity from large numbers of simple molecules. The **second law of thermodynamics** shows, however, that disorder inevitably tends to increase in the universe as time passes. Things run down, become disorganized, and break. **Entropy** is a measure of this disorder.

Living things are not exempt from the second law of thermodynamics, but they can delay that inevitable descent into disorganization because the transformation of energy in living things allows a temporary and local remission of the second law. Thus, living organisms might be defined as localized regions where the flow of energy results in *increased* order, that is, areas of great complexity and low entropy. Car engines use the energy in gasoline only to move. Organisms use energy in food to move, to maintain their highly complex organization, and to grow. Cars and word processors don't fix themselves; fish and people do. This sophisticated use of energy seems to be a basic attribute of living things.

The main source of energy for living things on Earth is the sun. Life prospers, becomes ever more complex, and evolves into millions of forms by accepting sunlight and radiating waste heat to the cold of space. As we will see in the next chapter, with a few exceptions organisms get their power directly or indirectly through the capture, storage, and transmission of energy from sunlight. Light energy is transformed into chemical energy and finally into heat as organisms temporarily forestall the disorderly fate decreed by the second law of thermodynamics.

A Working Definition of Life

A definition of *life* should incorporate most of these ideas: Living things contain matter in a highly organized, low-entropy state; they can capture, store, and transmit energy. Organisms are also capable of reproduction and change through time, and they adapt to their environment. The essential differences between living and nonliving systems appear to be in the way living things use energy and in living systems' ability to reproduce.

BIOGEOCHEMICAL CYCLES

Living things contain only a tiny fraction of Earth's total supply of elements. The atoms and small molecules that make up the biochemicals, and thus the bodies, of organisms move between the living and nonliving realms in **biogeochemical cycles.** Living organisms are supported and sustained by huge nonliving reserves, and there is a large-scale transport of elements between the reserves and the organisms themselves. Sometimes an environment has enough of a required element to sustain life; sometimes the element is in short supply.

You learned in Chapter 6 about the ocean's density stratification. The tropical and temperate ocean is usually highly stratified, with a warm, less dense layer of water—the *mixed zone*—separated from the cold, dense *deep zone* by a strong pycnocline (see Figure 6.12). In both ocean areas, the atoms and small molecules that make up the bodies of organisms may cycle rapidly for a time among predators, prey, scavengers, and decomposers in the mixed surface layer. When these organisms die, their bodies sink below the sunlit upper sea to beneath the pycnocline, where they are isolated from the rapid biological activity of the surface. Upwelling plays a critical role in returning these substances to the surface, if only for a short reprieve before their eventual incorporation into deep sediments from which only the very slow progress of tectonic cycles will liberate them.

As you read about the biogeochemical cycles described in the following sections, remember that the elements and small molecules that form the tissues of an organism *are always on the move.* They may cycle rapidly in and out of living things, or they may be trapped in Earth for great spans of time, but the nature of the cycles dictates what will live where, which creatures will be successful, and, ultimately, the very composition of the ocean and the atmosphere.

Carbon

The largest of all biogeochemical cycles is the global **carbon cycle** (**Figure 13.3**).

Because of its ability to form long chains to which other atoms can attach, carbon is considered the basic building block of all life on Earth. Carbon enters the atmosphere by the respiration of living organisms (as carbon dioxide), volcanic eruptions that release carbon from rocks deep in Earth's crust, the burning of fossil fuels, and other processes.

As we will see in the next chapter, large and small plants (and plantlike organisms) capture sunlight and use this energy to incorporate, or *fix,* CO_2 into organic molecules. Some of these molecules are used as food and some as structural components. When a marine animal eats one of these plants (or plantlike organisms), one of three things can happen to the carbon: (1) it can be incorporated into the animal's body for growth, (2) it can be

respired by the animal (taken apart to harvest the energy), or (3) it can be wasted, excreted back into the seawater as **dissolved organic carbon** (**DOC**). Typically about 45% of the carbon is used for growth, about 45% is used for respiration, and about 10% is lost as DOC. The end product of respiration is CO_2, a gas eventually lost to the atmosphere. The DOC is used by bacteria, which are in turn eaten by protozoans, which are eaten by zooplankton, which are then eaten by fish, in what is termed the *microbial loop.* Eventually, the organisms (or at least their hard parts containing calcium carbonate, $CaCO_3$) sink below the mixed layer and begin the long fall toward the seabed. Most of the carbon will be respired (turned into CO_2) by bacteria long before it hits the bottom, but a small percentage (less than 1%) will reach the sediments and be buried. Over geologic time, the carbonate sediments can be uplifted and weathered and the carbon returned to the biologically active upper sea.

Because of the large amount of carbon dioxide available in the ocean and because CO_2 from the atmosphere dissolves readily in seawater, marine organisms almost never suffer from a deficit of available carbon. For life in the sea, the critical bottlenecks lie elsewhere—mainly in the nitrogen, phosphorus, and iron cycles.

Nitrogen

Nitrogen is a critical component of proteins, chlorophyll, and nucleic acids. Like carbon, nitrogen may be found in the bodies of organisms, as a dissolved gas (N_2), and as dissolved

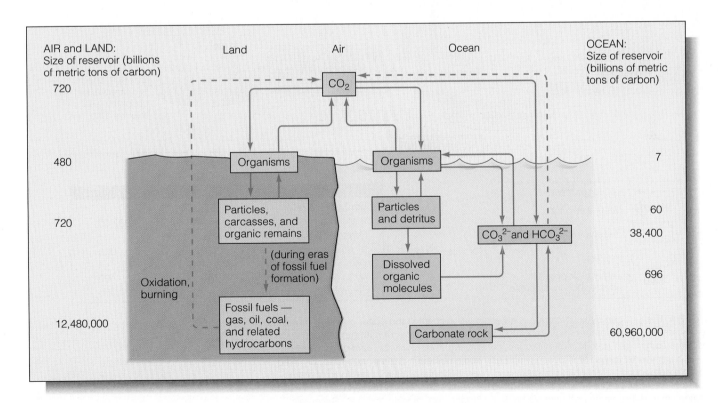

Figure 13.3 The carbon cycle. Reservoirs (boxes) and main transfer pathways (arrows) show the movement of carbon through living and nonliving systems. The numbers show the amount of carbon in each reservoir. CO_3^{2-} is the carbonate ion, HCO_3^- the bicarbonate ion.

organic matter (**dissolved organic nitrogen, DON**). It also exists as the dissolved inorganic ions nitrate (NO_3^-), nitrite (NO_2^-), and ammonium (NH_4^+).

We might expect nitrogen to be abundantly available in the ocean—Table 7.4 indicates that nitrogen accounts for 48% of the dissolved gas in seawater, by volume. But most organisms cannot use the free nitrogen in the atmosphere and ocean directly. It must first be bound with oxygen or hydrogen, or *fixed*, into usable chemical forms by specialized organisms, usually bacteria or cyanobacteria. Oceanic regions are thus frequently nitrogen-limited; the growth of plants and plantlike organisms is often held back because of a lack of available nitrogen. Readily usable forms of nitrogen enter the ocean primarily as nitrate in rivers and precipitation and by nitrogen-fixing organisms in the ocean.

Small oceanic plants use mostly nitrate, but many can also use nitrite and ammonium, and a few can use organic-nitrogen forms such as urea. After being assimilated by small plants, nitrogen is recycled by animals that consume the plants and then excrete primarily ammonium and urea. These reduced forms of nitrogen are then oxidized back into nitrate, via nitrite, by **nitrifying bacteria.** In the deep ocean, most of the nitrogen is in the form of nitrate. In anoxic sediments and certain low-oxygen regions of the ocean, **denitrifying bacteria** use nitrate in respiration and convert it back to nitrite and nitrogen gas, which is lost to the atmosphere. The other major loss occurs in the burial of nitrogen-containing organisms and debris in ocean sediments.

The **nitrogen cycle** is depicted in **Figure 13.4.**

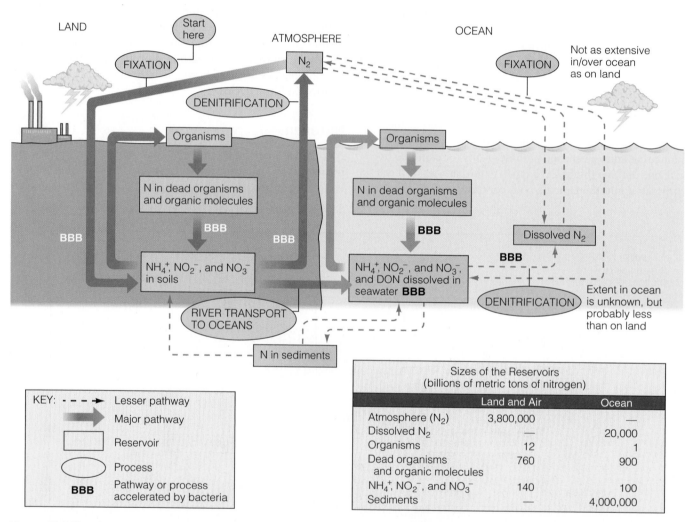

Figure 13.4 The nitrogen cycle. Nitrogen is an essential element in the construction of proteins, nucleic acids, and a few other critical biochemicals. The largest store of nitrogen resides in the atmosphere as nitrogen gas, N_2. This nitrogen, constituting 78% of the air, cannot be used by eukaryotic organisms (organisms possessing cells with a nucleus). It must be bound, or "fixed," into usable compounds by nitrifying bacteria and cyanobacteria, usually in the form of ammonium and nitrite ions, before it can be assimilated by plants. Denitrifying bacteria convert nitrate or ammonium in dead plants and animals back to nitrogen gas, closing the loop. Reservoirs (boxes) and main transfer pathways (arrows) of the nitrogen cycle are shown. The table shows the amount of nitrogen in each reservoir. NH_4^+ is the ammonium ion, NO_2^- the nitrite ion, and NO_3^- the nitrate ion. Estimates of dissolved organic nitrogen (DON) are still unclear.

Sizes of the Reservoirs (billions of metric tons of nitrogen)		
	Land and Air	Ocean
Atmosphere (N_2)	3,800,000	—
Dissolved N_2	—	20,000
Organisms	12	1
Dead organisms and organic molecules	760	900
NH_4^+, NO_2^-, and NO_3^-	140	100
Sediments	—	4,000,000

Phosphorus and Silicon

All organisms use phosphorus to link the parts of nucleic acids, and it is used in molecules that carry energy within the cell. Calcium phosphate is used in the formation of bones, teeth, and some shells. Several groups of marine organisms, including diatoms and radiolarians, make their skeletons from silica (silicon dioxide).

The phosphorus and silicon cycles in the ocean are relatively simple. Both elements enter the ocean in rivers and precipitation, and both are taken up by organisms near the ocean's surface. Phosphorus is generally released in organic form when organisms decompose, but at the pH of seawater this organic phosphorus is rapidly converted back to phosphate, which is then available to be reused by phytoplankton and bacteria. Silica in shells is released back to the water when organisms die and is then available to be reused as the ionized, dissolved form, silicate.

Both phosphorus and silicon cycle in three loops. The most rapid recycling occurs in the daily feeding, death, and decay of surface organisms. A slower loop occurs as the bodies of organisms fall below the pycnocline and phosphorus escapes downward into deep-ocean circulation. A few hundred years may pass before the phosphorus or silicon is again available in the sunlit surface waters where plants can take it up again. The longest loop begins with the phosphorus or silicon locked into rocks or shells that become marine sediments. Sediment subduction at converging plate margins and the

subsequent volcanic reemergence into the ocean of phosphorus- and silicon-containing compounds may take millions of years. The phosphorus cycle is shown in **Figure 13.5**.

Iron and Other Trace Metals

Iron is used in minute quantities in the reactions of photosynthesis, in certain enzymes crucial to nitrogen fixation, and in the structure of proteins. Other essential trace metals such as zinc, copper, and manganese are used by organisms in smaller quantities, primarily in enzymes. Although in absolute terms iron is required by organisms in very small quantities, its concentration in seawater relative to the concentration of the important compounds required for the production of organic matter (nutrients such as nitrogen and phosphorus) can sometimes be so low that phytoplankton growth is limited by the availability of iron. This may seem strange—after all, iron is one of the most abundant elements in Earth's crust. But iron is nearly insoluble in oxygenated seawater, and the little dissolved iron that is present is highly reactive, sticks to falling particles, and sinks to the bottom of the water column.

In general, the biogeochemical cycles of the trace metals follow the pattern we have described previously: uptake and recycling in the surface ocean, and regeneration—sometimes over long periods of time—at depth. However, much remains to be learned about the interactions between living organisms and trace metals. Iron and other trace metals exist in many

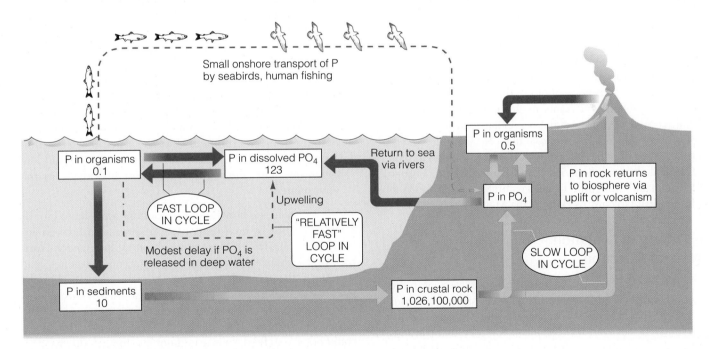

Figure 13.5 The phosphorus cycle. Phosphate is used by organisms in the molecules of genetics and in energy storage and release. Calcium phosphate is used in the formation of bones, teeth, and some shells. Phosphorus cycles in three loops, in which phosphorus moves fast, "relatively fast," or very slowly. The fast loop centers on the everyday activities of marine organisms in surface waters. The "relatively fast" loop involves returning phosphates to the surface when organisms (or parts of organisms) sink beneath the pycnocline. The slow loop is slow indeed—phosphorus-containing material falls to the seabed, is covered with sediment, is subducted, and millions of years later returns to the surface through volcanic eruptions.

chemical forms in seawater. Discovering what these forms are, how transformation between forms occurs, and how available the different forms are to marine organisms is a major focus of current research in trace metal biochemistry.

LIMITING FACTORS

Often, too much or too little of a single physical factor can adversely affect the function of an organism. We call that factor a **limiting factor,** a physical or biological necessity whose presence in inappropriate amounts limits the normal action of the organism. Imagine an ocean area in which everything is perfect for life—temperature, light, nutrients, proper pH, adequate CO_2—but there is *no available iron.* In that circumstance organisms could neither construct the enzymes necessary to convert nitrate to nitrite nor make amino acids essential to build proteins. Iron would be a limiting factor. If iron was present but light was absent, light would be the limiting factor. Sometimes *too much* of something—heat, for instance—can be limiting.

EVOLUTION AND LIFE IN THE OCEAN

Marine life has not accidentally capitalized on its rich fluid home. Earth's organisms did not arise in just a few thousand years. They have changed, generation by generation, over almost 4 billion years. The ability of living things to change through time to fit the physical and chemical environment, to become ever more efficient at extracting energy from their surroundings, to colonize virtually every location capable of sustaining them, and finally to investigate themselves by scientific logic has come about through **evolution.**

Evolution means change. Clothing styles evolve, systems of government evolve, our perceptions of the world evolve; *things change.* That animals and plants might be capable of change with the passage of time was not a popular idea in the nineteenth century. Indeed, the proposal that organisms do change with time began what has since been called a revolution in biology. The unlikely revolutionaries were Charles Darwin (**Figure 13.6**), a quiet and thoughtful English naturalist, and Alfred Wallace, an English biologist working in the Malay Archipelago.

The Theory of Evolution by Natural Selection

Why was evolution by **natural selection** controversial in the nineteenth century, and why is it capable of stirring passions even today? By the mid-1850s, Darwin and Wallace had independently discovered a mechanism for how living things might evolve—change—with the passage of time. Here are Darwin's main points:

1. In any group of organisms, more offspring are produced than can survive to reproductive age.
2. Random variations occur in all organisms. Some of these variations are inheritable; that is, they can be passed on to offspring.

Figure 13.6 Charles Darwin, codiscoverer of the principle of natural selection, ca. 1880. He is pictured between a Galápagos marine iguana (on the left) and the mainland iguana from which it presumably evolved (on the right). Darwin visited the Galápagos Islands off the Pacific coast of South America during his voyage aboard HMS *Beagle* in the early 1830s; his landmark book *On the Origin of Species* was published in 1859.

3. Some inheritable traits increase the probability that the organisms possessing them will survive. These are favorable traits.
4. Because bearers of favorable traits are more likely to survive, they are also more likely to reproduce successfully than bearers of unfavorable traits. Thus, favorable traits tend to accumulate in the population; they are *selected.*
5. The physical and biological (*natural*) environment itself does the selection. Favorable traits are retained because they contribute to the organism's success in its environment. These traits show up more often in succeeding generations if the environment stays the same. If the environment changes, other traits become favorable—and the organisms with those traits live most effectively in the environment.

It is easy to see how random variations are selected for or selected against by environmental pressures, but how do entirely new traits arise? They come about by spontaneous **mutation,** an inheritable change in an organism's genes (the structures that contain its assembly instructions). The vast majority of mutations are unfavorable, so the organisms possessing them are eliminated by other organisms or by the physical environment. For example, a tuna born with no eyes could not see to feed, so it would not live to reproductive age. But a tuna born with extraordinarily good eyesight might have more to eat than its cohorts, and being better nourished it might be especially effective in its reproductive efforts, spreading its genes far and wide. It is the accumulation of these favorable traits, either variations or mutations—and the elimination of unfavorable traits—that makes life possible in changing conditions on Earth (see **Box 13.1**).

The environment for life changes as time passes, and the changes are not always gradual. The biological history of Earth has been interrupted—catastrophically—at least six times in the last 450 million years. In these events, known as **mass extinctions,** a great many species died off simultaneously (in geological terms). Scientists are not certain of the causes of the mass extinctions, but leading candidates for a couple of these events include the collision of Earth with an asteroid or a comet.

Flocks of asteroids (**Figure a**) orbit the sun along with Earth and the other planets. Some of these asteroids have orbits that cross our own, and meetings are inevitable. Earth and its neighbors are pocked with impact craters (**Figure b**) as evidence of these meetings. The consequences for Earth of a collision with even a small asteroid are all but unimaginable. An asteroid only 10 kilometers (6 miles) in diameter would strike with an energy equivalent to the explosion of 500 billion tons of TNT (**Figures c** and **d**). More than 100 million metric tons of Earth's crust and mantle would be thrown into the atmosphere, obscuring the sun for decades and causing sulfur-rich acid rain that would poison the planet's surface. Concussive shock waves would shatter structures, crush large organisms, and trigger earthquakes for a radius of hundreds of kilometers. If the impact occurred in the Atlantic Ocean, say 1,600 kilometers (1,000 miles) east of Bermuda, the resulting tsunami would wash away the resort islands of the Caribbean and swamp most of Florida. Boston would be struck by a 100-meter (330-foot) wall of water. A hole more than 25 kilometers (16 miles) wide and perhaps 10 kilometers (6 miles) deep would mark the point of impact. Clouds of fine particles, accelerated to escape velocity, would travel around the sun in orbits that would intersect Earth's; a steady rain of fine debris might fall for tens of thousands of years.

These kinds of cataclysmic events have been disquietingly common in Earth's past. Where are the craters? As you may recall from Chapter 3's discussion of plate tectonics, much of the ocean floor has been recycled by the movement of lithospheric plates, so any undersea impact craters more than about 100 million years old have disappeared. The distortion and erosion of continents have obscured the outlines of ancient craters on land, although they are more readily visible from space than from the surface—especially if we know what to look for (as in Figure b). Evidence for one massive impact has been bolstered by the discovery of a thin, worldwide layer of iridium-rich continental rock dated at the boundary between the Cretaceous and Tertiary periods (see Appendix II). Iridium is rare on Earth but common in asteroids. The thin iridium-rich layer may have formed from the dust settling after a collision some 65 million years ago.

a What one asteroid looks like, courtesy of the *Galileo* spacecraft en route to study Jupiter. This asteroid would extend halfway from Washington, D.C. to Baltimore.

b What an impact crater looks like. Aerial view of Manicougan Crater, Quebec, where an asteroid struck Earth some 210 million years ago. The crater is about the size of Rhode Island. Asteroids are rocky, metallic bodies with diameters ranging from a few meters to 1,000 kilometers (600 miles). Most were swept into the planets during their formation, but about 6,000 large asteroids are still orbiting the sun in a belt between Mars and Jupiter. Unfortunately, the orbits of many dozens of others cross Earth's orbit.

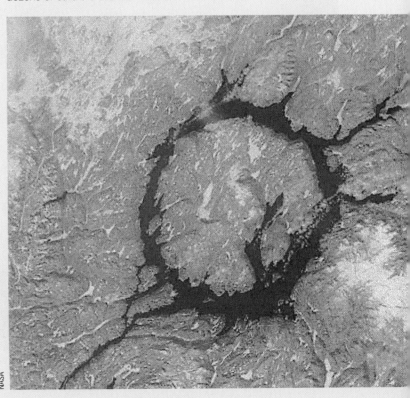

c An artist's conception of a large cometary nucleus, 13 to 16 kilometers (8 to 10 miles) across, striking Earth 65 million years ago. The cataclysmic release of energy is thought to have propelled shock waves and huge clouds of seabed, crust, and even mantle material across Earth, producing a time of cold and dark that contributed to the extinction of many species, including the dinosaurs. The resulting changes in the composition of the atmosphere may also have been toxic to animals.

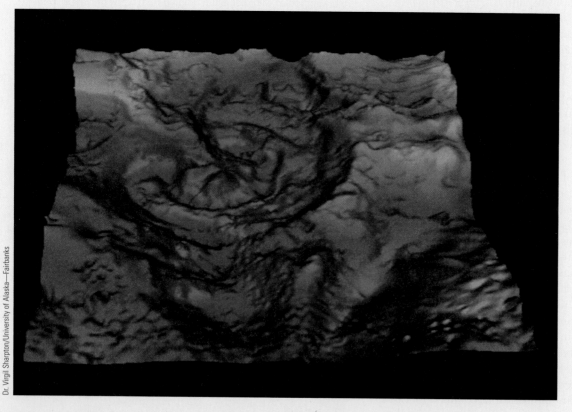

d A gravity map of the Chicxulub crater off Mexico's Yucatán coast, the largest impact crater yet found on Earth. The crater is at least 180 kilometers (113 miles) in diameter. This impact has been blamed for the mass extinctions at the end of the Cretaceous period some 65 million years ago.

Table a The Six Great Extinctions			
Geological Period in Which Extinction Occurred	Millions of Years Ago	Percent Marine Extinctions for:[a]	
		Families	Genera
Late Ordovician	435	27	57
Late Devonian	365	19	50
Late Permian	245	57	83
Late Triassic[b]	220	23	48
Late Cretaceous[b]	65	17	50
Late Eocene	35	2	16

Source: Sagan, C., and A. Druyan. 1985. *Comet* (New York: Random House).
[a]Rough estimate of percent of all families and genera of marine animals with hard parts (so we have fossil evidence of their existence) rendered extinct; numbers are given to the nearest 5 million years (see column 2).
[b]Mass extinctions for which asteroid or comet strikes may be responsible.

As can be seen in **Table a,** vast numbers of marine organisms have perished in mass extinctions. (Extinctions of land families and genera are thought to have been roughly comparable.) At the end of the Cretaceous period, about 65 million years ago, almost one of every five families, half the genera, and three-quarters of the species disappeared. Many scientists believe an asteroid impact triggered the mass extinction. Clearly these tumultuous interruptions do not represent biological business as usual. In a few instances, rocky visitors from space appear to have massively disrupted the environment for life on this planet. The animals, plants, bacteria, and single-celled organisms we see on Earth are descendants of the survivors.

Note that although mutations occur randomly, evolution by natural selection is anything but random. The natural environment winnows favorable mutations from unfavorable ones—hence the origin of the term *natural selection.* The process takes a great deal of time, but time is in abundant supply now that geologists have shown Earth to be about 4.6 billion years old.

This evolutionary viewpoint provided a new way of looking at life: An organism is a vessel holding a particular combination of traits, *testing* that combination. If the organism is successful it will reproduce, and the genes responsible for its good traits will continue in the population. If the organism is not successful that combination will be eliminated. There is no biological predeterminism. Organisms don't *want* to evolve, and individual organisms *don't* evolve. Rather, generation after generation, groups of individuals respond to environmental pressures by changing. The changes can be in shape, size, color, biochemistry, behavior, or any other aspect of the organism. Evolution by natural selection is the accumulation of these beneficial inheritable structural or behavioral traits, known as favorable **adaptations.** Organisms with favorable adaptations have more reproductive success than less well adapted organisms.

A **species** is a group of actually (or potentially) interbreeding organisms that is reproductively isolated from all other forms of living things. How do new species arise? One way is by physical isolation, such as that created when land animals or birds are rafted or blown from a mainland shore to an isolated oceanic island. Because the number of breeding animals within a species on an island may be small, evolutionary change may be rapid; that is, favorable traits may accumulate quickly in the population, and, generation after generation, the species will change relatively rapidly to suit its new habitat. In general, the smaller the reproducing population, the more rapid the rate of evolutionary change.

For example, on the Galápagos Islands off the coast of Ecuador, Darwin observed finches and marine lizards, most of which closely resembled their mainland South American ancestors (see again Figure 13.6). They were not the same species of birds and reptiles that occurred in mainland South America, however. This suggested to Darwin that isolation was a driving force of evolution.

Evolution, then, is the maintenance of life under changing conditions by continuous adaptation of successive generations of a species to its environment. It is a remarkable, beautiful, productive theory.

Evolution in the Marine Environment

Evolutionary theory contains important implications for marine science. The ocean has a much larger inhabitable volume than dry land, and as we will see in the discussion of physical factors that follows, it is often an easier place to live than either the terrestrial or freshwater realms. The ocean contains only one-fifth of the species known to science, but some scientists suggest that counting species is not a valid way of assessing the success of an environment. All major animal groups (known as *phyla*) are found in the sea, and one-third of them are exclusively marine. If plants and single-celled organisms are included, at least 80% of all phyla include marine species. Perhaps an examination of lifestyles (what an organism *does*) is more important in judging the biological diversity of an environment than what an organism *is*. There are more ways of "making a living" in the ocean than on land. Indeed, some important oceanic lifestyles have no terrestrial counterparts. For example, filter feeding—practiced by clams, barnacles, and some whales—relies on a relatively dense fluid matrix and is therefore not seen on land. Also, marine food chains tend to be more complex than terrestrial ones and contain more trophic levels. Marine life has evolved in countless ways to take advantage of nearly every scrap of energy, nearly every nook and cranny of space.

Since physical conditions in the open ocean are relatively uniform, large marine animals with similar lifestyles but different heritages eventually tend to look much the same. That is, similar conditions may result in coincidentally similar organisms. The shark (a fish), the ichthyosaur (an extinct marine reptile), the penguin (a bird), and the porpoise (a mammal) resemble one another in shape even though they are only remotely related, because the physics of rapid movement through water requires a similar streamlined shape (see **Figure 13.7**). Traits leading to this

Figure 13.7 Convergent evolution in sharks, ichthyosaurs, penguins, and dolphins. Selection for adaptations that permitted rapid swimming resulted in superficially similar shapes among these four kinds of vertebrates, even though they are only remotely related.

shape were independently selected by environmental conditions. These accumulated adaptations resulted in superficially similar animals, each derived from different and diverse stock. The process is known as **convergent evolution.**

Through these processes life and Earth change together. Life is tenacious; it has survived catastrophe and calm. In every instance, however, *the environment isn't right for the organisms; rather, the organisms are right for the environment.* One is reminded of author Pår Lagerkvist's observation, "Things need be as they are." Living things have adapted to the physical conditions of their environment, and they continue to increase in number, complexity, and efficiency.

CLASSIFICATION OF OCEANIC LIFE

The organizing principle of evolution by natural selection has given biologists a powerful tool for arranging organisms into logical groups by their ancestry. The study of biological classification is called **taxonomy** (*taxo* = to put in order, *onoma* = name).

Classification schemes have been around for as long as people have looked at living things. The Greek philosopher Aristotle proposed a system of classifying animals based on their exterior similarities, but his results were not very useful. Using his system we would place airline pilots, gliding squirrels, flying fish, and grasshoppers into the same group because each can fly! Such a system is an **artificial system of classification.** (Another artificial system of classification would be grouping books by jacket color, page size, or typeface.) By contrast, the **natural system of classification** for living organisms biologists use today relies on structural and biochemical similarities among organisms. We place all insects together regardless of their flying ability, just as we place all books by Melville together, all compositions of the Bach family together, and all sea stars together because each group has a common underlying natural origin. The groups are arranged *systematically*—that is, in some order that makes structural and evolutionary sense.

One of the first to classify groups of organisms into natural categories was the eighteenth-century Swedish naturalist Carl von Linné, or, as he called himself, **Carolus Linnaeus** (**Figure 13.8**). In his zeal to classify every aspect of the natural world, Linnaeus invented three supreme categories, or **kingdoms**: animal, vegetable, and mineral. Today's biologists leave the mineral kingdom to the geologists and have expanded Linnaeus's two living kingdoms to six. The names and characteristics of these six kingdoms are listed in **Figure 13.9** and **Table 13.2.**

Determining the placement of an organism into a kingdom requires a fundamental understanding of the nature of the organism and its evolutionary relationship to other organisms. Placement is determined by careful analysis of an organism's evolutionary heritage and genetic makeup. Five of the modern kingdoms are "natural"; that is, they include

Figure 13.8 Carolus Linnaeus—the father of modern taxonomy—in Laplander costume. (He went on a scientific expedition to Lapland in 1732.)

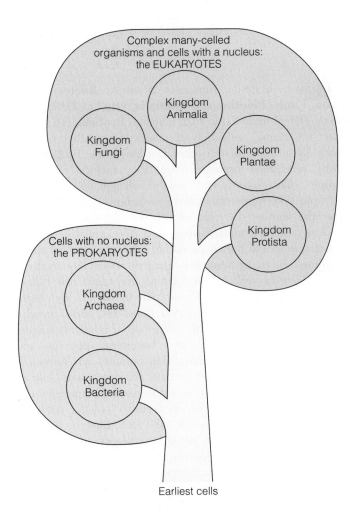

Figure 13.9 The hierarchy of kingdoms presumably evolved from a distant common ancestor. Bacteria and Archaea contain cells without nuclei or organelles, the **prokaryotes**. Fungi, protists, animals, and plants contain cells with nuclei and organelles, the **eukaryotes**.

Table 13.2 Classification of Organisms into Six Kingdoms

Group	Kingdom	Characteristics	Examples
Prokaryotes: Single-celled organisms lacking a nucleus and other internal structural subdivisions; feed by absorption, photosynthesis, chemosynthesis.	Bacteria	Single chromosome, asexual reproduction, extreme metabolic diversity, no nucleus or cytoskeleton.	Bacteria, cyanobacteria ("blue-green algae").
	Archaea	Superficially similar to bacteria, but with many different genes capable of producing different kinds of enzymes; often live in extreme environments.	*Methanococcus, Pyrolobus*, "extremophiles."
Eukaryotes: Single- or multicelled organisms possessing a nucleus and other internal structural subdivisions; feed by absorption, photosynthesis, or ingestion of particles.	Protista	Usually unicellular, sexual or asexual reproduction, great genetic diversity.	Diatoms and dinoflagellates, radiolarians and foraminifera, single- and multicellular marine algae (seaweeds).
	Fungi	Usually multicellular, sexual or asexual reproduction; release enzymes that break down organic material for absorption.	Molds, mushrooms, symbionts within lichens.
	Plantae	Multicellular photosynthetic autotrophs, sexual or asexual reproduction.	Mosses, ferns, flowering plants.
	Animalia	Multicellular heterotrophs, sexual or asexual reproduction.	Invertebrates, vertebrates.

organisms of a similar fundamental nature: **Bacteria, Archaea,** Fungi, **Plantae,** and **Animalia,** listed in Table 13.2. But one kingdom in Table 13.2, kingdom **Protista,** is a somewhat "unnatural" collection of diverse forms that do not clearly fit into any of the other five kingdoms. Protists are mostly unicellular organisms, but they include both autotrophs (like diatoms and most dinoflagellates) and heterotrophs (like foraminifera and radiolarians). Larger marine algae (the seaweeds) are also classified in kingdom Protista.

Kingdoms Protista, Plantae, and Animalia will be discussed at length in the chapters that follow, but special mention should be made here of the present renaissance in studies of Bacteria and Archaea (some archaeans are called "extremophiles" because of their ability to tolerate high temperatures, high methane concentrations, and anoxic conditions). The importance of this vast group of relatively simple one-celled creatures is only now becoming apparent—indeed, some researchers believe their biomass may be much greater than that of all other forms of life combined! Only in the last decade have researchers possessed the special tools and sterile techniques to enable them to look for Archaea deep within sediments or inside solid rock. Bacteria and Archaea have been found in astonishing places: in sediments 500 meters (1,640 feet) below the Pacific Ocean floor, and 1.5 kilometers (0.93 mile) beneath the Columbia Plateau in bare basaltic rock. Some biologists expect to find Archaea as deep as 5 kilometers (3.1 miles) below the solid surface.

Linnaeus's great contribution was a system of classification based on **hierarchy,** a grouping of objects by degrees of complexity, grade, or class. In this boxes-within-boxes approach, sets of small categories are nested within larger categories. Linnaeus devised names for the categories, starting with kingdom (the largest category) and passing down through phylum, class, order, family, and genus to species (the smallest category). In 1758 he published a catalog of all animals then known, his monumental *Systema Naturæ (The System of Nature)*. **Figure 13.10** shows the classification of a popular fish, Rex sole, using the Linnaean method. Note the nested arrangement of category-within-category, the categories becoming more specific with every downward step.

Linnaeus also perfected the technique of naming animals. The *genus* and *species* names—the names of the last two nested categories—constitute an organism's **scientific name.** *Octopus bimaculatus* is the scientific name of a common west coast octopus: *Octopus* is the generic name, *bimaculatus* the specific name. A closely related species, *Octopus dofleini,* is a larger animal that ranges to Alaska. *Octopus bimaculatus* and *Octopus dofleini* are not interfertile (they're not the same species), but, as their shared generic name suggests, they are closely related.

The advantage of a scientific name over a common name is immediately apparent to anyone trying to identify a shell found on the beach. The same shell may have many different common names in many different languages, but it will have *only one scientific name.* When you discover that name in a good key to shells, you can use it to find references that will tell you what is known about the animal, its lifestyle, its range, and its evolutionary history.

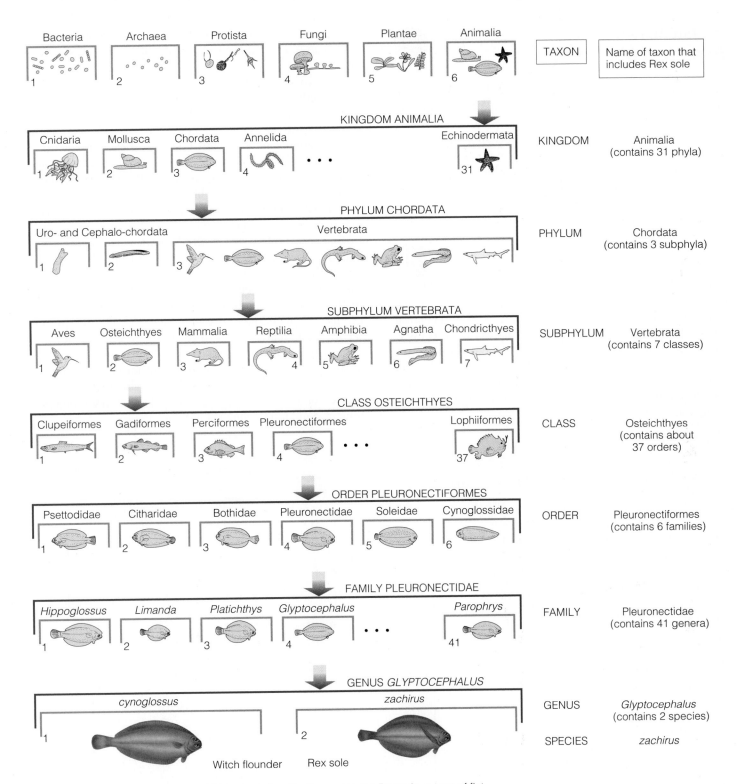

TAXON	Name of taxon that includes Rex sole

Bacteria 1 · Archaea 2 · Protista 3 · Fungi 4 · Plantae 5 · Animalia 6

KINGDOM ANIMALIA

Cnidaria 1 · Mollusca 2 · Chordata 3 · Annelida 4 · · · · Echinodermata 31

| KINGDOM | Animalia (contains 31 phyla) |

PHYLUM CHORDATA

Uro- and Cephalo-chordata 1, 2 · Vertebrata 3

| PHYLUM | Chordata (contains 3 subphyla) |

SUBPHYLUM VERTEBRATA

Aves 1 · Osteichthyes 2 · Mammalia 3 · Reptilia 4 · Amphibia 5 · Agnatha 6 · Chondricthyes 7

| SUBPHYLUM | Vertebrata (contains 7 classes) |

CLASS OSTEICHTHYES

Clupeiformes 1 · Gadiformes 2 · Perciformes 3 · Pleuronectiformes 4 · · · · Lophiiformes 37

| CLASS | Osteichthyes (contains about 37 orders) |

ORDER PLEURONECTIFORMES

Psettodidae 1 · Citharidae 2 · Bothidae 3 · Pleuronectidae 4 · Soleidae 5 · Cynoglossidae 6

| ORDER | Pleuronectiformes (contains 6 families) |

FAMILY PLEURONECTIDAE

Hippoglossus 1 · *Limanda* 2 · *Platichthys* 3 · *Glyptocephalus* 4 · · · · *Parophrys* 41

| FAMILY | Pleuronectidae (contains 41 genera) |

GENUS *GLYPTOCEPHALUS*

cynoglossus 1 · *zachirus* 2

Witch flounder · Rex sole

| GENUS | *Glyptocephalus* (contains 2 species) |
| SPECIES | *zachirus* |

Figure 13.10 The modern system of biological classification, using the Rex sole, a type of flat-fish *(Glyptocephalus zachirus),* as an example. Note the "boxes-within-boxes" approach, a hierarchy. Ellipses (. . .) indicate groups not shown for clarity.

PHYSICAL FACTORS AFFECTING MARINE LIFE

Marine organisms depend on the ocean's chemical composition and physical characteristics for life support. Any aspect of the physical environment that affects living organisms is called a **physical factor**. Living in the ocean has advantages over living on land—physical conditions in the sea are usually milder and less variable than physical conditions on land. The most important physical factors for marine organisms are light, temperature, salinity, dissolved nutrients, dissolved gases, acid–base balance, and hydrostatic pressure.

Light

As on land, sunlight is essential to most life in the sea. Marine plants (and some protists) convert inorganic matter to organic compounds by capturing light energy to form chemical bonds. As we will learn in the next chapter, this process is known as photosynthesis. On land, most photosynthesis proceeds at or just above ground level. But seawater, unlike soil, is relatively transparent, which allows photosynthesis to proceed to a much greater distance below the ocean surface than in the soil.

Water, however, is more transparent to some colors of light than others. In clear water, blue light penetrates to the greatest depth, while red light is absorbed near the surface. **Figure 13.11** shows the depths attained by light of various wavelengths (colors) in clear ocean water. Light energy absorbed by water turns to heat.

The depth to which light penetrates is also limited by the concentration and characteristics of particles in the water. These particles, which may include suspended sediments, dustlike bits of once-living tissue, or the organisms themselves, scatter and absorb light.

How far down does light penetrate? Near the coasts, conditions become uncomfortably dark for divers at depths of 30 to 40 meters (100 to 130 feet). The human eye is most sensitive to the blue wavelengths; in very clear tropical waters with a smooth surface and a high solar angle, human observers in submersibles have seen dark blue light at about 200 meters (660 feet). Photometers much more sensitive than the human eye have detected light at even greater depths—the present record is 590 meters (1,935 feet) in the tropical Pacific!

Light is important not only for photosynthesis. Some deep-water fish use even very dim light for body orientation, feeding, and predator avoidance. Light also triggers physiological rhythms in phytoplankton and determines breeding time in some multicellular organisms.

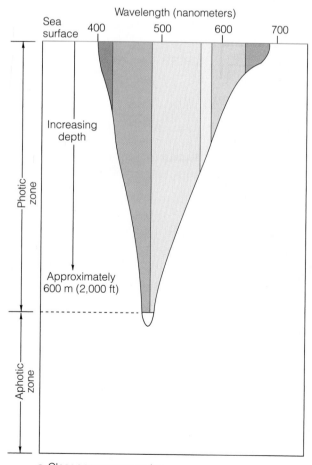

a Clear open ocean water

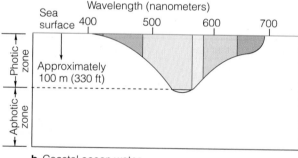

b Coastal ocean water

Figure 13.11 Penetration of light into the ocean. (**a**) In clear open ocean water, sensitive instruments can detect light to a depth of 600 meters (2,000 feet). (**b**) Because of the suspended particles often present in coastal waters, light cannot penetrate as far—about 100 meters (330 feet) is typical. The sunlit upper zone is called the *photic zone*. The dark ocean beneath is called the *aphotic zone*.

Temperature

The rate at which chemical reactions occur in a living organism is largely dependent on the molecular vibration we call heat. Since agitation brings reactants together, warmer temperatures increase the rate at which chemical reactions occur. Thus, an organism's **metabolic rate,** the rate at which reactions proceed within an organism, increases with temperature. The metabolic rate approximately doubles with a 10°C (18°F) temperature rise. The interior temperature of an organism is directly related to the rate at which it moves, reacts, and lives.

The great majority of marine organisms are "cold blooded," or **ectothermic** (*ektos* = outside, *therme* = heat), having an internal temperature that stays very close to that of their surroundings. A few complex animals—mammals and birds and some of the larger, faster fishes—are "warm blooded," or **endothermic** (*endon* = within), meaning they have a stable, high internal temperature.

In general, the warmer the environment of an ectotherm within its tolerance range, the more rapidly its metabolic processes will proceed. Tropical fish in a heated aquarium will therefore eat more food and require more oxygen than goldfish of the same size living in an unheated but otherwise identical aquarium. The tropical fish will generally grow faster, have a faster heartbeat, reproduce more rapidly, swim more swiftly, and live a shorter life. But you can't just crank the heater up another notch for even faster fish—eventually the little fellows will cook as their proteins distort. The upper limit of temperature that an ectotherm can tolerate is often not much higher than its optimum temperature. The lower limit is usually more forgiving because molecules are merely slowed.

Do endotherms also have narrow temperature requirements? Yes and no. Endotherms can tolerate a tremendous range of *external* temperature compared to ectotherms— think of a whale migrating from polar waters to the tropics, or an Emperor penguin incubating an egg at −51°C (−60°F)— but their internal temperatures vary only slightly. In our own case, consider the temperatures of places inhabited by humans in contrast to the narrow internal temperature range physicians consider normal. (An increase of only 8° is life threatening.) Sophisticated thermal regulation mechanisms make it possible for endotherms to live in a variety of habitats, but they pay a price: Their high metabolic rates make proportionally high demands on food supply and gas transport. But the benefit of having a biochemistry finely tuned to a single efficient temperature is worth the regulatory difficulties involved.

Ocean temperature varies with both depth and latitude. The average temperature of the world ocean is only a few degrees above freezing, with warmer water found only in the sunlit surface zones of the temperate and tropical ocean and in rare deep, warm chemosynthetic communities. Though temperature ranges of the ocean are considerable (**Figure 13.12**), they are usually much narrower than comparable ranges on land. Marine organisms have the definite advantage: They are almost never exposed to sustained tempera-

tures above 30°C (86°F), while some terrestrial species must tolerate long periods when temperature reaches or even exceeds 60°C (140°F). There are exceptions, however. A marine archaean, genus *Pyrolobus,* grows within hydrothermal vents at temperatures between 90°C and 113°C (194°F and 238°F). While that temperature exceeds the boiling point of water at sea level, it's far below boiling in the great pressure of its home depths within the deep vents. This organism is believed to be the most tolerant to heat of any, but others with similar tolerance may exist (see Questions from Students #5 at the end of this chapter).

Salinity

Cell membranes are greatly affected by the salinity of surrounding water. As we saw in Chapter 7, the salinity of seawater can vary in places because of rainfall, evaporation, runoff of water and salts from land, and other factors. Surface

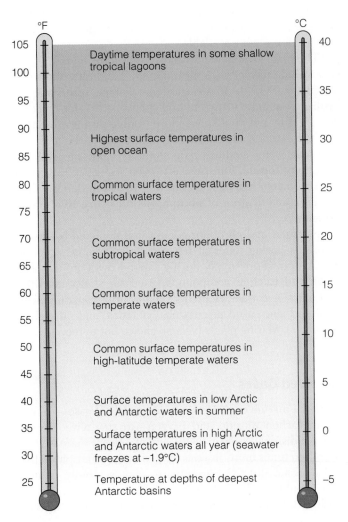

Figure 13.12 Temperatures of open marine waters capable of supporting life. Some isolated areas in the ocean, notably within and beneath hydrothermal vents, may contain specialized living organisms— extremophiles—capable of thriving at temperatures of 113°C (238°F).

salinity varies most, from lows of 6% or less (along the coast of the outer Baltic Sea in early summer) to year-round highs exceeding 40% in the Red Sea. Salinity is less variable with increasing depth, with the ocean typically becoming slightly saltier with depth.

Changing salinity can physically damage membranes, and concentrated salts can alter protein structure. Salinity can affect the specific gravity, the density, and therefore the buoyancy of an organism. Salinity is also important because it can cause water to enter or leave a cell through the membrane, changing the cell's overall water balance. Seawater is nearly identical in salinity to the interior of all but the most advanced forms of marine life; this means that maintaining salt balance, and therefore water balance, is easy for most marine species.

Dissolved Nutrients

A **nutrient** is a compound required for the production of organic matter.[1] Some nutrients help form the structural parts of organisms, some make up the chemicals that directly manipulate energy, and some have other functions. A few of these necessary nutrients are always present in seawater, but most are not readily available.

The main inorganic nutrients required by marine organisms include nitrogen (as nitrate, NO_3^-) and phosphorus (as phosphate, PO_4^{3-}). As any gardener knows, plants require fertilizer—mainly nitrogen and phosphates—for success. Ocean gardeners would have more trouble raising crops than their terrestrial counterparts, though, because the most fertile ocean water contains only about 1/10,000 the available nitrogen of topsoil. Phosphorus is even scarcer in the ocean, but fortunately less of it is required by living things, which contain about 1 atom of phosphorus for every 16 atoms of nitrogen.

Nitrogen and phosphorus are often depleted by marine organisms during times of fast growth and rapid reproduction. Also in short supply during rapid growth are dissolved silicon and calcium compounds (used for shells and other hard parts) and trace elements such as iron, copper, magnesium, and zinc (used in enzymes, vitamins, and other large molecules). Marine plants have no choice but to recycle these nutrients.

Dissolved Gases

Nearly all marine organisms require dissolved gases—in particular, carbon dioxide and oxygen—to stay alive. Oxygen does not dissolve easily in water, and as a result there is about a hundred times more gaseous oxygen in the atmosphere than in the ocean. But CO_2, essential to photosynthesis, is much more soluble and reactive in seawater than is oxygen (as **Table 13.3** shows). Although as much as a thousand times

---

[1]A broader definition of a *nutrient* includes all needed substances an organism obtains from its environment except oxygen, carbon dioxide, and water.

Table 13.3	Solubility of Gases in Seawater as a Function of Temperature (Salinity = 33‰)		
	Solubility (ml/l at atmospheric pressure)[a]		
Temperature	**N_2**	**O_2**	**CO_2**
0°C (32°F)	14.47	8.14	8,700.0
10°C (50°F)	11.59	6.42	8,030.0
20°C (68°F)	9.65	5.26	7,350.0
30°C (86°F)	8.26	4.41	6,660.0

Source: F. G. Walton-Smith. *CRC Handbook of Marine Science* (Cleveland, OH: CRC Press, 1974).
[a]Figures are given at saturation, the maximum amount of gas held in solution before bubbling begins.

more carbon dioxide than oxygen can dissolve in water, normal values at the ocean surface average around 50 milliliters per liter for carbon dioxide and 6 milliliters per liter for oxygen. At present the ocean holds about 60 times as much carbon dioxide as the atmosphere. Because of this abundance, marine plants almost never run out of it.

Deep water tends to contain more carbon dioxide than surface water. Why should this be? Table 13.3 also shows the relationship between water temperature and its ability to dissolve gases. Note that colder water contains more gas at saturation. You may recall that the deepest and densest seawater masses are formed at the surface in the cold polar regions, and, as we have seen, more CO_2 can dissolve in that low-temperature environment. The dense water sinks, taking its large load of CO_2 to the bottom, and the pressure at depth helps keep it in solution. CO_2 also builds in deep water because only heterotrophs (animals) live and metabolize there and because CO_2 is produced as decomposers consume falling organic matter. No photosynthetic organisms are present in the dark depths to use this excess CO_2 since there is insufficient sunlight to permit photosynthesis to occur.

Rapid photosynthesis at the surface lowers CO_2 concentrations and increases the quantity of dissolved oxygen. Oxygen is least plentiful just below the limit of photosynthesis because of respiration by many small animals at middle depths. (These relationships were shown in Figure 7.8.)

Low oxygen levels can sometimes be a problem at the ocean surface. Plants produce more oxygen than they use, but they produce it only during daylight hours. The continuing respiration of plants at night will sometimes remove much of the oxygen from the surrounding water. Oxygen depletion may lead to the death of the small drifting organisms in enclosed coastal waters, areas not flushed by currents or tides. Decomposition of their bodies by bacteria will further lower oxygen levels, leading to a broad die-off of marine life.

The greatest variability in levels of dissolved gas is found at the surface near shore. Less dramatic changes occur in the open sea.

Acid–Base Balance

The complex chemistry of Earth's life forms depends on precisely shaped enzymes, large protein molecules that speed up the rate of chemical reactions. Like heat, strong acids or bases distort the shapes of these vital proteins, and they lose their ability to function normally.

The acidity or alkalinity of a solution is expressed in terms of a *pH scale*, a logarithmic measure of the concentration of hydrogen ions in a solution. Recall (from Figure 7.9) that 7 on the pH scale is neutral, with smaller numbers indicating greater acidity and larger numbers indicating greater alkalinity.

Seawater is slightly alkaline; its average pH is about 8. The dissolved substances in seawater act to *buffer* pH changes, preventing broad swings of pH when acids or bases are introduced. The normal pH range of seawater is much less variable than that of soil—terrestrial organisms are sometimes limited by the presence of harsh alkali soils that damage cell components.

Though seawater remains slightly alkaline, it is subject to some variation. When dissolved in water, some CO_2 becomes carbonic acid. In areas of rapid plant growth, pH will rise because CO_2 is used by the plants for photosynthesis. And because temperatures are generally warmer at the surface, less CO_2 can dissolve in the first place. Thus, surface pH in warm productive water is usually around 8.5.

At middle depths and in deep water, more CO_2 may be present. Its source is the respiration of animals and bacteria. With cold temperatures, high pressure, and no photosynthetic plants to remove it, this CO_2 will lower the pH of water, making it more acid with depth. Thus, deep, cold seawater below 4,500 meters (15,000 feet) has a pH of around 7.5. This lower pH can dissolve calcium-containing marine sediments. A drop to pH 7 can occur at the deep-ocean floor when bottom bacteria consume oxygen and produce hydrogen sulfide.

Hydrostatic Pressure

Marine organisms are often subject to great pressure from the constant weight of water above them, but this **hydrostatic pressure** presents very little difficulty to them. In fact, the situation in the ocean is parallel to that on land. Land animals live in air pressurized by the weight of the atmosphere above them (1 kilogram per square centimeter, or 14.7 pounds per square inch, at sea level) without experiencing any problems. Indeed, atmospheric pressure is necessary for breathing, flight, and some other physical necessities of life.

Pressures inside and outside an organism are virtually the same, both in the ocean and at the bottom of the atmosphere. Thus, marine organisms do not need heavy shells to keep from being crushed by hydrostatic pressure. Great pressure does have some chemical effects: At high pressure, gases become more soluble, some enzymes are inactivated, and metabolic rates for a given temperature tend to be slightly higher. These effects are felt only at great depth, however. Unless marine organisms have gas-filled spaces in their bodies, a moderate change in pressure has little effect.

The Interplay of Factors

The physical factors we have surveyed work in concert to provide the physical environment for oceanic life. A change in one factor usually produces a change in others. For example, the ocean absorbs more solar radiation in spring, increasing surface temperature. The resulting stratification of the water column will allow phytoplankton and other plants to remain closer to the surface, conduct photosynthesis, multiply, exhaust the nutrients, die, and sink to the depths. If density becomes very high at the surface, the saline water will sink. It will be replaced by nutrient-rich water from below, stimulating new plant growth. The growing plants will absorb CO_2, causing the pH of the water to rise. And so it goes.

Figure 13.13 charts a number of physical factors at one location off southern California. The intermix of related factors—all changing with depth, all influencing (and being influenced by) living organisms—gives us a glimpse of the intricacy of the relationship of marine life to its physical environment.

ORGANISMS AND OCEAN TOGETHER

To the list of physical factors discussed previously we can add the concept of biological factors. A **biological factor** is a biologically generated aspect of the environment that affects living organisms. The most important biological factors for most marine organisms are the feeding (trophic) relationships we will explore in Chapter 14. Some other factors include crowding, metabolic wastes in the water, and the defense of territory. Biological factors operate in association with purely physical factors to shape the evolution of marine organisms.

Diffusion, Osmosis, and Active Transport

If a cube of dye is not disturbed in a container of still water, the dye molecules will eventually diffuse evenly throughout the water. This process, known as **diffusion,** might take weeks to complete (**Figure 13.14**). The energy to distribute the dye comes from heat, the random vibration of molecules: The warmer the water, the faster the diffusion. The net transfer of material in diffusion occurs from a region of high concentration to regions of lower concentration.

Diffusion is an important marine process. For example, minerals dissolve, and their components tend to diffuse randomly throughout a liquid environment. Liquids and gases can also diffuse through water from zones of high concentration to zones of lower concentration. Mass transport (the movement of substances in currents, for example) is more important than diffusion in moving dissolved substances over large distances. Diffusion is most important over small distances, particularly the distances within and between living cells.

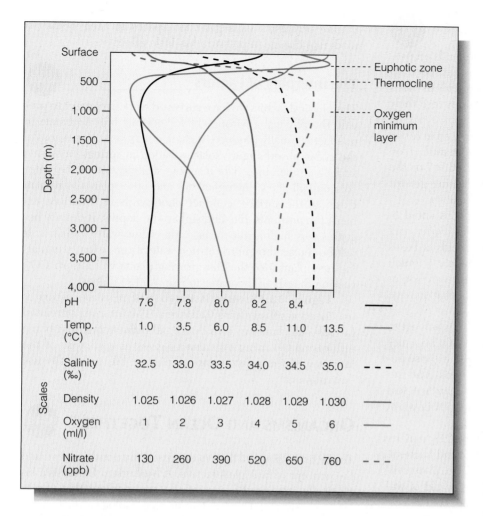

pH	7.6	7.8	8.0	8.2	8.4	
Temp. (°C)	1.0	3.5	6.0	8.5	11.0	13.5
Salinity (‰)	32.5	33.0	33.5	34.0	34.5	35.0
Density	1.025	1.026	1.027	1.028	1.029	1.030
Oxygen (ml/l)	1	2	3	4	5	6
Nitrate (ppb)	130	260	390	520	650	760

Scales

Figure 13.13 The variability of physical factors with depth, illustrated by a sampling off southern California.

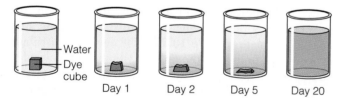

Figure 13.14 An example of diffusion. Molecules of dye gradually diffuse through the surrounding water as a dye cube dissolves. Random molecular movement spreads the dye away from the region of high concentration at the cube's surface. At the same time, water moves from its own region of high concentration (next to the dye cube) to areas of lower concentration (within the disintegrating dye cube itself). After many days, dye will be distributed evenly throughout the water. Stirring would greatly accelerate the mixing process, but this input of mechanical energy is not required if you don't mind waiting for diffusion alone to accomplish the task.

Cells are bounded by membranes, complex films through which certain selected substances can passively diffuse. Molecules will tend to diffuse across membranes from areas where they are highly concentrated to areas where they are less concentrated. For example, oxygen resulting from photosynthesis will diffuse through a membrane from inside a plant cell (a region of high oxygen concentration) to the outside (a region of lower oxygen concentration). Because biological membranes are selective and allow only certain kinds of small molecules to pass, they are considered *semipermeable.*

Diffusion of water through a membrane is called **osmosis** (*osmos* = a thrusting). In osmosis, water moves between two solutions of different water concentrations through a membrane permeable to water but not to salts. If the water outside a cell membrane contains less dissolved salt than the water inside, it will diffuse from the region of higher water concentration (outside the cell) to the region of lower water concentration (inside the cell), causing the cell to swell (as in Figure 13.16b). Salt is prevented by the nature of the membrane from moving outside to balance the situation. This is why human swimmers avoid getting fresh water in nasal membranes: Our body fluids are more saline than fresh water, so cells in our sinuses take up the water and swell painfully.

Most simple marine organisms have nearly the same concentration of dissolved substances in their body fluids as seawater does. They are almost **isotonic** to their fluid environment (*isos* = equal, *tonos* = strength) and so experience little net flow of water through their outer membranes. In fresh water a marine animal would be **hypertonic** (*hyper* = over) to its surroundings; water would move *into* the animal through its cell membranes. If the animal had no way to eliminate the water, its cells would burst. The same kind of marine animal moved to Utah's highly saline Great Salt Lake would be **hypotonic** (*hypo* = under), and water would flow *out of* its cells, resulting in dehydration and collapse. **Figure 13.15** summarizes these relationships. Because they are nearly isotonic to their surroundings, simple marine organisms have a big advantage over their freshwater counterparts; freshwater animals must expend large amounts of energy in excretory systems designed to transport water from their tissues back to the outside. Because of these specialized excretory mechanisms, very few freshwater and marine organisms can successfully change places.

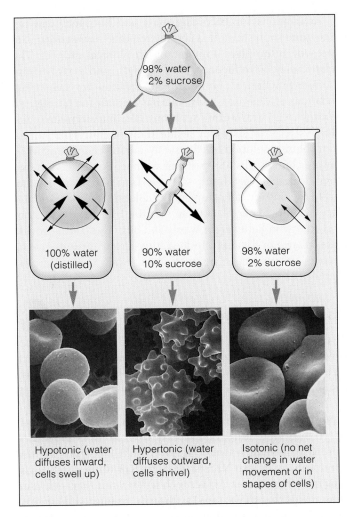

Figure 13.15 The effects of osmosis in different environments. The semipermeable bag at the top of the figure represents the concentration of water and dissolved solids typical in the cells of higher life forms. If the bag is placed into the solutions indicated, it will swell, shrink, or stay the same size. The cells pictured here are human red blood cells immersed in (from left to right) distilled water, concentrated seawater, and water of the same tonicity as human blood.

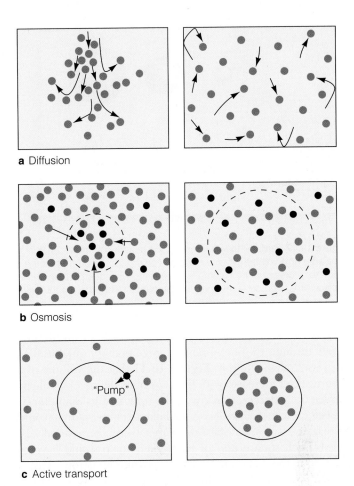

a Diffusion

b Osmosis

c Active transport

Figure 13.16 A comparison of the three main ways by which substances move into and out of cells. (**a**) In diffusion, molecules introduced into a container (left) become evenly distributed after a period of time (right). (**b**) In osmosis, the diffusion of water may cause a cell to swell. The blue dots represent water molecules, the black dots represent dissolved particles, and the arrows indicate the direction of water movement into the original cell. Under different conditions, water may move out of the cell, causing it to shrink. (**c**) Active transport enables a cell to accumulate molecules even when there are more inside the cell than outside. Cells may expel other molecules by the same process.

Active transport is the reverse of passive diffusion. In active transport dissolved substances are "pumped" through a membrane "uphill" from a region of low concentration to a region of high concentration. Active transport requires energy because this "uphill" movement defies the normal "downhill" functioning of the second law of thermodynamics. When a cell exports a finished product (the sugars made in photosynthesis, for example) through a membrane to a storage area in which this product is concentrated, it uses active transport. Active transport is a common process in living things, and much of any organism's energy is expended on facilitated movement "uphill," against the normal direction of flow.

Diffusion, osmosis, and active transport are all temperature-dependent, proceeding at a more rapid rate as temperature rises. These three processes are compared in **Figure 13.16.**

Surface-to-Volume Ratio

Though individual organisms are often large, the cells that comprise them are always small. The smaller the cell, the more efficiently materials can cross the outer cell membrane to be distributed through the interior. A small cell has enough surface area to permit the passage of oxygen (or carbon dioxide, or glucose, or wastes) at a rate sufficient to supply the metabolic needs of its internal machinery.

If a round cell grew only by expanding its volume, the surface area of its outer membrane would not increase at an

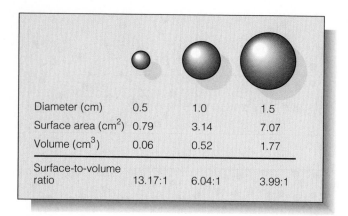

Diameter (cm)	0.5	1.0	1.5
Surface area (cm²)	0.79	3.14	7.07
Volume (cm³)	0.06	0.52	1.77
Surface-to-volume ratio	13.17:1	6.04:1	3.99:1

Figure 13.17 Surface-to-volume ratio: the relationship between surface area and volume when a sphere is enlarged. Notice that as the diameter increases, the volume is increasing more quickly than the surface area. The surface-to-volume ratio thus *decreases* as the cell increases in size.

adequate rate. The cell would soon be unable to shuttle materials through the membrane in large enough quantities to keep itself alive. Note in **Figure 13.17** that the volume of a spherical cell increases with the *cube* of its diameter, while its surface area increases only with the *square* of its diameter. If the cell grew to 4 times its original diameter, its volume would increase 64 times, but its surface area would increase only 16 times. Each square unit of outer membrane would have to serve four times as much interior volume as before. Past a certain point, the inward flow of nutrients and gases (and the outward flow of wastes) would not be fast enough to supply the cell's metabolic needs, and the cell would die. Large cells have an inefficient ratio of surface area to internal volume.

Instead, cells grow by dividing, not by limitless expansion. Diffusion, osmosis, and active transport can be accomplished more efficiently by cells with a high **surface-to-volume ratio**—that is, by small cells with a relatively large surface area and a relatively small volume.

Gravity and Buoyancy

The density of seawater is nearly the same as the average density of living material, so the weight of a marine organism is usually counterbalanced by its buoyancy. Seawater has a density of about 1.025 (pure water = 1.000). Some parts of a fish are heavier than seawater and some parts lighter, but overall a marine fish has a density of around 1.07—about 5% denser than the same volume of seawater. To counteract the mass of heavy muscle and bone, many swimming fishes have gas-filled bladders. Gas is much lighter (less dense) than seawater, so only a small volume of gas is required. Fishes from very deep water usually don't have gas bladders, but they do have lightweight bones and oily, watery flesh to lighten their mass and make swimming easier. At the other end of the scale, most members of the mackerel family are such strong swimmers that the heavy mass of their muscular bodies is of no apparent concern (see Figure 13.18d).

Adaptations for balancing mass and buoyancy can become quite sophisticated. The skeleton of a marine organism, if present, often plays a minor role in physical support. The skeleton may be reduced in size, strength, and mass because its major function is only to provide stiffness for swimming, predation, or defense. A few marine animals further lighten themselves by substituting solutions rich in ammonium chloride (NH_4Cl) for seawater inside their bodies. This ammonium chloride solution has the same osmotic characteristics as seawater but weighs less per unit volume. Some jellyfish and other floating invertebrates actively transport heavy sulfate ions to the outside, allowing lighter chloride ions to take their place within. Many planktonic (drifting) forms of life store food energy in lightweight waxes and oils. These also provide mass balance or flotation. Indeed, this trick contributes to the diatoms' great success (news of which awaits you in the next chapter).

Viscosity and Movement

Viscosity is a fluid's internal resistance to flow. Molasses flows much more slowly than water, so molasses has a much higher viscosity than water. Temperature has an effect on viscosity, and molasses in January flows very slowly indeed. Water has a much lower viscosity than molasses (and is *much* easier to swim through), but the viscosity of water is also affected by temperature. Seawater at 0°C (32°F) is about 25% more viscous than seawater at 20°C (68°F), a significant difference. Salinity also affects viscosity. Salt water is more viscous than fresh water. Air is much less viscous than seawater, but animals that move rapidly through air must still contend with its viscosity. Note that birds and other flying species are streamlined.

For marine organisms dealing with the viscosity of seawater, the main problem is **drag,** the resistance to movement of an organism induced by the fluid through which it swims. The amount of drag depends on the viscosity of the water and the speed, shape, and size of the moving organism.

Very small marine organisms are more directly affected by viscosity than larger ones. Swimming is difficult for tiny animals; the viscosity of water impedes their efforts at movement. Gradual sinking, however, can be a problem, and very small drifting organisms have plumes, hairs, ribbons, spines, and other extensions to increase their friction-producing area. Since warm water is less viscous than cold, warm-water species bear more of this ornamentation. Tropical species also tend to be smaller than related polar species. Why? Because a larger surface-to-volume ratio provides greater surface area—and hence slower sinking—for small drifting organisms in the less viscous warm water.

Large swimming animals have a different problem with viscosity: the chaotic water movement known as **turbulence.** Water's resistance to flow results in turbulence around the swimmer and in the swimmer's trailing wake, both of which tend to slow the animal. An animal can minimize this frictional loss by streamlining. As an example, **Figure 13.18** shows three objects of identical frontal area moving through water. Streamlined shape (c) will have only about 6% of the drag of shape (a), a disk. Note the teardrop shape of the fast-swimming tuna in Figure 13.18d.

Other unique adaptations have evolved in swimming species. Some fish secrete a small amount of friction-reducing mucus or oil onto their surface to minimize the formation of eddies, and some can tuck maneuvering fins into body recesses. Some fast-moving marine mammals have skins that passively indent to smooth out eddies as they form. In a sense, any reduction of drag is food saved, prey not needed, a greater ability to escape being eaten, and energy that can go to other concerns (such as reproduction).

Water Movement

The role of ocean currents in the horizontal distribution of marine organisms or their larvae is obvious, but there are subtler advantages of drifting with the currents. For example, some kinds of small drifting animals are known to swim each day through considerable *vertical* distances. Many reasons have been proposed to explain this behavior. One advantage to the organisms may be predator avoidance; they feed near the surface only during the relatively safe, dark night hours. Other benefits may include increased opportunity for feeding during the movement itself or for nutrient uptake in deeper water. Another possibility is that animals may be taking advantage of the differential speeds of currents with depth. Geostrophic currents move fastest at the surface and less rapidly as depth increases. Drifting animals grazing on primary producers at the surface could descend, drift at a different rate, and then reappear in a different surface water mass a day later. This tactic could prevent being "stuck" in one surface mass with few opportunities for feeding.

Other kinds of water movement that influence marine life are associated with the rise and fall of tides, wave action and wave shock, and density differences. In each case, organisms are directly influenced by movement that, in cases such as tidal height and wave shock, can be limiting factors.

CLASSIFICATIONS OF THE MARINE ENVIRONMENT

Scientists have found it useful to divide the marine environment into **zones,** areas with homogeneous physical features. Attempts to classify the oceanic realm began in France in the 1830s but were hampered by lack of knowledge about the physical and biological features of the ocean. The situation changed rapidly near the end of the 1800s with the invention of tools for measuring physical factors and taking samples at various depths. Now these divisions can be made on the basis of light, temperature, salinity, depth, latitude, water density, or almost any of the other physical dimensions we have discussed. Some classifications are more useful than others, however, and we will survey those classifications in this section.

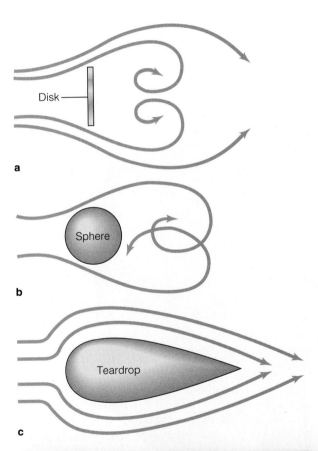

a

b

c

d

Figure 13.18 Turbulence and drag. At the same speed, with the same frontal area, shape (**a**) will have about 15 times more drag than shape (**c**). Shape (**b**) shows only a small improvement in drag over the disk. The fast-swimming yellowfin tuna seen in (**d**) is shaped like a teardrop to minimize turbulence and drag.

Classification by Light

The sunlit layer of water at the ocean's surface is called the **photic zone** (*photos* = light). At noon in the clear tropics the photic zone extends to a depth of approximately 600 meters (2,000 feet), but in mid-latitude water small organisms (and other small scattering particles) are more abundant, so light reaches down only to about 150 meters (500 feet). The upper part of the photic zone—where there is sufficient light for plant production by photosynthesis to exceed loss of carbohydrates through respiration—is called the **euphotic zone** (*eu* = good). The lower part of the photic zone is the **disphotic zone** (*dys* = difficult), a region where animals can see but there is insufficient light for productive photosynthesis. Below the dysphotic zone is the deepest, largest region of the open ocean, extending to the seabed—the **photic zone** (*a* = without), the dark zone that extends to the bottom.

Classification by Location

When the great English and American oceanographic institutions were founded, the need for a common, detailed nomenclature of position became urgent. Using a model from Harald Sverdrup as a point of departure, the National Academy of Sciences in the mid-1950s devised the scheme shown in **Figure 13.19**. It has withstood the test of time and is in common use today.

The primary division is between water and ocean bottom. Open water is called the **pelagic zone** (*pelagius* = of the sea) and is divided into two subsections: the **neritic zone** (*neritos* = shallow), near shore over the continental shelf, and the

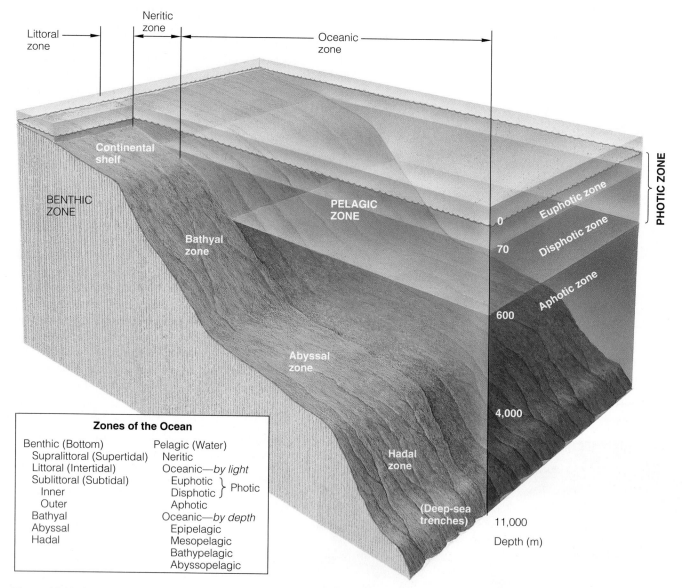

Zones of the Ocean

Benthic (Bottom)	Pelagic (Water)
Supralittoral (Supertidal)	Neritic
Littoral (Intertidal)	Oceanic—*by light*
Sublittoral (Subtidal)	Euphotic ⎱ Photic
Inner	Disphotic ⎰
Outer	Aphotic
Bathyal	Oceanic—*by depth*
Abyssal	Epipelagic
Hadal	Mesopelagic
	Bathypelagic
	Abyssopelagic

Figure 13.19 Classification of marine environments. This diagram is designed to show divisions and is somewhat exaggerated in indicating the proper proportions.

deep-water **oceanic zone,** beyond the continental shelf. The oceanic zone is further divided by depth into zones. The *epipelagic zone* (*epi* = atop) corresponds to the lighted photic zone. In the aphotic depths are layered the *mesopelagic* (*mesos* = in the middle), *bathypelagic* (*bathos* = depth), and *abyssopelagic* (*a* = without, *byssos* = bottom) zones. Abyssopelagic water is the water in the deep trenches.

Divisions of the bottom are labeled **benthic** (*benthos* = bottom) and begin with the intertidal **littoral zone** (*litoral* = of shore), the band of coast alternately covered and uncovered by tidal action. (The *supralittoral zone,* the splash zone *above* the high intertidal, is not technically part of the ocean bottom.) Past the littoral is the **sublittoral zone** (*sub* = below), which is further divided into inner and outer segments: The *inner sublittoral* is ocean bottom near shore, and the *outer sublittoral* is ocean floor out to the edge of the continental shelf.[2] The **bathyal zone** covers seabed on the slopes and down to great depths, where the **abyssal zone** begins. The **hadal zone** (*Hades* = underworld) is the deepest seabed of all, the trench walls and floors.

Classification by Behavior

Some biologists distinguish organisms by their behavior and swimming abilities. **Pelagic** organisms live suspended in seawater. They are immensely varied, but all have the common problems of maintaining their vertical position, producing or obtaining food, and surviving long enough to reproduce. They can be divided into two broad groups based on their lifestyle. The **plankton** drift or swim weakly, going where the ocean goes, unable to move consistently against waves or current flow. The **nekton** are pelagic organisms that actively swim. You'll meet a few of these organisms in the chapters that follow.

QUESTIONS FROM STUDENTS

1. **Why is nitrogen often a limiting factor in marine organisms? If air is 78% nitrogen, there should be lots of nitrogen around even if the solubility of nitrogen in the ocean is low.**

Yes, but marine plants cannot use dissolved atmospheric nitrogen (N_2) directly. The preferred forms of nitrogen for uptake are ammonium (NH_4^+) and nitrate (NO_3^-), an ion formed by the oxidation of ammonium and nitrite (NO_2^-). Nitrate runoff from soil is an especially rich source of this often limiting nutrient, which is why coastal water tends to support greater plankton populations than oceanic water. There is an important exception, however. Some kinds of bacteria and cyanobacteria can use dissolved atmospheric nitrogen, so they take N_2 directly from seawater (see again Figure 13.4). In general, their contribution to total oceanic productivity is

relatively low, however. In certain areas of the world (for example, the subtropical north Pacific Ocean), new evidence suggests that nitrogen fixation by very small organisms is a major source of nitrogen for primary productivity.

2. **If humans have a fluid much like seawater bathing our cells, why can't we drink seawater and survive?**

Human cells function in an environment hypotonic to seawater; that is, blood plasma is less saline than seawater. Drinking seawater therefore causes water to leave the intestinal walls, flood the intestine, and leave the body. There is a net loss via intestine or kidneys even if the seawater is diluted with fresh water before drinking. Moral: Never drink seawater in a survival situation at sea, and never dilute fresh water with seawater (in any proportion) to stretch your supply.

3. **Perfectly smooth body surfaces are best for attaining high speeds, right?**

Not always. Animals that move rapidly through fluids (water or air) are almost never perfectly smooth; their skin usually contains small projections such as hair or feathers or scales. Even small whales that appear glassy smooth have flexible skins that indent to smooth out flow irregularities. A small degree of surface roughness and "give" makes it more difficult for the fluid to adhere to the moving surface, thereby cutting drag.

This finding has been put to good use: Golf balls are covered by little dimples instead of being smooth, America's Cup yacht builders now incise patterns of tiny lines into their previously flawless hulls, and competitive swimmers no longer need to shave their bodies for maximum speed.

4. **How important was the revolution in biology sparked by Darwin's and Wallace's work?**

Very important. According to Ernst Mayr, a famous evolutionary biologist and philosopher of biology, "The Darwinian revolution was the most fundamental of all intellectual revolutions in the history of mankind. While such revolutions as those brought about by Copernicus, Newton, Lavoisier, or Einstein affected only one particular branch of science, or the methodology of science as such, the Darwinian revolution affected every thinking man. A world view developed by anyone after 1859 was by necessity quite different from any world view formed prior to 1859." How could it be otherwise? (See Mayr, 1988.)

5. **How can an organism, even something as relatively simple as a bacterium, possibly live at a depth of 1.5 kilometers in solid rock, or beneath hundreds of meters of marine sediment? What would it eat? How could it withstand the heat?**

These creatures, aptly called "extremophiles," have been isolated from the more familiar surface ecosystems for millions of years. They appear to have evolved exotic metabolisms, very slow reproductive rates, and the ability to make use of a broad variety of energy-rich molecules seeping down

---

[2]Researchers sometimes use the words *supertidal, intertidal,* and *subtidal* instead of the *littoral* terms.

from above or moving up from below. One recently discovered deep bacterium, appropriately named *Bacillus infernus*, cannot tolerate oxygen. As for temperature, their proteins must be strongly cross-linked, able to withstand the unraveling effects of great heat. Much remains to be learned about this vast underground biological economy!

6. Other than listing it as a kingdom, you didn't mention Fungi. Are there any marine fungi?

Though of great terrestrial importance, few fungi exist in the ocean. Most of those that do are confined to the intertidal zone, where they live in close association with marine algae. On land we would call these symbioses "lichens," but botanists are hesitant to categorize the marine equivalent with that word. A few types of fungi have been found in subtidal sediments, where they fill the same role as on land—decomposers of organic matter. Fungi are incapable of photosynthesis, and DNA studies have shown that they are more closely related to animals than to plants.

KEY CONCEPTS TO REVIEW

- Living things contain matter in a highly organized, low-entropy state. Living things can capture, store, and transmit energy; reproduce; and adapt to their environment over time.

- Matter cycles through living and nonliving systems in biogeochemical cycles.

- Working from what may have been a single instance of origin, evolution through natural selection has produced the great diversity of life on Earth. Living things change as time passes, capitalizing on available energy and becoming ever more efficient at its use.

- Life on Earth is notable for both unity and diversity: *diversity* because there are at least 5 million different species (kinds) of living things on Earth, *unity* because each species shares the same underlying mechanisms for basic life processes.

- Marine organisms are naturally classified by their physical characteristics and by the degree to which they resemble other organisms.

- Various physical factors affect the density, variety, and success of marine life. These factors include water's transparency, temperature, dissolved nutrients, salinity, dissolved gases, hydrostatic pressure, and acid–base balance.

- Biological factors that affect marine life include diffusion and osmosis, surface-to-volume ratio, and the interplay of organism and environment.

- A limiting factor is a physical or biological factor whose presence in an inappropriate amount limits the success of an organism.

- The environments populated by marine life may be classified by their physical or biological characteristics.

INTERNET STUDY RESOURCES

The Web site for this book contains helpful study aids. Log on to:

www.brookscole.com/053437557xs

and click on the Chapter-by-Chapter area. Choose Chapter 13 and select a resource:

- **Flash Cards** allows you to test your mastery of the Terms and Concepts to Remember for this chapter.
- **Tutorial Quizzes** provides a multiple-choice practice quiz.
- **Student Guide to InfoTrac College Edition** will lead you to Critical Thinking Projects that use InfoTrac College Edition as a research tool.

TERMS AND CONCEPTS TO REMEMBER

abyssal zone
active transport
adaptation
Animalia
aphotic zone
Archaea
artificial system of classification
Bacteria
bathyal zone
benthic zone
biogeochemical cycle
biological factor
carbon cycle
convergent evolution
denitrifying bacteria
diffusion
disphotic zone
dissolved organic carbon (DOC)
dissolved organic nitrogen (DON)
drag
ectotherm, ectothermic
endotherm, endothermic
energy
entropy
eukaryote
euphotic zone
evolution
hadal zone
hierarchy
hydrostatic pressure
hypertonic
hypotonic
isotonic

kingdoms
limiting factor
Linnaeus, Carolus (Carl von Linné)
littoral zone
mass extinction
metabolic rate
mutation
natural selection
natural system of classification
nekton
neritic zone
nitrifying bacteria
nitrogen cycle
nutrient
oceanic zone
osmosis
pelagic
pelagic zone
photic zone
physical factor
plankton
Plantae
prokaryote
Protista
scientific name
second law of thermodynamics
species
sublittoral zone
surface-to-volume ratio
taxonomy
turbulence
viscosity
zone

STUDY QUESTIONS

Review Questions

1. What evidence suggests that life on Earth originated in the ocean?

2. How are living things different from nonliving things? How can you differentiate between them?

3. Name and discuss five physical factors of the marine environment. How is each different in the ocean from physical factors on the land?

4. What is a biological factor? Give three examples.

5. How is an endotherm different from an ectotherm? Which are you? What are the advantages of a high, stable internal body temperature? The disadvantages?

6. How are diffusion, osmosis, and active transport related? Use these concepts to explain the problems a freshwater goldfish would have if it suddenly found itself in the ocean.

7. What problems do pelagic organisms encounter in maintaining their position in the water column and in swimming? How are these difficulties overcome?

8. How is the marine environment classified? Which scheme is most useful? Why, and to whom? Could you come up with an alternate classification?

9. How many living kingdoms did Linnaeus invent? How many do biologists use today? What are they? What major characteristics determine an organism's placement within each of them?

Critical Thinking Questions

1. The second law of thermodynamics states that entropy (disorganization) tends to increase with time. But living things tend to become *more* complex with time (embryos grow to adults, populations evolve). How can that be?

2. Can you suggest any ways humans might be altering biogeochemical cycles?

3. What is a limiting factor? Can you think of some examples not given in the text?

4. How is evolution by natural selection thought to work?

5. How would you define biological success? Does success depend on the size of an organism? Its beauty? The amount of space it controls? Its numbers?

6. How does a natural system of classification differ from an artificial system? Can you give an example of each? Was the hierarchy-based system invented by Linnaeus natural or artificial? What *is* a hierarchy-based system?

7. **InfoTrac College Edition Project** Explain how marine life can exist near undersea hydrothermal vents. What would such organisms lack, and what would they need and not need in order to exist in such an environment? Use InfoTrac College Edition to gather information for your answer

RESOURCES FOR FURTHER READING AND RESEARCH

The Web site for this book contains many ideas for further reading and research. Log on to:

 www.brookscole.com/053437557xs

and click on the Chapter-by-Chapter area. Choose Chapter 13 and select a resource:

- **References** lists the major books and articles consulted in writing this chapter, along with comments from the author about their content and reading level.

- **Hypercontents** takes you to an extensive list of sites with news, research, and images related to individual sections of the chapter.

- In **Student Guide to InfoTrac College Edition** scroll down to Suggested Readings from InfoTrac College Edition for brief descriptions of the articles listed and search hints for finding them in InfoTrac College Edition.

- **Regional InfoTrac College Edition** articles are organized into East Coast, West Coast, and Gulf Coast regions, allowing you to study oceanography on a more local level. You can also access Regional InfoTrac College Edition at www.localocean.com.

 For additional readings, go to InfoTrac College Edition, your online research library, at:

http://infotrac.thomsonlearning.com/

Primary Producers 14

THE FOREST BENEATH THE WAVES

The experience of diving in a kelp forest is a joy for any good swimmer interested in the marine environment, but for marine scientists it can be rapturous. The ocean often moves placidly in a kelp forest—the long seaweeds interfere with the circular movement of water molecules as the waves pass. Because kelp secretes a lubricant that smoothes a diver's way, visitors can glide easily between the closely entwined seaweeds, gently nudging the strands from their path. Sunlight filters between the fronds, sending illuminating shafts flickering into the deeper water below. Schools of fishes often move through the kelp, and countless animals—many too small to be seen—

Sunlight penetrates a kelp forest, one of the ocean's most beautiful and productive habitats.

nestle around its dark bases. The feeling is like that of being in a great grove of redwoods, and a diver's relative weightlessness allows nearly all parts of the forest to be explored. It is usually quiet here, peaceful and calm.

A biologist sees even more beauty in the scene and notes the rapid growth; the seaweeds are longer on this visit than a few weeks ago, and their stalks are thicker. Some worms have begun to make spiral tracings on the kelp surfaces. The animals on the seabed are different now and are arrayed in new patterns. A family of otters has moved to the area, attracted by the nutritious sea urchins scattered across the gravel-covered bottom. Schools of anchovetta flash overhead, swimming between the kelp blades in long, follow-the-leader schools, efficiently sieving the ocean for food. There are many more microscopic organisms living here than just the big seaweeds swaying in the currents: A gentle living haze crowds the water immediately ahead of the diver's faceplate. A flashlight beam reveals swarms of dust-size organisms in every direction. Countless millions of organisms drift unobserved, too small to reflect the light. Life abounds.

The ocean here fairly hums with metabolic activity. Food is being produced rapidly and in great quantity. Big seaweeds and microscopic algae are harnessing light from the sun to assemble food molecules as their forebears have done since the ocean was young. The temperate coastal ocean brims with life, nearly all of it dependent on the subtle light-driven biochemistry proceeding in this lovely, gently moving place.

14-1

WHAT TO WATCH FOR IN CHAPTER 14

Life runs on food, and some of the world's most important producers of food are found in the ocean. Understanding the term *primary producers* is central to this chapter: *Primary* indicates that food webs start with the organisms we will be discussing, and *producer* emphasizes that the organisms make glucose, an all-important food molecule. Primary producers are organisms that synthesize energy-rich organic compounds (food) from inorganic substances. If someone asks you "What is produced in primary productivity?" a safe answer is "The carbohydrate glucose."

Small, drifting photosynthetic producers—the phytoplankton—are responsible for most of the ocean's primary productivity. The larger marine producers we call seaweeds, and relatively simple organisms that depend on chemosynthesis, account for much of the rest. Many physical and biological factors influence marine primary productivity, the most important being the availability of light and inorganic nutrients. Worldwide oceanic productivity is roughly equivalent to land productivity, but a *much* smaller mass of producers is responsible for productivity in the ocean than on land—marine producers are much more efficient in assembling glucose molecules.

Phytoplankton—and zooplankton, the small, drifting or weakly swimming animals that consume them—are usually the first links in oceanic food webs. Plankton are most abundant along the coasts, in the upper sunlit layers of the temperate zone, and in areas of equatorial upwelling.

The larger multicellular marine algae informally known as seaweeds are classified by color (that is, pigment composition) into three large groups: green, brown, and red algae. Some forms of brown algae, which we call kelp, grow in great underwater forests. Note that not all large marine plantlike organisms are algae; some are sea grasses and mangroves.

In this chapter, we examine primary productivity and then turn our attention to the producers themselves. In the next, we investigate the animals that depend on producers (and each other) for food.

Primary **producers** are organisms that synthesize energy-rich organic compounds from inorganic substances. These energy-rich compounds—we know them as **food**—power the biological world. But food doesn't magically appear in the ocean. It must be constructed—produced—from simpler compounds by the careful manipulation of energy.

The Capture and Flow of Energy

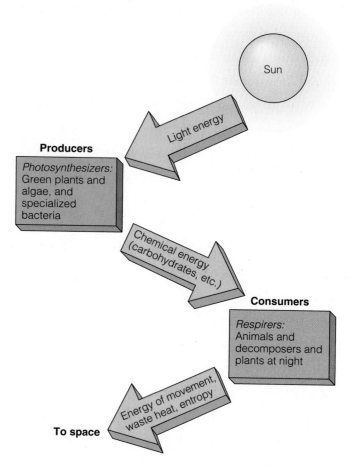

Most of the energy marine organisms need to make food comes directly or indirectly from the sun. The sun produces enormous quantities of energy, some in the form of visible light. Only about 1 part in 2,000 of the light that reaches Earth's surface is captured by organisms, but that "small" input of energy powers nearly all the growth and activity of living things. Light energy from the sun is trapped by chlorophyll in primary *producers* (certain bacteria, protists, and plants) and changed into chemical energy. The chemical energy is used to build simple carbohydrates (such as glucose) and other food molecules, which are then used by the producer or eaten by animals (or other organisms) called *consumers*. Because light energy is used to synthesize molecules rich in stored energy, the process is called **photosynthesis** (*photos* = light, *syn* + *tithenai* = to place together). Here is a general formula for photosynthesis:

$$\text{6 CO}_2 \; + \; \text{6 H}_2\text{O} \quad \xrightarrow{\textit{sunlight}} \quad \text{C}_6\text{H}_{12}\text{O}_6 \; + \; \text{6 O}_2$$

| 6 molecules of carbon dioxide | + | 6 molecules of water | (yields by photosynthesis) | 1 molecule of glucose (a carbohydrate) | + | 6 molecules of oxygen |

Carbohydrates such as glucose are made of atoms of carbon, hydrogen, and oxygen. Where did those atoms come from? Carbon arrives in the cell from surrounding water or air in the form of carbon dioxide. Water itself is also involved in making carbohydrates. Light energy is used to convert the energy-poor compounds CO_2 and H_2O into energy-rich food molecules such as carbohydrates. Oxygen is a by-product of these reactions. Notice in this formula that *a large molecule is formed from small ones.*

Photosynthesis is the predominant method of binding energy into carbohydrates, but there is another. **Chemosynthesis** is the production of usable energy directly from energy-rich inorganic molecules available in the environment rather than from the sun. It is employed by a few relatively simple forms of life, many of them widely distributed archaeans.

$$\text{6 CO}_2 + \text{6 H}_2\text{S} + \text{6 O}_2 + \text{6 H}_2\text{O} \rightarrow \text{C}_6\text{H}_{12}\text{O}_6 + \text{6 H}_2\text{SO}_4$$

| 6 molecules of carbon dioxide | + | 6 molecules of hydrogen sulfide | + | 6 molecules of oxygen | + | 6 H₂O molecules | (yields) | 1 molecule of glucose (a carbohydrate) | + | 6 molecules of sulfuric acid |

Figure 14.1 The flow of energy through living systems. At each step, energy is degraded (transformed into a less useful form).

As we will see in Chapter 16, some unusual forms of marine life depend on chemosynthesis. Overall, chemosynthetic production of food in the ocean is thought to be small compared to photosynthetic production.

The energy bound into carbohydrates by photosynthesis and chemosynthesis is released when food is used for growth, repair, movement, reproduction, and the other functions of organisms. The breakdown of food eventually produces waste heat, which flows away from Earth into the coldness of space. This one-way flow of energy is shown in **Figure 14.1.**

Obtaining Energy by Respiration

Primary producers make enough food through photosynthesis or chemosynthesis to meet their own needs, but animals must eat the photosynthesizers (or organisms that have consumed photosynthesizers) to gain an adequate supply of food.

The process used by plants and animals to disassemble food molecules and harvest energy is called cellular **respiration.**[1] A food molecule—shown here as the carbohydrate

---

[1] Respiration is a biochemical process and should not be confused with the mechanical process of breathing.

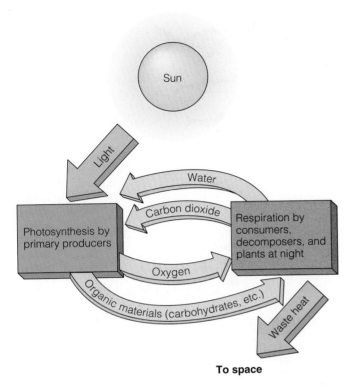

Figure 14.2 The flow of matter through living systems. Note that the beginning products for photosynthesis are the end products for respiration, and vice versa.

glucose—accompanied by oxygen is broken down into carbon dioxide and water. The process is shown here:

$C_6H_{12}O_6$	+	$6 O_2$	$\rightarrow$	$6 CO_2$	+	$6 H_2O$	+	chemical energy
1 molecule of glucose (a carbohydrate)		6 molecules of oxygen	(yields)	6 molecules of carbon dioxide	+	6 molecules water	+	energy

Respiration liberates the energy within the glucose molecule and makes it available for metabolism. CO_2 is released in the process. All organisms, *including the primary producers themselves,* carry out respiration. Notice in this formula that oxygen is being consumed and that *small molecules are formed from a large one*—the opposite of photosynthesis.

Now look at **Figure 14.2.** *The beginning products for photosynthesis are the end products for respiration, and vice versa.* The only loose ends are those of energy flow (Figure 14.1). The individual molecules of carbon dioxide, carbohydrates, oxygen, and water move back and forth in an endless cycle as energy flows through the system. Photosynthesis and respiration are opposite and complementary processes. All the free oxygen in the atmosphere and the ocean passes through organisms—in by respiration, out by photosynthesis—about every 2,000 years, and all the available carbon dioxide cycles through organisms in the reverse direction every 300 years. All the water in the ocean is broken down and re-formed by the chemistry of photosynthesis and respiration every 2 million years!

Feeding (Trophic) Relationships

Photosynthetic and chemosynthetic organisms can be called either primary producers or **autotrophs** (*auto* = self, *trophe* = nourishment) because they each make their own food. The bodies of autotrophs are rich sources of chemical energy for any organisms capable of consuming them. **Heterotrophs** (*hetero* = other, different) are organisms, such as animals, that must consume other organisms because they are unable to synthesize their own food molecules. Some heterotrophs consume autotrophs, and some consume other heterotrophs.

We can label organisms by their position in a "who eats whom" feeding hierarchy called a **trophic pyramid** (*trophos* = one who feeds). The primary producers shown at the bottom of the pyramid in **Figure 14.3** are mostly chlorophyll-containing photosynthesizers. The animal heterotrophs that eat them are called **primary consumers** (or herbivores), the animals that eat primary consumers are called **secondary consumers,** and so on to the **top consumer** (or top carnivore).

Note that the mass of consumers becomes smaller as energy flows toward the top of the pyramid. There are many small primary producers at the base, and very few large top consumers at the apex. Only about 10% of the energy from the organisms consumed is stored in the consumers as flesh, so each level is about one-tenth the mass of the level directly below. The rest of the energy is lost as waste heat as organisms live and work to maintain themselves.

Pyramids such as the one in Figure 14.3 can lead to the misconception that *one* kind of fish eats only *one* other kind of fish, and so on. Real communities are more accurately described as food webs, an example of which is shown in **Figure 14.4.** A **food web** is a group of organisms linked by complex feeding relationships in which the flow of energy can be followed from primary producers through consumers. Organisms in a food web almost always have some choices of food species.

PRIMARY PRODUCTIVITY

The synthesis of organic materials from inorganic substances by photosynthesis or chemosynthesis is called **primary productivity** (**Figure 14.5**). Primary productivity is expressed in *grams of carbon bound into organic material per square meter of ocean surface per year* ($gC/m^2/yr$). The organic material produced is usually glucose, a carbohydrate. The source of carbon for glucose is dissolved CO_2. *Phytoplankton*—small, drifting photosynthetic autotrophs you will meet in a moment—probably produce 90% to 96% of oceanic carbohydrates. Seaweeds—larger marine producers that we will also discuss later in this chapter—contribute from 2% to 5% of the ocean's primary productivity. Chemosynthetic organisms account for the rest. Though estimates vary widely, recent studies suggest that total ocean productivity ranges from 75 to 150 grams of carbon bound into carbohydrates per square meter of ocean surface per year (75 to 150 $gC/m^2/yr$). (For

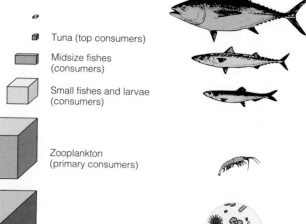

Trophic Level

A tuna sandwich 100 g (¼ pound)

5   For each kilogram of tuna,   Tuna (top consumers)

4   roughly 10 kilograms of midsize fish   Midsize fishes
must be consumed,   (consumers)

3   and 100 kilograms of small fish,   Small fishes and larvae
(consumers)

2   and 1,000 kilograms of small   Zooplankton
herbivores,   (primary consumers)

1   and 10,000 kilograms of primary   Phytoplankton
producers.   (primary producers)

Figure 14.3 A generalized trophic pyramid, showing the long and energetic history of a tuna sandwich. How many kilograms of primary producers are necessary to maintain 1 kilogram of tuna, a top carnivore? Using the trophic pyramid model shown here, you can see that 1 kilogram of tuna at the fifth trophic level (the fifth feeding step of the pyramid) is supported by 10 kilograms of midsize fish at the fourth, which in turn is supported by 100 kilograms of small fish at the third, who have fed on 1,000 kilograms of zooplankton (primary consumers) at the second, which have eaten 10,000 kilograms of phytoplankton (small autotrophs, primary producers) at the first. (These figures have been rounded off to illustrate the general principle. The actual measurements are difficult to make and quite variable.)

comparison, a well-tended alfalfa field produces about 1,600 gC/m²/yr.)

How does marine net productivity compare with terrestrial net productivity? Recent research suggests that the global net productivity in *marine* ecosystems is 35 to 50 billion metric tons of carbon bound into carbohydrates per year; global net *terrestrial* productivity is roughly similar at 50 to 70 billion metric tons per year.[2] However, the total producer **biomass** (the mass of living tissue) in the ocean is only 1 to 2 billion metric tons, compared with 600 to 1,000 billion metric tons of living biomass on land! As you can see in **Table 14.1** on page 353, marine producers are *much* more efficient in their production of food than their land-based counterparts.

The total mass of a primary producer is assumed to be about ten times the mass of the carbon it has bound into carbohydrates. Thus, a primary productivity of 100 gC/m²/yr represents the yearly growth of about 1,000 grams of primary producers for each square meter of ocean surface (see **Figure 14.6**). Between 35 and 50 billion metric tons of carbon is believed to be bound into carbohydrates in the ocean each year, so between 350 and 500 billion metric tons of marine producers is constructed annually. Each year this vast bulk is consumed by the metabolic activity of the pro-

ducers themselves and by the consumers that graze on them. The component atoms are reassembled by photosynthesis and chemosynthesis into carbohydrates in a continuous cycle.

It's no wonder that humans tend to overlook the importance of marine primary productivity—the organisms are mostly inconspicuous, but photosynthesis–respiration *cycling* (turnover time) is extraordinarily fast compared to terrestrial systems!

Measuring Primary Productivity

The rate at which autotrophs form organic material is not easy to measure. At first glance the problem might seem simple to solve: Collect all the organisms—large and small, autotrophs and heterotrophs—in an area and analyze their organic content and mass. Since all organisms are dependent on primary productivity, the mass of living tissue (biomass) in the area would seem directly proportional to productivity.

This approach has drawbacks. For example, a dense population of tiny drifting autotrophs (a high biomass) would interfere with light penetration, so the autotrophs would manufacture carbohydrates slowly and productivity would be low. In contrast, a sparse population of tiny drifting autotrophs (a low biomass) might encounter ideal conditions for photosynthesis and produce carbohydrates at a rapid rate. Small animals might immediately consume this production

---

[2] 1 billion metric tons = 1.1 billion tons.

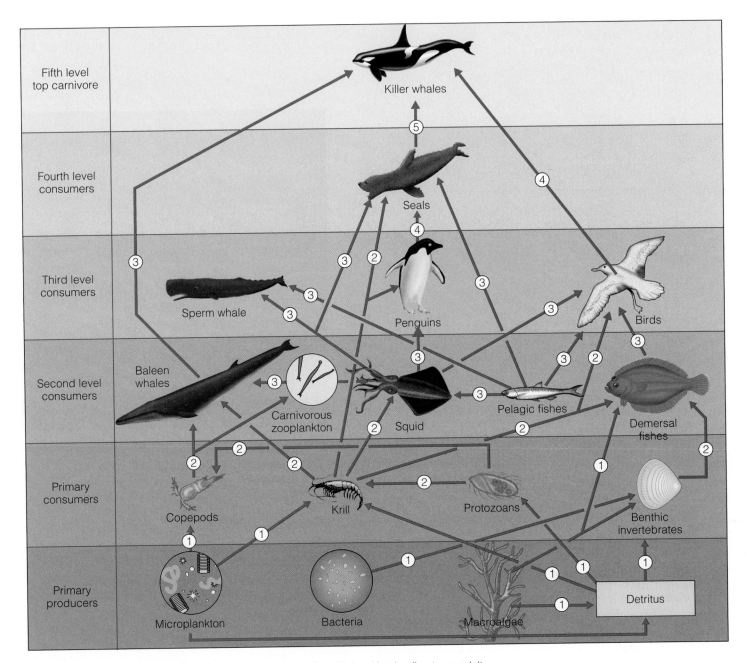

Figure 14.4 A simplified food web, illustrating the major trophic relationships leading to an adult killer whale. The arrows show the direction of energy flow; the numbers on each area represent the trophic level at which the organism is feeding. Note that the feeding relationships are more complex than one would assume from Figure 14.3.

and keep the biomass at low levels, but the productivity would be high.

What researchers need is a method of measuring the rate of productivity *directly*. Because we know the formula for photosynthesis (see page 348), measuring any one component of photosynthesis will tell us about the others. For example, from a measurement of carbon taken up by primary producers, we can calculate the rate of carbohydrate production.

This can be done by using atoms of carbon "tagged" by radioactivity. Though it is radioactive, carbon-14 (^{14}C) be-

haves chemically the same way as the much more common carbon-12 (^{12}C). And because it is radioactive, ^{14}C's progress through photosynthesis can be monitored. One of the ions formed when carbon dioxide dissolves in seawater is bicarbonate (HCO_3^-).[3] Carbon in bicarbonate is tagged, and known amounts of radioactive bicarbonate are added to two bottles of seawater. One bottle is exposed to light, and photosynthesis and respiration take place. The other bottle is

---

[3] This is shown in Figure 7.10.

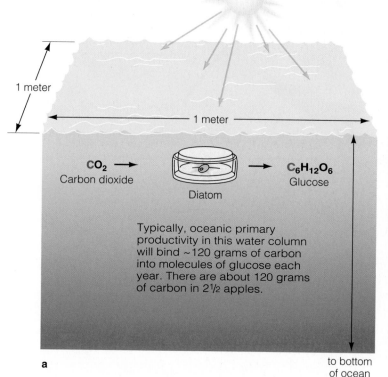

1 meter

1 meter

CO_2
Carbon dioxide

Diatom

$C_6H_{12}O_6$
Glucose

Typically, oceanic primary productivity in this water column will bind ~120 grams of carbon into molecules of glucose each year. There are about 120 grams of carbon in 2½ apples.

a

to bottom of ocean

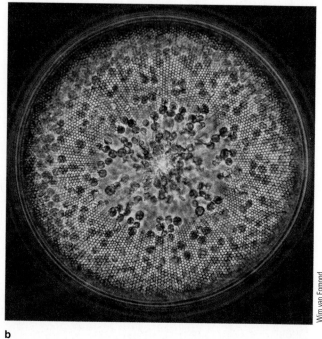

Wim van Egmond

b

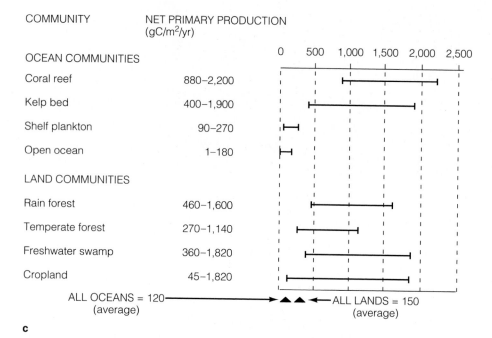

COMMUNITY	NET PRIMARY PRODUCTION (gC/m²/yr)	EQUIVALENT AMOUNT OF CARBON IN PUMPKINS AND APPLES/m²/yr (maximum values) (1 pumpkin = 20 apples)
OCEAN COMMUNITIES		
Coral reef	880–2,200	44.0
Kelp bed	400–1,900	38.0
Shelf plankton	90–270	5.4
Open ocean	1–180	3.6
LAND COMMUNITIES		
Rain forest	460–1,600	32.0
Temperate forest	270–1,140	22.6
Freshwater swamp	360–1,820	36.4
Cropland	45–1,820	36.4

0   500   1,000   1,500   2,000   2,500

ALL OCEANS = 120 (average) — ◀ ALL LANDS = 150 (average)

c

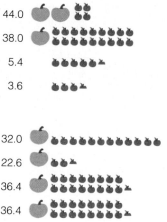

Figure 14.5 (**a**) Oceanic productivity—the incorporation of carbon atoms into carbohydrates—is measured in grams of carbon bound into organic material per square meter of ocean surface area per year (gC/m²/yr). (**b**) The diatom *Coscinodiscus,* an important marine primary producer. In a very bright light this single-celled plantlike organism would barely be visible to the unaided eye. (**c**) Net annual primary productivity in some marine and terrestrial communities. For visualization, an apple contains about 50 grams of carbon (50 gC). High productivities exceeding the equivalent of 20 apples/m²/yr are represented here by pumpkins, with each pumpkin equal to 20 apples. (*Source:* Milne, 1995.)

Table 14.1	Comparison of Global Net Productivity and Living Biomass in Marine and Terrestrial Ecosystems.		
Ecosystem	Net Primary Productivity (10^{15} grams/year)[a]	Total Plant Biomass (10^{15} grams)	Turnover Time (years)
Marine	35–50	1–2	0.02–0.06
Terrestrial	50–70	600–1,000	9–20

Source: Falkowski and Raven, 1997.
[a] 10^{15} grams is equivalent to 1 billion metric tons.

shielded from light, so only respiration occurs. The amount of radioactive carbon incorporated into carbohydrates is measured when the organisms are filtered out of the samples. The radioactivity is measured, and productivity is calculated from the following formula:

$$\text{Rate of production} = \frac{(R_L - R_D) \times M}{R \times t}$$

where R is the total radioactivity added to the sample, t is the number of hours of incubation, R_L is the radioactive count in the "light" bottle sample, R_D is the count in the "dark" sample, and M is the total mass of all forms of carbon dioxide in the sample (in milligrams of carbon per cubic meter). Productivity is expressed as the amount of carbon (in milligrams)

bound—or *fixed*—in new carbohydrates per volume of water (in cubic meters) per unit time (per hour). The rate of production varies from zero to as much as about 80 milligrams of carbon fixed into carbohydrates per cubic meter per hour. These data can be extrapolated to provide the amount of carbon fixed in the water column per square meter of surface per day ($gC/m^2/day$).

Researchers could also collect identical water samples from known depths, place the samples into pairs of identical transparent and opaque bottles, and then suspend an array of bottles from a buoy (**Figure 14.7**). The transparent bottles admit light; the opaque bottles block it. The difference in uptake of carbon in each pair of bottles over time gives an indication of the gross rate of productivity at each depth. Light bottle/dark bottle experiments with radioactive tracers may

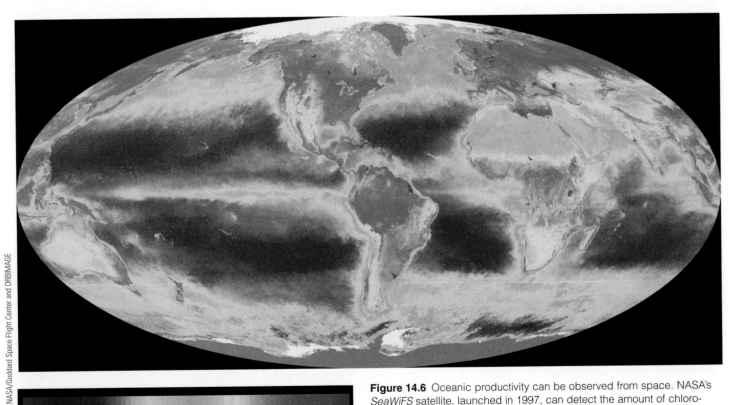

SeaWiFS Project, NASA/Goddard Space Flight Center and ORBIMAGE

.01   .02 .03   .05   .1   .2 .3   .5   1   2   3   5   10   15 20 30   50

Chlorophyll a Concentration mg/m3

Figure 14.6 Oceanic productivity can be observed from space. NASA's *SeaWiFS* satellite, launched in 1997, can detect the amount of chlorophyll in ocean surface water. Chlorophyll content allows an estimate of productivity. Red, yellow, and green areas indicate high primary productivity; blue areas indicate low. This image was derived from measurements made from September 1997 through August 1998.

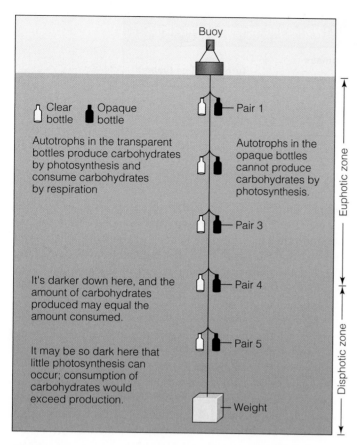

Figure 14.7 The light bottle/dark bottle technique for estimating marine primary productivity. Clear and opaque bottles are filled with water from the area to be studied. Pairs of the bottles are lowered to depths at which 100%, 50%, 25%, 10%, 5%, and 1% of surface illumination is present, and photosynthesis and respiration are allowed to occur. Carbohydrate production by autotrophs in the bottles is measured by radioactively tagged bicarbonate ions added to the seawater in each bottle. The difference in the quantity of carbohydrates in each pair of bottles over time gives an indication of the productivity at each depth. Bottle pair 4 is near the compensation depth, a depth at which production of carbohydrates equals consumption (see Figure 14.15). For an explanation of the euphotic and disphotic zones, please see Figure 14.14.

be conducted for various species and concentrations of producers at different light, nutrient, and temperature levels. Extrapolation from these experiments to other oceanic areas can then yield data on overall productivity.

Ocean conditions are complex and variable, however—rarely controlled. A new and promising method of gauging productivity may be the most effective and useful of all. Recent advances in remote sensing have made it possible to estimate the chlorophyll content of ocean water from orbiting satellites (Figures 14.6 and 14.16). Because the amount of chlorophyll present is directly related to the rate of photosynthesis, chlorophyll content can be a good indicator of productivity. As we have seen, however, the biomass of drifting autotrophs—the amount of phytoplankton in an area—is not always a true indicator of how rapidly substances are cycling through that biomass.

Factors That Limit Productivity

As you may recall, a *limiting factor* is a physical or biological necessity whose presence in inappropriate amounts limits the normal actions of an organism—in this case, production of carbohydrates. Photosynthetic autotrophs require four main ingredients to produce carbohydrates: water, carbon dioxide, inorganic nutrients, and sunlight. Obviously, water is not a limiting factor in the ocean. Carbon dioxide is almost never a limiting factor either, because of its high solubility in water and because of the large quantity of carbon dioxide dissolved in the ocean. So the two potential limiting factors in primary productivity are the availability of inorganic nutrients and sunlight.

Autotrophs require inorganic nutrients for two purposes: to construct the large organic molecules that make primary productivity possible and to construct their skeletons or protective shells. **Nonconservative nutrients** are nutrients that change in concentration with biological activity. After a period of rapid phytoplankton growth—a phenomenon known as a **plankton bloom**—ocean surface waters are often depleted of nonconservative nutrients such as nitrate, phosphate, iron, and silicate. These nutrients become part of the producers or of the animals that have eaten the producers. Unfortunately for the producers, valuable nutrients incorporated into the bodies of dead organisms tend to sink below the sunlit zone. The resulting low nutrient concentrations are the most important factors limiting the growth of marine producers. Photosynthetic productivity cannot continue unless the upwelling of deep water returns these nutrients to the surface.

The opportunity for exchange of nutrient-depleted surface water and nutrient-rich deep water is relatively high where there is little or no thermocline. In the Antarctic, the return of sunlight in the southern summer triggers tremendous phytoplankton blooms; Antarctic water sometimes becomes a rich planktonic soup as millions of diatoms compete for nutrients and sunlight. Indeed, because of the high nutrient levels from upwelling supplemented by iron suspended in glacial runoff from land, the summer waters between the Antarctic Convergence and the mainland of Antarctica are among the most productive on Earth. Phytoplankton blooms in the northern polar ocean are not usually as exuberant because the volume of upwelling water is much less and glaciers don't contribute as much rock dust to the ocean.

Because of the stability of the horizontal layers, upwelling is uncommon in the tropical ocean, except in the equatorial Pacific or in areas where currents impinge on interrupting islands or continents. The clear blue of the tropical ocean is the sure signature of an oceanic desert in which deep upwelling rarely occurs. When nutrients are available or are tightly recycled, as in shallow coral reefs, the tropical ocean explodes into vigorous productivity.

Lack of nutrients is the most common factor limiting primary productivity. If adequate nutrients are present, productivity depends on illumination. Too little light is obviously limiting for photosynthesizers; very little photosynthesis proceeds below 100 meters (330 feet), and no photosynthesizers are known to function below 268 meters (879 feet). Too much light can also be inhibiting, however. You might think that productivity would be greatest right at the ocean surface, where light is brightest. It isn't. Light there is often strong enough to overwhelm the photosynthetic chemistry of some photosynthesizers, especially phytoplankton.

Quantity is one important aspect of the light received by marine autotrophs. Quality, or color, is another. Chlorophyll is a green pigment and thus absorbs best in the red and violet wavelengths. Chlorophyll looks green because it reflects green light. But red and infrared light are effectively absorbed and converted into heat near the ocean surface; very little red light penetrates past 3 meters (10 feet).

PLANKTON

The phytoplankton we have been discussing are members of the plankton community. The organisms that constitute **plankton** (*planktos* = wandering) are as important as they are inconspicuous. Plankton drift or swim weakly, going where the ocean goes, unable to move consistently against waves or current flow.

The diversity of planktonic organisms is astonishing. There are giant, drifting jellyfish with tentacles 8 meters (25 feet) long; small but voracious arrowworms; many single-celled creatures that glow brightly when disturbed; mollusks with slowly beating flaps that resemble butterfly wings; crustaceans that look like microscopic shrimp; miniature, jet-propelled animals that live in jellylike houses and filter food from water; and shimmering, crystal-shelled algae. The only feature common to all plankton is their inability to move consistently laterally through the ocean. However, many can and do move vertically in the water column.

The plankton contain many different photosynthetic and chemosynthetic species and virtually every major group of animals. Thus, the term *plankton* is not a collective natural category like mollusks or algae, which would imply an ancestral relationship among the organisms; instead it describes a common ecological connection—a lifestyle. Members of the plankton community, also referred to as **plankters,** can and do interact with one another. There is grazing, predation, parasitism, and competition among members of this dynamic group. **Figure 14.8** suggests the rich diversity of plankton.

Collecting and Studying Plankton

The first large-scale, systematic study of plankton was carried out by biologists aboard the research vessel *Meteor* during the German Atlantic Oceanographic Expedition of 1925–26. Many of the tools and techniques they pioneered are still in use today. **Plankton nets** (**Figure 14.9**) of the type perfected aboard the *Meteor* are essential to plankton studies. These conical nets are customarily made of nylon or Dacron cloth woven in a fine interlocking pattern to assure consistent spacing between threads. The net is hauled slowly for a known distance behind a ship or cast to a set depth, then reeled in. Trapped organisms are flushed to the net's pointed end and carefully removed for analysis. Quantitative analysis of plankton requires both a count of the organisms and an estimate of the sampled volume of water.

Because very small plankton can slip through a plankton net, their capture and study require either concentration of water samples by centrifugation or entrapment by a filter. Filtration of water samples is currently the most common and widely used method used by biological oceanographers to collect plankton and bacteria for study and experimentation. Filters can be made of glass-fibers, polycarbonate membranes, or even matrices made of aluminum or silver. Some recent research suggests that *very* small producers, some of which are archaeans, may be responsible for at least as much primary productivity as their larger and better-known relatives!

Measurements of physical ocean conditions such as dissolved carbon dioxide and oxygen content, pH, temperature, and light intensity at the time and place of sampling are of special importance in interpreting the samples. Simultaneous sampling at many locations can be useful in pinpointing the often subtle interplay between species and physical conditions that complicates our understanding of plankton biology.

Phytoplankton: The Autotrophs

Autotrophic plankton that generate glucose by photosynthesis are generally called **phytoplankton** (*phyton* = plant). A huge, nearly invisible mass of phytoplankton drifts within the sunlit surface layer of the world ocean. Phytoplankton are critical to all life on Earth because of their great contribution to food webs and their generation of large amounts of atmospheric oxygen through photosynthesis. Planktonic autotrophs are thought to bind *at least* 35 billion metric tons of carbon into carbohydrates each year, at least 40% of the food made by photosynthesis on Earth! These easily overlooked, mostly single-celled, drifting photosynthesizers are much more important to marine productivity than the larger and more conspicuous seaweeds.

There are at least eight major types of phytoplankton, of which the most prominent are the diatoms and dinoflagellates. In the past decade, however, the small coccolithophores and silicoflagellates, and extremely small cyanobacteria and autotrophic picoplankton, have been found to be more important in world productivity than most researchers could have imagined.

DIATOMS The dominant and most productive photosynthetic organisms in the plankton—and in the world—are the **diatoms.** Diatoms evolved comparatively recently and began to dominate phytoplanktonic productivity in the Cretaceous period about 100 million years ago. Their abundance and photosynthetic efficiency further increased the proportion of free oxygen in Earth's atmosphere. More than 5,600 species of diatoms are known to exist. The larger species are barely

35 mm

5 mm

a

1 dolphins, *Delphinus*	26 mullet larva, *Mullus*
2 tropic birds, *Phaëthon*	27 sea butterfly, *Clione*
3 paper nautilus, *Argonauta*	28 copepods, *Calanus*
4 anchovies, *Engraulis*	29 assorted fish eggs
5 mackerel, *Pneumatophorus*, and sardines, *Sardinops*	30 stomatopod larva
6 squid, *Onykia*	31 hydromedusa, *Hybocodon*
7 *Sargassum*	32 hydromedusa, *Bougainvilli*
8 sargassum fish (*Histro*)	33 salp (pelagic tunicate), *Doliolum*
9 pilot fish (*Naucrates*)	34 brittle star larva
10 white-tipped shark, *Carcharhinus*	35 copepod, *Calocalaus*
11 pompano, *Palometa*	36 cladoceran, *Podon*
12 ocean sunfish, *Mola mola*	37 foraminifer, *Hastigerina*
13 squids, *Loligo*	38 luminescent dinoflagellates, *Noctiluca*
14 rabbitfish, *Chimaera*	39 dinoflagellates, *Ceratium*
15 eel larva, *Leptocephalus*	40 diatom, *Coscinodiscus*
16 deep sea fish	41 diatoms, *Chaetoceras*
17 deep sea angler, *Melanocetus*	42 diatoms, *Cerautulus*
18 lantern fish, *Diaphus*	43 diatom, *Fragilaria*
19 hatchetfish, *Polyipnus*	44 diatom, *Melosira*
20 "widemouth," *Malacosteus*	45 dinoflagellate, *Dinophysis*
21 euphausid shrimp, *Nematoscelis*	46 diatoms, *Biddulphia regia*
22 arrowworm, *Sagitta*	47 diatoms, *B. arctica*
23 amphipod, *Hyperoche*	48 dinoflagellate, *Gonyaulax*
24 sole larva, *Solea*	49 diatom, *Thalassiosira*
25 sunfish larva, *Mola mola*	50 diatom, *Eucampia*
	51 diatom, *B. vesiculosa*

b

Figure 14.8 (**a**) Facing page: Representative plankton and nekton of the subtropical Atlantic Ocean. (**b**) Key. Deep-sea fishes and the sargassum fish (8) are shown near their actual relative size. Surface-dwelling fishes are at least 20 times larger in life than shown in the figure. This stylized representation shows organisms to be much more crowded than they would be in real life.

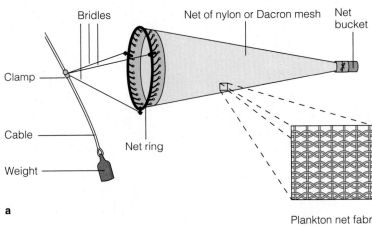

a

Plankton net fabric as seen under a microscope

b

Figure 14.9 Plankton nets. (**a**) The standard conical net is made of fine mesh and has a mouth up to 1 meter (3.3 feet) in diameter. The net is towed behind a ship for a set distance. The number of organisms present in the water can be estimated if the trapped organisms are counted and the volume of sampled water is known. (**b**) The net shown here has a somewhat coarser mesh because its target organisms, small shrimplike crustaceans known as *krill,* are relatively large.

Dennis Kelly

visible to the unaided eye. Most are round, but some are elongated, branched, or triangular.

Typical diatoms are shown in **Figure 14.10.** The name means "to cut through" (*dia* = through, *tomos* = to cut), which is a reference to the patterns of perforations through the diatom's rigid cell wall, or **frustule.** As much as 95% of the mass of the frustule consists of silica (SiO_2), giving this beautiful covering the optical, physical, and chemical characteristics of glass—an ideal protective window for a photosynthesizer. Magnification reveals that the frustule consists of two closely fitting halves, or **valves,** which fit together like a well-made gift box, the top valve adhering tightly over the lip of the bottom one. The pattern of perforations, slits, striations, dots, and lines on the surface of the valves is different for each diatom species.

Inside the diatom's tailored valves lies an extraordinary photosynthetic machine. Fully 55% of the energy of sunlight absorbed by a diatom can be converted into the energy of carbohydrate chemical bonds, one of the most efficient energy conversion rates known. Oxygen not needed in the cell's respiration is released through the perforations in the frustule into the water. Some oxygen is absorbed by marine animals, some is incorporated into bottom sediments, and some diffuses into the atmosphere. Most of the oxygen we breathe has moved recently through the glistening pores of diatoms.

For more effective light absorption, chlorophyll is accompanied in diatoms by other light-absorbing accessory pigments called **xanthophylls.** These yellow or brown pigments give most diatoms a yellow-green or tan appearance. Diatoms

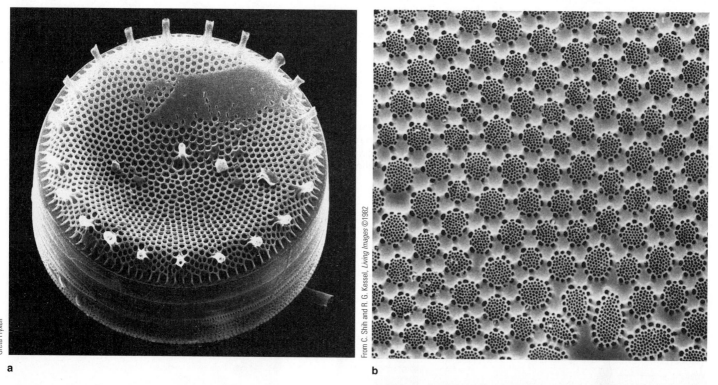

a

b

Figure 14.10 Diatoms. (**a**) The valve, or "shell," of the diatom *Coscinodiscus* as shown in a scanning electron micrograph of the beautiful silica frustule. (**b**) A closer view, showing the intricate perforations from which diatoms take their name. (**c**) A pennate (elongated) diatom. Each cell in this figure is about 15 micrometers (300 millionths of an inch) across.

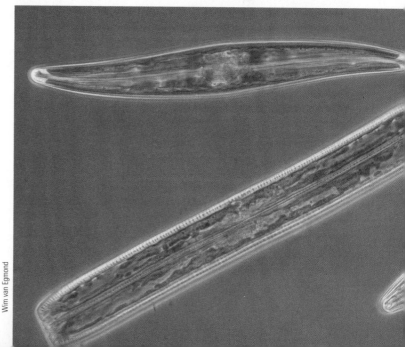

c

store energy as fatty acids and oils, compounds that are lighter than their equivalent volume of water and thus assist in flotation. As you might guess, flotation is a potential problem for diatoms because the mass of their heavy silica frustule seems at odds with their need to stay near the sunlit ocean surface. Oil floats, glass sinks, and a balanced amount of both reduces cell density and lightens the load. Not all diatoms need to float, however. Many nonplanktonic species lie on shallow bottoms, where light and nutrients are able to support photosynthesis. These benthic species are nearly always elongated (or *pennate*) in shape.

Like most single-celled organisms, diatoms reproduce by dividing in half and drifting apart (or, in the species of diatoms that form chains, remaining linked in long lines of cells). The cells may divide as rapidly as once each day. In most species the new valve is generated *within* the old one during division, so the average size of individuals in the population becomes smaller with time (**Figure 14.11**). Individuals will reach a minimum of about one-fourth the size of the original cell. When the cells become too small—or, more accurately, when the cells have too high a ratio of glass to living tissue—they become too heavy to float in spite of their buoyant oils. The problem is solved by sexual reproduction, which generates an **auxospore,** a naked cell without valves. If conditions are still favorable for growth, the auxospore will expand to the diatom's original size, form two thin new valves, and begin the cycle anew. If growing conditions are unsuitable, the auxospore will become dormant to await an opportunity for growth—which may be weeks or months away. When diatoms die, their valves fall to the seafloor to accumulate as layers of siliceous ooze (see Chapter 5).

DINOFLAGELLATES I remember as a child walking along a dark California beach where the sand in my footprints glowed after each step. Handfuls of sand thrown out over the wet beach surface burst into fans of blue light strong enough to cast a shadow, and swimmers emerged from the warm water with thousands of glowing points of light clinging to their bodies. The light I saw was caused by dinoflagellates that had multiplied rapidly in the nearshore waters off southern California. In the late 1950s, wastewater treatment was not as thorough as it is today, and the effluent was not pumped as far out to sea. The combination of warm, calm water and abundant nutrients often triggered such *blooms.* These displays are less common today, but on warm September or October nights after the northeast winds have blown the surface water out to sea and upwelling has brought nutrients to the shore, the sand sometimes glows, and waves can flash blue as they break. These dinoflagellates are *bioluminescent* (*bios* = life, *luminis* = light). **Bioluminescence** is the process by which energy from a chemical reaction is transformed into light energy. The imaginatively named compound *luciferin* is oxidized by the action of the enzyme *luciferase,* and in the process a blue-green light is emitted. The release of light is very efficient and thus is not accompanied by a release of heat, so the organism does not overheat in the oxidation process. (As is discussed in Questions

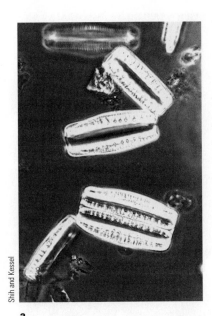

Shih and Kessel

a

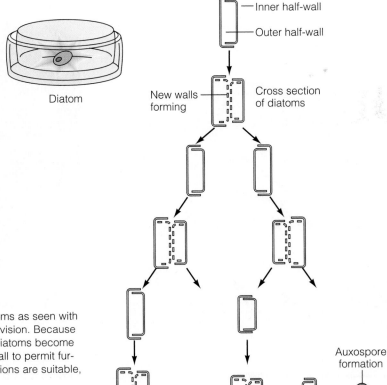

Diatom

Inner half-wall
Outer half-wall

New walls forming

Cross section of diatoms

Auxospore formation

b

Figure 14.11 How diatoms divide. (**a**) Dividing diatoms as seen with a light microscope. (**b**) Diatoms reproduce by cell division. Because a new inner shell is formed after each division, the diatoms become smaller with each generation. When they are too small to permit further reproduction, an auxospore forms. When conditions are suitable, it will germinate to produce a full-size diatom.

from Students #4, bioluminescence may have evolved as a sort of "intrusion alarm.")

Most **dinoflagellates** are autotrophs (**Figure 14.12**). A few species live within the tissues of other organisms (the zooxanthellae of coral animals you will meet in the next chapter, for example), but the great majority of dinoflagellates live free in the water, and only a few form chains. Most have two whiplike projections, called **flagella,** in channels grooved in their protective outer cell wall of cellulose. One flagellum drives the organism forward while the other causes it to rotate in the water. (Hence the name: *dino* = whirling, *flagellum* = whip.) Their flagella allow dinoflagellates to adjust orientation and vertical position to make the best photosynthetic use of available light, or to move vertically in the water column to obtain nutrients.

Dinoflagellates are widely distributed, solitary organisms that reproduce by simple fission; they rarely form colonies. During reproduction, the cellulose covering that surrounds most species splits, and the single cell divides in half. Each daughter subsequently replaces the missing portion of covering. Under favorable conditions the organisms can reproduce once a day, growing in number but not in size.

Some species of dinoflagellates can become so numerous that the water turns a rusty red as light reflects from the accessory pigments within each cell. These species are usually responsible for the phenomenon referred to as *harmful algal bloom* (or HAB) (see **Box 14.1**). During times of such rapid growth (usually in springtime), the concentration of microscopic planktonic organisms may briefly reach 6 to 8 million per liter (23 to 29 million per gallon).

COCCOLITHOPHORES AND SILICOFLAGELLATES

Coccolithophores are small, single-celled autotrophs covered with disks of calcium carbonate (coccoliths) fixed to the outside of their cell walls (**Figure 14.13**). Coccolithophores live near the ocean surface in brightly lit areas. The translucent covering of coccoliths may act as a sunshade to prevent absorption of too much light. In areas of high coccolithophore productivity, most notably in the Mediterranean and Sargasso Seas, their numbers occasionally become so great that the water appears milky or chalky. Coccoliths can also build seabed deposits of ooze—the famous White Cliffs of Dover in England consist largely of fossil coccolith ooze deposits uplifted by geological forces.

The **silicoflagellates** possess filamentous internal supporting structures of silica, the same material from which

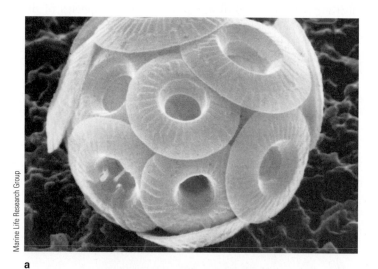
a

b

Figure 14.12 Two representatives of the genus *Ceratium*, a common dinoflagellate. Two flagella beat in opposing grooves in the armor-plated body: One groove is visible at the equator of the upper cell, between the two "horns." These specimens are about 0.5 millimeter (0.02 inch) across.

Figure 14.13 (**a**) Calcium carbonate plates in place on *Umbilicosphaera*, a coccolithophore. The tiny plates are 6 micrometers (120 millionths of an inch) across. (**b**) A bloom of coccolithophores in April 1998 turns the ocean surface turquoise in the Bering Sea, north of Alaska's Aleutian Islands.

HABs Can Be Trouble! · BOX 14.1

A harmful algal bloom (HAB) occurs when high concentrations of phytoplankton adversely affect the physiology of nearby organisms. The somewhat misleading name *red tide* is often used to describe patches of water turned a rusty red by the abundant growth of pigmented phytoplankton, usually certain dinoflagellate species (**Figure a**). But HABs don't always turn water red, nor are the organisms that cause them always visible. A number of factors are thought to contribute to HABs, including warm surface temperatures, reduced salinity, optimal nutrient and light conditions, and a mechanism of physically concentrating the dinoflagellates (such as gentle onshore winds).

Although the dinoflagellates responsible for most red tides are comparatively simple organisms, some have the ability to synthesize potent toxins as by-products of metabolism. Among the most effective poisons known, these toxins (if ingested) may affect nearby marine life (**Figure b**), or even humans by an indirect poisoning through the food chain. Some of the toxins are similar in chemical structure to the muscle relaxant curare, but they are tens of times more powerful.

One of these compounds, a neurotoxin (nervous system poison), affects between 10,000 and 50,000 individuals annually—most of them on tropical and subtropical islands where herbivorous fishes make up much of the diet. The fishes graze on seaweeds, inadvertently ingesting the dinoflagellates attached to the surface of the fronds. The toxin is stored in fat and so is most concentrated in the oldest and fattest (and thus most desirable) fish. Victims complain of abdominal pain, muscular aches, dizziness, anxiety, tingling in the hands and feet, and a curious inversion of the senses (hot seems cold, and vice versa); many die from the effects.

Clams, mussels, scallops, and oysters can also be dangerous, and shellfish poisonings are becoming increasingly common worldwide. These illnesses result when people consume invertebrates that filter seawater to obtain food. Paralytic, diarrhetic, neurotoxic, and amnesic shellfish poisonings have been reported by humans, but the filter-feeding invertebrate is insensitive or only marginally affected. A lethal dose of paralytic toxin kills by interfering with the normal function of heart muscle and affecting breathing. Neurotoxic compounds block communication between nerves and muscles; the first symptoms of intoxication are usually tingling lips and difficulty in swallowing. Diarrhetic shellfish poisoning causes diarrhea, nausea, and vomiting. Most mysterious is amnesic shellfish poisoning, a poorly understood disorder that results in the permanent loss of short-term memory. In 1987 three elderly people in eastern Canada died, and 107 were affected by amnesic poisoning. They ingested blue mussels (*Mytilus edulis*) contaminated with domoic acid, a

C.C. Lockwood

Red tides and toxic dinoflagellates. (**a**) Scanning electron micrograph of *Gymnodinium breve*, the dinoflagellate responsible for the outbreaks of red tides along the Florida coast. Other dinoflagellate species produce red tides in other parts of the world. (**b**) Portion of a fish kill that resulted from a dinoflagellate bloom.

During a red tide, the presence of millions of dinoflagellates turns seawater a brownish-red color.

Suisan Aviation Ltd., Japan

potent neurotoxin produced by the pennate diatom *Pseudonitzschia australis.* A 1998 *Pseudonitzschia* outbreak near California's Monterey Bay was responsible for the deaths of more than 400 sea lions and countless seabirds.

In areas where fisheries are not regulated by governmental agencies, people should avoid eating filter feeders during summer months when toxin-producing phytoplankton are most abundant. If shellfish from a particular area are unsafe, a governmental agency will issue an advisory, which may remain in effect for six weeks or more until the danger is past. Such warnings must be observed—a single clam can accumulate enough toxin to kill a human. There is no antidote; physicians can only treat the symptoms. Whales and seabirds are also at risk from dinoflagellate toxins.

People can be affected even if they don't eat seafood. Beachgoers in North Carolina complained of inflamed eyes and asthmalike symptoms when they inhaled dinoflagellates that had dried on the beaches and been blown inland by strong winds.

The number and severity of HABs appear to be increasing. Perhaps this is not surprising: Coastal waters receive industrial, agricultural, and domestic wastes, rich in nitrogen and other plant nutrients that stimulate algal growth. Also, the long-distance transport of algal species in the ballast water of cargo vessels can introduce alien species into coastal waters, where they may thrive in the absence of the organisms that naturally consume them. Australia has recently issued strict guidelines for discharging ballast in the country's ports. The Chinese government estimates $240 million in direct economic losses from 45 major HABs between 1997 and 1999. More than 75% of the stock of Hong Kong's fish farms were wiped out by one of these events. Seafood consumers should be aware of the source of their meal and the nature of poisoning symptoms.

diatoms construct their valves. One or two flagella are always present. Compared to diatoms, however, the overall structure and biochemistry of silicoflagellates seem primitive. Little is known about their worldwide distribution or abundance.

NANOPLANKTON AND PICOPLANKTON Most other types of phytoplankton are very small and so are called **nanoplankton** (*nanus* = dwarf) or **picoplankton** (*pico* = a trillionth). Recent evidence suggests that extraordinarily small picoplanktonic photosynthesizers may be much more numerous and contribute *much* more to overall plankton productivity than previously thought, especially in open ocean areas. **Box 14.2** explores this idea, a field of great interest to biological oceanographers.

The Euphotic Zone

As you may recall from Chapter 13, the sunlit layer of water at the ocean's surface is called the *photic zone* (see Figure 13.11). At noon in the clear tropics the photic zone may extend to a depth of around 600 meters (2,000 feet), but in midlatitude waters small organisms (and other small scattering particles) are more abundant, so light reaches down only to about 150 meters (500 feet). The upper part of the photic zone—where there is sufficient light for carbohydrate production by photosynthesis to exceed the loss of carbohydrates through respiration—is called the **euphotic zone** (**Figure 14.14**). The upper productive layer of ocean is a very thin

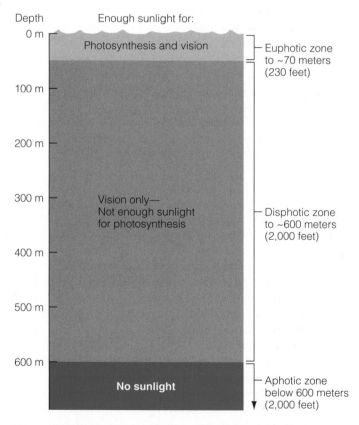

Figure 14.14 The relationship of the euphotic, disphotic, and aphotic zones to one another. (The euphotic zone statistic is for mid-latitude waters averaged through a year.)

In the mid-1970s biological oceanographers began to appreciate the contribution to oceanic productivity of very small phytoplankton termed picoplankton. Picoplankton, a word derived from the Latin term for "a millionth part," are extremely small—so tiny that they were long mistaken for fine particles of suspended sediments. These organisms are often too small to be resolved by light microscopes and slip undetected through all but the finest filters. Their size, typically about 0.2 to 2 micrometers (4 to 40 millionths of an inch) across, is made up for by their abundance: an astonishing 100,000 cells per milliliter (3 million cells per ounce) in tropical and subtropical oceanic waters previously thought to be all but devoid of phytoplankton.

Synechococcus is typical of this newly recognized type of organism. It was discovered when individual cells—little more than naked photosynthetic machines—fluoresced a bright orange when struck by ultraviolet light. Analysis of the fluorescence spectrum (living examples of *Synechococcus* are too small to study directly) revealed the presence of an odd chlorophyll variant previously known only in a rare mutant strain of corn. Structural details revealed by electron micrographs suggest that *Synechococcus* and its relatives are cyanobacteria.

Recent estimates suggest that picoplankton may account for up to 70% of all the photosynthetic activity in some parts of the open ocean! How could such a huge contribution to oceanic productivity have gone unnoticed for so long? In part because these autotrophs are so extremely small, and in part because they are efficiently grazed by microflagellates. Additionally, the products of their photosynthetic activity are closely cycled by even smaller heterotrophic bacteria in the immediate vicinity. Here is a complete micro-ecosystem—a community operating on the smallest possible scale—that manufactures and consumes particulate and dissolved carbon in amounts almost beyond comprehension, a sort of ecological black market below the "official economy" of the relatively huge diatoms and dinoflagellates. As if they weren't busy enough, these heterotrophic bacteria also decompose organic material spilled into the water when phytoplankton are eaten by zooplankton, turn soluble organic materials released by zooplankton back into inorganic nutrients, and process particulate organic matter in the water by releasing enzymes to break it down into dissolved organic matter, which they consume for their own growth. Biological oceanographers now believe that the greatest fraction of organic particles in the water column of the open ocean is composed of these metabolically active heterotrophic bacterial cells.

And what happens to these bacteria? Those that aren't consumed by the microflagellates may become infected by viruses, which cause the cells to burst. Those viruses that infect bacteria are referred to as *bacteriophages;* those infecting phytoplankton are termed *phycoviruses*. Viruses are extremely small (usually 20 to 250 nanometers in diameter) and are fundamentally different from other forms of life. Viruses have no metabolism of their own and must rely on a host organism for energy-requiring processes, including reproduction. Although we have known about viruses for some time, only in the late 1980s were their high abundances confirmed in a wide range of marine environments. How many are there? Between 10 and 100 million per milliliter of water—up to about 3 billion per ounce!

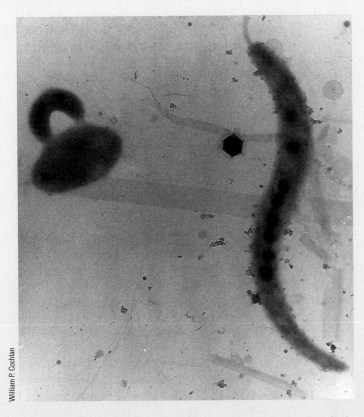

William P. Cochlan

Three bacteria—ellipsoidal, S-shaped, and C-shaped—from the Gulf of Bothnia (east of Sweden). The dark dots within the bacteria are viruses. The large hexagonal structure near the S-shaped bacterium is a relatively large virus with a diameter of 160 nanometers. The bacteria and viruses were photographed using a transmission electron microscope.

skin indeed; the water within this zone amounts to less than 2% of world ocean volume. Yet nearly all marine life depends on this fine illuminated band.

The lower part of the photic zone is the **disphotic zone,** a region where animals can see but there is insufficient light for productive photosynthesis. Below the disphotic zone is the deepest, largest region of the open ocean, extending to the seabed: the *aphotic zone,* the dark zone that extends to the bottom.

Compensation Depth

Remember that autotrophs respire as they photosynthesize—they use some of the carbohydrates and oxygen they produce. Carbohydrate production usually exceeds consumption, but not always. The deeper a phytoplankter's position, the less light it receives. At a certain depth, the production of carbohydrates and oxygen by photosynthesis through a day's time will exactly equal the consumption of carbohydrates and oxygen by respiration. This break-even depth is called the **compensation depth.** Compensation depth usually corresponds to the depth where about 1% of surface light penetrates (the lower limit of the euphotic zone). Compensation depth marks the bottom of the euphotic zone. In Figure 14.7, if bottle pair 4 is at compensation depth, the amount of carbohydrates produced in the transparent bottle will exactly equal the amount consumed, an indication of zero net productivity.

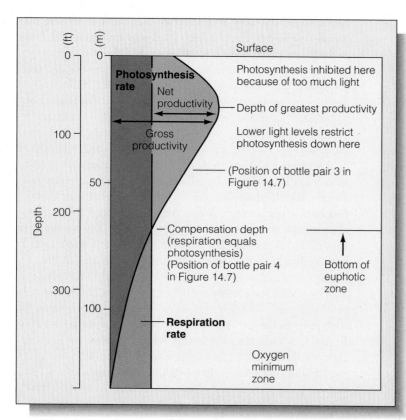

Figure 14.15 Compensation depth and its relationship to other aspects of productivity. Note the position of the bottom of the euphotic zone.

Figure 14.15 uses a chart to show compensation depth. Like the depth of greatest productivity, compensation depth changes with sun angle, turbidity, surface turbulence, and other factors. Remember that the compensation depth is always below the depth of greatest productivity and producers can still "make a profit" between the depth of greatest productivity and the compensation depth. If a producer slips below its compensation depth for more than a few days, however, it will consume its carbohydrate reserves and die.

Because of their greater efficiency, diatoms have a deeper compensation depth than dinoflagellates. Open tropical seas have the deepest potential compensation depths, but these regions are not generally productive because of nutrient deficiencies.

Global Distribution of Plankton Productivity

Where is phytoplankton productivity greatest? This question is among the most important in biological oceanography. Since phytoplankton form the base of nearly all oceanic food webs, the biological characteristics of any ocean area will depend heavily on the presence and success of phytoplankton.

With some exceptions, the distribution of phytoplankton corresponds to the distribution of macronutrients. Because of coastal upwelling and land runoff, nutrient levels are highest near the continents. Plankton are most abundant there, and productivity is highest. The water above some continental shelves sustains productivity in excess of 1 gC/m²/*day*! But what of the open ocean? Where is productivity greatest away from land?

IN THE TROPICS? Water circulating in the great tropical gyres has abundant sunlight and CO_2 but is generally deficient in surface nutrients because the strong thermocline discourages the vertical mixing necessary to bring nutrients from the depths. The tropical ocean away from land is therefore an oceanic desert nearly devoid of visible plankton. The typical clarity of tropical water underscores this point. In most of the tropics productivity rarely exceeds 30 gC/m²/yr, and seasonal fluctuation in productivity is low. But small cyanobacteria (and even tinier picoplanktonic producers) may be binding carbon at rapid rates, and heterotrophic bacteria can be using it as fast as it is made. Even though the standing crop (biomass) of these producers is low, the exceedingly high rate of "throughput" of carbon dioxide and carbohydrates in these communities makes their contribution to general productivity very high. This "black market economy" (discussed in Box 14.2) is not available to fishes and other larger consumers because the larger zooplankton, which are their prey, are unable to separate these astonishingly small organisms from the surrounding water. Bacteria simply utilize the carbon and shuttle the metabolic products back to the tiny producers.

Tropical coral reefs are exceptions to this general rule. Reef areas, which account for less than 2% of the tropical ocean surface, are productive places because autotrophic dinoflagellates live *within* the tissues of coral animals and don't drift as plankton. Nutrients are made available by coastal upwelling and by the coral's own metabolism. These nutrients are cycled tightly through the reef and are not lost to sinking.

IN THE POLAR REGIONS? At very high latitudes during the winter months, the low sun angle, the reduced light penetration due to ice cover, and weeks or months of darkness severely limit productivity. At the height of summer, however, 24-hour daylight, a lack of surface ice, and the presence of upwelled nutrients can lead to spectacular plankton blooms. Such blooms cannot last, however, because the nutrients that support them are not quickly recycled and the sun is above the critical angle for only a few weeks at best. The short-lived summer peak does not compensate for the long, unproductive winter months.

Average productivity at very high northern latitudes, averaging less than 25 gC/m²/yr, tends to be lower than production at high southern latitudes. The Arctic Ocean is almost surrounded by land masses, which limit water circulation and, therefore, nutrient upwelling, so nutrients are quickly depleted. The southern ocean, on the other hand, is enriched by water upwelling to replace sinking Antarctic Bottom Water. This rich mixture is stirred by the Antarctic Circumpolar Current. The Antarctic accounts for a much greater share of high-latitude production than the Arctic because nutrients rarely become depleted.

IN THE TEMPERATE AND SUBPOLAR ZONES? With the tropics generally out of the running because of nutrient deficiency and the north polar ocean suffering from slow nutrient turnover and low illumination, the prize for overall productivity in the world ocean goes to the temperate and southern subpolar zones. Thanks to the dependable light and moderate nutrient supply, the annual production in the nearshore temperate and southern subpolar ocean areas is the greatest of any open ocean area. Typical productivity in the temperate zone is about 120 gC/m²/yr. In ideal conditions southern subpolar productivity can approach 250 gC/m²/yr!

Figure 14.16 shows the levels of productivity in tropical, temperate, and polar ocean areas. Note that *nearshore* productivity is nearly always higher than *open ocean* productivity, even in the relatively productive temperate and south subpolar zones.

Curiously, the open ocean area with the greatest annual productivity is an exception to the general picture developed in this section. Slender, cold fingers of high productivity pointing west from South America and Africa along the equator are a result of wind-propelled upwelling due to Ekman transport on either side of the geographic equator.

SeaWiFS Project, NASA/Goddard Space Flight Center and ORBIMAGE

Figure 14.16 Measuring the concentration of chlorophyll in ocean water by satellite. This false-color image from the Coastal Zone Color Scanner aboard the *Nimbus 7* satellite represents average conditions in late spring in the Northern Hemisphere. It shows the concentration of chlorophyll in the upper layer of the ocean, with higher amounts indicated by green, orange, and red colors. Note the high phytoplankton concentrations induced by increased nutrient availability along the coasts and the equatorial upwelling west of Africa. The centers of the oceanic gyres contain relatively few phytoplankton, as shown by their purple hue.

Phytoplankton Productivity Through the Seasons

Figure 14.17 shows the relationship of phytoplankton biomass to season. The low, flat line representing annual tropical productivity contrasts with the high, thin peak representing the Arctic summer. The higher of the two peaks for the temperate zone indicates the plankton bloom of northern spring, caused by increasing illumination; the smaller temperate-zone peak, representing the northern fall bloom, is caused by nutrients mixed toward the surface with increasing storm activity and less thermal stratification.

Zooplankton: The Heterotrophs

Although this chapter emphasizes the producers, we will also spend a moment with the heterotrophic plankton, collectively called **zooplankton** (*zoion* = animal).[4] Zooplankters are the most abundant primary consumers of the ocean. They graze on the diatoms, dinoflagellates, and other phytoplankton at the bottom of the trophic pyramid in a way similar to cows

---

[4]More on these and other animals will be found in the next chapter.

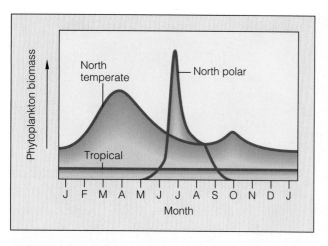

Figure 14.17 Variation in the biomass of phytoplankton by season and latitude. (The total area under each curve represents productivity.)

grazing on grass. The variety of zooplankton is surprising; nearly every major animal group is represented. Each is expert at the concentration of food from water. The most abundant zooplankters, accounting for about 70% of individuals, are tiny shrimplike animals called *copepods* (**Figure 14.18**). Copepods are crustaceans, a group that also includes crabs, lobsters, and shrimp.

Not all zooplankters are small. Many are in the 1- to 2-centimeter (½- to 1-inch) size range. The largest drifters are

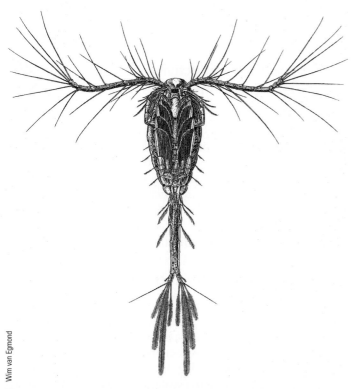

Wim van Egmond

Figure 14.18 A zooplanktonic copepod of genus *Calanus,* appropriately named after a mythical East Indian wanderer, is perhaps the most abundant and widely distributed animal in the ocean. The species shown here reaches a maximum size of about 0.5 millimeter (about 0.02 inch).

giant jellyfish of genus *Cyanea;* their bells may be more than 3.5 meters (12 feet) in diameter! We have a special term for plankton larger than about 1 centimeter (½ inch) across: **macroplankton** (*makros* = large).

Most zooplankton spend their whole lives in the plankton community, so we call them **holoplankton** (*holos* = entirely). But some planktonic animals are the juvenile stages of crabs, barnacles, clams, sea stars, and other organisms that will later adopt a benthic or nektonic (swimming) lifestyle. These temporary visitors are called **meroplankton** (*meros* = mixed). Most animal groups are represented in the meroplankton; even the powerful tuna serves a brief planktonic apprenticeship. These useful categories can be applied to phytoplankton as well as zooplankton. Holoplanktonic organisms are by far the more abundant forms of both phytoplankton and zooplankton.

A form related to amoebas is the **foraminifera** (**Figure 14.19**), which extend long protoplasmic filaments to snare food. Most foraminifera have calcium-carbonate shells. As we saw in Chapter 5, extensive white deposits of calcareous ooze have been built on the seabed from their skeletons. As is the case with some phytoplankton sediments, some of these layers have been uplifted and can be found on land.

No discussion of zooplankton would be complete without mention of *Euphausia superba,* a thumb-size shrimplike crustacean commonly called **krill** (**Figure 14.20**). Krill, primary consumers *extraordinaire,* graze on the abundant diatoms that prosper in the nutrient-rich open water of the southern polar ocean. In turn, krill are eaten in tremendous numbers by seabirds, squid, fishes, and whales.

Some 500 to 750 million metric tons (550 to 825 million tons) of krill inhabit the Antarctic Ocean. Krill comprise the greatest biomass of any single species, exceeding the total biomass of humanity by a factor of five! Krill travel in great schools that can extend over several square miles. They behave more like schooling fish than planktonic crustaceans—their primary swimming mode is horizontal, not vertical.

Plankton and Food Webs

Zooplankton and other animals eat phytoplankton, and still other animals eat these primary consumers. The mass of zooplankton is typically about 10% that of phytoplankton, which is reasonable because of the harvesting relationship that exists between them.

The zooplankters depend on phytoplankton for both food and oxygen. The decomposition of falling biological debris and the metabolic activity of zooplankton often create an **oxygen minimum zone** below the well-lighted surface zone: Oxygen is depleted by the animals there and not replaced by phytoplankton (see Figure 14.15). Some species of zooplankton and a number of small swimming animals make nightly migrations from the oxygen minimum zone toward the darkened surface layer to feed on the smaller organisms drifting there.

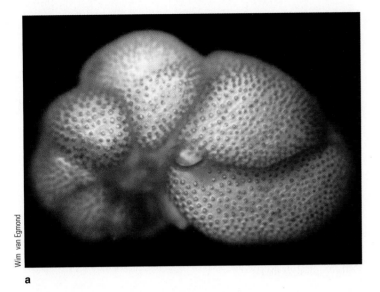

Wim van Egmond

Howard Spero

a

b

Figure 14.19 Foraminifera. The word means "bearers of windows." (**a**) Light streams through thin parts of the shells. (**b**) *Orbulina* feeding on a copepod (compare with Figure 14.18). The foram's spines are made of calcium carbonate and have a sticky layer on their surfaces. Zooplankton that bump into the spines stick to them and are subsequently digested.

It is interesting to note that the largest marine animals, such as whale sharks (fish) and baleen whales (mammals), do not expend their energy tracking down and attacking big animals. Instead, these largest of all feeders concentrate zooplankton from the water and consume them in vast quantities. The zooplankton they eat are usually not the primary consumers but the somewhat larger secondary consumers, which are usually crustaceans such as krill that have themselves fed on the microscopic primary consumers. In this way whales and other large filter feeders can harvest energy closer to the source, gaining the advantage of efficiency and quantity.

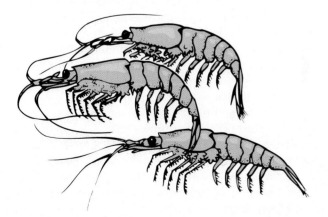

Figure 14.20 Krill (*Euphausia superba*). These shrimplike crustaceans, shown here approximately life-size, occur throughout the world ocean and are particularly numerous in Antarctic seas.

LARGER MARINE PRODUCERS

We now shift our attention from drifters to attached autotrophs. Relatively large attached autotrophs (the protists we call seaweeds) account for between 2% and 5% of the ocean's total primary productivity.

Most marine autotrophs, large or small, are algae. **Algae** is a collective term for autotrophs possessing chlorophyll and capable of photosynthesis but lacking vessels to conduct sap. The single-celled diatoms and dinoflagellates discussed earlier are classified as **unicellular algae.** *Seaweed* is the informal name for large, marine **multicellular algae.** A few species of marine plants are not algae but **angiosperms** (*angios* = covered, *sperma* = seed); these are flowering plants. The main groups of marine angiosperms are sea grasses and mangrove trees. Marine angiosperms are not considered seaweeds.

Algae are **nonvascular** organisms; that is, they do not have vessels. Algae require the same four ingredients for photosynthesis as vascular plants—carbon dioxide, sunlight, water, and inorganic nutrients—but in their case the ingredients are already present in one location, so vessels and flowing sap are not necessary. Either the alga is very small (perhaps a single cell) and lives in a moist spot on land where conditions are ideal, or it lives in water.

Multicellular marine algae occur in a great variety of sizes and shapes. Some types of algae form underwater forests; others grow in isolation. The largest can reach 62 meters (205 feet) in length. Their bodies are flexible, easily able to absorb shock, resistant to abrasion, streamlined to reduce water drag, and very strong. Their surfaces are often covered by a slick, mucilaginous material that lubricates them as they move, retards drying, and deters grazing animals. Algae do not grow below the euphotic zone because they all depend on photosynthesis to produce the energy-rich compounds necessary for life. Nearly 7,000 species of multicellular marine algae have been identified.

The marine lifestyle offers advantages. Large marine autotrophs suffer no droughts and nearly always have enough carbon dioxide for photosynthesis. Assuming suitable nutrient levels and a good foothold, only sunlight is required for productivity. Submersion in seawater brings the additional advantage of lightweight construction. A seaweed doesn't require strong support structures because it has nearly the same density as the surrounding seawater, so more of its bulk can be dedicated to photosynthesis. Indeed, productivity in some wave-washed seaweed beds may be the highest per unit area of any autotrophic community on Earth.

Some seaweeds' astonishingly high productivity is underscored by the fact that many species of large seaweeds are surprisingly leaky—carbohydrates and other products of photosynthesis diffuse from their blades and stipes like tea from a teabag. Up to half of all organic matter produced by these large plants can be lost in this way! The foam visible in surf near kelp beds is produced in part by these substances. Sea urchins and other heterotrophs can absorb these molecules directly through their skin or outer membranes, "feeding" on kelp plants that may be tens or even thousands of meters away.

Accessory Pigments

How can a few species of tropical marine photosynthesizers live at depths of over 250 meters (820 feet)? Most marine autotrophs have evolved specialized accessory pigments. **Accessory pigments** (or masking pigments) are light-absorbing compounds closely associated with chlorophyll molecules. Their presence in autotrophs greatly enhances photosynthesis because they absorb the dim blue light at depth and transfer its energy to the adjacent chlorophyll molecules. Accessory pigments may be brown, tan, olive green, or red; they give most marine autotrophs, especially seaweeds, their characteristic color. It is the combination of pigments in a photosynthesizer that determines its color and optimal depth distribution. The masking effect of accessory pigments is often so complete that an observer is unaware that the organism contains any chlorophyll, even when the seaweed is transparent. The absence of accessory pigments allows the bright green of chlorophyll to shine through.

Structure of Seaweeds

Common terms such as *leaf, stem,* and *root* are inappropriate for seaweeds because the definitions of those parts assume the presence of vessels. The structures in seaweeds that superficially resemble leaves are called **blades** (or fronds), the stemlike structures are called **stipes,** and the root-shaped jumble at the base is appropriately named a **holdfast. Gas**

bladders assist many species in reaching strongly illuminated surface water. Blades, stipes, and holdfast comprise the body of the organism, the **thallus.** These parts are labeled in **Figure 14.21.**

A thallus may be large or small, branching or tufted, a sheet or a filament, encrusting or elongated, rounded or pointed—algal variety is tremendous. Algal blades are symmetrically equipped with photosynthesizing tissue, absorb gases across their entire surface, and even participate in reproduction. Stipes are strong, photosynthesizing, shock-absorbing links tying the blades (or sheets or filaments) into a unit. The holdfast does not take up water and nutrients from the substrate as does the vascular root it superficially resembles, but it does anchor the seaweed in place and can provide incidental shelter for a rich variety of animal life. The gas bladders range in size from tiny grapelike bunches to single volleyball-size floats.

Classification of Seaweeds

Multicellular marine algae are classified in three divisions based on their pigments. The green algae, with their unmasked chlorophyll, are the **Chlorophyta** (*chloros* = green, *phyton* = plant); the brown algae, **Phaeophyta** (*phae* = tan, dusky); and the red algae, **Rhodophyta** (*rhodon* = rosy red).

THE CHLOROPHYTES The clear green color of chlorophytes is evidence of their lack of accessory pigments and suggests that they live at or near the surface, where red light is available. Land plants are thought to have evolved from chlorophyte ancestors. Only about 10% of the 7,000 species of Chlorophyta are marine, but some of those species are widely distributed and present in great numbers. Genus *Ulva* is a familiar, delicate, lettucelike edible seaweed able to tolerate the often impure waters of urban coasts. *Ulva* makes quick use of concentrated nutrients near sewage outfalls to grow and multiply. Most other species of green algae prefer clean water and are branched or threadlike. Some encrust hard surfaces and some live on sand, but none reaches the size of the large brown seaweeds, the phaeophytes.

THE PHAEOPHYTES Nearly all of the 1,500 living species of phaeophytes are marine. Some species of these largest of algae, which include the **kelps,** can reach lengths of 40 meters (132 feet)—the record length exceeds 60 meters (200 feet). To attain these dimensions the seaweed can grow at the spectacular rate of 50 centimeters (20 inches) per day! Part of the strategy of rapid growth is to reach bright surface water as soon as possible. Some brown algae are annuals; others live for up to seven years. The tan or brown color of phaeophytes comes from the accessory pigment *fucoxanthin,* which permits photosynthesis to proceed at greater depths than is possible for the unmasked chlorophytes. In ideal circumstances some larger brown algae can grow in water up to about 35 meters (115 feet) deep.

The Pacific Ocean's giant kelp forests, the world's largest, consist mostly of the magnificent genus *Macrocystis* (*macro* = large, *cyst* = bladder). The worldwide distribution of kelp is

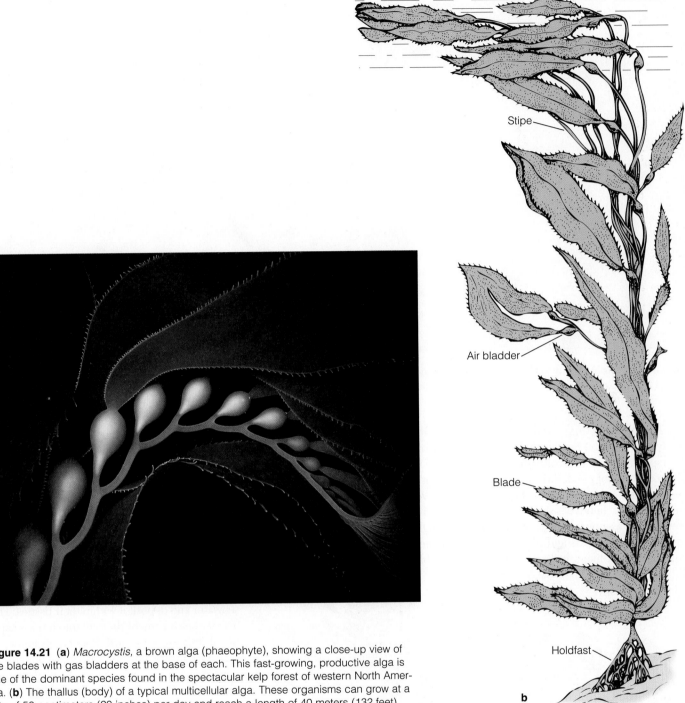

Stipe

Air bladder

Blade

Holdfast

© Bruce Hall

a

b

Figure 14.21 (**a**) *Macrocystis,* a brown alga (phaeophyte), showing a close-up view of the blades with gas bladders at the base of each. This fast-growing, productive alga is one of the dominant species found in the spectacular kelp forest of western North America. (**b**) The thallus (body) of a typical multicellular alga. These organisms can grow at a rate of 50 centimeters (20 inches) per day and reach a length of 40 meters (132 feet).

shown in **Figure 14.22.** Most brown algae live in temperate and polar habitats poleward of the 30° latitude lines, but a few live in the tropics.

THE RHODOPHYTES Most of the world's seaweeds are red algae; there are more rhodophyte species—about 4,000— than all other major groups of algae combined. Rhodophytes tend to be smaller and more anatomically and biochemically complex than phaeophytes. Rhodophytes excel in dim light

because of their sophisticated accessory pigments, reddish proteins called *phycobilins.* These compounds absorb and transfer enough light energy to power photosynthetic activity at depths where human eyes cannot see light. The record depth for a photosynthesizer is held by a small rhodophyte discovered in 1984 at a depth of 268 meters (879 feet) on a previously undiscovered seamount in the clear tropical Caribbean. The deepest rhodophytes grow very slowly and may be tens or even hundreds of years old.

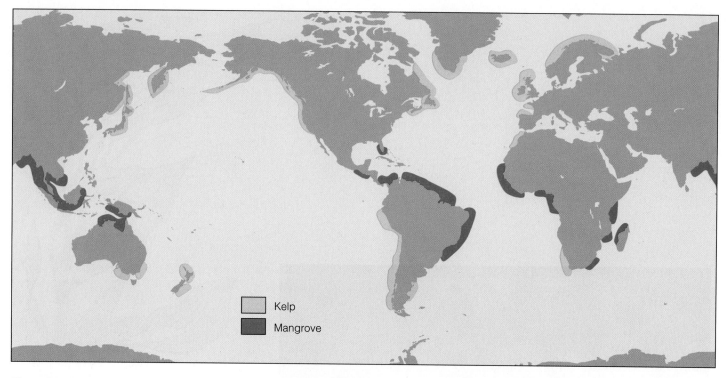

Figure 14.22 Distribution of kelp beds and mangrove communities worldwide.

Paradoxically, many red algae also live on rocks right at the water's surface. Surface light contains all colors of the visible spectrum, so the color-shifting capabilities of accessory pigments are not needed there. It is believed that the dark accessory pigments may act to shade the productive machinery from brilliant light. Most of the common surface-dwelling "reds" grow as a purple or pink film on rocks, shells, other seaweeds, and even glass and plastic. These coralline, or calcareous, algae look like a brilliant ceramic glaze, a bit of melting raspberry sherbet, or a knobby plastic membrane thrown carelessly over the surface. The name of one very common form, genus *Lithophyllum*, comes from the Latin terms for "stony" and "leaf," an apt description of an alga able to deposit calcium carbonate within its tissues (**Figure 14.23a**). Others grow in erect branching or bushy forms (**Figure 14.23b**). The gritty calcium in these seaweeds may deter grazing animals.

Another group of shallow rhodophytes is very important in the life of coral reefs. Like coral animals, encrusting coralline algae of genus *Lithothamnium* also remove large quantities of dissolved calcium carbonate from seawater and deposit it within their tissues. This activity helps cement the reef into a mass capable of resisting heavy surf. Some reefs in the Indian Ocean are not, as was once thought, of coral origin but rather were formed almost exclusively by the activity of coralline algae.

Seaweed Zonation

Zones or bands of dominant seaweeds are plainly visible on the sloping nearshore seafloor to the limit of light penetration. The vertical distribution of seaweeds depends in part on the pigments they possess. Chlorophytes lack accessory pigments and are rarely found more than 10 meters (33 feet) below the surface. Phaeophytes can operate from the surface down to about 30 meters (100 feet). Rhodophytes, as we have seen, can

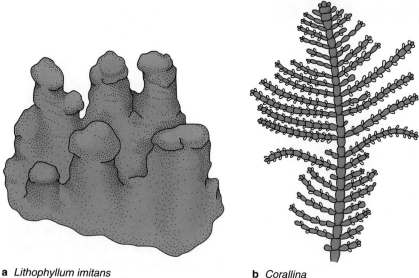

a *Lithophyllum imitans*   **b** *Corallina*

Figure 14.23 Two rhodophytes. *Lithophyllum* (**a**) is an encrusting species, *Corallina* (**b**) an erect species.

function from the surface to the photosynthetic record depth of 268 meters (879 feet).

Factors other than the response of pigments to light are also important in determining where these algae will prosper, especially in the intertidal zone. Wave shock, varying salinity and pH, the presence of grazers from land and sea, the drying effects of sun and wind, abrasion by waterborne particles of sand—all these factors and many more affect the location of intertidal producers. The resulting effects of these factors on marine organisms can be seen at low tide by any observant beachcomber. Look for bands of different marine algae (and angiosperms) the next time you go to the shore (see Figure 16.7).

Marine Angiosperms

14-24

Angiosperms are advanced vascular plants that reproduce with flowers and seeds. Most large land plants are angiosperms. Relatively few angiosperms live in water; the advantages conferred by vessels and roots are largely unnecessary in an aquatic environment. However, a few species of angiosperms have recolonized the ocean. All of these have descended from land ancestors, and all live in shallow coastal water. Angiosperms live at the surface, where the red light required for photosynthesis is abundant; they have no need for accessory pigments, and their chlorophyll is unmasked. The most conspicuous marine angiosperms are the sea grasses and the mangroves.

SEA GRASSES   Many people lump **sea grasses** in the informal seaweed group, but their resemblance to large marine algae is only superficial. These plants are not true grasses, but they do have leaves and stems, as well as roots capable of extracting nutrients from the substrate. Extensive stands of sea grasses are found on the coasts of North America, on the Atlantic coast of Europe, in Eastern Asia, in temperate Australia, and in South Africa. They form broad gray or green submerged meadows, which support extraordinarily rich communities of heterotrophs. The life cycle of sea grasses is much like that of other angiosperms, but their stringy pollen is distributed by flowing water rather than by insects or wind. About 45 species of sea grasses are known.

The most common sea grass is eelgrass, genus *Zostera,* a common inhabitant of the muddy shallows of calm bays and estuaries of the U.S. Atlantic and Pacific coasts. Similar habitats along the Gulf coast include stands of turtle grass (genus *Thalassia*) and manatee grass (genus *Syringodium*), which are angiosperms named after the animals that once shared their habitat and (in the case of the manatee) fed on them.

Perhaps the most beautiful sea grass is the vivid, emerald-green surf grass, genus *Phyllospadix*, with its seasonal flowers and fuzzy fruit (**Figure 14.24**). These hardy plants survive in the turbulent, wave-swept intertidal and subtidal zones of temperate East Asia and western North America.

Sea grasses can be remarkably productive. Some are capable of binding 1,000 gC/m^2/yr, three to five times the nearshore average for phytoplankton. One explanation for this efficiency is the advantage conferred by roots. Anaerobic bacteria in subtidal mud have been shown to bind dissolved nitrogen into nitrates easily available to these plants.

MANGROVES   Low, muddy coasts in tropical and some subtropical areas are often home to tangled masses of trees known as **mangroves.** These large, flowering plants are never completely submerged, but because of their intimate association with the ocean we consider them to be marine

Figure 14.24 The bright green sea grass *Phyllospadix* in a tide pool. Sea grasses are vascular plants, not seaweeds.

Tom Garrison

Figure 14.25 A mangrove (*Rhizophora*) growing in salt water in Everglades National Park, Florida. The tangled roots descending from the main branch are called *prop roots* or *stilt roots*. They provide anchorage for the mangrove, trap sediment, and provide protection for small organisms.

plants (**Figure 14.25**). They thrive in the sediment-rich lagoons, bays, and estuaries of the Indo-Pacific, tropical Africa, and the tropical Americas. Their distribution depends on temperature, currents, and rainfall (see again Figure 14.22).

The sediment in which mangrove trees live must be covered with brackish or salt water for part or all of the day. Many mangroves avoid taking up salt ions from seawater, or they selectively remove salt from sap with salt-excreting cells. The fine coastal mud they recolonize doesn't provide firm footing for these substantial plants, so an intricate network of arching prop roots is required for support. The strutlike prop roots are supplemented by many smaller roots equipped with breathing pores and air passages. Atmospheric oxygen is conducted by these passages to the portions of the plant submerged in oxygen-deficient mud.

The root system also traps and holds sediments around the plant by interfering with the transport of suspended particles by currents. Mangrove forests thus assist in the stabilization and expansion of deltas and other coastal wetlands. The root complex also forms an impenetrable barrier and safe haven for organisms around the base of the trees.

The mangrove communities of south Florida, among the world's most widespread, consist primarily of the red mangrove (genus *Rhizophora*). The spreading leaves of the tree protect the residents from the tropical sun, and the tangled roots keep out large predators and harbor huge numbers of fiddler crabs, worms, marine and terrestrial snails, fish, oysters, and red algae. Birds and insects inhabit the treetops, their droppings enriching the sediments below.

Mangrove seeds germinate on the trees. If the tide is out when the seed drops, the force of the fall will plant the seed in the muck, where it will continue to grow. If the tide is in,

the seed will drop into the water, suspend growth, float to a new location, and resume growth when it gets a foothold in the mud. The trees mature in 20 to 30 years and can attain a height of 8 to 10 meters (26 to 33 feet). New mangroves colonize mud and sandbars away from shore; they stabilize them and eventually create new land.

Commercial Importance

Chances are good that you have had some recent contact with marine autotrophs even if you haven't been in the ocean. The mucilaginous material that is so effective in making algal blades slick, in lowering friction, and in deterring grazers is also harvested and made into an important commercial product called *algin*. When separated and purified, its long, intertwining molecules are used to stiffen fabrics, make adhesives, suspend water and oil together in salad dressings, prevent the formation of gritty crystals in ice cream, clarify beer and wine, and manufacture shoe stains, soaps, and shaving creams. Fast-food restaurants are using carageenan, a similar seaweed extract, to replace some of the fat in newly popular healthier hamburgers. These substances also prevent fire-extinguishing foams from dispersing, permit chocolate milk to remain on the refrigerator shelf without separating, and keep the abrasives in liquid car waxes from settling to the bottom of the bottle. In biological laboratories bacteria are cultured on agar made from seaweed extracts. Very likely even the ink that forms the letters you are now reading has an algin or carageenan component!

You can find out more about the uses of marine plants in the discussion of marine resources in Chapter 17.

1. How does the productivity of seaweeds compare with the productivity of land plants?

Look again at Figure 14.5. Note that a rain forest, the most productive natural terrestrial community listed, is not as productive as coral reefs or kelp beds. Under ideal conditions, sea palms—a phaeophyte, genus *Postelsia*—can bind an astonishing 14,600 gC/m²/yr, the highest primary productivity recorded for any autotroph!

2. Do zooplankton have a compensation depth? If so, would it be above, below, or at the same level as the compensation depth of most phytoplankters?

The concept of compensation depth is meaningless for zooplankton. Compensation depth applies only to *autotrophs,* organisms capable of both photosynthesis and respiration. Since animals don't make their own food—aren't autotrophic—productivity in them can *never* equal consumption.

3. Isn't it too cold for seaweeds to survive at high latitudes?

No. As with phytoplankton, seaweed productivity can be high regardless of water temperature. Indeed, kelp in some Canadian bays grows most rapidly when water is near 0°C (32°F). The important limiting factors are the few hours of daylight in winter months, the relative lack of nutrients in the north polar ocean, and the abrasive grinding of ice in nearshore areas.

4. What's this about bioluminescence in dinoflagellates having evolved, possibly, as an "intrusion alarm"?

Some dinoflagellates flash when being attacked by zooplankton. The glow may startle the zooplankter and distract it from feeding. It may also alert larger organisms to the presence of that zooplankter and attract a predator toward the offending grazer. Behavioral studies suggest that dinoflagellate grazers react negatively to a dinoflagellate's blue flash, and perhaps this is why.

5. What proportion of total world productivity is achieved by chemosynthetic means?

An interesting and controversial question! In the marine environment, chemosynthesis is thought to occur primarily at hydrothermal vents associated with the 65,000-kilometer (41,000-mile) network of mid-ocean ridges. Specialized communities of organisms have evolved in the zones of warm or hot mineral-rich water percolating through the seafloor in these places. Once thought to be rare and isolated, vent communities have now been discovered in broadly separate locations and may be quite common to mid-ocean ridges. The trophic pyramid in these deep communities is based on chemosynthetic bacteria. Some biological oceanographers have suggested that if vent communities are abundant on nearly all ridges, if bacteria can grow at warmer temperatures than previously thought, or if archaeans really are living in vast numbers at great depths in sediments or solid seabed, chemosynthesis may account for a *much* larger proportion of total oceanic productivity than was previously thought. This is a very exciting area of marine research!

6. Phytoplanktonic organisms account for a significant percentage of world primary productivity. Why *are* phytoplankton so successful? Is there something about the planktonic lifestyle that lends itself to high efficiency?

Phytoplankton are successful because of their size. Because water supports them, small marine autotrophs don't require the elaborate support systems of large terrestrial plants. Their small size allows easy diffusion of required nutrients into their single cells and prompt transport of wastes out; vessels and sap are not needed. Nearly all of their tissue can be photosynthetic.

With transparent siliceous valves to protect them, diatoms are especially successful. The relatively recent evolution of diatoms was a high point in the development of marine autotrophs, and the additional oxygen these organisms contributed to the atmosphere has greatly influenced the success of life on land. More plants have led to more animals. After a few hundred million years of evolution, the planktonic lifestyle is elegantly tuned and interlocked into the complex and productive system we now observe in the world as a whole.

7. You mentioned that krill can swim horizontally over considerable distances. Doesn't that disqualify them for inclusion in the plankton?

Japanese researchers, using two ships equipped with side-scan sonar, tracked a large school of swimming krill for 14 days across 278 kilometers (172 miles). This new finding does indeed threaten krill's usual classification as zooplankton (animals unable to move consistently in one direction against waves or current flow). Traditions die hard, though, and this keystone of the Antarctic ecosystem will not likely lose its status as the premier zooplankter anytime soon.

- Energy flows through living systems, but matter is recycled among producers, consumers, and decomposers.

- Primary producers—autotrophs—are organisms that synthesize food (the carbohydrate glucose) from inorganic substances by photosynthesis and chemosynthesis. Heterotrophs (for example, animals) cannot produce glucose and must consume autotrophs or other heterotrophs to get food.

- Feeding (trophic) relationships resemble webs rather than chains. Consumers usually have a choice of food species.

- Primary productivity is expressed in grams of carbon bound into organic material per square meter of ocean surface per year (gC/m²/yr). Primary productivity occurs in the euphotic zone, which constitutes less than 2% of total ocean volume.

- The compensation depth is the depth above which photosynthetic production of carbohydrates exceeds consumption of carbohydrates by the producer itself.

- On a total annual basis, primary productivity is greatest in the temperate zone, near shore. The greatest productivity at any one time occurs around the Antarctic in local summer.

- Phytoplankton—drifting photosynthetic autotrophs—are among the world's most important primary producers. Two groups, the diatoms and the dinoflagellates, account for most oceanic productivity.

- Zooplankton—drifting heterotrophs—are the most abundant consumers in the ocean.

- Seaweeds, a form of multicellular marine algae, are efficient primary producers but collectively account for only a small percentage of total marine productivity.

INTERNET STUDY RESOURCES

The Web site for this book contains helpful study aids. Log on to:

 www.brookscole.com/053437557xs

and click on the Chapter-by-Chapter area. Choose Chapter 14 and select a resource:

- **Flash Cards** allows you to test your mastery of the Terms and Concepts to Remember for this chapter.
- **Tutorial Quizzes** provides a multiple-choice practice quiz.
- **Student Guide to InfoTrac College Edition** will lead you to Critical Thinking Projects that use InfoTrac College Edition as a research tool.

TERMS AND CONCEPTS TO REMEMBER

accessory pigment	diatom
algae	dinoflagellate
angiosperm	disphotic zone
autotroph	euphotic zone
auxospore	flagellum
bioluminescence	food
biomass	food web
blade	foraminifera
chemosynthesis	frustule
Chlorophyta	gas bladder
coccolithophore	heterotroph
compensation depth	holdfast

holoplankton	plankton net
kelp	primary consumer
krill	primary producer
macroplankton	primary productivity
mangrove	respiration
meroplankton	Rhodophyta
multicellular algae	sea grass
nanoplankton	secondary consumer
nonconservative nutrient	silicoflagellate
nonvascular, adj.	stipe
oxygen minimum zone	thallus
Phaeophyta	top consumer
photosynthesis	trophic pyramid
phytoplankton	unicellular algae
picoplankton	valve
plankter	xanthophyll
plankton	zooplankton
plankton bloom	

STUDY QUESTIONS

Review Questions

1. What do primary producers produce? How is productivity expressed?

2. How can primary productivity be measured? Which method is considered the most accurate?

3. What are plankton? How are plankton collected? How are members of the plankton community differentiated? How are zooplankton different from phytoplankton?

4. Describe the most abundant and important phytoplankters. Which are most efficient in converting solar energy to energy in chemical bonds?

5. What are algae? Are all algae seaweeds? How are seaweeds classified? Which seaweeds live at the greatest depths? Why?

6. Give examples of marine angiosperms. Are they vascular or nonvascular plants? Are they considered seaweeds?

7. Of what commercial importance are marine plants?

Critical Thinking Questions

1. What factors limit productivity? What methods have marine producers developed to cope with the lack of red light needed by chlorophyll for photosynthesis?

2. Is phytoplankton productivity highest at the ocean surface? What advantage would optimum productivity at a depth below the surface provide to phytoplankton?

3. What is compensation depth? What happens to phytoplankton below that depth? To zooplankton?

4. Where in the ocean is plankton productivity greatest? Why?

5. How does a nonvascular plant differ from a vascular plant? Why are most marine plants nonvascular?

6. **InfoTrac College Edition Project** Phytoplankton are the major primary producers of the ocean; all other life forms depend on them, directly or indirectly. Accordingly, many marine scientists are concerned about changes that can alter phytoplankton populations. What would be the effects of a drastic drop in the ocean's phytoplankton? Research your answer to this question using InfoTrac College Edition.

RESOURCES FOR FURTHER READING AND RESEARCH

The Web site for this book contains many ideas for further reading and research. Log on to:

 www.brookscole.com/053437557xs

and click on the Chapter-by-Chapter area.

Choose Chapter 14 and select a resource:

- **References** lists the major books and articles consulted in writing this chapter, along with comments from the author about their content and reading level.
- **Hypercontents** takes you to an extensive list of sites with news, research, and images related to individual sections of the chapter.
- In **Student Guide to InfoTrac College Edition** scroll down to Suggested Readings from InfoTrac College Edition for brief descriptions of the articles listed and search hints for finding them in InfoTrac College Edition.
- **Regional InfoTrac College Edition** articles are organized into East Coast, West Coast, and Gulf Coast regions, allowing you to study oceanography on a more local level. You can also access Regional InfoTrac College Edition at www.localocean.com.

 For additional readings, go to InfoTrac College Edition, your online research library, at:

http://infotrac.thomsonlearning.com/

Marine Animals 15

"... Silver-Shining, Swift, Strong, Streamlined ..."

The mild, supportive ocean brims with animal life. **Animals** are active multicellular organisms incapable of synthesizing their own food. Animals must obtain food from primary producers—or from other animals that have consumed primary producers—to ensure their own survival.

All animal species encounter the same basic set of problems—to find food, avoid predators, and reproduce—and all have ways to solve these problems. It is the *variety* of survival strategies—adaptations to their environment—that makes marine animals so interesting. Each species, from sponge to whale, has a unique set of adaptations. It may sound like a circular argument, but by definition each living species is successful precisely because it is living: Its continuously evolving suite of adaptations has brought it through the rigors of food finding, predator avoidance, and reproduction time and time again.

The nature of these adaptations often amazes us. Consider the members of the tuna family, the ocean's fastest and widest-ranging animals. The body of a tuna is dedicated to speed. Its fins retract into slots, its eyes form a smooth surface with the rest of the head, and it may consume as much as 25% of its mass in food each day. Indeed, a tuna uses so much energy that one of its greatest physiological problems is to avoid overheating! The biological equivalent of the legendary Flying Dutchman, these powerful fishes are fated to travel continuously. They depend on forward motion to pass oxygen-rich water over their gills. If they ever stopped they would suffocate, and their massive bodies would fall to the depths.

Yellowfin tuna and happy fisherman. Among the strongest and fastest of pelagic animals, yellowfin migrate across great distances. The record yellowfin was caught in 1977 off the coast of Mexico and weighed 176 kilograms (388 pounds).

Anthony R. Ramirez

Tuna and their relatives swim enormous distances and exhibit astonishing bursts of speed. Studies have shown that albacore tuna regularly migrate from the coast of California to Japan and back, a one-way trip of 8,500 kilometers (5,300 miles), moving at an average speed of not less than 26 kilometers (16 miles) per day. Tagged bluefin tuna have traveled at least 7,770 kilometers (4,830 miles) across the North Atlantic in 119 days—more than 65 kilometers (40 miles) each day. In reality, the bluefins must have traveled much farther, continually detouring from a straight-line course to hunt for food. The fastest tuna can maintain speeds of over 75 kilometers (47 miles) per hour, and the sailfish, a close relative, can rocket to 110 kilometers (68 miles) per hour for a short time.

These magnificent fishes are valued for their meat, especially in Japan, where they are highly prized for sashimi, the raw fish component of sushi. In the Tokyo fish market, an old and fat bluefin tuna sold for more than $80,000 in 1995, a record for a single fish! In 1989 a school of huge bluefin appeared off the California Channel Islands—the largest caught weighed 458 kilograms (1,008 pounds). Many lucky fishermen paid off home mortgages and boat loans in one week of heroic fishing. But it is the living animal that provides the greatest inspiration: Silver-shining, swift, strong, and streamlined, these silent nomads slip through the ocean more than a million miles in a lifetime.

Animals like the tuna teach us an important lesson: *No matter how improbable a structure or behavior seems, it must benefit the species in some way—or it would not be present.* In the long run, every shape, fin, and movement helps an organism to succeed.

The history of animals begins far back in the history of the ocean. The path leading to today's formidable array of animals began with the availability of food.

15-1

WHAT TO WATCH FOR IN CHAPTER 15

Animals are heterotrophs; that is, they cannot synthesize their own food and must ultimately depend on primary producers (autotrophs) for nutrition. Animals did not exist until the increasing levels of free oxygen in the atmosphere permitted them to metabolize food obtained from autotrophs. And remember, it was the photosynthetic autotrophs themselves that contributed huge quantities of oxygen to the environment. True multicellular animals arose between 900 and 700 million years ago, near the end of this "oxygen revolution." More than 10 million species of animals are now thought to exist on Earth. Their variety is astonishing—a tribute to millions of years of complex interplay among environment, producer, and consumer.

Our survey of marine animals follows the course of their evolution. As you read this chapter, note the increasing complexity of animals as we move from those groups (*phyla*) whose basic structure seems to have solidified relatively early in the history of animals to those groups that seem to have evolved more recently. Every marine animal has evolved effective adaptations for capturing prey, avoiding danger, maintaining thermal and fluid balance with its surroundings, and competing for space. Our survey of marine animals stresses these adaptations. In the next chapter we will investigate how these animals interact with one another and with their environment.

As you may recall from Chapter 1, the first organisms to evolve on Earth were probably tiny creatures adept at absorbing organic molecules that formed spontaneously in the ocean. Primitive life forms could use the energy stored in these food molecules for growth and reproduction. Competition for food increased as these organisms grew more numerous. Had photosynthesis not evolved, life would probably have died out when the supply of usable energy-rich molecules in the environment was exhausted. Using sunlight, the first simple autotrophs assembled their own food from inorganic molecules and then broke down the food to release energy. With the success and proliferation of simple autotrophs, and with the abundance of oxygen they provided, the way was clear for the evolution of animals.

THE ORIGIN OF ANIMALS

The first animal-like creatures were single-celled organisms. They began to prosper in the ocean during the **oxygen revolution,** a time of radical change in Earth's atmosphere. During the oxygen revolution, between about 2 billion and 400 million years ago, the activity of photosynthetic autotrophs changed the composition of the atmosphere from less than 1% free oxygen to its present oxygen-rich mixture of more than 20% (**Figure 15.1**). The growing abundance of free oxygen made aerobic respiration practical, speeding the disassembly of food molecules that early animals obtained by eating the autotrophs. Ozone derived from this oxygen blocked most of the sun's dangerous ultraviolet radiation from reaching Earth's surface, permitting life to survive at the surface of the ocean and, later, on land.

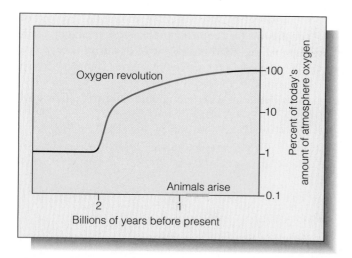

Figure 15.1 The oxygen revolution, a time of radical change in the composition of Earth's atmosphere, occurred between about 2 billion and 400 million years ago. The activity of photosynthetic autotrophs—mostly photosynthetic bacteria—resulted in a rapid rise in the amount of oxygen in the atmosphere, making the evolution of animals possible. Animals are thought to have arisen between 900 and 700 million years ago.

Animals grew in complexity as they became more abundant. Instead of drifting apart after reproduction, some dividing cells stuck together and formed colonies. True animals evolved as these colonies distributed labor among specialized cells, eventually increasing the degree of interdependence among cells within the colony. The colonies ceased to be simple aggregations of individuals and began to take on specific architectures for specific tasks.

A group of animals that shares similar architecture, level of complexity, and evolutionary history is known as a **phylum** (plural, *phyla; phylon* = tribe). No one knows how many phyla of animals may have developed during the time of rapid animal proliferation that occurred near the end of the oxygen revolution. Small but fascinating "snapshots" of early marine life are preserved in fossils found in the Ediacara Hills of Australia, in the Burgess Shale of British Columbia, and in the Chengjiang beds of southwestern China. These sites were once part of the warm, sediment-covered continental shelves of equatorial land masses. Animals inhabiting these places were abruptly buried—perhaps by turbidity currents—and their delicate features preserved. The animal in **Figure 15.2a** is about 600 million years old and shows evidence of a segmented body plan. The 530-million-year-old worm in **Figure 15.2b** is similar in outline to species alive today. Not all groups of early marine animals survived. Once abundant, trilobites like the one shown in **Figure 15.2c** have all perished. Indeed, most of the animals in these ancient fossil beds are extinct and may represent failed branches in animal evolution, unique designs that were not suited to later environmental conditions.

INTRODUCING THE INVERTEBRATES

Invertebrates are generally soft-bodied animals lacking a rigid internal skeleton for the attachment of muscles; but many invertebrates possess some sort of hard, protective outer covering, which can be continuous (like a snail shell) or segmented (like a lobster shell). More than 90% of all living and fossil animals are invertebrates. The category is convenient though somewhat artificial, containing as it does a vast variety of different organisms in many diverse phyla that have little in common. Invertebrates range in size from microscopic worms to giant squids. At least 33 invertebrate phyla are currently recognized, and almost every one of them has marine representatives. The nine most conspicuous invertebrate phyla are presented in this chapter in order of increasing complexity (for an overview, see **Table 15.1**).

Phylum Porifera

Sponges belong to the phylum **Porifera** (*porus* = holes, *ferre* = to bear), the most primitive true animals. Nearly all of the 10,000 species of these simple attached animals are marine. Sponges are widely distributed from intertidal zone to abyss and are found at all latitudes and in most benthic habitats. They range from the size of a bean to the size of a small

a

b

c

Figure 15.2 Ancient marine animals. (**a**) From Australia's Ediacara Hills, the fossil of a segmented marine animal about 600 million years old. (**b**) From the Burgess Shale of British Columbia, a marine worm about 530 million years old. In basic form, it resembles species alive today. (**c**) A 500-million-year-old trilobite fossil. Trilobites were once very abundant in the ocean, but all have died out.

automobile and come in a few basic shapes: branching, vase-like, and encrusting.

All sponges are **suspension feeders:** They strain plankton and tiny organic food particles from the surrounding water. A large sponge may filter more than 1,500 liters (400 gallons) of water each day. **Figure 15.3** shows a photograph and a cutaway diagram of a simple upright sponge. Water carrying food and oxygen enters at the incurrent pores and is swept toward the excurrent openings by flagellated collar cells. The sticky collars snare food particles drifting past, and digestion begins. The captured nutrients are distributed to other cells of the organism by wandering cells in the body of the sponge. Sponges have no digestive system, only individual digestive cells; they have no circulatory, respiratory, or nervous systems. Excretion and the movement of gases into and out of the animal occur by simple diffusion.

The flagellated cells of more elaborate sponges are concentrated in hundreds of tiny chambers. Their interiors resemble a Swiss cheese. A skeletal network of spicules (needles) of calcium carbonate or glassy silica prevents the internal chambers and canals from collapsing; a fibrous protein called *spongin* often serves the same purpose. Commercial natural bath sponges (not to be confused with the brightly colored synthetic sponges encrusting supermarket shelves) have been treated to retain only this spongin matrix; all the cells are gone.

Phylum Cnidaria

15-5

Jellyfish, sea anemones, and corals belong to the phylum **Cnidaria** (*knide* = nettle), which contains about 9,000 mostly marine species. (You may be more familiar with this phylum's old name, *Coelenterata*.) This group of carnivorous animals takes its name from the large, stinging cells called **cnidoblasts** (*knide* = nettle, *blastikos* = to shoot upward), deployed on tentacles that bend or retract toward the mouth. Each cnidoblast contains a capsule from which a coiled thread may be forcefully ejected (**Figure 15.4**). The thread can repel an aggressor or penetrate or entangle prey, often immobilizing the victim with a toxin. Then the prey—which may include the larger zooplankters and small fish—is drawn into the mouth, which leads to a saclike digestive cavity. The digested food is absorbed by cells of the inner layer and transported to other parts of the animal by migratory cells and

Table 15.1	Major Animal Phyla with Marine Examples
Phylum	**Marine Examples**
Invertebrates	
Porifera	Sponges
Cnidaria	Coral, jellyfish, sea anemones, siphonophores
Platyhelminthes	Flatworms, flukes, tapeworms
Nematoda	Roundworms
Annelida	Segmented worms
Mollusca	Chitons, snails, bivalves, squid, octopuses
Arthropoda	Crabs, shrimp, barnacles, copepods, krill
Echinodermata	Sea stars, sea urchins, sea cucumbers
Chordata[a]	Tunicates, salps, *Amphioxus*
Vertebrates	
Chordata	Fishes, reptiles, birds, mammals

[a]Phylum Chordata includes both vertebrate and invertebrate classes.

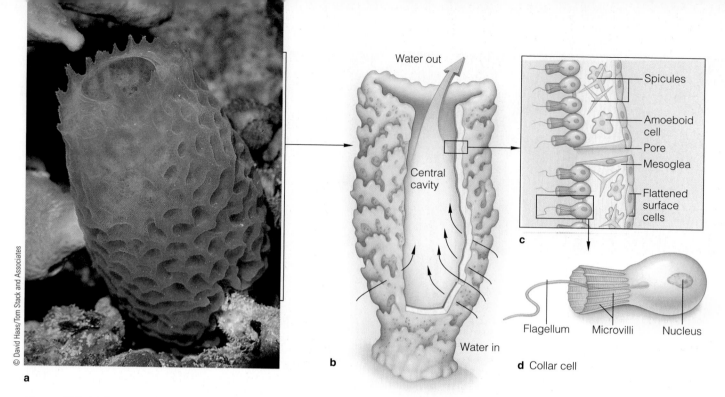

Figure 15.3 (**a**) An upright sponge. (**b**) The body plan of a simple sponge. (**c**) A section through a sponge body wall. (**d**) A type of collar cell.

diffusion. Because the digestive cavity has only one opening, indigestible bones and other wastes are eliminated through the mouth.

Members of this group are built of two layers of cells. The inner layer (the gastrodermis) is responsible for digestion and

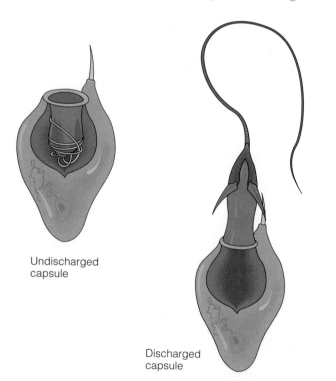

Undischarged capsule

Discharged capsule

Figure 15.4 A cnidoblast with stinging capsule undischarged (left) and discharged (right).

reproduction, the outer layer (the epidermis) for capture of prey and protection from attack. The layers are connected by a jelly-like mesoglea (*mesos* = middle, *glia* = glue). Cnidarians exhibit **radial symmetry;** that is, their body parts radiate from a central axis like the spokes of a wheel. Some, such as sea anemones and corals, attach to rocks or other objects; others, such as jelly-fish, swim freely in the water. No cnidarian possesses a definite head or concentration of sensory receptors, but a primitive network of nerves permits some species to respond to stimuli. These relatively simple animals depend on diffusion to move wastes and gases; they have no excretory or circulatory system.

Cnidarians occur in two forms: medusae and polyps. Jelly-fish are an example of the **medusa** body plan, named after a Greek mythological monster with a woman's face and hair that was a mass of writhing snakes. Medusae are predatory animals that swim by the rhythmic contraction of their bell-shaped bodies. They catch their prey with trailing tentacles armed with cnidoblasts. Some medusae are microscopically small, but the genus *Cyanea* can reach a bell diameter of 3.5 meters (12 feet), with tentacles 18 meters (59 feet) long! The sea wasp is a smaller but more dangerous species and is shown in **Figure 15.5.**

Sea anemones and corals are examples of the sedentary **polyp** (*polypous* = many-footed) body plan. Sea anemones have no skeleton and attach firmly to the substratum or burrow into it with a sticky basal disc on which they can slide slowly, like a snail. Other species such as the coral lack a basal disc; they contain a calcareous skeleton covered by living tissue and are permanently cemented in place.

Although they look like flowers, **corals**—along with sea anemones and jellyfish—are classified as cnidarians. Some

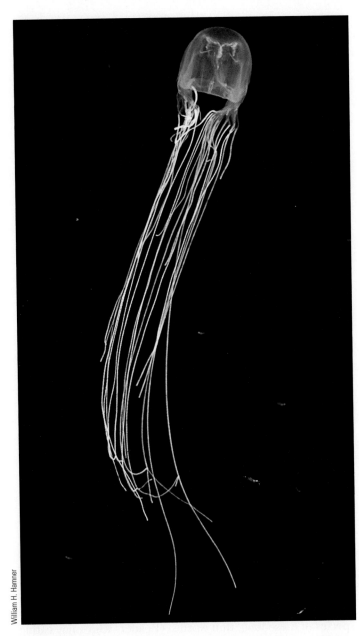

Figure 15.5 A sea wasp (*Chironix*), one of the most dangerous jellyfish. An inhabitant of tropical waters from Africa to northeastern Australia, it can kill a human within three minutes. The tentacles of a large specimen can be 15 meters (50 feet) long. *Chironix* has probably been responsible for more human deaths than sharks.

Tropical, reef-building corals are **hermatypic** (*hermatos* = mound builder). Their bodies contain masses of tiny, symbiotic dinoflagellates called **zooxanthellae** (*xanthos* = yellow). The single-celled zooxanthellae carry on photosynthesis, absorb waste products, grow, and divide within their coral host. The coral animals provide a safe and stable environment and a source of carbon dioxide and nutrients; the zooxanthellae reciprocate by providing oxygen, carbohydrates, and the alkaline pH necessary to enhance the rate of calcium carbonate deposition. The coral occasionally absorbs one of the resident dinoflagellates, "harvesting" the organic compounds for its own use.

Because of the needs of its zooxanthellae, hermatypic corals depend on light and warmth. Reef corals grow best in brightly lit water about 5 to 10 meters (16 to 33 feet) deep. They can form to depths of 90 meters (300 feet), but growth rates decline rapidly past the optimum 5- to 10-meter (16- to 33-foot) depth. In ideal conditions coral animals grow at a rate of about 1 centimeter (½ inch) per year. They prefer clear water because turbidity prevents light penetration and suspended inorganic particles interfere with feeding. The animals are protected from the harmful effects of bright sunlight by a mucous coating that contains an ultraviolet-blocking "suntan lotion."

Hermatypic corals also prefer water of average or slightly elevated salinity. Coral animals are highly susceptible to osmotic shock, and exposure to fresh water is rapidly fatal; thus, reefs growing in shallow water have a flat upper surface because rain is lethal. Fresh water and suspended sediments prevent reefs from forming near the mouths of rivers or in areas adjacent to islands or continents where rainfall is abundant.

The Cnidaria are a very successful invertebrate group, but their simple architecture would not serve more advanced organisms. Their blind-sac digestive system allows only one batch of food to be processed at a time; feeding opportunities arising during digestion cannot easily be accommodated. The lack of a distinct head with a concentration of sense organs is a drawback, as is the absence of circulatory, respiratory, and excretory systems. Having only two structural cell layers limits the complexity of systems and structures that can form within the organisms.

THE WORM PHYLA

A transition from relatively simple to more advanced organisms is made in the worm phyla, three of which are discussed here. The worm body plan exhibits **bilateral symmetry**, not radial symmetry; that is, each body has a left side and a right side that are mirror images of each other. Nearly all worms have some concentration of sensory tissue in what may be termed a head, and many have flow-through digestive systems and excretory systems to circulate fluids and eliminate waste. Some are efficient parasites, but most are free-living. A few burrow in cavities; others roam the seabed or lurk under rocks.

corals are solitary animals with bodies up to 30 centimeters (12 inches) in diameter, but most of the more than 500 species are ant-size organisms crowded into colonies called **coral reefs** (see Figures 12.26 and 12.27). The coral animals themselves construct the reefs by secreting hard skeletons of aragonite (a fibrous, crystalline form of calcium carbonate). The matrix of cup-shaped individual skeletons secreted by coral animals gives the colony its characteristic shape.

Tropical corals feed at night; at dawn the polyps retract into their skeletal cups to escape strong sunlight and predators. The anatomy of a coral polyp is shown in **Figure 15.6.**

Tentacles with
stinging cells

Epidermal
tissue

Mouth

Digestive
cavity

Interior
partition

Symbiotic
zooxanthellae

Mesenteric
filaments

Layers of
calcium
carbonate
forming a
skeleton

Pat Mason

a

b

Figure 15.6 (a) Close-up of hermatypic coral, showing expanded polyps. **(b)** Anatomy of a reef coral polyp, with enlarged detail showing a cross section of the outer covering and tissue. The symbiotic photosynthetic zooxanthellae are crucial to the survival of this type of coral.

The simplest worms are the well-named flatworms of phylum **Platyhelminthes** (*platys* = flat, *helmins* = worm). Some are parasites of vertebrates, such as the tapeworms of fish and marine mammals. However, most marine flatworms are free-living predators and scavengers; they can be found on the shady underside of intertidal rocks or sharing colonies or burrows with other animals. Few examples exceed 3 centimeters (1¼ inches) in length.

Flatworms are the most primitive organisms with a central nervous system. In some species a complex of nerve cells serving as a rudimentary brain connects the animal's simple nervous system to a pair of light-sensitive eyespots. The eyespots are small pigmented cups that can sense only the presence or absence of light—certainly a critical factor for an animal only a few cells thick that must avoid light to prevent overheating or detection.

Larger flatworms are necessarily thin because they lack a true respiratory and excretory system. Gases must be exchanged and wastes eliminated by diffusion through the animal's surface, so no cell can be more than a few cell diameters from the outside.

The first animals in our phylum survey that possess a flow-through digestive system (a digestive tract with mouth and anus rather than a digestive cavity) are members of the plentiful phylum **Nematoda** (*nematos* = thread). Also called

roundworms, these most successful of the worm phyla are present in nearly every imaginable terrestrial, aquatic, and marine habitat. Most of the 12,000 known species are free-living and microscopic, thriving in garden soil and marine sediments. Some make perfect parasites, however, and nearly all vertebrates and many invertebrates are parasitized by species of these long, thin worms. (Readers who enjoy eating raw fish, or sashimi, should read the information on fish parasites in **Box 15.1.**)

The nematodes' true claim to marine fame rests on their astonishing numbers in some soft sediments. A Dutch researcher investigating shallow subtidal mud in the North Sea reported 4,420,000 tiny individuals in a sample 1 meter square (10.8 square feet) by 4 centimeters (1.57 inches) deep! Swimmers and gardeners have encountered these unobtrusive worms by the thousands without ill effects. Free-living worms cannot become parasitic.

Members of the phylum **Annelida** (*annelus* = ring), which includes the familiar garden earthworm, are the most evolutionarily advanced worms. Their bodies are divided into a number of similar rings or segments. **Metamerism,** as this segmentation is called, is a convenient strategy for increasing the size of an animal simply by adding nearly identical units. We see evidence of metamerism in most higher animals. Each segment of an annelid can have its own circulatory, excretory, nerv-

People who enjoy sushi and sashimi (raw fish) run a slight risk of contracting herring worm disease or some other malady caused by marine parasites. More than 1,200 cases of humans being infested by marine parasites have been reported, mostly from Japan but also from Europe and North America. About 50 cases of parasitic marine worm infestations have been reported in the United States since 1980, probably because of an increase in the popularity of raw fish sushi, a Japanese delicacy.

The most dangerous parasites are nematodes of the genus *Anisakis*. These coiled worms—most often found in herring, red snapper, salmon, halibut, and some species of bass—are about 4 centimeters (1½ inches) long when unwound. They are usually coiled in balls the size of pinheads, either between groups of muscles or at the junction of muscles and organs. The worms look like tan or brown dots, often contrasting nicely with light-colored flesh.

Ingesting *Anisakis* can lead to much unpleasantness. Sometimes, the parasite can be felt in the throat within 24 hours of the meal; coughing may dislodge it. Worms that have been swallowed may cause intense abdominal pain, nausea, vomiting, fever, and diarrhea. Symptoms of an infestation may resemble those of a peptic ulcer. A person will become dangerously ill if the worm penetrates the intestine. Japanese physicians see these infestations rather frequently and recognize them, but American doctors often mistake the signs for an ulcer or appendicitis and treat them accordingly. In a recent issue of the *New England Journal of Medicine,* a surprised surgeon reported removing a stricken patient's apparently normal appendix. Just as he was closing the incision he saw a 4.2-centimeter (1.65-inch) nematode crawl onto the surgical drapes.

Sushi lovers can greatly lessen the danger of *Anisakis* infestations by patronizing specialty restaurants that employ experienced chefs. Sashimi from tuna or octopus is almost always free of the worms; salmon is potentially the most troublesome. Inspect each piece carefully, and *don't eat anything that moves!*

These parasites are killed by proper freezing or cooking. The fish must be frozen at −20°C (−4°F) for more than 60 hours or cooked for at least 5 minutes at 60°C (140°F). (Microwave cooks please note: Quick microwaving of fish does not raise all parts of the portion to the same high temperature.) You can safely enjoy a hot fish dinner if the fish has been uniformly heated.

15-7

Tom Garrison

Diners inspect sashimi, the raw component of sushi. Though rare, marine parasites can sometimes be spotted—and avoided—by a careful look.

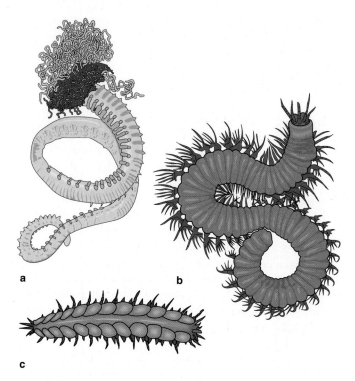

Figure 15.7 Marine annelids of the class Polychaeta. (**a**) *Amphitrite,* a sedentary tube worm that extends its sticky tentacles across the seafloor to capture food. (**b**) *Eunice,* an active predator. (**c**) *Serpula,* which lives in stony tubes on the sides of rocks and captures prey with its grasping mouthparts.

Figure 15.8 A tube-dwelling polychaete with its "feather duster" appendages deployed. A mucous coating on the "duster" catches food, and the appendage itself works as a gill for gas exchange.

ous, muscular, and reproductive systems, but some segments (such as those forming the head) are specialized for specific tasks.

The 5,400 species of class **Polychaeta** (*poly* = many, *chaetae* = bristles), the largest and most diverse class of annelids, are the most important marine annelids (**Figure 15.7**). Polychaetes are often brightly colored or iridescent worms with pairs of bristly projections extending from each segment. They range in length from about 1 to 15 centimeters (½ to 6 inches). Some polychaetes burrow through and devour sediments or move freely over the bottom in search of food; others construct fixed parchmentlike or calcareous tubes from which only parts of their heads emerge. Mobile polychaetes have well-developed heads with prominent sense organs, and they can function as efficient predators. Some skewer prey with their sharp mouthparts. Tube-dwelling polychaetes are a common sight to divers on most continental shelves. The "feather duster" tops of some of them (**Figure 15.8**) exchange gases and remove food from the water. A few tropical forms have biting jaws at their tips, a surprise for the unwary student! All can instantly retract their feathery gills when disturbed.

THE ADVANCED INVERTEBRATES
Phylum Mollusca

The conspicuous phylum **Mollusca** (*molluscus* = soft bodied) contains 80,000 species, second in size only to the huge phylum Arthropoda. The phylum Mollusca includes such di-

verse members as clams, snails, octopuses, and squid. Most mollusks are marine, and most have an external or internal shell. A few molluskan species possess acute sight and even considerable intelligence.

Mollusks and annelids probably shared a common origin—possibly a distantly ancestral segmented worm—and therefore share a few basic characteristics. Like annelids, mollusks are bilaterally symmetrical and generally have obvious heads, flow-through digestive tracts, and well-developed nervous systems. A few mollusks are segmented. Unlike annelids, however, some mollusks achieve great size, secrete beautifully fitted shells in which to take refuge, and exhibit great structural diversity.

We will briefly discuss three molluskan classes here: The class **Gastropoda,** the snails; the class **Bivalvia,** the clams, oysters, and mussels; and the class **Cephalopoda,** the nautiluses, octopuses, cuttlefish, and squid. Though greatly different in shape and habits, each class shows its link to a common ancient ancestor by sharing similar underlying parts. **Figure 15.9** shows some of the structural similarities in their body plans.

Gastropods (*gaster* = stomach, *pod* = foot) include the abalone, conch, limpet, and garden snail. Members of this largest class of mollusks usually inhabit relatively large shells, where they can seek refuge in case of danger. Some gastropods are grazers, some are suspension feeders, and some are predators. A few marine species—the pteropods and heteropods—are planktonic, but most marine gastropods are found wandering on rocky bottoms or other firm substrates.

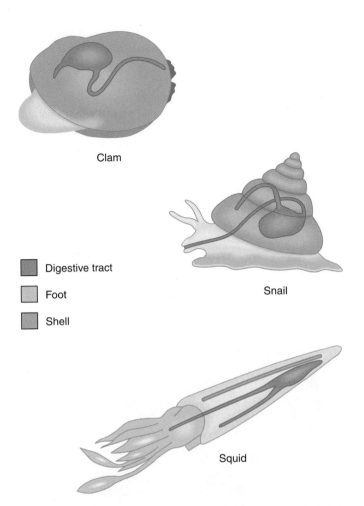

Clam

Snail

Squid

- ■ Digestive tract
- □ Foot
- ▨ Shell

Figure 15.9 The molluskan body plan as it has been modified in the major groups. (Source: From *Living Invertebrates*. V. & J. Pearse/ M. & R. Buchsbaum, 1987. The Boxwood Press. Adapted with permission of the publisher.)

Gastropod shells are often structures of great beauty (**Figure 15.10**). The animal adds to and enlarges the opening as it grows. The shell is frequently coiled to compress its mass and allow for easier maneuverability. A foot and head protrude from the shell while the snail moves about. The shell

BERMUDA

E II R

25¢

Bursa finlayi

Figure 15.10 Gastropod shells are often aesthetically pleasing. They have been prized by collectors for centuries and used to decorate many objects, postage stamps included.

Mark McMahon

Figure 15.11 A brightly colored reef nudibranch (*Ancula pacifica*) searches for food. The brilliant gill-like structures on its back assist in gas exchange. Although the nudibranch is usually in plain sight, its terrible taste seems to discourage other animals from eating it.

itself is secreted in three principal layers: a fibrous outer covering that may serve to distribute shock; a strong, crystalline layer of calcium carbonate ($CaCO_3$) to provide strength; and an inner layer of smooth $CaCO_3$ to provide nonabrasive surroundings for the resident. Not all gastropods have shells. Nudibranchs (*nudus* = naked, *branchia* = gills), also called *sea slugs,* are lovely (or even bizarre-looking) shell-less gastropods (**Figure 15.11**).

Although some snails use their foot to burrow, a gastropod foot cannot attach to sand or mud. The shell-and-foot configuration of gastropods is therefore not well-suited to life on most of the ocean floor, which is covered by sediments. Evolution of the bivalves (*bi* = two, *valv* = door) admitted mollusks to this rich sedimentary habitat. Animals enclosed in twin shells (clams, oysters, mussels, and scallops) are bivalves that surrender mobility for protection and gather food by suspension feeding rather than pursuit (see **Figure 15.12**). Burrowing species dig with a strong muscular foot, and extend their siphons to the surface to obtain water and eject wastes. In other species the foot has other functions: In mussels it secretes the tough threads that attach the organisms to wave-swept rocks—a neat turnabout trick by which bivalves can invade a gastropod habitat.

The most highly evolved mollusks are the magnificent cephalopods (*cephalon* = head, *pod* = foot), a group of marine predators containing nautiluses, octopuses, and squid. These well-named animals have a head surrounded by a foot divided into tentacles. The nautiluses retain a large coiled external shell, but squid have only a thin vestige of the shell within their bodies—and octopuses have none at all. Cephalopods can move by creeping across the bottom, by swimming with special fins, or by squirting jets of water from an interior cavity.

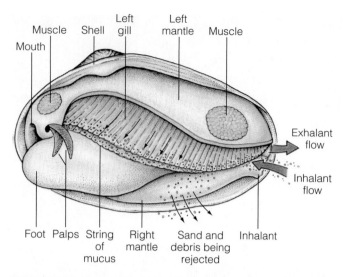

Figure 15.12 Bivalves are suspension feeders that make their living by filtering the water for edible particles. In this diagram (showing a bivalve with its left shell removed), water and tiny bits of food are swept into the animal by the movement of tracts of cilia on the gills. Food settles onto the gills and is then driven toward the mouth and swallowed.

Most cephalopods catch prey with stiff adhesive discs on their tentacles that function as suction cups, and they tear or bite the flesh with horny beaks. Nautiluses (**Figure 15.13**) are an open-ocean group that hunts at considerable depths, their strong shells buoyed with gas-filled chambers. Little is known of their natural history. Squid and octopuses are more advanced cephalopods. They can confuse predators by secreting clouds of ink. Some squid eject a kind of "dummy" of coagu-

lated ink that's a rough duplicate of their size and shape, and the squid is long gone by the time the attacker discovers the deception! At least one species of squid living below the euphotic zone produces a sparkling luminous ink instead of black ink (which would be ineffective in the darkness). Squid can grow to surprising sizes (**Figure 15.14**): The record length, including tentacles, is 18 meters (59 feet)! Most are much smaller.

Squid may be the largest invertebrates, but octopuses are the most intelligent. About as smart as puppies and with even better eyesight, some nearshore species of octopuses kept in captivity soon learn to recognize their keepers and forage at night through adjacent aquariums for tidbits. Octopuses use their visual acuity, ability to survive out of water for a short time, and intelligence to good advantage in their intertidal or subtidal homes, memorizing the positions of hiding places, escape holes, and good hunting locations.

Only about 450 species of cephalopods live today, a small percentage of the number of species known from fossils. The

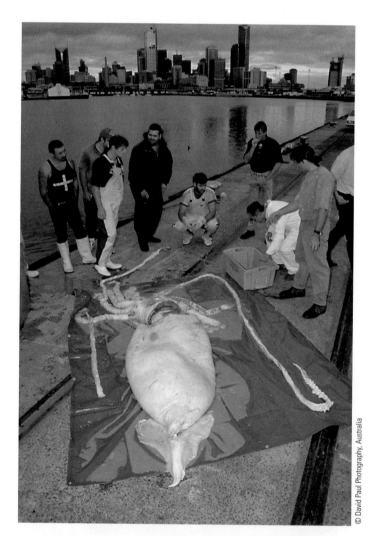

Figure 15.13 A chambered nautilus (*Nautilus*), from the only living cephalopod group with an external shell. The animal occupies the outermost chamber; the other chambers are partially filled with gas to provide buoyancy. Chambers are added as the animal grows.

Figure 15.14 This large squid, genus *Architeuthis*, was captured by commercial fishermen in New Zealand. It weighs 114 kilograms (250 pounds) and is 7.6 meters (25 feet) long. The world's largest preserved specimen, it was placed on display in New York City's American Museum of Natural History in October 1999.

small number of species suggests that sophistication is no guarantee of biological success.

Phylum Arthropoda

The phylum **Arthropoda** (*arthron* = joint, *pod* = foot)—a group that includes the lobsters, shrimp, crabs, krill, and barnacles—is a phylum of superlatives. Over a million species of arthropods are now known! Arthropods are the most successful of Earth's animal phyla, occupying the greatest variety of habitats, consuming the greatest quantities of food, and existing in almost unimaginable numbers. As you read in the last chapter, the krill, one type of zooplanktonic arthropod, comprise the greatest biomass of any single species on Earth. The arthropod body plan is a variation on the basic annelid theme; bodies show clear segmentation with a pair (or pairs) of appendages per segment. All are bilaterally symmetrical.

Arthropods have not achieved the nervous system development of cephalopod mollusks, nor do they have the advantages of intelligence or extraordinary eyesight. They do, however, exhibit three remarkable evolutionary advances that have led to their great success:

- An *exoskeleton*, a strong, lightweight, form-fitted external covering and support.
- *Striated muscle*, a quick, strong, lightweight form of muscle that makes rapid movement and flight possible.
- *Articulation*, the ability to bend appendages at specific points. The appendages of more primitive phyla can usually bend anywhere (like cooked spaghetti), but arthropod appendages bend at a joint. There are no ball-and-socket joints in arthropods; instead, each joint along an appendage moves through a different plane to ensure a full range of motion.

Most important of these advances is the **exoskeleton.** Unlike the often cumbersome shell of a gastropod, the exoskeleton of an arthropod fits and articulates like a finely tailored suit of armor. It is made in part of a tough, nitrogen-rich carbohydrate called **chitin,** which may be strengthened by calcium carbonate. Its three layers serve to waterproof the covering, tint it a protective color, and make it resilient and strong. Muscles within the animal are attached to the exoskeleton to move the appendages.

Such an arrangement sounds ideal, but the difficulties encountered by an organism with an exoskeleton are profound. How can muscle leverage be obtained? How can encrusting organisms be discouraged? How can ducts and feeding passages remain unblocked? And, perhaps most critically, how can the organism grow? That each of these problems has been solved is obvious by the group's overwhelming success, but the growth issue deserves a closer look.

We vertebrates grow by steadily adding length to the bones of our *internal* skeleton and bulk to our bodies. An *external* skeleton obviously limits growth and must be shed, or **molted,** at regular intervals. Arthropods do not have a steady growth pattern; instead, their external growth progresses in a

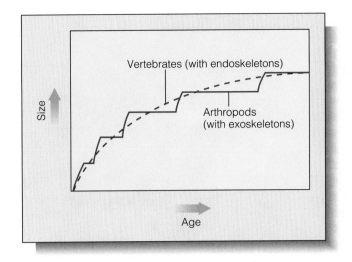

Figure 15.15 Generalized arthropod and vertebrate growth curves compared. The vertical segments of the arthropod curve represent molting periods. (Source: From James L. Sumich. *An Introduction to Biology of Marine Life,* 5/e. Copyright © 1992 Wm. C. Brown Communication, Inc. Dubuque, Iowa.)

series of steplike jumps (**Figure 15.15**) as the animal molts and replaces its exoskeleton. The arthropod grows without getting bigger between these jumps in size. An aquatic arthropod slowly substitutes body mass for water held in the tissues between molts. When molting, it suddenly takes on water from outside the body, expanding its tissues without growing in muscle mass. The shell splits and falls away, and, through a magnificently orchestrated sequence of glandular secretions, the animal quickly regenerates a new exoskeleton one size larger (**Figure 15.16**).

The largest class of arthropods, the class Insecta, is poorly represented in the sea: There is only one marine genus and five known open-ocean species, all of which are water striders. The class **Crustacea** (*crustaceus* = having a shell or rind), however, has 30,000 primarily marine species, including gill-breathing lobsters, crayfish, shrimp, crabs, water fleas, copepods, krill, amphipods, barnacles, and others. Their bodies usually have between 16 and 20 segments; the appendages may be specialized for sensing, food handling, walking, fighting, defense, and so forth. About 70% of all zooplankton are minute crustaceans, like that in Figure 14.18, which graze on diatoms and dinoflagellates. (Thumb- or bean-size planktonic crustaceans called *krill* constitute most of the food supply of the largest whales.) More familiar to most of us are the lobsters and crabs, including those we esteem for food. The largest crustacean is the giant king crab, which can reach a legspan of 3.6 meters (12 feet). The heaviest individual recorded, however, was a lobster caught off Chatham, Massachusetts, in 1949. (The beast weighed 22 kilograms, or 48 pounds, and reportedly served 10 people!)

The activity and importance of billions of crustaceans as scavengers, predators, parasites, grazers, and general participants in oceanic biology are hard to overestimate. Large or small, obvious or retiring, crustaceans dominate the world of marine animals.

a

b

Figure 15.16 A molting arthropod. (**a**) A Dungeness crab (*Cancer magister*) backing out of the exoskeleton (right) that it is abandoning. (**b**) Clear of the old exoskeleton, the soft-bodied crab takes in water and expands. It immediately begins to secrete a new exoskeleton. Note the obvious increase in the animal's size.

Phylum Echinodermata

15-10

The exclusively marine phylum **Echinodermata** (*echinos* = hedgehog, *derma* = skin) is an odd group sharply different from other members of the animal kingdom. The 6,000 species of echinoderms lack eyes or brains, have a radially symmetrical body plan based on five sections or projections (**Figure 15.17**), move slowly, and include only two known parasitic representatives.

Living echinoderms are divided into five classes: The four most familiar are the class **Asteroidea,** the sea stars; the class **Ophiuroidea,** the brittle stars; the class **Echinoidea,** the sea urchins and sand dollars; and the class **Holothuroidea,** the sea cucumbers.

Nearly all asteroids (*aster* = star, *oidea* = resembling), or sea stars, are star-shaped echinoderms with arms not completely delineated from the central disc (Figure 15.17a). The arms usu-

Figure 15.17 Pentamerous (five-sided) symmetry in two classes of echinoderms. At some time in their lives, all members of this phylum possess a radially symmetrical body plan based on five sections or projections. (**a**) A sea star; (**b**) a close-up of the five-part jaws centered on the underside of a sea urchin.

a

b

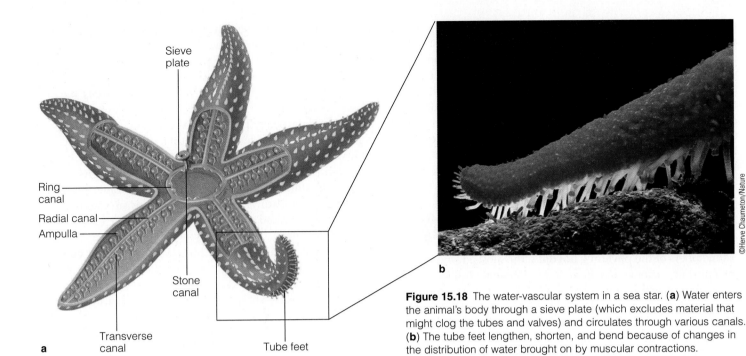

Figure 15.18 The water-vascular system in a sea star. (a) Water enters the animal's body through a sieve plate (which excludes material that might clog the tubes and valves) and circulates through various canals. (b) The tube feet lengthen, shorten, and bend because of changes in the distribution of water brought on by muscular contractions.

ally have spiny projections on top and delicate tube feet beneath. The tube feet work like suction cups and can grip objects; they also participate in gas exchange. Tube feet are part of a sea star's most striking feature: its unique **water-vascular system** (**Figure 15.18**), a complex of water-filled canals, valves, and projections used for locomotion and feeding. Operating like a hydraulic power system, the water-vascular system's plumbing can transmit forces generated by muscles on one side of the sea star to arms on the other side. Using this system, the tube feet of a sea star can grip a clam or mussel and exert a continuous pull (sometimes for hours) to force open its valves, even though one or more of the star's arms may tire. When the mussel finally opens, the sea star expels its stomach from its mouth, slips it between the mussel's shells, and digests the victim in place.

Delicate ophiuroids (*ophidion* = a snake) have long, slender arms. They are called *brittle stars* because of their unusual strategy for evading capture: If grasped by a predator, a brittle star will often detach its arm and escape, then later regenerate the arm. Ophiuroids are perhaps the most widely distributed benthic marine animals; a few species are found in great numbers on the deep, sedimented seabeds of the world ocean (**Figure 15.19**). Many ophiuroid species live beneath intertidal and subtidal rocks. Longitudinal grooves on the underside of each arm enable some species of brittle stars to locate food particles and transfer them by cilia to the mouth. Other species wave their arms through the water to capture plankton on sticky strands of mucus strung between adjacent arm spines.

Echinoids (*echinos* = hedgehog) are familiar to most coastal residents. The prickly appearance of a sea urchin or the smooth velvety surface of a sand dollar does not at first suggest the phylum's five-sided symmetry. Close observation, however, reveals just such an arrangement, overlain by a few bilaterally symmetrical features. (This can be seen in Figure 15.17b.) Urchins can feed either by taking bits of food into

the mouth with complex grasping, chewing jaws or by simply absorbing food molecules into the mucus layer that covers their bodies and flows toward the mouth.

Holothuroids (*holos* = entirely, *thur* = tubelike), the well-named sea cucumbers, also do not look five-sided at first glance. Most sea cucumbers feed by thrusting small, sticky appendages into the surrounding sediments, retracting the appendages, and eating the catch. The main claim to fame of certain nearshore species is an exuberant and seemingly suicidal defense mechanism: When annoyed, these holothurans

Figure 15.19 Ophiuroids (brittle stars) feed on edible particles in the surface layer of sediments on the continental slope off New England. They are among the most widely distributed of all benthic animals.

pressurize themselves and eviscerate violently, forcefully ejecting their digestive systems, respiratory apparatus, and gonads! The water around them churns with viscera, most of which are coated with the same sticky substance used to attach food to the feeding structures. A predator eating this material could glue his gill openings or esophagus shut. But of what value is a defense mechanism that results in the death of the defender? None, of course. The holothuran does not die but lies quietly for two to three weeks regenerating its insides from food reserves in its intact outer covering. After this brief sabbatical, it's as good as new!

THE CHORDATES

All members of the phylum **Chordata,** the most advanced animal phylum, at some time in their development possess a stiffening **notochord** (*notus* = back, *chorda* = cord), a tubular dorsal nervous system, and gill slits behind the oral opening. The notochord was critical in evolution. It permitted a more complex embryonic development by providing a rigid "scaffold" on which the developing embryo could be constructed, and it provided an internal mechanical foundation for skeletal and muscular development. About 5% of the 45,000 species of chordates lose their notochord as they develop; these species are called *invertebrate chordates*. The other 95% of chordates retain their notochord (or the vertebral column that forms around it) into adulthood; these are the familiar *vertebrate chordates* (such as fishes, reptiles, birds, and mammals).

Invertebrate Chordates

Two invertebrate chordates are of interest here. The **tunicates** (or sea squirts) are suspension feeders that superficially look and function like sponges. Their common name

comes from an extraordinarily strong and flexible tunic (outer covering). Close investigation reveals a body plan much different from that of the primitive sponges, however (**Figure 15.20**). Solitary or colonial, attached as adults or free-swimming, tunicates filter water with a special mucous plankton net capable of trapping a wide variety of microscopic food particles. The mucus is generated by a long glandular seam on one side of a basketlike interior structure (the pharynx); it is then driven around to the other side by tiny cilia and is collected and swallowed by an esophagus leading to a small stomach. Salps, related zooplanktonic forms, act almost like miniature jet engines—taking in water at one end, filtering it, and ejecting it from the other end to force themselves ahead. It stretches the imagination to consider these animals within the same phylum as seagulls or dolphins or people, but all chordate embryos share the same fundamental architecture.

Amphioxus (= sharp at both ends), another invertebrate chordate, is a small, semitransparent animal that buries itself in sand in shallow marine waters worldwide (**Figure 15.21**). It is a transitional form, an invertebrate with some vertebrate features. *Amphioxus* swims by undulating its body in a fishlike motion and feeds by combing the water for microscopic organisms. The animal's importance lies in its well-developed dorsal tubular nerve cord, which is very similar to the spinal cord of a vertebrate.

Vertebrate Chordates

Vertebrates (*vertebratus* = jointed) are the members of the phylum Chordata that possess backbones. The name refers to the segments of the backbone. About 95% of all chordates are vertebrates—nearly 50,000 species. Here are the familiar creatures found in drawings by generations of children—the

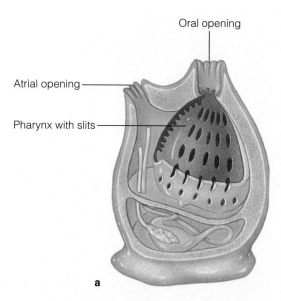

Oral opening

Atrial opening

Pharynx with slits

a

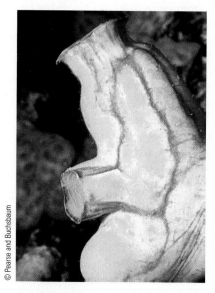

© Pearse and Buchsbaum

b

Figure 15.20 A tunicate. (**a**) Body plan showing the oral opening, through which water enters the animal's filtration system, and the atrial opening, from which it exits. (**b**) An adult tunicate. (Source for Figure 15.20a: From *Living Invertebrates.* V. & J. Pearse/M. & R. Buchsbaum, 1987. The Boxwood Press. Adapted with permission of the publisher.)

Figure 15.21 *Amphioxus*, an invertebrate chordate whose notochord and dorsal tubular nervous system make it a transitional species between invertebrates and vertebrates.

fishes, frogs, lizards, chickens, cats, and dogs most of us first think of when we hear the word *animal*.

Vertebrate Evolution and Classification

Like other chordates, vertebrates had a remote marine ancestor. The first chordates had the stiffening notochord that gives the phylum its name, but they lacked the backbone of true vertebrates. The line to modern vertebrates probably passed through an *Amphioxus*-like predecessor that lived in the ocean more than 500 million years ago. Besides having a backbone, vertebrate chordates differ from the invertebrate chordates by having an internal skeleton of calcified bone or cartilage (or both). This scaffold allows uninterrupted support during growth; it also protects vital organs and provides a foundation to which muscles may attach to permit the strength and rapid responses characteristic of active animals. The vertebrate skull, a special unit of the skeleton, provides secure housing for the brain, eyes, and other sense organs that have made the evolution of intelligence possible. Only the simplest vertebrates lack jaws. The central nervous system is partially enclosed within the backbone, which extends from the skull, and the pairs of nerves passing between the vertebral segments allow rapid and efficient communication between brain and body. The most abundant and successful vertebrates are the fishes, which exist in an extraordinary variety of forms and habitats. Least successful in the marine environment are the amphibians.

As is the case with any natural system of organization, vertebrate classification reflects our understanding of vertebrate evolution. **Figure 15.22** indicates the likely evolutionary relationship among vertebrate species. Note that all higher vertebrates appear to be derived from fishlike ancestors. In an architectural sense all higher vertebrates (including amphibians, reptiles, birds, and mammals) are highly modified, four-limbed, air-breathing fish!

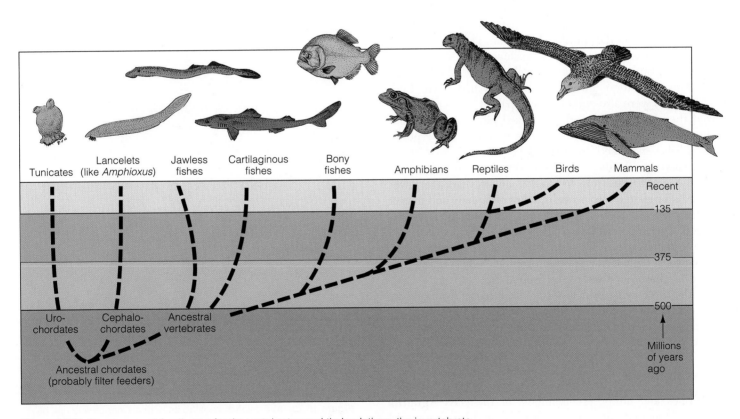

Figure 15.22 One proposed family tree for the vertebrates and their relatives, the invertebrate chordates.

Fishes are vertebrates that usually live in water and possess gills for breathing and fins for swimming. There are more species of fishes, and more individuals, than species and individuals of all other vertebrates *combined*—not a surprising fact considering the vast oceanic habitat the planet provides. Adult fishes range in length from less than 10 millimeters to over 20 meters (0.4 inch to 60 feet), and they weigh from about 0.1 gram to about 41,000 kilograms (0.004 ounce to 45 tons). Some fish are capable of short bursts of speed in excess of 120 kilometers (75 miles) per hour; some species hardly move at all.

Fishes live near the surface and at great depth, in warm water and cold, even frozen within ice or dried in balls of mud. Like other cold-blooded organisms, or **ectotherms,** the great majority of fishes are incapable of generating and maintaining a steady internal temperature from metabolic heat; thus, the internal body temperature of a fish is usually the same as that of the surrounding environment. About 40% of the 30,000-plus fish species live all or part of their lives in fresh water; 60% live exclusively in seawater. Fishes have evolved to fit almost every conceivable watery habitat, but they are most numerous on the bottom or in productive seawater over the continental shelves. Some species have a "sixth sense," an ability to detect small changes in the electrical field surrounding their bodies, that assists in the detection of prey or avoidance of predators. Some electric eels, catfish, and rays can use internally generated electricity for defense and offense.

The first fishes probably evolved in the ocean around 500 million years ago. These jawless animals were little more than motile sucking digestive tubes, but they did have the structural advantages of the chordate body plan. The earliest jawed fishes, with their grasping and crushing mouths, were far more successful at feeding on invertebrates with shells or exoskeletons than their jawless predecessors had been. Early jawed fishes were also equipped with paired fins, which stabilized their movements and minimized pitching and rolling when they attacked their prey. The numbers and types of jawed fishes increased dramatically beginning about 410 million years ago. By the end of the Devonian period, the so-called Age of Fishes from 408 to 360 million years ago, jawed fishes had radiated into a vast number of aquatic and marine habitats in which their dominance has remained unchallenged. The ancestral jawed fishes gave rise to cartilaginous fishes (animals whose skeleton is made of stiff cartilage) and bony fishes. Only a few jawless forms, such as lampreys and hagfishes, have persisted to the present day.

The three living fish classes are as different from one another as they are different from amphibians, reptiles, birds, and mammals. We will look briefly at each in turn, beginning with the jawless fishes and ending with the advanced bony fishes.

Class Agnatha

15-16

Hagfishes and lampreys—members of the class **Agnatha** (*a* = lacking, *gnathos* = jaw)—lack jaws and have no paired appendages to aid in locomotion. Their thick, snakelike bodies are pierced by gill slits and (in some species) the openings of slime glands. Their round, sucking mouths are surrounded by organs sensitive to touch and smell. Agnathan eyes are degenerate and covered by thick skin. The body ends in a flattened tail that undulates to provide forward motion. The skin consists of resilient fibrous layers currently in high demand in South Korea for the manufacture of "eelskin" leather goods sold in upscale leather shops and department stores. Fewer than 50 species of agnathans are known.

The pinkish, soft hagfishes live in colonies on continental shelf sediments, where they burrow for polychaete worms or scavenge for weak or dead organisms. They prefer feeding on the soft inner flesh and internal organs of their prey, which they abrade with a rasping tongue. If a piece of food is too large, a hagfish will tie its flexible body into a loose knot, pass the knot toward the head, and brace against the prey for a better tearing grip (**Figure 15.23**). A hagfish defends itself primarily by producing copious quantities of clinging slime from glands along the side of its body. This slippery, odorous mass deters all but the most determined would-be diner. The hagfish passes a knot down its body to shed the slime soon after danger has passed.

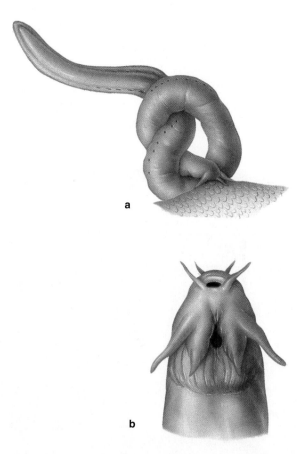

Figure 15.23 (a) A hagfish knotting its body to tear off a chunk of prey. **(b)** The head of a hagfish, showing the mouth and sensory appendages.

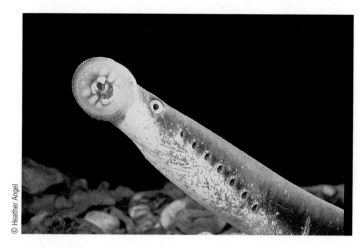

Figure 15.24 A lamprey pressing its toothed oral disc to the glass wall of an aquarium.

Lampreys (**Figure 15.24**), a group closely allied to hagfishes, possess a toothed, funnel-shaped mouth with which they can wear a passage into another vertebrate. Continuous sucking on this seeping wound provides nourishment for the lamprey. Prey are usually bony fish, but lampreys have been found feeding on several species of whales and porpoises. A lamprey will usually detach before it has killed the host. The largest hagfishes and lampreys can reach 1 meter (3.3 feet) in length, but most are smaller.

Class Chondrichthyes

Members of the group that includes sharks, skates, rays, and chimaeras have prowled the world ocean for at least 280 million years. Sharks have a history twice as long as that of di-

nosaurs. Fossilized teeth, fin spines, and sharks' eggs found in marine sediments tell us that today's sharks and rays are not much different from their ancient ancestors.

All members of the class **Chondrichthyes** (*chondros* = cartilage, *ichthys* = fish) have a skeleton made of a tough, elastic tissue called **cartilage,** the material that gives your ears and nose their shape. Though there is some calcification in the cartilaginous skeleton, true bone is entirely absent from this group. Unlike the more primitive agnathans, these fish have jaws with teeth, paired fins, and often active lifestyles. Sharks and rays tend to be larger than either agnathans or bony fishes, and except for some whales, sharks are the largest living vertebrates. Chimaeras, such as ratfishes, are comparatively rare forms found mostly at and below middle depths.

Only a small fraction of all fish species are members of class Chondrichthyes. About 350 species of sharks and 320 species of rays are known to exist. Nearly all are marine, though a few species inhabit estuaries and a very few are permanent inhabitants of fresh water. Although there are many exceptions, sharks tend to favor swimming through open water, whereas rays tend to be found on or near the bottom.

Skates and rays have a flattened appearance, with spreading pectoral fins attached to their bodies to form a triangular or rounded winglike shape. They glide through the water with a slow, rhythmic flapping of their fins. Neither sharks nor rays have gas bladders, and both are slightly negatively buoyant: They will slowly sink if they stop swimming. The skin of rays is usually smooth and greatly variable in color. (Sharks, in contrast, have a rough skin with thousands of tiny toothlike projections.) One family of rays carries a defensive barb at the base of the tail capable of inflicting a serious wound. Another ray family contains members measuring 7 meters (22 feet) across and weighing 1,700 kilograms (1.9 tons); an example is the giant manta (**Figure 15.25**). Still another produces a

Figure 15.25 A manta ray, sometimes called a *devilfish* because of the cartilaginous protuberances of the head, which are used to guide plankton into the mouth.

Figure 15.26 The great white shark, genus *Carcharodon,* a predator of seals, sea lions, and large fish. The great white shark is one of the most dangerous of sharks encountered by human swimmers. In this extraordinary sequence taken off the coast of South Africa, a large great white shark rockets from the ocean with a freshly caught sea lion in its jaws.

violent electric shock capable of stunning prey or disabling a human diver. The largest rays feed on plankton, but most smaller species crush mollusks and arthropods with smooth calcified plates in the mouth.

Sharks have an undeservedly bad reputation. More than 80% of shark species are less than 2 meters (6.6 feet) long as adults, and only a few of the remaining 20% are aggressive toward humans. Sharks don't hold grudges or behave in the malignant ways so vividly portrayed in popular novels and movies. Still, some species of sharks are indeed dangerous to humans, and the great white sharks in the genus *Carcharodon* (*karcharos* = sharp, *odontos* = tooth) are perhaps the most dangerous of all (**Figure 15.26**).[1] These swimmer's nightmares attain lengths of 7 meters (23 feet) and weigh up to 1,400 kilograms (3,000 pounds). (Reports of even larger great whites probably resulted from misidentification of large, harmless

species.) Great whites are not actually white, but rather grayish brown or blue above and creamy on the lower half. A dangerous relative, the mako shark, reaches a length of 4 meters (13 feet) and is even known to attack small boats. These and other predatory species, such as tiger sharks and hammerhead sharks, are attracted to prey by vibrations in the water, which they detect with sensitive organs arrayed in lines beneath the surface of their skin. Smell also plays an important role in hunting their prey, usually fish and marine mammals.

Though most famous, the so-called man-eaters are not the largest of shark species. This honor goes to the immense, warm-water whale sharks in the genus *Rhineodon* (*rhine* = a rasp, a reference to the fish's rough skin), which reach sizes in excess of 18 meters (60 feet) and 41,000 kilograms (90,000 pounds) (**Figure 15.27**). Whale sharks and their somewhat smaller relatives, the basking sharks, are docile and present little threat to people. These greatest of fishes swim slowly near the surface with their huge mouths open, feeding on plankton. They may filter as much as 2,200 cubic meters (2,500 tons) of water per hour through a fine mesh of gill rakers. Accumulated plankton are periodically backflushed into the mouth, where they are concentrated for swallowing.

---

[1]Some perspective: Worldwide, sharks are responsible for about six known human fatalities each year. Each year, more people are killed in the United States by dogs than have been killed by sharks in the last century. For every human killed by a shark, humans kill more than 16 million sharks, mostly for food and medicines.

Figure 15.27 A diver hitches a ride on a whale shark. Unless he is struck by the fish's tail as he dismounts, the diver is in no danger: Divers are not part of the diet of this type of shark.

Class Osteichthyes

15-18

The 27,000-plus species of bony fishes, members of class **Osteichthyes** (*osteum* = bone, *ichthyos* = fish), owe much of their great success to the hard, strong, lightweight skeleton that supports them. These most numerous of fish—and most numerous and successful of all vertebrates—are found in almost every marine habitat, from tide pools to the abyssal depths. Their numbers include the air-breathing lungfishes and lobe-finned coelacanths, whose ancient relatives broke from the path of fish evolution to establish the dynasties of land vertebrates.

About 90% of all living fishes are contained within the osteichthyan order **Teleostei** (*teleos* = perfect, *osteon* = bone), which contains the cod, tuna, halibut, perch, and other familiar species (**Figure 15.28**). Within this large category are different fishes with gas-filled swim bladders to assist in maintaining neutral buoyancy, independently movable fins for well-controlled swimming and communication, great speed for pursuit or avoidance of predators, highly effective camouflage, social organization, the ability to cluster together in defensive schools, orderly patterns of migration, and other advanced features. Their economic importance is great—some 86 million metric tons (94 million tons) of bony fishes is taken annually from the ocean to help satisfy the human demand for protein.

THE PROBLEMS OF FISHES

15-19

What problems are unique to a fish? Seawater may seem to be an ideal habitat, but living in it does present difficulties. Water is about a thousand times denser than air and a

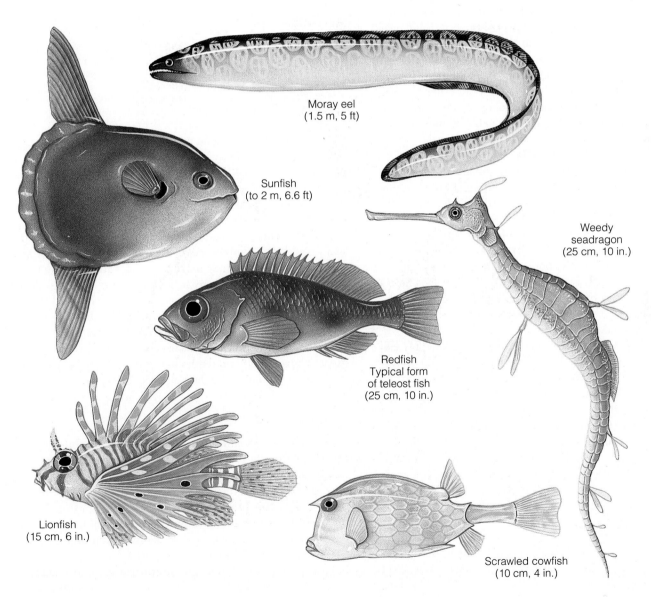

Figure 15.28 Some of the diversity exhibited by teleost (bony) fishes. These fishes are not drawn to scale.

Moray eel
(1.5 m, 5 ft)

Sunfish
(to 2 m, 6.6 ft)

Weedy
seadragon
(25 cm, 10 in.)

Redfish
Typical form
of teleost fish
(25 cm, 10 in.)

Lionfish
(15 cm, 6 in.)

Scrawled cowfish
(10 cm, 4 in.)

hundred times more viscous, and it impedes motion more effectively at low speeds. How can a fish best move through it? How can a fish maintain its vertical position in the water column? What about breathing? Can oxygen and carbon dioxide be exchanged efficiently underwater? What about osmotic balance? How can some species of fish move from fresh water to salt water? How is salt balance maintained across the thin gill membranes? How can predators be thwarted? Can group behavior, camouflage, or subterfuge be employed? There would seem to be many problems, but these most successful vertebrates have structures and behaviors to cope.

Movement, Shape, and Propulsion

15-20

Active fish usually have streamlined shapes that make their propulsive efforts more effective. A fish's resistance to movement, or *drag*, is determined by frontal area, body contour, and surface texture. Drag increases geometrically with increasing speed. Faster-swimming fish are therefore more highly modified to minimize the slowing effects of the dense, relatively sticky medium in which they live. Their tapering, torpedo-like body plan minimizes energy lost to drag.

A fish's forward thrust comes from the combined effort of body and fins. Muscles within slender, flexible fishes (such as eels) cause the body to undulate in S-shaped waves that pass down the body from head to tail in a snakelike motion. The eel pushes forward against the water much as a snake pushes against the ground (**Figure 15.29a**). This type of movement is not very efficient, however: The body must wave back and forth across a considerable distance, exposing a large frontal area to the water. The long body length (relative to width) needed to propagate the wave requires increased surface area, which increases drag. A more efficient swimming mechanism is shown in **Figure 15.29b.** More advanced fishes have a relatively inflexible body, which undulates rapidly through a shorter distance, and a hinged, scythelike tail to couple muscular energy to the water. The fish's body can be shorter and can face more squarely in the direction of travel, so the drag losses are lower.

How efficient are the best swimmers, and how fast can they swim? Estimates vary, but it is thought that in the fastest fish between 60% and 80% of muscle force delivered to the tail fin results in forward motion. Swordfish and marlin can reach 120 kilometers (75 miles) per hour in short bursts! Some of the fastest tuna (and a few swift sharks) sustain high speeds by maintaining their internal body temperature a few degrees above that of the surrounding water. The warmth permits them to oxidize food more rapidly and generates greater muscle power per unit of mass.

Maintenance of Level

15-21

The density of fish tissue is typically greater than that of the surrounding water, so fishes will sink unless their mass is offset by propulsive forces or by buoyant gas- or fat-filled bladders. Cartilaginous fishes have no **swim bladders** and must swim continuously to maintain their position in the water column.

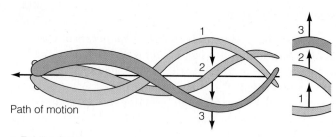

a Eel-like fishes

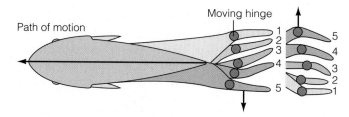

b Advanced fishes

Figure 15.29 How a fish's body shape affects its efficiency of movement. The undulating movement of eel-like fishes (**a**) requires more energy to generate forward motion than the hinged-tail movement of the more advanced fishes (**b**). (Source: From Pough et al., *Vertebrate Life*, 3/e. © 1989. Reprinted by permission of Pearson Eduction, Inc. Upper Saddle River, NJ 07458.)

Sharks generate lift with an asymmetrical tail and fins that act like airplane wings. Bony fishes that appear to hover motionless in the water usually have well-developed swim bladders just below their spinal columns. The volume of gas in these structures provides enough buoyancy to offset the animal's mass. The quantity of gas is controlled both by secretion and absorption of gas from the blood and by muscular contraction of the swim bladder to compensate for temporary changes in depth.

A swim bladder may make life easier for a slow-moving teleost, but the fastest bony fishes lack swim bladders. Why? Fast, powerful predators such as tuna, mackerel, and swordfish must be able to chase prey between depths. The expansion and compression of gas in the bladder would vary rapidly with changing depth, and the risk of rupture would be too great. These speedy predators have power to spare and don't seem inconvenienced by their slight negative buoyancy.

Gas Exchange

15-22

How can fish breathe underwater? **Gas exchange,** the process of bringing oxygen into the body and eliminating carbon dioxide, is essential to all animals. At first glance the task may seem more difficult for water breathers than for air breathers, but air-breathing animals add an extra step. We air breathers must first dissolve gases in a thin film of water in our lungs before they can diffuse across a membrane.

Fish take in water containing dissolved oxygen at the mouth, pump it past fine **gill membranes,** and exhaust it through rear-facing gill slits. The higher concentration of free oxygen dissolved in the water causes oxygen to diffuse through the gill membranes into the animal; the higher con-

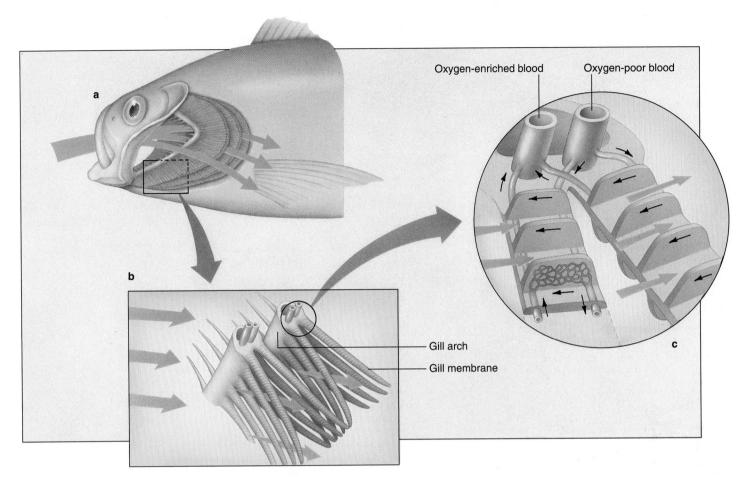

Figure 15.30 Cutaway drawing of a mackerel, showing the position of the gills (**a**). Broad arrows in (**b**) and (**c**) indicate the flow of water over the gill membranes of a single gill arch. Small arrows in (**c**) indicate the direction of blood flow through the capillaries of the gill filament in a direction opposite to that of the incoming water. This mechanism is called *countercurrent flow.*

centration of CO_2 dissolved in the blood causes CO_2 to diffuse through the gill membranes to the outside. The gill membranes themselves are arranged in thin filaments and plates efficiently packaged into a very small space (**Figure 15.30**). Water and blood circulate in opposite directions—in a countercurrent flow—which increases transfer efficiency.

An active fish like a mackerel requires so much oxygen and generates so much waste CO_2 that its gill surface area must be ten times its body surface area. (Sedentary fishes have proportionally less gill area.) With their large gill area and countercurrent flow, active fish extract about 85% of the dissolved oxygen in water flowing past their gills. Air-breathing vertebrates, by contrast, extract only about 25% of the oxygen from air entering their lungs.

Osmotic Considerations

15-23

Marine invertebrates and primitive vertebrates such as agnathans have an internal salt concentration nearly identical to that of seawater. The body fluids of more advanced vertebrates generally are only about one-third as saline as seawater; they are **hypotonic** to their oceanic surroundings. The

gill and intestinal membranes of fish are necessarily thin to allow passage of gases and food molecules, but this thinness also makes the passage of water, or **osmosis,** inevitable.

As we saw in Chapter 13, osmosis moves water across membranes from regions of high water concentration to regions of lower water concentration. Marine teleosts (bony fishes), with their higher relative internal concentration of water (and lower concentration of salts) per unit of fluid volume, continuously *lose* water to their environment; freshwater teleosts, with their lower relative internal concentration of water, constantly *absorb* water from their environment. If these fishes were incapable of **osmoregulation**—that is, if they had no active way of adjusting their internal salt concentration—they would quickly die of fluid imbalance.

Figure 15.31 summarizes the ways bony fishes cope with osmoregulatory difficulties. The skin of both freshwater and marine teleosts is nearly impermeable to water and salts. A freshwater fish (Figure 15.31a) does not drink water. It uses large kidneys to generate copious quantities of dilute urine to export the invading water, and it actively absorbs salts through its gills both from the surrounding water and from its own urine. A marine fish (Figure 15.31b), on the other hand,

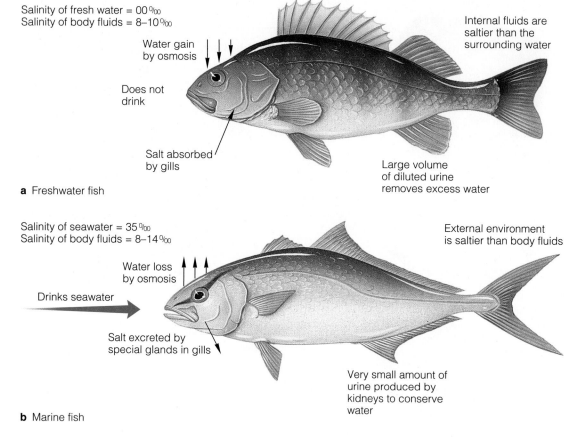

Salinity of fresh water = 00‰
Salinity of body fluids = 8–10‰

Water gain
by osmosis

Does not
drink

Salt absorbed
by gills

Internal fluids are
saltier than the
surrounding water

Large volume
of diluted urine
removes excess water

a Freshwater fish

Salinity of seawater = 35‰
Salinity of body fluids = 8–14‰

Water loss
by osmosis

Drinks seawater

Salt excreted by
special glands in gills

External environment
is saltier than body fluids

Very small amount of
urine produced by
kidneys to conserve
water

b Marine fish

Figure 15.31 Osmoregulation in freshwater and marine fishes.

makes only small quantities of urine, actively drinks seawater (some species drinking up to 25% of their body mass per day), and eliminates excess salts through special salt-secreting cells in the gills. Bony fish consume a substantial amount of energy in these essential osmoregulatory tasks.

Chondrichthyes (sharks, rays) employ a different strategy. Their internal fluids are supplemented with urea to produce a nearly neutral osmotic pressure across gill membranes. Although the proportion of dissolved solids in their body fluids is different from that in seawater, the quantity of dissolved solids is nearly identical, and the net flow of water into or out of the gills and intestinal membranes is minimal. Shark meat has an unpleasant, bitter taste (or tastes uncomfortably rich) unless it has been soaked in seawater to wash away the urea.

Feeding and Defense

Competitive pressure among the large number of fish species has caused a wonderful variety of feeding and defense tactics to evolve. Sight is very important to most fishes, enabling them to see their prey or avoid being eaten. Even some deep-water fishes that live below the photic zone have excellent eyesight for seeing luminous cues from potential mates or meals. Hearing is also well developed, as is the ability to detect low-frequency vibrations with the **lateral-line system.**

This mechanism consists of a series of small canals in the skin and bones around the eyes, over the head, and down the sides of the body. The canals are richly supplied with nerves and connect to the surface through tiny pores in the skin. The nerves report changes in current direction, water pressure, or sonic environment to the brain. Predatory sharks use their lateral-line systems to detect prey.

Defense measures are well advanced in fishes. Sea horses (and their relatives, the box fishes) depend on the simple expedient of armor plating for protection. Others, such as the puffer fish, inflate with water and erect bristly spines to become less attractive as a snack. Subtler means of offense or defense depend on trickery—looking like something you're not, or changing color to blend with the background. These kinds of **cryptic coloration** (*kryptos* = hidden) or camouflage may be active or passive. **Figure 15.32** shows an example of passive cryptic coloration. The kelp bass closely resembles its seaweed habitat. Small animals foraging nearby run the risk of being eaten, and the fish itself may escape the notice of a larger predator. An example of active cryptic coloration is shown in **Figure 15.33.** Here a turbot—a bottom dweller—has reproduced the pattern of the background on its own surface, and so blends inconspicuously into the surroundings. Actively swimming fish employ a different method of color blending called *countershading*. Their dark tops and

a

Figure 15.32 Passive cryptic coloration in a kelp fish. (**a**) The fish resembles a seaweed blade in shape, color, and pattern. (**b**) The fish nestles within the blades, nearly disappearing from view, moving with the blades as they sway in the water.

b

Figure 15.33 Active cryptic coloration in a turbot. The fish is able to blend into its surroundings because of chromatophores (skin cells that can change color). Though the fish is in plain view, the only obvious features are its round, black eyes.

silvery bottoms make the fish less obvious to predators above or below.

About a quarter of all bony fish species exhibit **schooling** behavior at some time during their life cycle. A fish school is a massed group of individuals of a single species and size class packed closely together and moving as a unit. There is no leadership in fish schools, and the movement of fish within them seems to be controlled automatically by direct interaction between lateral-line sensors and the locomotor muscles themselves. I can personally attest to the effectiveness of schooling as a means of defense. On a few diving trips I've noticed a large moving mass just beyond the limit of clear visibility. Is it a fish school, or is it a single large animal? Many predators might not stay around long enough to find out! Schools have the added benefits of reducing chance detection by a predator, providing ready mates at the appropriate time, and increasing feeding efficiency.

Perhaps the most effective way to avoid being eaten is simply to disappear. The surface-feeding flying fishes do this by accelerating rapidly, leaping into the air, and spreading their fins to act as wings. Refractive differences between air and water prevent the surprised predator from seeing where its intended meal has gone (**Figure 15.34**)!

AMPHIBIANS

Amphibians (frogs, salamanders, toads) are specialized vertebrates adapted to fresh water and to moist habitats on land. Only about 2,000 living species exist, and none is exclusively marine—although some large Southeast Asian frogs can tolerate water with a salinity of up to 28‰ for extended periods of time. Amphibians depend on the constant flow of water through their skins into their bodies to provide the fluid for the formation of urine to remove nitrogenous wastes. Placing an amphibian in seawater would cause water to flow through the skin in the opposite direction, dehydrating the animal.

Figure 15.34 Some fish avoid predators by disappearing completely—in one instant the predators see a potential meal; in the next it has left the ocean. This flying fish used specially modified pectoral fins to soar many meters across the surface. Some species lengthen the glide by extending a lower tail lobe into the water to reaccelerate.

Figure 15.35 A green sea turtle, species *Chelonia mydas*.

Their skin is too permeable to permit amphibians to colonize the marine environment.

MARINE REPTILES

Each of the three main groups within the class **Reptilia** has marine representatives: turtles, sea snakes and marine lizards (iguanas), and marine crocodiles. Like all reptiles, marine reptiles are ectothermic, breathe air with lungs, are covered with scales and a relatively impermeable skin, and are equipped with special **salt glands** to concentrate and excrete excess salts from body fluids. Except for one widely ranging species of turtle, all marine reptiles require the warmth of tropical or subtropical waters.

Sea Turtles

The best-known and most successful living marine reptiles are the eight species of sea turtles. Unlike land turtles, from which they evolved, sea turtles have relatively small, streamlined shells without enough interior space to retract the head or limbs. The shell provides an effective passive defense, and adult sea turtles have no predators except humans. Their forelimbs are modified as flippers and provide propulsive power; their hind limbs act as rudders. The two species of green sea turtles (*Chelonia*) are the most abundant and widespread of living species (**Figure 15.35**). Green turtles range over great distances looking for the marine algae, turtle grass, and other plants on which they feed. The largest living turtle is the carnivorous Atlantic leatherback, a streamlined animal with a soft, skin-covered "shell"; it reaches lengths in excess of 2 meters (6½ feet) and may weigh more than 600 kilograms (1,300 pounds). All sea turtles are endangered species.

Sea turtles are justly famous for navigating over remarkable distances, which in some cases have been progressively increased (as you may recall from Chapter 3) by seafloor spreading. Sea turtles return at two-, three-, or four-year intervals to lay eggs on the beaches on which they were hatched. Homing behavior can be a great advantage to an animal; if the parent survived its earliest childhood at this location, it will probably be a suitable place for hatching the next generation. The navigation of green turtles to tiny Ascension Island, an emergent point of the Mid-Atlantic Ridge between Brazil and Africa, has been extensively studied. Researchers have found that the turtles use solar angle (to find latitude), wind-wave direction, smell, and visual cues—first to find the island and then to discover the spot on the beach where they hatched perhaps 20 years before!

Marine Crocodiles

Fortunately, perhaps, there is only one living species of true marine crocodile (**Figure 15.36**). Marine crocodiles live in mangrove swamps and reef islands and on isolated mainland shores in the tropical western Pacific. They hunt in packs and sometimes move ashore to consume any large, warm-blooded animals they can find. They are extremely aggressive, extraordinarily fast in attack, and responsible for a few human deaths each year. Male marine crocodiles of northern Australia and Indonesia have attained lengths of 7 meters (23 feet) and may weigh over a ton! (The much smaller and less dangerous Florida and Louisiana crocodilians live in fresh, brackish, and salt waters.)

Sea Snakes

Sea snakes are probably the most evolutionarily advanced of the marine reptiles. Fifty species of sea snakes are known; most are native to the warm Indian Ocean and western Pacific areas. None lives in the Atlantic. They hunt among the coral heads or uneven rocky bottoms and rise at intervals to breathe. With the exception of one egg-laying species, sea

Figure 15.36 A marine crocodile. Marine crocs are the world's largest, reaching a length of 7 meters (23 feet). They hunt in packs and can swim long distances between tropical islands.

All birds lay eggs on land, and most incubate them and provide care for the young. Some seabirds may stay at sea for years, but all must eventually return to land to breed.

Only about 270 kinds of birds, about 3% of known bird species, qualify as seabirds. Most seabirds live in the Southern Hemisphere. Like the marine reptiles, seabirds have special salt-excreting glands in their heads to eliminate the excess salt taken in with their food. Salty brine from these glands may sometimes be seen dripping from the tip of their beaks. Marine birds are voracious feeders and are often found wherever the ocean teems with life. True seabirds generally avoid land unless they are breeding; they obtain virtually all their food from the sea and seek isolated areas for reproduction.

Seabirds may be divided into four groups: tubenoses (albatrosses, petrels), pelicans and their relatives, the gulls, and penguins.

snakes give birth in the water to living young and never come onto land. All sea snakes are highly venomous, with venoms that are among the most powerful animal poisons known.

Sea snakes are effective swimmers, with flattened tails. These generally slow-moving snakes lunge at passing small fish, grasp them in their teeth, and wait as a saliva-borne neurotoxic venom seeps into the wound. (Sea snakes do not have hollow injection fangs.) The paralyzed fish are then swallowed head first.

Though not particularly aggressive, these snakes are very dangerous to humans. Members of genus *Pelamis* are responsible for perhaps 20 fatalities every year, mainly among Asian fishermen who are scratched by the snake's teeth while removing it from a net. Sea snakes are most common in Southeast Asia, in the Gulf of California, and on the tropical Pacific coasts of Central and South America. There is concern that warming water will allow these unwelcome animals to extend their range to the southwestern coast of the United States.

MARINE BIRDS

Birds (class **Aves**) probably evolved from small, fast-running dinosaurs about 160 million years ago. Their reptilian heritage is clearly visible in their scaly legs and claws and in the configuration of their internal organs and skeletons. The success of the 8,600 living species of birds is due in large part to the evolution of feathers (derivatives of reptilian scales) used to insulate the body and provide aerodynamic surfaces for flight. Birds (and mammals) are **endotherms**: They generate and regulate metabolic heat to maintain a constant internal temperature that is generally higher than that of their surroundings.

Flying birds have light, thin, hollow bones without fatty insulation; they have forsaken the heavy teeth and jaws of reptiles for a lightweight beak. Their highly efficient respiratory system can accept great quantities of oxygen, and their large, four-chambered heart circulates blood under high pressure.

The Tubenoses

The 100 species of tubenoses of order Procellariiformes (*procella* = storm) are the world's most oceanic birds. Their prosaic common name does not convey any sense of their beauty and grace; it refers to the plumbing in their beak responsible for sensing air speed, detecting smells, and ducting saline water from the salt glands. Foraging for months across the ocean in conditions of strong winds and high waves, seeking no shelter during storms, being exposed to both tropical heat and polar sleet, soaring continually through air with a gliding efficiency exceeding that of the most perfectly built human sailplane—the beautiful albatrosses, petrels, and shearwaters are true masters of the sky.

The largest of the tubenoses are the magnificent wandering albatrosses (*Diomedea*), which reach a wingspan of 3.6 meters (12 feet) and a mass of 10 kilograms (22 pounds). (Ancestors of these great birds were even larger—some had wingspans in excess of 5.5 meters, or 18 feet!) The key to the albatross's success lies in its aerodynamically efficient wing (**Figure 15.37**), a very long, thin, narrow, cupped, and pointed structure ideal for high-speed soaring and gliding. This beautiful wing allows albatrosses to cover great distances in search of food with very little expense of energy, flying continuously for weeks or months at a time. Using the uplift from wind deflected by ocean waves to stay aloft, they soar in long, looping arcs. Albatrosses rarely flap their wings.

Satellite tracking data indicate that wandering albatrosses routinely cover 15,000 kilometers (9,300 miles) on foraging trips and reach speeds of 80 kilometers (50 miles) per hour. High speeds over long distances are their specialty. One bird was observed to travel 808 kilometers (502 miles) at an average speed of 56 kilometers (35 miles) per hour.

Albatrosses were once thought to locate food exclusively by sight, but recent research suggests that their extraordinarily acute sense of smell also plays a role in feeding. They can evidently find schools of fish by the odor of fish oil wafted

Figure 15.37 Wandering albatross (*Diomedea exulans*). This largest of all albatrosses can have a wingspan of 3.6 meters (12 feet) and may routinely cover 15,000 kilometers (9,300 miles) on foraging trips lasting up to eight months. They reach speeds of 80 kilometers (50 miles) per hour. Note the long, thin, cupped, pointed, narrow wing, a perfect structure for soaring. They breed every other year and spend more than 90% of their lives over the open sea.

tens of kilometers downwind. Tubenoses catch fish (and squid) by dipping their bill into the water during flight or during rare, brief stops on the surface when the water is calm. Albatrosses take shore leave only during breeding season. Chicks are hatched and raised on remote islands to which albatrosses regularly migrate from virtually any point over the world ocean. They are magnificent animals—no one who has seen one sweeping over the sea surface ever forgets the experience.

The Pelicans and Their Relatives

Birds in the order Pelecaniformes (*pelekan* = pelican)—which includes cormorants, frigate birds, and boobies in addition to pelicans—all have throat pouches and webbed feet. These large birds, commonly seen off tropical and subtropical coasts, don't spend as much time over the open sea as their tubenose relatives. Pelicans have broad, flat wings well adapted to slow flight. The wings are folded back when the pelican crashes into the water in pursuit of prey. In contrast, the aerobatic frigate birds can neither walk nor swim and so must feed on flying fish caught during flight, on small squid or fish obtained while hovering, or on regurgitated food stolen from other seabirds as they return to their rookeries from a day of hunting. These highly maneuverable birds are extraordinarily light and delicate; a frigate bird's skeleton weighs less than its feathers! The most oceanic of this group are the relatively small tropic birds,

which are among the few seabirds that regularly inhabit the relatively unproductive mid-oceanic gyres.

The Gulls

The order Charadriiformes (*charadra* = a cleft dweller) is a shorebird group with both marine and freshwater representatives. A few of the 115-plus species of gulls and terns are familiar to anyone who has spent time near the shore. Some gulls are found far inland at lakes or at dumps (where they can indulge their skill at scavenging), but most are found along the coasts eating nearly anything available. They are extraordinarily maneuverable and efficient flyers, buoyant swimmers, and—thanks to long legs at the middle of the body—surprisingly good runners.

Terns, smaller and finer in shape than gulls, are more oceanic and may travel to sea for prolonged periods in search of food. Most terns are excellent divers and plunge for their prey. The arctic tern has the most extensive migratory route of any bird, completing a 24,000-kilometer (15,000-mile) round-trip each year. It avoids winter altogether by moving from high northern latitude to high southern latitude at just the right time.

The Penguins

Penguins have completely lost the ability to fly, but they use their reduced wings to swim for long distances and with great maneuverability. The name of their order, Sphenisciformes (*sphenискos* = little wedge), refers to the shortness and shape of their wings. Their flightlessness makes it practical to have fatty insulation, greasy peglike feathers, stubby appendages, and large size and mass; indeed, such heat-conserving adaptations are critical to marine survival in very cold climates. Their neutral buoyancy is an advantage as they forage for food underwater. Emperor penguins, the largest of the living penguin species, may dive to depths of 265 meters (875 feet) and stay submerged for 10 minutes or more. Penguins feed on fish, large zooplankters, bottom-dwelling mollusks or crustaceans, and squid. A few of the 18 species spend two uninterrupted years at sea between breedings.

Penguins are native only to the Southern Hemisphere and range from the size of a large duck to a height of more than 1 meter (3.3 feet) and a mass exceeding 36 kilograms (80 pounds). They are thought to consume about 86% of all food taken by birds in the southern ocean—about 34 million metric tons (37 million tons) per year, mostly larger zooplanktonic crustaceans. The small Galápagos penguin lives a comparatively easy life fishing the cold, nutrient-rich Humboldt Current at the equator, but its Antarctic relatives lead what must surely be the most rigorous existence of any seabird. For example, Emperor penguins breed and incubate during the bitterly cold Antarctic winter (**Figure 15.38**). Unlike most other birds, Emperors do not establish a territory; instead they huddle together in tight crowds to conserve heat. The mob of penguins moves slowly around the breeding area as warm penguins from the center of the mob circulate to the

Figure 15.38 Emperor penguins and a maturing chick. One of the larger penguin species, Emperors subsist mainly on a diet of fish and squid.

outside to be replaced at the core by their chilled peripheral friends.

MARINE MAMMALS

The class **Mammalia** (*mamma* = breast), to which humans belong, is the most advanced vertebrate group. About 4,300 species of mammals are known. The three living groups of marine mammals are the porpoises, dolphins, and whales of order **Cetacea;** the seals, sea lions, walruses, and sea otters of order **Carnivora;** and the manatees and dugongs of order **Sirenia.**

Each order arose independently from land ancestors. They all exhibit the mammalian traits of being endothermic, breathing air, giving birth to living young that they suckle with milk from mammary glands, and growing hair at some time in their lives. Unlike other mammals, however, these extraordinary creatures have become adapted to life in the ocean and are dramatically different in appearance and function both from other modern forms and from the terrestrial precursors from which they branched only about 50 million years ago.

All marine mammals share four common features:

- Their *streamlined body shape* with limbs adapted for swimming makes an aquatic lifestyle possible. Efficient locomotion depends on minimum drag and maximum ability to transfer propulsive energy from the muscles to the water. Thin, stiff flippers and tail flukes situated at the rear of the animal drive it forward, and similarly shaped forelimbs act as rudders for directional control. Drag is reduced by a slippery skin or hair covering.

- They *generate internal body heat* from a high metabolic rate and conserve this heat with layers of insulating fat and, in some cases, fur. They also have relatively few surface capillaries, thus further reducing heat loss from the blood in the skin. Their large size gives them a favorable surface-to-volume ratio; with less surface area per unit of volume, they lose less heat through the skin. This is why there are no marine mammals smaller than a sea otter; a small mammal would lose body heat too rapidly. These adaptations are critical to animals living in cold water where food is readily available.

- The *respiratory system is modified* to collect and retain large quantities of oxygen. The air duct "plumbing" of marine mammals is typically much different from that of land mammals, and the lungs can be more thoroughly emptied before a fresh breath is drawn. The biochemistry of blood and muscle is optimized for the retention of oxygen during deep, prolonged dives. Some whales can stay submerged for 90 minutes.

- A number of *osmotic adaptations* free marine mammals from any requirement for fresh water. Unlike other marine vertebrates, the marine mammals do not have salt-excreting glands or tissues. They swallow little water during feeding (or at any other time), and their skin is impervious to water. This minimal seawater intake, coupled with their kidneys' abilities to excrete a concentrated and highly saline urine, permits them to meet their water needs with the metabolic water derived from the oxidation of food.

Order Cetacea

The 79 living species of cetaceans (*ketos* = whale) are thought to have evolved from an early line of ungulates—hooved land mammals related to today's horses and sheep—whose descendants spent more and more time in productive, shallow waters searching for food. Modern whales range in size from 1.8 meters (6 feet) to 33 meters (110 feet) in length and weigh up to 100,000 kilograms (110 tons). Their paddle-shaped forelimbs are used primarily for steering, and their hind limbs are reduced to vestigial bones that do not protrude from the body. They are propelled mainly by horizontal tail flukes that are moved up and down by powerful muscles at the animal's posterior end. A thick layer of oily blubber provides insulation, buoyancy, and energy storage. One or two nostrils are located at the top of the head and have special valves to prevent intake of water when submerged. Whales have large, deeply convoluted brains and are thought to form complex family and social groupings.

Modern cetaceans are further divided into two suborders. **Figure 15.39** shows representative whales in each division. Whales in the suborder **Odontoceti** (*odontos* = tooth), the toothed whales, are active predators and possess teeth to

Humpback whale

Bowhead whale

Right whale

Minke whale

Blue whale

Fin whale

Feeding on krill

Sei whale

Gray whale

Mysticetes (baleen whales)

Figure 15.39 Some representatives of the order Cetacea. Note the differences between Mysticeties and Odontoceties.

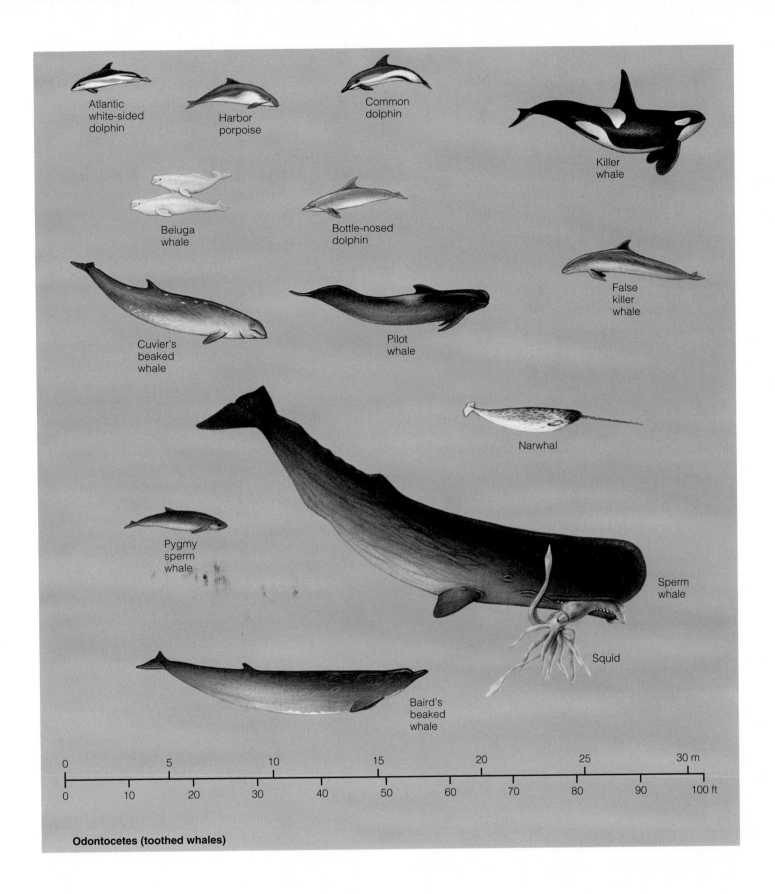

Atlantic white-sided dolphin

Harbor porpoise

Common dolphin

Killer whale

Beluga whale

Bottle-nosed dolphin

False killer whale

Cuvier's beaked whale

Pilot whale

Narwhal

Pygmy sperm whale

Sperm whale

Squid

Baird's beaked whale

0		5		10		15		20		25		30 m

0	10	20	30	40	50	60	70	80	90	100 ft

Odontocetes (toothed whales)

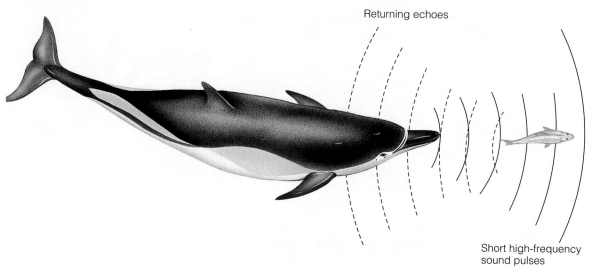

Returning echoes

Short high-frequency
sound pulses

Figure 15.40 Echolocation, used by toothed whales to locate and perhaps stun their prey.

subdue their prey. Toothed whales have a high brain-mass-to-body-mass ratio, and though much of their additional brain tissue is involved in formulating and receiving the sounds on which they depend for feeding and socializing, many researchers believe them to be quite intelligent. Smaller whales in this group include the orca (killer whale) and the familiar dolphins and porpoises of oceanarium shows. The largest toothed whale is the 18-meter (60-foot) sperm whale, which can dive to at least 1,140 meters (3,740 feet) in search of the large squid that provide much of its diet.

Toothed whales search for prey using **echolocation,** the biological equivalent of sonar; they generate sharp clicks and other sounds that bounce off prey species and return to be recognized (**Figure 15.40**). Reflected sound is also used to build a "picture" of the animal's environment and to avoid hitting obstacles while swimming at high speed. Odontocete whales are now thought to use sound offensively as well. Recent research indicates that some odontocetes can generate sounds loud enough to stun, debilitate, or even kill their prey. In one experiment dolphins produced clicks as loud as 229 decibels, equivalent to a blasting cap exploding close to the target organism. Sperm whales, it has been calculated, may generate sounds exceeding 260 decibels! (The decibel scale is logarithmic; compare this figure to the 130-decibel noise of a military jet engine at full power 20 feet away!) How this prodigious noise is generated is not yet known, nor do we know how such energy is radiated from the whale without damaging the organs that produce and focus it.

Whales in the suborder **Mysticeti** (*mystidos* = unknowable), the whalebone or baleen whales, have no teeth and are thought to have branched from the line leading to toothed whales early in whale evolution. Filter feeders rather than active predators, these whales subsist primarily on krill, a relatively large, shrimplike crustacean zooplankter obtained in productive polar or subpolar waters. They do not dive deep

but instead commonly feed a few meters below the surface. Their mouths contain interleaving triangular plates of bristly, hornlike **baleen** (**Figure 15.41**), used to filter the zooplankton from great mouthfuls of water or from mud scooped from a shallow seabed. The plankton are concentrated as water is expelled, then swept from the baleen plates by the whale's tongue, compressed to wring out as much seawater as possible, and swallowed through a throat not much larger in diameter than a grapefruit. A great blue whale, largest of all animals, requires about 3 metric tons (6,600 pounds) of krill each day during the feeding season—about 1 million calories per day. When nursing, a baby blue whale gains mass at a rate of about 4 kilograms (9 pounds) an hour! The short, efficient food chain from phytoplankton to zooplankton to whale provides the vast quantity of food required for whales' survival. A feeding blue whale with its mouth hugely distended with seawater and krill is shown in **Figure 15.42.**

Mysticeti is an excellent name for these odd and wonderful animals. We know comparatively little of their social structure, intelligence, sound-producing abilities, navigational skills, or physiology. We do know that humpback whales use complex songs in group communication (see **Box 15.2**) and that blue whales may use very low frequency sound to communicate over tremendous distances. Recent studies of the lenses of the eyes of bowhead whales suggest they may be more than 200 years old! Until recently our primary response to all whales has been to slaughter them in countless numbers for meat and oil, with little thought for their extraordinary abilities and assets. More of this depressing history may be found in the discussion of marine resources in Chapter 17.

Order Carnivora

The order Carnivora (*carnis* = flesh, *vorare* = to devour) includes land predators ranging from dogs and cats to bears and weasels, but the members of the carnivoran suborder **Pinni-**

Deep Voices : BOX 15.2

The largest vertebrates make the loudest sounds. No one is sure just how whales produce the rumbles, clicks, whistles, and groans familiar to generations of sailors, but acoustical biologists are making progress.

The sound-making apparatus of toothed (odontocete) whales serves two purposes: communication and feeding. The squeals and whistles associated with communication appear to be made in the same way sounds are produced by the stretched neck of a deflating rubber balloon. These noises are usually accompanied by streams of bubbles from the blow-hole. The extremely loud clicks, sounds important in feeding, are generated by an as-yet-unknown mechanism. Researchers do know that oil-filled chambers within the head of a toothed whale focus and direct the sounds once they are produced. These sounds are so powerful that they can debilitate prey.

Like all whales, baleen (mysticete) whales lack vocal cords, but they do have a constriction between the back of their larynx and the tube that carries air to their lungs. One theory suggests that the tissues of this constriction vibrate at low frequencies as air is forced around or across it. A sac extending off the laryngeal cavity—a device that probably plays a role in pressure regulation during diving—may resonate with and amplify these vibrations and direct them outside. Baleen whales don't expel air from their blowholes during vocalization, so the air used in making the sounds must move back and forth within a closed interior system.

A competing theory holds that sounds are made by direct muscular action. Nerves activated by the brain could

a Humpback whale.

cause muscles to twitch at sonic frequencies, and the sound generated would move through the skin and into the water. Perhaps both methods are used; sounds are often heard to overlap antiphonally, sometimes producing unbroken melodies lasting a few minutes. No comparative analysis of the theories has been possible because of the difficulty of closely observing vocalizing baleen whales. Even obtaining fresh and undamaged whale specimens is difficult.

The patterns of communicating and socializing sounds of whales have been extensively studied. Humpbacks, a baleen species (**Figure a**), are particularly melodious. Each of the populations of humpbacks—one in each of the major ocean basins—sings a unique song. Within each population, all humpbacks share the same song, a complex sequence of roars, ticks, ascending and descending scales, and booms. Each song lasts from 10 to 30 minutes and then repeats.

Changes occur in the songs over time, but the population's shared song is still recognizable even after many years have passed. Significant changes in the song don't occur during the six silent months when the humpbacks are visiting their cold-water feeding grounds (**Figure b**), but they do occur during the six months of breeding activity when the songs are in continuous use.

What do these repetitive songs tell us about the life of a humpback whale? So far, very little. But the complex syntax of the songs—complete with themes and variations, rhymes, and inverted phrases— suggests a rich shared heritage and extensive interactions among individuals in each population. One wonders how the noises of human civilization, from submarine sonar and seismic profilers to ships' propellers and engine exhausts, influence their symphonic display.

15-37

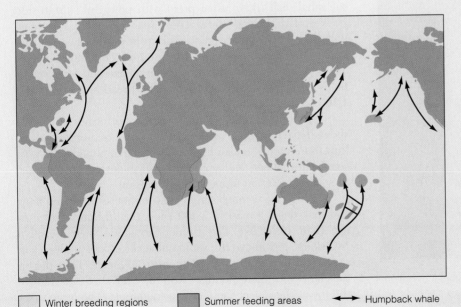

☐ Winter breeding regions   ■ Summer feeding areas   ↔ Humpback whale

b Migration routes of humpback whales showing winter breeding regions and summer feeding areas.

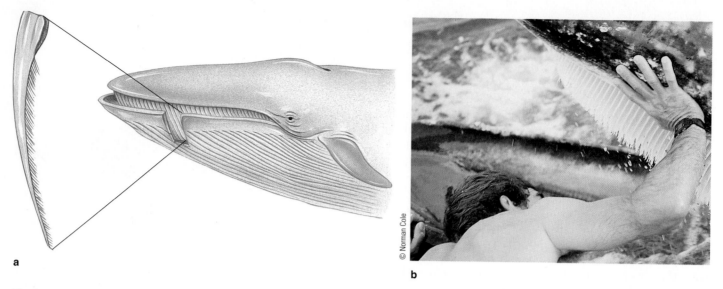

a

b

© Norman Cole

Figure 15.41 (**a**) A plate of baleen and its position in the jaw of a baleen whale. For clarity, the illustration shows an area of the mouth cut away. (**b**) A student and a gray whale, up close and personal. This gray whale uses its stiff, coarse baleen plates to sieve crustaceans from shallow bottom mud.

© Richard Sears/MICS-Photo

Figure 15.42 A blue whale surfaces from a dive with throat pleats distended with water and food. The whale will filter the water in its mouth through plates of baleen, leaving behind krill, which it swallows. For another view of a blue whale, see Figure 17.16.

pedia—the seals, sea lions, and walruses (see **Figure 15.43**)—are almost exclusively marine. Unlike the cetaceans, the gregarious pinnipeds (*pinna* = wing, *pedalis* = foot) leave the ocean for varying periods of time to mate and raise their young. They appear to have evolved from the same stock as modern bears.

True seals have a smooth head with no external ear flaps, the external part of the ear having been sacrificed to further streamline the body. They are covered with short, coarse hair without soft underfur. Seals are graceful swimmers that pursue small fish (their usual prey) with powerful side-to-side strokes of their hind limbs. These rear appendages are partially fused and always point back from the hind end of the body; thus, they are of very little use for locomotion on land. The elephant seal, named for its large size and long snout, holds the diving depth record for all air-breathing vertebrates: 1,560 meters (5,120 feet).

Sea lions, familiar to many as the performers in "seal" shows, have hind limbs with a greater range of motion and are thus more mobile on land. They have a streamlined head with small external ears and a pelt with soft underfur; unlike seals, they use their front flippers for propulsion.

Walruses are much larger than either seals or sea lions and may reach a mass of 1,800 kilograms (2 tons). Walruses use their tusks like sled runners to guide their sensitive "whiskers" just above the sediment surface, looking for clam siphons at depths of up to 91 meters (300 feet). They dig up the clams with their mouths, crush the clam shell and remove the meat, then eject the inedible fragments. The large tusks are also useful for hauling their heavy bodies onto ice floes.

The suborder **Fissipedia** (*fissus* = split, *pedalis* = foot) has many members (including cats, dogs, raccoons, and

Figure 15.43 Some representatives of the suborder Pinnipedia. (**a**) The California sea lion, the "seal" of seal shows. (**b**) Three juvenile harbor seals anxiously await lunch in a marine mammal care facility.

bears) and two truly marine representatives: sea otters and polar bears, relative newcomers to the marine environment.

Sea otters, active and pleasing creatures (**Figure 15.44**), are the smallest of marine mammals. Human demand for their fur, the densest and warmest of any animal, has caused their near-extermination. The modern population of the Pacific sea otter descends from a very few individuals accidentally overlooked by fur hunters of the late nineteenth and early twentieth centuries. Playful and intelligent, otters rarely exceed 120 centimeters (4 feet) in length; they eat voraciously, consuming up to 20% of their body mass in mollusks, crustaceans, and echinoderms each day. Sometimes they lie on their back in the water, balancing a rock on their chest. They hammer the shell of the prey against the rock until the shell cracks, then extract morsels of food with small, nimble fingers and roll over in the water to clean away the debris.

Figure 15.44 Two sea otters (*Enhydra lutris*) in a California kelp bed. Sea otters have the densest fur of any mammal.

Figure 15.45 A female polar bear with two cubs takes time out from seal hunting; the icy Arctic Ocean stretches beyond.

Polar bears (**Figure 15.45**) spend most of their lives stalking seals and stranded whales on the frozen northern polar ocean. They swim between ice floes with large, oarlike forepaws and can cross 100 kilometers (62 miles) of open water. The world's largest bears, male polar bears may grow to 2.5 meters (8.2 feet) and weigh 800 kilograms (1,800 pounds). They wander over enormous distances in search of food. One tagged bear moved 3,200 kilometers (2,000 miles) in a year, and bears have been seen in the vicinity of the pole. Apart from humans, polar bears have no enemies. Once an object of exploitation for their thick white fur, their numbers appear to be increasing under protection. As many as 40,000 are thought to exist today.

Figure 15.46 A manatee, or sea cow.

Order Sirenia

The bulky, lethargic, small-brained dugongs and manatees, collectively called sirenians (*siricis* = a mermaid) (**Figure 15.46**), are the only herbivorous marine mammals. Like the cetaceans, they appear to have evolved from the same ancestors as modern ungulates. They make their living grazing on sea grasses, marine algae, and estuarine plants in coastal temperate and tropical waters of North America, Asia, and Africa. Some species live in fresh water. The largest sirenians reach 4.5 meters (15 feet) in length and weigh 680 kilograms (1,500 pounds). They were first compared to mermaids by early Greeks who noted the female manatee's habit of resting in an upright position in the water and holding a suckling calf to her breast. Sirenians have been hunted extensively, and only about 10,000 individuals are now thought to exist worldwide. Even though they are protected now, many are killed or wounded each year in Florida by the propellers of powerboats.

The animals we have discussed in this chapter do not exist in isolation but interact with one another and with plants and other autotrophs in complex marine communities. In the next chapter we will turn our attention to these groupings of animals.

QUESTIONS FROM STUDENTS

1. **Representatives of the phylum Mollusca are both the largest and the most intelligent invertebrates. Yet mollusks are *not* considered the most successful nor the most highly evolved invertebrates. Why not?**

The most often used measure of success in biology is the *number* of species and individuals within a group. The number of arthropod species and individuals greatly exceeds the number of molluskan species and individuals—arthropods are more successful.

The question of evolutionary position—that is, which group is most highly evolved—is a matter of some controversy, but consider this automotive analogy. Modern cars possess a number of sophisticated technological features, such as electronic fuel injection, turbochargers, independent suspension, high-speed radial tires, antilock brakes, aerodynamic enhancements, nifty stereo systems, and so forth. All of these innovations *could* be fitted to a 1959 Cadillac, but the older car's primitive chassis design would limit the performance of the total package. Cephalopod mollusks are an *old* chassis design. Excellent eyes and relatively high intelligence have been added to the ancient chassis, but the physical limitations of musculature and the lack of a jointed skeleton do not allow the class to exploit the full potential of these "options." Though well represented in the fossil record, fewer than 500 species of cephalopods exist today. They may represent an evolutionary dead end.

2. If krill are the most successful *invertebrate*, what is the most widely distributed and successful *vertebrate*?

Probably small pelagic bony fishes of genus *Cyclothone*. These *bristlemouths* are found just below the euphotic zone in virtually all the world ocean away from shore. Bristlemouths feed on plankton and on organisms swimming into their visual field at the limit of light penetration. Some ichthyologists believe there are more living individuals of *Cyclothone microdon* than of any other single vertebrate species on Earth.

3. I've heard that sharks don't get cancer. Could we find out what prevents cancer in sharks and synthesize it for human use?

In fact sharks do develop cancer. The mistaken impression that shark cartilage may prevent or cure cancer in humans has contributed to sales of shark-derived supplements (derived mainly from hammerheads and spiny dogfish) worth more than $25 million in 1998. A large percentage of the 100 million sharks killed each year are taken for these ineffective medicines.

4. How do birds such as the arctic tern and wandering albatross navigate across the trackless ocean?

No one is certain. Experiments done at Cornell University with homing pigeons suggest that homing birds use a combination of magnetic and optical cues to return to their starting points. Even polarized light and the positions of certain stars might be involved. However it works, the behavior is not learned but is instinctive to the bird.

5. What's the difference between a dolphin fish and a dolphin mammal?

The name confusion began in Aristotle's time and has not abated. Dolphin mammals are small toothed whales; dolphin fish are teleosts (bony fish). The dolphin fish appears on restaurant menus and in seafood markets as mahi-mahi, dorado, and other names. Dolphin mammals are slaughtered in dismaying

numbers in association with tuna fishing by some non-U.S. fishing fleets, but none of their meat appears on U.S. plates.

6. OK, how is a porpoise different from a dolphin?

Dolphin and *porpoise* are the common names of two subtly different groups of odontocetes. *Porpoise,* as a term, refers to the group with smaller members, which have spade-shaped teeth, a triangular dorsal fin, and a smooth front end tapering to a point. *Dolphins* are usually larger and have an extended bottlelike jaw filled with sharp, round teeth. The small jumping whales in most oceanarium shows are dolphins, but killer whales are a species of large porpoise. To make matters even more complicated, the common dolphin seen in ocean-themed amusement parks, *Tursiops truncatus,* is often referred to as a porpoise, even by show announcers. This confusion between common names points out how useful scientific names can be: The real name of the animal, *Tursiops,* is clear and unambiguous.

7. If whales are so intelligent, why haven't they learned to avoid the whalers' catcher boats?

There are many theories, some of which are presented in an interesting and curious book called *Mind in the Waters* by Joan McIntyre (see bibliography). It may be that they are not as intelligent as was once thought. It may be that they have no conception of violence—a paradoxical thought considering the way the odontocetes make their living. These animals should perhaps be considered intelligent for their native environment but not be evaluated in relation to terrestrial organisms in a radically different environment with radically different problems.

8. If some seals and whales dive to great depths, why don't they get the bends?

The bends is a painful and occasionally fatal condition brought on by nitrogen gas—the most plentiful component of air—leaving solution and forming bubbles in the blood. The condition is analogous to what happens in a newly opened soda bottle. Human divers take a source of air with them to depth, breathe the air under pressure, and dissolve excess nitrogen gas in their blood. If they have been down long enough or deep enough, the release of pressure upon surfacing will be like taking the cap off the soda bottle. Whales don't use scuba tanks—they have no source of supplemental air at depth. They "tank up" at the surface by oxygenating their blood and tissues, but they have no excess gas to bubble from the blood at the end of their dives.

KEY CONCEPTS TO REVIEW

- Animals are active multicellular organisms incapable of synthesizing their own food. Animals must obtain food from primary producers—or from other animals that have consumed primary producers—to ensure their own survival.

- Animals arose near the end of the oxygen revolution—a time between about 2 billion and 400 million years ago during which the activity of photosynthetic autotrophs changed the composition of the atmosphere to its present oxygen-rich mixture.

- The invertebrates (animals without backbones) and vertebrates (animals with backbones) are grouped by similarities in external appearance and internal architecture into groups called phyla.

- All vertebrates are contained within the phylum Chordata. Vertebrates possess a bony spinal column as adults. The most abundant and diverse vertebrates are the fishes.

- Invertebrates are generally soft-bodied animals lacking a rigid internal skeleton, but many possess a hard outer protective covering. Almost every invertebrate phylum has marine representatives.

- Each marine phylum has developed a set of adaptations—characteristics that have evolved to allow animals to capture prey, avoid danger, maintain thermal and fluid balance with their surroundings, and compete for space.

endotherm
exoskeleton
Fissipedia
gas exchange
Gastropoda
gill membranes
hermatypic
Holothuroidea
hypotonic
invertebrate
lateral-line system
Mammalia
medusa
metamerism
Mollusca
molt
Mysticeti
Nematoda
notochord
Odontoceti
Ophiuroidea
osmoregulation

osmosis
Osteichthyes
oxygen revolution
phylum
Pinnipedia
Platyhelminthes
Polychaeta
polyp
Porifera
radial symmetry
Reptilia
salt glands
schooling
Sirenia
suspension feeder
swim bladder
Teleostei
tunicate
vertebrate
water-vascular system
zooxanthellae

INTERNET STUDY RESOURCES

The Web site for this book contains helpful study aids. Log on to:

www.brookscole.com/053437557xs

and click on the Chapter-by-Chapter area. Choose Chapter 15 and select a resource:

- **Flash Cards** allows you to test your mastery of the Terms and Concepts to Remember for this chapter.
- **Tutorial Quizzes** provides a multiple-choice practice quiz.
- **Student Guide to InfoTrac College Edition** will lead you to Critical Thinking Projects that use InfoTrac College Edition as a research tool.

TERMS AND CONCEPTS TO REMEMBER

Agnatha
animal
Annelida
Arthropoda
Asteroidea
Aves
baleen
bilateral symmetry
Bivalvia
Carnivora
cartilage
Cephalopoda
Cetacea

chitin
Chondrichthyes
Chordata
Cnidaria
cnidoblast
coral
coral reef
Crustacea
cryptic coloration
Echinodermata
Echinoidea
echolocation
ectotherm

STUDY QUESTIONS

Review Questions

1. What is an animal? How is an animal different from an autotroph? From a protist?

2. Which animal phylum is most successful? How is success defined? What structural advances contribute most to that phylum's immense success?

3. Which phylum contains the largest representatives? The most intelligent?

4. Which phyla exhibit radial symmetry? Bilateral symmetry? No symmetry?

5. What are the classes of living fishes? Which is considered most primitive? Most advanced? Which class has the largest individuals? Which is the most economically important?

6. How do fishes swim? What problems are associated with swimming? How do fishes overcome these problems? How do fishes make defensive use of color, shape, and schooling behavior?

7. How do fishes "breathe" underwater? Do marine fishes experience any unusual problems by being immersed in seawater?

8. How are seabirds different from land birds? Are there more species of seabirds than land birds?

9. What characteristics are shared by all marine mammals?

Critical Thinking Questions

1. When did the first true animals evolve? What atmospheric changes had to happen before animal life was possible? Are descendants of most of the early forms of animal life represented in the ocean today?

2. How can an arthropod grow within a "tailored" shell? How can an animal grow without getting bigger, or get bigger without growing?

3. Are all chordates vertebrates? Are all vertebrates chordates? What distinguishes a vertebrate?

4. There are seven living classes of vertebrates, but only six are marine. List the seven classes. Which class has no permanent marine representative? Why not?

5. How are odontocete (toothed) whales different from mysticete (baleen) whales? Which are the better known and studied?

6. **InfoTrac College Edition Project** The term *zooplankton* includes animals whose larval forms are small, such as crabs, squid, barnacles, and so on, as well as animals that remain small throughout their lives. Contrast the survival strategies of the two kinds of zooplankton: Would you expect the temporary, larval members of the zooplankton community to have different ways of surviving compared to the lifelong residents? Use InfoTrac College Edition to research this question.

RESOURCES FOR FURTHER READING AND RESEARCH

The Web site for this book contains many ideas for further reading and research. Log on to:

www.brookscole.com/053437557xs

and click on the Chapter-by-Chapter area. Choose Chapter 15 and select a resource:

- **References** lists the major books and articles consulted in writing this chapter, along with comments from the author about their content and reading level.
- **Hypercontents** takes you to an extensive list of sites with news, research, and images related to individual sections of the chapter.
- In **Student Guide to InfoTrac College Edition** scroll down to Suggested Readings from InfoTrac College Edition for brief descriptions of the articles listed and search hints for finding them in InfoTrac College Edition.
- **Regional InfoTrac College Edition** articles are organized into East Coast, West Coast, and Gulf Coast regions, allowing you to study oceanography on a more local level. You can also access Regional InfoTrac College Edition at www.localocean.com.

 For additional readings, go to InfoTrac College Edition, your online research library, at:

http://infotrac.thomsonlearning.com/

Marine Communities 16

THE RESOURCEFUL HERMIT

There is an enormous variety of organisms in marine communities, but let's spend a moment with one of the more entertaining ones in an intertidal community, the hermit crab. These small, pleasant relatives of edible crabs and lobsters have engaged the attention of generations of seaside visitors. The hermits rush around sand-swept rocks or the floors of tidal pools, withdrawing into their borrowed shells at the slightest sign of danger. Their activity appears random, but their fighting, snooping, hiding, probing, and scuffling are purposeful. Like all animals, hermit crabs must struggle to eat, avoid predators, and mate. Hundreds of structural and behavioral adaptations contribute to their success. Sensors on the hermit's antennae and mouth parts alert him to the presence of food; good eyesight, muscular coordination, and a tough form-fitting covering usually foil fast-moving predators; and brilliant blue bands around the tips of his legs may signal his availability for mating.

A hermit crab surveys his domain.

One unique behavioral adaptation shared by all hermit crabs involves the selection of a temporary home. The front parts of a hermit crab—mostly pincers, antennae, and mouth parts—are formidable, but its hindquarters are delicate and subject to attack. To protect its flank, a hermit searches for any enclosed portable object to climb into, usually an unoccupied snail shell. A hermit crab's borrowed or stolen shell seems a source of both inordinate pride and perpetual difficulty. No shell is ever completely satisfactory. A hermit crab will carefully inspect any substitute dwelling, occupied or not, and consider

414

whether to abandon its current digs in favor of the new candidate. House hopping proceeds fairly smoothly until the supply of suitably sized shells is exceeded by the number of potential occupants (as might happen when the crabs grow rapidly in times of abundant food). Then things get serious. Snails can be evicted from their self-made homes even before they're through with them—while the hermit gets a house *and* a meal in a single transaction. Two crabs may fight for hours or even days over one shell. Two others might simultaneously occupy the opposite ends of an abandoned worm tube, spending most of their day pulling each other in different directions. Sometimes two crabs swap shells at a moment's notice. Renters' remorse sets in almost immediately, and they're off to see if they can find something even more suitable. At times human observers can't resist laughing at the all-too-human goings-on.

WHAT TO WATCH FOR IN CHAPTER 16

The organisms you met in the last two chapters don't live alone. They are distributed throughout the marine environment in specific communities: groups of interacting producers, consumers, and decomposers that share a common living space. The types and variety of organisms found in a particular community depend on the physical and biological characteristics of that living space.

Any community is a dynamic place, growing and shrinking and changing its composition as residents respond to environmental fluctuations. As you'll see, the growth and distribution of organisms within communities depend on the often subtle interplay of physical and biological factors. The relative numbers of species and individuals in a community depend in part on whether its environment is relatively easy and free of stressors, or relatively hard and full of potential limiting factors. A few prominent marine communities are compared and contrasted in this chapter, and some representative adaptations of their residents are discussed. Compare and contrast conditions in these communities as you read. Why do organisms live where they do?

The chapter also discusses symbioses: intimate relationships of different species in ways that help or harm each other. Symbiotic relationships are very common in the ocean, and most forms of marine life are actively involved in them.

MARINE COMMUNITIES

A **community** is all the living things in a defined area. The organisms of a community will interact with and depend on one another, often in complex ways. Communities can be huge or surprisingly small. The largest marine community—and the most sparsely populated—lies within the uniform mass of permanently dark water between the sunlit surface and the deep bottom. Few animals live there because so little food is available, but those organisms that survive are among the strangest in the ocean. Opportunities for feeding in the deep open-ocean community are few and far between, so some animals are able to consume prey larger than themselves should the occasion arise. Because so few animals are present, mating is also a rare event—in a few species males and females become permanently bonded during their first encounter, the male burrowing into the female's body for a lifelong free ride.

In contrast, the smallest obvious marine communities may be those established against solitary rocks on an otherwise flat, featureless seabed. Drifting larvae will colonize the place; the established community can seem like an oasis of life and activity in an otherwise static sedimentary desert. Seaweeds will grow, worms will burrow, snails will scrape food from the hard surfaces, and small fishes will nestle among crevices. Hundreds of small plants and animals can live their lives within a meter of one another, interacting in a compact, solitary community with no similar environment available for thousands of meters. The larvae of the next generation drift away with little chance of finding a suitable place to carry on their lives. Microscopic communities also exist—an interacting set of populations can exist on a single grain of sand or on one decomposing fish scale.

ORGANISMS WITHIN COMMUNITIES

There are many different places to live and many different "jobs" for organisms within even a simple community. A **habitat** is an organism's "address" within its community, its physical *location*. Each habitat has a degree of environmental uniformity. An organism's **niche** (*nidus* = nest) is its "occupation" within that habitat, its relationship to food and enemies, an expression of what the organism is *doing*. For example, the small fishes living among the coral heads in a coral reef community share the same habitat, but each species has a slightly different niche. Each population in the community has a different "job" for which its shape, size, color, behavior, feeding habits, and other characteristics particularly suit it.

Biologists also survey the *range* of species in a community. The variety of species in a given area is an expression of **biodiversity** (biological diversity). Communities with high biodiversity—a coral reef for example—are characterized by complex interactions between and among residents.

The Influence of Physical and Biological Factors

A favorable balance of physical and biological factors is critical to each individual organism's success—and therefore to community success and longevity. The study of this balance and of the relationships of organisms and interactions within communities is called **ecology** (*oikos* = house, *logos* = study of). Marine ecologists are concerned with the types of organisms within marine communities, as well as their habitats, niches, distribution, numbers, and reactions to variations in their environment.

Physical and biological factors in the environment determine the location and composition of a community. As we saw in Chapter 13, physical factors such as temperature, pressure, and salinity affect the success of an organism: Seaweeds adapted to cold water usually cannot survive prolonged exposure to abnormally warm water, and lower-than-normal salinity can dangerously disrupt the fluid balance of most marine invertebrates. Biological factors are influences on an organism by members of its own population or other populations, in its own or other communities. Biological factors include crowding, predation, grazing, parasitism, shading from light, waste substances, and competition for limited oxygen.

A physical or biological factor that limits an organism's success in a community prevents that organism from feeding, growing, reproducing successfully, defending itself, sensing danger, or otherwise functioning successfully. As you may remember, some tropical fishes die in an unheated home aquarium because their physiology is optimized for a water temperature higher than the interior temperature of a typical house. If their natural tropical environment temporarily dropped to the temperature of a cool room, some of the fishes would become too sluggish to catch fast-moving food or avoid larger predators. Temperature would *limit* their success. Other fishes can adapt relatively easily to changes in temperature, however. Temperature would be a limiting factor to a sensitive tropical fish species, but not to a species with a wider temperature tolerance.

The prefixes *steno-* (meaning "narrow") and *eury-* (meaning "wide, broad") are sometimes used to describe those species, populations, or individuals that have narrow or wide tolerance to specific factors. **Stenothermal** species of tropical fish function better in a heated aquarium, but **eurythermal** fish do not need this extra attention. Likewise, **stenohaline** (*halos* = salt) marine organisms require a stable saline environment; they are unable to withstand relatively small fluctuations in salinity. **Euryhaline** species, however, can withstand a wide range of salinity and perhaps even tolerate some exposure to fresh water. Temperature and salinity are only two of the many physical factors that vary in a marine organism's native community. Light and pressure are examples of other such factors.

Figure 16.1 provides an idealized look at the tolerance of organisms to a varying physical factor such as temperature. A eurythermal organism's optimal range is wide (represented by a broad curve), but a stenothermal organism's optimal range will be narrow (which would be represented by a steeper, tighter curve). The same reasoning applies to the salinity tolerance of stenohaline or euryhaline species, or to the pressure tolerance of stenobaric (*baros* = pressure) or eurybaric ones.

An animal or plant is almost never exposed to fluctuations of only one physical or biological factor in its environment. An organism can die if subjected to changes in several environmental factors at once, even if each individual change is within its range of tolerance. Even small changes in temperature and salinity, survivable in themselves, might prove lethal if a particular species of tropical fish were simultaneously exposed to both.

Competition

The availability of resources such as food, light, and space in a community determines the number and composition of the populations of organisms within that community. Competition for the necessities of life may occur within the community between members of the *same* population or between members of *different* populations. Subtle swings in physical or biological factors may give one population the advantage for a time but then shift to favor another.

When members of the same population (all members of the same species) compete with each other, some individuals will be larger, stronger, or more adept at gathering food, avoiding enemies, or mating. These animals tend to prosper, forcing their less successful relatives to emigrate, fight, or die in the course of competition. The most successful organisms have the most surviving offspring, so useful inheritable variations are passed along in greater quantity to the next generation. As we saw in our discussion of evolution, this kind of competition continually fine-tunes a population to its environment.

When members of different populations compete, one population may be so successful in its "job" that it eliminates all competing populations. In a stable community, two populations cannot occupy the same niche for long. Eventually the more effective competitor overwhelms the less effective one. Extinction from this kind of head-to-head competition is probably uncommon, but restriction of a population because of competition between species is not. For example, the little barnacle *Chthamalus* lives on the uppermost rocks in many intertidal communities, while the larger barnacle *Balanus* lives on the lower rocks (see **Figure 16.2**). Planktonic larvae of both species can attach themselves to rocks anywhere in the intertidal zone and begin to grow. In the lower zone the more rapidly growing *Balanus* pushes the weaker *Chthamalus* off the rocks, while at higher positions *Balanus* cannot survive because it is less resistant to drying and exposure than the tough little *Chthamalus*. At the top and bottom of their distribution the two species do not compete for food or space. The competition at the intersection of their ranges prevents each species from occupying as much of the habitat as might otherwise be possible.

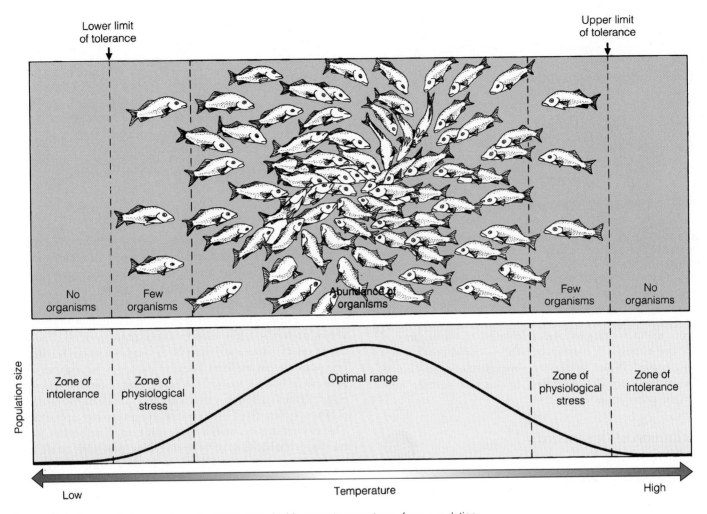

Figure 16.1 Range of tolerance to a physical factor—in this case, temperature—for a population of organisms.

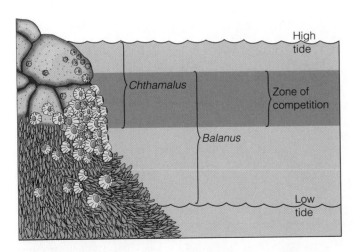

Figure 16.2 Competition between two species of barnacles prevents either from occupying as much of the intertidal zone as might otherwise be possible. The central zone of overlap indicates the area where *Chthamalus* (the smaller barnacle) and *Balanus* (the larger barnacle) compete for food and space. (Source: From Sylvia S. Mader, *Biology: Evolution, Diversity, and the Environment,* 2/e. © 1987 Wm. C. Brown Communications, Inc. Dubuque, Iowa.)

Growth Rate and Carrying Capacity in Communities

Organisms newly introduced into a favorable environment with no competitors for food or space will reproduce exponentially, tracing a J-shaped population growth curve (**Figure 16.3**). In nature very few populations reproduce at this maximal rate, however, because environmental conditions are rarely ideal and limiting factors in the environment quickly slow the rate of population growth. The sum of the effects of these limiting factors in the environment is called **environmental resistance**. Environmental resistance causes the actual population growth curve to be lower than the maximum potential growth curve. Figure 16.3 shows the growth rate in number of individuals over time for both potential and actual situations. Note that the curve resulting when limiting factors intrude is S-shaped; it gradually flattens toward an upper limit of the number of individuals in the population.

The final number of organisms oscillates around the **carrying capacity** of the environment for that species. The carrying capacity is the population size of each species that a community can support indefinitely under a stable set of

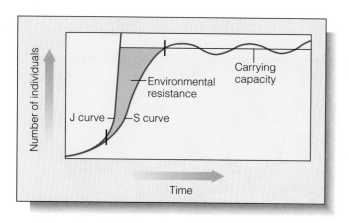

Figure 16.3 The J-shaped curve of population growth of a species is converted to an S-shaped curve when the population encounters environmental resistance. The physical or biological conditions responsible for the cessation of growth are called *limiting factors*.

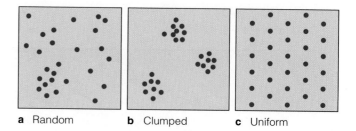

a Random   b Clumped   c Uniform

Figure 16.4 Random, clumped, and uniform population distribution patterns. The clumped pattern is most common in nature.

environmental conditions, and it changes if environmental conditions change. A marine population could *crash* if upwelling ceased, if new predators were introduced, if climate varied, if food supplies dwindled, or if a new parasite infiltrated the population.

Distribution of Organisms in a Community

As we have seen, physical and biological factors affect the number and positions of organisms in a community. The number of individuals per unit area (or volume) is known as the **population density.** Rare individuals have a much lower population density than dominant ones. In general, more different species exist in benign habitats where physical factors stay near optimal values (like a coral reef or rain forest), and fewer species exist in rigorous habitats where physical factors range to extremes (like a beach or a desert). That is, "easy" habitats typically have high biodiversity—they contain more species in more niches within a given area—and harsh habitats usually have a lower species diversity. Relatively few species can cope with the stressful environment of the polar ocean surface, for example, but many species have adapted to the relatively benevolent environment of a tropical reef.

Individual organisms are almost never distributed randomly throughout their habitat. A **random distribution** implies that the position of one organism in a community *in no way influences* the position of other organisms in the same community. Further, a truly random distribution (such as that shown in **Figure 16.4a**) indicates that conditions are precisely the same throughout the habitat, an extremely unlikely situation except possibly in the unvarying benthic communities of abyssal plains.

The most common pattern for distribution of marine organisms is small, patchy aggregations, or clumps. **Clumped distribution** (see **Figure 16.4b**) occurs when conditions for

growth are optimal in small areas because of physical protection (in cracks in an intertidal rock), nutrient concentration (near a dead body lying on the bottom), initial dispersal (near the position of a parent), or social interaction.

Uniform distribution, with equal space between individuals (**Figure 16.4c**), such as the arrangement of trees we see in orchards, is the rarest natural pattern of all. The distribution of some garden eels throughout their territories becomes almost uniform because each eel can extend from its burrow just far enough to hassle neighbor eels spaced at equal distance. There eventually is a break in the order of position—they don't line up row upon row like apple trees. The closest approximation to a uniform distribution in nature is probably that of breeding king penguins in an area bounded by a straight cliff. The birds nesting nearest the cliff arrange raised hillocks of stones one pecking-distance apart along the cliff face, the second row of birds is interspersed between birds in the first, and so on. For a few rows away from the cliff the pattern sometimes resembles a checkerboard (**Figure 16.5**).

Figure 16.5 King penguins incubating their eggs on South Georgia Island, a rare example of near-uniform population distribution.

CHANGE IN MARINE COMMUNITIES

Like the organisms that constitute them, communities change through time, but marine communities do not evolve as rapidly as terrestrial communities. The slow changes associated with seafloor spreading, climate cycles, atmospheric composition, or newly evolved species have shaped this generally slow evolution. As on land, the species, community composition, and location of a marine community are changed by the environmental factors to which members of the community are exposed. Communities themselves can gradually modify the physical aspects of their environment: A coral reef is an extreme example. The massive accumulation of coral and sediments within the reef can alter current patterns, influence ocean temperature, and change the proportions of dissolved gases.

But rapid changes can occur in marine communities. A natural catastrophe—a volcano erupting, a landslide that blocks a river, or the collision of an asteroid with Earth, for example—can disrupt a community. Similarly, human activities such as altering an estuary by damming a river, dumping excess nutrients into a nearshore area, or stressing organisms with toxic wastes can cause rapid, disruptive changes. Offshore communities change abruptly near new sewage outfalls, for example.

A stable, long-established community is known as a **climax community.** This self-perpetuating aggregation of species tends not to change unless disrupted by severe external forces such as violent storms, significant changes in current patterns, epidemic diseases, or an influx of great amounts of fresh water or pollutants. A disrupted climax community can be reestablished through the process of **succession,** the orderly changes of a community's species composition from temporary inhabitants to long-term inhabitants. Disruption makes the environment more hostile to the original species, but destruction of species in the original community leaves open habitats and niches. A few highly tolerant species will move into the area, eventually drawing in other species that depend on them. If the environment is permanently changed by the disruption, a climax community different from the previous one will be established.

EXAMPLES OF MARINE COMMUNITIES

There are many distinct marine communities. We briefly discussed the plankton community in Chapter 14, and here we survey some other important examples.

Rocky Intertidal Communities

Anyone who spends time at the shore, especially a rocky shore, is soon struck by a curious contradiction. Although the rocky shore looks like a very difficult place for organisms to make a living, the **intertidal zone**—the band between the highest high-tide and lowest low-tide marks—is one of Earth's most densely populated areas. Hundreds of species and individuals crowd this junction of land and sea.

The problems of living in the intertidal zone are formidable. The tide rises and falls, alternately drenching and drying out the animals and plants. **Wave shock,** the powerful force of crashing waves, tears at the structures and underpinnings of the residents. Temperature can change rapidly as cold water hits warm shells or as the sun shines directly on exposed organisms. In high latitudes ice grinds against the shoreline, and in the tropics intense sunlight bakes the rocks. Predators and grazers from the ocean visit the area at high tide, and those from land have access at low tide. Freshwater runoff can osmotically shock the occupants during storms. Annual movement of sediment onshore and offshore can cover and uncover habitats. Yet, astonishingly, the richness, productivity, and diversity of the rocky intertidal community—especially in the world's temperate zones—are matched by very few other places. There is intense competition for space. Life abounds.

One reason for the great diversity and success of organisms in the rocky intertidal zone is the large quantity of food available. The junction between land and ocean is a natural sink for living and once-living material. Minerals dissolved in water running off the land serve as nutrients for the inhabitants of the intertidal zone as well as for plankton in the area. The crashing surf and strong tidal currents keep nutrients stirred and ensure a high concentration of dissolved gases to support a rich population of autotrophs. Many of the larval forms and adult organisms of the intertidal community depend on plankton as their primary food source.

Another reason for the success of organisms here is the large number of habitats and niches available for occupation (see **Figure 16.6**). The habitats of intertidal animals and plants vary from hot, high, salty splash pools to cool, dark crevices. These spaces provide hiding places, quiet places to rest, attachment sites, jumping-off spots, cracks from which to peer to obtain a surprise meal, footing from which to launch a sneak attack, secluded mating nooks, or darkness to shield a retreat. The niches of the creatures in this community are varied and numerous—encrusting algae produce carbohydrates, snails scrape algae from rocks, hermit crabs scavenge for tidbits, and octopuses wait in ambush for likely meals.

The most obvious and important physical factor in intertidal communities is the rise and fall of the tides. Organisms living between the high-tide and low-tide marks experience very different conditions than do those residing below the low-tide line. Within the intertidal zone itself, organisms are exposed to varying amounts of emergence and submergence. For example, **Figure 16.7a** plots the number of hours of exposure to air in a California intertidal zone through six months' time versus the tidal height. Because some organisms can tolerate many hours of exposure while others are able to tolerate only a very few hours per week or month, the animals and plants sort themselves into three or more horizontal bands, or subzones, within the intertidal zone. Each distinct

Figure 16.6 A Pacific coast tide pool and intertidal shore. (**a**) A diagrammatic view. (**b**) Key.

a

b

1 bushy red algae, *Endocladia*
2 sea lettuce, green algae, *Ulva*
3 rockweed, brown algae, *Fucus*
4 iridescent red algae, *Iridea*
5 encrusting green algae, *Codium*
6 bladderlike red algae, *Halosaccion*
7 kelp, brown algae, *Laminaria*
8 Western gull, *Larus*
9 intrepid marine biologist, *Homo*
10 California mussels, *Mytilus*
11 acorn barnacles, *Balanus*
12 red barnacles, *Tetraclita*
13 goose barnacles, *Pollicipes*
14 fixed snails, *Aletes*
15 periwinkles, *Littorina*
16 black turban snails, *Tegula*
17 lined chiton, *Tonicella*
18 shield limpets, *Collisella pelta*
19 ribbed limpet, *Collisella scabra*
20 volcano shell limpet, *Fissurella*
21 black abalone, *Haliotis*
22 nudibranch, *Diaulula*
23 solitary coral, *Balanophyllia*
24 giant green anemones, *Anthopleura*
25 coralline algae, *Corallina*
26 red encrusting sponges, *Plocamia*
27 brittle star, *Amphiodia*
28 common sea star, *Pisaster*
29 purple sea urchins, *Strongylocentrotus*
30 purple shore crab, *Hemigrapsus*
31 isopod or pil bug, *Ligia*
32 transparent shrimp, *Spirontocaris*
33 hermit crab, *Pagurus,* in turban snail shell
34 tide pool sculpin, *Clinocottus*

Figure 16.7 The relationship between amount of exposure and vertical zonation in a rocky intertidal community. (**a**) A graph showing intertidal height versus hours of exposure. The 0.0 point on the graph, the tidal datum, is the height of mean lower low water. (**b**) Vertical zonation, showing four distinct zones. The uppermost zone (I) is darkened by lichens and cyanobacteria; the middle zone (II) is dominated by a dark band of the red alga *Endocladia;* the low zone (III) contains mussels and gooseneck barnacles; and the bottom zone (IV) is home to sea stars (*Pisaster*) and anemones (*Anthopleura*). The bands in the photograph correspond approximately to the heights shown in the graph. (Source for Figure 16.7a: From *Between Pacific Tides,* 5/e by Edward F. Ricketts, et al. Revised by David W. Phillips. Stanford Univeristy Press. © 1985 by the Board of Trustees of the Leland Stanford Junior University. Reprinted with permission.)

Many people know that John Steinbeck was awarded the 1962 Nobel prize for literature for his 1939 novel *The Grapes of Wrath*. Few know that this famous writer was an avocational marine biologist with a deep interest in marine intertidal communities. He was introduced to this rich habitat by Ed Ricketts, a real person who became the fictional character "Doc" in Steinbeck's popular novels *Cannery Row*, *Tortilla Flat*, and *Sweet Thursday*. These novels were set in Monterey, then an important northern California fishing town. Ed Ricketts ran a small commercial biological supply business on Cannery Row. When Steinbeck and Ricketts first met (in a dentist's office), each had heard of the other: Steinbeck knew that Ricketts was a curious character interested in invertebrate zoology and classical music, and Ricketts knew that Steinbeck was a promising newspaper reporter and budding author. They became fast friends and went on many collecting expeditions together. They even wrote a book, *The Sea of Cortez*, about their trip to collect marine invertebrates on the shore of the Gulf of California; it combines travel adventures, humor, vivid scientific description, and personal philosophy.

Ed Ricketts was ahead of his time as a biologist. His approach to intertidal life was community based—what we would today call an ecological emphasis. He attempted to combine all his experiences and observations in an environment into an integrated picture of the whole. Today's biologists can appreciate Ricketts's urge to bring together the physical and biological factors affecting each species within a community—to decipher, as he once wrote, "the Zen of a segment of shore."

In 1939, Ricketts published a landmark book, *Between Pacific Tides*. Unlike previous guides to seashore life, *Between Pacific Tides* was organized by community and not by organism type. " The treatment," he wrote in the preface to the first edition, "is ecological and inductive; that is, the animals are treated according to their most characteristic habitat, and in the order of their commonness, conspicuousness, and interest." The graceful writing and accurate observations quickly made it a classic, a book that has deeply influenced generations of marine scientists (whose labs are, not surprisingly, often filled with classical music). *Between Pacific Tides* is now in its fifth edition.

© Ed Ricketts, Jr.

Ed Ricketts at the Great Tide Pool, Pacific Grove, California.

Ed Ricketts died in an automobile accident in May 1948. Steinbeck wrote a foreword to the second edition, which Ricketts was preparing. The foreword concludes: "There are good things to see in the tide pools and there are exciting and interesting thoughts to be generated from the seeing. Every new eye applied to the peep hole which looks out at the world may fish in some new beauty and some new pattern, and the world of the human mind must be enriched by such fishing."

16-11

zone is an aggregation of animals and plants best adapted to the conditions within that particular narrow habitat. The zones are often strikingly different in appearance, even to a person unfamiliar with shoreline characteristics. This zonation is clearly evident in the rocky shore of **Figure 16.7b.**

For intertidal areas exposed to the open sea, wave shock is a challenging physical factor. Fist-size rocks have been thrown 100 meters (330 feet) into the air by the force of breaking waves. Large intertidal plants must be immensely strong, elastic, and slippery to avoid being shredded by wave energy. **Motile** (*motus* = moving) animals move to protective overhangs and crevices, where they cower during intense wave activity. Attached, or **sessile** (*sessilis* = sitting), animals hang on tightly, often gaining assistance from rounded or low-profile shells, which deflect the violent forces of rushing water around their bodies. Some sessile animals have a flexible foot that wedges into small cracks to provide a good hold; others, like mussels, form shock-absorbing cables that attach to something solid.

Desiccation (drying) by exposure to air and sunlight is another source of intertidal stress. Again, motile organisms have an advantage because they can move toward water left in tidal pools or muddy depressions by the retreating ocean. Attached animals and plants must await the water's return, huddled in low spots, moist pockets, or cracks in the rocks, or within tightly closed shells. Water trapped within a shell can keep gills moist for the needed exchange of gases. A protective mucous coating can retard evaporative water loss from exposed soft animal body parts or blades of seaweed. When the weather is too warm—or when the tide is out for an unusually long time—a deceptive calm settles over the zone, to be relieved only by the returning ocean water.

Seaweed Communities

The shelter and high productivity of a kelp forest can help provide a near-ideal environment for animals. When light and nutrient conditions are optimal, large algae can make carbohydrate molecules so rapidly and in such quantity that sugars leak from their tissues like tea from a tea bag. Resident animals such as sea urchins (**Figure 16.8**) are able to grow rapidly by collecting these molecules on their surfaces and transporting them directly into their bodies. As the algae weaken with age and productivity declines, the urchins' sharp teeth can gnaw at the thalluses. Other animals graze on the blades, nestle within the holdfasts, and consume kelp flakes and debris. Sea otters may eventually move in to eat the urchins. As with all other communities, the kelp forest changes as its inhabitants come and go and as time passes.

Sand Beach and Cobble Beach Communities

Some intertidal areas are sandy, some are muddy, and others consist of gravel or cobbles. (A few shores combine these elements within a small area.) The usual rigors of the intertidal zone are intensified for organisms surviving on loose sub-

© Bruce Hall

Figure 16.8 Sea urchins in a kelp bed. The urchins can absorb carbohydrates that leak from the algae, or gnaw the stipes and holdfasts with teeth. Too many urchins can destroy a kelp forest by releasing the kelp from its holdfasts.

strates. Indeed, it may surprise you to learn that in spite of its generally benign conditions the ocean contains what may well be the most hostile, rigorous, and dangerous environments for small living things on Earth: high-energy sand and cobble beaches.

As environments go, sand beaches don't seem particularly nasty places to us humans; many people consider the beach to be about the finest habitat around. Seals and sea lions spend a lot of time at the beach and seem to enjoy the experience as much as people do. In short, for organisms of about our size, the problems of living on a beach are manageable.

For smaller organisms, however, a beach is a forbidding place. Sand itself is the key problem. Many sand grains have sharp, pointed edges, so rushing water turns the beach surface into a blizzard of abrasive particles. Jagged grit works its way into soft tissues and wears away protective shells. A small organism's only real protection is to burrow below the surface, but burrowing is difficult without a firm footing. When the grain size of the beach is small, capillary forces can pin down small animals and prevent them from moving at all. If these organisms are trapped near the sand surface, they may be exposed to predation, to overheating or freezing, to osmotic shock from rain, or to crushing as heavy animals walk or slide on the beach.

As if this weren't enough, those that survive must contend with the difficulty of separating food from swirling sand and the dangers of leaving telltale signs of their position for predators or being excavated by crashing waves. A few can run for their lives—some larger beach-dwelling crabs depend on their good eyesight and sprinting ability to outrace onrushing waves!

To these horrors must be added the usual problems of intertidal life discussed earlier. Not surprisingly, very few species have adapted to wave-swept sandy beaches! Some of the successful ones are shown in **Figure 16.9.** The few that have done so—mostly small, fast-burrowing clams, sand crabs, sturdy polychaetes, and other minute worms—consume a rich harvest of plankton and organic particles washed onto the beach and filtered from the water by the uppermost layer of sand.

Cobble beaches are even more uninviting. The rounded rocks clack and bump together as waves pound the shore; most small animals are crushed. Except for nimble, insectlike "beach hoppers" and a few species of scavenging terrestrial insects, most loose rock-strewn shores are understandably void of anything much larger than microscopic organisms.

Surely the most difficult environment of all are the black sand beaches, derived from pulverized lava on tropical volcanic islands such as Hawaii. Besides all the other difficulties mentioned, we must add the lava's ability to store solar heat until temperatures approach 71°C (160°F) just below the surface of the sand. Almost nothing can tolerate these beaches for more than just a few minutes—including human feet!

Salt Marshes and Estuaries

Muddy-bottomed salt marshes are among the most interesting intertidal shores. Much of the high primary productivity of a salt marsh comes from sea grasses, mangroves, and other vascular plants that can prosper in a marine (or partly marine) environment.

As you may recall from Chapter 12, salt marshes often form in an **estuary,** a broad, shallow river mouth where fresh water and salt water mix (see Figures 12.29–12.31). Wave shock is usually reduced in estuaries—surf is blocked from estuaries by longshore bars or by twisting passages connecting to the ocean. The salinity of water within an estuary may vary with tidal fluctuations from seawater through **brackish** water (a mixture of salt water and fresh water) to fresh water. Many of the organisms living in estuaries are necessarily euryhaline. In areas near the river entrance, however, the water may be

a

b

Figure 16.9 Sand beach organisms. (**a**) Dime-size coquina clams (*Donax*) lie at the surface awaiting a ride up the beach on an incoming wave. They will bury themselves in the loose sediment, push up their siphons, and filter the water for food. When the tide retreats, they will again pop to the surface and allow the waves to take them back down the beach. (**b**) A sand crab (*Emerita*), beloved of all beach-going children and beginning lab students, attempts to bury itself in anticipation of an onrushing wave. It gleans food from passing water with its feathery antennae. (**c**) and (**d**) are examples of interstitial animals. These organisms are tiny enough to live in the spaces (or interstices) between sand grains, too small to be seen by the unaided eye. (**c**) A crustacean. (**d**) A tardigrade.

d

c

Figure 16.10 An estuarine marsh. Urban developers often destroy coastal marshes to build marinas or homes, but citizens near this marsh in Orange County, California, have recognized its natural value and have set it aside as a marine preserve.

almost fresh, whereas near the outlet it may be of oceanic salinity. These different salinities often lead to a distinct horizontal zonation of organisms. Temperature range is also potentially extreme, especially in the tropics or during the temperate zone summer when a receding tide abandons residents to the heat of the sun. Strong currents may move in estuaries as the tide rises and falls and the river flows. Flowing water serves the same purpose as waves in mixing nutrients and gases in the intertidal estuary community.

Estuarine marshes (such as the one shown in **Figure 16.10**) are richer and exhibit greater species diversity than marshes exposed only to seawater. Primary productivity in estuaries is often extraordinarily high because of the availability of nutrients, the great variety of organisms present, strong sunlight, and the large number of niches. Decomposition of fast-growing, salt-tolerant plants provides the raw material for the large and complex food webs and rapid nutrient turnover characteristic of these communities. The standing biomass (mass of living matter per unit area or volume) in a typical estuary is among the highest per unit of surface area of any marine community.

Estuarine organisms show unique adaptations to their rich and variable environment. Some estuarine plants trap fine silt particles at their roots, thus countering the erosive action of current flow. Small plants are often filamentous, bristling with tiny projections anchoring them to the substratum. Larger plants have extensive root systems to hold themselves in place and to colonize new areas. Most of the resident animals burrow into the muck, scurry rapidly across the surface, or hide in the vegetation. Clams and snails work their way through the substratum, obtaining food and shelter at the same time (**Figure 16.11**). Polychaete worms dig for targets of opportunity, and crabs dart after any interesting morsels. Since planktonic larvae would be washed out to sea, most es-

tuarine organisms produce nonplanktonic larvae, lay eggs on firm objects, or carry eggs on their bodies.

Estuaries are sometimes called marine nurseries because so many juvenile organisms are found there. This is especially true for fish. Many pelagic species spend their larval lives in the protective confines of an estuary taking advantage of the many feeding opportunities available. Most of the commercially exploited fish species on the North American Atlantic coast utilize estuaries as juvenile feeding grounds. The human interference of development and pollution is thus doubly stressful in estuaries, affecting both permanent residents and the sensitive larval stages of open-water animals.

You may also recall that estuaries are not permanent features—they are very sensitive to changes in sea level. Most Atlantic coast estuaries probably formed during the sea-level

Figure 16.11 Snails (*Cerithidea*) in an estuary search the surface mud for food.

a

Figure 16.12 (**a**) The coral reef habitat. (**b**) Key. (**c**) A coral reef in the Red Sea.

b

c

© Peter Scoones/Planet Earth Pictures

rise of the last 3,000 to 10,000 years. Estuaries are probably more common today than they were at the height of the last glaciation, when sea level was lower.

Coral Reef Communities

16-15

Tropical coral reefs typically form in areas of high wave energy; indeed, reef organisms preferentially build into high-energy environments in an attempt to be first in obtaining dissolved and suspended material in the water. In most reefs there is an approximate balance between construction and destruction. The reef consists of actively growing coral colonies and fragments of material of different sizes from coral boulders worn down to fine sand.

Corals are by no means the only participants in reef life, however; they may account for only about half of the biomass in these areas. Other reef residents include calcareous algae whose secretions help "cement" the reef together as well as a bewildering array of encrusting, burrowing, producing, and consuming creatures ranging upward in size from the microscopic. Some tunnel into the coral or shatter it in search of food, contributing to the erosion of the reef. Fierce competition exists among reef organisms for food, living space, protection from predators, and mates. The bright colors, protective camouflage, spines, and various toxins and venoms common to tropical organisms are probably related to the intense struggle for existence that goes on in these beautiful but deceptively calm-looking places. A typical reef scene is depicted in **Figure 16.12.** More than 1 million species are thought to inhabit the ocean's coral reef ecosystems!

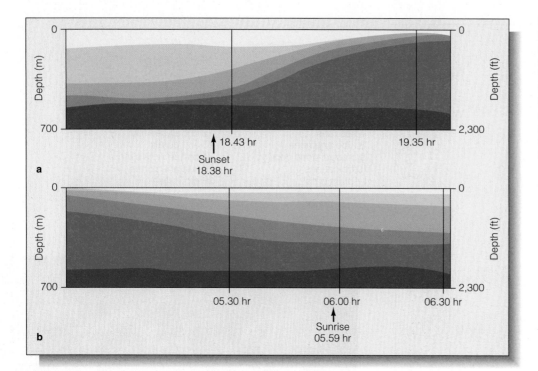

Figure 16.13 The movements of deep scattering layers, as recorded by an echo sounder. (**a**) Three distinct layers move toward the surface at sunset. (**b**) Before sunrise the layers move down again. This phenomenon is caused by organisms that migrate up and down with changing amounts of sunlight. (In both traces another scattering layer remains at a constant depth.)

The Open Ocean

16-16

About 83% of the ocean's total biomass is concentrated in its uppermost 200 meters (660 feet). Here we find most of the fishes and plankton. Of the ocean's biomass, only 0.8% is found below 3,000 meters (10,000 feet). Nearly all deep-ocean habitats are sparsely populated, but the few species of animals that have adapted to this impoverished place range through virtually all oceanic latitudes. There are no photosynthetic autotrophs in the deep ocean because there is no light. With the exception of rift communities (discussed more in a moment), consumers at great depths must depend on the productivity of the water column above.

A peculiar pelagic community lives at the uppermost limits of the permanent darkness. Named after its ability to reflect sound pulses and appear to echo sounders as a false bottom,

the **deep scattering layer** (**DSL**) is a relatively dense aggregate of fishes, squid, and other animals that usually migrate up and down in synchrony with daylight (**Figure 16.13**). Deep scattering layers—there is often more than one—are found in all ocean areas except the Arctic, and they are best developed in places of high surface productivity. The DSL is most pronounced during daylight hours, when members of the community congregate at the lowest limit of light penetration. At nightfall many of the organisms migrate to the surface to feed on plankton. Most residents of the DSL have large, sensitive eyes, which permit them to feed by detecting the faint shadows of prey above. Some members of the community have built-in luminescent organs that cast dim blue light downward; this light masks their own shadows, so they have less chance of being detected and eaten (see **Figure 16.14**).

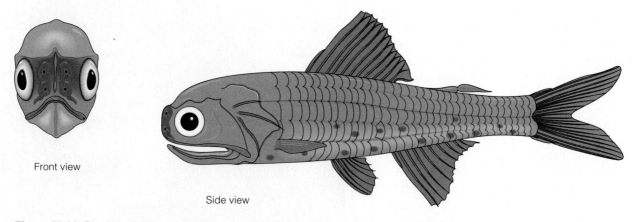

Front view

Side view

Figure 16.14 *Diaphus elucens*, a common mesopelagic fish with large light organs (here shaded blue) between the eyes and on the body. The light organs mask the fish's shadow and may identify it to potential mates. The fish is 5.5 centimeters (2 inches) long.

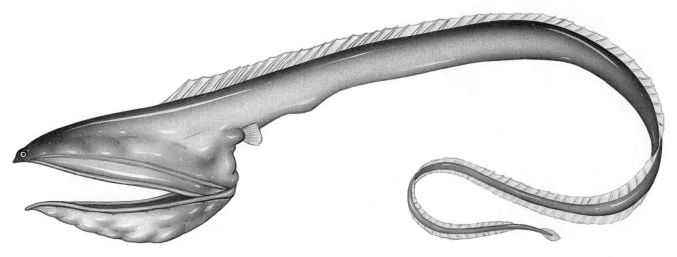

Figure 16.15 The deep-sea gulper (*Eurypharynx pelecanoides*), a bathypelagic species with a worldwide distribution beneath tropical waters. Its length is about 60 centimeters (24 inches). (Source: From J. C. Briggs, *Marine Zoogeography.* © 1974 by McGraw-Hill, Inc.)

Between the DSL and the bottom, in the bathypelagic zone, the ocean is nearly devoid of life. Rarely is food available in this zone; tiny crumbs of organic material are usually broken down by microbes before reaching mid-depths, and the bodies of large organisms continue their fall to the seafloor. But recent research has shown that the deep-water environment is "patchy"—nutrient-rich zones caused by sinking remnants of phytoplankton blooms or the falling excretions of large fish populations can temporarily enrich a water parcel to the benefit of deep residents.

The few animals in this vast middle volume of ocean below the DSL are among Earth's most bizarre. Gulper eels (**Figure 16.15**) have extendable jaws and a stomach capable of consuming prey larger than the eels themselves, an adaptation of great importance when one considers that a gulper eel may not encounter a feeding opportunity more often than once or twice a year! Bioluminescence, the biological production of light, is important here in both feeding and mate attraction. Some deep-swimming organisms attract their infrequent meals with a luminous lure (**Figure 16.16**). These animals also use patterns of glowing spots or lines to identify themselves to members of the same species, a necessary first step in mating. Some use flashes of light to dazzle or frighten potential predators.

The Deep-Sea Floor

Most of the deep-ocean floor is an area of endless sameness. It is eternally dark, almost always very cold, slightly hypersaline (to 36‰), and highly pressurized. Scientists once thought that such rigors would limit the extent of communities there. Not so. In the 1980s researchers investigating bottoms at depths between 1,500 and 2,500 meters (5,000 to 8,000 feet) found an average of nearly 4,500 organisms per square meter. There were 798 species recorded in twenty-one 1m² samples, and 46 of these species were new to science!

The feeding strategies of animals living on the deep-ocean floor are unique to this harsh environment. Tripod fish (**Figure 16.17**) are blind and use sensitive extensions of their fins and gill coverings to detect the movement of prey many meters away. Some organisms whose mouths blend with the natural contours of the ooze act as living caves into which small creatures crawl for protection. The predator need not even swallow to get the prey into its gut—back-pointing spines direct the victim along a one-way path to the stomach! Other species are capable of smelling large, sunken dead animals for many kilometers downcurrent, then spending weeks or months slowly following the scent to its source. The metabolic rate of organisms in cold water tends to be low, so most deep animals require relatively little food, move slowly, and

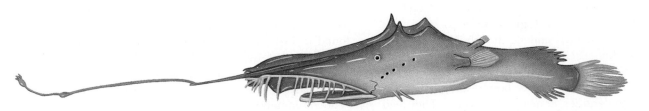

Figure 16.16 Some species of deep-sea anglerfishes have bioluminescent lures. *Lasiognathus saccostoma,* known only from the western Atlantic, has fishing equipment that consists of an extensible rod, filament, "float," and illuminated lure with little hooks. Its body is about 20 centimeters (8 inches) long. (Source: From J. C. Briggs, *Marine Zoogeography.* © 1974 by McGraw-Hill, Inc.)

Figure 16.17 A blind tripod fish, an abyssal benthic species. The long, curved projections are thought to aid in sensing the distant vibrations of prospective prey.

Charles D. Hollister, WHOI

live very long lives. Some may feed less than once in a year and may live to be hundreds of years old. Common deep benthic representatives are seen in **Figure 16.18.**

This deep uniform environment, though rigorous, holds benefits for those few species adapted to it. Perhaps the best-adapted large organisms are the ubiquitous ophiuroids (brittle stars) that inhabit sedimentary bottoms, sometimes at arm-tip to arm-tip densities, at nearly all latitudes. They feed by collecting tiny nutritive particles—some of which will have fallen through the water for many months—along grooves beneath their arms and transporting them to their mouths. As you may recall from Chapter 15, brittle stars are among the most widely distributed of Earth's animals (see Figure 15.19).

The organisms within deep pelagic and benthic communities share some curious adaptations. Gigantism is a common characteristic—individuals of representative families tend to be much larger in deep water than related individuals in the shallow ocean. Fragility is also common in the depths. Not only are heavy support structures unnecessary in the calm deep environment, but the relatively low water pH and high pressure discourage the deposition of calcium—and thus skeletal development. Some animals have slender legs or stalks to raise themselves above the sediment, and some come apart like warm gelatin at the slightest touch. Except for its influence on enzyme activity, hydrostatic pressure is not a problem for these animals. They live in balance with the great pressure; their internal pressure is precisely the same as that outside their bodies.

Recent research has shown that seabed communities are not confined to the uppermost layer of sediments. Populations of bacteria have been found as deep as 500 meters (1,640 feet) below the Pacific seafloor. For energy the bacteria depend on the metabolism of sulfur and methane compounds laid down in the sediments more than 4 million years ago!

Hydrothermal Vent and Cold Seep Communities

Biologists around the world were excited in 1977 when scientists in Woods Hole Oceanographic Institution's submersible *Alvin* discovered an entirely new type of marine community over 3,000 meters (10,000 feet) below the surface. They were searching the near-freezing bottom for the source of some unusually warm water, which had been detected by remote probes. What they found were jets of superheated water (to 350°C, 650°F) blasting from rift vents in the young oceanic ridge 350 kilometers (220 miles) north and east of the Galápagos Islands. Seawater had percolated into fractures in the active crust and had been warmed by heat from nearby magma chambers at a depth of 1 to 3 kilometers (0.6 to 1.9 miles). Convection moved this water back to the seafloor where it emerged as hot springs. The heated water dissolved minerals from the surrounding basaltic rock. As this water emerged and cooled, some inorganic sulfides precipitated, turning the water black. The term *black smoker* was quickly applied to these active vents (**Figure 16.19**).

Clustered around the vents were dense aggregations of large, previously unknown animals. Bottom water in the area was laden with hydrogen sulfide (H_2S), carbon dioxide, and oxygen, upon which specialized archaea and bacteria were found to live. These chemosynthesizers form the base of a food chain that extends to the animals. Large crabs, clams, sea anemones, shrimp, and unusual worms were found in this warm oasis. These communities were called hydrothermal vent communities.

Some of the *tube worms*, contained in their own long, parchmentlike tubes, measured 3 to 4 meters (10 to 13 feet) in length and were the diameter of a human arm. These unusual animals have been tentatively identified as pogonoph-

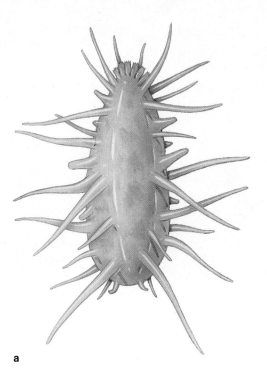

a

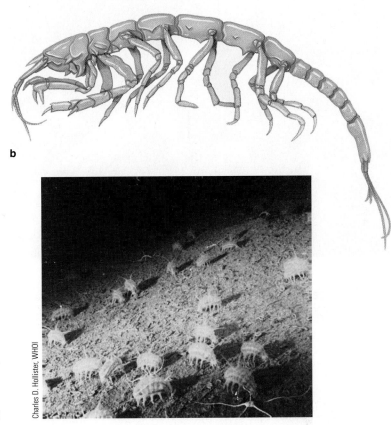

b

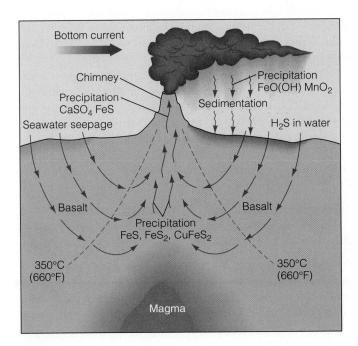

Charles D. Hollister, WHOI

c

Figure 16.18 Abyssal benthic animals. (**a**) *Oneirophanta,* a 10-centimeter (4-inch) holothuran found on the abyssal plains of the North Atlantic. (**b**) *Apseudes galatheae,* a blind, thumb-size crustacean found in the Kermadec Trench, north of New Zealand. (**c**) *Oneirophanta*—as in (**a**)—and brittle stars search for food on a continental slope at a depth of about 1,000 meters (3,300 feet). Brittle stars like these are among the world's most cosmopolitan organisms, found on nearly all deep sediments. (Source for Figure 16.18a & b: From J. C. Briggs, *Marine Zoogeography.* © 1974 by McGraw-Hill, Inc.)

orans, members of a small phylum of invertebrates also found in fairly shallow water. Three species of the newly and appropriately named genus *Riftia* (**Figure 16.20a**) have been identified so far. The tubes of these pogonophorans are flexible and capable of housing the length of the animal when it

retracts. The animals extend tufts of tentacles from the openings of their tubes. Feeding was something of a puzzle because these animals have no mouth, digestive tract, or anus. The trunks of the worms were found to contain large "feeding bodies" tightly packed with bacteria similar to those seen in the water and on the bottom near the hydrothermal vents. The worms' tentacles absorb hydrogen sulfide from the water and transport it to the bacteria, which then use the hydrogen sulfide as an energy source to convert carbon dioxide to organic molecules. The ultimate source of the worms' energy (and the energy of most other residents in this community) is

Figure 16.19 The path of water associated with a hydrothermal vent. Seawater enters the fractured seabed near an active spreading center and percolates downward where it comes in contact with rocks heated by a nearby magma chamber. The warmed water expands and rises in a convection current. As it rises, the hot water dissolves minerals from the surrounding fresh basalt. When the water shoots from a weak spot in the seabed, some of these minerals condense to form a "chimney" up to 20 meters (66 feet) high and 1 meter (3.3 feet) in diameter. As the water cools, metal sulfides precipitate out and form a sedimentary layer downcurrent from the vent. Bacteria in the sediment, in the surrounding water, and within specialized organisms make use of the hydrogen sulfide (H_2S) in the water to bind carbon into glucose by chemosynthesis. This chemosynthesis forms the base of the food chains of vent organisms. (This illustration is not to scale.)

a

b

a and b: Woods Hole Oceanographic Institution

Figure 16.20 Some large organisms of hydrothermal vents. (**a**) *Riftia,* large tube worms (pogonophorans) that contain masses of chemosynthetic bacteria in special interior pouches. (**b**) A vent field dominated by the giant white clam *Calyptogena magnifica.* Each clam is about the size of a man's shoe and contains chemosynthetic bacteria within its gill filaments.

this energy-binding process, called **chemosynthesis**, which replaces photosynthesis in the world of darkness.

The clams and shrimp of these vent communities are equally unusual. For example, the large white clam *Calyptogena magnifica* grows among uneven basaltic mounds (**Figure 16.20b**). Each the size of a shoe, the clams shelter the same kinds of bacteria as *Riftia.* Though the clam retains its filter-feeding structures, it too derives nutrition from the specialized bacteria embedded in the cells of its gill filaments. Small shrimp discovered at the vents in 1985 have been found to possess special organs that may allow them to sense heat from the vents. Such an adaptation would permit them to range away from the vents for food yet return to the warmth and richness of the home community.

Water streaming from the vents contains similar bacteria, but in this case bacteria adapted to extreme conditions (extremophiles). In one study, between 10^5 and 10^9 of these cells were found in each milliliter of ejected water! Chemosynthesis apparently occurs in the surrounding water, productivity that replaces the light-driven carbon-binding reactions of surface picoplankters in this dark world.

Hydrothermal vent communities have now been found off Florida and Louisiana, off California and Oregon, in the North Sea, east of Japan, and in at least 30 other locations. (See Figure 16.19 for more details.) Not all are on oceanic ridges, and not all are in areas spouting hot water. The slow upward percolation of cool, mineral-rich water seeping from beneath sediments encourages the growth of mats of chemosynthetic bacteria that are able to metabolize sulfur-containing compounds or methane. The methane, where present, appears to come from the decomposition of organic material in the sediment or underlying sedimentary rocks. Cool or cold vent communities dependent on these bacteria may be more widespread than communities near areas of superheated water, and they may be longer lived.

Cold seep communities are less dramatic but probably more widespread. Hypersaline seawater rich in minerals, hydrogen sulfide, and sometimes methane percolates upward through the seabed to emerge in broad fields at near-ambient temperatures. These areas are not always associated with the edges of tectonic plates, and about 25 large fields have been discovered. The dominant large organisms at cold seeps are bivalve mollusks, pogonophoran worms, and a few species of sponges—but the heroes are the chemosynthetic bacteria and archaea that form the base of the food chain in these deep, mysterious places.

Studies of vent and cold seep communities suggest many questions. Do hot vent and seep communities occupy the active central rift valleys of a significant percentage of the 65,000 kilometers (41,000 miles) of Earth's oceanic ridges? Did the hot or cold vents serve as the birthplace of life on Earth? Perhaps the deep vent and seep communities will prove to be more important to overall marine productivity than has been previously supposed. Marine biologists are eager to continue their explorations.

Whale Fall Communities

You may wonder how the specialized organisms that inhabit vent and seep communities disperse over the great distances between vent systems. The problem becomes more acute with the recent knowledge that the lifetimes of the vents themselves may be measured in tens of years at most. How are such unique organisms recruited? How are they dispersed?

The answer may lie in the "stepping stones" provided by the fallen bodies of whales. Even though humans have greatly diminished the numbers of living whales, researchers estimate that whale carcasses may be spaced at roughly 25-kilometer (16-mile) intervals across areas such as the North Pacific. Studies of fallen whale skeletons have shown the presence of sulfur-oxidizing chemosynthetic bacteria. As sulfide produced by these bacteria diffuses out of the bone, planktonic larvae of vent organisms might sense its presence, settle, grow, and reproduce. With luck, their offspring might drift to another whale fall and repeat the process. After many steps the organisms would reach a new or newly active vent.

Are Polar Marine Communities Different from Tropical Marine Communities?

Relatively few species have been able to adapt to the stern polar conditions. The harshness of the polar environment has favored the evolution of marine species that tend to move more slowly, develop and grow at slower rates, live longer lives, have fewer offspring, and attain larger sizes than their tropical counterparts. Organisms in polar communities generally have less elaborate and ornamented shapes, produce larval forms that more closely resemble the adult, store proportionally greater reserves of food, and evolve more slowly.

When growing conditions are optimal, the few species that can withstand polar environmental rigor frequently populate an area in extremely large numbers. Plankton communities are fine examples of this principle. The brief, intense plankton blooms of polar summer usually consist of only 1 or 2 species of primary producers, while a similar bloom in temperate waters may comprise from 10 to 20 species. Penguins are another example: Breeding grounds are often jammed with thousands of individuals of one species nesting as closely together as possible.

The situation is reversed in the tropics. Environmental conditions at low latitudes are not as intimidating, and a great many kinds of animals and plants have adapted to life in the warm zone. There are thousands of species of tropical fish, tropical birds, tropical plants, tropical insects, and tropical plankton, but you usually won't see many examples of any one species in a small area. Try remembering the idea this way: In tropical communities, if many organisms are present there will be *few* individuals of *many* species. In the polar regions, if many organisms are present there will be *many* individuals of *few* species.

The overall appearance of the resident species often differs with latitude. Whereas polar species are typically dull in color or pattern, tropical organisms are frequently conspicuous in coloration and behavior. Their bright colors and patterns are probably related to the need to differentiate among the diverse species for feeding and mating competition.

SYMBIOTIC INTERACTIONS AND DEPENDENCIES

Newcomers to biology are often surprised to learn that more than half of the animal species known to science are not free-living. Most are actively involved in close symbiotic relationships with at least one other life form in their community. These relationships are often intricate and sometimes quite strange.

Symbiosis (*sym* = with, *bios* = life) is the biologist's term for the co-occurrence of two species in which the life of one is closely interwoven with the life of the other. The symbiotic bond is often so strong that one organism is totally dependent on the other.

Symbioses

There are three general types of symbioses: mutualism, commensalism, and parasitism.

In **mutualism,** as the name implies, both the symbiont and the host—the larger organism with which the symbiont lives—benefit from the relationship. True mutualism is rare among marine organisms, but a few examples have been observed. One is the relationship between an anemone fish and its sea anemone. In this symbiosis a small, brightly colored anemone fish nestles within the tentacles of a sea anemone (**Figure 16.21**). The mechanism that permits the little fish to do this without being stung is not well understood, but biologists believe that the fish gradually desensitizes the stinging cells of the sea anemone by using mucous secretions from its own skin. In return for the sea anemone's

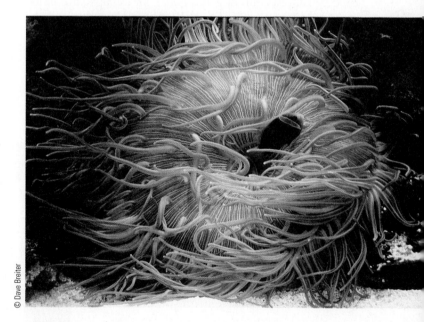

© Dave Breiter

Figure 16.21 Mutualism. Some species of sea anemones have a symbiotic relationship with anemone fish in which the fish receives protection from predators and the sea anemone receives scraps of food from the fish.

protection, the fish feeds the sea anemone scraps of food and may even lure prey within the sea anemone's reach.

Another example of mutualism is the relationship between certain cnidarians, such as coral animals, and the specialized dinoflagellates (known collectively as *zooxanthellae*) that live within their tissues. Both organisms benefit: The autotrophic dinoflagellates have a safe home and a ready source of carbon dioxide, and the animals have a handy, built-in source of carbohydrates. Without their resident dinoflagellates, the reef corals would be unable to deposit calcium, and the rich tropical reef communities we know today would not exist.

Perhaps the most striking examples of mutualism involve cleaning symbioses. In these relationships small organisms (usually fish or shrimp) establish a "cleaning station" at which they remove dead tissue and troublesome surface parasites from the skin, mouth, and gill coverings of larger animals (usually fish) visiting the site. The cleaner eats the dead tissue and parasites, so again both animals benefit. The cleaned animals frequently defend the cleaning station and its cleaners from attack by would-be predators.

In **commensalism,** the symbiont benefits from the association while its host neither benefits nor is harmed. For example, biologists once believed that the association between pilot fish and shark was mutualistic; the pilot fish was thought to guide the shark to a meal and in turn be permitted to dine on the scraps. We now know that the pilot fish is only an opportunistic commensal, taking what scraps it can from the shark's meal. (On the other hand, the mutualistic anemone fish/sea anemone partnership was thought to be commensal before the anemone fish was observed feeding the sea anemone.)

Some of the most curious commensals are those that enter their partner's habitat and, after a period of growth, cannot escape. Small pea crabs of genus *Fabia,* for example, live inside mussel shells, eating food particles that are brought inside by the normal feeding and respiratory actions of the mussel. After a time, the crab grows so large that it cannot leave the mussel because it is larger than the gap between the mussel's shells.

Parasitism is the most highly evolved and by far the most common symbiotic relationship. The parasite lives in (or on) the host for at least part of its life cycle and obtains food at the host's expense. For obvious reasons, parasites do not usually kill their hosts, but they can seriously affect the host organism by reducing its feeding efficiency, depleting its food reserves, reducing its reproductive potential, lowering its resistance to disease, or otherwise sapping its energy. The host–parasite relationship is finely balanced and extraordinarily delicate. The parasite must in some way be aware of the host's physical condition in order to avoid weakening the host so much that it dies. On the other hand, the parasite must take as much energy from the host as possible to ensure its own success.

All major phyla have parasitic marine representatives. However, the most widely distributed and successful group of parasitic marine animals is the roundworms of the phylum *Nematoda.* Like nearly all parasites, nematodes have a **species-specific relationship** with a host. A species-specific relationship is an exclusive relationship between two species; parasites can usually parasitize only one species of host. The reason for this interdependency is the delicacy of the biochemical feedback mechanisms that inform the parasite that its activity may be overstressing the host. The feedback responses are, by necessity, tailored to specific host–parasite pairs. The parasite will not usually survive if it settles in or on a host for which it is not specifically "programmed."

More than one species of parasite *can* infect a single host, however. Nematode worms and other parasites may comprise a significant percentage of the weight of many fishes. The parasitic burden of a normal, apparently healthy sea lion may exceed 2.3 kilograms (5 pounds) and 20 species! Parasites are usually small, but one species of nematode parasite of the fin whale reaches 7 meters (23 feet) in length and is the diameter of a pencil. And, by the way, parasites can have their own parasites!

These three categories (mutualism, commensalism, and parasitism) form a continuum in nature. There are few clearly mutualistic relationships that do not suggest at least a touch of commensalism, and few commensalistic relationships that do not hint of parasitism. The whale barnacles pictured in **Figure 16.22** glean most of their food directly from the ocean, but some of their nutrition is derived from

Figure 16.22 Whale barnacles. These arthropods can be buried up to 3 centimeters (1¼ inches) deep in the skin of certain whales. They derive part of their nutrition from the flesh and circulating fluids of the whale. Some individuals are up to 7.6 centimeters (3 inches) across.

© Norman Cole

the flesh and circulating fluids of their host. It's sometimes hard to tell where one kind of relationship stops and another begins.

Dependencies

Marine animals that can capture and eat only a single prey species are said to be *dependent* on that species. Though it is a controversial stance, some marine biologists consider such a specific feeding relationship to be a variety of symbiosis. These relationships are called **dependencies.** Dependencies differ from true symbioses because the lives of the species involved are not interwoven for extended periods of time.

One of the most extreme dependencies is that of a recently discovered species of copepod (a planktonic arthropod), which can eat only one particular size of one species of diatom because of the configuration of the copepod's mouthparts. Such a specific feeding dependency is unusual; most organisms are able to eat a broader range of food species.

1. If the rocky, sandy, or muddy intertidal zone represents such a challenging mix of environmental factors, why do so many organisms live there?

Difficulty in biology is a relative term. It may seem a circular argument, but wherever organisms live, conditions for life at that place are biologically tolerable, food is available, and environmental conditions are not so extreme as to preclude success. Organisms live in abundance where energy is available. Where there is food, sunlight, or biodegradable compounds, there is life. Natural selection has sorted out the ways that work in this zone from the ways that do not, and the adaptations that work give the organisms living in the intertidal zone's many niches access to a rich harvest of nutrients.

2. Is the species-specificity rule of parasitism ever broken?

Yes, sometimes with catastrophic results. As an example, the lung flukes that inhabit the respiratory tract of most sea lions will "abandon ship" if the sea lion is weakened or dying. A sea lion in this condition sometimes comes out of the water onto a beach to rest. If your pet dog should discover the animal there and sniff at its nose, some of the parasites could transfer from sea lion to dog and establish themselves in the dog's lungs. Because the dog is not the species-specific host for the lung parasites, the biochemical machinery that tells the parasite that it is weakening its host is missing. The dog may die a painful death in a few weeks from an overinfestation with lung flukes. The parasites, will, of course, also die. The interaction will have been a failure in all respects.

3. Could there be any huge undiscovered Godzilla-type sea monsters in the deep ocean?

Probably not, unless they can extract energy directly from water molecules! The deep pelagic feeding situation is simply not rich enough to support the energy needs of an active population of violent, aggressive, city-eating (metrophagous?) reptiles. Scientists never say never, but classic science fiction films aside, it doesn't look promising.

4. What are the historical foundations of ecology?

Many students confuse ecology with conservation efforts such as recycling or with environmental activism such as picketing large industrial polluters. But neither of these activities is ecology.

Ecology as a subdivision of natural science has a long history, which began with the writings of Pliny and other Romans who took an interest in a unified view of nature. Some European Renaissance scholars developed a similar outlook. In more recent times the French naturalists René-Antoine Réaumur (1663–1757) and Georges-Louis de Buffon (1707–88)—along with the great German zoologist and Darwinian Ernst Häckel, who coined the word *oekologie* in 1869—helped build the foundations of the modern science. Population studies were advanced by C. Elton in his 1927 book *Animal Ecology,* and modern biometric analyses of populations and communities were pioneered by the Australian H. G. Andrewartha and his colleagues and contemporaries. The important concept of primary productivity was developed by two American freshwater biologists—E. Birge and C. Juday—in the 1930s.

Modern ecology came of age in 1942 when Robert L. Lindeman, an American, began the study of ecological energetics. His work detailing the flow of energy through ecosystems was among the first to consider an ecosystem as a unified whole. Most modern ecological studies can be traced to his insight.

- A community is all the living things in a defined area. The organisms of a community will interact with and depend on one another, often in complex ways.

- An organism's habitat is its "address"; its niche is its "occupation."

- The distribution of organisms in communities is rarely random; clumped distribution is the most common.

- Though among Earth's most rigorous habitats, the high-energy rocky and sandy shores are heavily populated by diverse organisms able to take advantage of the abundant nutrients found there.

- Highly productive salt marshes and estuaries shelter a great variety of benthic life forms and serve as nurseries for some pelagic organisms.

- Coral reef communities are exceptions to the general rule that tropical oceans are unproductive. Closely cycled nutrients and zooxanthellae in coral organisms make for high productivity and high biodiversity.

- The open ocean below 3,000 meters (10,000 feet) is among the least densely populated habitats on Earth. Lack of food is the main limiting factor.

- The deep-sea floor is the ocean's most uniform habitat. It is populated by large numbers of mostly small, highly specialized organisms.

- Recently discovered deep-vent communities, near black smokers and atop cold seeps, depend on chemosynthesis by bacteria and archaeans for energy.

- Symbiosis is the co-occurrence of two species in which the life of one is closely interwoven with the life of the other. The most common symbiosis is parasitism.

INTERNET STUDY RESOURCES

The Web site for this book contains helpful study aids. Log on to:

www.brookscole.com/053437557xs

and click on the Chapter-by-Chapter area. Choose Chapter 16 and select a resource:

- **Flash Cards** allows you to test your mastery of the Terms and Concepts to Remember for this chapter.
- **Tutorial Quizzes** provides a multiple-choice practice quiz.
- **Student Guide to InfoTrac College Edition** will lead you to Critical Thinking Projects that use InfoTrac College Edition as a research tool.

TERMS AND CONCEPTS TO REMEMBER

biodiversity
brackish
carrying capacity
chemosynthesis
climax community
clumped distribution
commensalism
community

deep scattering layer (DSL)
dependency
desiccation
ecology
environmental resistance
estuary
euryhaline
eurythermal

habitat
intertidal zone
motile
mutualism
niche
parasitism
population density
random distribution

sessile
species-specific relationship
stenohaline
stenothermal
succession
symbiosis
uniform distribution
wave shock

STUDY QUESTIONS

Review Questions

1. How does a population differ from a community? A niche from a habitat?

2. Describe a typical growth curve for a population. What factors influence the shape and eventual height of the curve?

3. Which marine habitat is the most sparsely populated? Why?

4. Describe the residents of the deep ocean. How do inhabitants of the vent communities differ from other deep-floor organisms? What is the source of nutrition for vent communities?

5. What are the three types of symbioses? Give an example of each.

Critical Thinking Questions

1. What is a limiting factor? Give a few examples.

2. In what ways can members of the same population compete with one another? How might members of different populations compete? Contrast the results of these kinds of competition.

3. What factors influence the distribution of organisms within a community? How are these distributions described? Why is random distribution so rare?

4. What problems confront the inhabitants of the intertidal zone? How do you explain the richness of this zone in spite of these rigors?

5. Why must the host–parasite relationship be so finely balanced? What would be the result of an imbalance?

6. **InfoTrac College Edition Project** *Pfiesteria piscicida* is a species of toxic algae that kills fish and makes people sick. In 1997 major outbreaks of this species occurred in Atlantic coastal estuaries and bays. Is the rapid growth of *Pfiesteria* caused by human activity? If so, what needs to be done? Research this question using InfoTrac College Edition.

RESOURCES FOR FURTHER READING AND RESEARCH

The Web site for this book contains many ideas for further reading and research. Log on to:

www.brookscole.com/053437557xs

and click on the Chapter-by-Chapter area. Choose Chapter 16 and select a resource:

- **References** lists the major books and articles consulted in writing this chapter, along with comments from the author about their content and reading level.
- **Hypercontents** takes you to an extensive list of sites with news, research, and images related to individual sections of the chapter.

- In **Student Guide to InfoTrac College Edition** scroll down to Suggested Readings from InfoTrac College Edition for brief descriptions of the articles listed and search hints for finding them in InfoTrac College Edition.
- **Regional InfoTrac College Edition** articles are organized into East Coast, West Coast, and Gulf Coast regions, allowing you to study oceanography on a more local level. You can also access Regional InfoTrac College Edition at www.localocean.com.

For additional readings, go to InfoTrac College Edition, your online research library, at:

http://infotrac.thomsonlearning.com/

Marine Resources 17

AN ENDLESS SUPPLY?

World economies began a period of unprecedented growth in the decade following World War II. Leaders began to look to the sea to provide a greater proportion of the mineral and biological resources needed to sustain their economies and feed their citizens. The great ocean seemed an inexhaustible source of food, oil, and ore. It was comforting to think that as we learned more about the ocean, we would surely discover vast treasure troves available for the taking.

But memories were short. Resources that seem limitless can be depleted in a few generations. Consider, for example, the history of the North Atlantic's most valuable resources, its fisheries. For hundreds of years, Georges

A dangerous way to make a living: Crab fishermen work on deck in bad weather in the Bering Sea, Alaska. The 120-foot crab boats and crews are often buffeted by winter storms—winds over 100 miles per hour and seas to 50 feet are not unheard of. Large iron crab pots, some weighing more than 750 pounds, are moved by hand as rolling waves toss the boat like a cork in a bathtub. A crab fisher can make $20,000 per catch—for a successful trip lasting about six weeks. Commercial fishing is the most dangerous profession in the United States.

© Dan Parrett/Alaska Stock

Bank, one of the world's most productive fishing grounds, yielded seemingly endless supplies of cod, haddock, flounder, and halibut. American colonists considered halibut a trash fish, unfit for human consumption. By the 1830s, however, halibut had become fashionable and demand skyrocketed. Individual halibut boats fishing Georges Bank brought in ten tons of halibut per day! As the reproducing stock was removed, the fishery was rapidly and catastrophically depleted. Today, catches of halibut are so rare in the northwestern Atlantic that regulators don't bother to keep statistics on them.

More recently, on the other side of the North Atlantic, fishers of at least seven European nations compete for cod and haddock whose historic abundance will soon be little more than a memory. Fishing pressure is now so great that less than 1% of one-year-old cod remain in the ocean long enough to spawn. Each year 60% of all the cod present are swept into nets. The hunt on the Dutch continental shelf is so intense that every square meter of seabed is trawled, on average, once to twice each year! The 1992 collapse of the northwestern cod fishery cost 35,000 jobs (18,000 of them in Newfoundland alone). The northeastern fishery is now teetering on the same brink.

Marine fish are the only wildlife still hunted on a large scale. Properly managed, fish and other biological resources can be renewable resources, but the depleted minerals, oil, natural gas, and most other physical resources are nonrenewable. Because they are easier to exploit, physical resources on the land have many economic advantages over physical resources from the sea. Also, on land, valuable elements have been concentrated into ores by sedimentation, weathering, and other natural processes, and the variety of minerals that can be extracted far exceeds anything we have seen in the ocean. The most important physical resources are fluid resources such as petroleum (oil) and natural gas. These substances are probably as abundant in the continental shelves as on dry land, but obtaining them is very expensive and often dangerous.

With very few exceptions, the persistent prediction that the continued exploitation of the ocean will provide an increasing percentage of the material needs of a growing human population—oceanic riches for everyone—cannot be realized. As we will see, marine resources grow more valuable as they become scarcer.

17-1

WHAT TO WATCH FOR IN CHAPTER 17

Here we see two sides of humanity's use of the ocean. On the one hand, we find the ocean's resources useful, convenient, essential. On the other, we find we cannot exploit those resources without damaging their source. World economies are now dependent on oceanic materials, and we are unwilling to abandon or diminish their use until we see clear signs of severe environmental damage. As you'll see in this chapter and the next, humanity has embarked on an unintentional global experiment in marine resource exploitation and waste management. We adjust course only with hesitation as we rush into the unknown.

Marine resources include physical resources such as oil, natural gas, building materials, and chemicals; marine energy; biological resources such as seafood, kelp, and pharmaceuticals; and nonextractive resources such as transportation and recreation. The contribution of marine resources to the world economy has become so large that international laws now govern their allocation.

In spite of their abundance, marine resources provide only a fraction of the worldwide demand for raw materials, human food, and energy. Similar resources on land can usually be obtained more safely and at lower cost. The management of marine resources—especially biological resources—for long-term benefit has been largely unsuccessful.

CLASSES OF MARINE RESOURCES

We will discuss four groups of marine resources in this chapter:

- **Physical resources** result from the deposition, precipitation, or accumulation of useful substances in the ocean or seabed. Most physical resources are mineral deposits, but petroleum and natural gas, mostly remnants of once-living organisms, are included in this category. Fresh water obtained from the ocean is also a physical resource.
- **Marine energy resources** result from the extraction of energy directly from the heat or motion of ocean water.
- **Biological resources** are living animals and plants collected for human use.
- **Nonextractive resources** are uses of the ocean in place; transportation of people and commodities by sea, recreation, and waste disposal are examples.

Marine resources can be classified as either renewable or nonrenewable:

- **Renewable resources** are naturally replaced by the growth of marine organisms or by other natural processes.
- **Nonrenewable resources** such as oil, gas, and solid mineral deposits are present in the ocean in fixed amounts and cannot be replenished over time spans as short as human lifetimes.

PHYSICAL RESOURCES
Petroleum and Natural Gas

Offshore petroleum and natural gas generated nearly $233 billion in worldwide revenues in 1998. Here the ocean makes a significant contribution to current world needs: About 32% of the crude oil and 24% of the natural gas produced in 1998 came from the seabed. About a third of known world reserves of oil and natural gas lie along the continental margins. Major U.S. marine reserves are located on the continental shelf of southern California, off the Texas and Louisiana Gulf coast, and along the North Slope of Alaska. The deep-sea floors probably contain little or no oil or natural gas.

Oil is a complex chemical soup containing perhaps a thousand compounds, mostly hydrocarbons. Petroleum is almost always associated with marine sediments, suggesting that the organic substances from which it was formed were once marine. Planktonic organisms or soft-bodied benthic marine animals are the most likely candidates. Their bodies apparently accumulated in quiet basins where the supply of oxygen was low and there were few bottom scavengers. The action of anaerobic bacteria converted the original tissues into simpler, relatively insoluble organic compounds that were probably buried—possibly first by turbidity currents, then later by the continuous fall of sediments from the ocean above. Further conversion of the hydrocarbons by high temperatures and pressures must have taken place at considerable depth, probably 2 kilometers (1.2 miles) or more be-

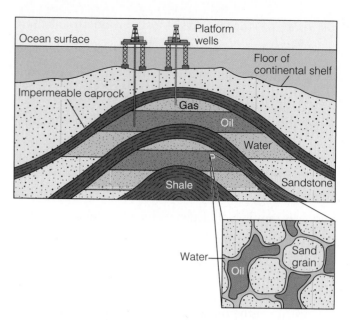

Figure 17.1 Oil and natural gas are often found together beneath a dome of impermeable caprock. Oil and gas are found not in great hollow reservoirs, but within pore spaces in rock (inset).

neath the surface of the ocean floor. Slow cooking under this thick sedimentary blanket for millions of years completed the chemical changes that produce oil.

If the organic material cooked too long or at too high a temperature, the mixture turned to methane, the dominant component of natural gas. Deep sedimentary layers are older and hotter than shallow ones and have higher proportions of natural gas to oil. Very few oil deposits have been found below a depth of 3 kilometers (1.8 miles). Below about 7 kilometers (4.4 miles), only natural gas is found.

Oil is less dense than the surrounding sediments, so it can migrate from its source rock through porous overlying formations. It collects in the pore spaces of reservoir rocks when an impermeable overlying layer prevents further upward migration of the oil (**Figure 17.1**). When searching for oil, geologists use sound reflected off subsurface structures to look for the signature combination of layered sediments, depth, and reservoir structure before they drill.

Drilling for oil offshore is far more costly than drilling on land because special drilling equipment and transport systems are required. Most marine oil deposits are tapped from offshore platforms resting in water less than 100 meters (330 feet) deep (**Figure 17.2a**). As oil demand (and therefore price) continues to rise, however, deeper deposits farther offshore will be exploited from larger platforms. The largest and heaviest platform is *Statfjord-B*, in position since 1981 northeast of the Shetland Islands in the North Sea. The tallest drilling platform is *Ursa*, a tension-leg structure deployed by the Shell Oil Company in 1998 (**Figure 17.2b**). *Ursa* floats in water more than 1,160 meters (3,800 feet) deep and is held in position off the Louisiana coast by 16 lengths of 81-centimeter (32-inch) steel cables anchored into the seabed and pulling against the semisubmerged platform's buoyancy (**Figure**

a

b

c

Figure 17.2 (**a**) An oil drilling platform in the relatively shallow water off southern California. (**b**) Shell Oil Company's tension-leg platform *Ursa,* deployed in 1,160 meters (3,800 feet) of water 108 kilometers (130 miles) southeast of New Orleans. (**c**) Platform *Mars,* deployed off Louisiana in 912 meters (2,933 feet) of water in 1996, is pictured here in this imaginary view in relation to the Houston skyline. Tension-leg platforms are held in position by steel cables anchored into the seabed, pulling against the semi-submerged platform's buoyancy. Twelve lateral cables anchored to the seabed (not shown) prevent sideways movement. Both the *Ursa* and *Mars* platforms are designed to withstand hurricane-force waves of 22 meters (71 feet) and winds of 225 kilometers (140 miles) per hour.

17.2c). At full production the 14 wells on platform *Ursa* are projected to produce 30,000 barrels of oil and 80 million cubic feet of natural gas per day. The total cost of platform *Ursa* exceeded $1.45 billion. Even larger tension-leg platforms are being planned.

Methane Hydrate

The largest known reservoir of hydrocarbons on Earth is not coal or oil, but methane-laced ice crystals—methane hydrate—in the sediments of some continental slopes. Little is known about their formation, but methane hydrates exist in thin layers 200 to 500 meters (660 to 1,650 feet) below the seafloor, where they are stable and long-lived. Sediment rich in methane hydrate looks like green Play-Doh. When brought to the warm, low-pressure conditions at the ocean surface, the sediment fizzes vigorously as the methane escapes. It burns vigorously if ignited.

Though abundant, exploitation of this resource would be very costly and dangerous. Even if engineers could bring the sediment to the surface before the methane disappeared, extracting the methane from the sediment and liquefying it for efficient use would be prohibitively expensive.

Could escape of methane from marine sediments have played a role in ancient climate change? Some researchers think so. Methane is a powerful greenhouse gas (see Chapter 18), and changes in ocean circulation that result in deep-ocean warming could release large quantities of methane. About 55 million years ago the deep ocean warmed by at least 4°C (7°F). The large-scale escape of methane from the seabed may have raised surface temperatures abruptly, melted surface ice, and lowered oxygen levels in the deep sea. If global warming continues today, will methane escaping from deep sediments exacerbate the problem?

Sand and Gravel

Sand and gravel are not very glamorous marine resources, but they are second in dollar value only to oil and natural gas. More than 1 billion metric tons (1.1 billion tons) of sand and gravel, valued at more than half a billion dollars, was mined offshore in 1998. Only about 1% of the world's total sand and gravel production is scraped and dredged from continental shelves each year, but the seafloor supplies about 20% of the sand and gravel used in the island nations of Japan and the United Kingdom. The world's largest single mining operation is the extraction of aragonite sand at Ocean Cay in the Bahamas. Sand is suction-dredged onto an artificial island and then shipped on specially designed vessels. This sand, about 97% calcium carbonate, is used in Portland cement, glass, and animal feed supplements and in the reduction of soil acidity.

Most of the exploitable U.S. deposits of marine sand and gravel are found off the coasts of Alaska, California, Washington, and the East Coast states from Virginia to Maine. Nearshore deposits are widespread, easily accessible, and used in concrete needed to construct buildings and highways. Offshore oil wells in Alaska are built on huge man-made

gravel platforms; the large quantities of gravel available at those locations make offshore drilling practical there.

Magnesium and Magnesium Compounds

Magnesium, the third most abundant dissolved element, precipitates from seawater, mainly in the form of magnesium chloride ($MgCl_2$) and magnesium sulfate ($MgSO_4$) salts. Magnesium metal (a strong, lightweight material used in aircraft and structural applications) can be extracted by chemical and electrical means from a concentrated brine of these salts. Worldwide, about half the production of metallic magnesium is derived from seawater; about 60% of U.S. production comes from a single seawater processing facility in Texas. The value of this metal to the U.S. economy was $320 million in 1998.

Magnesium compounds are also valuable. Magnesium salts are used in chemical processes, in foods and medicines, as soil conditioners, and in the lining of high-temperature furnaces. About 25% of the magnesium compounds produced in the United States are derived from seawater. In 1998 they were worth about $70 million to the U.S. economy.

Salts

As you may remember from Chapter 7, the ocean's salinity varies from about 3.3% to 3.7% by weight. When seawater evaporates, the remaining major constituent ions (see Table 7.1) combine to form various salts, including calcium carbonate ($CaCO_3$), gypsum ($CaSO_4$), table salt (NaCl), and a complex mixture of magnesium and potassium salts. Table salt makes up slightly more than 78% of the total salt residue.

Seawater is evaporated in large salt ponds in arid parts of the world (**Figure 17.3**). Operators can segregate the various salts from one another by shifting the residual brine from pond to pond at just the right time during the evaporation process. As we've seen, the magnesium salts are used as a source of magnesium metal and magnesium compounds. The potassium salts are processed into chemicals and fertilizers. Bromine (a useful component of certain medicines, chemical processes, and antiknock gasoline) is also extracted from the residue. Gypsum is an important component of wallboard and other building materials. About a third of the world's table salt is currently produced from seawater by evaporation. In North America, some of this salt is used for snow and ice removal. Salt is also used in water softeners, agriculture, and food processing. In 1998 the United States produced by evaporation about 4.7 million metric tons (5.2 million tons) of table salt with a value of about $145 million.

Manganese Nodules

In Chapter 5 you read about manganese nodules, the rounded black objects that litter the abyssal plains, particularly in the Pacific (see Figure 5.12). These slow-growing lumps were first seen in bottom samples taken by scientists aboard HMS *Challenger* in 1874. The iron, manganese, cop-

Figure 17.3 Salt evaporation ponds at the southern end of San Francisco Bay in California. Operators can segregate the various salts from one another by shifting the residual brine from pond to pond at just the right time during the evaporation process. The colors in the ponds are imparted by algae and other microorganisms that thrive at varying levels of salinity. In general, the highest-salinity ponds have a reddish cast.

per, nickel, and cobalt content of the nodules makes them particularly attractive to industrial nations lacking onshore sources of these crucial materials. The U.S. Bureau of Mines estimates that the richest deposits in the Pacific exceed 16 billion metric tons (17.6 billion tons), about 20 times more than all known terrestrial reserves and more than 2,000 years' worth of production at present rates of use.

Manganese nodules have been dredged from the seabed in small-scale trials, but no commercial mining ventures exist. Various recovery schemes have been proposed, including a system resembling a vacuum cleaner, but the difficulties of collecting large numbers of nodules from abyssal depths in excess of 4,000 meters (12,000 feet) have rendered these plans uneconomical—at least until prices for manganese, copper, and nickel rise as the terrestrial sources are consumed.

Phosphorite Deposits

17-9

Also discovered during the *Challenger* Expedition were irregular chunks of phosphorite, first collected from the continental rise off South Africa. Sedimentary phosphorite deposits, from which industrial chemicals and phosphate-rich agricultural fertilizers can be made, formed from the decaying remains of marine organisms that lived in areas of extensive upwelling. The richest deposits occur at depths between 30 and 300 meters (100 and 1,000 feet). Post-*Challenger* investigations have revealed rich deposits of this important material off the coasts of Florida, California, western South America, and western Africa. Even though phosphorite deposits occur in much shallower water than manganese nodules do, the cost of recovering the resource from the ocean greatly exceeds that of recovery from land. Much of the U.S. supply of phosphorus and phosphates comes from the strip-mining of ancient, uplifted shallow-water phosphorite deposits in central Florida.

Metallic Sulfides and Muds

17-10

The recent discovery of metal-rich sulfides around hydrothermal vents has spurred interest among economists as well as oceanographers. Heated seawater carrying large quantities of metals and sulfur leached from the newly formed crust pours out through vents and fractures. The metals—mainly zinc, iron, copper, lead, silver, and cadmium—combine with the sulfur and precipitate from the cooler surrounding water as mounds, coatings, and chimneys. While these deposits are certainly commercial-grade ores, they are neither large nor extensive. Also, they are subject to solution and oxidation on the seafloor and are not likely to be preserved in thick layers for long periods of time. Nevertheless, scientists are planning mineralogical studies of the fast-spreading rift zones along the East Pacific Rise near the mouth of the Gulf of California, the Galápagos Rift, and the Juan de Fuca Ridge off the coast of Oregon.

The Red Sea is another area where lithospheric plates are diverging and where molten material is close to the surface. Seawater seeping in through deep faults and fractures comes into contact with fresh, hot basalts and dissolves metals and salts from the rock. The recycled water emerges at temperatures of around 100°C (212°F) and is extremely saline, about 250‰ to 300‰ (average seawater is about 34‰). Though the solutions are hot, their great density causes them to stay on the floor of the Red Sea in deep, fault-bounded basins. Three great pools of hot water have been discovered at a depth of about 2,000 meters (6,600 feet). The metals precipitating within these basins produce muds rich in metal sulfides, silicates, and oxides of commercial concentration. Recovery of these ores is expected to begin early in this century.

Fresh Water

Only 0.017% of Earth's water is liquid, fresh, and available at the surface for easy use by humans. Another 0.6% is available as groundwater within half a mile of the surface. Unfortunately, much of this is polluted or otherwise unfit for human consumption. The fact that fresh, pure water often costs more per gallon than gasoline emphasizes its scarcity and importance. More than any other factor in nature, the availability of **potable water** (water suitable for drinking) determines the number of people who can inhabit any geographic area, their use of other natural resources, and their lifestyle.

Fresh water is becoming an important marine resource. Exploitation of that resource by **desalination,** the separation of pure water from seawater, is already under way, mainly in the Middle East, West Africa, Peru, Florida, Texas, and California. More than 1,500 desalination plants are currently operating worldwide, producing a total of about 13.3 billion liters (3.5 billion gallons) of fresh water per day. The largest desalination plant, in Saudi Arabia, produces about 114 million liters (30 million gallons) daily!

Several desalination methods are currently in use. *Distillation* by boiling is the most familiar; about three-fourths of the world's desalinated water is produced in this way. Distillation uses a great deal of energy, making it a very expensive process. *Freezing* is another effective but costly method of desalination; ice crystals exclude salt as they form, and the ice can be "harvested" and melted for use. Solar or geothermal power may bring down the cost of distillation or freezing, but more efficient, less energy-intensive mechanisms are being developed. Among these is *reverse osmosis desalination.* In this process seawater is forced against a semipermeable membrane at high pressure. Fresh water seeps through the membrane's pores while the salts stay behind. About a quarter of desalinated water is produced in this way. Reverse osmosis uses less energy per unit of fresh water produced than distillation or freezing, but the necessary membranes are fragile and costly.

One of the most elegant ideas for inexpensive desalination is illustrated in **Figure 17.4.** Other than the sunlight needed to make water vapor, this passive system requires no energy except what is required by pumps to move water through plastic-covered troughs and by small fans to keep the plastic sheets inflated.

Desalination, water conservation, and perhaps even iceberg harvesting will become more common as water becomes more polluted, scarcer, and more valuable.

MARINE ENERGY

The energy crises of 1973 and 1979 focused public attention on the need for unconventional sources of power. Those sources that are not consumed in use—solar power or wind

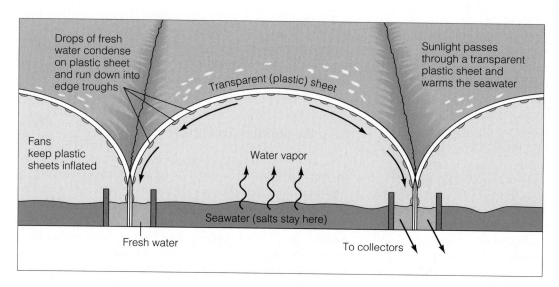

Figure 17.4 An effective low-energy, low-tech scheme for desalination. Water vapor condenses on the plastic film and runs down into collection troughs.

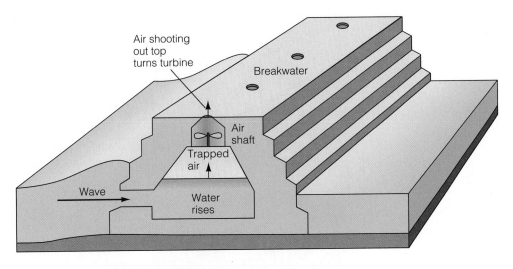

Figure 17.5 A device for using wave energy to generate power, one of several being tested around the world.

power, for example—are preferable to nonrenewable sources such as fossil fuels. Anyone who has watched the ocean knows that so restless a place must surely be rich in energy. The renewable energy is certainly there, but extracting it in useful form is not easy.

Waves and Currents

17-13

Waves are the most obvious manifestation of oceanic energy—ask any surfer about the energy in a wave. As you may remember from Chapter 10, wind waves store wind energy and transport it toward shore (see again Figure 10.19).

Many devices have been proposed to harness this energy; Japan, Norway, Britain, Sweden, the United States, and Russia have built small experimental plants to evaluate their effectiveness. One of these devices (shown in **Figure 17.5**) uses the rush of air trapped by waves entering breakwater caissons to power a generator. So far none of the experimental plants has produced power at a competitive price, but some designs show promise. There is, however, no free lunch. Extensive use of wave power generators along a shore could deprive that shore of its natural wave energy. Changing the wave patterns could alter longshore transport or disrupt the life cycles of marine organisms. (And some schemes would ruin waves for surfers and beachcombers.)

Ocean currents might also be harnessed. Huge, slowly turning turbines immersed in the Gulf Stream have been proposed, but their necessary size and complexity make them prohibitively expensive.

Thermal Gradient

17-14

The greatest potential for energy generation in the ocean lies in exploiting the thermal gradient between warm surface water and cold, deep water. As you may remember from Chapter 6, 1 calorie of heat will raise the temperature of 1 cubic centimeter of seawater by about 1°C (1.8°F). If we assume a temperature difference of 15°C (27°F) between surface wa-

ter and deep water, the total heat energy in each cubic meter of surface water (relative to deep water) is about *15 million calories*. If this energy were released in 1 second, it would generate 60 megawatts of power. Extracting the heat energy from 1,600 cubic meters of warm seawater per second would provide power equivalent to the full generating capacity of all the nuclear power plants in the United States!

How might the immense potential of thermal gradients be harnessed? The earliest proposal was made in the 1880s by a French physicist and inventor, Jacques d'Arsonval. His idea for generating electricity is shown in **Figure 17.6a.** Warm seawater would be pumped into the plant through openings near the ocean surface. This water would pass through heat exchangers and boil liquid ammonia—a liquid with a very low boiling point—into a pressurized vapor. This gas would turn a turbine, which would spin an electrical generator, and then pass into another heat exchanger cooled by water pumped from the depths. Here the ammonia would condense into a liquid, creating a vacuum that would draw more ammonia vapor through the turbine. The liquid ammonia would be pumped back to the first heat exchanger to repeat the cycle. A small plant based on this model was built in Cuba in 1930. It operated successfully for a short time before being destroyed in a tropical storm. A modern equivalent, the largest plant yet built to exploit the ocean's thermal gradient, produced 210 kilowatts of electrical power at Keahole Point, Kailua-Kona, Hawaii (**Figure 17.6b**). Cold water was pumped into the plant from a depth of 610 meters (2,000 feet). Having proved its design, the experimental plant was disassembled in 1997. These devices are called OTEC plants, for *Ocean Thermal Energy Conversion*.

Why has this promising technology not been more broadly exploited? Mainly because of the low efficiency of the OTEC process. The efficiency of heat-driven power generators depends on the *difference* in temperature between the hottest part of the system and the coldest. A fossil-fueled generating plant can be highly efficient because of the great difference in temperature between the flame and the water (or air) cooling the heat exchanger. A nuclear plant is less

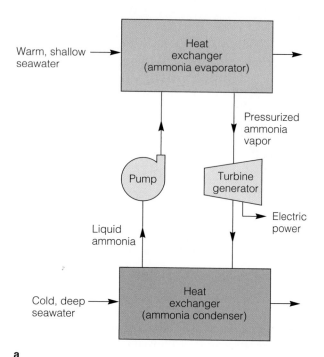

National Energy Laboratory of Hawaii

Figure 17.6 (**a**) Basic aspects of the thermal energy conversion system first proposed by Jacques d'Arsonval in the 1880s. (**b**) A 210-kilowatt OTEC (*Ocean Thermal Energy Conversion*) plant at Kailua-Kona, Hawaii. Having proved its design, the experimental plant was disassembled in 1997.

efficient because the reactor core cannot be as hot as a flame. The efficiency of an OTEC plant—with only a 15°C temperature difference between "hot side" and "cold side"—would be only about 2%. Therefore, huge amounts of warm and cold water would have to circulate through the plant. An OTEC plant with the same generating capacity as a single large nuclear power plant would need to process a continuous flow of water equal to five times the average flow of the Mississippi River—and that estimate doesn't include the power necessary to operate the OTEC plant's massive internal pumps.

There are other problems. An OTEC plant would need to be sited in the tropics, where warm water is layered over cold. Tropical cyclones common in these areas would wreak havoc with the plant's long pipes and delicate generators. Marine life stimulated by the upwelled, nutrient-rich cold water would grow in the surrounding ocean and foul the heat exchangers. Construction and maintenance costs would be astronomical, and the transmission of power to a distant shore presents special difficulties. Still, as energy becomes more and more costly, the OTEC option may prove practical.

BIOLOGICAL RESOURCES

Ancient kitchen middens (garbage dumps of bones and shells) found in many coastal regions demonstrate that humans have used the sea for thousands of years as a source of food and medicines. Now the human population threatens to outgrow its food supply. Contemporary food production and distribution practices are unable to satisfy the nutritional needs of all the world's 6 billion people, while starvation and malnutrition are major problems in many nations. Can the ocean help?

Compared to the production from land-based agriculture, the contribution of marine animals and plants to the human intake of all protein is small, probably around 4%. Although most of that protein comes from fish, marine sources account for only about 18% of the total *animal* protein consumed by humans. Fish, crustaceans, and mollusks contribute about 14.5% of the total; fish meal and by-products included in the diets of animals raised for food account for another 3.5%. About 85% of the annual catch of fish, crustaceans, and mollusks comes from the ocean, and the rest from fresh water.

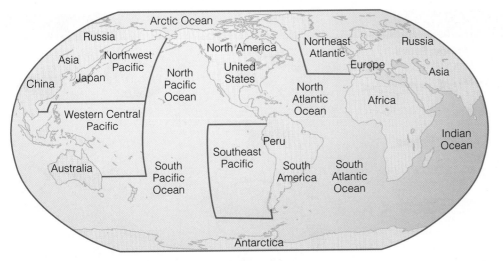

Figure 17.7 The top five marine fish harvesting nations and the top marine fishing areas in 1998 (Source: *National Geographic*, January 2001.)

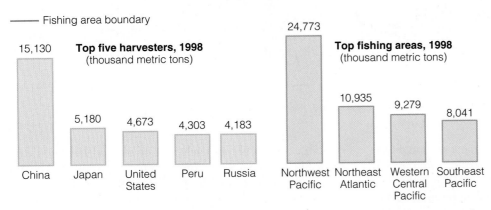

— Fishing area boundary

Top five harvesters, 1998
(thousand metric tons)

15,130 — China
5,180 — Japan
4,673 — United States
4,303 — Peru
4,183 — Russia

Top fishing areas, 1998
(thousand metric tons)

24,773 — Northwest Pacific
10,935 — Northeast Atlantic
9,279 — Western Central Pacific
8,041 — Southeast Pacific

The sea will probably not be able to provide substantially more food to help alleviate future problems of malnutrition and starvation caused by human overpopulation; indeed, population growth will likely absorb any resource increase. Nevertheless, these resources currently sustain a great many people.

Fishes, Crustaceans, and Mollusks   17-16

Fishes, crustaceans, and mollusks are the most valuable living marine resources. Commercial fishers took about 100.8 million metric tons (110.9 million tons) of these animals in 1997. The distribution of the 1998 catch is shown in **Figure 17.7.**

Of the thousands of species of marine fishes, crustaceans, and mollusks, fewer than 500 species are regularly caught and processed. The ten major groups listed in **Table 17.1** supply about 95% of the commercial marine catch. Note that the largest commercial harvest is of the herring and its relatives, which account for more than a fifth of the live weight of all living marine resources caught each year. **Figure 17.8** shows some of the most important commercial species.

Fishing is a big business, employing more than 15 million people worldwide. It is also the most dangerous job in the United States—commercial fishers suffer 155 deaths per 100,000 workers each year.[1] Though estimates vary widely, the

Table 17.1 World Commercial Catch of Marine Fishes, Crustaceans, and Mollusks, by Species Group, 1997	
Species Group	**Millions of Metric Tons, Live Weight**
Herring, sardines, anchovies	21.5
Jacks, mullets, sauries	10.9
Mollusks	15.5
Cods, hakes, haddocks	10.2
Redfish, basses, conger eels	7.4
Crustaceans	6.8
Tunas, bonitos, billfishes	4.8
Mackerel, snooks, cutlass fishes	5.2
Flounders, halibuts, soles	1.0
Miscellaneous marine fishes	17.5
Total for all marine sources, excluding marine mammals and aquatic plants	**100.8**

Source: U.S. Department of Commerce. Fisheries of the United States, 1999.

---

[1] Sebastian Junger's extraordinary 1997 book (and the 2000 movie) *The Perfect Storm* chillingly recalls these dangers.

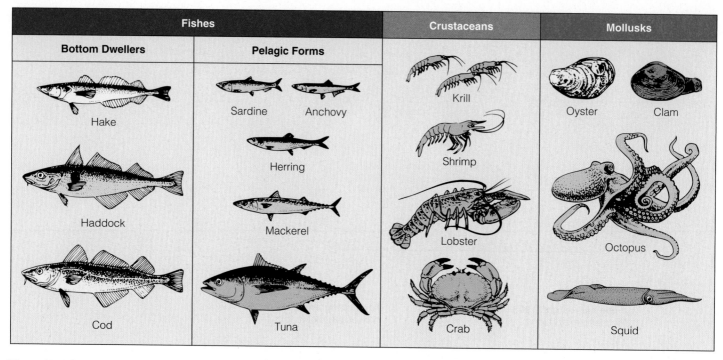

Figure 17.8 Some major types of commercially harvested fishes, crustaceans, and mollusks.

value in 1997 for the worldwide marine catch was thought to be around $92 billion. About half of the world's commercial marine catch is taken by the five countries shown in Figure 17.7. About 75% of the annual harvest is taken by commercial fishers who operate vast fleets working year-round, using satellite sensors, aerial photography, scouting vessels, and sonar to pinpoint the location of fish schools. Huge factory ships often follow the largest fleets to process, can, or freeze the animals on the run. Catching methods no longer depend on hooks and lines but rather on large trawl nets (**Figure 17.9**), purse seines, or gill nets. The living resources of the ocean are under furious assault: Between 1950 and 1997 the commercial marine fish catch increased more than fivefold.

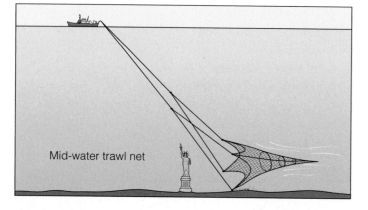

Figure 17.9 Stern trawler fishing. After sonar on the trawler finds the fish, they're captured by a trawl net more than 122 meters (400 feet) wide. Boards angled to the water flow keep the net's mouth open. The largest nets extend about 0.8 kilometer (½ mile) behind the towing vessel and are large enough to hold a dozen 747 jetliners.

The cost to obtain each unit of seafood has risen dramatically in spite of all this high-tech assistance. The increasing expense of fuel for the fishing fleets and processing plants, the cost of wages for the crews, and the greater distances that boats must cover to catch each ton of fish have all helped drive up the cost of seafood. In spite of greater efforts, the total marine catch leveled off around 1970 and remained surprisingly stable until 1980, when greater demand and increasing prices began to drive the tonnage upward again. Another plateau occurred from 1989 to 1992. Then, under renewed attack, ocean yields rose again. Despite new investments, world fisheries have almost certainly peaked, and we may look forward to a period of declining harvests. Since 1970 the world human population has grown, so the average *per capita* world fish catch has fallen alarmingly (see **Figure 17.10**).

Fishery Mismanagement

Can we continue to take huge amounts of food from the ocean year after year? The **maximum sustainable yield,** the maximum amount of each type of fish, crustacean, and mollusk that can be caught without impairing future populations, probably lies between 100 and 135 million metric tons (110 and 150 million tons) annually. As can be seen in Table 17.1, current yield now exceeds the lower figure. Fleets are obtaining fewer tons per unit of effort and are ranging farther afield in their urgent search for food. We may be perilously close to the catastrophic collapse of more fisheries. As of 1997, 30% of recognized marine fisheries were overexploited or already depleted, and 50% more were at their presumed limit of ex-

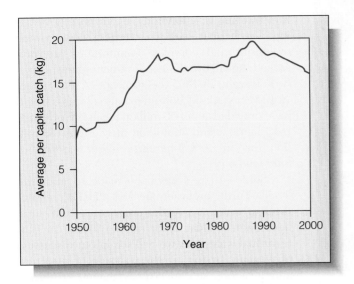

Figure 17.10 Average *per capita* world fish catch reached a peak in the late 1980s and has declined since. The decline is projected to continue in this century.

for capturing the remaining animals. When the impending catastrophe is obvious, governments will sometimes intervene to set limits or close a fishery altogether. In 1999 New England officials approved a plan to close a large section of the Gulf of Maine to fishing in an attempt to replenish once-abundant cod stocks. As many as 2,500 fishers and 700 boats were idled, and losses may exceed $21 million a year. Given the choice between immediate profit and a long-term sustainable fishery, fishers made their priorities clear by initiating a recall campaign against officials who had approved the ban!

Fishers sometimes abandon an exhausted fishery to search for new and profitable resources. New Zealand's orange roughy—a large, mild-tasting deep-water fish—became popular with chefs in the early 1980s. The orange roughy grows slowly, can live more than 100 years, and takes 25 to 30 years to become sexually mature. The use of precision locating equipment and fishing techniques developed for deep water brought the orange roughy to the brink of **commercial extinction** in just 13 years. (Commercial extinction is the depletion of a resource species to a point where it is no longer profitable to harvest.) With Pacific orange roughy no longer in dependable supply, in 1996 interest turned to the Patagonian toothfish, another large, slow-growing deep-water fish. Prices have risen to $7,000 per metric ton; boats from northern countries are in a race for the catch using 5,000 baited hooks each day on a 100-kilometer (62-mile) line. More than $300 million worth of toothfish was taken in 1996, and twice that value is projected for 1997. Few remain. The cycle repeats.

The intended organism is not the only victim. In some fisheries, **bykill**—animals unintentionally killed when desirable organisms are collected—exceeds target catch (**Figure 17.12**). In 1992 in the Bering Sea, fishers discarded 16 million red king crabs and kept only 3 million. Discarded creatures outnumber targeted shrimp by between 125% and 830%. In 1993 in the Gulf of Mexico, fishing for shrimp

ploitation. The U.S. National Marine Fisheries Service estimates that 45% of the fish stocks whose status is known are now **overfished**—so many fish have been harvested that there is not enough breeding stock left to replenish the species. Recent trends are seen in **Figure 17.11.**

Even when faced with evidence of overfishing, the fishing industry seldom follows a rational course. The industry's dominant motivating force is quick financial return, even if it means depleting a stock and disrupting the equilibrium of a fragile ecosystem. Long-term stability is forsaken for short-term profit. When the catch begins to drop, the industry increases the number of boats and develops more efficient techniques

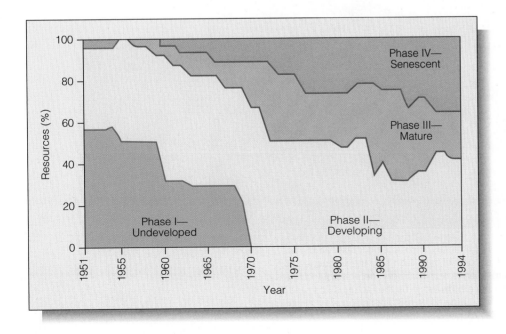

Figure 17.11 The percentages of world marine fish resources in different stages of development, 1951–94. Undeveloped = a low and relatively constant level of catches; developing = rapidly increasing catches; mature = a high and plateauing level of catches; senescent = catches declining from higher levels. (*Source:* Garcia, S. M., and R. Grainger, 1996. *FAO Fisheries Technical Paper* 359.)

Figure 17.12 Fish jam the rearmost part of a trawl net before it is hoisted from the water. Perhaps one-third of these fish are unwanted species—bykill—that will be thrown over the side.

© Bill Curtsinger Photography

tom trawling is, if anything, even more devastating. The habitat itself is disturbed; slow-growing organisms and complex communities are ransacked. Again, bykill exceeds target organisms by a wide margin. In 1995 the governor of Alaska said, "Last year's [Alaska bottom fishing] discards would have provided about 50 million meals." Worldwide bykill approached 30 million metric tons (33 million tons) in 1998, *a quantity nearly one-third of total landings!*

Some progress has been made to minimize bykill. Turtle exclusion devices (TEDs), chutes through which sea turtles are ejected from nets, have been mandated for shrimp fishing in U.S. territorial waters. As we will see in our discussion of whaling, by forcing the redesign of tuna nets and the adoption of special ship maneuvers during netting, the "dolphin-safe" tuna campaign, an adjunct to the 1972 Marine Mammals Protection Act, has spared the lives of hundreds of thousands of dolphins.

resulted in a bykill of 34 million red snappers and 2,800 metric tons (3,090 tons) of sharks. (The annual commercial catch of red snapper in the gulf is typically 3 million fish.) Albatrosses are endangered in the southern ocean because they take baited longline hooks cast for southern bluefin tuna. Bot-

There are other disruptive, mistargeted fishing techniques. One employs **drift nets**—fine, vertically suspended nylon nets as much as 7 meters (25 feet) high and 80 kilometers (50 miles) long. A deployed drift net is shown in **Figure 17.13.** Drift net technology was pioneered by a United Nations agency to help impoverished

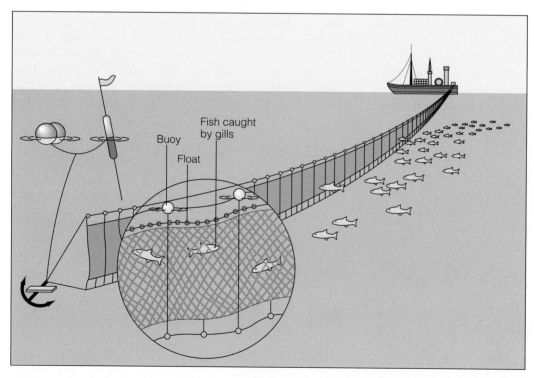

Buoy
Float
Fish caught by gills

Figure 17.13 Drift net fishing. Typically, a fleet of huge factory ships uses sophisticated electronic detection devices and aircraft to find large schools of fish. Each ship then launches 20 to 50 fast, small catcher boats to set hundreds to thousands of miles of drift nets, weighted to stay at the desired depth. After drifting overnight the nets are hauled in by the factory ships, and the catch is processed on board by canning or freezing. Large-scale drift net fishing was officially banned in 1993, but it continues in the Pacific and has begun in the Atlantic.

Figure 17.14 An unintentional but unavoidable consequence of drift net fishing.

Asian nations turn a profit from what had been subsistence fishing. Until 1993, Taiwanese, Korean, and Japanese vessels deployed some 48,000 kilometers (30,000 miles) of these "walls of death" each night—more than enough to encircle Earth! Drift nets caught the fish and squid for which they were designed, but they also entangled everything else that touched them, including turtles, birds, and marine mammals (**Figure 17.14**). The process has been compared to strip-mining—the ocean is literally sieved of its contents. An estimated 30 kilometers (18 miles) of these fine, nearly invisible nets was lost each night—about 1,600 kilometers (1,000 miles) per season. These remnants, made of non-biodegradable plastic, become *ghost nets,* continuing to entangle fish and other animals perhaps for decades (**Figure 17.15**).

Though large-scale drift net fishing is now banned, pirate netting continues in the Pacific and has begun in the Atlantic. Completely unrestrained by regulation, these outlaw fishermen operate where their activities cause maximum damage to valuable reproductive stock.

Figure 17.15 An abandoned plastic fishing net drapes the intertidal zone at remote Clarion Island, Revillagigedo Islands, Mexico. Before being stranded here, this net may have entangled fish and marine mammals for years.

"Madhouse Economics"

The fishing industry spent $124 billion to catch $70 billion worth of fish and other marine life in 1995. Government subsidies made up most of the $54 billion deficit. Artificial supports included fuel-tax exemptions, price controls, low-interest loans underwritten by government or state agencies, and grants of equipment. More than 1,600 large fishing vessels have been reflagged to maneuver around legal requirements in their original countries or to take advantage of subsidies. In an effort to preserve well-paying jobs, governments encourage construction of underutilized ships and research in new and often-destructive technology. The efficiency of the new ships reduces the number of fish available, thereby reducing the value of hulls and equipment on the market. Unable to sell their vessels, owners are forced to continue fishing to repay their loans. In an important 1995 paper in *Scientific American*, Carl Safina understandably described the situation as "madhouse economics."

Whaling

Since the 1880s, whales have been hunted to provide meat for human and animal consumption; oil for lubrication, illumination, industrial products, cosmetics, and margarine; bones for fertilizers and food supplements; and baleen for corset stays. **Figure 17.16** shows a whale being butchered for food and oil. An estimated 4.4 million large whales existed in 1900; today slightly more than 1 million remain. Eight of the 11 species of large whales once hunted by the whaling industry are commercially extinct. The industry pursued immediate profits despite obvious signs that most of the "fishery" was exhausted (**Figure 17.17**).

Substitutes exist for all whale products, but the harvest of most commercial species did not stop until whaling became uneconomical. In 1986 the International Whaling Commission, an organization of whaling countries established to manage whale stocks, placed a moratorium on the slaughter of large whales. Except for a suspiciously large harvest of whales taken by the Japanese for "scientific purposes," commercial whaling ceased in 1987. Fewer than 700 large whales were taken in 1988.

Now, however, the number of whales being taken is rising. What protection there is may have come too late to save some species from extinction—and protection may be only temporary. Under intense pressure from its major fishing industry, Norway resumed whaling in 1993. Japan never stopped. Their target, the minke whale, is the smallest and most numerous of the great whale species (see Figure 15.39).

Figure 17.16 The whaling industry has pushed most of the dozen or so species of great whales to the brink of extinction through overharvesting. This photograph shows a blue whale being butchered at the Coal Harbour Whaling Station, British Columbia, Canada, in 1963. Commercial whaling of great whales was banned in 1985 but began again in 1993.

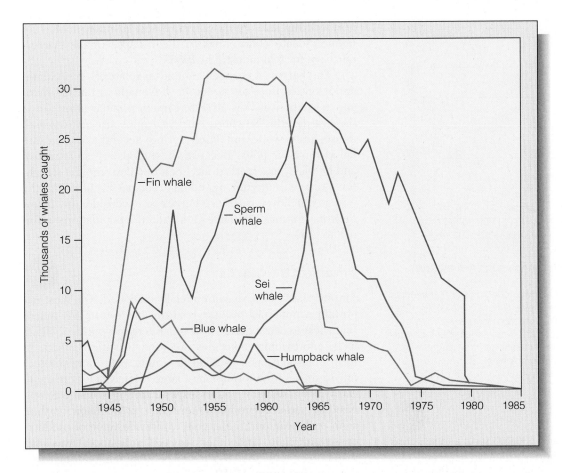

Figure 17.17 All five species included in this graph are commercially extinct, and the blue whale is still in danger of total extinction even though commercial blue whale hunting ended in 1964. At that time only about 1,000 blue whales were left, and experts fear that the population may have been too small to recover.

The meat and blubber of this whale is prized in Japan as an expensive delicacy. The minke whale population has been estimated at 1.2 million, a number that may withstand the present level of harvest. A decision to increase minke whaling, however, may doom the minke to the same fate as most other whale species. In 2000, Japan extended "scientific" whaling to sperm whales and Bryde's whales.

There is a glimmer of hope. In 1994 the International Whaling Commission voted overwhelmingly to ban whaling in about 21 million square kilometers (8 million square miles) around Antarctica, thus protecting most of the remaining large whales that feed in those waters. The sanctuary is often ignored. In their quest for minke (and other whales), Japanese and Norwegian whalers have entered the area and harpooned many animals. Between them, the Japanese and Norwegians caught 1,078 minke whales in 1999. Chile, Peru, and North Korea are considering joining the hunt. Pirate whalers based in other countries can also catch whales in the Antarctic and sell the flesh on the Japanese market. Conservation efforts can do some good, however. Although it was hunted nearly to extinction, the California gray whale has long been off limits to all but aboriginal hunters. Its numbers have grown, and it was removed from the endangered list in 1993.

Many more small whales have been killed than large ones, but not for food or raw materials. For reasons that are not well understood, dolphins (which are small whales) gather above schools of yellowfin tuna in the open sea. Fishermen have learned to find the tuna by spotting the dolphins. Nets cast to catch the tuna also entangle the air-breathing dolphins, and the mammals drown. The dead dolphins are simply pitched over the side as waste. More than 6 million small whales have been killed in association with tuna fishing since 1971 (see **Figure 17.18**).

Passage of the U.S. Marine Mammals Protection Act in 1972 brought a drop in the number of dolphin deaths. However, from 1977 to 1987 the number of tuna boats in the U.S. fleet declined from more than 100 boats to 34 boats, while the foreign fleet rose from fewer than 10 to more than 70 boats. Foreign fishermen do not always abide by the dolphin-saving provisions of the act, but the U.S. Congress voted in 1988 to ban importation of tuna not caught in accordance with new methods designed to reduce the kill. In 1990, American tuna processing companies agreed to buy tuna only from fishermen whose methods do not result in the death of dolphins. American commercial fishermen have agreed to comply, and conservationists hope that the foreign fleets will follow their lead.

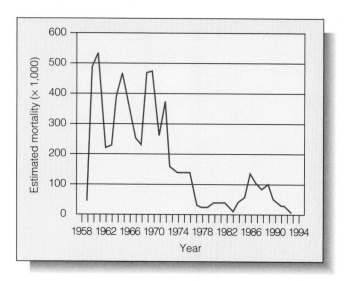

Figure 17.18 Estimates of the total mortality of dolphins of all species caused by the U.S. and foreign purse-seine fishery in the eastern Pacific, 1959–93. The Marine Mammals Protection Act, passed by Congress and signed by the president in 1972, was responsible for a dramatic decrease in dolphin bykill. "Dolphin-safe" tuna is caught in ways that do not drown the air-breathing dolphins.

Fur-Bearing Mammals

Worldwide, between 400,000 and 500,000 seals and sea lions are killed annually for fur. Eight species of seals and one species of sea lion are of economic importance. By the turn of the twentieth century a few of these species had been hunted nearly to extinction, but a system of hunting quotas has allowed most of the populations to recover. A provision of the 1972 U.S. Marine Mammals Protection Act protects all fur-bearing marine mammals in U.S. territory except for the

northern fur seal. Public pressure and an unfavorable economic climate contributed to the collapse of the Alaskan northern fur seal industry in 1986.

The harp seal, a species found in relatively great abundance on ice floes off the coast of Canada in the Labrador and Barents Seas, has attracted much popular attention in the last decade. Newborn harp seal pups are covered with dense, pure white fur. Each spring, several thousand are killed for their pelts. Because these pups are so attractive and helpless, much effort has been made to stop the slaughter. Harp seals are not an endangered species, however; the rate at which harp seals are harvested is considerably below their maximum sustainable yield, and the population continues to grow.

Botanical Resources

Marine plants are also commercially exploited. The most important commercial product is **algin,** made from the mucus that slickens seaweeds. When separated and purified, algin's long, intertwining molecules are used to stiffen fabrics; to form emulsions such as salad dressings, paint, and printer's ink; to prevent the formation of large crystals in ice cream; to clarify beer and wine; and to suspend abrasives. The U.S. seaweed gel industry produces more than $220 million worth of algin each year, and the annual worldwide value of products containing algin (and other seaweed substances) was estimated to be more than $41 billion in 1998. A harvesting barge is shown in **Figure 17.19.**

Seaweeds are also eaten directly. The Japanese consume 150,000 metric tons (165,000 tons) of *nori* each year; seaweed and seaweed extracts are also eaten in the United States, Britain, Ireland, New Zealand, and Australia. Their mineral content and fiber are useful in human nutrition.

Figure 17.19 A kelp cutter harvesting kelp. Metal shears about a meter below the surface trim the kelp, and a moving ramp brings it aboard.

Norm Cole

Aquaculture and Mariculture

17-22

Aquaculture is the growing or farming of plants and animals in any water environment under controlled conditions. Aquaculture production currently accounts for more than one-quarter of all fish consumed by humans. Most aquaculture production occurs in China and the other countries of Asia. In 1998, more than 28 million metric tons (31 million tons) of fish—mostly freshwater fish—was produced worldwide. By 2010, fish aquaculture could overtake cattle ranching as a food source.

Mariculture is the farming of *marine* organisms, usually in estuaries, bays, or nearshore environments or in specially designed structures using circulated seawater. Mariculture facilities are sometimes placed near power plants to take advantage of the warm seawater flowing from their cooling condensers.

Worldwide mariculture production is thought to be about one-eighth that of freshwater aquaculture. Several species of fish, including plaice and salmon, have been grown commercially, and marine and brackish-water fish account for 67% of the total production. Shrimp mariculture is the fastest-growing and most profitable segment, with an annual global value exceeding $6 billion in 1998.

In the United States, annual revenue from mariculture exceeds $150 million, with most of the revenue generated from salmon and oyster mariculture (**Figure 17.20**). About half of the oysters consumed in North America are cultured. Not all mariculture produces food; cultured pearls are an important industry in Japan, China, and northern Australia. Japan leads the world in mariculture.

Ranching, or open-ocean mariculture, is an interesting variation. Juveniles are grown in a certain area, released into the ocean, and expected to return when mature. The natural homing instinct of salmon makes salmon ranching quite success-ful—about 20% of the world supply of salmon is ranched. Yet only about 1 in 50 of ranched juveniles returns to the area of release. In Japan, recent attempts to extend ranching techniques to yellowtail and tuna have met with only limited success.

Mariculture is an expanding industry, growing at about 8% annually compared to a growing decline for the world marine fishery as a whole. Mariculture produces mostly luxury seafoods such as oysters and abalone, and it uses fish meal from so-called *trash fish* as feed. (Trash fish are edible but considered unappealing because they are bony or bad tasting.) As world population increases and the demand for protein grows among the world's millions of undernourished people, however, today's trash fish may become more acceptable as human food.

Drugs

17-23

The earliest recorded use of medicines derived from marine organisms appears in the *Materia Medica* of the emperor Shen Nung of China, 2700 B.C. Modern medical researchers estimate that perhaps 10% of all marine organisms are likely to yield clinically useful compounds. One such medicine is derived from a Caribbean sponge and is already in use: Acyclovir, the first antiviral compound approved for humans, has been fighting herpes infections of the skin and nervous system since 1982. A class of anti-inflammatory drugs known as pseudopterosins, developed by researchers at the University of California, has also been successful.

Newly discovered compounds are also being tested. A common bryozoan—a small encrusting invertebrate—has been found to produce a potent anticancer chemical that is now being tested in human volunteers. Extracts from 30% of all tunicate species investigated show antiviral and antitumor

Figure 17.20 Oyster mariculture in Puget Sound. Oyster larvae attached to old oyster shells are put in screen bags, which are placed on intertidal mud flats. The bags must be lifted monthly to clean out excess mud. Oysters are ready for harvesting in about three years.

Joseph W. Holliday

activity; one of these extracts, Didemnin-B, shows promise as a treatment for malignant melanoma, the deadliest form of skin cancer. Another tunicate derivative, Esteinascidin 743, has been found to be useful in the treatment of skin, breast, and lung cancers. A related compound found in the same organism, Aplidine, shows promise in shrinking tumors in pancreatic, stomach, bladder, and prostate cancers.

Cancer is not the only target. A compound derived from cyanobacteria stimulated the immune system of test animals by 225% and cells in culture by 2,000%; this drug may become useful in treating AIDS. Vidabarine, another antiviral drug developed from sponges, may attack the AIDS virus directly.

Other useful chemicals derived from marine organisms include Padan, a powerful new insecticide derived from an annelid worm, and a new family of steroids that may prove to be useful in the regulation of human growth. In the not-too-distant future, drugs from marine organisms may be among our most valuable marine resources.

NONEXTRACTIVE RESOURCES

Transportation and recreation are the main nonextractive resources the ocean provides. People have been using the ocean for transportation for thousands of years. Through most of this time the transport of cargo has produced far more revenue than the movement of passengers. At present, oil tankers ship the greatest gross tonnage of any type of cargo (over 275 million metric tons—303 million tons—annually). Oil accounts for about 53% of the total value of the world trade transported by sea; iron, coal, and grain make up another 24%.

Nearly half of the world's crude oil production is transported to market by ships. Tankers are needed because very few of the major oil-drilling sites are close to areas where the demand for refined oil products is highest. The largest tankers (**Figure 17.21**) are more than 430 meters (1,300 feet) long and 66 meters (206 feet) wide, and they carry more than 500,000 metric tons (3.5 million barrels) of oil.

Modern harbors are essential to transportation. Cargoes are no longer loaded and off-loaded piece by piece by teams of longshoremen. Today's harbors bristle with automated bulk terminals, high-volume tanker terminals (both offshore and dockside), containership facilities (see **Box 17.1**), roll-on–roll-off ports for automobiles and trucks, and passenger facilities required by the growing popularity of cruising. Most of this specialized construction has occurred since 1960. New Orleans is now the greatest North American port; nearly 210 million tons of cargo—most of it grain—passed through its docks in 1999. The world's largest container terminal is near

Figure 17.21 *Murex*, first of a series of five VLCCs (very large crude carriers) being built for the Shell Oil Company by Daewoo Heavy Industries, South Korea. These ships, first of a new generation of tankers, are of double-hulled design and will carry 2.15 million barrels of crude oil at a service speed of 28 kilometers (18 miles) per hour. Somewhat smaller than the largest tankers, *Murex* is 332 meters (1,089 feet) long and extends 22 meters (72 feet) below the surface when fully loaded. *Murex*, built to high standards of reliability and safety, is seen here unballasted on the day of her naming, 17 January 1995.

Hong Kong, China. In 1995, it became the first facility to move more than 1 million containers in a single month!

Transportation is sometimes combined with recreation. In the last decade, the cruise industry has experienced spectacular growth. Passengers on luxurious liners can enjoy a few relaxing days on the ocean crossing the North Atlantic, visiting tropical islands, or touring places accessible to the public only by ship. (Indeed, tourism is now the world's largest industry.) Ocean-related leisure pursuits, including sport fishing, surfing, diving, day cruising, sunbathing, dining in seaside restaurants, and just plain relaxing, contribute to the economy. In addition to being important producers of revenue, public aquariums and marine parks (such as Sea World) are centers of education, research, and captive breeding programs (**Figure 17.22**). Even public interest and curiosity about whales are a source of recreational revenue. In the United States in 1994, whale-watching trips generated an annual direct revenue of $122 million; indirect revenues including dolphin displays, whale artwork and books, and conservation group donations amounted to another $382 million.

And don't forget about real estate. As coastal real estate values will attest, people enjoy living near the ocean. By 2015 about 165 million Americans will be living in coastal areas on property valued in excess of $7 trillion.

The Law of the Sea

Prehistoric peoples living near the shore were the earliest users of marine resources. With the rise of nation-states and military establishments, the fight for control of marine resources began. Some nations assumed that the ocean belonged to all and endeavored to guarantee free right to passage and resources. Others decided that it belonged to none and tried to control access to ports and resources by force. In 1604 Hugo Grotius, a learned Dutch jurist, wrote *De Jure Praedae* (*On the Law of Prize and Booty*), a treatise justifying the action of a Dutch admiral who had successfully defended Dutch trading rights in a dispute with Portugal. One chapter of this work, in which Grotius defended free ocean access for all nations, was reprinted in 1609 under the title *Mare Liberum* (*A Free Ocean*). *Mare Liberum* formed the basis for all modern international **laws of the sea.**

About a century later, in 1703, the concept of territorial seas adjacent to land was recognized. A country's seaward boundary was set at about 5 kilometers (3 miles)—the distance a cannonball could be fired from shore. This 5-kilometer (3-mile) limit stood until 1945.

The United Nations and the International Law of the Sea

After World War II the technology became available to search for oil and natural gas on continental shelves. After U.S. oil companies found rich deposits beyond the 5-kilometer (3-mile) limit off Louisiana, President Harry Truman issued a proclamation annexing the physical and biological resources of the continental shelf contiguous to the United States. Other nations rushed to make similar claims.

The United Nations then became involved. A committee of the General Assembly began to formulate policy, which was later presented at the First United Nations Conference on the Law of the Sea in 1958 in New York. Twenty-four years of effort by delegates from many interested nations resulted in the 1982 Draft Convention on the Law of the Sea. In April 1982 the United Nations adopted the convention by a vote of

Figure 17.22 Public aquariums and marine parks are important resources.

Containers, World Economics, and Your Shoes

BOX 17.1

a A representative of a U.S. manufacturer of high-tech apparel inspects and approves a shipment of goods being made in a factory in Shenzhen, China.

Tom Garrison

Look at the labels of your clothing and shoes. Mine seem to come from everywhere—the Dominican Republic, Malaysia, Sri Lanka, Israel, Thailand, Bangladesh, and especially China. Our local supermarket stocks watermelon in the middle of winter, canned clams from Korea, fresh tuna from Thailand, and New Zealand apples. The citizens of Hong Kong consume more California oranges per capita than any other people in the world. The electronics store at the local mall is brimming with DVD players and MP3 recorders, nearly all produced in the Far East.

This trade revolution is possible because of a revolution in shipping: the *container*. A container is a large, strong, internationally standardized packing box for cargo. Goods can be safely stowed and transported in a secure, locked space that is resistant to damage, theft, and pilferage. A container

is designed to be easily lifted and moved and for the most efficient use of space. It may be carried by any type of transportation—road, rail, or sea.

Before the widespread advent of containers (about 25 years ago), a large cargo ship could take two weeks to load, and the irregularly packaged cargo could easily be damaged or simply disappear. Now, a modern containership can be loaded in two days and can cross the Pacific from China to Los Angeles in 16 days.

Standardization of container sizes makes loading the ships efficient. The basic unit, set by the International Organization of Standardization (ISO), is a 20-foot container. With dimensions of 20 feet in length, 8 feet in width, and 8 feet 6 inches in height, a container can be loaded with 15 to 20 metric tons of cargo. A larger-size container 40 feet in length has become more popular in the last decade; it can be loaded with up to 30 metric tons of cargo. A 45-foot container is used for special purposes. Additionally there are standardized designs for general-purpose containers, hard-top containers, flat-rack containers, bulk containers, and refrigerated containers. These efficiencies mean that imported goods are transported more quickly and at lower cost than ever before.

Specialized ships (containerships), some of which are about as large as the U.S. Navy's *Nimitz*-class aircraft carriers, are designed to transport up to 4,100 standard containers at a time! They operate safely with a small crew and are driven by efficient diesel or gas turbine engines.

A container cycle begins when a U.S. importer's representative approves the goods for transport (**Figure a**). The containers are loaded and locked at a factory or production facility, then moved by truck (**Figure b**) or rail to a container terminal (**Figure c**) where cranes load them onto a containership (**Figure d**). Once unloaded, cleared through customs, and transported by rail and truck to the importer's central warehouse, the containers are unlocked and opened, and their contents are distributed. Container shipping is efficient, swift, and sure. Clothing made in Hong Kong can be worn in Kansas City less than a month after manufacture.

Containers that have carried finished goods to the United States return to containerships and factories carrying raw materials (plastic pellets, fabric, computer chips, fruit, or even recycled cardboard). The cycle repeats.

International trade by sea is growing at an explosive rate. Look around you on the highways, and you'll see shipping containers on trucks and trains going everywhere. Almost all of them are part of the international container cycle.

17-25

Tom Garrison

b Sealed safely in a container, the apparel is transported from factory to port through specialized economic zones in China.

Tom Garrison

c The clothing leaves from the busiest container port in the world. In 1995, this state-owned facility on the Pearl River estuary in southeastern China became the first to transfer 1 million containers in one month.

Port Authority of NY and NJ

d *Regina Maersk,* one of the world's largest containerships, enters the port of New York.

130 to 4, with 17 abstentions. (The United States, Turkey, Venezuela, and Israel voted against the convention.) By 1988 more than 140 countries had signed all or most parts of the treaty. It is now legally binding, but signatories have selectively chosen to respect or ignore its individual provisions.

Here are some important features of the 1982 Draft Convention:

- **Territorial waters** are defined as extending 12 miles (18.2 kilometers) from shore. A nation has the right to jurisdiction within its territorial waters. Straits used for international navigation are excluded from a nation's territorial waters in that any vessel has the right to innocent passage.
- The 200 nautical miles (370 kilometers) from a nation's shoreline constitute its **exclusive economic zone** (**EEZ**). Nations hold sovereignty over resources, economic activity, and environmental protection within their EEZs.
- All ocean areas outside the EEZs are considered the **high seas.** In tradition, the high seas are common property to be shared by the citizens of the world. An International Seabed Authority was established to oversee the extraction of mineral resources from the deep sea.
- The values of protecting the ocean and preventing marine pollution were endorsed.
- Subject to some conditions, the freedom of scientific research in the ocean was encouraged.

The convention places about 40% of the world ocean under the control of the coastal countries, within the EEZs. The resources of the remaining 60%, the high seas, are to be shared by the citizens of the world.

The United States Exclusive Economic Zone

The United States did not sign the 1982 Draft Convention for a variety of reasons. One reason was the concern that private enterprise would be deprived of profits if it were made to share high seas resources with other countries. Instead, the United States unilaterally claimed sovereign rights and jurisdiction over all marine resources within its own 200-nautical-mile region, which it called the **United States Exclusive Economic Zone.** The proclamation—similar in most ways to the 1982 United Nations treaty but lacking the provision of shared high seas resources—was signed by President Ronald Reagan on 10 March 1983. The United States EEZ brings within national domain over 10.3 million square kilometers (4 million square miles) of continental margins, an area 30% larger than the land area of the United States and a region of diverse geological and oceanographic settings (see **Figure 17.23**).

The first step in exploring this new region has been to map the surface of the seafloor, a project that is still going on. The first bottom surveys were conducted off the Pacific coast of the United States because of the energy and mineral resources known to exist there. Massive deposits of various metallic sulfides are being explored at the hydrothermal vents on the Gorda and Juan de Fuca Ridges. More than a hundred previously unknown volcanoes have been mapped within the U.S. EEZ off the U.S. Pacific coast. Huge faults, submarine landslides, seamounts, and details of the spreading crest of the oceanic ridges are a few of the features that have been discovered so far. Knowledge of the modern tectonic setting for ore formation has been valuable in locating new ore deposits on land.

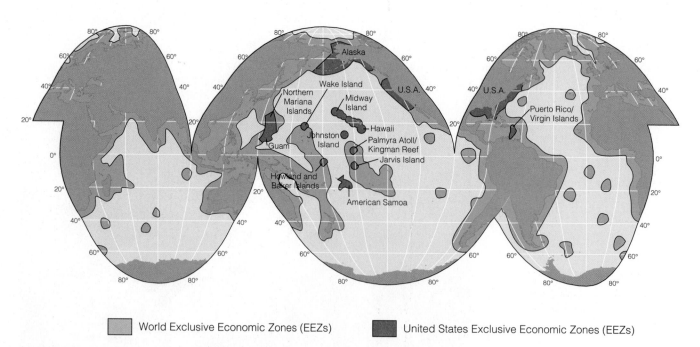

Figure 17.23 The U.S. Exclusive Economic Zone (EEZ) is shown here in red. Note how much of the EEZ is associated with distant possessions. Other EEZs are shown in darker blue.

Manganese nodules and crusts have been discovered within the U.S. EEZ off the Atlantic and Pacific coasts, Hawaii, and the Pacific Island territories. Cobalt, present in the manganese crusts, is being studied to determine how the deposit is formed and how it can be retrieved economically.

The EEZ project also includes research into meteorology, accurate weather forecasting, environmental studies, effects of plate movement on the ocean floor, effects of mining on seafloor organisms, and geohazards such as submarine landslides and earthquakes. It is hoped that this new era of marine exploration and research will lead to informed decisions about offshore activities that will benefit not only the United States but also other maritime nations. Maybe the treasure trove is really there, waiting for these new developments in marine technology and international cooperation to tap the riches of the sea.

QUESTIONS FROM STUDENTS

1. When will we run out of oil?

By the time the world's crude oil supplies have been substantially depleted, the total amount of oil extracted is expected to range from 1.6 to 2.4 trillion barrels. If this estimate is correct, consumption could continue at the present level until some time between 2025 and 2040, at which time it would drop quite rapidly. But, in fact, we will never run *completely* out of oil—there will always be some oil within Earth to reward great effort at extraction. The days of unlimited burning of so valuable a commodity are nearing an end, however. Future civilizations will surely look back in horror at the fact that their ancestors actually burned something as valuable as lubricating oil.

2. Where are petroleum geologists looking for new reserves of oil?

In 2000, a team of researchers at the U.S. Geological Survey raised its previous estimate of the world's remaining crude oil reserves by 20% to a total of 649 billion barrels. The largest reserves of undiscovered oil are believed to lie in existing fields in the Middle East; northeast of Greenland's continental shelf; in areas in Alaska, Siberia and the Caspian Sea; and in the deltas of the Niger and Congo in Africa. The shelf surrounding Antarctica may contain the greatest reserves of all. (The difficulties of Antarctic extraction would be extraordinary, indeed!)

3. How does the sea salt I see in the health food stores differ from regular table salt?

Unlike the producers of regular table salt, the producers of commercial sea salt do not move the evaporating brine from pond to pond to isolate the different precipitates. Sea salt therefore contains the ocean's salts in their natural proportions, with NaCl making up 78% of the mix. Sea salt has a slightly bitter taste from the potassium and magnesium salts present. Some people believe that the variety of minerals in sea salt is beneficial, but most people obtain adequate amounts of these minerals from a normal diet.

4. What about treasure as a marine resource?

Florida entrepreneur Mel Fisher has recovered nearly half a billion dollars in gold and archaeological treasures from sunken ships along the U.S. southeastern coast. (He believes there's more where that came from.) In 1981, a British salvage company recovered $80 million in gold ingots from a British warship sunk in 1942 off Murmansk. Private consortia have plundered the hulk of the RMS *Titanic*—one can now purchase bits of coal salvaged from the wreck, and other artifacts will not be far behind. The German battleship *Bismarck* is being considered for exploitation. Even priceless archaeological treasures are being excavated and sold to the highest bidder. Treasure is not always economic—the scholarly value of an ancient shipwreck is beyond monetary calculation.

5. The problems of overfishing and bykill concern me. What seafood should I eat, and what should I avoid?

Eat sparingly, if at all:
Sharks, especially shark-fin soup
Swordfish
Bluefin tuna (most canned tuna is OK)
Orange roughy
Grouper
Toothfish

OK to eat:
Pacific cod
Atlantic striped bass (taken offshore)
Atlantic mackerel
Atlantic herring
Farm-raised shrimp
Ranched salmon
Canned tuna with "dolphin safe" labels

(*Sources:* Nash, M. "The Fish Crisis." *Time*, 11 August 1997, p 67; National Marine Fisheries Statistics fax-on-demand service.)

KEY CONCEPTS TO REVIEW

- Marine resources may be categorized as physical, biological, and nonextractive. They may be renewable or nonrenewable. The contribution of marine resources to the world economy has become so large that international laws now govern their allocation.

- Pollution is a potential side effect of resource extraction and use.

- Though expensive to obtain, offshore oil and natural gas presently provide nearly a third and a quarter (respectively) of world needs. Sand and gravel are second in dollar value.

- Except in a few locales, marine energy production is more experimental than practical.

- Though fishery output continues to increase, *per capita* use of oceanic biological resources has fallen since about 1970. Biological resources are often taken without regard to replenishment time.

- Worldwide bykill may amount to a third of total biological resource utilization.

- Whaling continues despite efforts to limit it.

- Aquaculture—much of it freshwater—accounts for over one-quarter of all fish consumed by humans.

- Promising drugs are being extracted from marine resources; some are approved for human use.

- World economics strongly depends on marine transport. Most international cargo is transported in standard containers aboard containerships.

- International agreements, collectively termed Laws of the Sea, govern resource allocation and trade.

INTERNET STUDY RESOURCES

The Web site for this book contains helpful study aids. Log on to:

 www.brookscole.com/053437557xs

and click on the Chapter-by-Chapter area. Choose Chapter 17 and select a resource:

- **Flash Cards** allows you to test your mastery of the Terms and Concepts to Remember for this chapter.
- **Tutorial Quizzes** provides a multiple-choice practice quiz.
- **Student Guide to InfoTrac College Edition** will lead you to Critical Thinking Projects that use InfoTrac College Edition as a research tool.

TERMS AND CONCEPTS TO REMEMBER

algin	marine energy resources
aquaculture	maximum sustainable yield
biological resources	nonextractive resources
bykill	nonrenewable resources
commercial extinction	overfishing
desalination	physical resources
drift net	potable water
exclusive economic zone	renewable resources
(EEZ)	territorial waters
high seas	United States Exclusive
Laws of the Sea	Economic Zone
mariculture	

STUDY QUESTIONS

Review Questions

1. Distinguish between physical and biological resources, and between renewable and nonrenewable resources.

2. What are the three most valuable physical resources? How does the contribution of each to the world economy compare to the contribution of that resource derived from land?

3. What are the sources of metals mined or extracted (or potentially mined or extracted) from the sea?

4. Distinguish between mariculture and aquaculture. Does freshwater or seawater aquaculture provide more human food worldwide? What are the prospects for mariculture?

5. What is a nonextractive resource? Give some examples.

Critical Thinking Questions

1. How are oil and natural gas thought to be formed? How can these substances be extracted from the seabed? Why are the physical characteristics of the surrounding rock important?

2. What method of ocean energy extraction has been most practical? Which has the greatest potential? Why has the latter not been exploited on a large scale?

3. Does the ocean provide a substantial percentage of all protein needed in human nutrition? Of all animal protein? What is the most valuable biological marine resource? The fastest-growing fishery?

4. What are the signs of overfishing? How does the fishing industry often respond to these signs? What is the usual result? What is bykill?

5. What are the advantages and disadvantages of a proclaimed EEZ? Do you feel the United States was justified in proclaiming its own EEZ separate from the provisions of the 1982 United Nations Convention?

6. **InfoTrac College Edition Project** Offshore oil drilling requires large drilling and storage structures. However, these structures can become obsolete, and their disposal can become problematic. Oil companies' proposed solution has been to sink the structures; one example is Shell Oil's Brent spar oil storage buoy off the coast of Scotland. This solution has been strongly opposed by environmental groups. Do you think that sinking is an acceptable method for disposing of obsolete oil rigs? Why or why not? Do some research using InfoTrac College Edition to back up your position.

RESOURCES FOR FURTHER READING AND RESEARCH

The Web site for this book contains many ideas for further reading and research. Log on to:

 www.brookscole.com/053437557xs

and click on the Chapter-by-Chapter area. Choose Chapter 17 and select a resource:

- **References** lists the major books and articles consulted in writing this chapter, along with comments from the author about their content and reading level.
- **Hypercontents** takes you to an extensive list of sites with news, research, and images related to individual sections of the chapter.

- In **Student Guide to InfoTrac College Edition** scroll down to Suggested Readings from InfoTrac College Edition for brief descriptions of the articles listed and search hints for finding them in InfoTrac College Edition.
- **Regional InfoTrac College Edition** articles are organized into East Coast, West Coast, and Gulf Coast regions, allowing you to study oceanography on a more local level. You can also access Regional InfoTrac College Edition at www.localocean.com.

 For additional readings, go to InfoTrac College Edition, your online research library, at: http://infotrac.thomsonlearning.com/

Environmental Concerns

A Cautionary Tale

Let me tell you a story.

Easter Island was home to a culture that rose to greatness amid abundant resources, attained extraordinary levels of achievement, and then died suddenly and alone, terrified, in the empty vastness of the Pacific. The inhabitants had destroyed their world.

Europeans first saw Easter Island (Rapa Nui) in 1722. The Dutch explorer Jacob Roggeveen encountered the small volcanic speck on Easter morning while on a scouting voyage. The island, dotted with withered grasses and scorched vegetation, was populated by a few hundred skittish, hungry, ill-clad Polynesians who lived in caves. During his one-day visit, Roggeveen was amazed to see more than 200 massive stone statues standing on platforms along the coast. At least 700 more statues were later found partially com-

Giant statues look to the horizon on Easter Island (Rapa Nui). The civilization that built the statues had all but destroyed its world when the first Europeans arrived in 1722.

pleted, lying in the quarries as if they had just been abandoned by workers. Roggeveen immediately recognized something inexplicable: "We could not comprehend how it was possible that these people, who are devoid of thick heavy timber for making machines, as well as strong ropes, nevertheless had been able to erect such images." The islanders had no wheels, no powerful animals, and no resources to accomplish this artistic, technological, and organizational feat. And there were too few islanders—600 to 700 men and about 30 women.

When Captain James Cook visited in 1774, the islanders paddled out to his ships in canoes "put together with manifold small planks . . . cleverly stitched together with very fine twisted threads." Cook noted the natives "lacked the knowledge and materials for making tight the seams of the canoes, they are accordingly very leaky, for which reason they are compelled to spend half the time in bailing." The canoes held only one or two people, and they were less than 3 meters (10 feet) long. Only three or four canoes were seen on the entire island, and Cook estimated the human population at less than 200. Within 50 years, by the time of Cook's visit, nearly all the statues had been overthrown, tipped into pits dug for them, often onto a spike placed to shatter their faces upon impact.

Archaeological research on Easter Island has revealed a chilling story. Only 165 square kilometers (64 square miles) in extent and the easternmost outpost of Polynesia, Easter Island is one of the most isolated places on Earth. Voyagers from the Marquesas first reached Easter Island about A.D. 350, possibly after being blown off course by a series of storms. They found a miniature paradise. In its fertile volcanic soils grew dense forests of palms, daisies, grasses, hauhau trees, and toromino shrubs. Large oceangoing canoes could be built from the long, straight, buoyant Easter Island palm and strengthened with rope made from the hauhau trees. Toromino firewood cooked the fish and dolphins caught by the newly arrived fishermen, and forest tracts were cleared to plant crops of taro, bananas, sugarcane, and sweet potatoes. The human population thrived.

By the year 1400 the population had blossomed to between 10,000 and 15,000 people. Gathering, cultivating, and distributing the rich bounty for so many inhabitants required complex political control. As numbers grew, however, stresses began to be felt—the overuse and erosion of agricultural land caused crop yields to decline, and the nearby ocean was stripped of benthic organisms so the fishermen had to sail greater distances in larger canoes. As resources became inadequate to support the growing population, those in power appealed to the gods. They redirected community resources to carve worshipful images of unprecedented size and power. The people cut down more large trees for ropes and rolling logs to place the heads on huge carved platforms.

By this time the island's seabirds had been consumed, and no new birds came to nest. The rats that had hitchhiked to the island on the first canoes were raised for food. Palm seeds were prized as a delicacy. All available land was under cultivation. As resources continued to shrink, wars broke out over dwindling food and space. By about 1550 no one could venture offshore to harpoon dolphins or fishes because the palms needed to construct seagoing canoes had all been cut down. The trees used to provide rope lashings were extinct, their wood burned to cook what food remained. The once-lush forest was gone. Soon the only remaining ready source of animal protein was being utilized: The people began to hunt and eat each other.

Central authority was lost, and gangs arose. As tribal wars raged, the remaining grasses were burned to destroy hiding places. The rapidly shrinking population retreated into caves from which raids were launched against enemy forces. The vanquished were consumed, their statues tipped into pits and destroyed. Around 1700 the human population crashed to less than a tenth of its peak numbers. No statues stood upright when Cook arrived.

Even if the survivors had wanted to leave the island, they could not have done so. No suitable canoes existed; none could be made. What might the people have been thinking as they chopped down the last palm? Generations later, their successors died, wondering what the huge stone statues had been looking for.

As you read this chapter, you may be struck by similarities between Easter Island and Earth. Here we will survey the pollutants, the habitat destruction, the mismanagement of living resources, and the global changes that are currently stressing the planet's environment. It's a depressing list, but at the end you'll find a glimmer of hope. After all, the Easter Islanders had no books, and no histories of other doomed societies. We might be able to learn from their mistakes.

WHAT TO WATCH FOR IN CHAPTER 18

Our species has always exercised its capacity to consume resources and pollute its surroundings, but only in the last few generations have our efforts affected the ocean and atmosphere on a planetary scale. The introduction into the biosphere of unnatural compounds (or natural compounds in unnatural quantities) has had, and will continue to have, unexpected detrimental effects. The destruction of marine habitats and the uncontrolled harvesting of the ocean's living resources have also disturbed delicate ecological balances. We have embarked on a time of inadvertent global experimentation and find ourselves in difficult situations for which solutions do not come easily.

This chapter may surprise you in unexpected ways. For example, did you know that Americans dump much more used engine oil into storm drains and sewers than was spilled by the grounding of the tanker *Exxon Valdez*? Or that pollution control and the damage caused by introduced species cost the U.S. economy about $300 billion every year? Your personal health may be jeopardized by pollutants in the ocean and atmosphere. Ours is the last generation to live as all others have lived—"pollute and move on" won't work anymore.

This is an unpleasant chapter—the only one I don't enjoy working on. Nonetheless, I urge you to read it carefully and decide what actions to take. After you've finished, read the Afterword that follows. You'll find some hope there.

MARINE POLLUTION

We define **marine pollution** as the introduction into the ocean by humans of substances or energy that changes the quality of the water or affects the physical, chemical, or biological environment. It is not always easy to identify a pollutant; some materials labeled as pollutants are produced in large quantities by natural processes. For example, a volcanic eruption can produce immense quantities of carbon dioxide, methane, sulfur compounds, and oxides of nitrogen. Excess amounts of these substances produced by human activity may cause global warming and acid rain. For this reason we need to distinguish between *natural pollutants* and *human-generated pollutants*.

We have long used the sea as a dump for our wastes. About three-quarters of the pollution entering the ocean comes from human activities on land (**Figure 18.1**). The ocean's great volume and relentless motion dissipate and distribute natural and synthetic substances, but its ability to absorb is not inexhaustible.

No one knows to what extent we have contaminated the ocean. By the time the first oceanographers began widespread testing, the Industrial Revolution was well under way and changes had already occurred. Traces of synthetic compounds have now found their way into every oceanic corner. It is sad to consider that we will never know what the natural ocean was like nor what remarkable plants and animals may have vanished as a result of human activity. Our limited knowledge of pristine conditions is gleaned from small seawater samples recovered from deep within the polar ice pack and from tiny air bubbles trapped in glaciers. There are few undisturbed habitats left to study, and few marine organisms are completely free of the effects of ocean pollutants.

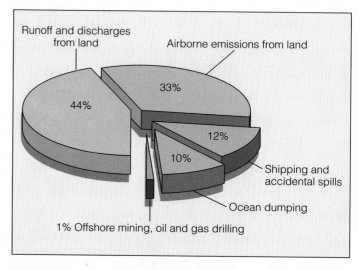

Figure 18.1 Sources of marine pollution. (Source: Joint Group of Experts on the Scientific Aspects of the Marine Environment, *The State of the Marine Environment,* UNEP Regional Seas Reports and Studies No. 115, Nairobi: U.N. Environment Programme, 1990).

Characteristics of a Pollutant

18-3

A **pollutant** causes damage by interfering directly or indirectly with the biochemical processes of an organism. Some pollution-induced changes may be instantly lethal; other changes may weaken an organism over weeks or months, alter the dynamics of the population of which it is a part, or gradually unbalance the entire community.

In most cases an organism's response to a particular pollutant will depend on its sensitivity to the combination of *quantity* and *toxicity* of that pollutant. Some pollutants are toxic to organisms in tiny concentrations. For example, the photosynthetic ability of some species of diatoms is diminished when chlorinated hydrocarbon compounds are present in parts-per-trillion quantities. Other pollutants may seem harmless, as when fertilizers flowing from agricultural land stimulate plant growth in estuaries. Still other pollutants may be hazardous to some organisms but not to others. For example, crude oil interferes with the delicate feeding structures of zooplankton and coats the feathers of birds, but it is biodegradable and simultaneously serves as a feast for certain bacteria. Indeed, many pollutants are ultimately **biodegradable**—that is, able to be broken down by natural processes into simpler compounds.

Pollutants also vary in their *persistence;* some reside in the environment for thousands of years, while others last only a few minutes. Some pollutants break down into harmless substances spontaneously or through physical processes (like the shattering of large molecules by sunlight). Sometimes pollutants are removed from the environment through biological activity. For example, some marine organisms escape permanent damage by metabolizing certain hazardous substances to harmless ones. Many other pollutants resist attack by water, air, sunlight, or living organisms, however, because the synthetic compounds of which they are composed resemble nothing in nature.

The ways in which pollutants are changing the ocean and the atmosphere are often difficult for researchers to determine. Environmental impact cannot always be predicted or explained. As a result, marine scientists vary widely in their opinions about what pollutants are doing to the ocean and atmosphere and what to do about it. Environmental issues are frequently emotional, and media reports tend to sensationalize short-term incidents (such as oil spills) rather than focus on more serious, long-term problems (such as atmospheric changes or the effects of long-lived chlorinated hydrocarbon compounds).

Oil Pollution

18-4

Oil is a natural part of the marine environment. Oil seeps have been leaking large quantities of oil into the ocean for millions of years. The amount of oil entering the ocean has increased greatly in recent years, however, because of our growing dependence on marine transportation for petroleum products, offshore drilling, nearshore refining, and street runoff carrying waste oil from automobiles (**Figure 18.2**).

In the late 1990s people were using nearly 6.8 billion metric tons (7.5 billion tons) of crude oil each year, slightly more than half of which was transported to market in large tankers. In 1998 about 6 million metric tons (6.6 million tons) of oil entered the world ocean. Natural seeps accounted for only about 10% of this annual input, about 600,000 metric tons. About a third of the total was associated with marine transportation. Some of this oil was not spilled in well-publicized tanker accidents but was released during the loading, discharging, and flushing of tanker ships. Between 150,000 and 450,000 marine birds are killed each year as a result of the release of oil from tankers (**Figure 18.3**).

Much more oil reaches the ocean in runoff from city streets or as waste oil dumped down drains, poured into dirt, or hidden in trash destined for a landfill. Every year more than 908 million liters (about 240 million gallons) of used motor oil—about 22 times the volume of the *Exxon Valdez* spill—finds its way to the sea (**Figure 18.4**). Used motor oil is much more toxic than crude oil or newly refined oil because it has developed carcinogenic and metallic components from the heat and pressure within internal combustion engines. It's no wonder that a sea surface completely free of an oil film is quite rare.

It is difficult to generalize about the effects that a concentrated release of oil—an oil spill from a tanker, coastal storage facility, or drilling platform—will have in the marine

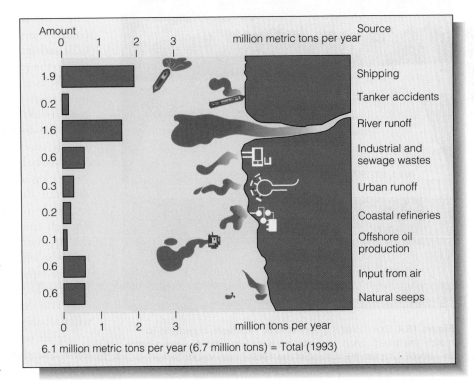

Figure 18.2 Sources of ocean oil pollution.

Reuters/Santiago Armas/Archive Photos

Figure 18.3 Birds coated with crude oil. Such birds die unless the oil is removed with a detergent solution. Many will die of cold or toxic effects even when the oil is removed; more will die because of suppressed immune systems. Up to half a million seabirds die each year from the effects of oil pollution. These birds died in a diesel oil spill in the Galápagos Islands in January, 2001.

Tom Garrison

Figure 18.4 Storm drains empty directly into rivers, bays, and ultimately the ocean. Each year, about 240 million gallons of used motor oil—about 22 times the volume of the 1989 *Exxon Valdez* tanker spill—is dumped down drains, poured into dirt, or concealed in trash headed for landfills. Much of this finds its way to the ocean.

environment. The consequences of a spill vary according to several factors: its location and proximity to shore; the quantity and composition of the oil; the season of the year, currents, and weather conditions at the time of release; and the composition and diversity of the affected communities. Intertidal and shallow-water subtidal communities are most sensitive to the effects of an oil spill.

Spills of *crude* oil are generally larger in volume and more frequent than spills of refined oil. Most components of crude oil do not dissolve easily in water, but those that do can harm the delicate juvenile forms of marine organisms even in minute concentrations. The remaining insoluble components form sticky layers on the surface that prevent free diffusion of gases, clog adult organisms' feeding structures, kill larvae, and decrease the sunlight available for photosynthesis. Even so, crude oil is not highly toxic, and it is biodegradable. Though crude oil spills look terrible and generate great media attention, most forms of marine life in an area recover from the effects of a moderate spill within about five years. For example, the 908 million liters (240 million gallons) of light crude oil released into the Persian Gulf during the 1991 Gulf War (see **Table 18.1**) dissipated relatively quickly and will probably cause little long-term biological damage.

Spills of *refined* oil, especially near shore where marine life is abundant, can be more disruptive for longer periods of time. The refining process removes and breaks up the heavier components of crude oil and concentrates the remaining

Table 18.1   Largest Marine Oil Spills Since 1960

Rank	Date	Incident and Location	Size of Spill (millions of gallons)
1	26 Jan 1991	Discharged into Persian Gulf during Gulf War	240.0
2	3 June 1979	Well spill into the Bay of Campeche, Mexico	140.0
3	19 July 1979	Tanker *Atlantic Empress,* wreck and subsequent tow, Caribbean NE of Trinidad and Tobago	83.8
4	4 Feb 1983	Well spill at Nowruz, Iran, Persian Gulf	80.0
5	6 Aug 1983	Fire aboard tanker *Castillo de Bellver,* South Africa	78.5
6	16 March 1978	Tanker *Amoco Cadiz* aground off Portsall, Brittany, France	68.7
7	10 Nov 1988	Tanker *Odyssey,* breakup NE of St. Johns, Newfoundland, Canada	43.1
8	11 April 1991	Tanker *Haven,* Mediterranean Sea off Genoa, Italy	42.3
9	18 Mar 1967	Tanker *Torrey Canyon* aground at Land's End, southwestern England	38.2
10	19 Dec 1972	Tanker *Sea Star,* Gulf of Oman, Oman	37.9
•			
•			
•			
46	24 March 1989[a]	Tanker *Exxon Valdez,* Prince William Sound, Alaska	11.0

Source: International Oil Spill Database, Cutter Information Corporation.

[a] The *Exxon Valdez* spill of March 1989 was the third worst in U.S. history. But when ranked with other massive marine oil spills involving both shipping and well accidents worldwide, it falls to number 46.

lighter, more biologically active ones. Components added to oil during the refining process also make it more deadly. Spills of refined oil are of growing concern because the amount of refined oil transported to the United States has risen dramatically through the 1980s and 1990s. A 1969 spill of refined oil at West Falmouth, Massachusetts, killed countless marine organisms and caused grave environmental damage that is still apparent in the form of reduced species diversity and fertility. In a similar accident in Mexico, the tanker *Tampico* went aground and dumped her load of refined gasoline into an unspoiled Baja California bay. The area still shows biological scars after nearly three decades.

The volatile components of any oil spill eventually evaporate into the air, leaving the heavier tars behind. Wave action causes the tar to form into balls of varying sizes, which then fall to the seafloor, where they may be assimilated by bottom organisms or incorporated into sediments. Bacteria will eventually decompose these spheres, but the process may take years to complete, especially in cold polar waters. This oil residue—especially if derived from refined oil—can have long-lasting effects on seafloor communities. The fate of spilled oil is summarized in **Figure 18.5.**

The methods used to contain and clean up an oil spill sometimes cause more damage than the oil itself. Detergents used to disperse oil are especially harmful to living things. Cleanup of the 1967 *Torrey Canyon* accident off the southern coast of England, one of the first large tanker accidents, did much more environmental damage than the 100,000 metric tons (110,000 tons) of crude oil released. Some resort beaches in the south of England were closed for two seasons, not because of oil residue but because of the stench of decaying marine life killed by the chemicals used to make the shore look clean.

Even the more sophisticated methods that were used in dealing with the *Exxon Valdez* disaster, the second worst oil spill in U.S. history (and the 46th worst spill ever; see Table 18.1), seem to have done more harm. When the supertanker *Exxon Valdez* ran aground in Alaska's Prince William Sound on 24 March 1989, more than 41 million liters (almost 11 million gallons; 29,000 metric tons) of Alaskan crude oil—about 22% of her cargo—escaped from the crippled hull (**Figure 18.6**). Only about 17% of this oil was recovered by a work crew of more than 10,000 people using containment booms, skimmer ships, bottom scrapers, and absorbent sheets. About 35% of the oil evaporated, 8% was burned, 5% was dispersed by strong detergents, and 5% biodegraded in the first five months. The rest of the oil, some 30% of the spill, formed oil slicks on Prince William Sound and

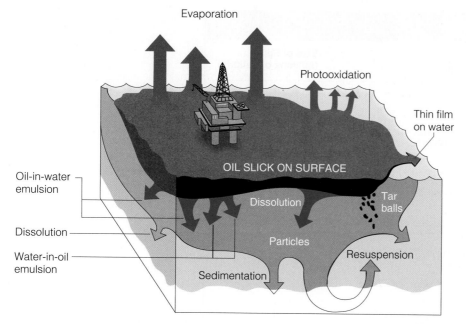

Evaporation

Photooxidation

Thin film on water

Oil-in-water emulsion

OIL SLICK ON SURFACE

Dissolution

Tar balls

Dissolution

Particles

Water-in-oil emulsion

Resuspension

Sedimentation

Figure 18.5 The fate of oil spilled at sea. Smaller molecules evaporate or dissolve into the water beneath the slick. Sunlight and atmospheric oxygen cause oxidation of the surface layer. Within a few days, water motion coalesces the oil into tar balls and semisolid emulsions of water-in-oil and oil-in-water. Tar balls and emulsions may persist for months after formation. Although bonding with sediment particles may sink oil droplets, suspended oil does not enter benthic subtidal sediments in large quantities. If crude oil is left undisturbed, bacterial activity will eventually consume it. Refined oil, however, can be more toxic, and natural cleaning processes take a proportionally longer time to complete.

fouled more than 450 kilometers (300 miles) of coastline (**Figure 18.7**).

Recent analysis of the affected parts of Prince William Sound shows the cleaned areas to be in generally worse shape than areas left alone. Most of the small animals that make up the base of the food chain in these areas were cooked by the 65°C (150°F) water used to blast oil from between the rocks. Others were smothered when the high-pressure jets rearranged sand and mud. It appears that an overambitious cleanup program can be counterproductive.

Sylvia Earle, chief on-site scientist of the National Oceanic and Atmospheric Administration (NOAA), has said, "Sometimes the best, and ironically the most difficult, thing to do in the face of an ecological disaster is to do nothing." The biological cost of the spill will not be known for many years. The cost of the cleanup has exceeded $3.5 billion. The legal costs could be even higher: In 1994 a federal jury awarded $5 billion in punitive damages to 14,000 plaintiffs including fishermen, native villagers, and fish processors. Exxon appealed the verdict.

Figure 18.6 Cleaning up the March 1989 *Exxon Valdez* oil spill, Prince William Sound, Alaska.

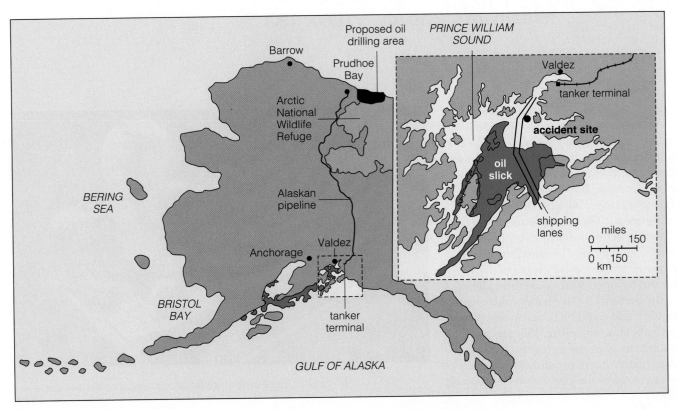

Figure 18.7 Site of the *Exxon Valdez* spill. About 41 million liters (11 million gallons) of crude oil fouled 450 kilometers (300 miles) of Alaskan coastline.

Of course the best way to deal with oil pollution is to prevent it from happening in the first place. Tanker design is being modified to limit the amount of oil intentionally released in transport. Legislation is being considered that would limit new tanker construction to stronger, double-hull designs (although shipping experts are uncertain that even a double-hull design could have prevented the *Exxon Valdez* spill). Perhaps most important, crew testing and training are being upgraded.

Heavy Metals

Small quantities of heavy metals are capable of causing damage to organisms by interfering with normal cell metabolism. Mercury and lead are two of the dangerous heavy metals being introduced into the ocean. Human activity releases about 5 times as much mercury and 17 times as much lead as is derived from natural sources, and incidents of poisoning, major cause of brain damage and behavioral disturbances in children, have increased dramatically over the last two decades (see **Box 18.1**). Lead particles from industrial wastes, landfills, and gasoline residue reach the ocean through runoff from land during rains, and the lead concentration in some shallow-water, bottom-feeding species is increasing at an alarming rate. Consumers should be wary of seafoods taken near shore in industrialized regions.

Copper, another heavy metal, is so effective in killing marine organisms that it has long been used in marine an-

tifouling paints. The ship *Pac Baroness,* a freighter carrying 21,000 metric tons (23,000 tons) of finely powdered copper, sank in 448 meters (1,480 feet) of water off the coast of central California after a collision in 1987. Her toxic cargo was scattered over the seabed, and a plume of copper-tainted water has been detected 40 kilometers (24 miles) downcurrent from the wreck. Marine life has been significantly disrupted in the area, which is a major fishing zone for Dover sole and rock cod. A similar incident off Holland in 1965 killed more than 100,000 fish and destroyed commercial mussel beds.

Tributyl tin, an antifouling paint additive banned in the United States in the late 1980s, is still accumulating in dolphins. Though the compound has been removed from store shelves, it persists in the food chain. Tributyl tin is a potent immunosuppressor and may damage the animals' ability to fight off bacterial and viral diseases. Tin accumulations are unusually high in small whales involved in strandings, and biologists suspect a relationship may exist.

Canadian and Norwegian researchers have found that coal combustion, electric utilities, steel and iron manufacturing, fuel oils, fuel additives, and incineration of urban refuse are the major sources of oceanic and atmospheric contamination by heavy metals. They now believe that the toxicity of heavy metals is a greater health problem than the toxicity of all organic and radioactive wastes. Wellness-conscious consumers see fish as a safe and healthful food. But with the

For the people of Minamata, Japan, who were poisoned by mercury released into the ocean from a nearby factory, heavy metal pollution is a continuing horror story. Between 1953 and 1960 more than a hundred people who ate shellfish taken from Minamata Bay were afflicted by a form of mercury poisoning now called Minamata disease (**Figure a**). Their symptoms included kidney damage, neuromuscular deterioration, birth defects, insanity, and eventually death.

The source of the mercury was a plastics plant that was dumping waste mercuric chloride into the bay. Bacterial action changed this chemical into a form that could be taken up and concentrated by benthic organisms. Villagers who ate clams and oysters, especially pregnant women and young children, were seriously affected.

The discharge of mercury has been stopped, but the bay (which had been a source of food for the poor) is still completely unusable for fishing and clamming, even after more than 30 years. This situation will not change for generations; mercury is not easily removed from sediments and has a long residence time.

a A mother bathes her daughter, a victim of Minamata disease.

In 1997, the last surviving victims of the poisoning ended a 40-year battle for compensation. Each accepted a cash settlement of $24,200 from the Chisso Corporation.

ocean still receiving heavy metal–contaminated runoff from the land, a rain of pollutants from the air, and the fallout from shipwrecks, we can only wonder how much longer most seafood will be safe to eat.

Synthetic Organic Chemicals

Heavy metals are not the only poisons that may contaminate seafood. Many different synthetic organic chemicals also enter the ocean and become incorporated into its organisms. Some of these synthetic organic substances are listed in **Table 18.2.** Ingestion of even small amounts of these compounds can cause illness or even death. Rachel Carson's alarming discussion of the effects of these compounds in her 1962 book *Silent Spring* began the environmental movement in the United States (**Figure 18.8**).

Halogenated hydrocarbons—a class of synthetic hydrocarbon compounds that contain chlorine, bromine, or iodine—are used in pesticides, flame retardants, industrial solvents, and cleaning fluids. The concentration of **chlorinated hydrocarbons**—the most abundant and dangerous halo-

genated hydrocarbons—is so high in the water off New York State that officials have warned women of childbearing age and children under 15 not to consume more than half a pound of local bluefish a week. (They are told *never* to eat striped bass caught in the area.) One administrator for the U.S. Environmental Protection Agency has written, "Anyone who eats the liver from a lobster taken from an urban area is living dangerously."

The level of synthetic organic chemicals in seawater is usually very low, but some organisms at higher levels in the food chain can concentrate these toxic substances in their flesh. This **biological amplification** is especially hazardous to top carnivores in a food web.

The damage caused by biological amplification of DDT, a chlorinated hydrocarbon pesticide, is particularly instructive. In the early 1960s, brown pelicans began producing eggs with thin shells containing less than normal amounts of calcium carbonate. The eggs broke easily, no chicks were hatched, and the nests were eventually abandoned. The pelicans were disappearing. The trail led investigators to DDT. Plankton absorbed DDT from the water, fishes that fed on

Table 18.2	Synthetic Organic Chemicals That Have Been Detected in the Ocean
Name	**Major Health Effects**
Aldicarb (Temik)	High toxicity to the nervous system
Benzene	Chromosomal damage, anemia, blood disorders, and leukemia
Carbon tetrachloride	Cancer; liver, kidney, lung, and central nervous system damage
Chloroform	Liver and kidney damage; suspected cancer
Dioxin	Skin disorders, cancer, genetic mutations
Ethylene dibromide (EDB)	Cancer and male sterility
Polychlorinated biphenyls (PCBs)	Liver, kidney, and lung damage
Trichloroethylene (TCE)	In high concentrations, liver and kidney damage, central nervous system depression, skin problems, and suspected cancer and mutations
Vinyl chloride	Liver, kidney, and lung damage; lung, cardiovascular, and gastrointestinal problems; cancer and suspected mutations

Source: Miller, 1997.

these microscopic organisms accumulated DDT in their tissues, and the birds that fed on the fishes then ingested it (**Figure 18.9**). The whole food chain was contaminated, but because of biological amplification the top carnivores were most strongly affected. A chemical interaction between DDT and the birds' calcium-depositing tissues prevented the formation of proper eggshells. DDT was eventually banned for use in the United States, and the pelican and osprey populations are recovering.

Biological amplification of other chlorinated hydrocarbons has also affected other species. **Polychlorinated biphenyls** (**PCBs**), fluids once widely used to cool and insulate electrical devices and to strengthen wood or concrete, may be responsible for the behavior changes and declining fertility of some populations of seals and sea lions on islands off the California coast. PCBs have also been implicated in a deadly viral epidemic among dolphins in the western Mediterranean. Ingestion of the chemical may have severely weakened the dolphins' immune systems and made it impossible for them to fight the infection—up to 10,000 dolphins died during the summer of 1990. Even more alarming to biologists is the recent discovery that the nearshore dolphins off U.S. coasts are intensely contaminated. The concentration of chlorinated hydrocarbons in these animals exceeds 6,900 parts per million (ppm), a concentration high enough to disrupt the dolphins' immune systems, hormone production, reproductive success, neural function, and ability to stave off cancers.[1] This level vastly exceeds the 50 ppm limit the U.S. government considers hazardous for animals and the 5 ppm considered the maximum acceptable level for humans! Investigations are continuing, as are probes into the effects of dioxin and other synthetic organic poisons accumulating in the oceanic sink.

Marine animals are even being killed by chemicals used during illicit drug trafficking. In 1997, 42 dolphins and at least three large whales were found dead off Mexico's west coast, victims of the cyanide-based chemicals used by drug merchants to mark ocean drop-off sites. Ships approach the shore at night, drop bales of illegal drugs overboard, mark them with toxic luminescent compounds, and retreat into international waters. Pick-up crews in small boats then retrieve the drugs. Squid and fishes are drawn to the light, encounter the poisonous chemicals, and die. They are in turn eaten by larger animals, with disastrous results.

Figure 18.8 Rachel Carson, author of the influential 1962 book *Silent Spring* on the misuse of synthetic pesticides. This work is credited with beginning the environmental movement in the United States.

---

[1]If you drag a beached porpoise into the ocean, you could theoretically receive a $10,000 fine for improper disposal of polluted materials!

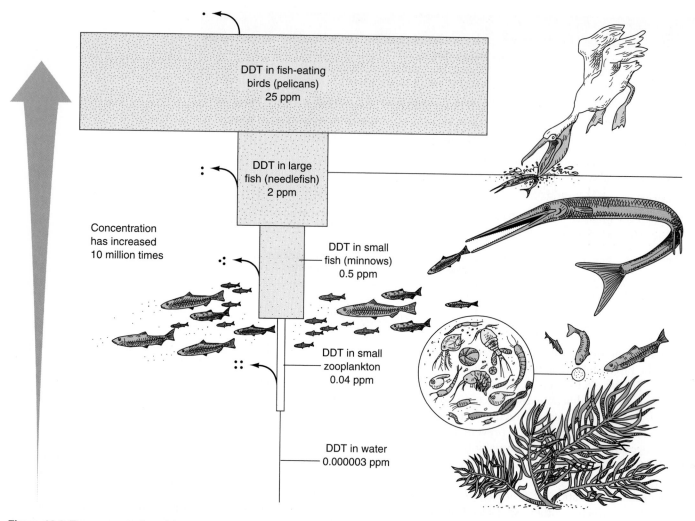

Figure 18.9 The concentration of the pesticide DDT in the fatty tissues of organisms was biologically amplified approximately 10 million times in this food chain of an estuary adjacent to Long Island Sound, near New York City. Dots represent DDT, and arrows show small losses of DDT through respiration and excretion.

Within the figure:

DDT in fish-eating
birds (pelicans)
25 ppm

DDT in large
fish (needlefish)
2 ppm

Concentration
has increased
10 million times

DDT in small
fish (minnows)
0.5 ppm

DDT in small
zooplankton
0.04 ppm

DDT in water
0.000003 ppm

Figure 18.10 indicates the locations of U.S. coastal areas degraded by synthetic and processed organic chemicals. Production of such chemicals subject to biological amplification in food chains currently exceeds 91 million metric tons (100 million tons) each year. Even greater expanses of ocean will be affected if the predicted 1% of that production reaches the sea.

Eutrophication

Not all pollutants kill organisms. Some dissolved organic substances act as nutrients or fertilizers that speed the growth of marine autotrophs, causing eutrophication. **Eutrophication** (*eu* = good, well; *trophos* = feeding) is a set of physical, chemical, and biological changes that take place when excessive nutrients are released into the water. Too much fertility can be as destructive as too little. Eutrophication stimulates the growth of some species to the detriment of others, destroying the natural biological balance of an ocean area. The

extra nutrients come from wastewater treatment plants, factory effluent, accelerated soil erosion, or fertilizers spread on land. They usually enter the ocean from river runoff and are particularly prevalent in estuaries.

Eutrophication is occurring at the mouths of almost all the world's rivers. The Mississippi River provides an example. In 1999 a zone of low or no oxygen covered 20,000 square kilometers (7,700 square miles) of the continental shelf south and west of the Mississippi delta (**Figure 18.11**). Fishes, shrimp, and other animals suffocated if they could not flee. The primary cause of this "dead zone" is the nitrogen fertilizer washed into the gulf from millions of farm fields across the Mississippi Basin. A spring drought in 2000 reduced the nutrient flow that fuels the algal blooms that deplete oxygen—the low-oxygen zone shrank to just 4,400 square kilometers (1,700 square miles) in July 2000. As river flow increases, however, so will the size of the "dead zone." In the future, government officials hope to reduce fertilizer runoff to minimize the problem.

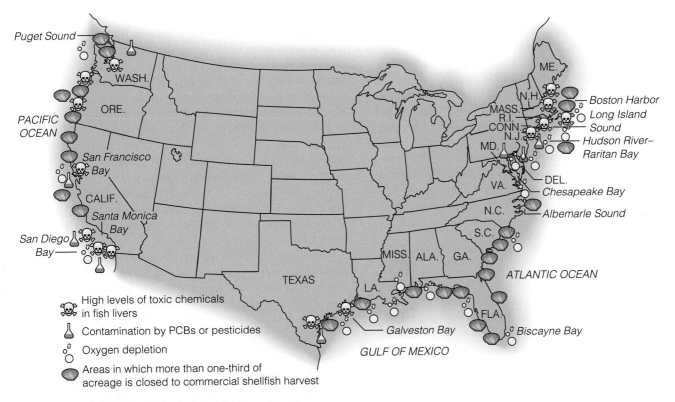

High levels of toxic chemicals in fish livers

Contamination by PCBs or pesticides

Oxygen depletion

Areas in which more than one-third of acreage is closed to commercial shellfish harvest

Figure 18.10 Chemical assault on major U.S. coastal areas.

The most visible manifestations of eutrophication are the red tides, yellow foams, and thick green slimes of vigorous plankton blooms (see Box 14.1). These blooms typically consist of one dominant phytoplankter that grows explosively, overwhelming other organisms. Huge numbers of algal cells can choke the gills of some animals and (at night, when sunlight is unavailable for photosynthetic oxygen production) deplete the free oxygen content of surface water (note the symbols for oxygen depletion in Figure 18.10).

In nearshore waters, low oxygen is now thought to cause more mass fish deaths than any other single agent, including oil spills. It is a leading threat to commercial shellfisheries.

Toxic substances released from the abundant algae can sicken or kill other species. In 1978 a phytoplankton bloom was linked to the catastrophic die-off of fish near the German coast of the North Sea. In 1988 a similar fate befell dolphins off the Atlantic coast of the United States. As many as 1 million fluke and flounder were killed in the summer of 1989

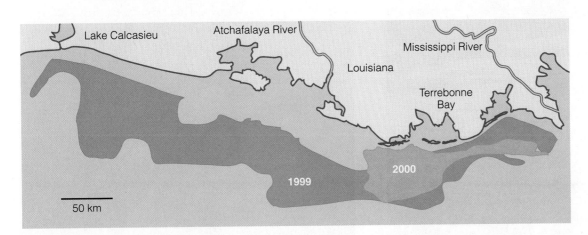

Figure 18.11 The "dead zone," an area of anoxic conditions on the continental shelf off southern Louisiana. Algal blooms fueled by nutrients running off farmlands and into the Mississippi River deplete the oxygen in this area. The "dead zone" shrank in 2000 because of decreased river flow during a spring drought. Unless nutrient runoff is limited, the zone will grow. (Source: *Science, 289,* August 4, 2000, p. 723. Reprinted with permission.)

when an algal bloom depleted the oxygen in New Jersey's Raritan Bay. During some of these blooms, beaches were covered with foul-smelling foam more than 1 meter (3.3 feet) thick.

These exceptional algae blooms appear to be increasing in number and intensity. There is little mention of foam events before about 1930, but since 1978 there has been at least one every year. A similar pattern has been reported for red tides. The threat is especially great in countries experiencing rapid economic growth. After two decades of breathless development, severe harmful algal blooms (HABs) have begun to plague China's southeastern coastline. In 1998 a massive bloom wiped out up to 75% of Hong Kong's fish farms, wiped out many local fishers. The Chinese government estimates $240 million in direct economic losses from 45 HAB incidents from 1997 to 1999.

Solid Waste

18-9

Not all pollutants enter the ocean in a dissolved state; much of the burden arrives in solid form. Some solid waste is ultimately biodegradable, but plastic, which now makes up almost 8% of the solid waste stream, is not. Scientists estimate that some kinds of synthetic materials—plastic six-pack holders, for example—will not decompose for about 400 years!

Americans generate 120 million metric tons of plastic waste each year, about 500 kilograms (1,100 pounds) per person. By the year 2000 plastics will account for more than 10% of all solid waste. Since the ocean is treated as the ultimate dump, some of this waste plastic finds its way to the sea. In 1997 more than 4,500 volunteers scoured the New Jersey–New York coast to collect debris. More than 75% of the 209 tons recovered was plastic (paper and glass accounted for another 15%). In isolated areas the dunes of plastic debris can build to surreal proportions (**Figure 18.12**).

A 1987 survey by the Woods Hole Oceanographic Institution found that each square mile of ocean surface off the northeastern coast of the United States has more than 46,000 pieces of plastic floating on the surface. This material includes ropes, fishing line and nets, plastic sheeting and bags, and granules of broken plastic cups. A staggering 100,000 marine mammals and 2 million seabirds die *each year* after ingesting or being caught in plastic debris! Sea turtles mistake plastic bags for their jellyfish prey and die from intestinal blockages. Seals and sea lions starve after becoming entangled in nets (**Figure 18.13**) or muzzled by six-pack rings. The same kinds of rings strangle fish and seabirds. Adding ingredients to plastics that would hasten their decomposition would add only 5% to 7% to their cost, but this price increase is currently unacceptable to industry.

What should we do with plastic and other solid wastes such as glass and paper, disposable diapers, scrap metal, building debris, and all the rest? Dumping it into the ocean is clearly unacceptable, yet places to deposit this material are becoming scarce. In 1999 the average New Yorker threw away more than

Figure 18.13 Sea lions (seen here) and seals die by the hundreds each year after becoming entangled in plastic debris, especially discarded and broken fishing nets.

Figure 18.12 Plastic mounds on an isolated shore of Niihau, one of the Hawaiian islands.

a ton of waste annually. California's Los Angeles and Orange Counties generate enough solid waste to fill Dodger Stadium every eight days. Transportation of waste to sanitary landfills becomes more expensive as nearby landfills reach capacity.

Is incineration the answer? We currently burn about 10% of our trash. By contrast, Sweden burns half its solid wastes. Unfortunately, even with air pollution control devices, incinerators still emit great quantities of tiny particles, heavy metal residue, and immense amounts of carbon dioxide. Dumping of the residual ash is also a problem—ash from Philadelphia's garbage incinerators has been turned away by states as far away as South Carolina, and one shipload of the stuff was refused entry at an African port.

Is recycling the answer? The Japanese currently recycle about 50% of their solid waste and are importing even more; scrap metal and waste paper headed for Japan are the two biggest exports from the Port of New York. Americans are buying back their own refuse in the form of appliances, automobiles, and the cardboard boxes that hold their televisions and compact disc players. Massachusetts and California have set a goal of recycling 25% of their waste; the city of Seattle is now approaching 30%. The direct savings to consumers, as well as the environmental rewards to ocean and air, will be significant.

The *best* solution is a combination of recycling and reducing the amount of debris we generate by our daily activities. We will soon have no other choice.

Sediment

Runoff from mining, farming, forestry, and other land uses often contains large amounts of sediment. This material can cloud the water, impede photosynthesis, and clog the gills of marine organisms. Coastal ecosystems have been smothered and buried by soils and sands washed into the ocean after strong rains. The problem is particularly apparent in areas of rapidly growing human populations and in the vicinity of coastal mining and dredging operations. In Louisiana, the coastal marshes are vanishing as sediments from upstream bury them at the rate of 130 to 155 square kilometers (50 to 60 square miles) a year. The sediments come from poor agricultural practices and uncontrolled commercial development. Sediments washing from local rivers have killed 75% of Costa Rica's Caribbean reefs. Philippine reefs are similarly endangered. By the early 1980s about 75% of the reefs surrounding the Philippines had been damaged beyond recovery by the effects of sedimentation and erosion. The economic cost of the loss of Philippine reefs is tremendous; more than 100,000 jobs and $80 million in fish catches are lost each year. Because of lowered reef productivity, 5 million Filipinos are unable to catch the fish and crustaceans they require for proper nutrition. Half of all children living in coastal regions of the Philippines are malnourished.

Sewage

About 98% of sewage is water. Sewage treatment separates the fluid component from the solids, treats the water to kill disease organisms and reduce the levels of nutrients, and re-

leases it into a river or the ocean. The remaining **sewage sludge** is a semisolid mixture of organic matter containing bacteria and viruses, toxic metal compounds, synthetic organic chemicals, and other debris. It is digested, thickened, dried, and shipped to landfills, burned to generate electricity, or dumped into the ocean. The liquid effluent wanders from its outlet and circulates with currents, but sludge and other insoluble residues may stay near the outfall or dump site for years. The amount of wastewater and sewage sludge increased by 60% in the 1990s.

Until 1992 almost 2 billion liters (half a billion gallons) of partially treated sewage poured into Boston Harbor every day. A new sewage treatment plant opened in that year, 22 years after the deadline mandated by the U.S. Clean Water Act. About the same amount of treated sewage pours from outfall pipes extending about 5 miles out to sea from the coast at Los Angeles. Treatment plants in southern California are sometimes overwhelmed after heavy rainstorms, and raw sewage enters the ocean in large quantity. Rain or shine, areas around the entrance to San Diego Harbor are often so contaminated with sewage from the city of Tijuana, Mexico, that anyone who enters the water runs the risk of bacterial or viral infection.

Not only human sewage is involved. In June 1995, 26 million gallons of hog feces and urine spilled into a narrow stream feeding North Carolina's New River estuary. Thousands of fish died from direct effects, and hundreds of thousands more perished as algae grew and consumed oxygen.

Sludge may be an even greater problem than raw sewage. The water just south of Long Island, New York, has been one of the most intensely used ocean dumping sites in the world. The area is covered with sludge, which creates an oxygen-poor environment in which few animals can survive. During storms some of this material washes up on local beaches, contaminates shellfish beds, and routinely causes disease outbreaks among people consuming raw oysters and clams from the area. After a public outcry, a new dumping site was selected beyond the edge of the continental shelf. Ten million tons of wet sludge produced by treatment plants in New York and New Jersey since March 1986 has been dropped there by huge barges. The plume from this sludge now contaminates the Gulf Stream.

Liquid effluent and sludge also contribute to eutrophication. As we've seen, some forms of algae thrive in nutrient-rich wastewater, where they multiply to prodigious numbers, secrete toxins, or deplete oxygen in the surrounding water and cause the death of fishes and crustaceans.

Waste Heat

Shoreside electrical generating plants use seawater to cool and condense steam. The seawater is returned to the ocean about 6°C (10°F) warmer, a difference that may overstress marine organisms in the effluent area. **Figure 18.14** shows how widely the heated water spreads out around a coastal power installation. Some recent power plant designs minimize environmental impact by pumping colder water from

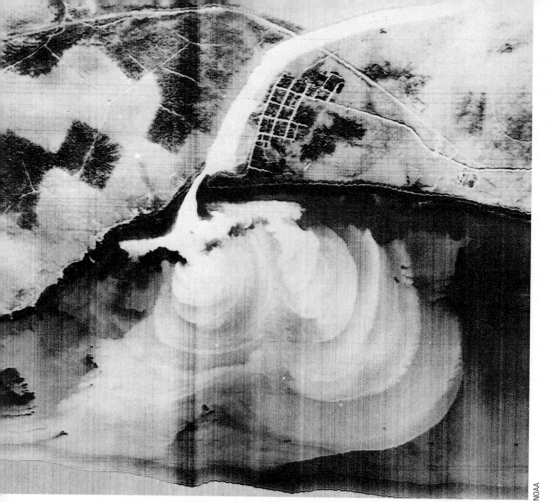

Figure 18.14 Thermal effluent from power plants and sewage outfalls, shown in a thermograph of the Quinault River as it enters the Pacific at Taholah, Washington State. In this infrared photograph, white areas indicate warmer water.

NOAA

farther offshore, warming it to the temperature of the seawater surrounding the plant site, and then releasing it. This method minimizes the impact on the surrounding communities, but it still shocks those eggs, larvae, plankton, and other organisms that are sucked through the power plant with the cooling water.

Introduced Species

Several thousand species are in transit every day in the ballast water of tankers and other ships. Juvenile forms of marine organisms can easily hitch rides across otherwise insurmountable oceanic barriers and set up housekeeping at distant shores. These foreign organisms sometimes outcompete native species and reduce biological diversity in their new habitats. New marine diseases can also be introduced in this way. Even canals and fishery enhancement projects can introduce potentially destabilizing new species.

The Chinese mitten crab (genus *Eriocheir*, **Figure 18.15**) and a common decorative aquarium seaweed (genus *Caulerpa*) are good examples of the destruction exotic marine species can wreak. Probably brought to the western United States and the Atlantic coast of Europe as larval forms in ships' ballast water, the mitten crab is an energetic crustacean that burrows into riverbanks and levees, causing them to col-

lapse. Sometimes reaching densities of 30 animals per square meter (2.7 per square foot), the crabs clog water supply pipes and, if frustrated in migration, wander onto streets and into people's backyards and homes. The aquarium alga may be even more of a menace. It spreads through fragmentation. After escaping into the Mediterranean in 1984, this Caribbean native now chokes more than 4,050 hectares (10,000 acres) of ocean floor off Spain, France, Italy, and Croatia. It has recently turned up in southern California bays, where it is outcompeting local organisms and greatly reducing biodiversity.

The problems caused by introduced species will grow. There are about 250 known exotic species in California's San Francisco Bay, the most invaded estuary in the world—more inadvertent global experimentation.

The Costs of Pollution

In 1998 government and industry in the United States spent about $220 billion on the control of atmospheric, terrestrial, and marine pollution—an average of $800 per U.S. citizen. This figure was equivalent to about 1.6% of the gross national product, or 2.7% of capital expenditures by U.S. business. That same year the United States lost 4% of its gross national product through environmental damage. Clearly the financial costs of pollution will continue to increase.

Figure 18.15 Chinese mitten crabs, exotic (introduced) species that burrow into riverbanks and levees, causing them to collapse.

But there are other costs. Failure to control pollution will eventually threaten our food supply (marine and terrestrial), destroy whole industries, produce a greater disparity between have and have-not nations, and cause a decline in the health of all the planet's citizens. To these costs must be added the aesthetic costs of an ocean despoiled by pollution; few of us look forward to sharing the beach with oiled birds, jettisoned diapers, or clumps of medical waste.

HABITAT DESTRUCTION

The pollution processes we have discussed don't affect individual organisms alone. They influence whole habitats, especially the most complex and biologically sensitive shallow-water habitats.

Bays and Estuaries

The hardest-hit habitats are estuaries, the hugely productive coastal areas at the mouths of rivers where fresh water and seawater meet. Pollutants washing down rivers enter the ocean at estuaries, and estuaries often contain harbors, with

their potential for oil spills. As little as 1 part of oil for every 10 million parts of water is enough to seriously affect the reproduction and growth of the most sensitive bay and estuarine species. Some of the estuaries along Alaska's Prince William Sound, site of the 1989 *Exxon Valdez* accident, were covered with oil to a depth of 1 meter (3.3 feet) in places. The spill's effects on the $150-million-a-year salmon, herring, and shrimp fishery will be felt for years to come.

Other Pacific coast habitats have been polluted in different ways. Bottom sediments in Seattle's Elliott Bay are contaminated with a poisonous mix of lead, arsenic, zinc, cadmium, copper, and PCBs. Tumors on the livers of English sole, which dwell on the sediments, have been linked to these compounds. Pollutants are so abundant in southern San Francisco Bay that clams and mussels contain near-lethal concentrations of heavy metals. Birds migrating from Central America to the Arctic Circle run a risk of being poisoned by stopping there to feed. There is also a risk to humans who eat ducks shot in this area because of the high concentrations of pollutants they contain. Similar warnings apply to all fish caught in and near southern California estuaries between Santa Barbara and Ensenada, Mexico.

Estuaries and bays along the U.S. Gulf Coast, one of the most polluted bodies of water on Earth, are also being severely stressed. About 40% of the nation's most productive fishing grounds, including its most valuable shrimp beds, are found in the gulf. Nearly 60% of the gulf's oyster and shrimp harvesting areas—about 13,800 square kilometers (5,300 square miles)—are either permanently closed or have restrictions placed on them because of rising concentrations of toxic chemicals and sewage. Half of Galveston Bay, once considered the second most productive estuary in the United States, is off limits to oyster fishers because of sewage discharges.

Estuaries along the Atlantic coast are also threatened. From southern Florida to central Georgia, more than 325 square kilometers (125 square miles) of sea grasses (which act as nurseries for a great variety of marine life) has been killed by a virus. Scientists speculate that pollutants from urban and agricultural runoff have weakened the plants' resistance. Fishermen to the north in Chesapeake Bay have been mystified by a sudden decline in the abundance of fish and crustaceans, a change marine scientists attribute to increasing pollution in the area. Lobstermen in New England have noticed an alarming increase in the incidence of tumors in lobsters' tail and leg joints. Could these changes also be caused by toxic wastes? The beluga whale population of Canada's St. Lawrence estuary collapsed in the early 1980s. High levels of PCBs, DDT, and heavy metals—substances biologically amplified in the whales' food—were blamed for the tumors, ulcers, respiratory ailments, and failed immune systems discovered during the autopsies of 72 dead whales.

People also "develop" estuaries into harbors and marinas. In the 1960s and 1970s California led the world in the acreage of bays and estuaries filled in for recreational marinas. Harbors grew smaller as more of the area was filled for docks and storage facilities. One hundred and fifty years ago, San Francisco Bay covered 1,131 square kilometers (437 square miles).

Today only 463 square kilometers (179 square miles) remains—the rest has been filled in. Filling of estuaries is just as threatening to the natural reproductive cycles of shrimp and fish as is poisoning by toxic wastes.

Some states control the development of coastal regions. A citizens' initiative passed by Californians in 1972 limited development of that state's coastal zone. Massachusetts laws make it illegal to fill any marsh or estuarine region, even areas that are privately owned. Similar legislation is pending in a few other coastal states.

Coral Reefs

Pollution may also be damaging coral reefs. Marine biologists have been baffled by recent incidents of coral bleaching—corals expelling their symbiotic zooxanthellae—in the Caribbean and tropical Pacific. The 1989 bleaching event in the Caribbean is the most recent ecological disruption in a series that includes a massive fish kill in 1980, a die-off of about 95% of the individuals of one sea urchin species in 1983–84, and a slowly spreading coral malady known as white-band disease. Biologists do not know for certain what caused these disturbances, but increasing pollution is thought to be partially responsible.

Some chemical pollution is intentional. Especially damaging to tropical reefs has been the practice of using cyanide to collect tropical fish. Fishermen squirt a solution of sodium cyanide over the reef to stun valuable species. Many fish die; those that survive are sent to collectors all over the world. At the same time the invertebrate populations of the sensitive coral reef communities are decimated.

Not all coral reef pollutants are chemicals. Fishermen in Indonesia and Kenya dynamite the reefs to kill fish that hide among the coral branches. Reefs throughout the world are mined for construction material, for ornamental pieces, or for their calcium carbonate (to make plaster and concrete). About 27% of the world's coral reefs have already disappeared. Researchers fear 70% of the reefs close to large population centers will disappear in the next 50 years.

Other Habitats

Between 1963 and 1977, about half of India's extensive mangrove forests were cut down. About a third of Ecuador's mangroves have been converted to ponds used in shrimp mariculture. Agricultural expansion is expected to wipe out all Philippine mangroves in ten years. Not even the calm, cold communities of the abyssal plains are safe from disruption—imagine the effect manganese nodule mining will have on the delicate organisms of the deep bottom.

GLOBAL CHANGES

The ocean and the atmosphere are extensions of each other, and human activity has changed the atmosphere as it has changed the ocean. Pollutants injected into the air can have global consequences for the ocean and for all Earth's inhabitants. Potentially the most destructive atmospheric problems are depletion of the ozone layer, global warming, and acid rain.

Ozone Layer Depletion

Ozone is a molecule formed of three atoms of oxygen. Ozone occurs naturally in the atmosphere. A diffuse layer of ozone mixed with other gases—the **ozone layer**—surrounds the world at a height of about 20 to 40 kilometers (12 to 25 miles).

Seemingly harmless synthetic chemicals released into the atmosphere—primarily **chlorofluorocarbons** (**CFCs**) used as cleaning agents, refrigerants, fire-extinguishing fluids, spray-can propellants, and insulating foams—are converted by the energy of sunlight into compounds that attack and partially deplete Earth's atmospheric ozone. Ozone levels in the stratosphere have decreased by about 3% over parts of the United States since 1969. A 4% drop has been noted over Australia and New Zealand, and a 50% decrease has been observed near the North and South Poles (**Figure 18.16**). The amount of depletion varies with latitude (and with the seasons) because of variations in the intensity of sunlight. In September 2000 the largest ozone depletion ever recorded was detected over Antarctica. About 11 million square kilometers (4 million square miles) was affected.

This decline in ozone alarms scientists because stratospheric ozone intercepts some of the high-energy ultraviolet radiation coming from the sun. Ultraviolet radiation injures living things by breaking strands of DNA and unfolding protein molecules. Species normally exposed to sunlight have evolved defenses against average amounts of ultraviolet radiation, but increased amounts could overwhelm those defenses. Land plants such as soybeans and rice would be subjected to sunburn that decreases their yields. Even plankton in the uppermost 2 meters (6.5 feet) of ocean would be affected: Recent research in fact indicates an alarming decrease in phytoplankton primary productivity of between 6% and 12% in the coastal waters around Antarctica. In 1997, biologists working at the U.S. Palmer Research Station in Antarctica confirmed lethal ultraviolet damage to seastar embryos drifting near the sea surface.

Our own species would not escape: A 1% decrease in atmospheric ozone would probably be accompanied by a 5% to 7% increase in human skin cancer (**Figures 18.17** and **18.18**). According to medical researchers' estimates, the 4% ozone depletion over Australia and New Zealand will cause at least a 20% increase in human skin cancers over the next two decades. Strong ultraviolet light can also suppress the immune system and cause eye cataracts. The quantity of photochemical smog shrouding our cities will also increase as ozone levels fall.

In June 1990, representatives of 53 nations agreed to ban major production and use of ozone-destroying chemicals by the year 2000. Legislation is being proposed to phase out CFC production in the United States, and a sense of great urgency surrounds ongoing research to find safe substitutes for these substances. Researchers hope to see a decrease in the size of polar ozone holes by 2005 or 2010.

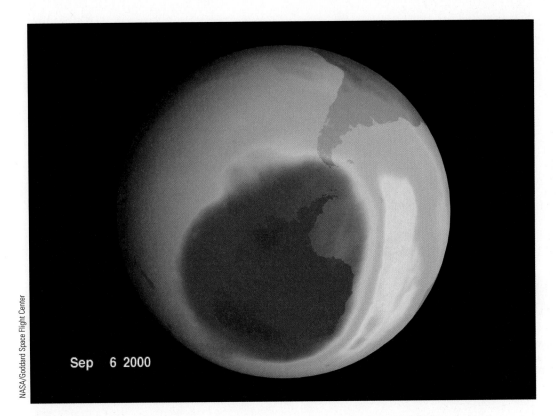

Sep 6 2000

Figure 18.16 A computer-generated view of the largest ozone "hole" (ozone depletion area) yet observed, 6 September 2000. The hole covers 28.3 million square kilometers (11 million square miles)—about three times larger than the land mass of the United States. These observations reinforce concern about the fragility of Earth's atmospheric ozone.

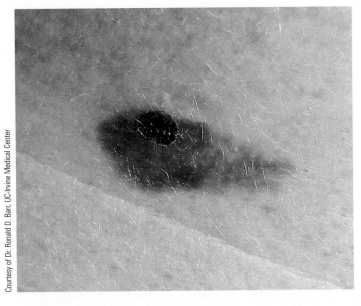

Figure 18.17 A malignant melanoma in an early stage. Left untreated, this most dangerous form of skin cancer could spread and cause the death of the patient. Skin cancer appears to be associated with exposure to the sun. If you spend much time in the sun you increase your risk of skin cancer. Be alert for changes in moles, or for any pigmented lesion that is asymmetrical, has an irregular border, is an unusual color, grows in size, or begins to itch. If you notice any of these characteristics, see a physician.

Figure 18.18 If ozone depletion continues, the "perfect tan" may have to be none at all. © The New Yorker Collection 1989. Stuart Leeds from cartoonbank.com. All rights reserved.

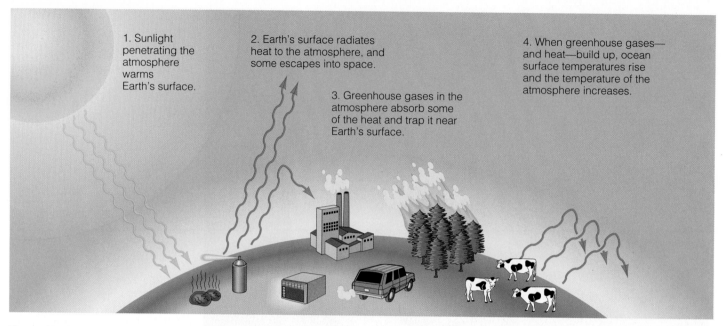

1. Sunlight penetrating the atmosphere warms Earth's surface.

2. Earth's surface radiates heat to the atmosphere, and some escapes into space.

3. Greenhouse gases in the atmosphere absorb some of the heat and trap it near Earth's surface.

4. When greenhouse gases—and heat—build up, ocean surface temperatures rise and the temperature of the atmosphere increases.

Figure 18.19 How the greenhouse effect works. Since the beginning of the Industrial Revolution, emissions of carbon dioxide, methane, and other greenhouse gases have increased. Most researchers believe these gases have contributed to a general warming of Earth's atmosphere and ocean.

Global Warming

The surface temperature of Earth fluctuates slowly over time. The global temperature trend has been generally upward in the 18,000 years since the last ice age, but the *rate* of increase has recently accelerated. This rapid warming is probably the result of an enhanced **greenhouse effect,** the trapping of heat by the atmosphere. Glass in a greenhouse is transparent to light but not to heat. The light is absorbed by objects inside the greenhouse, and its energy is converted into heat. The temperature inside a greenhouse rises because the heat is unable to escape. On Earth **greenhouse gases**—carbon dioxide, water vapor, methane, CFCs, and others—take the place of glass. Heat that would otherwise radiate away from the planet is absorbed and trapped by these gases, causing a rise in surface temperature. **Figure 18.19** shows this mechanism.

The greenhouse effect is necessary for life; without it, Earth's average atmospheric temperature would be about $-18°C$ ($0°F$). Earth has been kept warm by natural greenhouse gases. The sources of these gases are volcanic and geothermal processes, the decay and burning of organic matter, and respiration and other biological processes. The removal of these gases by photosynthesis and absorption by seawater appears to prevent the planet from overheating. But the human demand for quick energy to fuel industrial growth, especially since the beginning of the Industrial Revolution, has injected unnatural amounts of new carbon dioxide into the atmosphere from the combustion of fossil fuels. Carbon dioxide is now being produced at a greater rate than it can be absorbed by the ocean. **Figure 18.20** shows how much carbon dioxide has increased in the atmosphere since about 1750.

The atmosphere's carbon dioxide content is now rising at the rate of 0.4% each year. Though CFCs are now declining, methane levels continue to increase. (The primary sources of methane include intestinal gases from cows and termites, and the decay of refuse and vegetation cleared from land.)

There has been a $5°C$ ($9°F$) rise in global temperature from the end of the last ice age until today. Carbon dioxide and other human-generated greenhouse gases produced since 1880 are thought to be responsible for about half of that

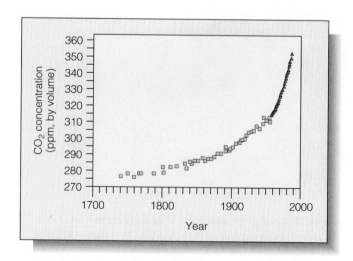

Figure 18.20 The rise in atmospheric carbon dioxide since 1750. Measurements up to about 1960 (■) came from samples of air trapped in the Antarctic ice; later measurements (▲) were made directly at Mauna Kea, Hawaii (after Watson et al., 1990).

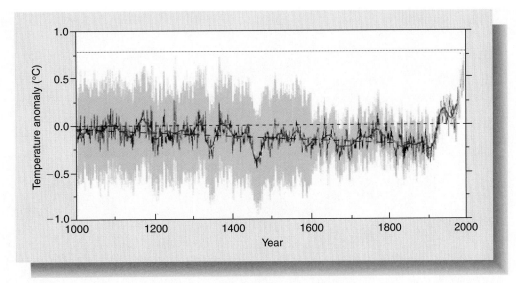

Figure 18.21 The year 2000 was the warmest on record. Temperature records preserved in tree rings and glacial ice have revealed a large and abrupt warming unique in the past 1,000 years. In April 2000, a report by the United Nations Panel on Climate Change concluded: ". . . there has been a discernible human influence on global climate." (Source: "Northern Hemisphere Millennial Temperature Reconstruction" from *Geophysical Research Letters*, vol. 26, no. 6, p. 759–762 by Michael E. Mann, Raymond S. Bradley, & Michael K. Hughes. Copyright © 1999 American Geophysical Union. Reprinted by permission.)

increase. A new survey of 300 borehole temperatures on four continents has confirmed that Earth is getting warmer and that the rate of warming has been increasing since 1900. Subsurface temperatures confirm that the average global temperature has increased 0.4 to 0.8°C (0.7 to 1.4°F) over the last century (**Figure 18.21**). The year 2000 was the warmest on record. A consensus of experts meeting in Shanghai in January 2001 confirmed that if greenhouse emissions continue to rise, the average should increase another 5.8°C (10.5°F) during the twenty-first century.[2]

---

[2] This estimate is based on projections of data, not on models.

The south polar ice sheets are already shrinking in the warmth. Imagine the effect of a significant rise in sea level on the harbors, coastal cities, river deltas, and wetlands where one-third of the world's people now live. As **Figure 18.22** suggests, the cost to society would be enormous.

Unfortunately, it will be exceedingly difficult to curtail our production of carbon dioxide and methane. In the last hundred years, industrial production has increased fiftyfold; we have burned roughly 1 billion barrels of oil, 1 billion metric tons of coal, and 10 billion cubic meters of natural gas. Carbon dioxide is a major product of combustion for these hydrocarbon compounds. The world's energy demand is projected to increase 3.5 times between now and the year 2025, with carbon dioxide emissions 65% higher than today (see **Figure 18.23**).

Reprinted by permission of the Los Angeles Times Syndicate

Figure 18.22 For island nations such as the Maldives, even a small rise in sea level could spell disaster. Strung out across 880 kilometers (550 miles) of the Indian Ocean, 80% of this island nation of 263,000 people lies less than 1 meter (3.3 feet) above sea level. Of the country's 1,180 islands, only a handful would survive the median estimate of sea level rise by 2100. Most of the population lives in fishing villages on low islands like the one shown here, where the effects of this century's sea-level rise of 10 to 25 centimeters (4 to 10 inches) have already been felt.

C. Mayhew & R. Simmon/NASA

Figure 18.23 Earth at night, 27 November 2000. Human-made light highlights developed or populated areas on the surface. This composite image emphasizes the present extent of industrialization and resource use.

At a meeting in Kyoto, Japan, in 1997, leaders and representatives of 160 countries established carbon dioxide emission targets for each developed country. The United States, for example, would reduce its carbon dioxide emissions to 7% below 1990 levels by 2012. This is thought to be economically untenable, and the Kyoto Protocol has not been ratified by the U.S. Senate. In any case, those levels would only slow the eventual effects on Earth's climate.

Alternatives to fossil fuel must be found if we are to maintain world economies and prevent an increase in global temperature with all its uncertainties. The only alternative source of energy that currently produces significant amounts of power is **nuclear energy,** which now generates about 17% of the electricity produced in the United States. Despite much publicity to the contrary, these pressurized water reactors have good records of dependable power production and safety.[3] The problem with nuclear power lies not so much in the everyday operation of the reactors but in disposing of the nuclear wastes they produce. By 1992 about 55,000 highly radioactive spent fuel assemblies were in temporary storage in deep pools of cooled water; they must be stored for 10,000 years before their levels of radioactivity will be low enough to pose no environmental hazard. Radioactive substances emit **ionizing radiation,** a form of energy able to penetrate and permanently dam-

age cells. It is mandatory that artificial sources of ionizing radiation be isolated from the environment.

Will citizens of industrial countries (and countries wishing to become industrialized) agree to lessen the danger of increased global warming by slowing their economic growth, decreasing their dependence on fossil fuel combustion, and developing safe alternative sources of energy? Some insight may be gained from the behavior of ranchers and industrialists in the rain forests of New Guinea, the Philippines, and Brazil. The Amazon rain forest of Brazil is being burned at a rate of about 12 square kilometers (almost 5 square miles) *each hour,* acreage equivalent to the area of West Virginia every year. Huge stands of trees that should be nurtured to absorb excess carbon dioxide are being destroyed. The cleared land is used for farms, cattle ranches, roads, and cities. The priorities are clear.

MARINE SANCTUARIES

Beginning in 1972, the federal government has established a dozen national marine sanctuaries (**Figure 18.24**). They vary in size but collectively cover about 46,000 square kilometers (18,000 square miles) of coral reefs, whale migration corridors, undersea archaeological sites, deep canyons, and zones of extraordinary beauty and biodiversity.

Despite their name, these sanctuaries are not off-limits to commercial fishing, trawling, or dredging. In May 2000, Pres-

---

[3] The Soviet reactor at Chernobyl that exploded in 1986 was of a much different design.

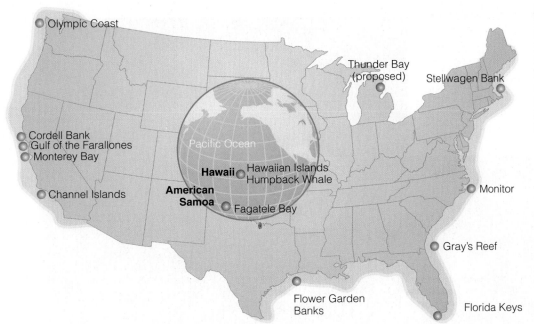

Figure 18.24 The U.S. marine sanctuaries. Since 1972, a dozen national marine sanctuaries have been designated in the coastal waters of the United States and American Samoa. Despite their name, most sanctuaries are not off-limits to commercial exploitation. (Source: *Popular Science,* September 2000.)

ident Bill Clinton issued an executive order directing federal agencies to establish a national framework for managing sanctuaries, wildlife refuges, and other protected areas that together cover about 1% of U.S. territorial waters. The new framework is to include ecological reserves where "consumptive uses" of marine resources will be prohibited. As of May 2001, the framework had not been applied.

What Can Be Done?

In a pivotal paper published in 1968, biologist Garrett Hardin examined what he termed "The Tragedy of the Commons." Hardin's title was suggested by his study of societies in which some agricultural areas were held *in common*—that is, were jointly owned by all residents. Citizens of these societies owned small homes, plots of land, and perhaps a cow that was put to pasture on the commons. Each farmer *kept* the milk and cheese given by his cow but *distributed* the costs of cow ownership—overgrazing of the commons, cow excrement, fouled drinking water, and so on—among all the citizens. This arrangement worked well for centuries because wars, diseases, and poaching kept the numbers of people and cows well below the carrying capacity of the land. But eventually political stability and relative freedom from disease allowed the human and cow populations to increase. Farmers pastured more cows on the commons and gained more benefits. Soon the overstressed commons could no longer sustain the growing number of cows, and the area held in common was ruined. Eventually no cows could survive there.

The lesson applies to our present situation. Hardin noted that in our social system each individual tends to act in ways that maximize his or her material gain. Each of us gladly keeps the *positive* benefit of work but willingly distributes the *costs* among all. For example, this morning I drove to my college office; the benefit to me was one trip to my office. A cost of this short drive was the air pollution generated by the fuel combustion in my car's engine. Did I route the exhaust fumes through a hose to a mask held tightly over my nose and mouth? That is, did I reserve the environmental costs of my actions for my own use, just as I had reserved for myself the benefit of my ride to work? No. I shared those fumes with my fellow Californians, just as you shared your morning's sewage with your fellow citizens, or just as the factory down the road shared its carbon dioxide with all the world. Indeed, the world itself is our commons. The modern tragedy of the commons rests on these kinds of actions.

The carrying capacity of the whole Earth-commons may already have been exceeded. Births now exceed deaths by about three people per second, 10,400 per hour. *Each year* there are 95 million more of us, a total equal to nearly one-third of the population of the United States. The number of people has tripled in the twentieth century and is expected to double again before reaching a plateau sometime in the twenty-first century. Another billion humans will join the world population in the next ten years, 92% of them in third world countries (**Figure 18.25**).[4] One-fifth of the world's people already suffer from abject poverty and hunger.

This exploding population is not content with using the same proportion of resources used today. Citizens of the

---

[4] In the United States alone, the population is growing by the equivalent of four Washington D.C.s every year, another New Jersey every 3 years, another California every 12.

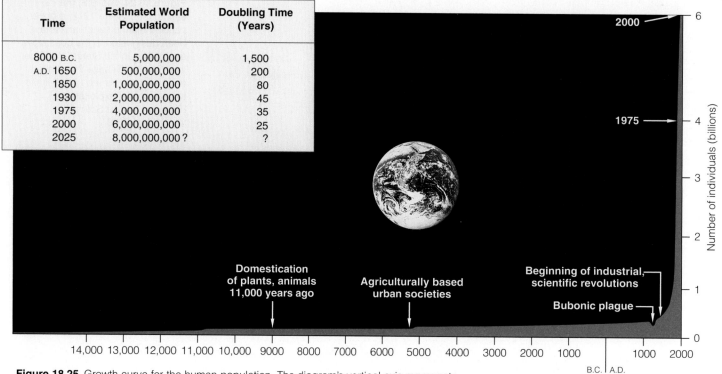

Time	Estimated World Population	Doubling Time (Years)
8000 B.C.	5,000,000	1,500
A.D. 1650	500,000,000	200
1850	1,000,000,000	80
1930	2,000,000,000	45
1975	4,000,000,000	35
2000	6,000,000,000	25
2025	8,000,000,000 ?	?

Figure 18.25 Growth curve for the human population. The diagram's vertical axis represents world population, in billions. (The slight dip between the years 1347 and 1351 represents the 25 million deaths in Europe from the bubonic plague.) The growth pattern over the past two centuries has been exponential, sustained by agricultural revolutions, industrialization, and improvements in health care. The list in the blue box tells how long it took for the human population to double in size at different times in history. The number of people on Earth now exceeds 6 billion. The last billion was added in only 12 years.

world's least-developed countries are influenced by education and advertising to demand a developed-world standard of living. They look with justifiable envy on the United States, a country with 5% of the planet's population that consumes 35% of its raw-material resources and 25% of its energy while generating 30% of industry-related carbon dioxide. Can the world support a population whose expectations are rising as rapidly as their numbers? In Garrett Hardin's words, "We can maximize the number of humans living at the lowest possible level of comfort, or we can try to optimize the quality of life for a smaller population." The burgeoning human population is the greatest environmental problem of all. The last Easter Islanders would have understood completely.

We cannot expect science to solve the problem for us. Most of the decisions and necessary actions fall outside pure science in the areas of values, ethics, morality, and philosophy. *The solution to environmental problems, if one exists, lies in education and action.* Each of us is obliged to become informed on issues that affect Earth, its ocean, and its air—to learn the arguments and weigh the evidence. Once informed, we must act in rational ways. Chaining yourself to an oil tanker is not rational, but selecting well-designed, long-lasting, recyclable products made by responsible companies with minimal environmental impact (and encouraging others to do so) certainly is.

Obvious answers and quick solutions are often misleading; a great deal of research and work is needed to give reliable insight into the many difficult questions that confront us. The present trade-off between financial and ecological considerations is often strongly tilted in favor of immediate gain, short-term profit, and immediate convenience. Education may be the only way to modify these destructive behaviors. Garrett Hardin suggests that absolute freedom in a commons brings ruin to all.

Humanity is part of the natural world, not its master. We may be able to learn to live in harmony with this small, beautiful, blue world. We need not relive the Tragedy of the Commons on a planetary scale. True convenience and true progress depend on the preservation of open space, serious and sustained attention to population control, conversion to a steady-state economy instead of one that must grow to stay alive, business incentives for preservation, the use of renewable resources, and, above all, *public education on environmental issues.* We must ask ourselves difficult questions: What is the optimal quality of life? How can I achieve balance between my material needs and the needs of Earth? What do I want to leave for my children? How can I reserve the quiet, renewing ocean for myself, for all species, and for the future? Our cities are crowded and our tempers are short. Times of turbulent change lie before us. The trials ahead will be severe.

Figure 18.26 A clean, calm corner of the world ocean.

Each of us, individually, needs to take a stand. We must preserve the sunsets and fog; the waves to ride; the cold, clean spray on our faces—our one world ocean (**Figure 18.26**). Margaret Mead summarized our potential for making a difference: "Never doubt that a small group of thoughtful, committed citizens can change the world. Indeed it is the only thing that ever has." We need to start now.

QUESTIONS FROM STUDENTS

1. What can an individual do to minimize his or her impact on the ocean and atmosphere?

Remember that Earth and all life forms are interconnected. There are no true consumers, only users: Nothing can truly be thrown away (there is no "away"). We must abandon the pollute-and-move-on ethic that has guided the actions of most humans for thousands of years, and we must work toward a society more in harmony with the fundamental rhythms of life that sustain us. *Our task is not to multiply and subdue Earth.* It may not be too late to change our ways. We need to act individually to effect change. We should think globally and act locally.

2. Is pollution always a bad thing?

Some forms of pollution bring temporary benefits. For example, some of Florida's once-endangered manatees are thriving at the warm outfalls of coastal power stations. On a larger scale, if increased greenhouse warming does develop, some computer models indicate increased rainfall, longer growing seasons, and increased crop yields over broad latitude bands in the temperate zones of both hemispheres. On the whole, though, the less human intervention in complex natural systems, the better.

3. What are the most dangerous threats to the environment, overall?

The underlying causes of the problems discussed in this chapter are (1) human population growth and (2) a growth-dependent economy. Stanford professor Paul Ehrlich said recently, "Arresting global population growth should be second in importance on humanity's agenda only to avoiding nuclear war." The present world population, now above 6 billion, seems doomed to reach at least 10 billion before leveling off. And what if everybody wants the same number of cows?

4. What role does public perception play in pollution issues?

A large role, indeed! Until recently, relatively small-scale but highly visible insults have claimed most of the public's attention and have driven us to action. The messy breakup of the oil tanker *Torrey Canyon* off the southern coast of England in 1967 galvanized world opinion and set the stage for the present environmental movement. More recently, in the summer of 1988 beachgoers were horrified to discover that more than 80 kilometers (50 miles) of northern New Jersey and Long Island beaches had been temporarily closed because of medical debris littering the shore. Some of the dozens of vials of blood, syringes, stained bandages, and surgical sutures tested positive for the viruses that cause AIDS and hepatitis B. Similar incidents occurred in Rhode Island and Massachusetts.

Appalling and visible though such incidents are, their long-term effects on the ocean as a whole are negligible. Public attention has recently turned to issues of larger consequence. The drought and heat of the past few summers brought terms like *greenhouse effect* and *ozone layer* to local newspapers and dinner table conversation. Weekend fishermen worry about eating their catch. They wonder about the unseen threats as much as the obvious ones. Perceptions are changing.

KEY CONCEPTS TO REVIEW

- A pollutant causes damage by interfering directly or indirectly with the biochemical processes of an organism.

- Spills of crude oil are generally larger in volume and more frequent than spills of refined oil. Given equal volumes, refined oil and used oil are more damaging than crude oil to the marine environment.

- Though it generated much media attention, the *Exxon Valdez* oil spill was only the 46th worst spill since 1960. Americans flush more oil down drains every year than was spilled by the *Exxon Valdez.*

- Synthetic organic chemicals and heavy metals are persistent environmental poisons; they can sometimes be concentrated to toxic levels by biological amplification through food chains.

- Uncontrolled algal growth through artificially high nutrient levels—eutrophication—can be dangerous to organisms and community diversity.

- Exotic (introduced) species are becoming more of a problem as marine transportation grows, because they may be carried to new locations inadvertently.

- Human activity is destroying habitats. Coral reefs are especially hard hit, with an estimated 70% of the world's reefs at risk.

- Depletion of the atmosphere's ozone layer exposes surface organisms to higher levels of ultraviolet radiation, and consequently to a greater risk of cancer.

- Changes in the composition of the atmosphere can influence global climate, ocean temperatures, and sea level. Experts estimate that at least half of recent global warming is caused by human activities.

- The solution to environmental problems, if one exists, lies in education and rational action.

INTERNET STUDY RESOURCES

The Web site for this book contains helpful study aids. Log on to:

www.brookscole.com/053437557xs

and click on the Chapter-by-Chapter area. Choose Chapter 18 and select a resource:

- **Flash Cards** allows you to test your mastery of the Terms and Concepts to Remember for this chapter.

- **Tutorial Quizzes** provides a multiple-choice practice quiz.
- **Student Guide to InfoTrac College Edition** will lead you to Critical Thinking Projects that use InfoTrac College Edition as a research tool.

TERMS AND CONCEPTS TO REMEMBER

biodegradable
biological amplification
chlorinated hydrocarbons
chlorofluorocarbons (CFCs)
eutrophication
greenhouse effect
greenhouse gases
ionizing radiation

marine pollution
nuclear energy
ozone
ozone layer
pollutant
polychlorinated biphenyls
  (PCBs)
sewage sludge

STUDY QUESTIONS

Review Questions

1. What is pollution? What factors determine how dangerous a pollutant is?

2. What is eutrophication? How can "good eating" be hazardous to marine life?

3. What parts of the marine environment are hardest hit by human activities?

4. What synthetic chemicals appear to be causing depletion of Earth's protective ozone layer? What is the likely result?

5. How is waste heat dangerous to marine life?

6. What introduced species affect your area? Are they dangerous? Costly?

7. What was the largest oil spill? How does this compare with routine introduction of oil into the marine environment by intentional release and leakage from landfills?

Critical Thinking Questions

1. Why is refined oil more hazardous to the marine environment than crude oil? Which is spilled more often? What happens to oil after it enters the marine environment?

2. What heavy metals are most toxic? How do these substances enter the ocean? How do they move from the ocean to marine organisms and people?

3. Few synthetic organic chemicals are dangerous in the very low concentrations in which they enter the ocean. How are these concentrations increased? What can be the outcome when these substances are ingested by organisms in a marine food chain?

4. What is the greenhouse effect? Is it always detrimental? What gases contribute to the greenhouse effect? Why do

most scientists believe that Earth's average surface temperature will increase over the next few decades? What may result?

5. What is the Tragedy of the Commons? Do you think Garrett Hardin was right in applying the old idea to modern times? What will you do to minimize your negative impact on the ocean and atmosphere?

6. How might global warming or a decrease in stratospheric ozone *directly* affect the ocean?

7. **InfoTrac College Edition Project** Debris and pollution in the ocean can be a deadly problem for marine life as well as an unsightly and dangerous addition to beaches. In some cases, however, cleanup may do more harm than good. Explore the question of marine pollution and its cleanup. In what cases should we emphasize prevention rather than remediation? Can there be a positive side to disposing of manufactured materials in the sea? Find information for your answer using InfoTrac College Edition.

 Earth Systems Today CD Question

1. Go to "Weather and Climate" and then to "Ozone Hole." Watch the 1996 to 2000 ozone-hole animation showing the seasonal changes in the size of the ozone hole. What can you imply from past data for the future of ozone-hole size? Should mid-latitude dwellers be concerned?

RESOURCES FOR FURTHER READING AND RESEARCH

The Web site for this book contains many ideas for further reading and research. Log on to:

 www.brookscole.com/053437557xs

and click on the Chapter-by-Chapter area. Choose Chapter 18 and select a resource:

- **References** lists the major books and articles consulted in writing this chapter, along with comments from the author about their content and reading level.
- **Hypercontents** takes you to an extensive list of sites with news, research, and images related to individual sections of the chapter.
- In **Student Guide to InfoTrac College Edition** scroll down to Suggested Readings from InfoTrac College Edition for brief descriptions of the articles listed and search hints for finding them in InfoTrac College Edition.
- **Regional InfoTrac College Edition** articles are organized into East Coast, West Coast, and Gulf Coast regions, allowing you to study oceanography on a more local level. You can also access Regional InfoTrac College Edition at www.localocean.com.

For additional reading, go to InfoTrac College Edition, your online research library, at: http://infotrac.thomsonlearning.com/

Afterword

The marine sciences are at the threshold of a new age. The recent revolutions in biology and geology are being assimilated, and the road ahead seems clearer. A revolution in the design of sampling devices, robot submersible vehicles, and data processing has brought new vigor to oceanography. Satellite-borne sensors can provide in an instant data that would have taken years to collect using surface ships. Shipboard technology has become so sophisticated that Wyville Thomson or Fridtjof Nansen would hardly recognize our sensors or sampling devices.

The tools may be different, but the spirits of those who use them remain the same. Today's marine scientists are like all the men and women who have gone before: We want to know about the ocean. We eagerly search our mailboxes for journals bearing the latest research news, watch television listings for any new ocean shows, inspect new samples with the enthusiasm of little kids, and share our insights with anyone at the drop of a hat. I am personally delighted that you have traveled with me this far. Those of us who enjoy an oceanographic background (and this now includes you) look at Earth with greater understanding than we did before we began. The whole concept of an ocean world appeals to us, gives us profound pleasure, and sobers us with a deep sense of responsibility. In no other field of science do so many ideas interweave to form so rich a tapestry.

Our journey together is over, but before we go our separate ways, I have four last ideas to share:

John Wilson Cramer IV

- *Change* has been a recurrent theme of this book. Earth's climate has changed with time, as has its atmospheric composition, its ocean chemistry, the size and positions of its continents, and its life forms. Earth may seem a calm and stable home, but it is really a violent place for inhabitation by such seemingly delicate objects as living things. Even so, life and the ocean have grown old together. The story of Earth is the story of change and chance; its history is written in the rocks, the water, and the genes of the millions of organisms that have evolved here. We are survivors.

  That survival may now be in question. Change is now progressing at an unnatural rate, and these human-induced changes are imposing stress on natural systems. What we do *with* and *to* the ocean is literally of planetary consequence. In the last century we have devel-

oped the physical, chemical, and biological machinery to destroy or rejuvenate the world ocean and all of its life. A painful time of inadvertent global experimentation lies just ahead.

All of us who love the colors and textures of this small wet world need to act to moderate the negative effects of the looming environmental crisis. In Chinese, the written character for the word *crisis* has two components: danger and opportunity. Informed citizens will express their concern, discuss this concern with others, and act whenever possible to minimize the threats and take advantage of new opportunities. Intelligence and beauty must triumph; we have no other rational alternative.

- Appreciation of the ocean doesn't come exclusively from the realm of science. Philosophers, artists, composers, and poets have had much to say about the sea. Read Homer's description of the ocean in *The Odyssey* (try Books IV, X, and XI). See how Lord Byron's feeling for the ocean colors his poetry (see, for instance, *Childe Harold's Pilgrimage*, stanzas 183 and 184). Read modern poet Robinson Jeffers's powerful *Continent's End.* Share Prospero's marine magic in Shakespeare's *The Tempest.* Find some of the evocative woodcuts of Rockwell Kent and the impressionistic ocean paintings of English artist J.M.W. Turner. Listen to Benjamin Britten's *Four Sea Interludes* from *Peter Grimes,* and Ralph Vaughan-Williams's *Sea Symphony* and *Sinfonia Antarctica* (but take care not to blow out your sound system). Read the ocean novels of Herman Melville and Jack London, and try reading the journals and accounts of the famous explorers and scientists you have met in this book. Sit on a quiet beach at night with the stars of the Milky Way shining softly overhead. The pervasive inspiration of the wave-breathing ocean is never far away.

- Don't let your involvement stop here. Lifelong learning is the truest joy, a pleasure that does not diminish with age, a source of wisdom and calm. We can learn much about patience, hope, and optimism from the ocean. We can learn much about the world—and about ourselves—by looking for the oceanic connections among things. I hope your interest in learning about the ocean has just been kindled. There is much good in the world. Go and add to it.

- Share your insight with family and friends. *Use* your new knowledge—make it your own. You don't need to be a college professor to talk to people about the beauty, history, and future of the ocean. My wife has always been patient and receptive of my oceanic tilt. Our children are my tolerant built-in students. Our daughter, son-in-law, and son are participants in Box 15.1 and Box 17.1. Along with your children, they will inherit the world.

APPENDIX I

Measurements and Conversions

Other than the United States, only two countries in the world—Liberia and Myanmar—do not use metric measurements. The metric system, a contribution of the French Revolution, conquered Europe along with Napoleon. It is based on a decimal system, a system familiar to Americans because of our decimal money system: 10 cents to a dime, 10 dimes to the dollar.

The first move toward a rational system of measurement was made in 1670 by Gabriel Mouton, the vicar of St. Paul's Church in Lyon, France. Instead of the then-prevalent measurement system based on the width of the king's hand, or the length of his outstretched right arm, or the weight of a particular basket of stones kept in the palace, Mouton suggested a length measure based on the arc of 1 minute of longitude, to be subdivided decimally. Other measurements would follow from this unit of length. His proposal contained the three major characteristics of the metric system: using Earth itself as a basis for measurement, subdividing decimally (by 10s), and using standard prefixes (*kilo, centi, milli,* and so on). These ideas were debated for 125 years before being implemented by a commission appointed by Louis XVI in one of his last official acts before being imprisoned during the French Revolution. One ten-millionth of the distance from the North Pole to the equator (on the line of longitude passing through Paris) was selected as the standard unit of length, the meter. A new unit of weight was derived from the weight of 1 cubic meter of pure water. Temperature was to be based on pure water's boiling and freezing points. A list of prefixes for decimal multiples and submultiples was proposed. In 1795 a firm decision was made to establish the system throughout France, and in 1799 the metric system was implemented "for all people, for all time."

At first, people objected to the changes, but the government insisted that old measurements be included side by side with the equivalent new (metric) ones. In everyday competition, the advantages of the metric system proved decisive; in 1840, it was declared a legal monopoly in France. The French public had been won over to the new, simple, rational system of measurement. All of Europe—and, eventually, virtually all other countries—followed.

Not the United States, however. Though Ben Franklin proposed that the country convert in the eighteenth century, the people of the United States have continued to insist that the metric system—now known as the Système International (SI)—is too difficult to learn and work with. The federal government has urged conversion to metric units to increase opportunities for international trade. In August 1988, President Ronald Reagan signed the Omnibus Trade and Competitiveness Act. This act amended the 1975 Metric Conversion Act, stating that by 1992 all federal agencies must, wherever feasible, use the metric (SI) system in their purchases, grants, and other business. (It should be noted that Canada began to convert in 1970 and has been metric since 1980.)

The government may be making the change, but the public clings tenaciously to inches, pints, and pounds. Why? Is it really simpler to add $\frac{1}{16}$ of an inch, $\frac{1}{32}$ of an inch, and $\frac{3}{8}$ of an inch to cut a bookshelf to length? Can you remember how many cups to a quart? How many pints to a gallon? How many miles to a league?[1] The reason we continue to use the old English Imperial system (which, of course, the English have long since abandoned) is because it is familiar to us. We know how long 5 inches is, and how much a quart is, and what 72° Fahrenheit represents. Perhaps by following the French example—by having measurements expressed everywhere in American *and* metric measurements—we may be able to make a complete conversion within a generation or two. That's why American and metric measurements are used together throughout this book. The process has already begun, of course: You use 35mm film, 2-liter soft drink containers, 750-milliliter wine bottles, 100-watt light bulbs—and you might run a 10-K (10-kilometer) race on Saturday.

The conversion factors listed here will give you an idea of how American and metric (SI) units are equivalent. Don't panic—the system is as rational and logical as it has always been. Note that 1 meter equals 100 centimeters and that 1 centimeter equals 10 millimeters. (See the table showing multiples and submultiples for an explanation of the relationship between prefixes like *centi* and *milli*.) Note that 2.54 centimeters equals 1 inch. (See the table of conversion factors if you wish to convert from one system to the other.) Some numerical oceanographic data are included in supplemental tables.

---

[1] A mile is 5,280 feet, and a league is 5,280 yards. Captain Nemo, in his fictional submarine *Nautilus,* traveled a distance greater than twice the circumference of Earth in Jules Verne's fantasy *20,000 Leagues Under the Sea,* a task made even more formidable by his total dependence on Imperial units of measurement!

Scientific Notation

Multiples and Submultiples

	Name	Common Prefixes
$10^{12} = 1,000,000,000,000$	trillion	tera
$10^9 = 1,000,000,000$	billion	giga
$10^6 = 1,000,000$	million	mega
$10^3 = 1,000$	thousand	kilo
$10^2 = 100$	hundred	hecto
$10^1 = 10$	ten	deka
$10^{-1} = 0.1$	tenth	deci
$10^{-2} = 0.01$	hundredth	centi
$10^{-3} = 0.001$	thousandth	milli
$10^{-6} = 0.000001$	millionth	micro
$10^{-9} = 0.000000001$	billionth	nano
$10^{-12} = 0.000000000001$	trillionth	pico

Conversion Factors

Area

1 square inch (in.²)	6.45 square centimeters
1 square foot (ft²)	144 square inches
1 square centimeter (cm²)	0.155 square inch 100 square millimeters
1 square meter (m²)	10^4 square centimeters 10.8 square feet
1 square kilometer (km²)	247.1 acres 0.386 square mile 0.292 square nautical mile

Mass

1 kilogram (kg)	2.2 pounds 1,000 grams
1 metric ton	2,205 pounds 1,000 kilograms 1.1 tons
1 pound	16 ounces 454 grams 0.45 kilograms
1 ton	2,000 pounds 907.2 kilograms 0.91 metric ton

Pressure

1 atmosphere (sea level)	760 millimeters of mercury at 0°C 14.7 pounds per square inch 33.9 feet of water (fresh) 29.9 inches of mercury 33 feet of seawater

Length

1 micrometer (μm)	0.001 millimeter 0.0000349 inch
1 millimeter (mm)`	1,000 micrometers 0.1 centimeter 0.001 meter
1 centimeter (cm)	10 millimeters 0.394 inch 10,000 micrometers
1 meter (m)	100 centimeters 39.4 inches 3.28 feet 1.09 yards
1 kilometer (km)	1,000 meters 1,093 yards 3,281 feet 0.62 statute mile 0.54 nautical mile
1 inch (in.)	25.4 millimeters 2.54 centimeters
1 foot (ft)	30.5 centimeters 0.305 meter
1 yard	3 feet 0.91 meter
1 fathom	6 feet 2 yards 1.83 meters
1 statute mile	5,280 feet 1,760 yards 1,609 meters 1.609 kilometers 0.87 nautical mile
1 nautical mile	6,076 feet 2,025 yards 1,852 meters 1.15 statute miles
1 league	15,840 feet 5,280 yards 4,804.8 meters 3 statute miles 2.61 nautical miles

Volume

1 cubic inch (in.³)	16.4 cubic centimeters
1 cubic foot (ft³)	1,728 cubic inches 28.32 liters 7.48 gallons
1 cubic centimeter (cc; cm³)	1 millimeter 0.061 cubic inch
1 liter	1,000 cubic centimeters 61 cubic inches 1.06 quarts 0.264 gallon
1 cubic meter (m³)	10^6 cubic centimeters 264.2 gallons 1,000 liters
1 cubic kilometer (km³)	10^9 cubic meters 10^{15} cubic centimeters 0.24 cubic mile

Temperature

$$°C = \frac{(°F - 32)}{1.8}$$

$$°F = (1.8 \times °C) + 32$$

$$°K = °C + 273.2$$

	°F	°C
Water boils	212	100

100°C = 212°F
(boiling point of water)

40°C = 104°F
(heat wave conditions)

37°C = 98.6°F
(normal body temperature)

30°C = 86°F
(very warm—almost hot)

20°C = 68°F
(a mild spring day)

10°C = 50°F
(a warm winter day)

0°C = 32°F
(freezing point of water)

		160	80
			60
Body temperature	98.6		37
		80	20
Water freezes	32		0
		0	−20
		−40	−40

Some Familiar Metric Approximations

Measurement	Metric Unit	Approximate Size of Unit
Length	millimeter	diameter of a paper clip wire
	centimeter	a little more than the width of a paper clip (about 0.4 inch)
	meter	a little longer than a yard (about 1.1 yards)
	kilometer	somewhat farther than ½ mile (about 0.6 mile)
Mass (Weight)	gram	a little more than the mass (weight) of a paper clip
	kilogram	a little more than 2 pounds (about 2.2 pounds)
	metric ton	a little more than a short ton (about 2,200 pounds)
Volume	milliliter	five of them make a teaspoon
	liter	a little larger than a quart (about 1.06 quarts)
Pressure	kilopascal	atmospheric pressure is about 100 kilopascals

Source: U.S. Metric Board Report.

Time

1 hour	3,600 seconds
1 day	24 hours 1,440 minutes 86,400 seconds
1 calendar year	31,536,000 seconds 525,600 minutes 8,760 hours 365 days

Speed

1 statute mile per hour	1.61 kilometers per hour 0.87 knot
1 knot (nautical mile per hour)	51.5 centimeters per second 1.15 miles per hour 1.85 kilometers per hour
1 kilometer per hour	27.8 centimeters per second 0.62 mile per hour 0.54 knot

Numerical Oceanographic Data

Equivalences in Concentration of Seawater

Seawater with 35 grams of salt per kilogram of seawater	3.5 percent 35 parts per thousand (‰) 35,000 parts per million (ppm)

Speed of Sound

Velocity of sound in seawater at 34.85 parts per thousand (‰)	4,945 feet per second 1,507 meters per second 824 fathoms per second

Area, Volume, and Depth of the World Ocean

Body of Water	Area (10^6 km²)	Volume (10^6 km³)	Mean Depth (m)
Atlantic Ocean	82.4	323.6	3,926
Pacific Ocean	165.2	707.6	4,282
Indian Ocean	73.4	291.0	3,963
All oceans and seas	361	1,370	3,796

APPENDIX II

Geological Time

As we saw in Chapter 2, astronomers and geologists have determined that Earth originated about 4.6 billion years ago. They have divided Earth's age into eras, roughly corresponding to major geological and evolutionary changes that have taken place, as shown in the chart in **Figure 1**. Note that the time spans of the different eras are not shown to scale; if they were, the chart would run off the page.

Recall that life began fairly soon after Earth's crust, atmosphere, and ocean formed. One way to conceive of the time span over which life evolved is to imagine it as a 24-hour clock, with life originating at midnight (**Figure 2**). In this scheme, invertebrates with hard parts (which make good fossils) became abundant about 4:30 P.M., and animals began to leave the ocean for land about 8 P.M. Our closest human ancestors (*Homo sapiens*) apeared about 2 seconds before midnight, agriculture began only ¼ second before midnight, and the Industrial Revolution has been around for $\frac{7}{1000}$ of a second.

Era	Period	Epoch	Millions of Years Ago (Ma)
CENOZOIC	Quaternary	Recent	0.01
		Pleistocene	1.65
	Tertiary	Pliocene	5
		Miocene	24
		Oligocene	37
		Eocene	58
		Paleocene	66
MESOZOIC	Cretaceous	Late	
		Early	98
	Jurassic		144
	Triassic		208
PALEOZOIC	Permian		245
	Carboniferous		286
	Devonian		360
	Silurian		408
	Ordovician		438
	Cambrian		505
PROTEROZOIC			570
ARCHEAN			2,500

Source: Geological Time Scale, Decade of North American Geology, Geological Society of America, 1983.

Figure 1 A geological time scale.

Times of Major Geological and Biological Events

1.65 Ma to present. Major glaciations. Modern humans emerge and begin what may be greatest mass extinction of all time on land, starting with ice age hunters.

65–1.65 Ma. Unprecedented mountain building as continents rupture, drift, collide. Major climatic shifts; vast grasslands emerge. Major radiations of flowering plants, insects, birds, mammals. Origin of earliest human forms.

65 Ma. Asteroid impact? Mass extinction of all dinosaurs and many marine organisms.

135–65 Ma. Pangaea breakup continues, broad inland seas form. Major radiations of marine invertebrates, fishes, insects, dinosaurs. Origin of flowering plants.

181–135 Ma. Pangaea breakup begins. Rich marine communities. Major radiations of dinosaurs.

205 Ma. Asteroid impact? Mass extinction of many organisms in seas, some on land; dinosaurs, mammals survive.

240–205 Ma. Recovery, radiations of marine invertebrates, fishes, dinosaurs. Gymnosperms the dominant land plants. Origin of mammals.

240 Ma. Mass extinction. Nearly all species in seas and on land perish.

280–240 Ma. Pangaea, worldwide ocean forms; shallow seas squeezed out. Major radiations of reptiles, gymnosperms.

360–280 Ma. Tethys Sea forms. Recurring glaciations, Major radiations of insects, amphibians. Spore-bearing plants dominate. Gymnosperms present. Origin of reptiles.

370 Ma. Mass extinction of many marine invertebrates, most fishes.

435–360 Ma. Laurasia forms, Gondwana moves north. Vast swamplands, early vascular plants. Radiations of fishes continue. Origin of amphibians.

435 Ma. Glaciations as Gondwana crosses South Pole. Mass extinction of many marine organisms.

500–435 Ma. Gondwana moves south. Major radiations of marine invertebrates, early fishes.

550–500 Ma. Landmasses dispersed near equator. Simple marine communities. Origin of animals with hard parts.

700–550 Ma. Supercontinent Laurentia breaks up; widespread glaciations.

2,500–570 Ma. Oxygen present in atmosphere. Origin of aerobic metabolism. Origin of protistans, algae, fungi, animals.

3,800–2,500 Ma. Origin of photosynthetic bacteria.

4,600–3,800 Ma. Formation of Earth's crust, early atmosphere, oceans. Chemical evolution leading to origin of life (anaerobic bacteria).

4,600 Ma. Origin of Earth.

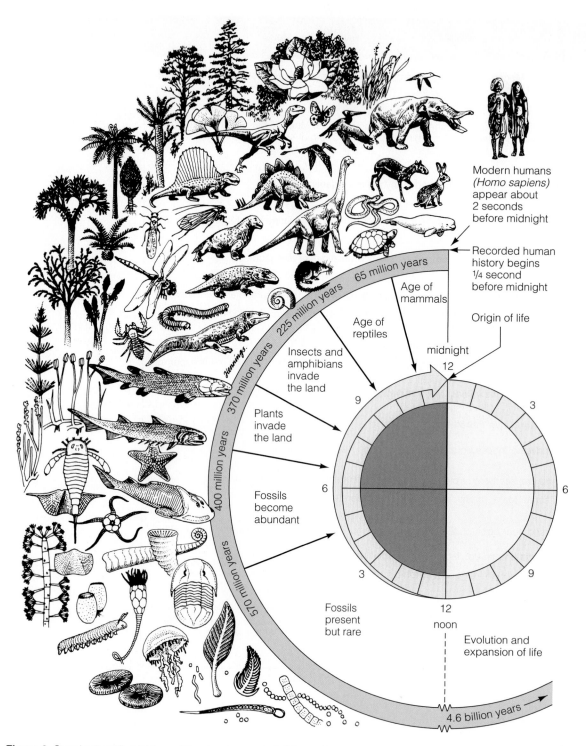

Modern humans
(Homo sapiens)
appear about
2 seconds
before midnight

Recorded human
history begins
¼ second
before midnight

Origin of life

Age of
mammals

midnight

Age of
reptiles

12

Insects and
amphibians
invade
the land

9

3

Plants
invade
the land

65 million years

225 million years

370 million years

Fossils
become
abundant

6

6

400 million years

570 million years

3

9

Fossils
present
but rare

12
noon

Evolution and
expansion of life

4.6 billion years

Figure 2 Greatly simplified history of the development of different forms of life on Earth through biological evolution, compressed to a 24-hour time scale.

Latitude and Longitude, Time, and Navigation

The ocean is large and easy to get lost in. A backyard, like that shown in **Figure 1,** is smaller, but we can still be lost in it if we don't have a frame of reference. Note that the yard is framed by a fence. We can refer to this frame to establish our position—in this case, at the intersection of perpendicular lines drawn from fence posts 2 and C. Many towns are arranged in this way: Fourth and D Streets intersect at a precise spot based on the municipal frame of reference.

But the World Is Round: Spherical Coordinates

If the world were flat, a simple scheme of rectangular coordinates would serve all mapping purposes—a rectangle, like the yard in Figure 1, has four sides from which to measure. A sphere has no edges, no beginnings or ends, so what shall we use as a frame of reference for Earth? Because Earth turns, the poles—the axis of rotation—are the only absolute points

of reference. We can draw an imaginary line equidistant from the North and South Poles, a line that *equates* the globe into northern and southern halves: the equator. Other lines, drawn parallel to the equator, further divide the sphere north and south of the equator. These lines, or parallels, are lines of **latitude** (**Figure 2**).

We can further subdivide Earth by drawing lines at regular intervals through both poles. Note that unlike the parallels, these lines, called meridians, are all equally long. Meridians are lines of **longitude** (**Figure 3**).

If you travel north from the equator, you can count the parallels (lines of latitude) that cross your path to find out how far you have gone. Likewise, if you travel east from a reference meridian, you can count the meridians (lines of longitude) that cross your path to find out how far you have gone. Just as a football player on the field knows his distance from the goal line by the yard lines that cross his run, so you know how far north or east you have gone by the lines that have crossed your path.

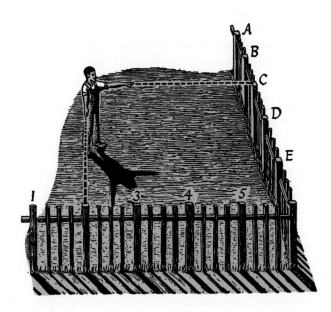

Figure 1

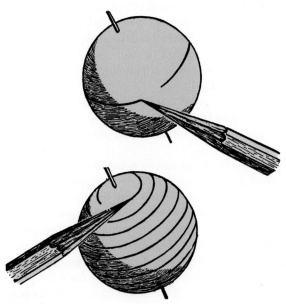

Figure 2

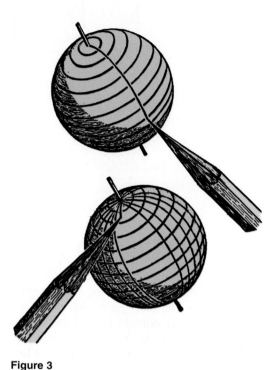

Figure 3

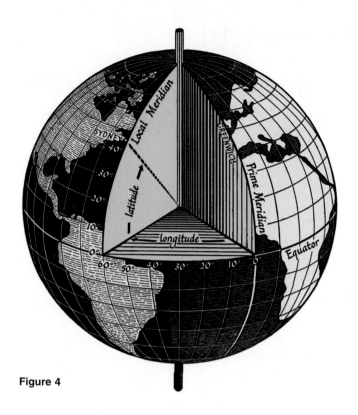

Figure 4

Because there are no continuous lines of fence posts on the spherical Earth, our reference frame for latitude and longitude must be marked from the equator and poles by some other means. This is done by degrees.

Why Degrees?

Degrees measure fractions of a circle. We need to know what fraction of Earth's circumference separates us from the equator and from the reference meridian to have a definite idea of our location.

Babylonian astronomers first divided the circle into 360 degrees (°). Why 360? The moon cycles around Earth every 30 days. It takes about 12 months ("moonths") to make a year. Thus, $30 \times 12 = 360$, the number of days they supposed was in a year. Circles were divided the same way. As we saw in Chapter 1, the Greek librarian Hipparchus applied this division to the surface of Earth.

In **Figure 4**, we have marked the position of Sydney, Canada. A line drawn from Sydney to the center of Earth intersects the plane of the equator at an angle of 46° to the north. That is its latitude.

The reference meridian, the meridian from which all others are marked, is known as the prime meridian. Unlike the equator, there is no earthly reason why the prime meridian should pass through any particular place. It passes through Greenwich, England, because an international agreement signed in 1884 decreed it so. The meridian on which Sydney, Canada, lies intersects the plane of the prime meridian at an angle of 60°. The angular distance of Sydney from the prime meridian is 60° to the west. That is its longitude. So its position is 46°N 60°W.

We can do this for each hemisphere. A line drawn to the center of Earth from Sydney, Australia, intersects the plane of the equator at an angle of 34° south latitude. Sydney, Australia, lies 151° east of the prime meridian. Thus, its position is 34°S 151°E. (Note that the greatest possible longitude is 180°; once you pass 180°, the line opposite the prime meridian, you begin to come around the other side of Earth, and the angle to Greenwich decreases.)

What Does Time Have to Do with This?

Meridians are often numbered from the prime meridian in 15° increments. Earth takes 24 hours to complete a 360° rotation. Divide 360° by 24 hours and you get 15, the number of degrees the sun moves across the sky in 1 hour. Meridians on a globe are often spaced to represent 1 hour's turning of Earth toward or away from the sun, toward or away from the moon.

You can use this fact to find your east-west position, your longitude. Imagine that you have a radio that can tell you the precise time of noon at Greenwich.[1] If your local noon comes *before* Greenwich noon, you are east of Greenwich. For instance, if the sun is highest in your sky at 10 A.M. Greenwich time, you are 2 hours before—30° east—of Greenwich. Earth

---

[1]Any shortwave radio will do. Tune it to 2.5, 5, 10, 15, or 20 mHz for radio stations WWV (Colorado) or WWVH (Hawaii). These stations broadcast time signals giving a measure of coordinated universal time, an international time standard based on the time at Greenwich. For a telephone report of coordinated universal time, call WWV at (303) 499-7111, or go to www.time.gov on the Internet.

must turn 2 more hours before the sun will shine directly above the Greenwich meridian. If your local noon is *after* Greenwich noon, you are west of Greenwich. Suppose that the sun is at high noon and your chronometer, set at Greenwich time, says 6 P.M. That means that Earth has been turning 6 hours since noon at Greenwich, and 6 hours times 15° per hour is 90°. That's your longitude relative to Greenwich: 90°W.

Navigation

Longitude is half the problem. To find latitude and obtain a position, we need to measure the angle north or south of the equator. But we can't use the time difference between local noon and Greenwich noon to determine latitude because the sun moves from east to west and we want to measure north-south position. Instead, we use the angle of the North Star above the horizon. Polaris, the current North Star, lies almost exactly above the North Pole. If we were standing at the North Pole, the North Star would appear almost directly overhead; ideally, the angle from the horizon to the star would be 90°, the same as the latitude of the North Pole (**Figure 5**). At Sydney, Canada, the angle from the horizon to the star would be about 46°—again, the same as the latitude. What would the angle of Polaris be at the equator, 0° latitude? (If you enjoyed your high school geometry course, you might try to prove that the angle from the horizon to Po-

laris is equal to the latitude at any position in the Northern Hemisphere.)

Polaris is not visible in the Southern Hemisphere, so how can we find south latitude? By finding the angle above the horizon of other stars. In practice, navigators in both hemispheres use a sextant to measure angles from the horizon to selected stars, planets, the moon, and the sun. The time of the observation is carefully noted. The navigator takes these readings to his or her stateroom, consults a series of mathematical tables, does some relatively simple calculations to compensate for observational errors, and comes up to the pilothouse with the vessel's latitude and longitude, accurate (in the best of circumstances) to within ½ mile, marked on a small slip of paper. The daily results are always entered into the ship's log.

New Tricks

Discovering position by measuring the angular positions of heavenly bodies—celestial navigation—is a dying art. Global positioning satellites, loran-C, inertial platforms, radar, and other electronic wonders have largely replaced the romance of a navigator standing on the bridge squinting through a sextant. The slip of paper has been supplanted by the glow of backlit liquid-crystal readouts or a chart with an X marking the ship's position, accurate to within 50 feet, feeding out of a slot. Still, when the power fails, the human navigator becomes the most popular person on board.

APPENDIX IV

Maps and Charts

It is easier to draw a diagram to show someone how to get to a place than to describe the process in words. For centuries, travelers have made special diagrams—maps and charts—to jog their own memories and to show others how to reach distant destinations. A **map** is a representation of some part of Earth's surface, showing political boundaries, physical features, cities and towns, and other geographical information. A **chart** is also a representation of Earth's surface, but it has been specially designed for convenient use in navigation. It is intended to be worked on, not merely looked at. A **nautical chart** is primarily concerned with navigable water areas. It includes information such as coastlines and harbors, channels, obstructions, currents, depths of water, and the positions of aids to navigation.

Any flat map or chart is necessarily a distortion of the spherical Earth. If we roll a flat sheet of paper around a globe to form a cylinder, the paper will contact the globe only along one curve. Let's assume that it's the equator. If the lines of latitude and longitude on the globe are covered with ink, only the equator will contact the paper and print an exact replica of itself. Unroll the cylinder, and that part of the new map will be a perfect representation of Earth. To include areas north and south of the equator, we will have to "throw them forward" onto the paper; we need to *project* them in some way.

Now imagine our globe to be a translucent sphere. If we place a bright light at its center, we can project the lines of latitude and longitude onto the rolled paper cylinder (**Figure 1**). Careful tracing of these lines will result in a map, but the areas away from the equator will be distorted: The farther from the equator, the greater the distortion. A useful modification of this projection—one that does not distort high latitudes as dramatically—was devised by Gerhardus Mercator, a Flemish cartographer who published a map of the world in 1569. Though landmasses and ocean areas are not depicted as accurately in a Mercator projection as they would be on a globe, such a map is still useful because it enables mariners to steer a course over long distances by plotting straight lines.

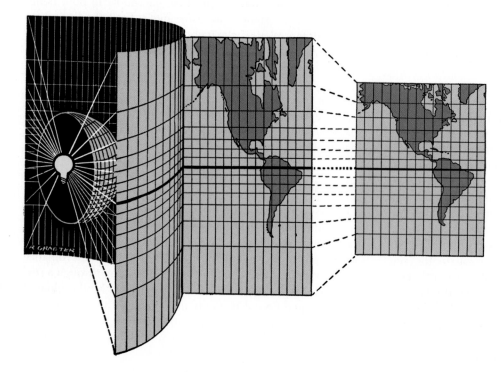

Figure 1 Central projection of a globe upon a cylinder, and a modified map structure, the Mercator, made to the same scale along the equator.

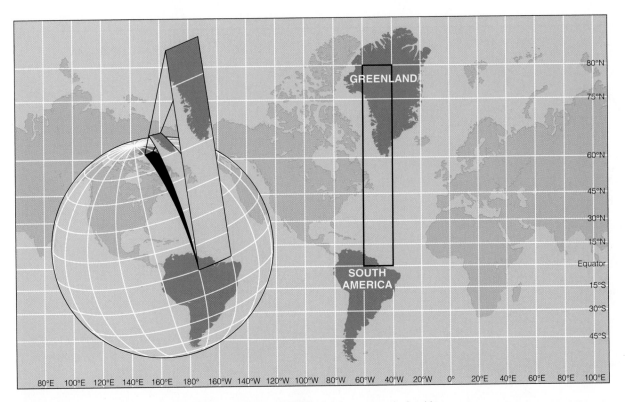

Figure 2 A gore of the globe peeled and projected according to the scheme devised by Gerhardus Mercator. This is the projection used on modern sailing charts. Note that this projection's distortion at high latitudes makes Greenland and South America appear about the same size. Next time you're near a globe, check their real sizes.

The distortion in Mercator projections has led generations of schoolchildren to believe that Greenland is the same size as South America (**Figure 2**). Mercator charts can distort our perceptions of the ocean as well: The area of the continental shelves at high latitudes, the amount of primary productivity in the polar regions, and the importance of ocean currents at the northerly or southerly extremes of an ocean basin may be exaggerated if presented in Mercator projection. The projection used in this book—a further modification of the Mercator projection known as the Miller projection—was chosen for its more accurate representation of surface area at high latitudes.

Mapmakers have invented other projections, each with advantages and disadvantages for particular uses. Some are conical projections: a flat sheet of paper wrapped into a cone with its edge touching the globe at a line of latitude north (or south) of the equator and the point of the cone above the North (or South) Pole. Conical projections do not distort high-latitude areas in the same way a Mercator projection does and, if drawn for the ocean area in which a mariner is sailing, can be used to draw great circle routes as straight lines. However, the distortions inherent in a conical projection prevent it from being used to represent more than about one-third of the globe on a single sheet of paper. Other projections, like the point-contact projection shown in **Figure 3,** try to minimize distortion around a specific location. All map and chart projections are distorted in some way; a sphere cannot be flattened onto a plane without deformation. Marine scientists

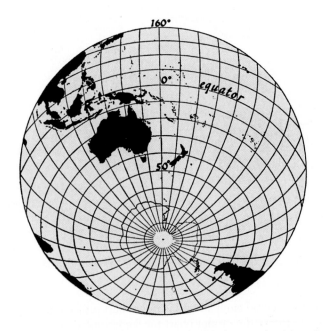

Figure 3 A Lambert equal-area projection (a type of point-contact projection), centered at 50°S 160°E.

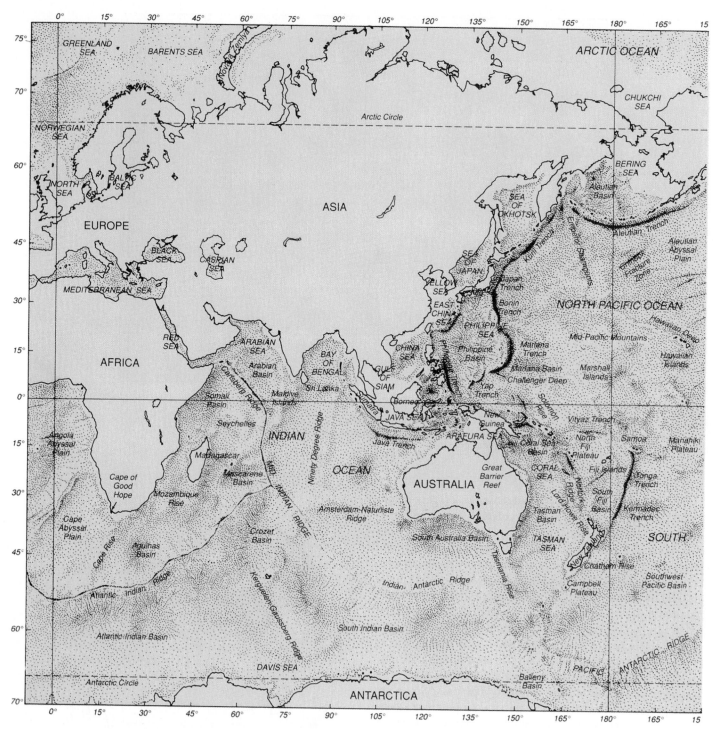

Figure 4 A map of the world, with many oceanic features labeled.

necessarily become familiar with various chart projections and are careful to use the proper chart for the intended purpose.

Figure 4 is a Mercator projection of the world. On it are indicated areas of interest discussed in this book.

For Further Study

Herring, T. 1996. "The Global Positioning System." *Scientific American*, 274 (no. 2): 44–50.

Krause, G., and M. Tomczak. 1995. "Do Marine Scientists Have a Scientific View of the Earth?" *Oceanography* 8 (no. 1): 11–16. Chart distortions often distort our interpretation of data, as this well-illustrated paper demonstrates.

Wilford, J. N. 1998. "Revolution in Mapping." *National Geographic* 193 (no. 2): 6–39. The usual thorough treatment of a rapidly changing topic.

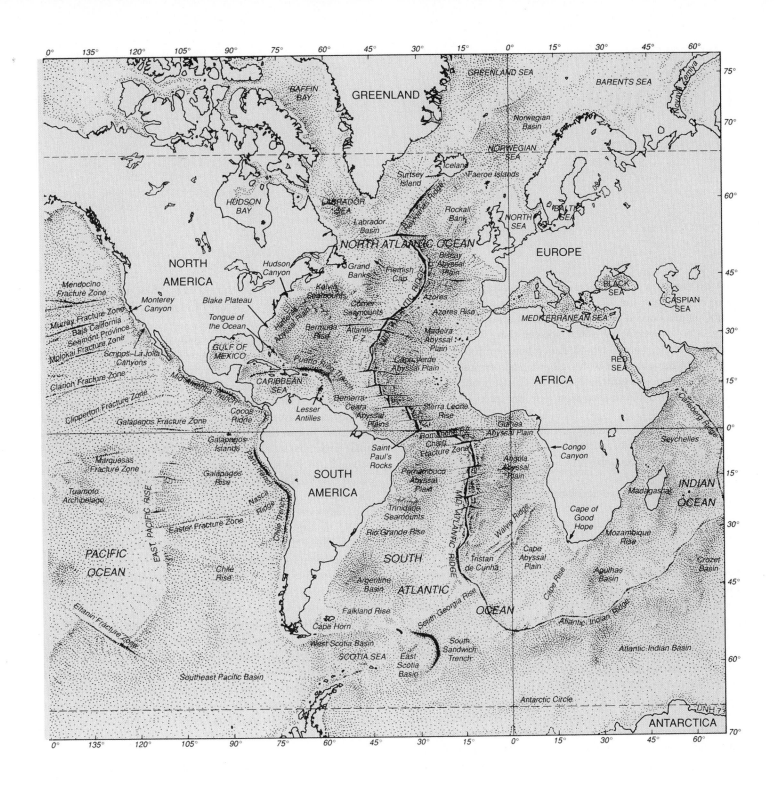

APPENDIX V
The Coriolis Effect

Newton's first law states that an object in motion will naturally move in a straight line, but an object at rest on Earth's surface is moving in a curved path. Because the motion is not straight, it must be experiencing some acceleration, which may be written as v^2/R (R equals the distance between Earth's center and that of the object; it is, for all practical purposes, Earth's radius). The force is written $F = v^2/R$. This centrifugal inertia varies with latitude and is written, per unit mass, as

Force = mass $\times$ (velocity)^2 $\times$ radius at that latitude
$\times$ cosine of latitude
$$F = Mv^2R\phi(\cos \phi)$$

Thus, the centrifugal acceleration is

$$F/M = v^2R\phi(\cos \phi)$$

The Coriolis effect (F_c) (per unit mass) is a function of an object's velocity across Earth's surface (horizontal velocity v) and the vertical component of Earth's rotational velocity. It is mathematically stated as $F_{CF} = 2wv \sin \phi$, where w is Earth's angular velocity and ϕ is the latitude. This deflecting force is always at right angles to the velocity and, of course, performs no work. The Coriolis effect is greatest at the poles (90° latitude) and zero at the equator (0° latitude). Mathematically, this is due to the fact that the sine is a maximum (1.00) at 90° and zero at 0°. In physical terms, however, the Coriolis effect vanishes at the equator because there the vertical component of Earth's rotation vanishes; the vertical component is greatest at the poles. When an object is at rest, its centrifugal force and deflecting force are equal; but when the object has a hor-

izontal motion across Earth's surface, it has, depending on the direction, an excess or deficit of centrifugal inertia, which is termed the *Coriolis effect*.

It is by no means a foregone conclusion that the deflection due to the Coriolis effect on a moving object will be greater at higher latitudes, because the velocity of the object can vary. One would expect a large deflection if such an object moves very rapidly. The Coriolis effect would indeed be large, and the distance traversed by the object would also be large per unit time, but in that time period Earth would rotate only a small angle distance, and the object's actual deflection would be small. In the same period of time, a slow-moving object would cover a short distance along its path of travel (velocity = distance/time, so distance = velocity $\times$ time), producing a large deflection but a small Coriolis acceleration.

While the deflection force is not a physical force, it is necessarily real to an observer within a rotating system and must be taken into consideration when describing motion, especially for movements of fluids such as water or of air, which move at relatively low velocities and experience appreciable deflection over long distances.

In addition, the equation $C_F = 2wv \sin \phi$ actually expresses the magnitude of the Coriolis acceleration and would, for a unit mass, numerically equal the Coriolis effect. The product of the object's mass and the Coriolis acceleration is the Coriolis effect, $C_F = ma$ (Newton's second law). For objects in *vertical* motion, the Coriolis acceleration is greatest at the equator and least at the poles.

Source: Ingmanson and Wallace, 1995, courtesy of Wadsworth Publishing Co.

Taxonomic Classification of Marine Organisms

Exclusively nonmarine phyla generally have been omitted, along with most extinct phyla and classes.

KINGDOM BACTERIA: Single-celled prokaryotes with a single chromosome that reproduce asexually and exhibit high metabolic diversity.

KINGDOM ARCHAEA: Superficially similar to bacteria, but with genes capable of producing different kinds of enzymes. Often live in extreme environments.

KINGDOM PROTISTA: Eukaryotic single-celled, colonial, and multicellular autotrophs and heterotrophs.

    PHYLUM CHRYSOPHYTA. Diatoms, coccolithophores, silicoflagellates.

    PHYLUM PYRROPHYTA. Dinoflagellates, zooxanthellae.

    PHYLUM CRYPTOPHYTA. Some "microflagellates"; cryptomonads.

    PHYLUM EUGLENOPHYTA. A few "microflagellates"; mostly freshwater.

    PHYLUM ZOOMASTIGINA. Nonphotosynthesizing flagellated protozoa.

    PHYLUM SARCODINA. Amoebas and their relatives.

        Class Rhizopodea. Foraminiferans.
        Class Actinopodea. Radiolarians.

    PHYLUM CILIOPHORA. Ciliated protozoa.

    PHYLUM CHLOROPHYTA. Multicellular green algae.

    PHYLUM PHAEOPHYTA. Brown algae, kelps.

    PHYLUM RHODOPHYTA. Red algae, encrusting and coralline forms.

KINGDOM FUNGI: Fungi, mushrooms, molds, lichens; mostly land, freshwater, or highest supratidal organisms; heterotrophic.

KINGDOM PLANTAE: Photosynthetic autotrophs.

    DIVISION ANTHOPHYTA. Flowering plants (angiosperms). Most species are freshwater or terrestrial. Marine eelgrass, manatee grass, surfgrass, turtle grass, salt marsh grasses, mangroves.

KINGDOM ANIMALIA: Multicellular heterotrophs.

    PHYLUM PLACOZOA. Amoeba-like multicellular animals.

    PHYLUM MESOZOA. Wormlike parasites of cephalopods.

    PHYLUM PORIFERA. Sponges.

    PHYLUM CNIDARIA. Jellyfish and their kin; all are equipped with stinging cells.

        Class Hydrozoa. Polyplike animals that often have a medusa-like stage in their life cycle, such as Portuguese man-of-war.
        Class Scyphozoa. Jellyfish with no (or reduced) polyp stage in life cycle.
        Class Cubozoa. Sea wasps.
        Class Anthozoa. Sea anemones, coral.

    PHYLUM CTENOPHORA. "Sea gooseberries," comb jellies; round, gelatinous, predatory, common.

    PHYLUM PLATYHELMINTHES. Flatworms, tapeworms, flukes; many free-living predatory forms, many parasites.

    PHYLUM NEMERTEA. Ribbon worms.

    PHYLUM GNATHOSTOMULIDA. Microscopic, wormlike; live between grains in marine sediments.

    PHYLUM GASTROTICHA. Microscopic, ciliated; live between grains in marine sediments.

    PHYLUM ROTIFERA. Ciliated; common in fresh water, in plankton, and attached to benthic objects.

    PHYLUM KINORYNCHA. Small, spiny, segmented, wormlike; live between grains in marine sediments; all marine.

PHYLUM ACANTHOCEPHALA. Spiny-headed worms; all parasitic in vertebrate intestines.

PHYLUM ENTOPROCTA. Polyplike, small, benthic suspension feeders.

PHYLUM NEMATODA. Roundworms. Common, free-living, parasitic.

PHYLUM BRYOZOA. Common, small, encrusting colonial marine forms.

PHYLUM PHORONIDA. Shallow-water tube worms; suspension feeders; a few centimeters long; all marine.

PHYLUM BRACHIOPODA. Lampshells; bivalve animals, superficially like clams; scarce, mainly in deep water.

PHYLUM MOLLUSCA. Mollusks.
    Class Monoplacophora. Rare deep-water forms with limpetlike shells.
    Class Polyplacophora. Chitons.
    Class Aplacophora. Shell-less, sand burrowing.
    Class Gastropoda. Snails, limpets, abalones, sea slugs, pteropods.
    Class Bivalvia. Clams, oysters, scallops, mussels, shipworms.
    Class Cephalopoda. Squid, octopuses, nautiluses.
    Class Scaphopoda. Tooth shells.

PHYLUM ARTHROPODA.
    Subphylum Crustacea. Copepods, barnacles, krill, isopods, amphipods, shrimp, lobsters, crabs.
    Subphylum Chelicerata. Horseshoe crabs, sea spiders.
    Subphylum Uniramia. Insects, centipedes, millipedes; one genus and five species in the ocean.

PHYLUM PRIAPULIDA. Small, rare, wormlike, subtidal.

PHYLUM SIPUNCULA. Peanut worms; all marine.

PHYLUM ECHIURA. Spoon worms.

PHYLUM ANNELIDA. Segmented worms; includes polychaetes such as feather duster worms and some oligochaete deep-sea bristle worms.

PHYLUM TARDIGRADA. "Water bears"; tiny, eight-legged animals with the ability to survive long periods of hibernation.

PHYLUM PENTASTOMA. Tongue worms; parasites of vertebrates.

PHYLUM POGONOPHORA. Beard worms; no digestive system; deep-water tube worms; all marine.

PHYLUM ECHINODERMATA. Spiny-skinned, benthic, radially symmetrical, most with a water-vascular system.
    Class Asteroidea. Sea stars.
    Class Ophiuroidea. Brittle stars, basket stars.
    Class Echinoidea. Sea urchins, sand dollars, sea biscuits.
    Class Holothuroidea. Sea cucumbers.
    Class Crinoidea. Sea lilies, feather stars.
    Class Concentricycloidea. Sea daisies.

PHYLUM CHAETOGNATHA. Arrowworms; stiff-bodied, planktonic, predaceous, common.

PHYLUM HEMICHORDATA. Acorn worms; unsegmented burrowers.

PHYLUM CHORDATA.
    Subphylum Urochordata. Sea squirts, tunicates, salps.
    Subphylum Cephalochordata. Lancelets, *Amphioxus*.
    Subphylum Vertebrata.
        Class Agnatha: Jawless fishes: lampreys, hagfishes; cartilaginous skeleton.
        Class Chondrichthyes. Sharks, skates, rays, sawfish, chimaeras; cartilaginous skeleton.
        Class Osteichthyes. Bony fishes.
        Class Amphibia. Frogs, toads, salamanders; no marine species.
        Class Reptilia. Sea snakes, turtles, one species of crocodile.
        Class Aves. The birds.
            Order Sphenisciformes. Penguins.
            Order Procellariiformes. Albatrosses, petrels.
            Order Charadriiformes. The gulls.
            Order Pelecaniformes. The pelicans.
        Class Mammalia. Warm-blooded, with hair and mammary glands.
            Order Cetacea. Whales, porpoises, dolphins.
            Order Sirenia. Manatees.
            Order Carnivora. Two marine families.
                Suborder Pinnipedia. Seals, sea lions, walruses.
                Suborder Fissipedia. Sea otters.
            Order Primates. One family that regularly enters the ocean.
                Family Hominidae. Humans.

APPENDIX VII

Calculating the Tide-Generating Force

We can calculate the strength of the tide-generating force as follows: Let the distance between the center of Earth and that of the moon be R, let the radius of Earth be r, and let m be the mass of the moon divided by the mass of Earth. The average gravitational attraction, which is equal to the centrifugal acceleration, is then proportional to m/R^2, because the gravitational attraction is proportional to the mass and inversely proportional to the square of the distance.

Consider a point on Earth closest to the moon. The distance of this point from the center of the moon is $R - r$ and so the gravitational attraction is proportional to

$$\frac{m}{(R - r)^2}$$

The tide-generating force, which is equal to the difference between the gravitational attraction and the centrifugal force, is proportional to

$$\frac{m}{(R - r)^2} - \frac{m}{R^2} = \frac{mR^2 - m(R - r)^2}{(R - r)^2 R^2}$$

Because r is small compared to R, the equation is approximately equal to $2mr/R^3$. In contrast, in the same units, the gravitational attraction of Earth is $1/r^2$. Thus, the ratio of the tide-generating force to the acceleration of gravity is

$$\frac{Tidal\ force}{a_g} = \frac{2mr^3}{R^3} = 1.176 \times 10^{-7}$$

Because $m = 1/81.45$, $R = 60/34r$.

APPENDIX VIII

Working in Marine Science

Working in the marine sciences is wonderfully appealing to many people. They sometimes envision a life of diving in warm, clear water surrounded by tropical fish, or descending to the seabed in an exotic submersible outfitted like Captain Nemo's fictional submarine in *20,000 Leagues Under the Sea*, or living with intelligent dolphins in a marine life park. Then reality sets in. There are rewards from working in the marine sciences, but they tend to be less spectacular than the first dreams of students looking to the ocean for a life's work.

A marine science worker is paid to bring a specific skill to a problem. If that problem lies in warm, tropical water or in a marine park, fine. But more likely the problem will yield only to prolonged study in an uncomfortable, cold, or dangerous environment. The intangible rewards can be great; the physical rewards are often slim. Having said that, let me add that no endeavor is more interesting or exciting, and few are more intellectually stimulating. Doing marine science is its own reward.

Training for a Job in Marine Science

Marine science is, of course, science. And science requires mathematics—you need math to do the chemistry, physics, measurements, and statistics that lie at the heart of science. Your first step in college should be to take a math placement test, enroll in an appropriate math class, and spend time doing math. *Math is the key to further progress in any area of marine science.*

With your math skills polished, start classes in chemistry, physics, and basic biology. Surprisingly, except for one or two introductory marine science classes, you probably won't take many marine science courses until your junior year. These introductory classes will be especially valuable because a balanced survey of the marine sciences can aid you in selecting an appealing specialty. Then, with a good foundation in basic science, you can begin to concentrate in that specialty.

Other skills are important too. The ability to write and speak well is crucial in any science job. Also critical is computer literacy. Expertise in photography or foreign languages or the ability to field-strip and rebuild a diesel engine or hydraulic winch will put you a step above the competition at hiring time. Certification as a scuba diver is almost mandatory; you can never have too much diving experience. (Remember, though, diving is only a tool, a way to deliver an informed set of eyes and an educated brain to a work site.) You should be in good health. Indeed, good aerobic fitness is essential in most marine science jobs; stamina is often a crucial factor in long experiments under difficult conditions at sea. It is also desirable to be physically strong—marine equipment is heavy and often bunglesome. And it helps greatly if you are not prone to seasickness.

Deciding what school to attend will depend on your skills. Readers of this book will probably be enrolled in a general oceanography course in a college or university. The first step would be to discuss your interests with your professor (or his or her teaching assistants). You'll need to attend a four-year college or university to complete the first phase of your training. If you're attending a two-year institution, picking a specific transfer institution can come later, but keep a few things in mind: No matter where you take your first two years of training, you need thorough preparation in basic science. You should attend an institution with strengths in the area of your specialty (such as geology, biology, and marine chemistry). And you should be reasonable in your expectations of acceptance if you're a transfer student (that is, don't try for Stanford or Yale with a B average).

Another thing: Most marine scientists have completed a graduate degree (a master's degree or doctorate). Most graduate students hold teaching or research assistantships (that is, they get paid for being grad students). In all, progress to a final degree is a long road, but the journey is itself a pleasure.

If the thought of four or more years of higher education doesn't appeal to you, does that mean there's no hope? Not at all. Many students begin a program with the goal of becoming a marine technician, animal trainer at a marine life park, marina or boatyard employee or manager, or crew member on a private yacht. Those jobs don't always require a bachelor's degree. Jobs at Sea World and other marine theme parks do require athletic ability, extreme patience, public speaking skills, a love of animals, and, usually, diving experience. Few positions are available, but there is some turnover in the ranks of junior trainers, and being hired is certainly possible.

Becoming a marine technician is an especially attractive alternative to the all-out chemistry-physics-math academic route. For every highly trained marine scientist, there are perhaps five technical assistants who actually do the experiments, maintain the equipment, work daily with organisms, and build special apparatus. Marine technicians tend to spend more time at hands-on tasks than marine scientists. Most of these folks (including the author of the letter that ends this appendix) have the equivalent of a two-year technical degree, usually from a community college.

Don't quit your job, burn your bridges, leave your family, sell your possessions, and dedicate yourself monklike to marine science. Do some investigation. Nothing is as valuable as *actually going out and talking to people who do things that you'd like to do*. Ask them if they enjoy their work. Is the pay OK? Would they start down the same road if they had it to do all over again? You may decide to expand your involvement in marine science in a more informal way, by becoming a volunteer; joining the Sierra Club, Audubon Society, Greenpeace, or other environmental group; working for your state's fish and game office as a seasonal aide; or attending lectures at local colleges and universities.

If you decide to continue your education, don't be discouraged by the time it will take. Have a general view of the big picture, but proceed one semester at a time. Again, remember that the educational journey is itself a great pleasure. Don Quixote reminds us of the joys of the road, not the inn.

The Job Market

Marine science is very attractive to the general public. People are naturally drawn to thoughts of working in the field. Unfortunately, there aren't a great many jobs in the marine sciences. But there will always be some jobs, and people will fill them. Those people will be the best prepared, most versatile, and most highly motivated of those who apply. Perhaps not surprisingly, marine biology is the most popular marine science specialty. Unfortunately, it is also the area with the smallest number of nonacademic jobs. Museums, aquariums, and marine theme parks employ biologists to care for animals and oversee interpretive programs for the public. A few marine biologists are employed as monitoring specialists by water management agencies like sanitation districts, which discharge waste into the ocean. Electrical utilities that use seawater to cool the condensers in power-generating plants almost always have a handful of marine biologists on staff to watch the effects of discharged heat on local marine life and to write the reports required by watchdog agencies. State and federal agencies employ marine biologists to read and interpret those documents and to set standards. Relatively small businesses, like private shipyards, agricultural concerns, and chemical plants, can't afford their own staff biologists, so private consulting firms staffed by marine biologists and other specialists have arisen to assist in the preparation of the environmental impact reports required of businesses under various legislation.

There are more jobs in physical oceanography: marine geology, ocean engineering, and marine chemistry and physics. Thousands of marine geologists work for oil and mineral companies; indeed, with the increasing emphasis on offshore resources, the market for these people may be increasing. Marine engineers are needed to design, construct, and maintain offshore oil rigs, ships, and harbor structures. Marine chemists are hard at work figuring ways to stop corrosion and to extract chemicals from seawater. Physicists are vitally interested in the transmission of underwater sound and light, in the movement of the ocean, and in the role the ocean plays in global weather and climate. Economists, lawyers, writers, and mathematicians also work in the marine science field.

Many biological and physical oceanographers are teachers and professors. Indeed, there are nearly as many marine scientists employed in the academic world as there are in private industry and government. If you like the idea of teaching, you might consider this avenue. The demand for science teachers at all educational levels is already great and is expected to increase.

Four factors will be significant in influencing your employability:

1. *Experience.* Employers are favorably impressed by experience, especially work experience related to the duties of the position for which you are applying. Volunteer work counts.
2. *Grades.* Good grades are important, especially for positions in government agencies. A grade point average of 3.0 or higher in all college work increases your chances of employment and should give you a higher starting salary.
3. *Geographical availability.* Don't restrict yourself geographically. Not everyone can work in Hawaii or California, but four out of ten marine scientists work in just three states: California, Maryland, and Virginia.
4. *Diversification.* Again, mastery of more than one specialty gives you an employment edge. Being a plankton connoisseur and also able to repair a balky computer while ordering in-port supplies over a radiotelephone in Spanish makes a lasting impression.

Report from a Student

Students in marine science programs graduate, get jobs, and move on. One of the pleasures of being a professor is hearing from them. One of our former students, an employee of the Marine Science Institute at the University of California, Santa Barbara, recently reported his activities as part of a team using the submersible *Alvin* to investigate plumes of warm water issuing from hydrothermal vents along the southern Juan de Fuca Ridge. The nature of his work—and his enthusiasm for it—is clearly evident in this excerpt. Dan Dion writes:

> The buoyant plume experiment wasn't going very well. The chemistry dives were pushed back because of technical difficulties and poor weather (rough seas cut two dives). The first two buoyant plume dives ended in failure. The first one because of mechanical/electrical problems, the second because of a computer crash. Everyone worked around the clock to get things in order for dive 2440. I was scheduled to go down with John Trefrey, from the Florida Institute of Technology. Cindy Van

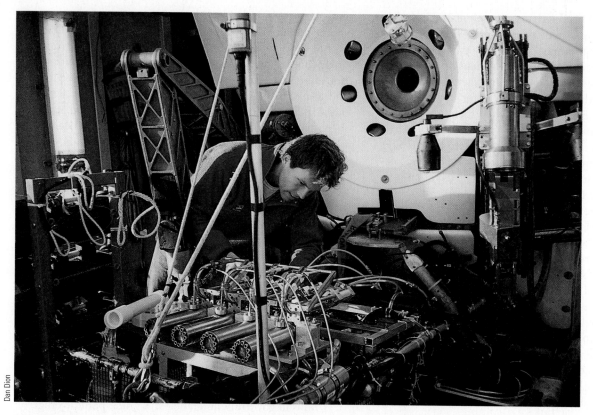

Figure 1 Dan Dion attaching hydraulic actuators to water-sampling bottles, in preparation for a dive by *Alvin*. The bottles were part of a sampling program that included measurements of water conductivity, transmissivity, temperature, and iron and manganese ion content near a hydrothermal vent. This information was later merged with data from transponder navigation to obtain a three-dimensional map of plume structure and chemistry.

Dover was our pilot. We launched *Alvin* right on schedule at 0800, and descended from the glacier blue water into the bioluminescent snowstorm of the euphotic zone. During the hour and a half descent we listened to the music of Enya in the soft light of the sub as we busily prepared ourselves for the experiment: booting up the computers, loading film in the cameras, tapes in the recorders, etc. We had three laptop computers to deal with in the cramped spaces of the sub. I was in charge of two of them, one that plotted our in-sub navigation (from transponders), and the other that controlled and recorded data from the [continuous temperature-depth-conductivity probe]. The third laptop was connected to the chemical analyzer and John was in control of that. All of the instruments were operating perfectly. I periodically saved the computer file in the event of another crash. We reached the bottom right on target; Monolith Vent was in sight, 2261 meters below the surface. We did a video survey of the vent, especially a chimney that was rapidly growing back after the geologists had decapitated it just a few days earlier. We ascended to 55 meters-off-bottom and began our drive-throughs. To me, the navigator, it was the ultimate video game. From the computer screen I would guide *Alvin* through a dark abyss, calling out headings that would maneuver us into a "lawn mower" pattern crisscrossing the plume. It was quite visible, and even beautiful; wispy, intricate patterns of "smoke" which seemed to dance like graceful ghosts. We completed passes at 35, 20, 10, and 5 meters above the bottom, then one last one at 45 meters. Eight hours of sub time went by so quickly! Our dive was a huge success; in addition to all the sam-

Figure 2 Dan Dion enters the hatch of *Alvin* to visit hydrothermal vents 2,261 meters (7,416 feet) below the surface off the coast of Oregon.

People in Marine Science

The late nineteen nineties saw a significant change in the profile of people involved in oceanographic research in the United States. Some of these changes are summarized here.

1. The number of oceanographers employed in all sectors increased from 1,130 in 1973 to 2,460 in 1989.
2. The percent of oceanographers who consider teaching as their primary activity decreased from 21% in 1973 to 11% in 1989.
3. The percentage of oceanographers who consider research as their primary activity remained stable at 40%.
4. In 1989, a majority of Ph.D. oceanographers (60%) were employed at educational institutions.
5. The distribution of oceanographers by discipline (biological, chemical, geological, physical) has remained almost constant.
6. The percentage of women oceanographers increased from 4% in 1973 to 11% in 1989. In 1992, 30% of students at major oceanographic institutions were women.
7. The two largest oceanographic institutions are the Scripps Institution of Oceanography (affiliated with the University of California, San Diego) and the Woods Hole Oceanographic Institution. In 1970 the faculty at these two institutions constituted 40% of the field. In 1990 it constituted only 25%.

Source: "Workshop Report, National Science Foundation Conference on Undergraduate Programs in Ocean Sciences," Galveston, Texas, 1992.

ples we obtained, we generated over 25 megabytes of data. I used everything I learned . . . from computer skills to navigation and marlinspike seamanship (and, of course, chemistry!).

Marine science is equipment training sessions, long cruises, seminars and lectures, visiting experts, hot sand volleyball games, and chilly labs with classical music. Marine science is a long and demanding road; but it is, quite honestly, great fun. Captain Nemo and his sub never had it this good!

Oceanography Programs in North America

Here is a list of the largest and most comprehensive marine science teaching and research institutions in North America.

Bedford Institute of Oceanography, Dartmouth, Nova Scotia, Canada
Dalhousie University, Halifax, Nova Scotia, Canada
Louisiana State University, Baton Rouge, LA 70803
Massachusetts Institute of Technology, Cambridge, MA 02139
McGill University, Montreal, Canada
Mississippi State University, Starkville, MS 39759
North Carolina State University, Raleigh, NC 27695
Old Dominion University, Norfolk, VA 23529
Oregon State University, Corvallis, OR 97331-2134
State University of New York, Stony Brook, NY 11794-5001
Texas A & M University, College Station, TX 77843-3146
University of Alaska, Fairbanks, AK 99775-5040
University of British Columbia, Vancouver, BC, Canada
University of California, San Diego, La Jolla, CA 92093-0232*
University of Connecticut, Groton, CT 06340
University of Delaware, Newark, DE 19716
University of Florida, Gainesville, FL 32611-0409
University of Georgia, Athens, GA 30602
University of Hawaii, Honolulu, HI 96822
University of Maine, Orono, ME 04469-5715
University of Maryland, College Park, MD 20742
University of Michigan, Ann Arbor, MI 48109-2099
University of New Hampshire, Durham, NH 03824
University of North Carolina, Wilmington, NC 28403
University of Rhode Island, Narragansett, RI 02882-1197
University of South Carolina, Columbia, SC 29208
University of Southern California, Los Angeles, CA 90089-1231
University of Virginia, Charlottesville, VA 22903
University of Washington, Seattle, WA 98105
Woods Hole Oceanographic Institution, Woods Hole, MA 02543

*The Scripps Institution of Oceanography is affiliated with the University of California, San Diego.

FOR MORE INFORMATION

Organizations

1. Write to the following organizations:

American Society of Limnology and Oceanography A nonprofit, professional scientific society that seeks to promote the interests of limnology and oceanography and related sciences and to further the exchange of information across the range of aquatic science disciplines. ASLO Business Office, 5400 Bosque Boulevard, Suite 680, Waco, TX 76710, 817/399-9635, e-mail: business@aslo.org

Association for Women in Science A nonprofit association dedicated to achieving equity and full participation for women in science, mathematics, engineering, and technology. 1200 New York Ave., Suite 650 NW, Washington, DC 20005, 202/326-8940, e-mail: awis@awis.org

Center for Marine Conservation A nonprofit membership organization dedicated to protecting marine wildlife and its habitats and to conserving coastal and ocean resources. 1725 DeSales St., NW, Washington, DC 20036

The Cousteau Society A nonprofit environmental education organization dedicated to the protection and improvement of the quality of life for present and future generations. 870 Greenbrier Circle, Suite 402, Chesapeake, VA 23320

Marine Advanced Technology Training Center For training of marine technicians—specialists in the deployment and maintenance of tools used in marine science. Marine Advanced Technology Training Center, Monterey Peninsula College, 980 Fremont St., Monterey, CA 93940, www.marinetech.org

Marine Technology Society An international, interdisciplinary society devoted to ocean and marine engineering science and policy. 1828 L St., NW, Suite 906, Washington, DC 20036-5104, 202/775-5966, fax: 202/429-9417

National Marine Educators Association An organization that brings together those interested in the study and enjoyment of the world of water—both fresh and salt. Web site: www.marine-ed.org

National Oceanic and Atmospheric Administration A government agency that guides use and protection of our oceans and coastal resources, warns of dangerous weather, charts the seas and skies, and conducts research to improve our understanding and stewardship of the environment. NOAA Correspondence Unit, 1305 East-West Highway, #8624, Silver Spring, MD 20910

National Science Foundation An independent agency of the federal government established in 1950 to promote and advance scientific progress in the United States. The National Science Foundation, 4201 Wilson Blvd., Arlington, VA 22230, 703/292-5111

Oceanic Engineering Society An organization that promotes the use of electronic and electrical engineers for instrumentation and measurement work in the ocean environment and the ocean/atmosphere interface. For info, contact: http://auv.tamu.edu/oes

The Oceanography Society A professional society for scientists in the field of oceanography. 5912 Le May Rd., Rockville, MD 20851, 301/881-1101, fax: 301/881-1102, e-mail: info@tos.org

Office of Naval Research A component of the U.S. Navy, ONR plans, fosters, and encourages scientific research and technology development in recognition of their paramount importance to national security. ONR Public Affairs Office, Ballston Centre Tower One, 800 North Quincy Street, Arlington, VA 22217, e-mail: onrpao@onr.navy.mil

The Society for Marine Mammalogy A professional organization that supports the conservation of marine mammals and the educational, scientific, and managerial advancement of marine mammal science. Daniel K. Odell, SMM Education Committee, Sea World, 7007 Sea World Drive, Orlando, FL 32821-8097, 407/363-2662, fax: 407/345-5397, e-mail: odell@pegasus.cc.ucf._edu

Publications

2. Read one or more of the following publications:

    Careers in Oceanography and Marine-Related Fields—available from The Oceanography Society

    Education and Training Programs in Oceanography and Related Fields—$6, available from the Marine Technology Society

    Strategies for Pursuing a Career in Marine Mammal Science ($1.50 for students, $3 for others)—Allen Press, P.O. Box 1897, Lawrence, KS 66044-8897, 800/627-0629

    Taking the Initiative: Report on a Leadership Conference for Women in Science and Technology—available from the Assoc. for Women in Science

    You Can Teach Science (high school version)—available from the Office of Public Information, National Science Teachers Assoc., 1840 Wilson Blvd., Arlington, VA 22201, 703/243-7100

3. Consult the catalog of any college or university offering a marine science curriculum.

4. Send for a copy of *Ocean Opportunities: A Guide to What the Oceans Have to Offer*. The pamphlet is available from The Marine Technology Society, 1828 L Street, NW, Suite 906, Washington, DC 20036-5104, 202/775-5966.

5. See Wunsch, Carl. 1993. "Marine Science in the Coming Decades." *Science* 259 (no. 5093, 15 January): 296–7. An expert oceanographer's glimpse into the future.

6. See Yentsch, C. M., and C. J. Snyderman. 1992. *The Woman Scientist: Meeting the Challenges for a Successful Career*. New York: Plenum. Two marine scientists examine the facets of surviving and succeeding in academia and research science.

7. See American Meteorological Society. 1992. *Curricula in the Atmospheric, Oceanic, and Related Sciences*. Boston: American Meteorological Society. Source of current information on marine science programs in the United States, Canada, and Puerto Rico.

Note: The organizations and publications listed here are but a sampling of the resources that are available to help you learn more about careers in the marine sciences. Each contact you make in your search for information will lead you to more contacts and more information.

Glossary

absolute dating Determining the age of a geological sample by calculations of radioactive decay and/or its position in relation to other samples.

absorption Conversion of sound or light energy into heat.

abyssal hill Small sediment-covered inactive volcano or intrusion of molten rock less than 200 meters (650 feet) high, thought to be associated with seafloor spreading. Abyssal hills punctuate the otherwise flat abyssal plain.

abyssal plain Flat, cold, sediment-covered ocean floor between the continental rise and the oceanic ridge at a depth of 3,700 to 5,500 meters (12,000 to 18,000 feet). Abyssal plains are more extensive in the Atlantic and Indian Oceans than in the Pacific.

abyssal zone The ocean between about 4,000 and 5,000 meters (13,000 and 16,500 feet) deep.

accessory pigment One of a class of pigments (such as fucoxanthin, phycobilin, and xanthophyll) present in various photosynthetic plants and that assist in the absorption of light and the transfer of its energy to chlorophyll. Also called *masking pigment*.

accretion An increase in the mass of a body by accumulation or clumping of smaller particles.

acid A substance that releases a hydrogen ion (H^+) in solution.

acid rain Rain containing acids and acid-forming compounds such as sulfur dioxide and oxides of nitrogen.

acoustical tomography A technique for studying ocean structure that depends on pulses of low-frequency sound to sense differences in water temperature, salinity, and movement beneath the surface.

active margin Continental margin near an area of lithospheric plate convergence. Also called *Pacific-type margin*.

active sonar A device that generates underwater sound from special transducers and analyzes the returning echoes to gain information of geological, biological, or military importance.

active transport The movement of molecules from a region of low concentration to a region of high concentration through a semipermeable membrane at the expense of energy.

adaptation An inheritable structural or behavioral modification. A favorable adaptation gives a species an advantage in survival and reproduction. An unfavorable adaptation lessens a species' ability to survive and reproduce.

adhesion Attachment of water molecules to other substances by hydrogen bonds. Wetting.

Agnatha The class of jawless fishes: hagfishes and lampreys.

ahermatypic Describing coral species lacking symbiotic zooxanthellae and incapable of secreting calcium carbonate at a rate suitable for reef production.

air mass A large mass of air with nearly uniform temperature, humidity, and density throughout.

algae Collective term for nonvascular plants possessing chlorophyll and capable of photosynthesis. (Singular, alga.)

algin A mucilaginous commercial product of multicellular marine algae. Widely used as a thickening and emulsifying agent.

alkaline Basic. See also *base*.

alternation of generations A reproductive cycle in which a plant alternates between sexual and asexual stages.

amphidromic point A "no-tide" point in an ocean caused by basin resonances, friction, and other factors around which tide crests rotate. About a dozen amphidromic points exist in the world ocean. Sometimes called a *node*.

angiosperm A flowering vascular plant that reproduces by means of a seed-bearing fruit. Examples are sea grasses and mangroves.

angle of incidence In meteorology, the angle of the sun above the horizon.

animal A multicellular organism unable to synthesize its own food and often capable of movement.

Animalia The kingdom to which multicellular heterotrophs belong.

Annelida The phylum of animals to which segmented worms belong.

Antarctic Bottom Water The densest ocean water (1.0279 g/cm^3), formed primarily in Antarctica's Weddell Sea during Southern Hemisphere winters.

Antarctic Circle The imaginary line around the Earth, parallel to the equator at 66°33'S, marking the southernmost limit of sunlight at the June solstice. The Antarctic Circle marks the northern limit of the area within which, for one day or more each year, the sun does not set (around 21 December) or rise (around 21 June).

Antarctic Circumpolar Current The current driven by powerful westerly winds north of Antarctica. The largest of all ocean currents, it continues permanently eastward without changing direction.

Antarctic Convergence Convergence zone encircling Antarctica between about 50° and 60°S, marking the boundary between Antarctic Circumpolar Water and Subantarctic Surface Water.

Antarctic Ocean An ocean in the Southern Hemisphere bounded to the north by the Antarctic Convergence and to the south by Antarctica.

aphelion The point in the orbit of a satellite where it is farthest from the sun; opposite of *perihelion*.

aphotic zone The dark ocean below the depth to which light can penetrate.

apogee The point in the orbit of a satellite farthest from the main body; opposite of *perigee*.

aquaculture The growing or farming of plants and animals in a water environment under controlled conditions. Compare *mariculture*.

Arctic Circle The imaginary line around Earth, parallel to the equator at 66°33'N, marking the northernmost limit of sunlight at the December solstice. The Arctic Circle marks the southern limit of the area within which, for one day or more each year, the sun does not set (around 21 June) or rise (around 21 December).

Arctic Convergence Convergence zone between Arctic Water and Subarctic Surface Water.

Arctic Ocean An ice-covered ocean north of the continents of North America and Eurasia.

Arthropoda The phylum of animals that includes shrimp, lobsters, krill, barnacles, and insects. The phylum Arthropoda is the world's most successful.

artificial system of classification A method of classifying an object based on attributes other than its reason for existence, its ancestry, or its origin. Compare *natural system of classification*.

Asteroidea The class of the phylum Echinodermata to which sea stars belong.

asthenosphere The hot, plastic layer of the upper mantle below the lithosphere, extending some 350 to 650 kilometers (220 to 400 miles) below the surface. Convection currents within the asthenosphere power plate tectonics.

atmospheric circulation cell Large circuit of air driven by uneven solar heating and the Coriolis effect. Three circulation cells form in each hemisphere. See also *Ferrel cell*; *Hadley cell*; *polar cell*.

atoll A ring-shaped island of coral reefs and coral debris enclosing, or almost enclosing, a shallow lagoon from which no land protrudes. Atolls often form over sinking, inactive volcanoes.

atom The smallest particle of an element that exhibits the characteristics of that element.

ATP Adenosine triphosphate, the compound that acts as the immediate source of energy for all life on Earth. The energy stored in ATP is provided directly by photosynthesis or by respiration of glucose.

authigenic sediment Sediment formed directly by precipitation from seawater. Also called *hydrogenous sediment*.

autotroph An organism that makes its own food by photosynthesis or chemosynthesis.

auxospore A naked diatom cell without valves. Often a dormant stage in the life cycle following sexual reproduction.

Aves The class of birds.

backshore Sand on the shoreward side of the berm crest, sloping away from the ocean.

backwash Water returning to the ocean from waves washing onto a beach.

bacteria Single-celled prokaryotes, organisms lacking membrane-bound organelles.

baleen The interleaved, hard, fibrous, hornlike filters within the mouth of baleen whales.

barrier island A long, narrow, wave-built island lying parallel to the mainland and separated from it by a lagoon or bay. Compare *sea island*.

barrier reef A coral reef surrounding an island or lying parallel to the shore of a continent, separated from land by a deep lagoon. Coral debris islands may form along the reef.

barycenter The center of mass and center of rotation of the Earth–moon system, 1,700 kilometers (1060 miles) inside Earth.

basalt The relatively heavy crustal rock that forms the seabeds, composed mostly of oxygen, silicon, magnesium, and iron. Its density is about 2.9 g/cm³.

base A substance that combines with a hydrogen ion (H⁺) in solution.

bathyal zone The ocean between about 200 and 4,000 meters (700 and 13,000 feet) deep.

bathybius Thomas Henry Huxley's name for an artifact of marine specimen preservation he thought was a remnant of the "primeval living slime."

bathymetry The discovery and study of submerged contours.

bathyscaphe Deep-diving submersible designed like a blimp, which uses gasoline for buoyancy and can reach the bottom of the deepest ocean trenches. From the Greek *batheos* ("depth") and *skaphidion* ("a small ship").

bay mouth bar An exposed sandbar attached to a headland adjacent to a bay and extending across the mouth of the bay.

beach A zone of unconsolidated (loose) particles extending from below water level to the edge of the coastal zone.

beach scarp Vertical wall of variable height marking the landward limit of the most recent high tides. Corresponds with the berm at extreme high tides.

benthic zone The zone of the ocean bottom. See also *pelagic zone*.

berm A nearly horizontal accumulation of sediment parallel to shore. Marks the normal limit of sand deposition by wave action.

berm crest The top of the berm; the highest point on most beaches. Corresponds to the shoreward limit of wave action during most high tides.

big bang The hypothetical event that started the expansion of the universe from a geometric point. The beginning of time.

bilateral symmetry Body structure having left and right sides that are approximate mirror images of each other. Examples are crabs and humans. Compare *radial symmetry*.

biodegradable Able to be broken by natural processes into simpler compounds.

biodiversity The variety of different species within a habitat.

biogenous sediment Sediment of biological origin. Organisms can deposit calcareous (calcium-containing) or siliceous (silicon-containing) residue.

biogeochemical cycle Natural processes that recycle nutrients in various chemical forms from the nonliving environment to living organisms and then back to the nonliving environment.

biological amplification Increase in concentration of certain fat-soluble chemicals such as DDT or heavy-metal compounds, in successively higher trophic levels within a food web.

biological factor A biologically generated aspect of the environment, such as predation or metabolic waste products, that affects living organisms. Biological factors usually operate in association with purely physical factors such as light and temperature.

biological resource A living animal or plant collected for human use. Also called a *living resource*.

bioluminescence Biologically produced light.

biomass The mass of living material in a given area or volume of habitat.

biosynthesis The initial formation of life on Earth.

Bivalvia The class of the phylum Mollusca that includes clams, oysters, and mussels.

Bjerknes, Vilhelm (1862–1951) Pioneering Norwegian physicist and discoverer of the nature and formation of extratropical cyclones, which cause most mid-latitude weather.

blade Algal equivalent of a vascular plant's leaf. Also called a *frond*.

bond See *chemical bond*.

brackish Describing water intermediate in salinity between seawater and fresh water.

breakwater An artificial structure of durable material that interrupts the progress of waves to shore. Harbors are often shielded by a breakwater.

buffer A group of substances that tends to resist change in the pH of a solution by combining with free ions.

buoyancy The ability of an object to float in a fluid by displacement of a volume of fluid equal to it in mass.

bykill Animals unintentionally killed when desirable organisms are collected.

caballing Mixing of two water masses of identical densities but different temperatures and salinities, such that the resulting mixture is denser than its components.

calcareous ooze Ooze composed mostly of the hard remains of organisms containing calcium carbonate.

calcium carbonate compensation depth The depth at which the rate of accumulation of calcareous sediments equals the rate of dissolution of those sediments. Below this depth, sediment contains little or no calcium carbonate.

calorie The amount of heat needed to raise the temperature of 1 gram (0.035 ounce) of pure water by 1°C (1.8°F).

capillary wave A tiny wave with a wavelength of less than 1.73 centimeters (0.68 inch), whose restoring force is surface tension; the first type of wave to form when the wind blows.

Carnivora The order of mammals that includes seals, sea lions, walruses, and sea otters.

carrying capacity The size at which a particular population in a particular environment will stabilize when its supply of resources—including nutrients, energy, and living space—remains constant.

cartilage A tough, elastic tissue that stiffens or supports.

cartographer A person who makes maps and charts.

catastrophism The theory that Earth's surface features are formed by catastrophic forces such as the biblical flood. Catastrophists believe in a young Earth and a literal interpretation of the biblical account of Creation.

celestial navigation The technique of finding one's position on Earth by reference to the apparent positions of stars, planets, the moon, and the sun.

cell The basic organizational unit of life on this planet.

Cephalopoda The class of the phylum Mollusca that includes squid, octopuses, and nautiluses.

Cetacea The order of mammals that includes porpoises, dolphins, and whales.

CFCs See *chlorofluorocarbons*.

***Challenger* Expedition** The first wholly scientific oceanographic expedition, 1872–76. Named for the steam corvette used in the voyage.

chart A map that depicts mostly water and the adjoining land areas.

chemical bond An energy relationship that holds two atoms together as a result of changes in their electron distribution.

chemical equilibrium In seawater, the condition in which the proportion and amounts of dissolved salts per unit volume of ocean are nearly constant.

chemosynthesis The synthesis of organic compounds from inorganic compounds using energy stored in inorganic substances such as sulfur, ammonia, and hydrogen. Energy is released when these substances are oxidized by certain organisms.

chitin A complex nitrogen-rich carbohydrate from which parts of arthropod exoskeletons are constructed.

chiton A polyplacophoran mollusk.

chlorinated hydrocarbons The most abundant and dangerous class of halogenated hydrocarbons, synthetic organic chemicals hazardous to the marine environment.

chlorinity A measure of the content of chloride, bromine, and iodide ions in seawater. We may derive salinity from chlorinity by multiplying by 1.80655.

chlorofluorocarbons (CFCs) A class of halogenated hydrocarbons thought to be depleting Earth's atmospheric ozone. CFCs are used as cleaning agents, refrigerants, fire-extinguishing fluids, spray-can propellants, and insulating foams.

chlorophyll A pigment responsible for trapping sunlight and transferring its energy to electrons, thus initiating photosynthesis.

Chlorophyta Green algae.

Chondrichthyes The class of fishes with cartilaginous skeletons: the sharks, skates, rays, and chimaeras.

Chordata The phylum of animals to which tunicates, Amphioxus, fishes, amphibians, reptiles, birds, and mammals belong.

chromatophore A pigmented skin cell that expands or contracts to affect color change.

chronometer A very consistent clock. It doesn't need to tell accurate time, but its rate of gain or loss must be constant and known exactly so that accurate time may be calculated.

clamshell sampler Sampling device used to take shallow samples of the ocean bottom.

classification A way of grouping objects according to some stated criteria.

clay Sediment particle smaller than 0.004 millimeter in diameter; the smallest sediment size category.

climate The long-term average of weather in an area.

climax community A stable, long-established community of self-perpetuating organisms that tends not to change with time.

clockwise Rotation around a point in the direction that clock hands move.

clumped distribution Distribution of organisms within a community in small, patchy aggregations, or clumps; the most common distribution pattern.

Cnidaria The phylum of animals to which corals, jellyfish, and sea anemones belong.

cnidoblast Type of cell found in members of the phylum Cnidaria that contains a stinging capsule. The threads that evert from the capsules assist in capturing prey and repelling aggressors.

coast The zone extending from the ocean inland as far as the environment is immediately affected by marine processes.

coastal cell The natural sector of a coastline in which sand input and sand outflow are balanced.

coastal upwelling Upwelling adjacent to a coast, usually induced by wind.

coccolithophore A very small planktonic alga carrying discs of calcium carbonate, which contributes to biogenous sediments.

cohesion Attachment of water molecules to each other by hydrogen bonds.

colligative properties Those characteristics of a solution that differ from those of pure water because of material held in solution.

Columbus, Christopher (1451–1506) Italian explorer in the service of Spain who discovered islands in the Caribbean in 1492. Although traditionally credited as the discoverer of America, he never actually sighted the North American continent.

commensalism A symbiotic interaction between two species in which only one species benefits and neither is harmed.

commercial extinction Depletion of a resource species to a point where it is no longer profitable to harvest the species.

community The populations of all species that occupy a particular habitat and interact within that habitat.

compass An instrument for showing direction by means of a magnetic needle swinging freely on a pivot and pointing to magnetic north.

compensation depth The depth in the water column at which the production of carbohydrates and oxygen by photosynthesis exactly equals the consumption of carbohydrates and oxygen by respiration. The break-even point for autotrophs. Generally a function of light level.

compound A substance composed of two or more elements in a fixed proportion.

condensation theory Premise that stars and planets accumulate from contracting, accreting clouds of galactic gas, dust, and debris.

conduction The transfer of heat through matter by the collision of one atom with another.

conservative constituent An element that occurs in constant proportion in seawater. For example, chlorine, sodium, and magnesium.

constructive interference The addition of wave energy as waves interact, producing larger waves.

consumer A heterotrophic organism.

continental crust The solid masses of the continents, composed primarily of granite.

continental drift The theory that the continents move slowly across the surface of Earth.

continental margin The submerged outer edge of a continent, made of granitic crust. Includes the continental shelf and continental slope. Compare *ocean basin*.

continental rise The wedge of sediment forming the gentle transition from the outer (lower) edge of the continental slope to the abyssal plain. Usually associated with passive margins.

continental shelf Gradually sloping submerged extension of a continent, composed of granitic rock overlain by sediments. Has features similar to the edge of the nearby continent.

continental slope The sloping transition between the granite of the continent and the basalt of the seabed. The true edge of a continent.

contour current A bottom current made up of dense water that flows around (rather than over) seabed projections.

convection Movement within a fluid resulting from differential heating and cooling of the fluid. Convection produces mass transport or mixing of the fluid.

convection current A single closed-flow circuit of rising warm material and falling cool material.

convergence zone The line along which waters of different density converge. Convergence zones form the boundaries of tropical, subtropical, temperate, and polar areas.

convergent evolution The evolution of similar characteristics in organisms of different ancestry; the body shape of a porpoise and a shark, for instance.

convergent plate boundary A region where plates are pushing together and where a mountain range, island arc, and/or trench will eventually form. Often a site of much seismic and volcanic activity.

Cook, James (1728–1779) Officer in the British Royal Navy who led the first European voyages of scientific discovery.

coral Any of over 6,000 species of small cnidarians, many of which are capable of generating hard calcareous (aragonite, $CaCO_3$) skeletons.

coral reef A linear mass of calcium carbonate (aragonite and calcite) assembled from coral organisms, algae, mollusks, worms, and so on. Coral may contribute less than half of the reef material.

core The innermost layer of Earth, composed primarily of iron, with nickel and heavy elements. The inner core is thought to be a solid 6,000°C (11,000°F) sphere, the outer core a 5,000°C (9,000°F) liquid mass. The average density of the outer core is about 11.8 g/cm^3, and that of the inner core is about 16 g/cm^3.

Coriolis, Gaspard Gustave de (1792–1843) The French scientist who in 1835 worked out the mathematics of the motion of bodies on a rotating surface. See *Coriolis effect*.

Coriolis effect The apparent deflection of a moving object from its initial course when its speed and direction are measured in reference to the surface of the rotating Earth. The object is deflected to the right of its anticipated course in the Northern Hemisphere and to the left in the Southern Hemisphere. The deflection occurs for any horizontal movement of objects with mass and has no effect at the equator.

cosmogenous sediment Sediment of extraterrestrial origin.

counterclockwise Rotation around a point in the direction opposite to that in which clock hands move. Also called *anticlockwise*.

countercurrent A surface current flowing in the opposite direction from an adjacent surface current.

covalent bond A chemical bond formed between two atoms by electron sharing.

crest See *wave crest*.

crust The outermost solid layer of Earth, composed mostly of granite and basalt; the top of the lithosphere. The crust has a density of 2.7–2.9 g/cm^3 and accounts for 0.4% of Earth's mass.

Crustacea The class of phylum Arthropoda to which lobsters, shrimp, crabs, barnacles, and copepods belong.

cryptic coloration Camouflage. May be active (under control of the animal) or passive (an unalterable color or shape).

current Mass flow of water. (The term is usually reserved for horizontal movement.)

cyclone A weather system with a low-pressure area in the center around which winds blow counterclockwise in the Northern Hemisphere and clockwise in the Southern Hemisphere. Not to be confused with a tornado, a much smaller weather phenomenon associated with severe thunderstorms. See also *extratropical cyclone; tropical cyclone*.

Darwin, Charles (1809–1882) English biologist. Co-discoverer (with Alfred Russell Wallace) of evolution by natural selection.

deep scattering layer (DSL) A relatively dense aggregation of fishes, squid, and other mesopelagic organisms capable of reflecting a sonar pulse that resembles a false bottom in the ocean. Its position varies with the time of day.

deep-water wave A wave in water deeper than one-half its wavelength.

deep zone The zone of the ocean below the pycnocline, in which there is little additional change of density with increasing depth. Contains about 80% of the world's water.

degree An arbitrary measure of temperature. One degree Celsius (°C) = 1.8 degrees Fahrenheit (°F).

delta The deposit of sediments found at a river mouth, sometimes triangular in shape (hence the name after the Greek letter).

density The mass per unit volume of a substance, usually expressed in grams per cubic centimeter (g/cm^3).

density curve A graph showing the relationship between a fluid's temperature or salinity and its density.

density stratification The formation of layers in a material, with each deeper layer being denser (weighing more per unit of volume) than the layer above.

dependency A feeding relationship in which an organism is limited to feeding on one species or, in extreme cases, on one size phase of one species.

deposition Accumulation, usually of sediments.

depositional coast A coast in which processes that deposit sediment exceed erosive processes.

desalination The process of removing salt from seawater or brackish water.

desiccation Drying.

destructive interference The subtraction of wave energy as waves interact, producing smaller waves.

diatom Earth's most abundant, successful, and efficient single-celled phytoplankton. Diatoms possess two interlocking valves made primarily of silica. The valves contribute to biogenous sediments.

diffusion The movement—driven by heat—of molecules from a region of high concentration to a region of low concentration.

dinoflagellate One of a class of microscopic single-celled flagellates, not all of which are autotrophic. The outer covering is often of stiff cellulose. Planktonic dinoflagellates are responsible for "red tides."

dispersion Separation of wind waves by wavelength (and therefore wave speed) as they move away from the fetch (the place of their formation). Dispersion occurs because waves with long wavelengths move more rapidly than waves with short wavelengths.

disphotic zone The lower part of the photic zone, where there is insufficient light for photosynthesis.

dissolution The dissolving by water of minerals in rocks.

disthermal zone The zone of stable temperature below the thermocline.

disturbing force The energy that causes a wave to form.

diurnal tide A tidal cycle of one high tide and one low tide per day.

divergent evolution Evolutionary radiation of different species from a common ancestor.

divergent plate boundary A region where plates are moving apart and where new ocean or rift valley will eventually form. A spreading center forms the junction.

doldrums The zone of rising air near the equator known for sultry air and variable breezes. See also *intertropical convergence zone (ITCZ)*.

downwelling Circulation pattern in which surface water moves vertically downward.

drag The resistance to movement of an organism induced by the fluid through which it swims.

drift net Fine, vertically suspended net that may be 7 meters (25 feet) high and 80 kilometers (50 miles) long.

drumlin A streamlined hill formed by a glacier.

DSL See *deep scattering layer*.

dynamic theory of tides Model of tides that takes into account the effects of finite ocean depth, basin resonances, and the interference of continents on tide waves.

earthquake A sudden motion of Earth's crust resulting from waves in Earth caused by faulting of the rocks or by volcanic activity.

eastern boundary current Weak, cold, diffuse, slow-moving current at the eastern boundary of an ocean (off the west coast of a continent). Examples include the Canary Current and the Humboldt Current.

ebb current Water rushing out of an enclosed harbor or bay because of the fall in sea level as a tide trough approaches.

Echinodermata The phylum of exclusively marine animals to which sea stars, brittle stars, sea urchins, and sea cucumbers belong.

Echinoidea The class of the phylum Echinodermata to which sea urchins and sand dollars belong.

echolocation The use of reflected sound to detect environmental objects. Cetaceans use echolocation to detect prey and avoid obstacles.

echo sounder A device that reflects sound off the ocean bottom to sense water depth. Its accuracy is affected by the variability of the speed of sound through water.

ecology Study of the interactions of organisms with one another and with their environment.

ectoderm The outermost layer of cells in a developing embryo.

ectotherm An organism incapable of generating and maintaining steady internal temperature from metabolic heat and therefore whose internal body temperature is approximately the same as that of the surrounding environment. A cold-blooded organism.

eddy A circular movement of water usually formed where currents pass obstructions, or between two adjacent currents flowing in opposite directions, or along the edge of a permanent current.

EEZ See *exclusive economic zone*.

Ekman spiral A theoretical model of the effect on water of wind blowing over the ocean. Because of the Coriolis effect, the surface layer is expected to drift at an angle of 45° to the right of the wind in the Northern Hemisphere and 45° to the left in the Southern Hemisphere. Water at successively lower layers drifts progressively to the right (N) or left (S), though not as swiftly as the surface flow.

Ekman transport Net water transport, the sum of layer movement due to the Ekman spiral. Theoretical Ekman transport in the Northern Hemisphere is 90° to the right of the wind direction.

electron A tiny negatively charged particle in an atom responsible for chemical bonding.

element A substance composed of identical atoms that cannot be broken into simpler substances by chemical means.

El Niño A southward-flowing nutrient-poor current of warm water off the coast of western South America, caused by a breakdown of trade-wind circulation.

endoderm The innermost layer of cells in a developing embryo.

endotherm An organism capable of generating and regulating metabolic heat to maintain a steady internal temperature. Birds and mammals are the only animals capable of true endothermy. A warm-blooded organism.

energy The capacity to do work.

ENSO Acronym for the coupled phenomena of El Niño and the Southern Oscillation. See also *El Niño; Southern Oscillation*.

entropy A measure of the disorder in a system.

environmental resistance All the limiting factors that act together to regulate the maximum allowable size, or carrying capacity, of a population.

epicenter The point on Earth's surface directly above the focus of an earthquake.

epipelagic zone The lighted, or photic, zone in the ocean.

equator See *geographical equator; meteorological equator*.

equatorial upwelling Upwelling in which water moving westward on either side of the geographical equator tends to be deflected slightly poleward and replaced by deep water often rich in nutrients. See also *upwelling*.

equilibrium theory of tides Idealized model of tides that considers Earth to be covered by an ocean of great and uniform depth capable of instantaneous response to the gravitational and inertial forces of the sun and the moon.

Eratosthenes of Cyrene (276–192 B.C.) Greek scholar and librarian at Alexandria who first calculated the circumference of Earth about 230 B.C.

erosion A process of being gradually worn away.

erosional coast A coast in which erosive processes exceed depositional ones.

estuary A body of water partially surrounded by land where fresh water from a river mixes with ocean water, creating an area of remarkable biological productivity.

euphotic zone The upper layer of the photic zone in which net photosynthetic gain occurs. Compare *photic zone*.

euryhaline Describing an organism able to tolerate a wide range in salinity.

eurythermal Describing an organism able to tolerate wide variance in temperature.

eurythermal zone The upper layer of water, where temperature changes with the seasons.

eustatic change A worldwide change in sea level, as distinct from local changes.

eutrophication A set of physical, chemical, and biological changes brought about when excessive nutrients are released into water.

evaporite Deposit formed by the evaporation of ocean water.

evolution Change. The maintenance of life under constantly changing conditions by continuous adaptation of successive generations of a species to its environment.

excess volatiles A compound found in the ocean and atmosphere in quantities greater than can be accounted for by the weathering of surface rock. Such compounds probably entered the atmosphere and ocean from deep crustal and upper mantle sources through volcanism.

exclusive economic zone (EEZ) The offshore zone claimed by signatories to the 1982 United Nations Draft Convention on the Law of the Sea. The EEZ extends 200 nautical miles (370 kilometers) from a contiguous shoreline. See also *United States Exclusive Economic Zone*.

exoskeleton A strong, lightweight, form-fitted external covering and support common to animals of the phylum Arthropoda. The exoskeleton is made partly of chitin and may be strengthened by calcium carbonate.

experiments Tests that simplify observation in nature or in the laboratory by manipulating or controlling the conditions under which observations are made.

extratropical cyclone A low-pressure mid-latitude weather system characterized by converging winds and ascending air rotating counterclockwise in the Northern Hemisphere and clockwise in the Southern Hemisphere. An extratropical cyclone forms at the front between the polar and Ferrel cells.

fault A fracture in a rock mass along which movement has occurred.

Ferrel, William (1817–1891) The American scientist who discovered the mid-latitude circulation cells of each hemisphere.

Ferrel cell The middle atmospheric circulation cell in each hemisphere. Air in these cells rises at 60° latitude and falls at 30° latitude. See also *westerlies*.

fetch The uninterrupted distance over which the wind blows without a significant change in direction, a factor in wind wave development.

Fissipedia The carnivoran suborder that includes sea otters.

fjord A deep, narrow estuary in a valley originally cut by a glacier.

flagellum A whiplike structure used by some small organisms and gametes to move through the environment. (Plural, flagella.)

float method A method of current study that depends on the movement of a drift bottle or other free-floating object.

flood current Water rushing into an enclosed harbor or bay because of the rise in sea level as a tide crest approaches.

flow method A method of current study that measures the current as it flows past a fixed object.

food General term for organic molecules capable of providing energy to heterotrophs when combined with oxygen during biochemical respiration.

food web A group of organisms associated by a complex set of feeding relationships in which the flow of food energy can be followed from primary producers through consumers.

foraminiferan One of a group of planktonic amoeba-like animals with a calcareous shell, which contributes to biogenous sediments.

forced wave A progressive wave under the continuing influence of the forces that formed it.

Forchhammer's principle See *principle of constant proportions*.

foreshore Sand on the seaward side of the berm, sloping toward the ocean, to the low tide mark.

fracture zone Area of irregular, seismically inactive topography marking the position of a once-active transform fault.

Franklin, Benjamin (1706–1790) Published the first chart of an ocean current in 1769.

free wave A progressive wave free of the forces that formed it.

freezing point The temperature at which a solid can begin to form as a liquid is cooled.

fringing reef A reef attached to the shore of a continent or island.

front The boundary between two air masses of different density. The density difference can be caused by differences in temperature and/or humidity.

frontal storm Precipitation and wind caused by the meeting of two air masses, associated with an extratropical cyclone. Generally, one air mass will slide over or under the other, and the resulting expansion of air will cause cooling and, consequently, rain or snow.

frustule Siliceous external cell wall of a diatom consisting of two interlocking valves fitted together like the halves of a box.

fucoxanthin A brown or tan accessory pigment found in many species of brown algae and some species of diatoms.

fully developed sea The theoretical maximum height attainable by ocean waves given wind of a specific strength, duration, and fetch. Longer exposure to wind will not increase the size of the waves.

galaxy A large rotating aggregation of stars, dust, gas, and other debris held together by gravity. There are perhaps 50 billion galaxies in the universe and 50 billion stars in each galaxy.

gas bladder In multicellular algae, an air-filled structure that assists in flotation.

gas exchange Simultaneous passage, through a semipermeable membrane, of oxygen into an animal and carbon dioxide out of it.

Gastropoda The class of the phylum Mollusca that includes snails and sea slugs.

geographical equator 0° latitude, an imaginary line equidistant from the geographical poles.

geostrophic Describing a gyre or current in balance between the Coriolis effect and gravity; literally, "turned by Earth."

gill membrane The thin boundary of living cells separating blood from water in a fish's (or other aquatic animal's) gills.

Global Positioning System (GPS) Satellite-based navigation system that provides a geographical position—longitude and latitude—accurate to less than 1 meter.

GPS See *Global Positioning System.*

granite The relatively light crustal rock—composed mainly of oxygen, silicon, and aluminum—that forms the continents. Its density is about 2.7 g/cm³.

gravimeter A sensitive device that measures variations in the pull of gravity at different places on Earth's surface.

gravity wave A wave with wavelength greater than 1.73 centimeters (0.68 inch), whose restoring forces are gravity and momentum.

greenhouse effect Trapping of heat in the atmosphere. Incoming short-wavelength solar radiation penetrates the atmosphere, but the outgoing longer-wavelength radiation is absorbed by greenhouse gases and reradiated to Earth, causing a rise in surface temperature.

greenhouse gases Gases in Earth's atmosphere that cause the greenhouse effect; include carbon dioxide, methane, and CFCs.

groin A short, artificial projection of durable material placed at a right angle to shore in an attempt to slow longshore transport of sand from a beach. Usually deployed in repeating units.

group velocity Speed of advance of a wave train; for deepwater waves, half the speed of individual waves within the group.

Gulf Stream The strong western boundary current of the North Atlantic, off the Atlantic coast of the United States.

guyot A flat-topped, submerged inactive volcano.

gyre Circuit of mid-latitude currents around the periphery of an ocean basin. Most oceanographers recognize five gyres plus the Antarctic Circumpolar Current.

habitat The place where an individual or population of a given species lives. Its "mailing address."

hadal zone The deepest zone of the ocean, below a depth of 5,000 meters (16,500 feet).

Hadley, George (1685–1768) A London lawyer and philosopher who worked out the overall scheme of wind circulation in an effort to explain the trade winds.

Hadley cell The atmospheric circulation cell nearest the equator in each hemisphere. Air in these cells rises near the equator because of strong solar heating there and falls because of cooling at about 30° latitude. See also *trade winds.*

half-life Time required for one-half of all the unstable radioactive nuclei in a sample to decay.

halocline The zone of the ocean in which salinity increases rapidly with depth. See also *pycnocline.*

Harrison, John (1693–1776) British clockmaker who invented the modern chronometer in 1760.

heat A form of energy produced by the random vibration of atoms or molecules.

heat budget An expression of the total solar energy received on Earth during some period of time and the total heat lost from Earth by reflection and radiation into space through the same period.

heat capacity The heat, measured in calories, required to raise 1 gram of a substance 1 degree Celsius. The input of 1 calorie of heat energy raises the temperature of 1 gram of pure water by 1 degree.

Henry the Navigator (1394–1460) Prince of Portugal who established a school for the study of geography, seamanship, shipbuilding, and navigation.

hermatypic Describing coral species possessing symbiotic zooxanthellae within their tissues and capable of secreting calcium carbonate at a rate suitable for reef production.

heterotroph An organism that derives nourishment from other organisms because it is unable to synthesize its own food molecules.

hierarchy Grouping of objects by degrees of complexity, grade, or class. A hierarchical system of nomenclature is based on distinctions within groups and between groups.

high-energy coast A coast exposed to large waves.

high seas That part of the ocean past the exclusive economic zone, which is considered common property to be shared by the citizens of the world. About 60% of the ocean area.

high tide The high-water position corresponding to a tidal crest.

holdfast A complex branching structure that anchors many kinds of multicellular algae to the substrate.

holoplankton Permanent members of the plankton community. Examples are diatoms and copepods. Compare *meroplankton*.

Holothuroidea The class of the phylum Echinodermata to which sea cucumbers belong.

horse latitudes Zones of erratic horizontal surface air circulation near 30°N and 30°S latitudes. Over land, dry air falling from high altitudes produces deserts at these latitudes (for example, the Sahara).

hot spot A surface expression of a plume of magma rising from a stationary source of heat in the mantle.

hurricane A large tropical cyclone in the North Atlantic or eastern Pacific, whose winds exceed 118 kilometers (74 miles) per hour.

hydrogen bond Relatively weak bond formed between a partially positive hydrogen atom and a partially negative oxygen, fluorine, or nitrogen atom of an adjacent molecule.

hydrogenous sediment Sediment formed directly by precipitation from seawater. Also called *authigenic sediment*.

hydrostatic pressure The constant pressure of water around a submerged organism.

hydrothermal vent Spring of hot, mineral- and gas-rich seawater found on some oceanic ridges in zones of active seafloor spreading.

hypertonic Referring to a solution having a higher concentration of dissolved substances than the solution that surrounds it.

hypothesis A speculation about the natural world that may be verified or disproved by observation and experiment.

hypotonic Referring to a solution having a lower concentration of dissolved substances than the solution that surrounds it.

hypsographic curve A graph of the area of Earth's surface above any given elevation or depth above or below sea level.

ice age One of several periods (lasting several thousand years each) of low temperature during the last million years. Glaciers and polar ice were derived from ocean water, lowering sea level at least 100 meters (328 feet). (See Appendix II, "Geological Time.")

iceberg Large mass of ice floating in the ocean that was formed on or adjacent to land. Tabular icebergs are tablelike or flat; pinnacled icebergs are castellated, or jagged. Southern icebergs are often tabular; northern icebergs are often pinnacled.

ice cap Permanent cover of ice. Formally limited to ice atop land, but informally applied also to floating ice in the Arctic Ocean.

ice floe A mass of firm sea ice floating as a unit.

ice pack A large, floating expanse of broken ice masses pressed and frozen together.

inlet A passage giving the ocean access to an enclosed lagoon, harbor, or bay.

insolation rate The amount of solar energy reaching Earth's surface per unit time.

interference Addition or subtraction of wave energy as waves interact. Also called *resonance*. See also *constructive interference; destructive interference*.

intermediate-depth water wave A wave moving through water deeper than $\frac{1}{20}$ but shallower than $\frac{1}{2}$ its wavelength. Also called a *transitional wave*.

internal wave A progressive wave occurring at the boundary between liquids of different densities.

intertidal zone The marine zone between the highest high-tide point on a shoreline and the lowest low-tide point. The intertidal zone is sometimes subdivided into four separate habitats by height above tidal datum, typically numbered 1 to 4, land to sea.

intertropical convergence zone (ITCZ) The equatorial area at which the trade winds converge. The ITCZ usually lies at or near the meteorological equator. Also called the *doldrums*.

invertebrate Animal lacking a backbone.

ion An atom (or small group of atoms) that becomes electrically charged by gaining or losing one or more electrons.

ionic bond A chemical bond resulting from attraction between oppositely charged ions. These forces are said to be "electrostatic" in nature.

ionizing radiation Fast-moving particles or high-energy electromagnetic radiation emitted as unstable atomic nuclei disintegrate. The radiation has enough energy to dislodge one or more electrons from atoms it hits to form charged ions, which can react with and damage living tissue.

island arc Curving chain of volcanic islands and seamounts almost always found paralleling the concave edge of a trench.

isostatic equilibrium Balanced support of lighter material in a heavier, displaced supporting matrix. Analogous to buoyancy in a liquid.

isotonic Referring to a solution having the same concentration of dissolved substances as the solution that surrounds it.

ITCZ See *intertropical convergence zone*.

Jason-1 A follow-on satellite mission to *TOPEX/Poseidon*.

kelp Informal name for any species of large phaeophyte.

kingdom The largest category of biological classification. Five kingdoms are presently recognized.

knot A speed of 1 nautical mile per hour. See also *nautical mile*.

krill *Euphausia superba*, a thumb-size crustacean common in Antarctic waters.

lagoon A shallow body of seawater generally isolated from the ocean by a barrier island. Also, the body of water enclosed within an atoll, or the water within a reverse estuary.

land breeze Movement of air offshore as marine air heats and rises.

La Niña An event during which normal tropical Pacific atmospheric and oceanic circulation strengthens and the surface temperature of the eastern South Pacific drops below average values. Usually occurs at the end of an ENSO event. See also *ENSO*.

latent heat of fusion Heat removed from a liquid during freezing (or added to a solid during thawing) that produces a change in state but not a change in temperature. For pure water, 80 calories per gram at 0°C (32°F).

latent heat of vaporization Heat added to a liquid during evaporation (or released from a gas during condensation) that produces a change in state but not a change in temperature. For pure water, 585 calories per gram at 20°C (68°F).

lateral-line system A system of sensors and nerves in the head and midbody of fishes and some amphibians that functions to detect low-frequency vibrations in water.

latitude Regularly spaced imaginary lines on Earth's surface running parallel to the equator.

law A large construct explaining events in nature that have been observed to occur with unvarying uniformity under the same conditions.

Law of the Sea Collective term for laws and treaties governing the commercial and practical use of the ocean.

Library of Alexandria The greatest collection of writings in the ancient world, founded in the third century B.C. by Alexander the Great. Could be considered the first university.

light Electromagnetic radiation propagated as small, nearly massless particles that behave like both a wave and a stream of particles.

limiting factor A physical or biological environmental factor whose absence or presence in an inappropriate amount limits the normal actions of an organism.

Linnaeus, Carolus Carl von Linné (1707–1778). Swedish "father" of modern taxonomy.

lithification Conversion of sediment into sedimentary rock by pressure or by the introduction of a mineral cement.

lithosphere The brittle, relatively cool outer layer of Earth, consisting of the oceanic and continental crust and the outermost, rigid layer of mantle.

littoral zone The band of coast alternately covered and uncovered by tidal action; the intertidal zone.

longitude Regularly spaced imaginary lines on Earth's surface running north and south and converging at the poles.

longshore bar A submerged or exposed line of sand lying parallel to shore and accumulated by wave action.

longshore current A current running parallel to shore in the surf zone, caused by the incomplete refraction of waves approaching the beach at an angle.

longshore drift Movement of sediments parallel to shore, driven by wave energy.

longshore trough Submerged excavation parallel to shore adjacent to an exposed sandy beach. Caused by the turbulence of water returning to the ocean after each wave.

low-energy coast A coast only rarely exposed to large waves.

lower mantle The rigid portion of Earth's mantle below the asthenosphere.

low tide The low-water position corresponding to a tidal trough.

low tide terrace The smooth, hard-packed beach seaward of the beach scarp on which waves expend most of their energy. Site of the most vigorous onshore and offshore movement of sand.

lunar tide Tide caused by gravitational and inertial interaction of the moon and Earth.

macroplankton Animal plankters larger than 1 to 2 centimeters ($\frac{1}{2}$ to 1 inch). An example is the jellyfish.

Magellan, Ferdinand (c. 1480–1521) Portuguese navigator in the service of Spain who led the first expedition to circumnavigate Earth, 1519–22. He was killed in the Philippines.

magma Molten rock capable of fluid flow. Called *lava* above ground.

magnetometer A device that measures the amount and direction of residual magnetism in a rock sample.

Mammalia The class of mammals.

mangrove Large flowering shrub or tree that grows in dense thickets or forests along muddy or silty tropical coasts.

mantle The layer of Earth between the crust and the core, composed of silicates of iron and magnesium. The mantle has an average density of about 4.5 g/cm³ and accounts for about 68% of Earth's mass.

map A representation of Earth's surface, usually depicting mostly land areas. See also *chart*.

mariculture The farming of marine organisms, usually in estuaries, bays, or nearshore environments or in specially designed structures using circulating seawater. Compare *aquaculture*.

marine energy resource Any resource resulting from the direct extraction of energy from the heat or movement of ocean water.

marine pollution The introduction by humans of substances or energy into the ocean that changes the quality of the water or affects the physical and biological environment.

marine science The process (or result) of applying the scientific method to the ocean, its surroundings, and the life forms within it. Also called *oceanography* or *oceanology*.

masking pigment See *accessory pigment*.

mass A measure of the quantity of matter.

mass extinction A catastrophic, global event in which major groups of species perish abruptly.

Maury, Matthew (1806–1873) "Father" of physical oceanography. Probably the first person to undertake the systematic study of the ocean as a full-time occupation, and probably the first to understand the global interlocking of currents, wind flow, and weather.

maximum sustainable yield The maximum amount of fish, crustaceans, and mollusks that can be caught without impairing future populations.

mean sea level The height of the ocean surface averaged over a few years' time.

medusa Free-swimming body form of many members of the phylum Cnidaria.

membrane A complex structure of proteins and lipids that forms boundaries around and within the cell. It is usually semipermeable, allowing some kinds of molecules to pass through but not others.

meroplankton The planktonic phase of the life cycle or organisms that spend only part of their life drifting in the plankton.

mesoderm The middle layer of cells in a developing embryo.

mesosphere The rigid inner mantle, similar in chemical composition to the asthenosphere.

metabolic rate The rate at which energy-releasing reactions proceed within an organism.

metamerism Segmentation; repeating body parts.

Meteor **Expedition** German Atlantic expedition begun in 1925; the first to use an echo sounder and other modern optical and electronic instrumentation.

meteorological equator The irregular imaginary line of thermal equilibrium between hemispheres. It is situated about 5° north of the geographical equator, and its position changes with the seasons, moving slightly north in northern summer. Also called the *thermal equator*.

meteorological tide A tide influenced by the weather. Arrival of a storm surge will alter the estimate of a tide's height or arrival time, as will a strong, steady onshore or offshore wind.

metrophagy Tendency for large reptiles to eat entire cities.

microtektite Translucent oblong particles of glass, a component of cosmogenous sediment.

Milky Way The name of our galaxy. Sometimes applied to the field of stars in our home spiral arm, which is correctly called the Orion arm.

mineral A naturally occurring inorganic crystalline material with a specific chemical composition and structure.

mixed layer See *surface zone*.

mixed tide A complex tidal cycle, usually with two high tides and two low tides of unequal height per day.

mixing time The time necessary to mix a substance through the ocean, about 1,600 years.

mixture A close intermingling of different substances that still retain separate identities. The properties of a mixture are heterogeneous; they may vary within the mixture.

molecule A group of atoms held together by chemical bonds. The smallest unit of a compound that retains the characteristics of the compound.

Mollusca The phylum of animals that includes chitons, snails, clams, and octopuses.

molt To shed an external covering.

monsoon A pattern of wind circulation that changes with the season. Also, the rainy season in areas with monsoon wind patterns.

moon tide See *lunar tide*.

moraine Hills or ridges of sediment deposited by glaciers.

motile Able to move about.

multicellular Consisting of more than one cell.

multicellular algae Algae with bodies consisting of more than one cell. Examples are kelp and *Ulva*.

mutation A heritable change in an organism's genes.

mutualism A symbiotic interaction between two species that is beneficial to both.

Mysticeti The suborder of baleen whales.

nanoplankton Very small members of the plankton community. Examples are coccolithophores and silicoflagellates.

Nansen bottle A water-sampling instrument perfected early in this century by the Norwegian scientist and explorer Fridtjof Nansen.

natural selection A mechanism of evolution that results in the continuation of only those forms of life best adapted to survive and reproduce in their environment.

natural system of classification A method of classifying an organism based on its ancestry or origin.

nautical chart A chart used for marine navigation.

nautical mile The length of 1 minute of latitude, 6,076 feet, 1.15 statute miles, or 1.85 kilometers. (See Appendix I.)

neap tide The time of smallest variation between high and low tides occurring when Earth, moon, and sun align at right angles. Neap tides alternate with spring tides, occurring at two-week intervals.

nebula Diffuse cloud of dust and gas.

nekton Drifting organisms.

Nematoda The phylum of animals to which roundworms belong.

neritic Of the shore or coast. Refers to continental margins and the water covering them, or to nearshore organisms.

neritic sediment Continental shelf sediment consisting primarily of terrigenous material.

neritic zone The zone of open water near shore, over the continental shelf.

niche Description of an organism's functional role in a habitat. Its "job."

node The line or point of no wave action in a standing pattern. See also *amphidromic point*.

nodule Solid mass of hydrogenous sediment, most commonly manganese or ferromanganese nodules and phosphorite nodules.

nonconservative constituent An element whose proportion in seawater varies with time and place, depending on biological demand or chemical reactivity. An element with a short residence time. For example, iron, aluminum, silicon, trace nutrients, dissolved oxygen, and carbon dioxide.

nonconservative nutrient A compound or ion needed by autotrophs for primary productivity and that changes in concentration with biological activity.

nonextractive resource Any use of the ocean in place, such as transportation of people and commodities by sea, recreation, or waste disposal.

nonrenewable resource Any resource that is present on Earth in fixed amounts and cannot be replenished.

nonvascular Describing photosynthetic autotrophs without vessels for the transport of fluid. Examples are algae.

nor'easter (northeaster) Any energetic extratropical cyclone that sweeps the eastern seaboard of North America in winter.

North Atlantic Deep Water Cold, dense water formed in the Arctic that flows onto the floor of the North Atlantic ocean.

notochord Stiffening structure found at some time in the life cycle of all members of the phylum Chordata.

nuclear energy Energy released when atomic nuclei undergo a nuclear reaction such as the spontaneous emission of radioactivity, nuclear fission, or nuclear fusion. About 17% of the electrical power generated in the United States is provided by the nuclear fission of uranium in civilian power reactors.

nutrient Any needed substance that an organism obtains from its environment except oxygen, carbon dioxide, and water.

ocean (1) The great body of saline water that covers 70.78% of the surface of Earth. (2) One of its primary subdivisions, bounded by continents, the equator, and other imaginary lines.

ocean basin Deep-ocean floor made of basaltic crust. Compare *continental margin*.

oceanic crust The outermost solid surface of Earth beneath ocean floor sediments, composed primarily of basalt.

oceanic ridge Young seabed at the active spreading center of an ocean, often unmasked by sediment, bulging above the abyssal plain. The boundary between diverging plates. Often called a mid-ocean ridge, though less than 60% of the length exists at mid-ocean.

oceanic zone The zone of open water away from shore, past the continental shelf.

oceanography The science of the ocean. See also *marine science*.

oceanus Latin form of *okeanos*, the Greek name for the "ocean river" past Gibraltar.

Odontoceti The suborder of toothed whales.

oolite sand Hydrogenous sediment formed when calcium carbonate precipitates from warmed seawater as pH rises, forming rounded grains around a shell fragment or other particle.

ooze Sediment of at least 30% biological origin.

ophiolite An assemblage of subducting oceanic lithosphere scraped off (obducted) onto the edge of a continent.

Ophiuroidea The class of the phylum Echinodermata to which brittle stars belong.

orbit In ocean waves, the circular pattern of water particle movement at the air–sea interface. Orbital motion contrasts with the side-to-side or back-and-forth motion of pure transverse or longitudinal waves.

orbital inclination The 23°27' "tilt" of Earth's rotational axis relative to the plane of its orbit around the sun.

orbital wave A progressive wave in which particles of the medium move in closed circles.

osmoregulation The ability to adjust internal salt concentration.

osmosis The diffusion of water from a region of high water concentration to a region of lower water concentration through a semipermeable membrane.

Osteichthyes The class of fishes with bony skeletons.

outgassing The volcanic venting of volatile substances.

overfishing Harvesting so many fish that there is not enough breeding stock left to replenish the species.

oxygen minimum zone A zone in which oxygen is depleted by animals and not replaced by phytoplankton.

oxygen revolution The time span, from about 2 billion to 400 million years ago, during which photosynthetic autotrophs changed the composition of Earth's atmosphere to its current oxygen-rich mixture.

ozone O_3, the triatomic form of oxygen. Ozone in the upper atmosphere protects living things from some of the harmful effects of the sun's ultraviolet radiation.

ozone layer A diffuse layer of ozone mixed with other gases surrounding the world at a height of about 20 to 40 kilometers (12 to 25 miles).

Pacific Ring of Fire The zone of seismic and volcanic activity that encircles the Pacific Ocean.

paleoceanography The study of the ocean's past.

paleomagnetism The "fossil," or remanent, magnetic field of a rock.

Pangaea Name given by Alfred Wegener to the original "protocontinent." The breakup of Pangaea gave rise to the Atlantic Ocean and to the continents we see today.

Panthalassa Name given by Alfred Wegener to the ocean surrounding Pangaea.

parasitism A symbiotic relationship in which one species spends part or all of its life cycle on or within another, using the host species (or food within the host) as a source of nutrients. The most common form of symbiosis.

partially mixed estuary An estuary in which an influx of seawater occurs beneath a surface layer of fresh water flowing seaward. Mixing occurs along the junction.

passive margin Continental margin near an area of lithospheric plate divergence. Also called *Atlantic-type margin*.

passive sonar A device that detects the intensity and direction of underwater sounds.

PCBs See *polychlorinated biphenyls*.

pelagic Of the open ocean. Refers to the water above the deep-ocean basins, sediments of oceanic origin, or organisms of the open ocean.

pelagic sediment Sediments of the slope, rise, and deep-ocean floor that originate in the ocean.

pelagic zone The realm of open water. See also *benthic zone*.

perigee The point in the orbit of a satellite where it is closest to the main body; opposite of *apogee*.

perihelion The point in the orbit of a satellite where it is closest to the sun; opposite of *aphelion*.

period See *wave period*.

Phaeophyta Brown multicellular algae, including kelps.

photic zone The thin film of lighted water at the top of the world ocean. The photic zone rarely extends deeper than 200 meters (660 feet). Compare *euphotic zone*.

photon The smallest unit of light energy.

photosynthesis The process by which autotrophs bind light energy into the chemical bonds of food with the aid of chlorophyll and other substances. The process uses carbon dioxide and water as raw materials and yields glucose and oxygen.

pH scale A measure of the acidity or alkalinity of a solution. Numerically, the negative logarithm of the concentration of hydrogen ions in an aqueous solution. A pH of 7 is neutral; lower numbers indicate acidity, and higher numbers indicate alkalinity.

phycobilin A reddish accessory pigment found in red algae.

phylum One of the major groups of the animal kingdom whose members share a similar body plan, level of complexity, and evolutionary history (see Appendix VI). (Plural, phyla.) (The major groups of the plant kingdom are called divisions.)

physical factor An aspect of the physical environment that affects living organisms, such as light, salinity, or temperature.

physical resource Any resource that has resulted from the deposition, precipitation, or accumulation of a useful nonliving substance in the ocean or seabed. Also called a *nonliving resource*.

phytoplankton Plantlike, usually single-celled members of the plankton community.

picoplankton Extremely small members of the plankton community, typically 0.2 to 2 micrometers (4 to 40 millionths of an inch) across.

Pinnipedia The carnivoran suborder that contains the seals, sea lions, and walruses.

piston corer A seabed-sampling device capable of punching through up to 25 meters (80 feet) of sediment and returning an intact plug of material.

planet A smaller, usually nonluminous body orbiting a star.

plankter Informal name for a member of the plankton community.

plankton Drifting or weakly swimming organisms suspended in water. Their horizontal position is to a large extent dependent on the mass flow of water rather than on their own swimming efforts.

plankton bloom A sudden increase in the number of phytoplankton cells in a volume of water.

plankton net Conical net of fine nylon or Dacron fabric used to collect plankton.

Plantae The kingdom to which multicellular vascular autotrophs belong.

plate One of about a dozen rigid segments of Earth's lithosphere that move independently. The plate consists of continental or oceanic crust and the cool, rigid upper mantle directly below the crust.

plate tectonics The theory that Earth's lithosphere is fractured into plates, which move relative to each other and are driven by convection currents in the mantle. Most volcanic and seismic activity occurs at plate margins.

Platyhelminthes The phylum of animals to which flatworms belong.

plunging wave Breaking wave in which the upper section topples forward and away from the bottom, forming an air-filled tube.

polar cell The atmospheric circulation cell centered over each pole.

polar front Boundary between the polar cell and the Ferrel cell in each hemisphere.

polar molecule A molecule with unbalanced charge. One end of the molecule has a slight negative charge, and the other end has a slight positive charge.

polar ocean areas Zones to the south of the Antarctic Convergence and to the north of the Arctic Convergence.

polar regions Earth's cold area poleward of either the Arctic or Antarctic Circle.

pollutant A substance that causes damage by interfering directly or indirectly with an organism's biochemical processes.

Polychaeta The largest and most diverse class of phylum Annelida. Nearly all polychaetes are marine.

polychlorinated biphenyls (PCBs) Chlorinated hydrocarbons once widely used to cool and insulate electrical devices and to strengthen wood or concrete. PCBs may be responsible for the changes in and declining fertility of some marine mammals.

Polynesia A large group of Pacific islands lying east of Melanesia and Micronesia and extending from the Hawaiian Islands south to New Zealand and east to Easter Island.

polynya A gap in polar pack ice at which liquid water contacts the atmosphere.

polyp One of two body forms of Cnidaria. Polyps are cup-shaped and possess rings of tentacles. Coral animals are polyps.

Polyplacophora The class of the phylum Mollusca that includes the chitons.

poorly sorted sediment A sediment in which particles of many sizes are found.

population A group of individuals of the same species occupying the same area.

population density The number of individuals per unit area.

Porifera The phylum of animals to which sponges belong.

potable water Water suitable for drinking.

precipitate (1) A solid substance formed in an aqueous reaction. (2) The process by which a solute forms in and falls from a solution. The falling of water or ice from the atmosphere.

precipitation Liquid or solid water that falls from the air and reaches the surface as rain, hail, or snowfall.

pressure Force per unit area.

prey An organism consumed by a predator.

primary coast Coast on which terrestrial influences dominate. See also *secondary coast*.

primary consumer Initial consumer of primary producers. The consumers of autotrophs; the second level in food webs.

primary forces The forces that induce and maintain water flow in ocean current systems: thermal expansion, wind friction, and density differences.

primary producer An organism capable of using energy from light or energy-rich chemicals in the environment to produce energy-rich organic compounds. An autotroph.

primary productivity The synthesis of organic materials from inorganic substances by photosynthesis or chemosynthesis. Expressed in grams of carbon bound into carbohydrate per unit area per unit time (g $C/m^2/yr$).

Prince Henry the Navigator Established a center at Sagres, Portugal, for the study of marine science and navigation in the mid-1450s.

principle of constant proportions The proportions of major conservative elements in seawater remain nearly constant, though total salinity may change with location. Also called *Forchhammer's principle*.

progressive wave A wave of moving energy in which the wave form moves in one direction along the surface (or junction) of the transmission medium (or media).

Protista The kingdom of single-celled nucleated organisms to which protozoa, diatoms, and dinoflagellates belong. Also called *Protoctista*.

proton A positively charged particle at the center of an atom.

protostar Tightly condensed knot of material that has not yet attained fusion temperature.

protozoa Multiphyletic animal group within the kingdom Protista. Protozoa include amoebas, paramecia, foraminiferans, and radiolarians.

pteropod Small planktonic mollusk with a calcareous shell, which contributes to biogenous sediments.

P wave Primary wave. A compressional wave associated with an earthquake and that can move through both liquid and rock.

pycnocline The middle zone of the ocean in which density increases rapidly with depth. Temperature falls and salinity rises in this zone.

radial symmetry Body structure in which the body parts radiate from a central axis like spokes from a wheel. An example is a sea star. Compare *bilateral symmetry*.

radioactive decay The disintegration of unstable forms of elements, which releases subatomic particles and heat.

radiolarian One of a group of usually planktonic amoeba-like animals with a siliceous shell, which contributes to biogenous sediments.

radiometric dating The process of determining the age of rocks by observing the ratio of unstable radioactive elements to stable decay products.

random distribution Distribution of organisms within a community whereby the position of one organism is in no way influenced by the positions of other organisms or by physical variations within that community. A very rare distribution pattern.

reef A hazard to navigation. A shoal, a shallow area, or a mass of fish or other marine life.

refraction Bending of light or sound waves as they move at an angle other than 90° between media of different optical or acoustical densities. See also *wave refraction*.

refractive index The degree of refraction from one medium to another expressed as a ratio. The higher the ratio (refractive index), the greater the bending of waves between media.

refractometer A compact optical device that determines the salinity of a water sample by comparing the refractive index of the sample to the refractive index of water of known salinity.

relative dating Determining the age of a geological sample by comparing its position to the positions of other samples.

renewable resource Any resource that is naturally replaced on a seasonal basis by the growth of living organisms or by other natural processes.

Reptilia The class of reptiles, including turtles, crocodiles, iguanas, and snakes.

residence time The average length of time a dissolved substance spends in the ocean.

respiration Release of stored energy from chemical bonds in food; carbon dioxide and water are formed as by-products. (Respiration is a biochemical process and is not the same as the mechanical process of breathing.)

restoring force The dominant force trying to return water to flatness after formation of a wave.

reverse estuary An estuary along a coast in which salinity increases from the ocean to the estuary's upper reaches because of evaporation of seawater and a lack of freshwater input.

Rhodophyta Red multicellular algae.

Richter scale A logarithmic measure of earthquake magnitude. A great earthquake measures above 8 on the Richter scale.

rip current A strong, narrow surface current that flows seaward through the surf zone and is caused by the escape of excess water that has piled up in a longshore trough.

rogue wave A single wave crest much higher than usual, caused by constructive interference.

salinity A measure of the dissolved solids in seawater, usually expressed in grams per kilogram or parts per thousand by weight. Standard seawater has a salinity of 35‰ at 0°C (32°F).

salinometer An electronic device that determines salinity by measuring the electrical conductivity of a seawater sample.

salt gland Specialized tissue responsible for concentration and excretion of excess salt from blood and other body fluids.

salt wedge estuary An estuary in which rapid river flow and small tidal range cause an inclined wedge of seawater to form at the mouth.

sand Sediment particle between 0.062 and 2 millimeters in diameter.

sandbar A submerged or exposed line of sand accumulated by wave action.

sand spit An accumulation of sand and gravel deposited downcurrent from a headland. Sand spits often curl at their tips.

saturation State of a solution in which no more of the solute will dissolve in the solvent. The rate at which molecules of the solute are being dissolved equals the rate at which they are being precipitated from the solution.

scattering The dispersion (or "bounce") of sound or light waves when they strike particles suspended in water or air. The amount of scatter depends on the number, size, and composition of the particles.

schooling Tendency of small fish of a single species, size, and age to mass in groups. The school moves as a unit, which confuses predators and reduces the effort spent searching for mates.

science A systematic way of asking questions about the natural world and testing the answers to those questions.

scientific method The orderly process by which theories explaining the operation of the natural world are verified or rejected.

scientific name The genus and species name of an organism.

sea Simultaneous wind waves of many wavelengths forming a chaotic ocean surface. Sea is common in an area of wind wave origin.

sea breeze Onshore movement of air as inland air heats and rises.

sea cave A cave near sea level in a sea cliff cut by processes of marine erosion.

sea cliff Cliff marking the landward limit of marine erosion on an erosional coast.

seafloor spreading The theory that new ocean crust forms at spreading centers, most of which are on the ocean floor, and pushes the continents aside. Power is thought to be provided by convection currents in Earth's upper mantle.

sea grass Any of several marine angiosperms. Examples are *Zostera* (eelgrass) and *Phyllospadix* (surfgrass). Sea grasses are not seaweeds.

sea ice Ice formed by the freezing of seawater.

sea island Island whose central core was connected to the mainland when sea level was lower. Rising ocean separates these high points from land, and sedimentary processes surround them with beaches. Compare *barrier island*.

sea level The height of the ocean surface. See also *mean sea level*.

seamount Circular or elliptical projection from the seafloor, more than 1 kilometer (0.6 mile) in height, with a relatively steep slope of 20° to 25°.

sea power The means by which a nation extends its military capacity onto the ocean.

Seasat First satellite dedicated to oceanic research, launched in 1978.

SEASTAR Satellite capable of measuring the distribution of chlorophyll at the ocean surface, a measure of marine productivity.

seaweed Informal term for large marine multicellular algae.

secondary coast Coast dominated by marine processes. See also *primary coast*.

secondary consumer Consumer of primary consumers.

secondary forces The forces that influence the direction and nature of water flow in ocean current systems: the Coriolis effect, gravity, friction, and the shape of ocean basins.

second law of thermodynamics Disorder (entropy) in a closed system must increase over time. If disorder decreases, it does so at the expense of energy. Because the universe as a whole may be considered a closed system, it follows that an increase in order in one part must result in a decrease in order in another.

sediment Particles of organic or inorganic matter that accumulate in a loose, unconsolidated form.

seiche Pendulum-like rocking of water in an enclosed area; a form of standing wave that can be caused by meteorological or seismic forces, or that may result from normal resonances excited by tides.

seismic Referring to earthquakes and the shock of earthquakes.

seismic sea wave Tsunami caused by displacement of earth along a fault. (Earthquakes and seismic sea waves are caused by the same phenomenon.)

seismic wave A low-frequency wave generated by the forces that cause earthquakes. Some kinds of seismic waves can pass through Earth. See also *P wave*; *S wave*.

seismograph An instrument that detects and records earth movement associated with earthquakes and other disturbances.

semidiurnal tide A tidal cycle of two high tides and two low tides each lunar day, with the high tides of nearly equal height.

sensible heat Heat whose gain or loss is detectable by a thermometer or other sensor.

sessile Attached. Nonmotile; unable to move about.

sewage sludge Semisolid mixture of organic matter, microorganisms, toxic metals, and synthetic organic chemicals removed from wastewater at a sewage treatment plant.

shadow zone (1) The wide band at Earth's surface 105° to 143° away from an earthquake in which seismic waves are nearly absent. P waves are absent because they are refracted by Earth's liquid outer core; S waves are absent from this band and the zone immediately opposite the earthquake site because they are absorbed by the outer core. (2) In sonar, the volume of ocean from which sound waves diverge and in which a submarine may hide.

shallow-water wave A wave in water shallower than $\frac{1}{20}$ its wavelength.

shelf break The abrupt increase in slope at the junction between continental shelf and continental slope.

shore The place where ocean meets land. On nautical charts, the limit of high tides.

side-scan sonar A high-resolution sound-imaging system used for geological investigations, archaeological studies, and the location of sunken ships and airplanes.

siliceous ooze Ooze composed mostly of the hard remains of silica-containing organisms.

silicoflagellate A tiny single-celled phytoplankter with a siliceous skeleton.

silt Sediment particle between 0.004 and 0.062 millimeter in diameter.

Sirenia The order of mammals that includes manatees, dugongs, and the extinct sea cows.

slack water A time of no tide-induced currents that occurs when the current changes direction.

sofar Sound *fixing* and *ranging*. An experimental U.S. Navy technique for locating survivors on life rafts, based on the fact that sound from explosive charges dropped into the layer of minimum sound velocity can be heard for great distances. See also *sofar layer*.

sofar layer Layer of minimum sound velocity in which sound transmission is unusually efficient for long distances. Sounds leaving this depth tend to be refracted back into it. The sofar layer usually occurs at mid-latitude depths around 1,200 meters (4,000 feet).

solar nebula The diffuse cloud of dust and gas from which the solar system originated.

solar system The sun together with the planets and other bodies that revolve around it.

solar tide Tide caused by the gravitational and inertial interaction of the sun and Earth.

solstice One of two times of the year when the overhead position of the sun is farthest from the equator. The time of the solstice is midway between equinoxes.

solute A substance dissolved in a solvent. See also *solution*.

solution A homogeneous substance made of two components, the solvent and the solute.

solvent A substance able to dissolve other substances. See also *solution*.

sonar Sound *navigation* and *ranging*.

sound A form of energy transmitted by rapid pressure changes in an elastic medium.

sounding Measurement of the depth of a body of water.

Southern Oscillation A reversal of airflow between normally low atmospheric pressure over the western Pacific and normally high pressure over the eastern Pacific. The cause of El Niño. See also *El Niño*.

speciation The formation of new species. Charles Darwin suggested that this is accomplished through isolation and natural selection.

species Any group of actually or potentially interbreeding organisms reproductively isolated from all other groups and capable of producing fertile offspring. (Note: The word *species* is both singular and plural.)

species diversity Number of different species in a given area.

species-specific relationship An exclusive relationship between two species. Parasites are usually species-specific; that is, they can usually parasitize only one species of host.

spilling wave A breaking wave whose crest slides down the face of the wave.

spreading center The junction between diverging plates at which new ocean floor is being made. Also called *spreading zone*.

spring tide The time of greatest variation between high and low tides occurring when Earth, moon, and sun form a straight line. Spring tides alternate with neap tides throughout the year, occurring at two-week intervals.

standing wave A wave in which water oscillates without causing progressive wave forward movement. There is no net transmission of energy in a standing wave.

star A massive sphere of incandescent gases powered by the conversion of hydrogen to helium and other heavier elements.

state An expression of the internal form of matter. Water exists in three states: solid, liquid, and gas. A solid has a fixed volume and fixed shape, a liquid has a fixed volume but no fixed shape, and a gas has neither fixed volume nor fixed shape.

stenohaline Describing an organism unable to tolerate a wide range in salinity.

stenothermal Describing an organism unable to tolerate wide variance in temperature.

stipe Multicellular algal equivalent of a vascular plant's stem.

Stokes drift A small net transport of water in the direction a wind wave is moving.

storm Local or regional atmospheric disturbance characterized by strong winds often accompanied by precipitation.

storm surge An unusual rise in sea level as a result of the low atmospheric pressure and strong winds associated with a tropical cyclone. Onrushing seawater precedes landfall of the tropical cyclone and causes most of the damage to life and property.

stratigraphy The branch of geology that deals with the definition and description of natural divisions of rocks. Specifically, the analysis of relationships of rock strata.

subduction The downward movement into the asthenosphere of a lithospheric plate.

subduction zone An area at which a lithospheric plate is descending into the asthenosphere. The zone is characterized by linear folds (trenches) in the ocean floor and strong deep-focus earthquakes. Also called a *Wadati–Benioff zone*.

sublittoral zone The ocean floor near shore. The inner sublittoral extends from the littoral (intertidal) zone to the depth at which wind waves have no influence; the outer sublittoral extends to the edge of the continental shelf.

submarine canyon A deep, V-shaped valley running roughly perpendicular to the shoreline and cutting across the edge of the continental shelf and slope.

subsidence Sinking, often of tectonic origin.

Subtropical Convergence Convergence zone marking the boundary between Central Water and either Subarctic or Subantarctic Surface Water. The northern Subtropical Convergence lies at about 45°N in the Pacific and 60°N in the Atlantic; the southern Subtropical Convergence lies at 40° to 50°S.

succession The changes in species composition that lead to a climax community.

sun tide See *solar tide*.

supernova The explosive collapse of a massive star.

supralittoral zone The splash zone above the highest high tide; not technically part of the ocean bottom.

surf The confused mass of agitated water rushing shoreward during and after a wind wave breaks.

surface current Horizontal flow of water at the ocean's surface.

surface-to-volume ratio A physical constraint on the size of cells. As a cell's linear dimensions grow, its surface area does not increase at the same rate as its volume. As surface-to-volume ratio falls, each square unit of outer membrane must serve an increasing interior volume.

surface zone The upper layer of ocean in which temperature and salinity are relatively constant with depth. Depending on local conditions, the surface zone may reach to 1,000 meters (3,300 feet) or be absent entirely. Also called the *mixed layer*.

surf beat The pattern of constructive and destructive interference that causes successive breaking waves to grow, shrink, and grow again over a few minutes' time.

surf zone The region between the breaking waves and the shore.

surging wave A wave that surges ashore without breaking.

suspension feeder An animal that feeds by straining or otherwise collecting plankton and tiny food particles from the surrounding water.

sverdrup (sv) A unit of volume transport named in honor of oceanographer Harald U. Sverdrup: 1 million cubic meters of water flowing past a fixed point each second.

swash Water from waves washing onto a beach.

S wave Secondary wave. A transverse wave associated with an earthquake and that cannot move through liquid.

swell Mature wind waves of one wavelength that form orderly undulations of the ocean surface.

swim bladder A gas-filled organ that assists in maintaining neutral buoyancy in some bony fishes.

symbiosis The co-occurrence of two species in which the life of one is closely interwoven with the life of the other; mutualism, commensalism, or parasitism.

synoptic sampling Simultaneous sampling at many locations.

taxonomy In biology, the laws and principles covering the classification of organisms.

tektite A small, rounded, glassy component of cosmogenous sediments, usually less than 1.5 millimeters (1/20 inch) in length. Thought to have formed from the impact of an asteroid or meteor on the crust of Earth or the moon.

Teleostei The osteichthyan order that contains the cod, tuna, halibut, perch, and other species of bony fishes.

telepresence The extension of a person's senses by remote sensors and manipulators.

temperate zone The mid-latitude area between the Tropic of Cancer and the Arctic Circle and between the Tropic of Capricorn and the Antarctic Circle.

temperature The response of a solid, liquid, or gas to the input or removal of heat energy. A measure of the atomic and molecular vibration in a substance, indicated in degrees.

temperature–salinity (T-S) diagram A graph showing the relationship of temperature and salinity with depth.

terrane An isolated segment of seafloor, island arc, plateau, continental crust, or sediment transported by seafloor spreading to a position adjacent to a larger continental mass. Usually different in composition from the larger mass.

terrigenous sediment Sediment derived from the land and transported to the ocean by wind and flowing water.

territorial waters Waters extending 12 miles from shore and in which a nation has the right to jurisdiction.

thallus The body of an alga or other simple plant.

theory A general explanation of a characteristic of nature consistently supported by observation or experiment.

thermal equator See *meteorological equator*.

thermal equilibrium The condition in which the total heat coming into a system (such as a planet) is balanced by the total heat leaving the system.

thermal inertia Tendency of a substance to resist change in temperature with the gain or loss of heat energy.

thermocline The zone of the ocean in which temperature decreases rapidly with depth. See also *pycnocline*.

thermohaline circulation Water circulation produced by differences in temperature and/or salinity (and therefore density).

thermostatic property A property of water that acts to moderate changes in temperature.

tidal bore A high, often breaking wave generated by a tide crest that advances rapidly up an estuary or river.

tidal current Mass flow of water induced by the raising or lowering of sea level owing to passage of tidal crests or troughs. See also *ebb current*; *flood current*.

tidal datum The reference level (0.0) from which tidal height is measured.

tidal range The difference in height between consecutive high and low tides.

tidal wave The crest of the wave causing tides. Another name for a tidal bore. Not a tsunami or seismic sea wave.

tide Periodic short-term change in the height of the ocean surface at a particular place, generated by long-wavelength progressive waves that are caused by the interaction of gravitational force and inertia. Movement of Earth beneath tide crests results in the rhythmic rising and falling of sea level.

tombolo Above-water bridge of sand connecting an offshore feature to the mainland.

top consumer An organism at the apex of a trophic pyramid, usually a carnivore.

TOPEX/Poseidon Joint French–U.S. satellite carrying radars that can determine the height of the sea surface with unprecedented accuracy. Other experiments in this five-year program include sensing water vapor over the ocean, determining the precise location of ocean currents, and determining wind speed and direction.

tornado Localized, narrow, violent funnel of fast-spinning wind, usually generated when two air masses collide. Not to be confused with a cyclone. (The tornado's oceanic equivalent is a waterspout.)

trace element A minor constituent of seawater present in amounts of less than 1 part per million.

trade winds Surface winds within the Hadley cells, centered at about 15° latitude, that approach from the northeast in the Northern Hemisphere and from the southeast in the Southern Hemisphere.

transform fault A plane along which rock masses slide horizontally past one another.

transform plate boundary Places where crustal plates shear laterally past one another. Crust is neither produced nor destroyed at this type of junction.

transitional wave See *intermediate-depth water wave*.

transverse current East-to-west or west-to-east current linking the eastern and western boundary currents. An example is the North Equatorial Current.

trench An arc-shaped depression in the deep-ocean floor with very steep sides and a flat sediment-filled bottom coinciding with a subduction zone. Most trenches occur in the Pacific.

trophic level A feeding step within a trophic pyramid.

trophic pyramid A model of feeding relationships among organisms. Primary producers form the base of the pyramid; consumers eating one another form the higher levels, with the top consumer at the apex.

tropical cyclone A weather system of low atmospheric pressure around which winds blow counterclockwise in the Northern Hemisphere and clockwise in the Southern Hemisphere. Originates in the tropics within a single air mass, but may move into temperate waters if water temperature is high enough to sustain it. Small tropical cyclones are called *tropical depressions*, larger ones *tropical storms*, and great ones *hurricanes*, *typhoons*, or *willi-willis*, depending on location.

tropical ocean area Warm central zone of the ocean equatorward of the Subtropical Convergences where surface temperature is nearly always above 20°C (69°F).

Tropic of Cancer The imaginary line around Earth, parallel to the equator at 23°27'N, marking the point where the sun shines directly overhead at the June solstice.

Tropic of Capricorn The imaginary line around Earth, parallel to the equator at 23°27'S, marking the point where the sun shines directly overhead at the December solstice.

tropics The area between the Tropic of Cancer and the Tropic of Capricorn.

trough See *wave trough*.

tsunami Long-wavelength shallow-water wave caused by rapid displacement of water. See also *seismic sea wave*.

tunicate A type of suspension-feeding invertebrate chordate.

turbidite A terrigenous sediment deposited by a turbidity current; typically, coarse-grained layers of nearshore origin interleaved with finer sediments.

turbidity current An underwater "avalanche" of abrasive sediments thought responsible for the deep sculpturing of submarine canyons and a means of transport for sediments accumulating on abyssal plains.

turbulence Chaotic fluid flow.

ultraplankton Extremely small plankton, smaller than nanoplankton.

undercurrent A current flowing beneath a surface current, usually in the opposite direction.

unicellular Consisting of a single cell.

unicellular algae Algae with bodies consisting of a single cell. Examples are diatoms and dinoflagellates.

uniform distribution Distribution of organisms within a community characterized by equal space between individuals (the arrangement of trees in an orchard). The rarest natural distribution pattern.

uniformitarianism The theory that all of Earth's geological features and history can be explained by processes occurring today and that these processes must have been at work for a very long time.

United States Exclusive Economic Zone The region extending seaward from the coast of the United States for 200 nautical miles, within which the United States claims sovereign rights and jurisdiction over all marine resources.

United States Exploring Expedition The first U.S. oceanographic research voyage, launched in 1838.

upwelling Circulation pattern in which deep, cold, usually nutrient-laden water moves toward the surface. Upwelling can be caused by winds blowing parallel to shore or offshore.

$V = \sqrt{gd}$ Relation between velocity (V), the acceleration due to gravity (g), and water depth (d) for shallow-water waves.

$V = L/T$ Relation between velocity (V), wavelength (L), and period (T) for deep-water waves; velocity increases as wavelength increases. Typically measured in meters per second.

valve In diatoms, each half of the protective silica-rich outer portion of the cell. The complete outer covering is called the *frustule*.

vascular plant Plant having vessels for transport of fluid through leaves, stems, and roots. Examples are sea grasses, mangroves, and maple trees.

velocity Speed in a specified direction.

Vertebrata The subphylum of the phylum Chordata that includes animals with segmented backbones.

vertebrate A chordate with a segmented backbone.

Vikings Seafaring Scandinavian raiders who ravaged the coasts of Europe around A.D. 780–1070.

viscosity Resistance to fluid flow. A measure of the internal friction in fluids.

voyaging Traveling (usually by sea) with a specific purpose.

Wadati–Benioff zone See *subduction zone.*

water mass A body of water identifiable by its salinity and temperature (and therefore its density) or by its gas content or another indicator.

water vapor The gaseous, invisible form of water.

water-vascular system System of water-filled tubes and canals found in some representatives of the phylum Echinodermata and used for movement, defense, and feeding.

wave Disturbance caused by the movement of energy through a medium.

wave crest Highest part of a progressive wave above average water level.

wave-cut platform The smooth, level terrace sometimes found on erosional coasts that marks the submerged limit of rapid marine erosion.

wave diffraction Bending of waves around obstacles.

wave frequency The number of waves passing a fixed point per second.

wave height Vertical distance between a wave crest and the adjacent wave troughs.

wavelength The horizontal distance between two successive wave crests (or troughs) in a progressive wave.

wave period Time it takes for successive wave crests to pass a fixed point.

wave reflection The reflection of progressive waves by a vertical barrier. Reflection occurs with little loss of energy.

wave refraction Slowing and bending of progressive waves in shallow water.

wave shock Physical movement, often sudden, violent, and of great force, caused by the crash of a wave against an organism.

wave steepness Height-to-wavelength ratio of a wave. The theoretical maximum steepness of deep-water waves is 1:7.

wave train A group of waves of similar wavelength and period moving in the same direction across the ocean surface. The group velocity of a wave train is half the velocity of the individual waves.

wave trough The valley between wave crests below the average water level in a progressive wave.

weather The state of the atmosphere at a specific place and time.

Wegener, Alfred (1880–1930) German scientist who proposed the theory of continental drift in 1912.

well-mixed estuary An estuary in which slow river flow and tidal turbulence mix fresh and salt water in a regular pattern through most of its length.

well-sorted sediment A sediment in which particles are of uniform size.

westerlies Surface winds within the Ferrel cells, centered around 45° latitude, that approach from the southwest in the Northern Hemisphere and from the northwest in the Southern Hemisphere.

western boundary current Strong, warm, concentrated, fast-moving current at the western boundary of an ocean (off the east coast of a continent). Examples include the Gulf Stream and the Japan (Kuroshio) Current.

westward intensification The increase in speed of geostrophic currents as they pass along the western boundary of an ocean basin.

West Wind Drift Current driven by powerful westerly winds north of Antarctica. The largest of all ocean currents, it continues permanently eastward without changing direction. See *Antarctic Circumpolar Current*.

Wilson, John Tuzo (1908–1993) Canadian geophysicist who proposed the theory of plate tectonics in 1965.

wind The mass movement of air.

wind duration The length of time the wind blows over the ocean surface, a factor in wind wave development.

wind-induced vertical circulation Vertical movement in surface water (upwelling or downwelling) caused by wind.

wind strength Average speed of the wind, a factor in wind wave development.

wind wave Gravity wave formed by transfer of wind energy into water. Wavelengths from 60 to 150 meters (200 to 500 feet) are most common in the open ocean.

world ocean The great body of saline water that covers 70.78% of Earth's surface.

xanthophyll A yellow or brown accessory pigment that gives some marine autotrophs a yellow or tan appearance.

zone Division or province of the ocean with homogeneous characteristics.

zooplankton Animal members of the plankton community.

zooxanthellae Unicellular dinoflagellates that are symbiotic with coral and that produce the relatively high pH and some of the enzymes essential for rapid calcium-carbonate deposition in coral reefs.

Credits

This page constitutes an extension of the copyright page. We have made every effort to trace the ownership of all copyrighted material and to secure permission from copyright holders. In the event of any question arising as to the use of any material, we will be pleased to make the necessary corrections in future printings. Thanks are due to the following authors, publishers, and agents for permission to use the material indicated.

TEXT AND ILLUSTRATION CREDITS

Chapter 1

Figure 1.7a–e *The Universe Explained*, Colin A. Ronan, ed. New York: H. Holt, © 1994. **Figure 1.15** C. J. Allègre and S. H. Schneider, "The Evolution of the Earth," *Scientific American* (October 1994). Reprinted by permission of Ian Worpole.

Chapter 3

Figure 3.5 from *The Earth: An Introduction to Physical Geology*, 4/e, by Edward J. Tarbuck and Frederick K. Lutgens. © 1993 Macmillan College Publishing. Reprinted by permission of Pearson Education, Inc., Upper Saddle River, NJ 07458. **Figure 3.6** from *The Earth's Dynamic Systems*, 9/e, by Hamblin & Christiansen. Reprinted by permission of Pearson Education, Inc., Upper Saddle River, NJ 07458. **Box 3.1c** B. Bolt, "Earthquakes and Geological Discover," Scientific American Books, © 1993. **Figure 3.13** from *The Earth's Dynamic Systems*, 9/e, by Hamblin & Christiansen. Reprinted by permission of Pearson Education, Inc., Upper Saddle River, NJ 07458. **Figure 3.28** from *Earth's Dynamic Systems*, 8/e by Hamblin & Christiansen, © 1998. Reprinted by permission of Pearson Education, Inc., Uppder Saddle River, NJ 07458 **Figure 3.34** reprinted by permission of Dr. Roger Billham, University of Colorado.

Chapter 4

Figure 4.17 from Taylor & Francis International Scientific and Educational Publishers. **Figure 4.22a** Aluminum Company of America. **Figure 4.29** reprinted by permission of Chet Raymo.

Chapter 5

Figure 5.14c-d from *Exploration of the Oceans* by John G. Weihaupt. Copyright © 1979 John G. Weihaupt. **Figure 5.20** A. G. Fischer, *Geological History of the Western North Pacific*, vol. 168, p. 1210, June 5, 1970. © 1970 AAAS. Reprinted by permission.

Chapter 6

Figure 6.4 from Stahler, A. N., *Physical Geology*, published by Addison Wesley Longman, Inc. © 1981 by Arthur N. Stahler. Reprinted by permission of Pearson Education, Inc. **Figure 6.11** from G. P. Kuiper, ed., *The Earth as a Planet*, © 1954 The University of Chicago Press. Reprinted by permission.

Chapter 8

Figure 8.3 from Hamblin & Christiansen, *Earth's Dynamic Systems*, 9/e, p. 214. Reprinted by permission of Pearson Education, Inc., Upper Saddle River, NJ 07458. **Figure 8.9** CALVIN AND HOBBES. © 1990 Universal Press Syndicate. Reprinted with permission of Universal Press Syndicate. **Figure 8.13** *Meteorology*, 5/e by Moran and Morgan. © 1997 Prentice-Hall, Inc. **Table 8.2** from *Earth in Crisis: An Introduction to the Earth Sciences*, 2/e, Thomas L. Burrus, Herbert J. Spiegel, 1980. C. V. Mosby Co. Reprinted by permission of Thomas L. Burrus. **Figure 8.15a–b** reprinted by permission of Alan D. Iselin.

Chapter 9

Figure 9.5a from *Laboratory Exercises in Oceanography*, 3/e by Pipkin, Gorslin, Casey and Hammond. © 1987 by W. H. Freeman and Company. **Figure 9.5b** NASA 217: From M. Grant Gross, Oceanography: A View of the Earth, 5/e, p. 173, © 1990. Prentice-Hall. Reprinted with permission. **Box 9.2** written by Denise Smythe-Wright, Southampton Oceanography Centre, U.K.

Chapter 11

Figure 11.6 research by Philip L. F. Liv, Seung Nam Seo, and Sung Bum Yoon, and Civil and Environmental Engineering, Cor-

nell University. Visualization by Catherine Devine, Cornell Theory Center. Used by permission.

Chapter 12

Figure 12.6 used by permission of Dr. William Galloway, University of Texas, Austin and the Houston Geological Survey. **Figure 12.31** used by permission of Glen Wheless.

Chapter 15

Figure 15.9 from *Living Invertebrates*, V. & J. Pearse/M. & R. Buchsbaum, 1987. The Boxwood Press. Adapted with permission of the publisher. **Figure 15.15** from James L. Sumich, *An Introduction to Biology of Marine Life*, 5/e. Copyright © 1992 Wm. C. Brown Communication, Inc., Dubuque, Iowa. **Figure 15.20a** from *Living Invertebrates*, V. & J. Pearse/M. & R. Buchsbaum, 1987, The Boxwood Press. Adapted with permission of the publisher. **Figure 15.29** from Pough, et al., *Vertebrate Life*, 3/e © 1989. Reprinted by permission of Pearson Education, Inc. Upper Saddle River, NJ 07458.

Chapter 16

Figure 16.2 from Sylvia S. Mader, *Biology: Evolution, Diversity, and the Environment*, Second Edition. © 1987 Wm. C. Brown Communications, Inc. Dubuque, Iowa. **Figure 16.7** from *Between Pacific Tides*, 5/e by Edward F. Ricketts, et al. Revised by David W. Phillip. Stanford University Press, © 1985 by the Board of Trustees of the Leland Stanford Junior University. Reprinted with permission. **Figure 16.15** from J. C. Briggs, *Marine Zoogeography*. © 1974 by McGraw-Hill, Inc. **Figure 16.16** from J. C. Briggs, *Marine Zoogeography*. © 1974 by McGraw-Hill, Inc. **Figure 16.18a & b** from J. C. Briggs, *Marine Zoogeography*. © 1974 by McGraw-Hill, Inc.

Chapter 17

Figure 17.7 from *National Geographic*, January 2001

Chapter 18

Figure 18.11 from *Science*, 289 (August 4, 2000), p. 723. Reprinted by permission.

Figure 18.21 "Northern Hemisphere Millennial Temperature Reconstruction" from *Geophysical Research Letters*, Vol. 26, No. 6, p. 759-762 by Michael E. Mann, Raymond S. Bradley, and Michael K. Hughes. Copyright © 1999 American Geophysical Union. **Figure 18.24** from *Popular Science* (September 2000).

Appendixes

Appendix III, Figures 1-4 and **Appendix IV, Figures 1, 3** from David Greenhood, *Down to Earth: Mapping for Everybody*. **Appendix VIII, Figures 1 & 2 and excerpt pp. 511-513** reprinted by permission of Dan Dion.

PHOTO CREDITS

Chapter 1

Chapter opening photo courtesy of NASA. **Figure 1.2b** Tim Fitzharris/Index Stock Imagery/PictureQuest. **Figure 1.3a** Philip Richardson/Woods Hole Oceanographic Institution. **Figure 1.3b** Tom Garrison. **Figure 1.3c** Dan Burton Photography. **Figure 1.3d** Jay Dickman. Box 1.1 Slocum, J. 1899. *Sailing Alone Around the World*. Reprint. New York: Sheridan House, 1972. **Figure 1.5** Courtesy of ESO. **Figure 1.6** Image from TRACE (Transition Region and Coronal Explorer), a mission of the Stanford-Lockheed Institute for Space Research (a joint program of Lockheed-Martin Advanced Technology Center's Solar and Astrophysics Laboratory and Stanford's Solar Observatories Group), and part of the NASA Small Explorer program. Figure 1.8 NASA. **Figure 1.9a** © William K. Hartmann. **Figures 1.9b & 1.10** Hubble Space Telescope/STScI. **Figure 1.12** © William K. Hartmann. **Figure 1.13** Excerpted from *The History of Earth*. Copyright © 1991 by William Hartmann and Ron Miller. Used by permission of Workman Publishing Co., Inc., New York. All Rights Reserved. **Figure 1.14a** © William K. Hartmann. **Figure 1.14b** © Don Dixon. **Figure 1.17a** Tom Garrison. **Figure 1.17b** Woods Hole Oceanographic Institution. **Figure 1.18** Jeffrey Bada/Department of Earth and Space Sciences. **Box 1.2a–c** NASA/Robert Sullivan. **Box 1.2d & e** NASA/JPL/Malin Space Science Systems. **Box 1.2f** Malin Space Science Systems/MGS/JPL/NASA. **Box 1.2g** Bruce Macintosh, Seran Gibbard, Don Gavel, and Claire Max/Lawrence Livermore National Laroratory.

Chapter 2

Chapter opening photo Dan Dion. **Figure 2.1** Reprinted with permission from Doubleday, a division of Bantam, Doubleday, Dell Publishing Group, Inc. from *Sailing Ships* by Bjorn Landstrom, 1978. **Figure 2.4** Tom Garrison. **Figure 2.5** © 1991 Herbert Kawainui Kane. **Figure 2.6** © Larus Karl Ingason. **Figure 2.7** © 1991 Herbert Kawainui Kane. **Figure 2.9** James Ford Bell Library, University of Minnesota **Figure 2.10** The Granger Collection, New York. **Figure 2.12** © National Maritime Museum, Greenwich. **Figure 2.13** © 1991 Herbert Kawainui Kane. **Figure 2.15a, c** Tom Garrison. **Figure 2.15b** © National Maritime Museum, London. **Figure 2.16** Tom Garrison. **Figure 2.17** © National Maritime Museum, London. **Figure 2.18** United States Naval Academy Museum. **Figures 2.19 & 2.20** Library of Congress. **Figure 2.22** Reprinted by permission of the Royal Geographical Society, London, archives from the journal of Lt. Pelham Aldrich aboard *H.M.S. Challenger*, 1872–1875. **Figure 2.24** Deborah Day/Scripps Institute of Oceanography. **Figure 2.25** Hulton Deutsch Collection Limited/Liaison Agency. **Figure 2.26** Alda Amundsen, SPRI. **Figure 2.27** Scott Polar Research Institute. **Figure 2.28** Ted Delaca, University of Alaska. **Box 2.2** AP/Wide World Photos **Figure 2.29** Deep Sea Drilling Program. Texas A&M University. **Figures 2.30a & b** Tom Garrison **Figure 2.31** NASA.

Chapter 3

Chapter opening photo Roger Ressmeyer/CORBIS. **Figure 3.1** © Hans-Peter Bunge. **Box 3.1a** National Portrait Gallery, London. **Box 3.1b** Copyright, National Survey & Cadastre, Denmark. **Figure 3.7** WILBUR GARRETT/NGS Image Collection. **Figure 3.8** Dorothy L. Stout. **Figure 3.10** Alfred Wegener Institute for Polar and Marine Research. **Figure 3.17d** V. Courtillot. **Figure 3.25a** Fundemental Photographs. **Figures 3.27 & 3.30** Dr. Peter Sloss, NOAA/National Geophysical Data Center.

Chapter 4

Chapter opening a photo Dr. Don Walsh, International Maritime, Inc. **Chapter opening b** Naval Historical Center. **Figure 4.1** Deborah Day/Scripps Institute of Oceanography. **Figure 4.4a** U.S. Department of the Navy. **Figure 4.4c** Karen Marks, NOAA. **Box 4.1a** NUYTCO Research. **Box 4.1c** JAMSTEC. **Box 4.1d**

Ted Delaca, University of Alaska. **Box 4.1e** JAMSTEC. **Figures 4.12, 4.13 & 4.15** William Haxby, Lamont/Coluumbia. **Figure 4.18** Society for Sedimentary Geology. **Figure 4.19** U.S. Department of the Navy. **Figure 4.22a** Alcoa, Inc. **Figure 4.22b** Lamont-Doherty, Earth Observatory of Columbia University. **Figure 4.22c** Charles D. Hollister **Box 4.2b** Dr. Jian Lin, Woods Hole Oceanographic Institution. **Figure 4.26** Charles D. Hollister **Figure 4.27** Deborah Smith, Woods Hole Oceanographic Institution. **Figure 4.30** Dr. Peter Sloss, NOAA/National Geophysical Data Center.

Chapter 5

Chapter opening photo from *The History of Earth*. Copyright © 1991 by William Hartmann and Ron Miller. Used by permission of Workman Publishing Co., Inc., New York. All Rights Reserved . **Figures 5.1, 5.2, & 5.3** Charles D. Hollister, Woods Hole Oceanographic Institution. **Figure 5.6a** USGS. **Figure 5.6b** Department of Geology, University of Delaware. **Figure 5.6c** NASA. **Figure 5.7** Department of Geology, University of Deleware. **Figure 5.8a** Dr. Howard Spero. **Figure 5.8b** Wim van Egmond. **Figure 5.8c** Roger Witmer. **Figure 5.10a** Wim van Egmond. **Figure 5.10b** Roger Witmer. **Figures 5.11a & b** Susumi Honjo. **Box 5.1** from *Face of the Deep* by Bruce C. Heezen and Charles D. Hollister, 1971, Oxford University Press. Reprinted by permission of Woods Hole Oceanographic Institution. **Figures 5.12a & b** Unocal Research. **Figures 5.14a & 5.15a** Tom Garrison. **Box 5.2a & b** Dr. James Broda, Woods Hole Oceanographic Institution. **Figures 5.16 & 5.17** Deep Sea Drilling Project, Texas A&M University. **Figure 5.21** NASA/JPL/Malin Space Science Systems.

Chapter 6

Chapter opening photo © C. Sims/SuperStock, Inc. **Figure 6.7** © Chigmaroff/Davison, SuperStock, Inc. **Figure 6.17** Gunther Konnen, by permission of Bob Greenler. **Figures 6.18c & d** © Bruce Hall. **Figure 6.23b** EG&C Ocean Products. **Figure 6.23c** WESMAR.

Chapter 7

Chapter opening painting by Jacques Louis David, 1788. The Metropolitan Museum of Art, Purchase, Mr. and Mrs. Charles Wrightsman Gift, in honor of Everett Fahy, 1977. Copyright © 1989 by the Metropolitan Museum of Art. **Figure**

7.4 Woods Hole Oceanographic Institution. **Figure 7.5a** Tom Garrison. **Figure 7.5c** Dr. William P. Cochlan, Hancock Institute for Marine Sciences, University of Southern California. **Figure 7.6** David Breiter. **Figure 7.7** Tom Garrison.

Chapter 8

Chapter opening b photo AP/Wide World Photos. **Figure 8.1b** © Craig Tuttle/The Stock Market. **Figure 8.14** NASA. **Figure 8.19** NASA and Tom Garrison. **Figure 8.20** © Bettmann Archive/CORBIS. **Figure 8.21** NASA. **Figure 8.23** JPL/NASA. **Figure 8.24b** NOAA/National Weather Service. **Box 8.1a** SeaWiFS Project, NASA/Goddard Space Flight Center and ORIBIMAGE. **Box 8.1b** AP/Wide World Photos. **Figure 8.25** Jim Bell, Cornell University Astronomy Department.

Chapter 9

Chapter opening photo NASA. **Figures 9.7b & 9.8a** NASA/JPL. **Box 9.1** © Bettmann Archive/CORBIS. **Figure 9.11** O. Brown, R. Evans, M. Carle/University of Miami Rosenstiel School of Marine and Atmospheric Science. **Figure 9.16a** Mobil Oil Corporation. **Figures 9.17b & d** NASA. **Figures 9.18a–e** TOPEX/Poseidon Team, CNES, NASA. **Figure 9.26b** Tom Garrison. **Figure 9.26c** Woods Hole Oceanographic Institution. **Figure 9.26f** Webb Research. **Box 9.2** D. Smythe-Wright.

Chapter 10

Chapter opening photo © Ron Romanosky/Creation Captured. **Figure 10.9** Tom Garrison. **Box 10.1a** Raychel Ciemma. **Figure 10.12** NASA. **Figure 10.13** Office of NOAA Corps Operations. **Figure 10.15** U.S. Navy. **Box 10.2** Dr. W. C. Stillwell. Reprinted by permission of Suzanne M. Stillwell. **Figure 10.17** from Erich Lessing, Art Resource, NY. Crane, Walter (1845-1915). Neptune's Horses. Oil on Canvas. Neue Pinakothek, Munich, Germany. Reprinted by permission of Art Resource. **Figures 10.19a & b** © Dennis Junor/Creation Captured. **Figure 10.20b** University of Hawaii Press. **Figure 10.21** Tom Garrison. **Figure 10.22** University of Hawaii Press. **Figure 10.25b** NASA.

Chapter 11

Chapter opening photo D. Millsap. **Chapter opening photo inset** USGS. **Figure 11.1** AP/Wide World Photos. **Figures 11.2a & b** Thames Barrier Visitors Centre. **Figure 11.4** William F. Haxby and Lincoln F. Pratson. **Figures 11.7a–c** USGS. **Figure 11.8** Bishop Museum, Hawaii. **Figure 11.9** Kyodo Photo Service. **Figure 11.11a** Tom Garrison. **Figure 11.11b** AP/Wide World Photos. **Figure 11.12 (both)** Scott Walking Adventures, Halifax, Nova Scotia. **Figure 11.31a & b** © Jeff Greenberg/Photo Researchers, Inc. **Figure 11.32** F. W. Rowbotham. **Box 11.1a** Arthur Rackham. **Box 11.1b** BØDØ Arrangement, Norway. **Figure 11.33** Cabrillo Marine Aquarium. **Figure 11.34** Photothèque EDF.

Chapter 12

Chapter opening photo a © Billy E. Barnes/Stock, Boston/PictureQuest. **Chapter opening photo b** AP/Wide World Photos. **Figure 12.1a & b** Courtesy of William Haxby, Lamont/Columbia. **Figure 12.2** NASA. **Figure 12.3** © Bob Clemenz Photography. **Figure 12.4** USGS. **Figure 12.5** NASA. **Figure 12.8** John S. Shelton. **Figure 12.9** Dr. George P. Lafaker, USGS. **Figure 12.10** NASA. **Figure 12.11** © Ron Romanosky. **Figure 12.12b** © SuperStock, Inc. **Figures 12.16a & b** © Gerald G. Kuhn. **Figures 12.17a & c** Dr. William Wallace. **Figure 12.17b** Tom Garrison. **Figure 12.17d** © Dante de la Rosa. **Figure 12.18** © Carr Clifton Photography. **Figure 12.19** © John S. Shelton. **Figure 12.20b** NASA. **Figure 12.23** © Jack Mertz. **Figure 12.24** NASA. **Figure 12.25a & b** © Gerald G. Kuhn. **Figure 12.25c** © W. Demetrakas, O. C. Camera. **Figure 12.26** Great Barrier Reef Marine Park Authority. Queensland, Australia. **Figure 12.27d** © Douglas Faulkner, Science Source/Photo Researchers, Inc. **Figure 12.28** © Brian Parker/Tom Stack and Associates. **Figure 12.32** © John S. Shelton. **Figures 12.33a & b** The Fairchild Aerial Photography Collection at Whittier College. **Box 12.1** Peter Van Elk. **Figure 12.35** AP/Wide World Photos.

Chapter 13

Chapter opening photo © Peter Scoones/Planet Earth Pictures. **Figure 13.1** © Linda E. Tway. **Figure 13.2** Raychel Ciemma. **Box 13.1a & b** NASA. **Box 13.1c** © Don Davis. **Box 13.1d** Dr. Virgil Sharpton/University of Alaska-Fairbanks. **Figure 13. 7 shark** Courtesy of Frank Lane Picture Agency Limited. **Figure 13. 7 penguin** Courtesy of SuperStock, Inc. **Figure 13. 7 dolphin** Courtesy of Ed Degginger. **Figure 13. 18d** © Timothy Reese. **Figure 13. 19** L. K. Townsend

Chapter 14

Chapter opening photo © Ralph A. Clevenger/CORBIS. **Figure 14.5b** © Wim van Egmond. **Figure 14.6** SeaWiFS Project, NASA/Goddard Space Flight Center and ORIBIMAGE. **Figure 14.9b** Dennis Kelly, Orange Coast College. **Figure 14.10a** Greta Fryxell. **Figure 14.10b** From C. Shih and R. G. Kessel, *Living Images* © 1982 Sudbury, MA: Jones & Bartlett Publishers. www.jbpub.com. Reprinted with permission. **Figure 14.10c** © Wim van Egmond. **Figure 14.11a** Shih and Kessel. **Figure 14.12** Tom Garrison. **Figure 14.13a** Marine Life Research Group. **Figure 14.13b** SeaWiFS Project, NASA/Goddard Space Flight Center and ORIBIMAGE. **Box 14.1a** Florida Department of Environmental Protection, Florida Marine Research Institute, St. Petersburg. **Box 14.1b** C. C. Lockwood. **Box 14.1 bottom** Suisan Aviation Ltd., Japan. **Box 14.2** Dr. William P. Cochlan, Hancock Institute for Marine Sciences/University of Southern California. **Figure 14.16** SeaWiFS Project, NASA/Goddard Space Flight Center and ORIBIMAGE. **Figures 14.18 & 14.19a** © Wim van Egmond. **Figure 14.19b** Courtesy of Dr. Howard Spero. **Figure 14.21a** © Bruce Hall. **Figure 14.24** Tom Garrison. **Figure 14.25** National Park Service.

Chapter 15

Chapter opening photo Anthony R. Ramirez. **Figure 15.2a & b** The South Australian Museum. **Figure 15.2c** Dr. Chip Clark. **Figure 15.3a** © David Hass/Tom Stack and Associates. **Figure 15.3b-d** Raychel Ciemma. **Figure 15.5** William H. Hamner. **Figure 15.6a** Pat Mason. **Box 15.1** Tom Garrison. **Figure 15.8** © Bruce Hall. **Figure 15.11** © Mark McMahon. **Figure 15.12** Raychel Ciemma. **Figure 15.13** © 1992 Sea World of California, Inc. **Figure 15.14** © David Paul Photography, Australia. **Figures 15.16a & b** Washington State Department of Fisheries. **Figures 15.17a & b** © Bruce Hall. **Figure 15.18b** © Herve Chaumeton/Nature. **Figure 15.19** Charles D. Hollister, Woods Hole Oceanographic Institution. **Figure 15.20b** © Pearse and Buchsbaum. **Figure 15.21** © Herve Chaumeton/Nature. **Figure 15.24** © Heather Angel. **Figure 15.25** © Amos Nachoum/CORBIS. **Figure 15.26 series** Chris Fallows/www.apexpredators.com. **Figure 15.27** © James D. Watt/Planet Earth Sciences. **Figures 15.32a & b** and

15.33 © Bruce Hall. **Figure 15.34** © OSF, Debbie Stone/Animals Animals. **Figure 15.36** © Valerie Taylor. **Figure 15.37** © Johnny Johnson/Animals Animals. **Figure 15.38** © SuperStock, Inc. **Figure 15.41b** © Norman Cole. **Figure 15.42** © Richard Sears/MICS-Photo. **Figure 15.43a** © Bruce Hall. **Figure 15.43b** © Courtesy of Friends of the Sea Lion. **Figure 15.44** Courtesy of Friends of the Sea Otter/ Richard Bucich. **Figure 15.45** © Leonard Rue/Visuals Unlimited. **Figure 15.46** Florida Department of Natural Resources.

Chapter 16

Chapter opening photo Tom Garrison. **Figure 16.5** © Eric and David Hosking. **Figure 16.7b** Reprinted from *Between Pacific Tides*, Fifth Edition, by Edward F. Ricketts et al. Revised by David W. Phillips, with permission of the publishers, Stanford Univrsity Press. © 1985 by the Board of Trustees of the Leland Stanford Junior University. **Box 16.1** © Ed Ricketts, Jr. **Figure 16.6a** © 1972 Darwen Hennings. **Figure 16.8** © Bruce Hall. **Figure**

16.9a Tom Garrison. **Figure 16.9b** © Stan Elms/Visuals Unlimited. **Figures 16.10 & 16.11** Tom Garrison. **Figure 16.12a** © Darwen Hennings. **Figure 16.12c** © Peter Scoones/Planet Earth Pictures. **Figures 16.17 & 16.18c** Charles D. Hollister, Woods Hole Oceanographic Institution. **Figure 16.20 a & b** Woods Hole Oceanographic Institution. **Figure 16.21** © Dave Breiter. **Figure 16.22** © Norman Cole.

Chapter 17

Chapter opening photo © Dan Parrett/ Alaska Stock. **Figure 17.2a** © Bruce Hall. **Figures 17.2b & c** Shell Oil Company. **Figure 17.3** Courtesy of Cargill Salt, Newark, California. **Figure 17.6b** National Energy Laboratory of Hawaii. **Figure 17.12** © Bill Curtsinger Photography. **Figure 17.14** © Tom Campbell's Photographic. **Figure 17.15** Anthony R. Ramirez. **Figure 17.16** Dan Stanford. **Figure 17.19** Norman Cole. **Figure 17.20** Joseph W. Holliday. **Figure 17.21** Shell Oil Company. **Figures 17.22 &**

Box 17.1a–c Tom Garrison. **Box 17.1d** Port Authority of NY & NJ.

Chapter 18

Chapter opening photo Hiroyuki Matsumoto/Black Star Publishing/Picture-Quest. **Figure 18.3** Reuters/Santiago Armas/Archive Photos. **Figure 18.4** Tom Garrison. **Figure 18.6** 1989 Tony Dawson Photography. **Box 18.1** © W. Eugene Smith/Black Star. **Figure 18.12** © Ken Sakamoto/Black Star Photo. **Figure 18.13** © Tom Campbell's Photographic. **Figure 18.14** NOAA. **Figure 18.15** State of California. **Figure 18.16** NASA/Goddard Space Flight Center. **Figure 18.17** Courtesy of Dr. Ronald D. Barr, UC-Irvine Medical Center **Figure 18.18** The New Yorker Collection 1989 Stuart Leeds from cartoonbank.com. All rights reserved. **Figure 18.22** Reprinted by permission of the Los Angeles Times Syndicate. **Figure 18.23** C. Mayhew and R. Simmon/NASA. **Figure 18.26** © 2001 Stone/Mark Lewis.

Afterword

Page 490 Tom Garrison.

Index

Fishes (*continued*)
madhouse economics, 452
movement, shape, and propulsion,
395–396, *396*
osmoregulation, 397–398, *398*
per capita catch, *449*
Fissipedia, 408–409
Fissurella, 420–421
Fjord estuaries, *309*, 310
Fjords, 291, *292*, 308–309, *309*
Flagella, 360, *360*
Flatworms, 382
Float method, 230
Flood currents, 281
Florence, Hurricane, *200*
Florida, 100, *102*, 291
Flow method, 230
Fluids, *149*
Fluorine, *173*
Flying fishes, 399, *400*
Folger, Tim, 41, *42*
Food webs, 349, *351*, 366–367
Foraminifera, 129, *129*, 141, 356–357, *367*
Forbes, Edward, 20, 43
Forced waves, 240, 269
Forchhammer, Georg, 175
Forchhammer's principle, 175, 177
Foreshore, 300, *300*
Fossil fuels, 484
Fossils, 15–16, *16*, 66, 378, *379*
Fracture zones, 109, *109*
Fragilaria, 356–357
Fram, R/V, 45
Franklin, Benjamin, 41, *42*, 493
Free waves, 240, 269
Freezing point, 148, 171
Fremont, Capt. John C., 40
Frigate birds, 402
Fringing reefs, 306, *307*
Frogs, 399
Frontal storms, 198
Fronts, 198
Frustules, 358
Fucoxanthin, 368
Fucus, 420–421
Fully developed sea, 245, *245, 247*
Fungi, 332, *332*, 344

G

Gaia hypothesis, 319–320
Galápagos Islands, 111, 218, *326*, 329, *468*
Galaxies, 7–8, *8, 16*
Galileo spacecraft, 17, *327*
Galloway, William, 293, *294*
Galveston, Texas, 262, 304n, 312, 479
Ganymede, 17
Gas bladders, 368, *369*
Gases, definition, *149*
Gases, dissolved, 178–179, *179, 180, 336*, 336
Gas exchange, 396–397, *397*
Gastropoda, 384
Geneva, Lake, 263, *263*, 277
Geological time, *496–498*
Georges, Hurricane, *202*

Georges Bank, 438–439
Geosat satellite, 95, *95*
Geostrophic gyres, 212–214, *218*
Ghost nets, 451, *451*
Gibraltar, *120, 208, 227, 228*
Gigantism, 430
Gill membranes, 396–397, *397*
Glaciation, 103, *104*, 128, 294, *294*
Global Positioning System (GPS), 51, 54
Global warming. *See* Climate change
GLOBEC (Global Ocean Ecosystem
Dynamics), 51
Glomar Challenger, R/V, 49, *49*, 72, 120–121,
134, 137
Glossary of terms, 515–534
Glossopteris fossils, 66
Glyptocephalus zachirus, 333
Gold, *173*, 182
Gonyaulax, 356–357
Gorgonians, *122*
GPS (Global Positioning System), 51, 54
Grand Canyon, 141
Granite, 58–59, 125, *147*, 148
Gravel resources, 442
Gravitational energy, 10, 272, *272*
Gravity, buoyancy and, 340
Gravity waves, 240
Gray whales, *404, 408*, 453
Great Barrier Reef (Australia), *306*,
306–307
Great white sharks, *394*
Greeks, ancient ocean voyaging, 25, *26*
Greenhouse effect, 188n, *482*, 482–483
Greenhouse gases, *18*, 442, 482, *482*
Greenland Current, 219
Greenwich meridian, 28, *28*, 38, *40*, 500
Groins, 314, *314*
Grotius, Hugo, 457
Group velocity, 244
Growth rate, population, 417–418, *418*,
485–487, *486*
Grunion runs, 284, *284*
Gulf of California, *295*
Gulf of Mexico
coast, 312
deltas, 293
low energy regime, 297
pollution, 479
salt deposits, *103*
tidal patterns, 277, *277, 278*
Gulf Stream
discovery, 41, *42*
drift measurements, 215
effect on coastal climate, 234
power generation from, 234, 445
water flow in, *210*, 214–215, *216, 217, 218*
Gulls, 402, *420–421*
Gulper eels, 429, *429*
Guyot, Arnold, 84
Guyots, 84, *84*, 113–114
Gymnodinium breve, 361
Gypsum, 121, 133, 442
Gyres, 209–212, *210*, 219. *See also* Ocean
currents

H

HAB (harmful algal blooms), 360–362, *361,*
475–476
Habitats, 415, 418, 479–480
Häckel, Ernst, 435
Hadal zone, 342, 343
Haddock fishing, 439
Hadley, George, 193
Hadley cells, *193*, 193–194
Hagfishes, 392, *392*
Haldane, J. B. S., 13
Halibut fishing, 439
Haliotis, 420–421
Halley, Edmund, 35
Halocline, *155*, 156
Halosaccion, 420–421
Hansa Carrier, 230
Hantzschia, 284
Hardin, Garrett, 485–486
Harmful algal blooms (HAB), 360–362, *361,*
475–476
Harp seals, 454
Harrison, John, 38, *39*
Hastigerina, 129, *356–357*
Hatchetfish, *356–357*
Haven, 469
Hawaiian Islands, 30, 36, 53, 82, *82, 83*
Hawkbill, USS, *5*, 47, 99
Heard Island experiment, *163*, 163–164
Heat
budgets, *188*, 188–189, *189*
capacity, 147, *147, 152*, 171
global transfer of, 189, *189, 230*
phase changes, 149, *151*
sensible, 149
temperature and, 147, 150, *150*, 165
waste, 477–478, *478*
Heat budgets, *188*, 188–189, *189*
Heat capacity, 147, *147, 152*, 171
Heavy metals pollution, 471–472, *472*
Hemigrapsus, 420–421
Henry the Navigator, Prince of Portugal, *33*,
33–34
Henslow, John, 42
Henson, Matthew, 46
Herjulfsson, Bjarni, 31
Hermatypic corals, 381, *382*
Hermit crabs, *414*, 414–415, *420–421*
Herodotus, 293
Herring worm disease, 383
Hess, Harry, 68, 84, 141
Heteropods, 384
Heterotrophs, 349
Hierarchy of classification, 332
High seas, 460
High tides, 272, *272, 273, 274*
Hilo Bay, HI, 266, *267*
Himalayas, 77, *77, 79*
Hipparchus, 27, 500
Histro, 356–357
Holdfasts, 368
Hollow Earth theory, 40
Holoplankton, 366
Holothuroidea, 388–390, *431*

Medical debris, 487
Mediterranean Deep Water, 226, *227, 228*
Mediterranean Sea, *120,* 121, 226, *227,* 293
Medusae, 380
Meiji Seamount, *83*
Melanomas, 480, *481*
Melosira, 356–357
Melville, R/V, 57–58, 82
Mercator, Gerhardus, 502
Mercator projections, *502,* 502–503, *503, 504–505*
Mercury (element), *173,* 472, *472*
Mercury (planet), 9
Meroplankton, 366
Metabolic rates, 335
Metallic sulfides and muds, 443–444, 460
Metals pollution, 471–472, *472*
Metamerism, 382
Meteor, R/V, 93, 355
Meteor Expedition, 47
Meteorites, 20. *See also* Asteroid impacts; Asteroids
Meteorological tides, 283
Methane
   escape from sediments, 442
   hydrothermal vents, 432
   ocean of, 166
   in primitive atmosphere, 14, *14*
   as resource, 442
Methane hydrate, 442
Metric system, 493–495, *494, 495*
Metrophagy, 435
Michelangelo, 250
Microbial loops, 323
Micronesian stick charts, 30, *30*
Microtektites, 127, *127*
Mid-Atlantic Ridge
   discovery of, 39, 47, 68, 93
   Iceland, *110,* 110–111, *111,* 115
   models of, *74, 97, 107, 108, 109*
   paleomagnetism, *80*
   process, 75
   rate of spreading, 75n
   shape of, *69*
   volcanoes on, *113*
Migrations of peoples, 54
Milford Sound (New Zealand), *292*
Milky Way galaxy, 7
Miller, Stanley, 13–15, *15*
Miller projection, 503
Mind in the Waters, 411
Minerals, Earth's crust, 125
Minimata disease, 472, *472*
Mining, deep-sea, 133, 442–443. *See also* Marine resources
Minke whales, *404,* 452–453
Mississippi River delta, *293,* 474, *475*
Mitch, Hurricane, 203, *203*
Mixed layer, 154, 322
Mixed tides, *277, 277, 278*
Mixing times, 177
Mixtures, 170
MLLW (mean lower low water), 279, 286
Mola mola, 356–357
Molecules, 146

Mollusca, 384–387, *385, 386,* 410–411, 447–448, *448*
Molting, 387, *387, 388*
Monsoons, 196, *197*
Moon (Earth), 11, *12,* 270–273, *272, 273, 274*
Moraines, 294
Moray eels, *395,* 426–427
Morley, Lawrence, 79
Motile animals, 423
Mottez, J., 248
Mountain building, 61, 77, *77, 86*
Mouton, Gabriel, 493
M-reflector, 120–121, 133
Mud waves, 106, *106*
Mullet larvae, *356–357*
Mullus, 356–357
Multibeam systems, 93–94, *94*
Multicellular algae, 367–368. *See also* Seaweeds
Murex, 456
Muricid snails, *426–427*
Murray, Sir John, 43, 124
Musée Océanographique, 49
Mussels, 361, *420–421*
Mutations, 326, 329
Mutualism, *433,* 433–434
Mysticetes, *404,* 406–407
Mytilus, 420–421
Mytilus edulis, 361

N

Nanoplankton, 362
Nansen, Fridtjof, 39, *45,* 175
Nansen bottles, 175, *176*
Narwhals, *405*
National Maritime Museum, Greenwich, England, 38, *39*
National Oceanic and Atmospheric Administration (NOAA), 50
Natural gas resources, *440,* 440–442, *441. See also* Methane
Natural selection, 65, *326,* 326–329
Natural system of classification, 331
Naucrates, 356–357, 434
Nautical charts, 502
Nautilus, USS, 46
Nautiluses, *356–357,* 385–386, *386*
Navigation
   birds, 411
   celestial, 27, 38, *501,* 501
   GPS, 51, 54, 501
   Polynesian, *29,* 29–31, *30, 31, 254*
Nazca Plate, 72, 75–76, *76,* 97
Neap tides, 275, *275*
Nebulae, 7–9
Nekton, 343, *356–357*
Nelson, Adm. Horatio, 44
Nematoda, 382–383, 434
Nematoscelis, 356–357
Nemo, Capt., 493, 510
Neptune, 9
Neptune's Horse, 251
Neritic sediments, 127
Neritic zone, 342, *342*

Netherlands, 261, 313, *313*
New Guinea, 30, 53
New River estuary, 477
Newton, Sir Isaac, 269, 276
Newton's laws of motion, 506
Neyland, Robert, 48
Niches, 415
Nike shoe experiment, 230–231
Nile River delta, *293*
Niskin bottles, 175, *176*
Nitrifying bacteria, 324
Nitrogen
   in atmosphere, *173, 179, 186*
   cycle, 323–324, *324*
   dissolved, 336, *336,* 343
   as nutrient, *321,* 336, 338, 343
   in seawater, *173, 179, 179*
NOAA (National Oceanic and Atmospheric Administration), 50
Nobel Peace Prize, 46
Noctiluca, 356–357
Nodes, 263, *263,* 277
Nonconservative constituents, 178
Nonconservative nutrients, 354
Nonvascular organisms, 367
Nor-easters, 198–200, *200*
Nori, 454
North American Plate, *104*
North Atlantic currents, 210, 212, 213, 214, 218. *See also* Gulf Stream
North Atlantic Deep Water, 226, 228
North Equatorial Current, *210, 213*
Northern Madeira Abyssal Plain, *113*
Notochords, 390
Nuclear energy, 484
Nuclear fusion, in stars, 8, 8–10
Nuclear submarines, 46, 98, 99
Nuclear wastes, 484
Nudibranchs, 385, *385, 420–421, 426–427*
Nurdles, 316
Nutrients
   dissolved, 336
   limiting, 353–354
   nitrogen, *321,* 336, 338, 343
   nonconservative, 354
   phosphorus, *321,* 336

O

Ocean
   density structure, 154–156, *155*
   as evidence of life, 19
   formation of, 2, 10–13, *16*
   origin of term, 20, 25
   outside of Earth, 17–19
   size statistics, 2, 4
   solitude, 6
   surface height, 50–51, 95, 164, 212, *212,* 235
   thermal inertia, 151–154, *153*
   volume, 182, 290
Ocean basins. *See also* Ocean floors
   abyssal plains and hills, 112–113, *113*
   cross-sections, 96, *96, 101*
   floor topography, *96, 108, 109, 116–117*
   guyots, 84, *84,* 113–114